NUCLEAR SAFEGUARDS TECHNOLOGY 1978

VOL. II

The following States are Members of the International Atomic Energy Agency:

AFGHANISTAN
ALBANIA
ALGERIA
ARGENTINA
AUSTRALIA
AUSTRIA
BANGLADESH
BELGIUM
BOLIVIA
BRAZIL
BULGARIA
BURMA
BYELORUSSIAN SOVIET SOCIALIST REPUBLIC
CANADA
CHILE
COLOMBIA
COSTA RICA
CUBA
CYPRUS
CZECHOSLOVAKIA
DEMOCRATIC KAMPUCHEA
DEMOCRATIC PEOPLE'S REPUBLIC OF KOREA
DENMARK
DOMINICAN REPUBLIC
ECUADOR
EGYPT
EL SALVADOR
ETHIOPIA
FINLAND
FRANCE
GABON
GERMAN DEMOCRATIC REPUBLIC
GERMANY, FEDERAL REPUBLIC OF
GHANA
GREECE
GUATEMALA
HAITI
HOLY SEE
HUNGARY
ICELAND
INDIA
INDONESIA
IRAN
IRAQ
IRELAND
ISRAEL
ITALY
IVORY COAST
JAMAICA
JAPAN
JORDAN
KENYA
KOREA, REPUBLIC OF
KUWAIT
LEBANON
LIBERIA
LIBYAN ARAB JAMAHIRIYA
LIECHTENSTEIN
LUXEMBOURG
MADAGASCAR
MALAYSIA
MALI
MAURITIUS
MEXICO
MONACO
MONGOLIA
MOROCCO
NETHERLANDS
NEW ZEALAND
NICARAGUA
NIGER
NIGERIA
NORWAY
PAKISTAN
PANAMA
PARAGUAY
PERU
PHILIPPINES
POLAND
PORTUGAL
QATAR
ROMANIA
SAUDI ARABIA
SENEGAL
SIERRA LEONE
SINGAPORE
SOUTH AFRICA
SPAIN
SRI LANKA
SUDAN
SWEDEN
SWITZERLAND
SYRIAN ARAB REPUBLIC
THAILAND
TUNISIA
TURKEY
UGANDA
UKRAINIAN SOVIET SOCIALIST REPUBLIC
UNION OF SOVIET SOCIALIST REPUBLICS
UNITED ARAB EMIRATES
UNITED KINGDOM OF GREAT BRITAIN AND NORTHERN IRELAND
UNITED REPUBLIC OF CAMEROON
UNITED REPUBLIC OF TANZANIA
UNITED STATES OF AMERICA
URUGUAY
VENEZUELA
VIET NAM
YUGOSLAVIA
ZAIRE
ZAMBIA

The Agency's Statute was approved on 23 October 1956 by the Conference on the Statute of the IAEA held at United Nations Headquarters, New York; it entered into force on 29 July 1957. The Headquarters of the Agency are situated in Vienna. Its principal objective is "to accelerate and enlarge the contribution of atomic energy to peace, health and prosperity throughout the world".

Printed by the IAEA in Austria
July 1979

PROCEEDINGS SERIES

NUCLEAR SAFEGUARDS TECHNOLOGY 1978

PROCEEDINGS OF A SYMPOSIUM
ON NUCLEAR MATERIAL SAFEGUARDS
ORGANIZED BY THE
INTERNATIONAL ATOMIC ENERGY AGENCY
AND HELD IN VIENNA, 2–6 OCTOBER 1978

In two volumes

VOL.II

INTERNATIONAL ATOMIC ENERGY AGENCY
VIENNA, 1979

NUCLEAR SAFEGUARDS TECHNOLOGY 1978
IAEA, VIENNA, 1979
STI/PUB/497
ISBN 92–0–070179–5

FOREWORD

This was the fourth time that the IAEA had held an international symposium on the general theme of nuclear material safeguards, and the timeliness of this symposium, held in Vienna from 2 to 6 October 1978, was evident to all. World interest in international safeguards, noticeably increased by the ongoing International Nuclear Fuel Cycle Evaluation (INFCE), was high. So the topics of research and development in the field of safeguards, which had anyway been growing in importance, were significantly expanded. There was a clear need for researchers to come together in a common technical forum to discuss and evaluate their results and their expectations, and the date of the symposium had the combined advantages of being late enough to allow time for significant accomplishments to be reported and yet early enough to produce a meaningful input to INFCE.

The organization of a symposium such as this one always presents serious problems. International safeguards is itself a broad subject, extending from fundamental questions of how to design facilities to make them more easily safeguardable or how to design safeguards approaches to optimize effectiveness while minimizing such factors as intrusiveness or cost, to the more detail questions of how to measure specific material quantities or how to analyse statistically the resulting materials accountancy data. All these topics are important to international safeguards, and all have a rightful claim to a place in any broad-based symposium on nuclear material safeguards.

With all signs indicating that the symposium would be a large one, nobody should have been surprised that there were over 275 participants. Only rarely were there fewer than 100 listeners in the meeting room, and lively discussions were held throughout. The full texts of the papers, together with the discussions, are published here in two volumes.

The foreword of the proceedings of the last symposium in 1975 argued that "it cannot be claimed that all the problems of international safeguards have been solved". It may be disappointing to some to learn that, three years later, some of those problems still have not been solved. Technical problems, however, are noted for their capability of self-proliferation. Some problems have been solved, and significant progress has been achieved on a number of others. Those that still remain will be tackled with determination and confidence.

EDITORIAL NOTE

The papers and discussions have been edited by the editorial staff of the International Atomic Energy Agency to the extent considered necessary for the reader's assistance. The views expressed and the general style adopted remain, however, the responsibility of the named authors or participants. In addition, the views are not necessarily those of the governments of the nominating Member States or of the nominating organizations.

Where papers have been incorporated into these Proceedings without resetting by the Agency, this has been done with the knowledge of the authors and their government authorities, and their cooperation is gratefully acknowledged. The Proceedings have been printed by composition typing and photo-offset lithography. Within the limitations imposed by this method, every effort has been made to maintain a high editorial standard, in particular to achieve, wherever practicable, consistency of units and symbols and conformity to the standards recommended by competent international bodies.

The use in these Proceedings of particular designations of countries or territories does not imply any judgement by the publisher, the IAEA, as to the legal status of such countries or territories, of their authorities and institutions or of the delimitation of their boundaries.

The mention of specific companies or of their products or brand names does not imply any endorsement or recommendation on the part of the IAEA.

Authors are themselves responsible for obtaining the necessary permission to reproduce copyright material from other sources.

CONTENTS OF VOL. II

ADVANCED MATERIALS CONTROL CONCEPTS AND SYSTEMS
(Session VIII, Part 1)

THORIUM/URANIUM-233 FUEL CYCLES (Session VIII, Part 2)

SPENT FUEL REPROCESSING (Sessions IX and X)

Session VI and Session VII (Part 1)

DESTRUCTIVE AND NON-DESTRUCTIVE MEASUREMENT TECHNOLOGY *(cont.)*

Chairman (Session VI): V.M. GRYAZEV (USSR)
Chairman (Session VII): H.L. KRINNINGER (Federal Republic of Germany)

Rapporteur summary: *NDA equipment/techniques for safeguards applications*
Papers IAEA-SM-231/30, 69, 78, 79, 94, 100 were presented by
N. BARON as Rapporteur

Rapporteur summary: *Reference materials and calibration techniques*
Papers IAEA-SM-231/24, 68, 84, 92, 120 were presented by
M. CUYPERS as Rapporteur

METHODES ET REFERENCES PROPOSEES POUR LE CONTROLE DES MATIERES NUCLEAIRES

M. DESPRES, G. MALET, J. MOREL,
H. GOENVEC, J. LEGRAND
Laboratoire de métrologie
des rayonnements ionisants,
CEA, Centre d'études nucléaires de Saclay,
Gif-sur-Yvette, France

Abstract–Résumé

TECHNIQUES AND STANDARDS PROPOSED FOR SAFEGUARDING NUCLEAR MATERIALS.

Non-destructive analytical techniques are the ones best suited to in-situ safeguarding of nuclear materials. Gamma spectrometry occupies an important place among them and the following three applications of that technique are described: (a) Checking the ^{235}U enrichment of the uranium oxide (UO_2) pellets by means of a set of standards covering the range from 0.7 to 9.6%, especially designed for this purpose. It is shown that in the case of an infinitely thick sample the photon emission rate at 186 keV per unit surface, referred to the enrichment, is a constant equal to $(3.19 \pm 0.04)\ \gamma \cdot s^{-1} \cdot cm^{-2}$, which constitutes a direct measurement of the enrichment. (b) Determination of the isotopic composition of plutonium using an intrinsic 40-cm^3 germanium detector in the 120–450 keV energy range. A thorough examination of the sources of error shows that the accuracy attainable is 1–5%, except in the case of ^{240}Pu. Furthermore, there is a relationship between the $^{239}Pu/^{241}Pu$ ratio and the ^{242}Pu content which makes it possible to determine the latter, something which cannot be done by gamma spectrometry. (c) Determination of the uranium or plutonium concentration by differential absorptiometry using two radioactive sources spanning the discontinuity K of the element being measured. Exact knowledge of the mass absorption coefficients and the cell thickness provides a means of measuring the concentration directly. These three techniques have been tested in the laboratory as well as under actual operating conditions (measurements made through a stainless steel container).

METHODES ET REFERENCES PROPOSEES POUR LE CONTROLE DES MATIERES NUCLEAIRES.

Les méthodes non destructives d'analyse sont les mieux adaptées au contrôle in situ des matières nucléaires. Parmi celles-ci, la spectrométrie γ tient une place importante. Trois applications de cette technique sont présentées: a) Le contrôle de l'enrichissement en ^{235}U de pastilles d'oxyde d'uranium (UO_2) à l'aide d'un jeu de références dans la gamme de 0,7 à 9,6%, réalisé spécialement à cet effet. On montre que pour un échantillon d'épaisseur infinie, le taux d'émission photonique à 186 keV par unité de surface rapporté à l'enrichissement est une constante égale à $(3{,}19 \pm 0{,}04)\ \gamma \cdot s^{-1} \cdot cm^{-2}$, ce qui constitue une mesure directe de l'enrichissement. b) La détermination de la composition isotopique du plutonium à l'aide d'un détecteur au germanium intrinsèque de 40 cm^3 en utilisant le domaine d'énergie de 120 à 450 keV. Un examen approfondi des causes d'erreurs révèle que la précision que l'on peut atteindre est de 1 à 5%, sauf pour le ^{240}Pu. En outre il existe une relation entre le rapport $^{239}Pu/^{241}Pu$ et la teneur en ^{242}Pu, ce qui permet une détermination de ce dernier, non accessible par la spectrométrie γ. c) La détermination de la concentration en uranium ou en plutonium par une méthode d'absorptiométrie différentielle à l'aide de deux sources radioactives encadrant la discontinuité K de l'élément à doser. La connaissance précise des coefficients d'absorption massique et de l'épaisseur de la cellule permet une mesure directe de la concentration. Ces trois méthodes ont été testées non seulement en laboratoire mais aussi dans des conditions réelles d'exploitation (mesures faites à travers un conteneur en acier inoxydable).

I INTRODUCTION

Le contrôle des matières nucléaires de base implique en général la détermination de deux paramètres : la quantité et la qualité représentée par la composition isotopique de l'élément à contrôler.

Ces deux grandeurs peuvent être déterminées par des méthodes non destructives. La spectrométrie γ est désormais très développée car elle présente l'avantage d'être une technique simple et rapide. Nous nous sommes intéressés à la détermination :

- de la teneur en Uranium-235 sur des oxydes UO_2
- de la composition isotopique sur des oxydes de plutonium
- de la concentration de solutions nitriques d'U et de Pu par absorptiométrie différentielle.

II DETERMINATION DE LA TENEUR EN URANIUM 235 PAR SPECTROMETRIE γ

II.1 Généralités

Le contrôle du taux d'enrichissement en Uranium-235 sous forme solide est effectué selon la méthode dite de l'épaisseur infinie, fondée sur l'exploitation du pic d'absorption totale d'énergie 186 keV. Cette méthode comparative nécessite un jeu de références de pastilles d'oxyde uranium (UO_2) que nous avons réalisé dans la gamme 0,7 à 9,6 %.

II.2 Réalisation d'étalons UO_2

Des lots homogènes de pastilles UO_2 pour six teneurs différentes, à savoir 0,7 - 1,4 - 2,8 - 5,1 - 6,2 - 9,6 % ont été constitués; ce travail a été complété par un contrôle de pureté par spectrométrie γ, avec rejet éventuel de tout échantillon anormal.

L'étude de la dispersion des mesures, obtenues en comptage intégral au moyen d'une chaîne équipée d'un détecteur à scintillation INa (Tl) à puits, a permis de tester l'homogénéité de chaque lot.

Deux pastilles de chaque lot ont été prélevées en vue d'un contrôle de la teneur isotopique par spectrométrie de masse. Ces analyses en cours actuellement sont faites par deux laboratoires différents : Le Bureau Central des Mesures Nucléaires (B.C.M.N.) et le Laboratoire Central d'Analyse et de Contrôle (L.C.A.C./C.E.A.).

Le conditionnement des pastilles a été conçu pour éviter toute détérioration physique et faciliter leur identification. Ces pastilles sont placées dans un gainage de duralinox à fond mince de 0,5 mm d'épaisseur.

II.3 Mesure du taux d'émission

II.3.1 Principe

Le principe des mesures de l'enrichissement de l'uranium par spectrométrie γ est fondé sur la comparaison des différentes intensités du pic d'énergie 186 keV, sachant qu'elles sont proportionnelles à l'enrichissement. Les échantillons peuvent être de nature différente, cependant, la comparaison n'est crédible qu'en préservant les conditions d'expérience.

Dans ce qui suit, on appelle échantillon d'épaisseur infinie tout échantillon d'épaisseur telle qu'une augmentation sensible de cette dernière est sans effet sur l'émergence de la raie γ 186 keV. Pour un faisceau collimaté perpendiculaire à un échantillon homogène, le taux de comptage enregistré sous le pic d'absorption totale peut, d'après [1] et après simplification, s'écrire :

$$C_\infty = E.\alpha.\beta.\gamma.\delta$$

E : enrichissement en ^{235}U

$\alpha = I/\mu_n$: facteur caractéristique des constantes physiques
(I : taux d'émission γ à 186 keV pour 1 gramme d'^{235}U
μ_n : coefficient d'atténuation massique de l'uranium)

β : facteur de géométrie

γ : paramètre lié à la composition de l'échantillon

δ : atténuation d'écran.

II.3.2 Certification d'un taux de comptage en géométrie collimatée

L'objet de cette certification est de garantir le taux de comptage d'un étalon UO_2 pour une géométrie collimatée donnée. Les conditions d'expérience étant préalablement définies, tout utilisateur peut, par comparaison, déterminer l'enrichissement d'un échantillon quelconque.

a) Appareillage

La chaîne de mesure est équipée d'un détecteur à scintillation de type INa (Tl) de 25 mm de diamètre et de 25 mm d'épaisseur. Un collimateur conique en plomb est posé directement sur la fenêtre d'entrée ; ses dimensions, épaisseur : 10,00 ± 0,01 mm, diamètres 3,00 ±0,01 et 9,00 ± 0,01 mm, ont été choisies pour permettre des variations de la distance source-détecteur de l'ordre de 4 mm.

b) calcul de la surface du pic 186 keV

A cause de la résolution importante que présente tout détecteur à scintillation. Les pics d'énergies 144, 163, 194, 202 et 205 keV dus à l'^{235}U et à ses descendants interfèrent à la base du pic 186 keV et, par conséquent, ne permettent pas une estimation aisée du fond. Le calcul simple mais reproductible consiste à intégrer les contenus des canaux dans la zone 160 - 220 keV, et déduire le fond par une droite dont la pente et l'ordonnée sont définies à parti[r] des moyennes des contenus situés près des bornes d'intégration.

c) Résultats et commentaires

Six pastilles de chaque lot ont été mesurées en géométrie collimatée. L'essentiel des résultats est récapitulé dans le résumé ci-après ; les taux de comptage mentionnés se rapportent à l'ensemble des échantillons qui constitue le lot et sont affectés d'une erreur qui représente l'erreur statistique calculée à 1 σ. La figure 1 montre la variation du taux de comptage avec l'enrichissement, et les erreurs statistiques attendues pour des durées de mesure différentes.

Enrichissement nominal en % ^{235}U	Taux de comptage en géométrie collimatée ($c.s^{-1}$)	Taux de comptage rapporté à l'enrichissement ($c \cdot s^{-1}$)
0,7	1,171 ± 0,015	1,64 ± 0,02
1,4	2,21 ± 0,02	1,56 ± 0,02
2,8	4,10 ± 0,03	1,47 ± 0,01
5,1	7,48 ± 0,03	1,46 ± 0,01
6,2	9,10 ± 0,03	1,47 ± 0,01
9,6	14,08 ± 0,04	1,47 ± 0,01

Le taux de comptage rapporté à l'enrichissement, constant à partir d'une teneur de 2,8 %, est affecté d'un écart systématique d'autant plus grand que l'enrichissement devient faible, expliquant ainsi l'allure de la courbe en figure 1. Cet écart systématique est du à une modification de la courbure du fond situé sous le pic avec l'enrichissement comme le montre la figure 2. D'autres procédés de calcul de la surface du pic donnent des résultats semblables et montrent que le choix des bornes est fonction de l'enrichissement. La méthode de mesure d'un taux de comptage en géométrie collimatée étant comparative et faisant appel à l'utilisation d'étalons d'enrichissement connu, l'écart systématique précédent se trouve néanmoins compensé.

II.3.3 Mesure du taux d'émission γ surfacique

a) Principe

Cette mesure nécessite l'emploi de spectromètre γ à détecteur Ge-Li ou germanium intrinsèque, étalonné en efficacité absolue d'absorption totale à 186 keV pour un échantillon ponctuel situé à une distance donnée (> 5 cm) du détecteur. La pastille de référence est placée à la distance d'étalonnage ; un diaphragme de plomb, d'épaisseur suffisante pour atténuer totalement le rayonnement de 186 keV, est ajusté au diamètre de la pastille afin de s'affranchir de la détection des photons émis par la face cylindrique.

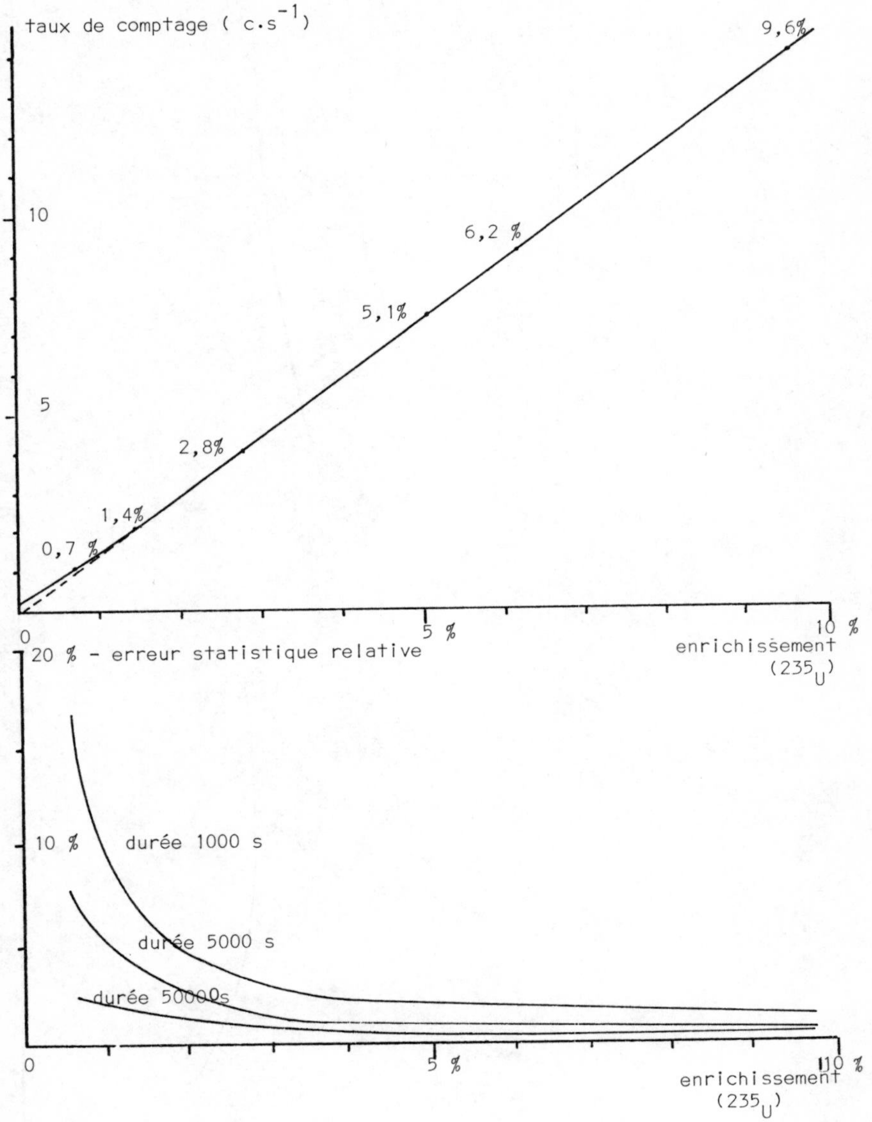

FIG.1. Taux de comptage et erreur statistique en fonction de l'enrichissement (géométrie collimatée – détecteur INa(Tl) 25 mm × 25 mm).

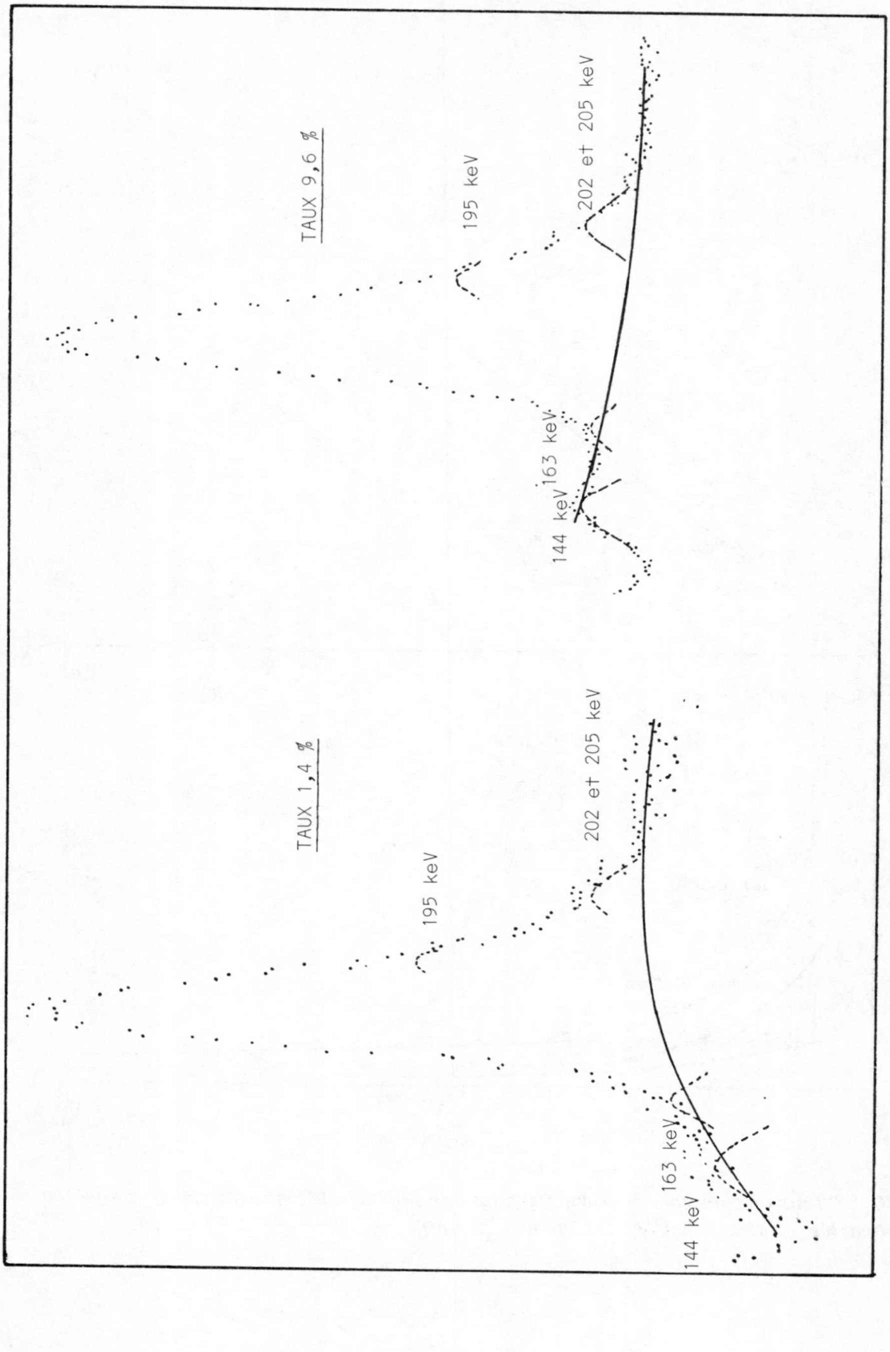

FIG.2. Evolution de la région du pic 186 keV avec l'enrichissement.

En admettant, dans un premier temps, que l'efficacité est constante pour une variation limitée de la distance source-détecteur, le taux de comptage enregistré sous le pic d'énergie 186 keV est :

$$C_{186} = \varepsilon ff_{186} . E . \frac{I}{\mu_n} . A . \gamma . \delta$$

où εff_{186} est l'efficacité d'absorption totale à l'énergie 186 keV.

A est la surface visible de l'échantillon et les autres paramètres sont ceux définis au paragraphe II.3.1.

Le taux d'émission photonique, rapporté à la surface de l'échantillon, s'écrit alors :

$$N_{186} = \frac{C_{186}}{\varepsilon ff_{186} . A . \delta} = \frac{E . I . \gamma}{\mu_n}$$

b) Causes d'erreurs

Sur l'expression précédente, il convient de faire quelques remarques. L'échantillon mesuré est différent d'un dépôt ponctuel, si bien que l'efficacité doit être corrigée des pertes, imputables à l'étalement de l'échantillon Δ1, à son épaisseur Δ 2 et à ses défauts de planéité Δ 3. Le diaghragme a pour inconvénient d'absorber partiellement certains photons émis en périphérie Δ4 ; d'autre part, comme toute mesure en spectrométrie γ , le taux de comptage à 186 keV est affecté par les pertes dues aux effets d'empilements Δ 5 et par la présence de ^{226}Ra et des descendants dans le bruit de fond Δ 6. Par conséquent, le taux d'émission photonique surfacique s'écrit donc :

$$N_{186} = \frac{E . I . \gamma}{\mu_n} = \frac{C_{186}}{\varepsilon ff_{186} . A . \delta .} \times 1 / \prod_{i=1}^{6} \Delta_i$$

c) Aspect expérimental

Les chaînes de spectrométrie γ 1 et 3 ont été utilisées (voir tableau I). La réponse en efficacité à 186 keV est entachée d'une incertitude de 1,6 % pour le détecteur intrinsèque et de 0,8 % pour le détecteur Ge-Li ; les distances source-détecteur déterminées par la relation de l'angle solide, corrigée des variations de géométrie sont respectivement de 5,85 cm et 11,9 cm.

L'exploitation des pics présents dans le spectre est effectuée selon deux procédés. Dans le premier cas, la surface est déduite de l'intégration d'une courbe de Gauss, superposée à une parabole pour tenir compte du fond et ajustée aux données selon le critère des moindres carrés. Pour le deuxième procédé, la surface est obtenue de la somme des contenus des canaux après déduction linéaire du fond.

d) Résultats

Les résultats obtenus sur une pastille de chaque lot figurent dans le tableau II. Ils représentent les taux d'émission photonique

TABLEAU I. CARACTERISTIQUES DES CHAINES DE SPECTROMETRIE γ UTILISEES

	Détecteur 1	Détecteur 2	Détecteur 3
Type	planaire intrinsèque Ge	coaxial intrinsèque Ge	Ge(Li) coaxial
Dimensions	3cm2 x 1 cm	41 cm3	83 cm3
Tension de polarisation	- 2000 volts	+ 2500 volts	+ 4000 volts
Préamplificateur	opto pulsé	normal refroidi	normal refroidi
Amplificateur	Canberra 1413	Canberra 1413	Elscint CAV-N 3
Codeur	C44B (100 MHz)	C 44 B (100 MHz)	CT 103 (200 MHz)
Analyseur	Didac 4000 canaux	Didac 4000 canaux	Plurimat 20 Intertechnique
Sortie des résultats	bande perforée	bande perforée	-
Résolution en énergie	200 eV à 5,9 keV 520 eV à 122 keV	0,80 keV à 122 keV 1,8 keV à 1332 keV	1,0 keV à 122 keV 2,0 KeV à 1332 keV
Reproduction des conditions géométriques	comparateur	-	banc optique microscopique

TABLEAU II. VARIATION DU RAPPORT N_{186}/E EN FONCTION DE L'ENRICHISSEMENT NOMINAL E

Taux d'enrichissement nominal %	Diamètre cm	Taux d'émission photonique surfacique N_{186} $\gamma.s^{-1}.cm^{-2}$			Rapport (N_{186}/E)
		Détecteur planaire	Détecteur coaxial	Moyenne	
0,7	1,300 ± 0,001	222 ± 6	229 ± 4	227 ± 3	3,19 ± 0,05 10^4
1,4	1,300 ± 0,001	453 ± 12	445 ± 7	447 ± 6	3,16 ± 0,05 10^4
2,8	0,803 ± 0,001	893 ± 26	884 ± 15	886 ± 13	3,18 ± 0,05 10^4
5,1	0,805 ± 0,001	1618 ± 44	1625 ± 28	1623 ± 24	3,17 ± 0,05 10^4
6,2	0,805 ± 0,001	1990 ± 50	1991 ± 32	1991 ± 27	3,21 ± 0,05 10^4
9,6	0,920 ± 0,001	3045 ± 80	3057 ± 46	3054 ± 40	3,20 ± 0,05 10^4

à 186 keV ramenés à l'unité de surface ($\gamma.s^{-1}.cm^{-2}$). Ces résultats tiennent compte des corrections décrites antérieurement et sont entachés d'une erreur totale qui représente la somme des erreurs aléatoires, calculées pour un niveau de confiance de 68 % et des erreurs systématiques additionnées linéairement. Le taux d'émission photonique surfacique adopté pour chaque teneur résulte de la moyenne pondérée des valeurs obtenues avec l'une et l'autre instrumentation. Il est intéressant de rapporter le taux d'émission photonique surfacique (N_{186}) à l'enrichissement (E) (Tableau II). En effet, ce paramètre est constant et caractérise la détection des photons γ d'énergie 186 keV pour tout échantillon d'uranium d'épaisseur infinie dans la gamme d'enrichissement étudiée. En dissociant les erreurs dues à l'étalonnage en efficacité, au traitement du spectre et à l'enrichissement, des autres erreurs considérées alors indépendantes, la meilleure valeur du taux d'émission surfacique rapporté à l'enrichissement est :

$$N_{186}/E = (3,19 \pm 0,04).10^{4}\ \gamma.s^{-1}.cm^{-2}$$

Cette dernière grandeur est égale au rapport $I.\gamma/\mu_n$ (cf II.3.1) ; en adoptant $\mu_n = 1,53\ cm^2.g^{-1}$ [2], affecté d'une erreur de 5 %, comme période T = 7,038 ± 0,007).10^8ans [3] correspondant à une activité de (7,99 ± 0,02).10^4 Bq pour 1 gramme d'uranium 235 et en tenant compte de la contribution de la raie γ d'énergie 182,4 keV, d'intensité relative de 0,65 %, il est possible de calculer l'intensité absolue de la raie γ d'énergie 185,7 keV. Le résultat trouvé : I_a = 0,62 ± 0,04 entaché d'une erreur importante imputable à la méconnaissance du coefficient μ_n, recouvre néanmoins la valeur 0,576 ± 0,023 publiée récemment [4].

II.4 Conclusion

De cette étude, il ressort que deux procédés peuvent être employés pour le contrôle en épaisseur infinie de la teneur en uranium 235. Le premier, fondé sur la comparaison d'échantillons au moyen d'un appareillage sommaire et d'une collimation adéquate, est de mise en oeuvre relativement facile ; cependant, il nécessite un étalonnage préalable de l'instrumentation avec des références de taux d'enrichissement connu, alors que le second, délicat puisqu'il exige de maîtriser tous les phénomènes qui régissent la réponse d'une instrumentation plus complexe, permet de s'en affranchir.

L'utilisation de détecteur au germanium intrinsèque d'efficacité appréciable et d'excellent pouvoir de résolution donne une troisième détermination de l'enrichissement en uranium 235, non décrite ici, qui implique l'analyse détaillée du spectre γ avec étalonnage interne, d'une façon analogue aux déterminations de composition isotopique du plutonium.

III DETERMINATION PAR SPECTROMETRIE γ DE LA COMPOSITION ISOTOPIQUE DU PLUTONIUM

La détermination par spectrométrie γ de la composition isotopique du plutonium a déjà été exposée par de nombreux auteurs [5-6-7-8]. Il n'existe malheureusement pas de méthodes universelles. Le choix des raies servant à l'analyse dépend de la composition isotopique, de la matrice et de la concentration du plutonium. Le mode de dépouillement et de calcul diffère selon les auteurs :

- déconvolution du spectre dans la région 90 - 110 keV

- emploi de jeu d'équations prenant en compte l'ensemble du spectre découpé en régions d'intérêt

- emploi de pics isolés représentatifs de chaque isotope.

Enfin, le Pu-242 n'ayant pas de raies γ caractéristiques intenses ne peut être déterminé de cette façon et doit donc être obtenu différemment.

III.1 Application aux oxydes de plutonium

III.1.1 Conditions expérimentales

Des séries d'expérience ont été faites sur des oxydes obtenus après retraitement de combustibles de réacteurs type PWR ou graphites gaz avec les chaînes 1 et 2 (tableau I). A titre d'exemple deux spectres sont représentés sur les figures 3 et 4. Les meilleures conditions dans notre cas sont :

- emploi du détecteur 2 (volume 41cm3)
- écran de cadmium
- temps d'analyse 1 heure pour 100 mg.

La figure 4 montre que le spectre obtenu à travers un container en inox est parfaitement exploitable, ce qui est important, car l'analyse directe du plutonium est possible à travers une canalisation (à partir d'une certaine concentration) ou dans son conditionnement initial.

III.1.2 Exploitation des spectres

Pour chaque isotope on choisit un ou plusieurs pics d'énergie connue dont on détermine la surface. De l'ensemble des rapports de deux isotopes P et P' on déduira la composition isotopique.

Ces rapports massiques peuvent se mettre sous la forme :

$$\frac{P}{P'} = \frac{S}{S'} \times \frac{T1/2}{T'1/2} \times \frac{\varepsilon'ERT}{\varepsilon ERT} \times \frac{I'\gamma}{I\gamma} \times \frac{M}{M'}$$

dans laquelle :

S et S' sont les surfaces des pics d'absorption totale utilisés

Iγ et I'γ les intensités γ absolues de ces mêmes pics

T1/2 et T'1/2 les valeurs des périodes correspondant aux deux isotopes

εERT et ε'ERT les efficacités relatives totales de détection

M et M' étant les masses atomiques des deux isotopes.

III.1.2.1 Choix des raies γ utilisées - Détermination de la réponse en efficacité

Après détermination des abscisses et des surfaces des pics présents dans le spectre, on sélectionne les raies γ du Pu-239 ou

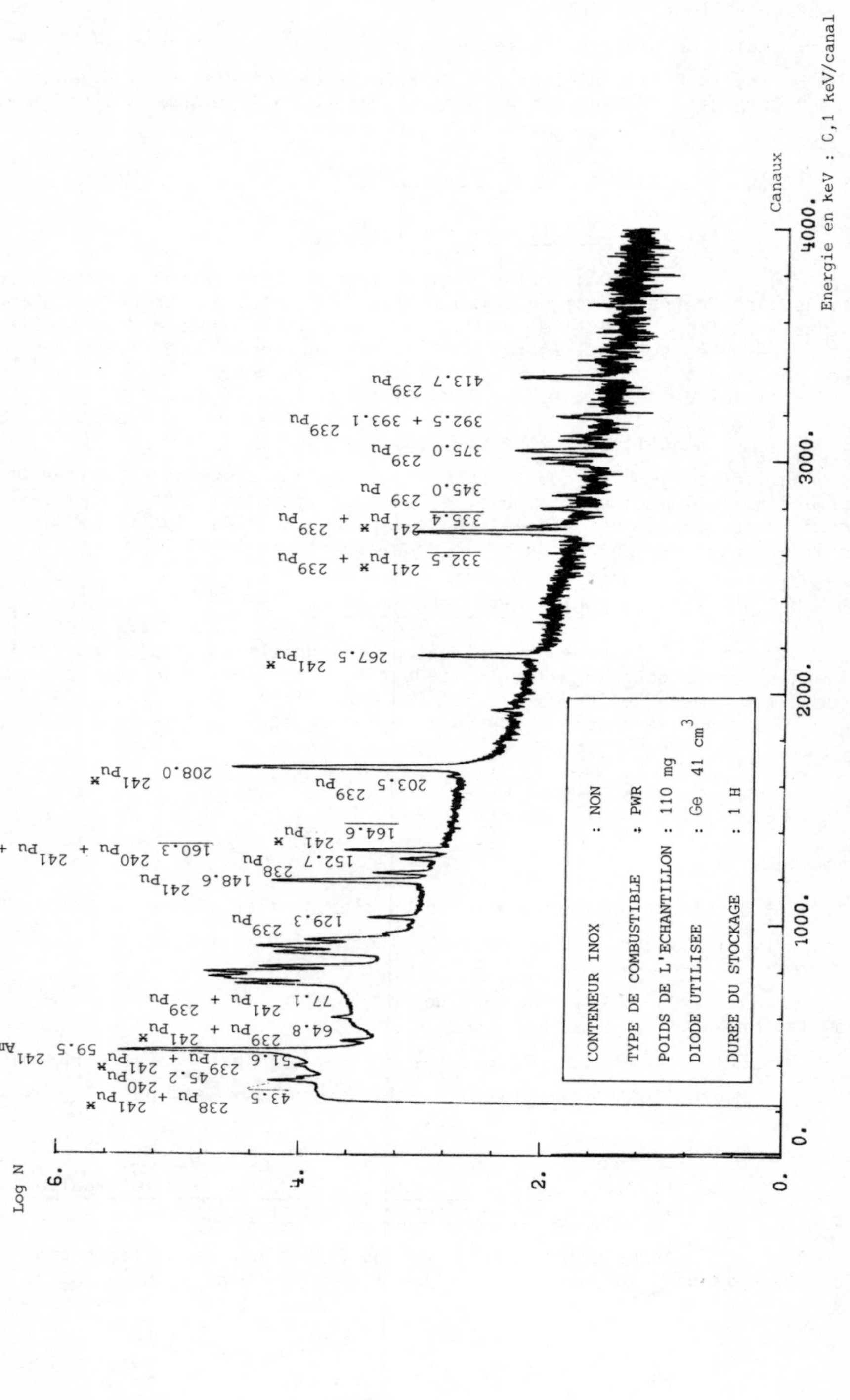

FIG.3. Spectre γ de PuO_2 sans conteneur inox.

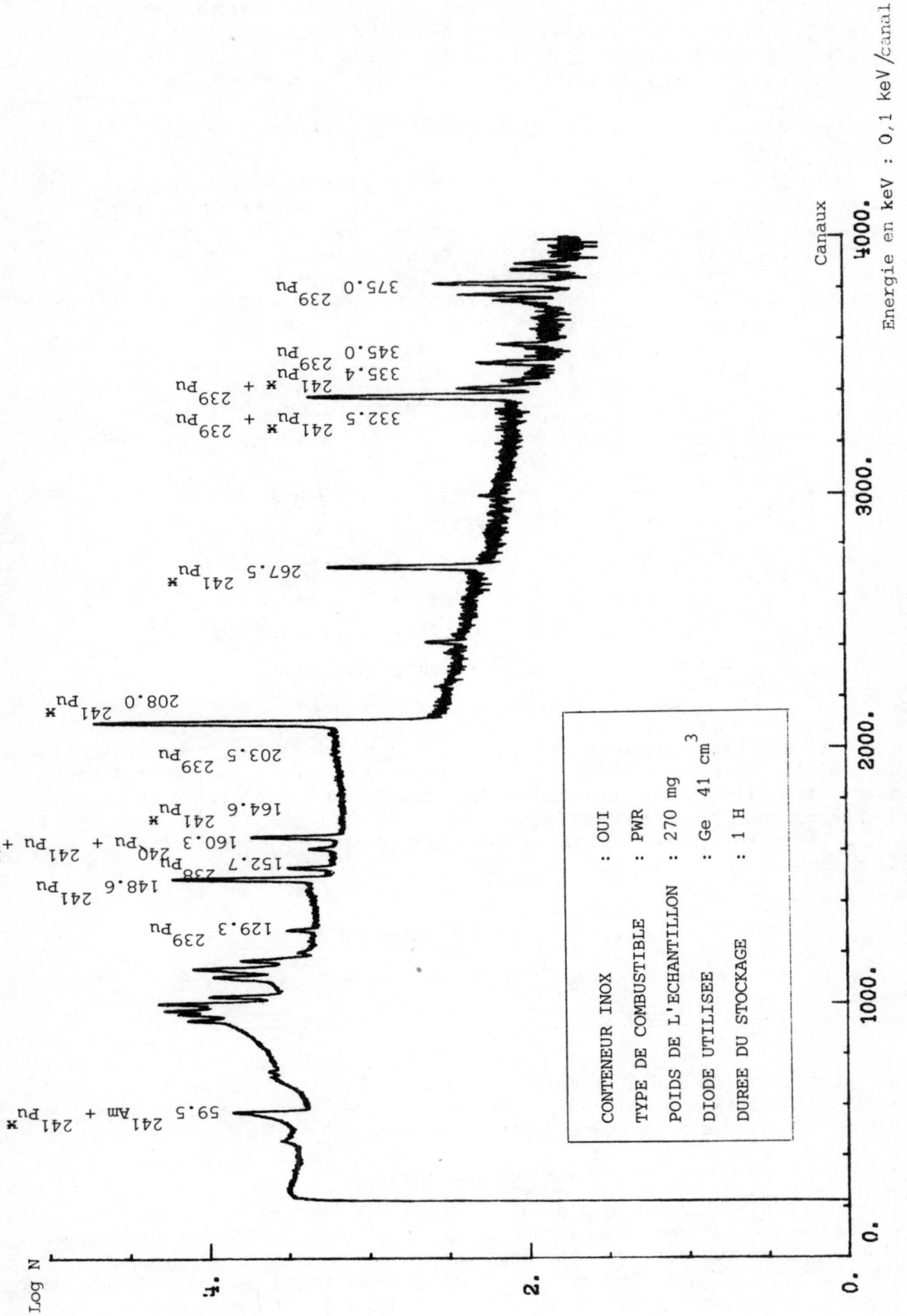

FIG.4. Spectre γ de PuO_2 avec conteneur inox.

du Pu-241 les plus intenses présentant un minimum d'interférences avec d'autres nuclides pour la détermination de la réponse en efficacité totale relative. En particulier pour le Pu-239, ces pics sont ceux d'énergies : 39, 52, 78, 129, 144, 203, 255, 341 + 345, 375, 380 + 383, 392 + 393, 411 + 414, 423 keV

et pour le Pu-241, en équilibre avec l'U-237, ceux d'énergies : 65, 149, 165, 208, 267.

Il convient de remarquer que certains pics de faible énergie (40 à 80 keV) ne sont pas toujours employés en raison de leur faible proportion, de certaines interférences notables, ou de l'absorption importante du conteneur. L'efficacité d'absorption totale relative, ajustée dans un système de coordonnées logarithmiques, tient compte de l'efficacité du détecteur pour nos conditions géométriques, de l'absorption des différents écrans et de l'autoabsorption de la source.

III.1.2.2 Constantes utilisées

Les périodes utilisées [3] recommandées par l'AIEA sont :

Pu-238	-	87,74 ±	0,09 ans
Pu-239	-	24110 ±	30 ans
Pu-240	-	6553 ±	8 ans
Pu-241	-	14,7 ±	0,4 ans
Pu-242	-	(3,76 ±	$0,02).10^5$ ans
Am-241	-	432,6 ±	0,6 ans

- Intensités absolues des raies γ : il n'existe pas, comme pour les périodes, de valeurs connues avec suffisamment de précision, en raison du nombre très limité d'études faites sur les isotopes du plutonium et du désaccord que présentent certains auteurs.

Les valeurs que nous avons adoptées sont en majeure partie extraites du rapport UCRL 52 139 [9].

III.1.2.3 Analyse des causes d'erreurs

Les principales causes d'erreurs sont :

. La méconnaissance des constantes fondamentales

Les valeurs les plus récentes présentent parfois un écart qui peut aller de 1 à 10 % suivant les nuclides. Ceci met un accent tout particulier sur le besoin en précision au niveau des intensités γ absolues. Une telle étude est en cours dans notre laboratoire.

. Le traitement des spectres

Dans le cas de pics isolés sur un fond monotone, la surface est entachée d'une erreur de quelques pour mille. Par contre, il n'en est plus toujours de même pour les doublets et nous pensons que même avec un traitement mathématique particulièrement élaboré, la déconvolution de multiplets donne une erreur importante (quelques pour cent).

TABLEAU III. ESTIMATION DES PRINCIPALES CAUSES D'ERREURS

	I_γ	T 1/2	Surface	Etalonnage erreur de modèle	Interférences	Erreur statistique	Erreur totale
Pu-238	3 %	0,1 %	≤ 0,5%	0,5 %	-	≤ 1 %	5 %
Pu-239	2 %	0,1 %	≤ 0,5%	0,5 %	-	≤ 1 %	4 %
Pu-240	2 %	0,1 %	< 0,5 %	0,5 %	8 %	≤ 5 %	16 %
Pu-241	2 %	2,7 %	≤ 0,5 %	0,5 %	-	< 0,5 %	6 %

. La précision de l'étalonnage

Le modèle mathématique choisi pour décrire la réponse en efficacité intervient peu sur la précision de l'étalonnage : les deux principales causes d'erreurs étant dans le cas présent, les constantes et les erreurs aléatoires sur la surface.

. La détermination des interférences

Dans le cas des Pu-238, 239, 241, les corrections d'interférences entraînent des erreurs négligeables. Par contre, dans le cas où l'on utilise pour la détermination du Pu-240 le pic à 160,3 keV, celui-ci contient du Pu-239 (161,5 keV) et du Pu-241 (160,0 keV) de façon significative (plusieurs pour cent).

. Les erreurs statistiques de comptage

Les erreurs statistiques de comptage, pour une durée d'analyse de 1 heure et des taux d'impulsions modérés (< 2000 c/s) n'entraînant pas de déformations notables du spectre par les effets d'empilement, sont relativement importantes (quelques pour cent) dans les régions 40 - 80 keV et 120 - 400 keV.

Le tableau III fait apparaître l'importance des erreurs précédemment décrites. Leur somme arithmétique représente l'erreur totale à laquelle on peut s'attendre pour une durée d'analyse de l'ordre de l'heure.

III.1.3. Le problème du Pu-242

Si dans les combustibles de la filière graphite-gaz, le Pu-242 peut être négligé, il n'en est pas de même dans les combustibles PWR. Malheureusement, celui-ci ne peut être déterminé à partir du rayonnement γ. Pour estimer sa proportion, plusieurs solutions ont été proposées dont la meilleure nous semble celle de Dragnev [5] qui suggère, après étalonnage préalable, d'employer le rapport Pu-239/Pu-240 pour déterminer le Pu-242.

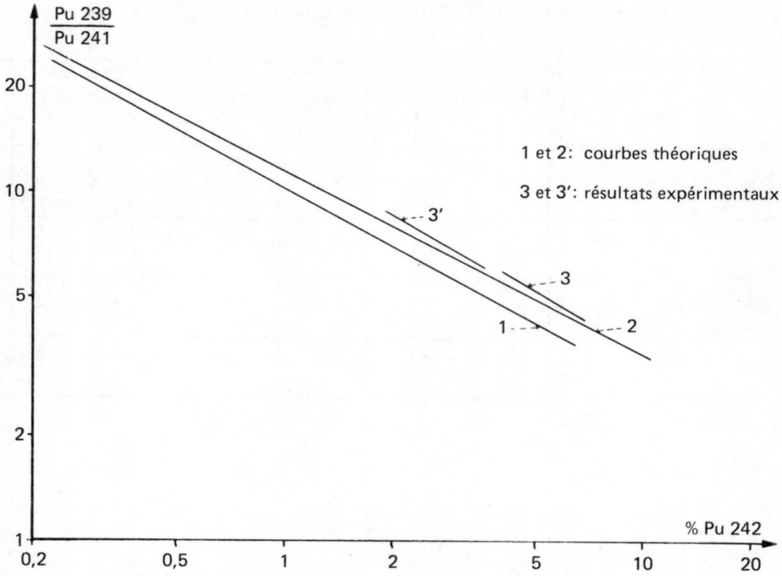

FIG. 5. Variation du rapport $^{239}Pu/^{241}Pu$ en fonction du % du ^{242}Pu.

L'avantage d'une telle détermination réside dans le fait que pour un réacteur donné et dans un certain domaine de précision (5 à 10 %), ainsi que nous l'avons vérifié à partir de documents internes et de valeurs expérimentales, les résultats obtenus semblent être indépendants du taux d'irradiation, de l'enrichissement initial (en première approximation) et pour le rapport Pu-239/Pu-240 du temps de refroidissement. Par contre, l'inconvénient de cette méthode est que ce rapport est celui qui risque d'être entaché de la plus grande erreur. Aussi, nous en avons essayé d'autres et en particulier le rapport Pu-239/Pu-241. Les résultats peuvent de la même façon être mis sous la forme d'une relation log-log (Fig. 5). Les valeurs expérimentales confirment l'existence d'une telle relation. Il faut cependant remarquer que dans ce cas une correction doit éventuellement être faite (refroidissement).

La méthode est donc relativement simple. Il suffit de construire la courbe correspondant au combustible à traiter à l'aide de quelques points obtenus par spectrométrie de masse. De celle-ci, tracée en coordonnée log-log, on déduira la teneur en Pu-242. Il est à remarquer que ceci n'est pas une gêne puisque de telles déterminations sont toujours nécessaires pour le calcul du bilan à l'entrée et à la sortie des usines. Enfin une étude systématique de ces corrélations est en cours de façon à essayer d'affiner les résultats obtenus pour le Pu-242.

III.1.4 Résultats obtenus

Nos essais ont porté principalement sur des oxydes en provenance de réacteurs PWR et graphite-gaz. Les résultats sont les suivants pour les PWR :

PWR	Pu-238/Pu-241	Pu-239/Pu-241	Pu-240/Pu-241
Spectrométrie de masse	0,122	4,999	2,175
Spectrométrie γ moyenne sur 3 résultats	0,117	4,905	2,209

Du rapport Pu-239/Pu-241 on tire la valeur du Pu-242 soit 5,1 % (si on emploie le rapport Pu-239/Pu-240 on trouve 5, 8 % soit un écart de 10 %).

La composition isotopique est alors

	Spectrométrie masse	Spectrométrie γ	Ecart
Pu-238	1,396	1,35	- 3,3 %
Pu-239	57,058	56,57	- 0,85 %
Pu-240	24,819	25,47	+ 2,5 %
Pu-241	11,413	11,53	+ 1 %
Pu-242	5,312	5,08	- 4,4 %

Les résultats obtenus peuvent être considérés comme satisfaisants compte tenu des précisions attendues énoncées précédemment.

III.2 Cas de solutions nitriques de Plutonium

Des essais préliminaires ont été faits sur des solutions nitriques de plutonium de 10 à 300g.l $^{-1}$.Ces solutions provenaient de dissolution de PuO_2 correspondant à celui des essais précédents ; les résultats sont identiques à ceux obtenus précédemment. Il en a été de même sur des solutions contenant outre le plutonium, de l'uranium et des produits de fission de l'ordre de quelques mCi.l^{-1}.

III.3 Conclusion

De ces séries d'expériences, nous pouvons tirer les enseignements suivants :

- Malgré le seul pic complexe (interférence avec Pu-239, Pu-241, Am-241) utilisable pour la mesure du Pu-240 à 160,3 keV, la région 120-450 keV nous semble être la plus appropriée parce que la plus simple d'exploitation pour la détermination par spectrométrie γ de la composition isotopique du plutonium. Dans les faibles énergies (inférieuresà 90 keV), il est très difficile d'obtenir une courbe de réponse en efficacité relative précise et la forte absorption du rayonnement rend impossible la mesure à travers un conteneur inox. La région située aux alentours de 600 keV implique des temps de mesure trop longs (dans nos conditions d'expérience, le pic à 642 keV du Pu-240 est difficilement décelable au bout de 15 heures de mesure). Quant à la zone 90-110 keV, elle nécessite une mise en oeuvre de moyens trop importants pour une précision qui reste à démontrer dans le cas d'une installation in situ.

- Les intensités γ absolues ont fait l'objet de trop peu de déterminations pour être connues à mieux que 1 %.

- Enfin, si ce que nous venons d'exposer s'applique particulièrement bien au contrôle des matières nucléaires, l'emploi de cette méthode peut s'appliquer aussi au contrôle des extractions par comptage neutronique lors du retraitement du combustible, la connaissance de la teneur en Pu-240 étant indispensable pour accéder à la concentration en plutonium.

IV DETERMINATION DE LA CONCENTRATION EN URANIUM ET EN PLUTONIUM

IV.1 Introduction

Diverses méthodes physiques d'analyse de la concentration en plutonium et en uranium sont déjà couramment employées. Citons pour mémoire la fluorescence X, la spectrométrie γ, l'absorptiométrie γ, l'interrogation neutronique active, la densitométrie ...

Ces méthodes demandent l'emploi d'une courbe d'étalonnage faite à l'aide d'étalons mesurés par d'autres méthodes et placés le plus souvent dans des conditions de géométrie et d'états physique et chimique identiques aux échantillons. De plus, certaines ne sont pas sélectives (dosage de U + Pu), ou trop sélectives (dosage du Pu-239 seul).

En se référant aux travaux de T.R. Canada et al. [10], nous envisageons ici l'application de la méthode d'absorptiométrie différentielle à la mesure de la concentration en plutonium ou en uranium d'une solution ou d'un solide homogène sans étalonnage systématique. En contrôle continu, une telle procédure est prohibitive.

En effet, dans l'hypothèse d'une géométrie collimatée et dans le cas où l'effet d'absorption par la matrice est négligeable, la concentration d'une solution de plutonium ou d'uranium peut être déterminée au moyen de la relation :

$$\rho_n = \frac{\mathrm{Log}\ \frac{T1}{T2}}{x.\ \Delta\ \mu n}$$

où T_1 et T_2 sont les transmissions pour des énergies E_1 et E_2 situées de part et d'autre et proches de l'énergie de liaison K, $\Delta\mu n$, la différence des coefficients d'absorption massique en $cm2.g^{-1}$ aux énergies E_1 et E_2 de U ou Pu, x l'épaisseur de l'échantillon en cm.

IV.1.1 Choix des sources pour le calcul du $\Delta\mu_n$

La discontinuité K de l'uranium est située à 115,6 keV, celle du plutonium à 121,8 keV.

Trois radionucléides présentent des transitions d'énergies proches de 115,6 keV ou de 121,8 keV. Il s'agit de l'Yb-169 avec une transition γ intense de 109,8 keV et du Se-75 à 121,1 keV, encadrant la discontinuité de l'uranium ; le Se-75 et le Co-57 à 122,1 keV encadrant celle du plutonium. En l'absence de ce dernier, il est possible d'utiliser pour l'uranium le cobalt à la place du sélénium en raison de la transition γ très intense qu'il présente sans interférences.

La différence des coefficients d'absorption massique pour l'uranium est $\Delta\mu_U$ = 2,81 cm2 g^{-1}, celle du plutonium est $\Delta\mu_{Pu}$ = 3,21 $cm^2.g^{-1}$ [2].

IV.1.2 Influence de la matrice

La correction de matrice peut être calculée. Dans le cas de solutions nitriques et pour des acidités de 0,1 N - 2 N et 5 N, les courbes de la figure 6 présentent la correction de matrice en fonction de la concentration de l'élément considéré et montrent que dans certains cas celle-ci est négligeable.

IV.1.3 Choix de l'épaisseur de l'échantillon

Pour des concentrations inférieures à 100 $g.l^{-1}$, il faut utiliser des échantillons relativement épais : quelques centimètres afin d'obtenir un rapport de transmissions significatif ; par contre, pour des concentrations plus importantes, des épaisseurs de l'ordre du centimètre sont suffisantes pour obtenir des mesures correctes, c'est-à-dire des variations du taux de comptage de 100 à 2.000 $c.s^{-1}$.

IV.1.4 Précision

L'erreur systématique relative, associée à la concentration s'écrit :

$$\left|\frac{\Delta\rho_n}{\rho_n}\right| = \left|\frac{\Delta T_1/T_2}{T_1/T_2} \cdot \frac{1}{\text{Log } T_1/T_2}\right| + \left|\frac{\Delta x}{x}\right| + \left|\frac{\Delta(\Delta\mu)}{\Delta\mu}\right|$$

Cette erreur totale dépend en partie de l'erreur sur le coefficient d'absorption massique et de l'incertitude sur l'épaisseur de l'échantillon.

- Influence de l'épaisseur : Il faut utiliser un conteneur à faces parallèles, d'épaisseur interne bien connue (< 0,001 cm) rempli totalement de la solution à doser.

- La valeur des coefficients d'absorption massique est sujette à deux causes principales d'erreurs : le choix des valeurs tabulées et la fonction choisie pour l'interpolation ; pour réduire la première, il est préférable de vérifier dans la région d'intérêt quelques valeurs issues de tables récentes, et pour la seconde de faire un ajustement par la méthode des moindres carrés.

- Le rapport des transmissions est affecté d'une part, d'erreurs systématiques imputables au calcul de la surface des pics d'absorption totale et aux effets d'empilements, et d'autre part aux erreurs statistiques de comptage. A priori, d'après l'expression précédente, la mesure de la concentration est d'autant plus précise que le rapport des transmissions est élevé. Cependant, cette dernière condition oblige de travailler avec des fortes activités entraînant des déformations notables du spectre tout en dégradant la statistique de comptage en présence de l'échantillon. Le meilleur compromis semble être un rapport des transmissions compris entre 4 et 10 mais ne peut malheureusement être toujours obtenu.

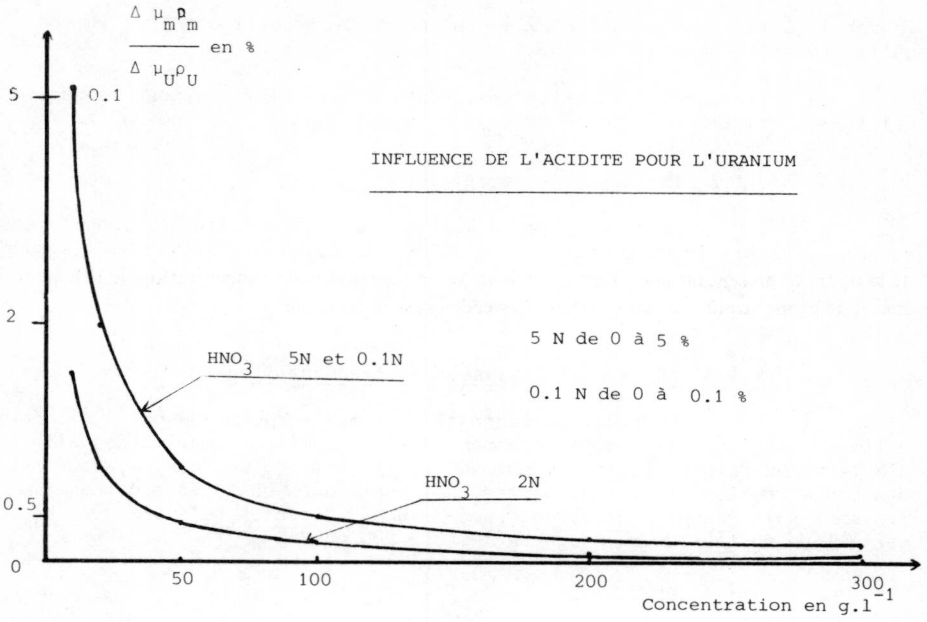

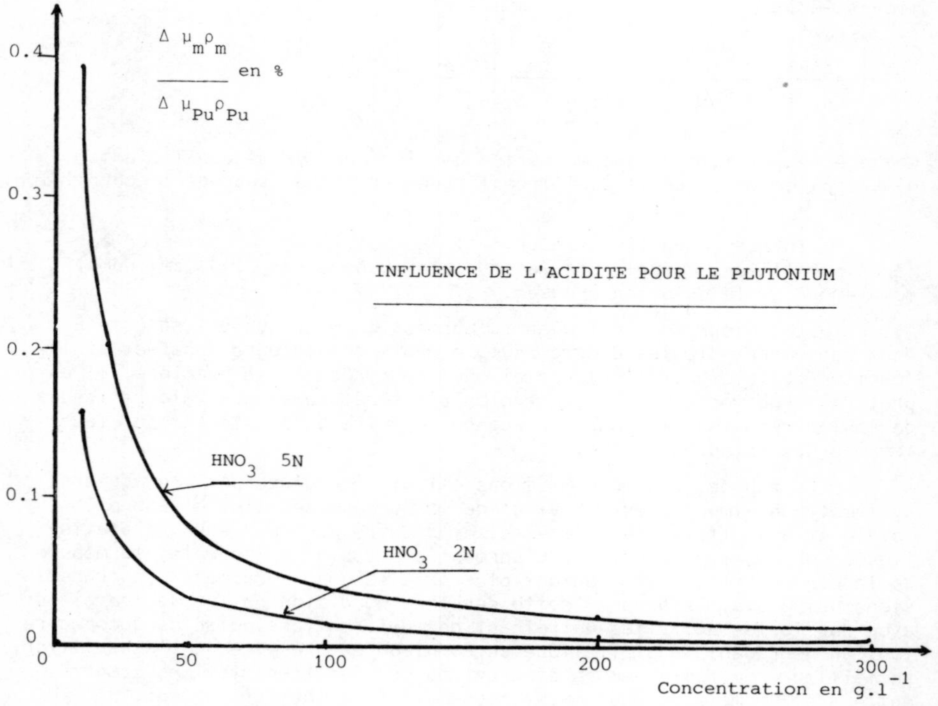

FIG.6. Influence de l'acidité pour l'uranium et le plutonium.

IV.2 Résultats expérimentaux

IV.2.1 Appareillage

Les expériences ont été effectuées au moyen de la chaîne n° 2 **(tableau I)**. La nécessité de travailler en faisceau parallèle conduit à utiliser deux collimateurs en plomb, de même ouverture (Ø 10 cm). Leur épaisseur, de 2 cm est choisie pour atténuer les rayonnements parasites d'énergie supérieure à 122 keV que l'on détecte notamment avec le Se-75 et l'Yb-169. La disposition des collimateurs, de la source d'émission γ et de l'échantillon est adaptée d'une part, pour réduire le rayonnement γ propre à l'échantillon, et d'autre part pour rendre le faisceau parallèle. Un collimateur intermédiaire peut être utilisé pour obtenir des taux de comptage comparables relevés avec ou sans échantillon. Les sources d'émission γ, de l'ordre du millicurie, sont placées dans un "château de plomb" d'épaisseur 1,5 cm, permettant de protéger l'expérimentateur, et de même ouverture que les diaphragmes précédents. Un écran de cadmium, de l'ordre du millimètre s'avère parfois nécessaire afin d'atténuer totalement les émissions γ ou X de faible énergie (<100 keV)

IV.2.2 Mesure de l'uranium

Les caractéristiques des échantillons et les résultats de l'analyse sont résumés ci-après. Les épaisseurs des échantillons sont différentes afin d'obtenir un rapport des transmissions favorable. Les mesures ont une durée de 600 s ; le taux d'impulsion est inférieur à 2000 $c.s^{-1}$.

échantillon dosé par chimie g l^{-1}	acidité N	épaisseur cm	$\frac{T1}{T2}$	concentration calculée g l^{-1}	écart/chimie %
354	1,7	2,5 ± 0,1	8,91	311	- 12,4
291	1,9	3,0 ± 0,1	8,19	249	- 14,8
234	2,13	3,5 ± 0,1	7,73	208	- 11,5
175	2,35	4,0 ± 0,1	5,4	152	- 12,0

Un écart systématique par défaut par rapport à la concentration réelle de l'uranium est observé et peut être imputable soit à la détermination de l'épaisseur de l'échantillon, soit au $\Delta\mu_h$ (les tables donnent une incertitude sur les μ_n de ± 10 %).

IV.2.3 Mesure du plutonium

IV.2.3.1 Cas de l'oxyde de plutonium

L'expérience est faite sur un oxyde de plutonium mis dans deux boîtiers d'acier inox + nickel de 6,4 mm d'épaisseur. L'épaisseur de l'oxyde est déterminée à partir de la masse connue, néanmoins, un contrôle fait apparaître des écarts d'absorption de l'ordre de 5 %. Les conditions expérimentales sont analogues à celles de l'uranium.

Masse superficielle calculée : 0,70 g cm^{-2}

Masse superficielle mesurée : 0,66 ± 0,07 $g.cm^{-2}$

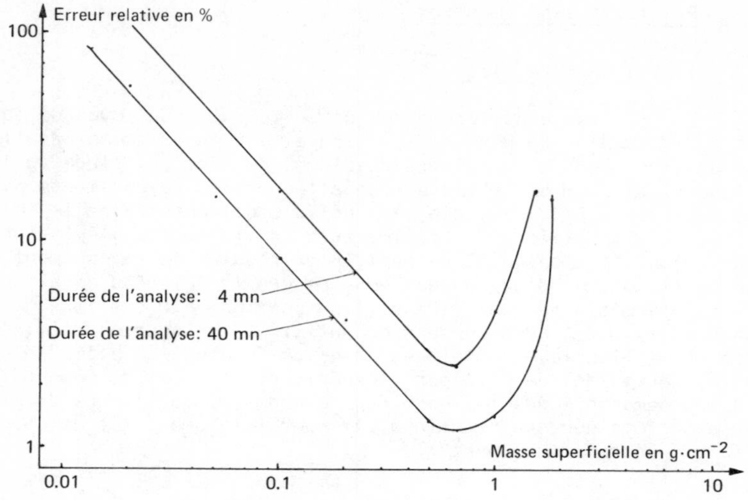

FIG. 7. Mesure de la concentration du plutonium. Erreur relative en fonction de la masse superficielle.

L'incertitude de mesure peut être due dans ce cas, en plus des erreurs précitées, aux défauts d'homogénéité en épaisseur de l'échantillon.

IV.2.3. 2 Cas de solutions de plutonium

Différentes solutions nitriques de plutonium sont analysées, le résumé suivant récapitule leurs caractéristiques et les résultats.

échantillon dosé par chimie $g.l^{-1}$	acidité N	épaisseur cm	$\frac{T1}{T2}$	concentration calculée $g\ l^{-1}$	écart/chimie %
223,2	5	1,9 ±0,1	3,55	210,5	- 5,7
111,6	5	"	1,89	104,5	- 6,4
44,64	5	"	1,28	42,3	- 5,2
22,32	2	"	1,15	22,1	- 1,0

De la même façon que précédemment, un écart systématique par défaut est observé. Elle est imputable aux mêmes causes. Il est à remarquer que pour cette détermination, nous ne pouvions pas être dans les meilleures conditions.

IV.2.4 Recherche des conditions optimales

A partir des précédentes mesures, il a été possible d'estimer les comptages et erreurs associés pour des quantités différentes de plutonium (il en serait de même pour l'uranium). La figure 7 fait apparaître l'erreur propre à la mesure en fonction de la masse superficielle à doser, cela pour des durées de comptage différentes à savoir 40 et 4 mn. Le minimum d'erreur est obtenu pour une masse superficielle de 0,6g·cm^{-2}, ce qui revient à dire que les conditions d'expérience sont bien adaptées et confirment que le meilleur rapport des transmissions est compris entre 4 et 10.

Pour des durées de stockage assez brèves, de 1 à 5 mn, l'erreur propre à la mesure est de quelques pour cents dans le domaine 0,4 à 1 gcm^{-2}, par contre, elle est plus importante pour des masses superficielles inférieures.

V CONCLUSION

Un étalonnage ne s'avère pas nécessaire si l'on connaît précisément d'une part, l'épaisseur de l'échantillon, et d'autre part, les coefficients d'absorption massiques de l'uranium ou du plutonium aux énergies (encadrant la discontinuité K) des raies γ utilisées émises par les sources.

L'incertitude de la méthode doit être de l'ordre du pour cent pour des durées de comptage modérées (quelques minutes) dans des limites définies sous réserve que :

- l'erreur sur l'épaisseur soit minimisée (quelques ‰) par l'utilisation d'un conteneur calibré à faces parallèles
- la différence des coefficients d'absorption massiques $\Delta\mu$ soit très bien connue (< 1 %)
- que l'erreur sur l'acidité soit négligeable (quelques ‰) ou que cette acidité soit connue.

Des études sont entreprises pour préciser le domaine d'application afin de déterminer la teneur en plutonium et en uranium dans une solution contenant un mélange U-Pu, et d'analyser l'influence des produits de fission.

En parallèle, un prototype permettant le contrôle en ligne dans une usine de retraitement des combustibles irradiés est en cours d'étude et de réalisation (conditionnement de l'échantillon, automatisme, traitement).

REMERCIEMENTS

Les auteurs tiennent à remercier leurs collègues, Mmes Imbert, Thomas, Bac, MM. Vatin et Dalmazzone pour la réalisation des références UO_2, M. Thiry pour la détermination de la concentration en uranium.

Enfin, ils remercient la Cogema - Etablissement de La Hague et tout spécialement le Chef du Service Laboratoire pour son amicale collaboration.

REFERENCES

[1] KULL, L.A., GINAVEN, R.O., GLANCY, J.E., A simple gamma-spectrometric technique for measuring isotopic abundances in nuclear materials, At. Energy Rev. **14** 4 (1976) 681.

[2] VEIGELE, W.M.J., Photon cross sections from 0,1 keV to 1 MeV for elements Z = 1 to Z = 94, Atomic Data Tables **5** (1973) 51–111.

[3] First Coordinated Research Meeting on the Measurement of Transactinium Isotope Nuclear Data, INDC (NDS)–96/N (June 1978).
[4] HARRY, R.J.S., AALDIJK, J.K., BRAAK, J.P., «Gamma-spectrometric determination of isotopic composition without use of standards», Safeguarding Nuclear Materials (Proc. Symp. Vienna, 1975) Vol.2, IAEA, Vienna (1975) 235.
[5] DRAGNEV, T.N., Intrinsic self-calibration of non destructive gamma spectrometric measurements, IAEA STR-60 (Internal Report); J. Radioanal. Chem. **36** (1977) 401.
[6] BANHAM, M.F., The Determination of the Isotopic Composition of Plutonium by Gamma Ray Spectrometry, AERE-R-8737 (1977).
[7] NESSLER, I., Détermination de la composition isotopique du plutonium par spectrométrie gamma, Centre d'Ispra, communication privée.
[8] GUNNINK, R., EVANS, J.E., In-Line Measurement of Total and Isotopic Plutonium Concentration by Gamma Ray Spectrometry, UCRL-52220 (1976).
[9] GUNNINK, R., EVANS, J.E., CRIDDLE, L., A Reevaluation of the Gamma Ray Energies and Absolute Branching Intensities of U-237, Pu-238, 239, 240, 241 and Am-241, UCRL-52139 (1976).
[10] CANADA, T.R., PARKER, J.L., REILLY, T.D., Total plutonium and uranium determination by gamma ray densitometry, Trans. Am. Nucl. Soc. **22** (1975) 140.

DISCUSSION

V.M. GRYAZEV *(Chairman):* Do you calibrate your apparatus using standard samples with different isotopic contents and, if so, how often?

G. MALET: In the K-edge densitometric technique, the result does not change with isotopic content because K-edge is a function of plutonium only. Thus, no calibration is necessary for a change in isotopic composition. However, it is better to perform monthly checks on the whole device using standard solutions to avoid instrument failures.

A. KEDDAR: How did you determine the percentage of ^{242}Pu using gamma spectrometry?

G. MALET: A calibration curve was constructed for a given reactor on the basis of mass spectrometry data (^{239}Pu/^{241}Pu as a function of the percentage of ^{242}Pu). The samples from the same reactor were analysed by gamma spectrometry. From the ^{239}Pu/^{241}Pu ratio obtained in this way, the percentage of ^{242}Pu can be derived using the calibration curve.

NON-DESTRUCTIVE ELEMENT AND ISOTOPE ASSAY OF PLUTONIUM AND URANIUM IN NUCLEAR MATERIALS*

H. EBERLE, P. MATUSSEK, H. OTTMAR, I. MICHEL-PIPER
Kernforschungszentrum Karlsruhe GmbH,
Institut für Angewandte Kernphysik,
Karlsruhe, Federal Republic of Germany

M.R. IYER, P.P. CHAKRABORTY
Bhabha Atomic Research Centre,
Health Physics Division,
Trombay, Bombay, India

Abstract

NON-DESTRUCTIVE ELEMENT AND ISOTOPE ASSAY OF PLUTONIUM AND URANIUM IN NUCLEAR MATERIALS.

Non-destructive assay techniques based on the analysis of X-rays and isotopic gamma rays have been employed for element and isotope analyses of uranium and plutonium materials. The selected techniques of analysis include high-resolution gamma spectroscopy for plutonium isotopic analysis of small and bulk plutonium samples, energy-dispersive X-ray fluorescence analysis of K X-rays for plutonium enrichment measurements in mixed U-Pu fuels, and K-edge gamma densitometry for the special nuclear material assay in solutions. The intrinsic calibration approach has been adopted for evaluating the parameter of interest from the various measurements. The results obtained from the non-destructive intrinsic calibration measurements compare reasonably well with the known reference values from destructive analyses. Some suggestions for the proper adjustment and choice of published atomic and nuclear data entering into the procedure of intrinsic calibration are given.

1. INTRODUCTION

In nuclear material accountancy measurements, usually both the element and isotope composition of the nuclear fuels represent parameters of interest to be determined for an estimation of the fissile content. Proven methods from analytical chemistry, and more recently also from NDA technology, are available to perform the necessary analyses. In the non-destructive assay, techniques based on the spectrometry of X-rays and isotopic gamma rays are employed for the analyses, because both radiations provide unique signatures for the elemental and isotopic characterization of the nuclear fuel. Some of these techniques have been utilized in the present investigations for the elemental and isotopic assay of plutonium and uranium in nuclear

---

* Part of the work performed within the framework of the Indo-Federal German Collaboration Agreement.

feed and product materials. In particular, we will report on (i) the plutonium isotopic analysis using high-resolution gamma spectrometry, (ii) the plutonium concentration determination in mixed U-Pu fuel materials by X-ray fluorescence analysis and (iii) on special nuclear material concentration measurements in solution using the K-edge gamma densitometry technique.

In most NDA applications, comparison measurements relative to specially prepared physical standards are necessary in order to determine quantitatively the parameter of interest from the unknown samples. An attractive feature common to the NDA techniques reported in this paper is that they allow to determine the parameter of interest without the use of external standards. The importance of the intrinsic calibration approach applicable to high-resolution gamma and X-ray spectrometry has been recently pointed out in an Advisory Group Meeting convened by the IAEA |1|. This paper describes some valuable experiences obtained in the determination of element and isotope concentrations using intrinsic calibration methods.

2. PRINCIPLE OF THE INTRINSIC CALIBRATION TECHNIQUE

Basically, the intrinsic calibration method applied to gamma-ray and X-ray measurements makes use of fundamental nuclear and atomic parameters such as gamma branching intensities, isotopic half-lives and photon cross-sections in order to derive element and isotope concentrations, or ratios of these concentrations, from measured spectral responses. The method might be therefore also called the 'Fundamental Parameters Technique'. The basis of the technique is best illustrated for the specific measurement examples discussed in this paper, viz. plutonium isotopic ratio measurements, plutonium concentration analysis in mixed U-Pu fuels, and measurement of special nuclear material concentration in solution.

In the plutonium isotopic analysis the isotopic ratio $N(i)/N(k)$ of the two isotopes i and k are deduced from the relation

$$\frac{N(i)}{N(k)} = \frac{I(i)}{I(k)} \times \frac{A(k)}{A(i)} \times \frac{RE(k)}{RE(i)}$$

where I, A and RE are measured peak intensity, gamma activity (in photons/g) and relative detection efficiency for gamma rays with energies $E(i)$ and $E(k)$ from isotopes i and k, respectively. Nuclear data entering into the activity ratio calculation are the isotopic half-lives and absolute gamma branching intensities. Intrinsic calibration in this context means the determination of dependence of the relative gamma detection efficiency, i.e. the ratio of peak areas to branching intensities, upon gamma-ray energy. This calibration relation changes with self absorption S of the sample and its cladding, but also with intrinsic detector efficiency, geometric properties of the arrangement (mainly collimator), and possibly electronic properties of the system (if, e.g., pulse shape discrimination is utilized for improved resolution). All these effects depend upon the gamma-ray energy E; if the above effects are denoted by e (for intrinsic efficiency of the detector), c (for collimator effects) and d (for pulse shape discrimination), then the calibration for determination of the *shape* of the curve is given by

$$RE(E) = S(E) \times \{ c(E)\ d(E)\ e(E) \} \quad vs\ E$$

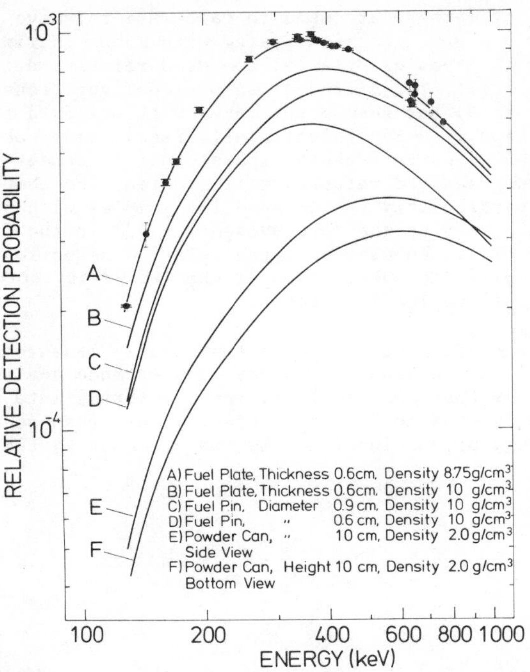

FIG. 1. Computed relative gamma-detection efficiency for different sample configurations. Intrinsic detector efficiency of a coaxial diode, gamma attenuation from typical cladding materials and from 1-mm Cd absorber are included in the calculations. For case (A) experimental points are given for comparison.

or, if the term in brackets which is typical for the detection system is called $\varepsilon(E)$,

$$RE(E) = S(E) \times \varepsilon(E) \text{ vs } E.$$

The second term $\varepsilon(E)$ will usually remain constant, or not change its shape as long as the apparatus remains unchanged; but $S(E)$ will be different unless samples of identical shape and density are measured. Two practical approaches for the determination of the relative detection efficiency can be distinguished:

i) Information about the energy dependence of $S \cdot \varepsilon$ is taken from the measured spectrum itself, i.e. in the case of plutonium isotopic analysis, for example, from measured peak ratios of ^{239}Pu and the corresponding intensity data. We have found from measurements on the plutonium reference materials NBS 948 and NBS 949d that the ^{239}Pu branching intensities reported by Gunnink et al. |2|, with a few minor exceptions, provide an internally consistent set of data |3|.

ii) Computational methods are used to calculate relative detection efficiency for a given detection system with known intrinsic efficiency $\varepsilon(E)$. Fig. 1 gives examples of computed relative detection efficiencies, $S(E) \cdot \varepsilon(E)$, for different sample configurations. The intrinsic efficiency $\varepsilon(E)$ has been established with standard photon sources. The comparison with the relative efficiency values determined from the measured spectrum (case A) shows close agreement between the computed and measured relative efficiencies. The computational approach is particularly recommended for samples with defined characteristics and geometry as they are present, e.g., in the fuel production area. An *a priori* knowledge of the relative detection efficiency would greatly simplify the evaluation of the different isotopic ratios from an intrinsic calibration measurement.

A similar procedure as in the isotopic ratio measurement is employed in the Pu/U ratio evaluation from X-ray fluorescence analyses. The only difference is that the gamma activity ratio entering into the isotopic ratio determination has to be substituted by the cross-section ratio for fluorescence X-ray production from the two elements by the excitation radiation, i.e.,

$$Pu/U = \frac{I_{Pu}}{I_U} \cdot \frac{(\sigma_K\ \omega_K f_K)_U}{(\sigma_K\ \omega_K f_K)_{Pu}} \cdot \frac{RE_U}{RE_{Pu}}$$

The atomic data required for the intrinsic calibration measurement are the K-shell photo absorption cross section for the exciting energy E_o, $\sigma_K(E_o)$, the fluorescence yield ω_K(i.e. the probability that a K-shell vacancy will result in a K X-ray emission) and the relative branching intensity f_K of the individual X ray. The relative branching intensities of X rays from the *same* element are also used to establish the relative detection efficiency *RE*.

Finally, the evaluation of the special nuclear material concentration from an intrinsic calibration measurement using absorption edge densitometry just only requires the knowledge of the total photon cross-section differences, $\Delta\mu = \mu(E_1) - \mu(E_2)$ of the transmitting radiations, E_1 and E_2, above and below the absorption edge energy for the element of interest. With the known value of $\Delta\mu$, the special nuclear material concentration ρ is then obtained from the measured ratio of transmissions $R = T(E_1)/T(E_2)$ according to the relation

$$\rho = \frac{\ln R}{\Delta\mu \cdot D},$$

where D is the thickness of the solution sample.

In principle, all the nuclear and atomic data required for the different analyses are available from published literature. However, because of uncertainties associated with these data, their use in intrinsic calibration measurement analyses would in many cases limit the accuracy ultimately attainable from the measurements. Effort should therefore be undertaken to better define the required atomic and nuclear data. A part of this work aimed at the testing of the validity of the published data, and, where possible, to suggest improvements of these.

3. PLUTONIUM ISOTOPIC ANALYSIS BY GAMMA-RAY SPECTROMETRY

The plutonium isotopic analysis by gamma-ray spectrometry experiences increasing attention in NDA safeguards measurements for several reasons. First, the technique provides a fast 'fingerprint' method for the characterization of the plutonium isotopic composition, which in many cases aids in a fast verification of the identity of the fuel. The major interest in the technique, however, relates to other NDA techniques designed for quantitative measurements on bulk plutonium samples, *viz.* neutron coincidence counting and calorimetry. Both techniques require supplementary information on the plutonium isotopic composition for the interpretation of their instrumental response in terms of total plutonium. In principle, gamma spectrometry offers a means to obtain this supplementary information from an independent NDA measurement. The attractiveness of the technique also relies on the fact that the plutonium isotopic composition can be derived from an intrinsically calibrated measurement without making use of physical standards.

Most of the practical experiences with the gamma-spectrometric plutonium isotopic analysis have been obtained so far from measurements on low-burnup plutonium in solutions |4|. Only limited experiences with the technique are available from isotopic measurements on solid samples of arbitrary shape, and from analyses on higher burnup plutonium fuel materials. In the following we will briefly comment on some experimental results from plutonium isotopic analyses on samples which also included this type of material.

3.1 Small Sample Analyses

High-resolution gamma spectroscopy has been used to record gamma spectra from a set of small plutonium samples (0.1 - 0.5 g) of widely different isotopic compositions, covering the range from low burnup to high burnup plutonium. The samples included the isotopic standards NBS 946, 947 and 948. The purpose of the measurements was to test different procedures of analysis for the determination of isotopic ratios, and to investigate the sensitivity of the different analyses to varying isotopic distributions. For the isotopic ratio evaluations we adopted the most general approach of analysis based on the intrinsic calibration procedure, which equally applies to plutonium isotopic composition measurements of any sample configuration. Published half-lives and gamma branching intensities |2,5| have been used to convert observed peak intensity ratios (corrected for relative detection efficiency) into isotopic ratios. For the peak area determination both the peak fitting method and the more simple (and practicable) channel summation method have been employed. The computer codes GAUSSFIT |6| and SAMPO |7| used for the peak fitting involved linear and parabolic background approximations, while in the channel summation method (CHSUM) the background has been approximated by a smooth step-like function.

In the following analyses spectral regions above 125 keV are used, which are probably the only accessible ones in arbitrary solid sample analyses. Since in the plutonium gamma spectrum, gamma rays from ^{241}Pu happen to be energetically close to gamma rays from the other plutonium isotopes, it appears advantageous to determine primarily the isotopic ratios as 238/241, 239/241 and 240/241. The following results have been obtained from the various isotopic ratio evaluations:

TABLE I . RESULTS OF Pu-238/Pu-241 AND Pu-238/Pu-239 RATIO MEASUREMENTS

Sample	Pu-238/Pu-241				Pu-238/Pu-239		
	Destructive Analysis(DA)	NDA/DA			Destructive Analysis(DA)	NDA/DA	
		GSFIT	SAMPO	CHSUM		GSFIT	SAMPO
NBS948	0.0295	1.030	1.020	1.087	1.149-04	1.052	1.014
NBS947	0.0827	0.976	0.964	0.967	3.724-03	0.968	0.978
NBS946	0.0786	0.977	0.975	0.973	2.829-03	0.981	0.969
75% Fissile	0.0416	0.971	0.975	0.972	2.299-03	0.977	0.968
70 % Fissile	0.1188	0.974		0.960	1.739-02	0.982	
Mean Values NDA/DA (Excluding NBS 948)		0.975	0.971	0.968		0.977	0.972

TABLE II. RESULTS OF Pu-239/Pu-241 RATIO MEASUREMENTS

Sample	Destr. Anal.(DA)	NDA/DA						
		164/161 keV	148/144 keV		208/203 keV			332-422 keV
		GSFIT	GSFIT	SAMPO	GSFIT	SAMPO	CHSUM	CHSUM
NBS948	255.3	0.992	0.991	0.999	1.019	1.005	0.987	1.021
NBS946	27.90	0.938	0.995	0.943	1.010	1.002	0.982	0.973
NBS947	22.33	1.035	0.991	0.986	1.021	1.003	0.961	0.971
75 % Fissile	18.16	0.986	1.019	0.941	1.045	0.984	0.953	0.969
70 % Fissile	6.86	1.073	0.992		1.003	0.950	0.932	0.972

$^{238}Pu/^{239}Pu$ and $^{238}Pu/^{241}Pu$ Ratios. These ratios have been determined using the gamma rays 152 keV(^{238}Pu), 148 keV (^{241}Pu) and 144 keV (^{239}Pu). Table I summarizes the results obtained from the different analyses. The ^{238}Pu abundances in the measured samples ranged between 0.01 and 1.09 %. The generally consistent results suggest that (i) the ^{238}Pu abundance in NBS 948 is about 6 % higher than listed, and (ii) that the actual branching intensity of the ^{238}Pu gamma line at 152 keV is about 3 % lower than reported by Gunnink et al. |2|. If the ^{238}Pu gamma branching intensity is adjusted accordingly, the gamma results would have come in close agreement to the chemical results. We believe that gamma spectrometry favourably competes with alpha spectrometry for the ^{238}Pu assay in higher burnup plutonium.

While the gamma spectrometry provides rapidly improving accuracy with increasing ^{238}Pu abundance, alpha spectrometry does not |8|. These features are particularly important for calorimetric measurements on higher fissile plutonium, which require a very accurate knowledge of the ^{238}Pu abundance.

^{239}Pu/^{241}Pu Ratio. Different gamma lines have been chosen to investigate the evaluation of this isotopic ratio. The results are listed in Table II. The pairs of gamma lines in the energy region between 144 and 208 keV are best suited for isotopic analyses on lower burnup plutonium as becomes evident from a comparison of the low burnup and high burnup plutonium spectra presented in Fig. 2. The intense ^{241}Pu gamma rays more or less swamp the ^{239}Pu gamma-ray signatures in this energy region when ^{241}Pu is present at higher concentrations. In this case, the higher energy gamma rays in the 330 to 420 keV region provide a better choice for ^{239}Pu/^{241}Pu ratio measurements. These gamma rays have been successfully employed for ^{239}Pu, ^{241}Pu and ^{241}Am abundance measurements on samples with higher ^{241}Pu content (^{239}Pu/^{241}Pu $\lesssim$ 50). We have chosen a multigroup analysis, using counts from the energy regions 328-339, 339-347, 365-378, 378-385 and 421-425 keV. If ^{241}Am is present at higher concentrations, also the ^{241}Am line at 419 keV is included in the analysis. The net peak counts from these energy regions as determined by the channel summation method have been used to form a set of linear equations, relating the peak counts to the unknown isotopic abundances. The matrix of linear equations has then been solved for the relative ^{239}Pu, ^{241}Pu and ^{241}Am abundances by the method of least-squares. This procedure provided consistent ^{239}Pu/^{241}Pu ratios for the indicated range of application, i.e. for ^{241}Pu abundances $\gtrsim$1 %. The observed bias of about 3 % of the gamma results listed in Table II possibly indicates small systematic errors in the branching intensities of the gamma rays involved. Once this bias has been recognized, it can be taken into account in future analyses. The ^{239}Pu/^{241}Pu ratios determined from the multigroup analysis have proved very insensitive to the ^{241}Am content in the samples. The ^{241}Am concentration itself is determined accurately from the multigroup analysis for ratios ^{239}Pu/^{241}Am $\lesssim$ 100. The analysis approach is also applicable to plutonium isotopic analyses on mixed U-Pu fuels because there are no interfering gamma rays from uranium in the 330 to 450 keV energy region.

^{240}Pu/^{241}Pu Ratio. Only the ^{240}Pu gamma ray at 160.3 keV could be used for the ^{240}Pu determination in the small sample analyses. This ^{240}Pu gamma-ray signature has to be extracted from an unresolved peak complex. The results listed in Table III indicate that the ^{240}Pu abundance cannot be reliably determined from intrinsically calibrated measurements using this spectral region. Large systematic errors can occur in the gamma analyses depending on the method used for the peak and background determination. For larger sample measurements, the 640 keV region will provide a better choice for the ^{240}Pu analysis |9|. An example of measurement from this energy region is given in Section 3.2.

We conclude from the present measurements that the available nuclear data, with a few minor exceptions, are well established in order to permit accurate isotopic analyses from intrinsically calibrated measurements, provided proper analysis procedures are employed. Otherwise significant biases can arise from the analyses which by far exceed the precisions calculated from counting statistics. Those biases are best recognized from interlaboratory comparison experiments. The ESARDA working group on techniques and standards for NDA analysis has recently started such an intercomparison excercise for gamma-spectrometric plutonium isotopic analysis.

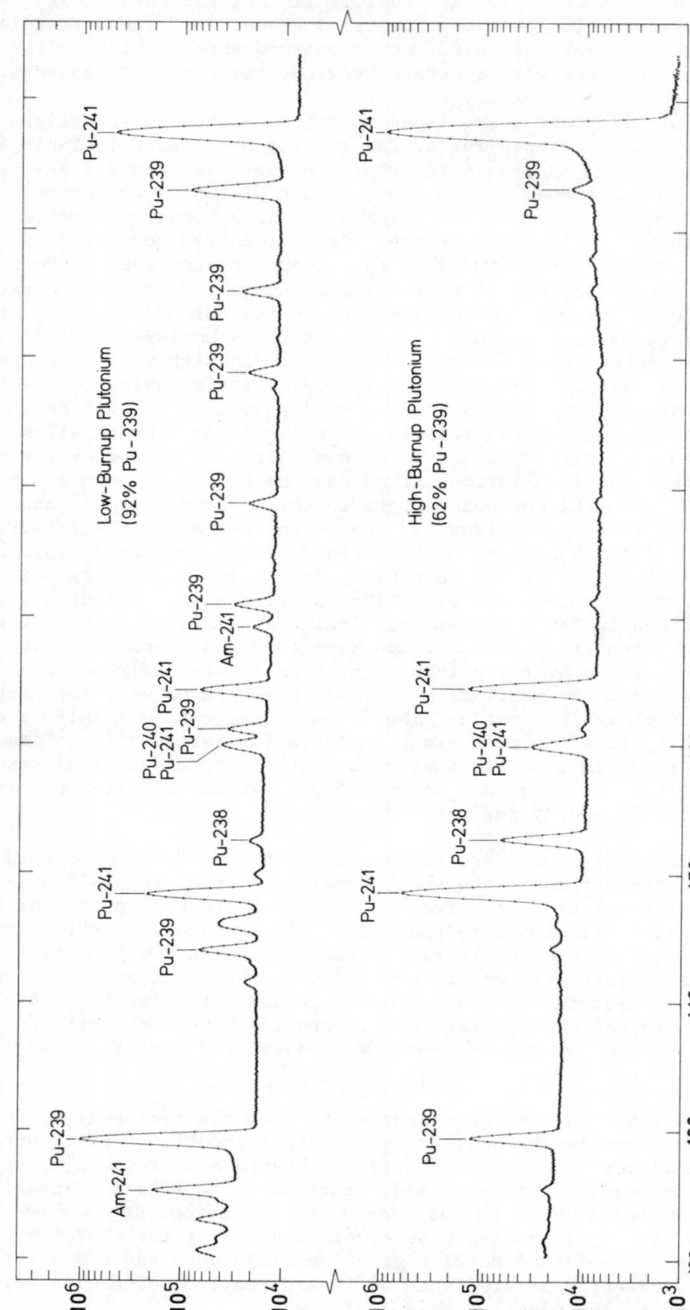

FIG.2. High-resolution gamma spectra from aged low burnup plutonium and recently separated high burnup plutonium.

TABLE III. RESULTS OF Pu-240/Pu-241 RATIO MEASUREMENTS

Sample	Destructive Analysis (DA)	NDA/DA			
		160[a]/164		160.3/164	160.3/160.0
		GSFIT	CHSUM	GSFIT	GSFIT
NBS948	22.14	0.967	0.930	0.988	0.77
NBS947	5.418	0.928	0.905	1.038	1.13
NBS946	4.068	0.928	0.914	1.063	1.15
75 % Fissile	6.440	0.981	0.945	1.021	1.13
70 % Fissile	2.557	0.955	0.951	1.079	1.06

[a] 160.3 keV intensity extracted from total 160 keV peak complex.

3.2 Bulk Sample Measurements

An important near-term application of the non-destructive plutonium isotopic analysis by gamma spectrometry would be to provide the isotopic information required for the interpretation of neutron coincidence measurements, which are now being increasingly employed for the assay of bulk plutonium samples. Here the main parameter of interest is the ^{240}Pu fraction in the fuel. To arrive at this information, it would be generally sufficient to know the abundances of the major isotopes ^{239}Pu, ^{240}Pu and ^{241}Pu. The abundances of the remaining two isotopes, ^{238}Pu and ^{242}Pu, could then be estimated fairly well from this information without introducing a significant error into the ^{240}Pu assay.

In order to obtain useful isotopic information on the major isotopes in a reasonable counting time, the measurements and data analysis procedures must be carefully optimized with respect to the different measurement conditions. For isotopic measurements on bulk samples, the higher energy gamma rays in the 330 to 450 keV region and in the 640 keV region can be advantageously employed as long as no fission product contaminations are present. While keeping the total detector count rate at a tolerable level (20 000 cps), the speed and precision of the ^{239}Pu, ^{240}Pu and ^{241}Pu analysis from the higher energy gamma rays will significantly improve if the intense low-energy X rays and gamma rays are removed by appropriately filtering the gamma spectrum. Table IV presents results obtained from isotopic measurements on a mixed U-Pu fuel plate which contained about 30 g of medium burnup plutonium (75 % fissile). Ten repetitive measurements, each 40 minutes long, have been performed on the sample with a medium-sized coaxial detector (V=18 ccm) at a total detector count rate of 15 000 cps. The low-energy gamma intensity was greatly reduced using absorbers of 2 mm Pb and 1 mm Cd. The above mentioned multigroup analysis in the 332 to 422 keV region using the channel summation method was employed for the ^{239}Pu/^{241}Pu ratio determination. The measured ratio again turns out to be about 3 % lower than the book values as in the small sample analyses, while the standard deviation of the ratios from the

TABLE IV. RESULTS OF ISOTOPIC RATIO MEASUREMENTS FROM A MIXED-OXIDE FUEL PLATE (10 RUNS, 40 MIN EACH).

	$^{239}Pu/^{240}Pu$	$^{239}Pu/^{241}Pu$	$^{239}Pu/^{241}Am$
Mean Value	3.239	31.68	44.35
Standard Deviation (%)	4.6	0.47	1.04
NDA/DA	1.002	0.970	Mean value from gamma rays 125/129, 169/171, 419/422, 659/662: 43.80

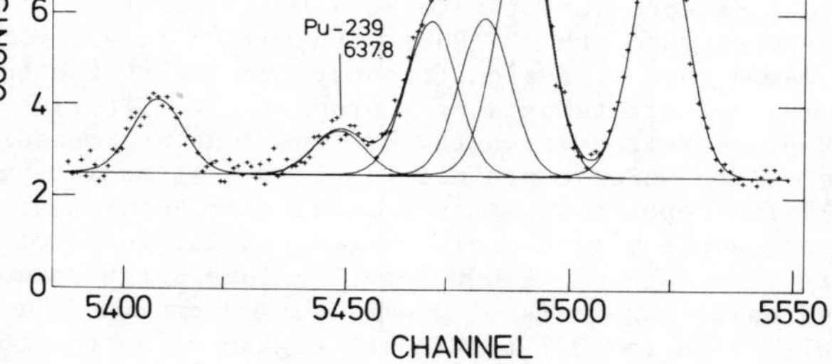

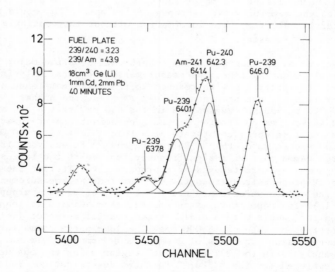

FIG. 3. $^{239}Pu/^{240}Pu$ ratio evaluation from 642.3/646 keV gamma rays.

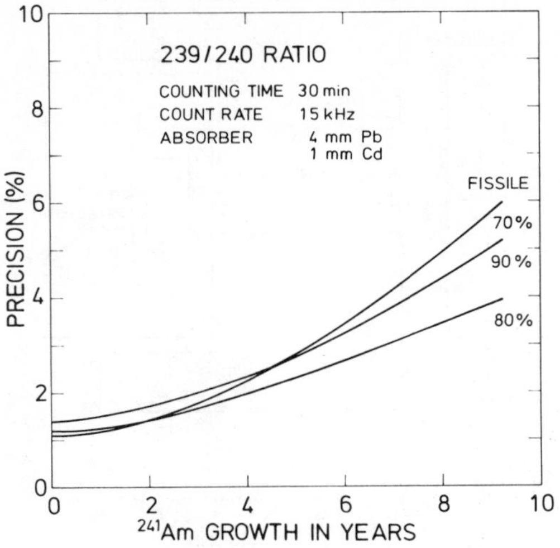

FIG.4. Expected precisions from $^{239}Pu/^{240}Pu$ ratio measurements for different types of material using the 640-keV region.

10 consecutive measurements is less than 0.5 %. For the $^{239}Pu/^{240}Pu$ ratio determination the peak fitting method has been applied to the 640 keV region, as shown in Fig. 3. A standard deviation of about 4 % was obtained for this ratio. If there is no fission product contamination present in the plutonium material, the precision and accuracy attainable for the $^{239}Pu/^{240}Pu$ ratio determination from this spectral region will be largely limited by the ^{241}Am content in the sample, as can be seen from Fig. 3.

In Fig. 4 we have plotted the approximate precisions that can be expected for the $^{239}Pu/^{240}Pu$ ratio determination from the 640 keV region for different types of materials, assuming the experimental conditions indicated in the figure. The measurements would require sample masses of $\gtrsim$ 10 g. For a 30 min assay, a precision in the order of 1 % is attainable if ^{241}Am growth is not more than approximately 2 years.

4. PLUTONIUM CONCENTRATION ANALYSIS IN MIXED U-Pu FUELS

In principle, the same analysis procedure as in the plutonium isotopic analysis can be applied to plutonium concentration measurements in mixed U-Pu fuel materials when isotopic gamma rays are used for the analysis. But the spectrum analysis becomes more complicated in this case, because both the ^{235}U and ^{238}U gamma rays interfere with plutonium gamma rays. In addition, the isotopic abundances of the gamma emitting uranium and plutonium isotopes should be known for the evaluation of elemental Pu/U ratios from isotopic gamma-ray intensities.

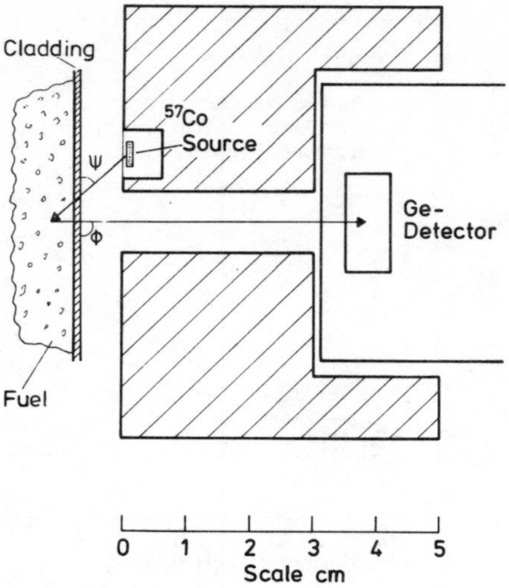

FIG.5. Experimental set-up for XRF measurements on mixed U-Pu fuel samples.

An alternate approach to elemental concentration measurements of mixed U-Pu fuel materials is the energy-dispersive analysis of fluorescent K X rays. The X-ray fluorescence (XRF) analysis offers the advantage that it can also be intrinsically calibrated, but does not require any *a priori* information about isotopic abundances, as it is the case in the isotopic gamma-ray analyses. In order to investigate the benefits of the XRF analysis technique for elemental concentration measurements, we applied the technique to plutonium concentration measurements in various mixed U-Pu sample materials, like fuel plates, fuel pins and solutions. For this purpose uranium and plutonium K X rays have been excited using the 122.06 keV gamma rays from a 10 mCi ^{57}Co source. The geometry of the experimental set-up used for the analyses is shown in Fig. 5. The fluorescent spectra obtained from a mixed-oxide fuel pin are shown in Fig. 6. The directly measured fluorescent spectra contain some interfering X rays and isotopic gamma rays from the natural decay. With a sufficiently intense ^{57}Co source (activity 50 to 100 mCi), the decay-generated X-ray intensity is about to be a few percent of the induced fluorescent intensity for most of the applications. For quantitative measurements the interfering radiations from the natural decay must be removed from the measured fluorescent spectrum. This is best accomplished by subtracting the self-radiation spectrum from the fluorescent spectrum as illustrated in Fig. 6.

Two other corrections are necessary before converting for the measured peak intensity ratio of I_{Pu}/I_U from a pair of uranium and plutonium X rays into the elemental ratio Pu/U: (i) the interelement enhancement correction applied to the observed uranium fluorescent intensity and (ii) correction for the relative detection efficiency. The latter one can be established from measured uranium X-ray intensities together with their known branching intensities. Fig. 7 shows the relative detection efficiency curves

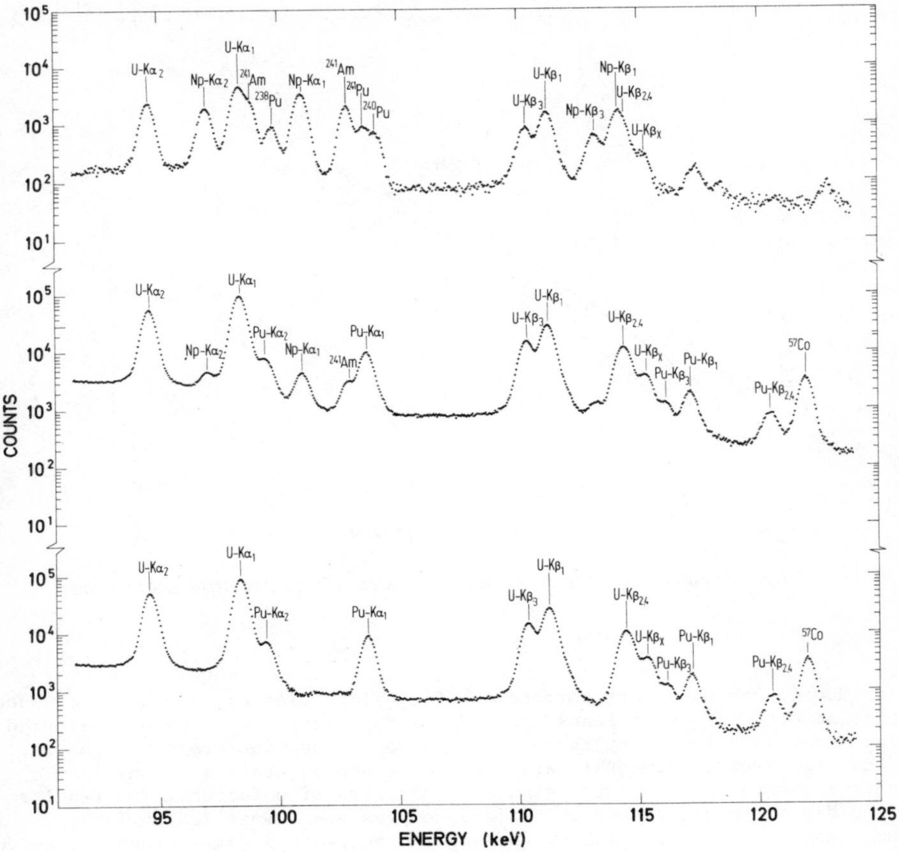

FIG. 6. Fluorescent K X-ray spectra obtained from a mixed-oxide fuel pin (Pu/U = 0.078). Top to bottom: passive spectrum (^{57}Co source shielded), XRF spectrum superimposed on passive spectrum, and net XRF spectrum after subtraction of passive spectrum.

for two different experimental configurations. The correction for the secondary enhancement of the uranium fluorescent intensity which is due to fluorescence induced by plutonium K_{β} X rays can be calculated |10| using known photon cross sections. Parameters which enter into the calculations are the average inclination of the incident and emergent radiations with respect to the sample surface (angles ψ and ϕ, see Fig. 5). Fig. 8 presents the results of computations for secondary fluorescence effect. The computations assumed infinite thickness of the sample for the X-ray radiation. For a plutonium enrichment of 30 % the fluorescent uranium intensity due to interelement enhancement is approximately 4 % of the total observed fluorescent intensity. The enhancement effect slightly depends on the geometry. The relative variation of the effect with X-ray energy is found to be almost negligible.

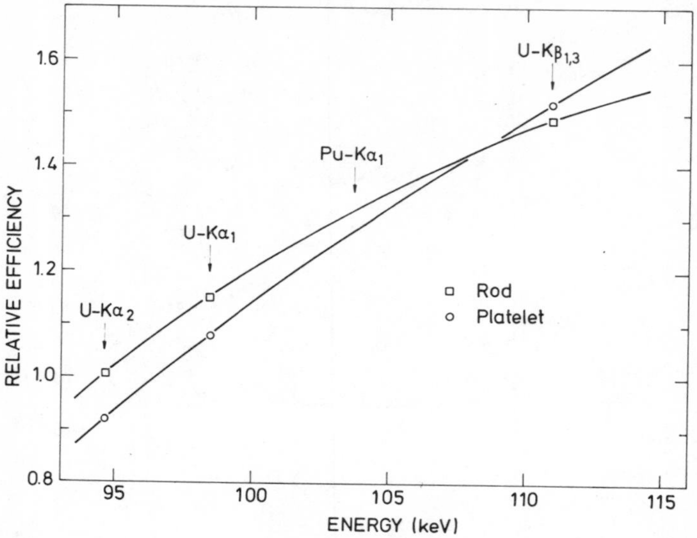

FIG. 7. Relative detection efficiency for K X-rays from two different samples.

Having the necessary corrections for self-radiation, secondary enhancement and detection efficiency applied, the Pu/U ratio can then be evaluated from X-ray peak intensity ratios using X-ray production cross section ratios calculated from fundamental parameters. The photon cross sections from Ref. |11| have been used to calculate the conversion factors. The results of Pu/U ratio measurements on various samples are summarized in Table V. In the last column of the Table, the applied percentage corrections for secondary enhancement are given. The solution sample was transparent for the X rays, thus no secondary fluorescence occured in this sample. The data in the Table show that the Pu/U ratios from the intrinsically calibrated X-ray fluorescence analyses are systematically lower than the reported book values. We do not yet have a plausibe explanation for this observation. The analyses of the solid samples have been performed on different sections of the fuel material, and generally consistent results have been obtained from the different measurements of the same sample. Probably the most critical factor entering into the data evaluation is the relative detection efficiency correction, which could be established from 3 experimental data points only (cf. Fig. 7). Computational methods are in progress to reassure the experimentally determined relative detection efficiency curves. In general, we found that the XRF analysis of induced K X rays provides a fast non-destructive method for plutonium concentration measurements in mixed U-Pu fuels.

5. CONCENTRATION MEASUREMENTS OF SPECIAL NUCLEAR MATERIALS IN SOLUTION

The non-destructive measurement of the concentration of special nuclear materials like uranium and plutonium in solution is best accomplished by gamma absorptiometry. To be element specific, the technique usually requires

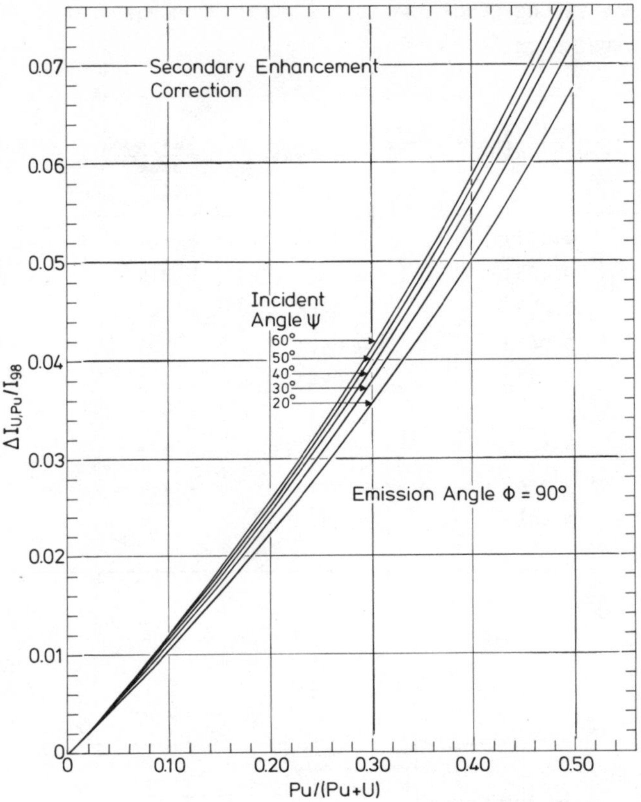

FIG. 8. Secondary enhancement $(\Delta I/I)_{U,Pu}$ of U K_{α_1} fluorescent intensity due to excitation by plutonium K_β X-rays for different incident angles ψ of the primary ^{57}Co radiation.

gamma transmission measurements at different photon energies in order to discriminate between the attenuation caused by the element of interest and that which is due to matrix materials. Gamma-ray absorptiometry which makes use of the total absorption cross-section discontinuity at either the L- or K-shell absorption edge of the special nuclear materials appears most advantageous for an unbiased concentration determination of the particular element of interest. Both discrete gamma rays closely bracketing the absorption edge and continuous photon beams are suitable for absorption-edge densitometry. We used both types of radiations for the special nuclear material concentration analysis using K-edge absorptiometry. For the sake of convenience the measurements have been performed on uranium solutions, but in principle there are no difficulties in applying the techniques also to the plutonium concentration assay.

Convenient gamma sources providing abundant discrete gamma rays which closely bracket the K-absorption edge of uranium are not available. We used therefore fluorescence radiation from uranium in conjunction with the

TABLE V. X-RAY FLUORESCENCE ANALYSIS RESULTS FROM Pu/U RATIO MEASUREMENTS.

Sample	Pu/U Weight Ratio		$\frac{\text{XRF-Book}}{\text{Book}}$ (%)	Enhancement Effect (%)
	Book Value	XRF Analysis		
Rod	0.0239	0.0235	- 1.7	0.27
Rod	0.0251	0.0242	- 3.6	0.29
Platelet	0.0299	0.0286	- 4.2	0.34
Rod	0.0432	0.0421	- 2.4	0.48
Rod	0.0777	0.0743	- 4.3	0.86
Solution	0.315	0.291	- 7.8	0.
Platelet	0.363	0.330	- 9.1	3.5
Platelet	0.366	0.340	- 7.1	3.6
Platelet	0.404	0.387	- 4.2	3.8

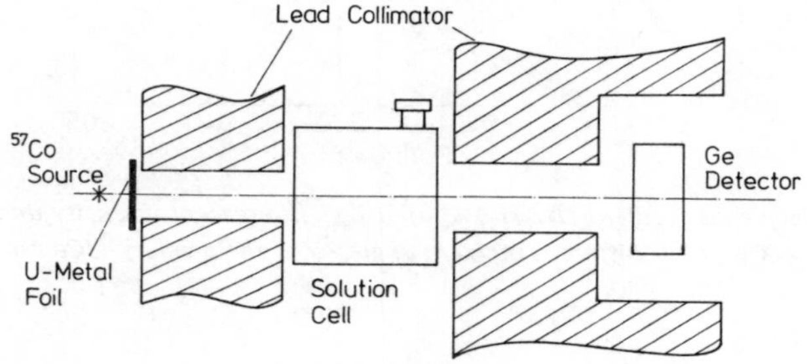

FIG. 9. Experimental set-up for K-edge densitometry using fluorescent X-rays and ^{57}Co gamma-rays.

122.06 keV gamma rays from a ^{57}Co source for uranium concentration measurements. The experimental set-up is shown in Fig. 9. The thickness of the fluorescent uranium foil (D=300 μm) is so chosen that it is still transparent for the ^{57}Co gamma radiation, and also thick enough to provide sufficient fluorescent intensity. This approach has the advantage that only a single photon source is required for the transmission measurements. The same technique can also be applied to plutonium concentration analyses. Either fluorescent uranium or plutonium K_β X rays in conjunction with the ^{57}Co radiation could be used in this case. Fig. 10 shows the intensity distribution of the transmitted radiations for different uranium concentrations.

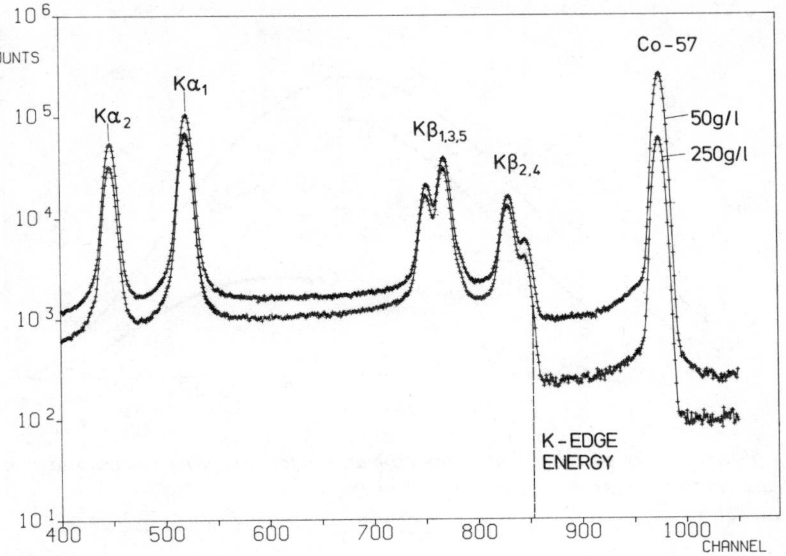

FIG.10. K-edge gamma absorptiometry with discrete X-rays and gamma rays.

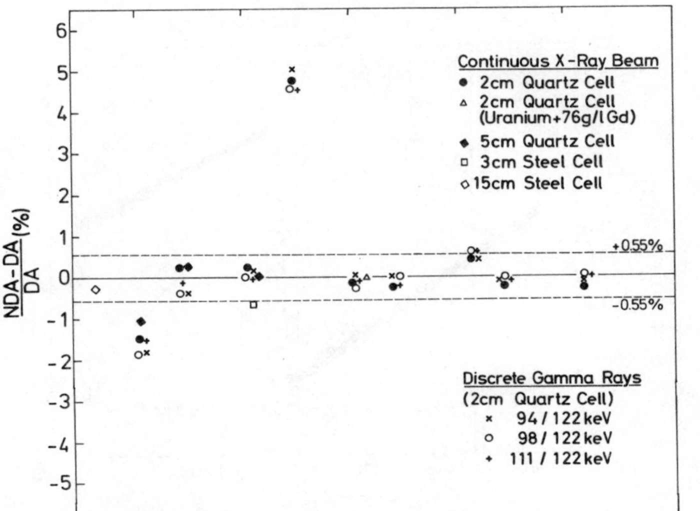

FIG.11. Uranium concentration analysis from K-edge gamma absorptiometry.

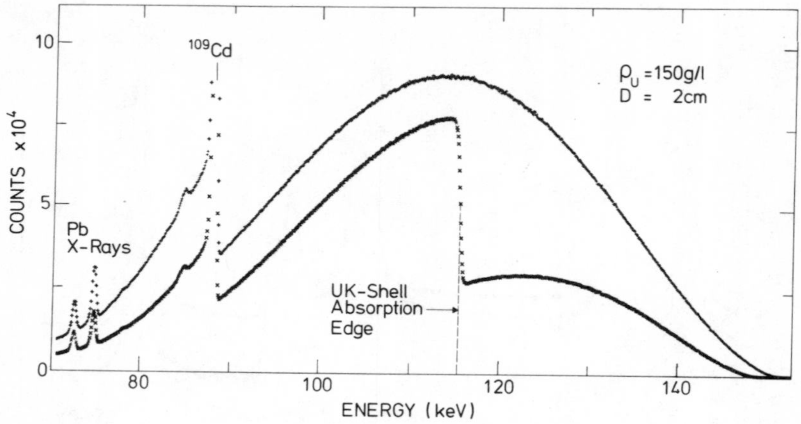

FIG.12. Filtered X-ray beam from an X-ray generator before and after the transmission through a 2-cm-thick uranium solution ($\rho_U = 150$ g/ℓ).

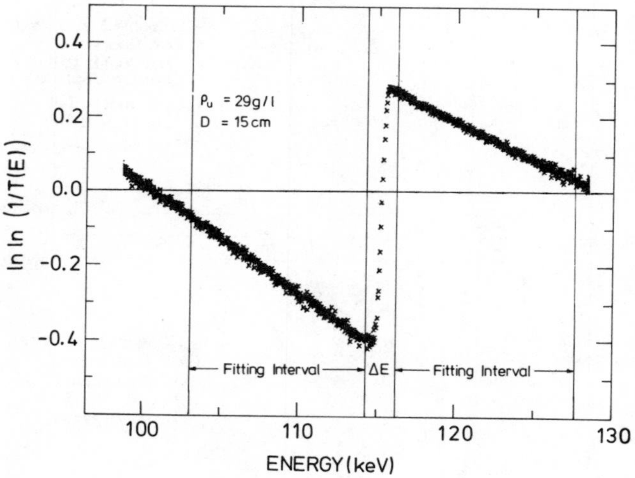

FIG.13. Energy dependence of the function ℓn ℓn (1/T(E)), where T(E) describes the transmission through the heavy element bearing solution, measured with respect to the transmission spectrum through a blank nitric acid solution.

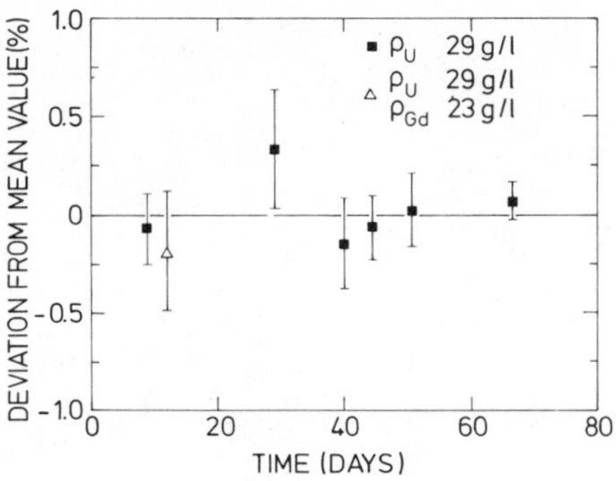

FIG.14. Reproducibility of the concentration analysis obtained from repetitive measurements on the same sample covering a two-month period.

The ^{57}Co gamma rays and the uranium K_β X rays (centroid energy $\bar{E}_{K\beta}$ = 111.90 keV) are sufficiently close to the uranium K-absorption edge energy (E_K=115.61 keV) so that the influence of matrix elements to the transmission ratio is greatly reduced. The transmission ratio evaluation requires a small correction due to fluorescent uranium X rays produced in the measured sample. This secondary uranium X-ray intensity is easily determined for a given uranium concentration and ^{57}Co gamma-ray intensity by removing the primary intensity of uranium X rays from the transmission spectrum. For the highest measured uranium concentration (ρ_U=260 g/ℓ) the secondary uranium X-ray intensity produced in the sample was measured to be less than 2 % of the transmitted primary X-ray intensity.

Fig. 11 summarizes the results obtained from measurements on prepared uranium standard solutions. The accuracy of the chemical analysis is estimated to be ± 0.3 % as indicated by the dashed lines in the Figure. The precision of the results from the various transmission measurements ranges between 0.15 and 0.25 %. All of the NDA analyses including the transmission measurements with the continuous X-ray beam described below provide internally consistent results. When normalized to the chemical values, the observed relative differences between the non-destructive and chemical results do not exceed the uncertainties assigned to the measurements except for two samples with large deviations. No satisfactory explanations could be found for such deviations.

The K-edge densitometry measurements conducted with the continuous X-ray beam used an X-ray generator for the production of the continuous photon energy distribution across the absorption edge. Fig. 12 shows the tailored X-ray beam before and after the transmission through a solution containing uranium (the ^{109}Cd gamma line served for digitally stabilizing the electronic gain). The spectra are measured with a high-resolution planar Ge detector (FWHM = 550 eV at 122 keV) with a slope of about 70 eV/channel.

The accurate concentration analysis requires an accurate determination of the ratio of the transmission, $R = T(E_1)/T(E_2)$, below and above the

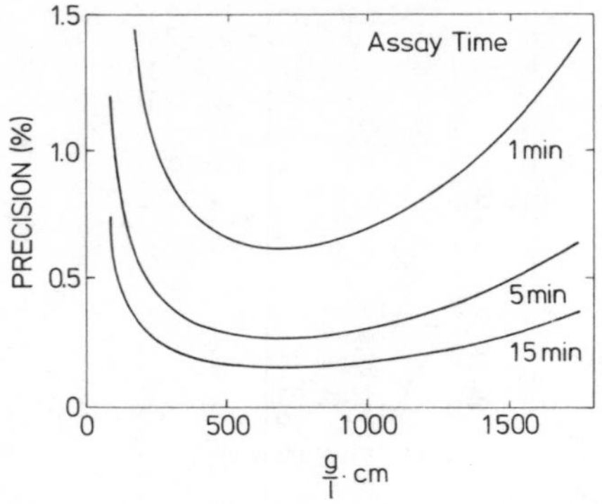

FIG.15. Precision on concentration assay attainable with K-edge absorptiometry for given analysis time intervals, plotted versus the product of concentration (g/ℓ) times sample thickness (cm).

absorption edge energy. The quantitative element concentration, ρ, is determined from this ratio by the relation

$$\rho = \frac{\ln R}{\Delta\mu \cdot D},$$

where Δμ and D denote the change of the total photon cross-section at the K-absorption edge and the sample thickness, respectively. We have tested several analysis procedures for determining the ratio of the transmissions at the absorption edge energy. An advantageous analysis approach which effectively utilizes most of the information provided by the continuous transmitted gamma beam has been worked out and implemented into a mini-computer-based analysis system. This approach consists of fitting on both sides of the absorption edge the experimentally determined transmission values, $T(E)$, to the function

$$\ln \ln (1/T(E)) = \ln(\mu(E) \cdot \rho \cdot D).$$

This function shows a nearly linear dependence upon the gamma energy E, as shown in Fig. 13. The transmission values, $T(E)$, are measured with respect to a reference spectrum which is once taken with good counting statistics through a blank nitric acid solution. This reference spectrum is permanently kept in the computer memory. The values $\ln \ln (1/T(E))$ are fitted by the least-squares method either to a linear or parabolic function on both sides of the absorption edge, using the fitting intervals as shown in Fig. 13. The transmission jump at the absorption edge energy is then calculated by extrapolating the fitted functions to the K-edge energy. Typically 150 to 200 channels are used as fitting intervals on both sides of the absorption edge, which corresponds to energy ranges of approximately 10 to 15 keV. The gap between the fitting intervals across the absorption edge, ΔE, depends on the detector resolution. This gap can be as narrow as 1.5 to 2 keV when high-resolution Ge detectors (FWHM 500-600 eV) are used for the transmission measurements.

TABLE VI. COMPARISON OF EXPERIMENTAL AND THEORETICAL TOTAL PHOTON CROSS SECTION DIFFERENCES

Element	Energies (keV)	Total Photon Cross Section Differences $\Delta\mu$ (cm^2/g)			
		Storm and Israel /12/	Veigele /11/	McMaster et al. /13/	Experimental This work
U	122.06/111.03	2.772	2.837	2.870	2.832 ± 0.014
U	122.06/98.44	2.274	2.339	2.363	2.323 ± 0.014
U	122.06/94.66	2.097	2.145	2.162	2.127 ± 0.012
U	115.61 (K-Edge Jump)	3.501	3.605	3.651	3.578 ± 0.02
Pu	121.80 (K-Edge Jump)	3.193	3.223	3.397	3.22 ± 0.03[a]

[a] Value taken from Ref. /14/.

This procedure of analyses has experimentally proved to give reproducible results which are highly insensitive to matrix properties. Fig. 14 shows the results obtained from the same sample on different occasions within a 2-month period. The results suggest an accuracy of better than 0.3 %. The insensitivity of this method of analyses to matrix effects may be demonstrated by the results obtained from uranium solutions with artificial gadolinium contamination (cf. Figs. 11 and 14).

The precision and accuracy of the special nuclear material concentration analysis attainable from the continuous X-ray beam measurement in a given analysis time is largely limited by the count rate that can be tolerated in the electronics of the gamma detection system without deteriorating appreciably the energy resolution. Fig. 15 shows the precisions attainable in different analysis times, plotted as a function of the product of heavy element concentration (in units of g/ℓ) times the sample thickness in cm. The precision values are calculated assuming that the X-ray intensity from the X-ray generator is adjusted to give always a fixed count rate of 10 kHz, and that 50 % of this count rate is covered in the fitting intervals. The curves illustrate that under optimum conditions, i.e., choosing the optimum sample thickness for a given heavy element concentration, precisions of less than 0.2 % can be achieved in an assay time of 15 min. We successfully applied the K-edge densitometry to concentration measurements in the range between ~30 to 300 g/ℓ, using a solution thickness ranging from 2 cm to 15 cm.

An accurate concentration analysis from intrinsic calibration measurement is possible, provided the total photon cross-section differences, $\Delta\mu$, of the transmitting radiations above and below the absorption edge are known with sufficient accuracy. We used the present measurements on the standard reference solutions to determine these data for the radiations involved in the analyses. Table VI compares the experimentally determined values of $\Delta\mu$ for different energies with those obtained from theoretical values published in compilations for photon cross sections. The results show that the photon cross sections compiled by Veigele |11| are close to our experimental values.

6. CONCLUSIONS

The present investigations have shown that fairly accurate information about the elemental and isotopic composition of nuclear fuels can be obtained from non-destructive measurements which do not necessitate recourse to reference materials or physical standards. In some applications dispensing with external standards means, however, that more effort has to be spent with the data evaluation. Care should be taken that no systematic errors occur from the individual approach of analysis adopted for the data evaluation. An initial test of the procedure of analysis performed on well characterized samples would help to recognize and eliminate possible biases in the NDA measurements.

The nuclear data necessary for the isotopic ratio analysis from intrinsic calibration measurements, i.e. absolute gamma-branching intensities and isotopic half-lives, appear well established within about 3%. Some larger deviations are observed from published compilations of total photon cross sections, which enter into the evaluation of the special nuclear material concentration in solution from absorption edge densitometry measurements. The present state-of-the-art of the gamma densitometry techniques permits to determine very accurately the required parameters from calibration measurements performed on standard reference solutions. Once these parameters are established, accurate concentration analyses are possible without using any reference standards.

REFERENCES

|1| INTERNATIONAL ATOMIC ENERGY AGENCY, Advisory Group Meeting on the Use of Physical Standards in Inspection and Measurements of Nuclear Materials by Non-Destructive Techniques, IAEA Document AG-112 (1977).

|2| GUNNINK, R., EVANS, J.E., PRINDLE, A.L., A reevaluation of the gamma-ray energies and absolute branching intensities of ^{237}U, ^{238,239,240,241}Pu and ^{241}Am, Rep. UCRL-52139 (1976).

|3| EBERLE, H., IYER, M.R., OTTMAR, H., Branching intensities of ^{239}Pu gamma rays, in Rep. KFK 2686 (1978) 95-98.

|4| GUNNINK, R., A system for plutonium analysis by gamma-ray spectrometry, Pt 1: Techniques for analysis of solutions, Rep. UCRL-51577 (1974).

|5| American National Standard ANSI N 15.22 (1975).

|6| HAASE, V., Methoden zur Analyse von Gamma-Spektren, Rep. KFK 730 (1968).

|7| ROUTTI, J.T., SAMPO: A FORTRAN IV program for computer analysis of gamma spectra from Ge(Li) detectors and other spectra with peaks, Rep. UCRL-19452 (1969).

|8| BEYRICH, W., SPANNAGEL, G., Preliminary report on the evaluation of the AS-76 interlabtest, Kernforschungszentrum Karlsruhe (July 1978), unpublished.

|9| DRAGNEV, T.N., SCHÄRF, K., Non-destructive gamma spectrometry measurement of $^{239}Pu/^{240}Pu$ and $^{240}Pu/Pu$ ratio, Int. J. Appl. Radiat. Isotopes 26 (1975) 125-29.

|10| SPARKS, C.J., jr., Quantitative X-ray fluorescent analysis using fundamental parameters, Advances in X-Ray Analysis 19 (1976) 19-52.

|11| VEIGELE, W.M.J., Photon cross sections from 0.1 keV to 1 MeV for elements Z=1 to Z=94, Atomic Data Tables 5 (1973) 51-111.

|12| STORM, E., ISRAEL, H., Photon Cross Sections from 1 keV to 100 MeV for elements Z=1 to Z=100, Nuclear Data Tables 7 6 (1970).

|13| McMASTER, W.H., DEL GRANDE, N.K., MALLETT, J.H., HUBBELL, J.H., Compilation of X-ray cross sections, Rep. UCRL-50174 R, Section 2 (1969).

|14| CANADA, T.R., BEARSE, R.C. TAPE, J.W., An accurate determination of the plutonium K-absorption edge energy using gamma-ray attenuation, Rep. LA-6675-PR (1977) 22-24.

DISCUSSION

R. SHER: Do you have any experience using L X-rays instead of K X-rays in your X-ray fluorescence technique?

H. OTTMAR: We did not use L X-rays because we wanted to stress the non-destructive character of the technique. Non-destructive determination of the plutonium enrichment in clad fuels is only possible when the more penetrating K X-rays are employed.

A. KEDDAR: What measuring time was adopted to get good statistics for ^{239}Pu and ^{240}Pu at the 642 and 646 keV energy lines and to avoid interference from ^{241}Am?

H. OTTMAR: An assay time of 30 minutes proved to be sufficient for fairly accurate determination of the $^{239}Pu/^{240}Pu$ ratio from the higher energy gamma rays at 642/646 keV when the assayed samples contained at least 5 – 10 g of plutonium. The major interference affecting the attainable precision does come from ^{241}Am. May I refer you to Fig.4 in our paper, where we indicate how the ^{241}Am growth affects the $^{239}Pu/^{240}Pu$ ratio analysis.

F.V. FRAZZOLI: Could you give an indication of the precision attainable with your XRF instrumentation and the measuring time required?

H. OTTMAR: The precision attainable depends, of course, on several factors, for example the strength of the excitation source and the plutonium concentration in the sample material under assay. With the present instrumentation which uses a 10-mCi ^{57}Co source, the plutonium concentration in mixed-oxide fuels containing a few per cent of plutonium can be determined with a precision of 1% in less than five minutes' counting time.

T.N. DRAGNEV: Do you use X-ray fluorescence analysis for determining the oxygen to heavy metal ratios as well as the Pu/U ratios?

H. OTTMAR: In the X-ray fluorescence analysis described here we measure only the heavy element ratio, not the metal-to-oxygen ratio.

T.N. DRAGNEV: You indicated three methods of determining the Pu/U ratios. Can you compare the merits of the passive gamma-spectrometric and X-ray fluorescence methods?

H. OTTMAR: I should mention that we also investigated the analysis of gamma rays from uranium and plutonium isotopes for the Pu/U ratio determination. We found, however, that this approach requires much more elaborate data analysis and longer counting times than X-ray fluorescence analysis.

A PORTABLE NEUTRON COINCIDENCE COUNTER FOR ASSAYING LARGE PLUTONIUM SAMPLES

M.S. KRICK, M.L. EVANS, N. ENSSLIN,
C. HATCHER, H.O. MENLOVE, J.L. SAPIR,
J.E. SWANSEN
Los Alamos Scientific Laboratory,
Los Alamos, United States of America
M. De CAROLIS, A. RAMALHO
Department of Safeguards,
International Atomic Energy Agency, Vienna

Abstract

A PORTABLE NEUTRON COINCIDENCE COUNTER FOR ASSAYING LARGE PLUTONIUM SAMPLES.

A portable high-level neutron coincidence counter (HLNCC) has been developed at the Los Alamos Scientific Laboratory (LASL) for the assay of plutonium by inspectors of the International Atomic Energy Agency (IAEA). The counter is designed to measure the effective ^{240}Pu mass in plutonium samples which may have a high plutonium content. The term "high-level" refers to the high neutron count-rates produced by large (several kilograms) PuO_2 or plutonium metal samples. The counter measures coincident fission neutrons in the presence of a random neutron background. Total plutonium content is calculated from the plutonium isotopic composition. Correction procedures for removing non-linearities in the counter response due to multiplication effects in the samples are being developed for plutonium metal and PuO_2 samples. The detector consists of 18 ^3He proportional counters embedded in six polyethylene slabs, which form an hexagonal well. Top and bottom end-plugs can be used to form a closed sample-counting cavity. The detector weighs approximately 35 kg. A portable electronics package featuring shift-register coincidence counting electronics was designed for use with the detector; it includes all the electronic subsystems required for the operation of the counter except for the preamplifiers, which are mounted on the side of the detector. The electronics package is interfaced to a Hewlett-Packard HP-97 programmable calculator and to standard data communications devices.

1. Introduction

A portable, high-level neutron coincidence counter (HLNCC) has been developed at the Los Alamos Scientific Laboratory (LASL) for the assay of high-content plutonium samples (∿1 g to several kg) by inspectors of the International Atomic Energy Agency (IAEA). The term "high level" refers to the high neutron count rates produced by large (several kg) PuO_2 or plutonium metal samples. Figure 1 shows the neutron detector, the electronics package, and the interfaced Hewlett-Packard HP-97

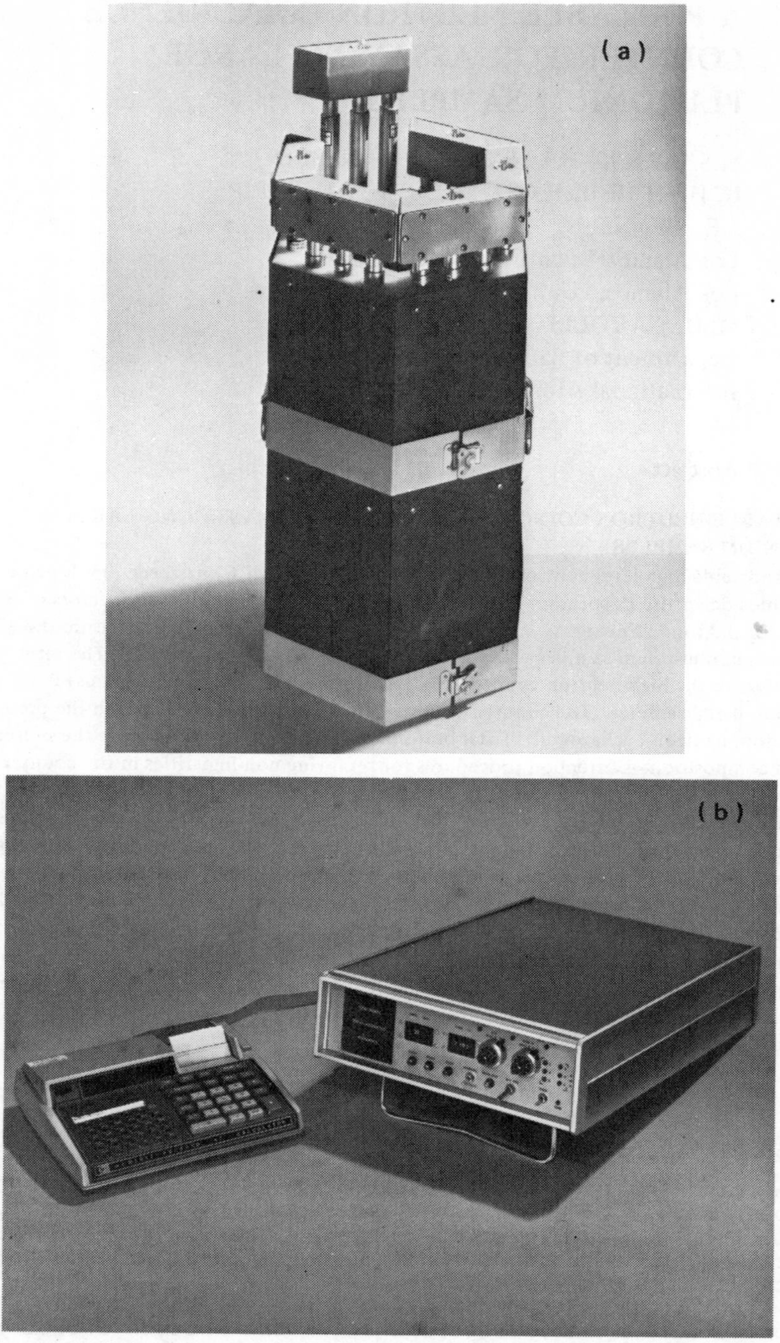

FIG.1. (a) HLNCC hexagonal neutron detector with one section of ^{3}He tubes partially removed. (b) HLNCC electronics package and HP-97 calculator.

programmable calculator. Three of these systems have been supplied by LASL to the IAEA for evaluation by the development and inspection sections.

The HLNCC assays plutonium-bearing material by detecting coincident fission neutrons from the plutonium in the presence of a random neutron background originating principally from (α,n) reactions in the material. The fission neutrons are primarily due to the spontaneous fission of the even-mass plutonium isotopes (^{238}Pu, ^{240}Pu, and ^{242}Pu) and to multiplication of (α,n) or spontaneous fission neutrons. The effective ^{240}Pu content of a sample is the mass of ^{240}Pu which would give the same corrected response to the measurement system as the actual ^{238}Pu, ^{240}Pu and ^{242}Pu content of the sample. If corrections can be made for dead time, multiplication, matrix, and geometric effects, the HLNCC assay determines the effective ^{240}Pu content of the sample. The total plutonium content is then calculated from the plutonium isotopic composition.

2. Neutron Detector

The coincidence counter basically consists of a sample-counting cavity surrounded by ^{3}He proportional counter neutron detectors embedded in cadmium-lined polyethylene slabs. Neutrons are absorbed by the polyethylene or cadmium, leak out the sides or ends of the detector, or are thermalized by the polyethylene and are captured by the ^{3}He proportional counters, which produce detector pulses via the ^{3}He(n,p)^{3}H reaction. The fraction of neutrons emitted in the detector which produces detectable pulses from the ^{3}He counters is the absolute detector efficiency and is approximately 12% for this counter.

The HLNCC is an example of a thermal neutron detector because most of the detected neutrons are thermalized in the polyethylene before being captured in the ^{3}He tubes. The die-away time τ of the detector is the average lifetime of a neutron in the detector and is approximately 32 μs for the HLNCC. The probability for a neutron to survive to time t following its emission is $\sim e^{-t/\tau}$.

Because the HLNCC is designed for field assay by IAEA inspectors, restrictions were placed on its weight and size to facilitate transportation and handling. To give flexibility in the physical configuration to accommodate a wide variety of sample containers, the counter is fabricated as six separate slabs that form an hexagonal well. The width of the well (18 cm minimum) accepts most PuO_2 sample cans, fast critical assembly fuel drawers, and some fuel rod assemblies; the slabs can be further separated to accept larger containers or, alternatively, two slabs can be used in a sandwich configuration to measure small samples or fuel rods. Three 2.5-cm-diam by 50-cm-long ^{3}He tubes (pressurized to 4 atm) are embedded in each section of the hexagon.

The HLNCC detector has three 0.4-mm-thick cadmium liners. The outside liner shields the detector from low-energy room background neutrons; the liner on the inside of the well prevents low-energy neutrons from returning to the sample after scattering in the detector and then causing neutron multiplication. The middle cadmium liner lowers the detector die-away time and reduces the sensitivity to hydrogenous materials in the sample or container.

The HLNCC is frequently used to assay PuO_2 in cans. Because the fill height of the PuO_2 in the cans varies with the mass and density in the can, it is desirable to have the detector's counting efficiency insensitive to these height variations. The dependence of the efficiency on axial position can be greatly reduced by placing plugs at both ends of the well to reflect into the detector neutrons that would otherwise escape through the ends. The effect of the end plugs is greater for the HLNCC than for most neutron well counters because the HLNCC is under-moderated and hence the polyethylene in the plugs has a large effect. The HLNCC units have 7.6-cm-thick end plugs with cadmium liners available for applications in which a flat central response is desired.

Design calculations for well counters show that in the low-efficiency range (<10%) the efficiency is directly proportional to the weight and die-away time (or the CH_2 thickness). The present detector weighs approximately 35 kg, including 7 kg for the end plugs. This particular efficiency and weight was chosen to give a total neutron counting rate of roughly 80,000 counts/s for a 2-kg PuO_2 sample (20% ^{240}Pu). The detector measures approximately 65 cm high by 35 cm wide.

3. Coincidence Counting

Neutrons originating from (α,n) reactions in the sample, from external sources, or from different fissions are uncorrelated in time (i.e., random), whereas neutrons emitted by a fissioning nucleus are time correlated. This correlation is exploited by the coincidence circuitry of the counter to distinguish between the coincident fission neutrons and the random background neutrons.

Figure 2 is a conceptual diagram showing the relative probability for detecting a neutron at time t following the detection of another neutron for a fissioning source (or a fissioning-plus-random source). Neutrons from a particular fission are detected at different times due to the detector die-away time. The probability for detecting another neutron from the same fission decreases approximately exponentially with time, whereas the probability for detecting a random neutron is constant with time (neglecting detector recovery time).

To distinguish correlated neutron events from random events (including neutrons from different fissions), two equal time periods are sampled by the coincidence circuit after a neutron has been detected (refer to Fig. 2). The circuit has a predelay, typically 4 μs, after the first neutron has been detected to allow the amplifiers to recover for detecting subsequent neutrons. The first time gate is then opened for typically 32 μs to detect the other neutron (or neutrons) associated with the fission event that gave rise to the first neutron. The neutrons detected during the first gate can be due

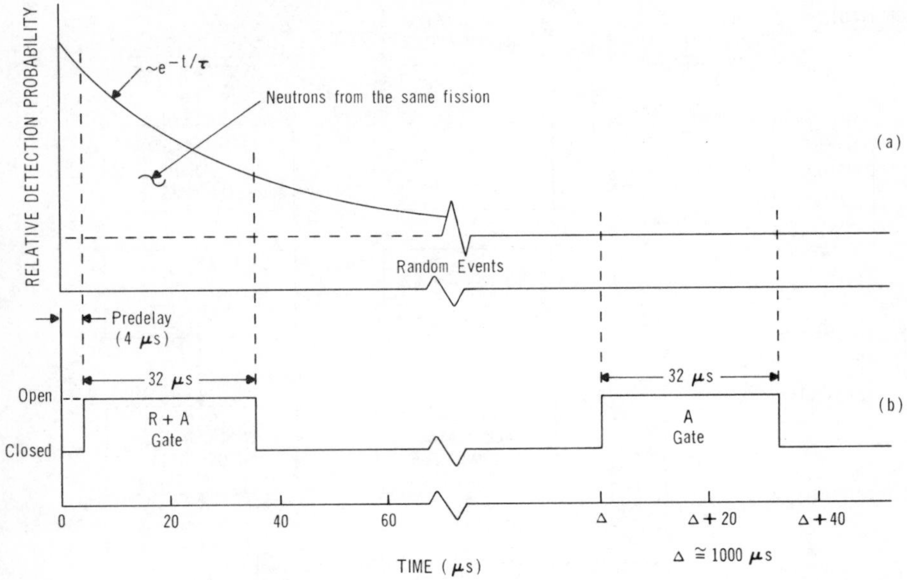

FIG.2. (a) Simplified relative neutron detection probability distribution.
(b) Coincidence gate timing diagram.

to either time-correlated spontaneous fission events or random neutron events. As a result, the counts accumulated during this gate are called the real-plus-accidental (R+A) events.

Following a long delay to ensure that any time correlation is removed, a second time gate is opened for the same duration as the first. The long delay must be at least several times longer than the detector die-away time and is 1024 μs in the HLNCC. This second gate counts only random events and is called the accidental (A) gate. Since the two gates are the same length, the net difference in counts received during the two gates is the net real (R) coincidence count and is related to the spontaneous fission rate in the sample.

4. Electronics

Each segment of the hexagonal detector uses a separate electronic channel for preamplification, amplification, and discrimination; the three ^3He counters in each segment are connected in parallel. Multiple analog channels are required to reduce the system dead time for high count rate applications. The six preamplifiers and the high voltage distribution components are contained in a box which mounts to the side of the detector. The electronics package contains all other electronic subsystems required to operate the HLNCC; these

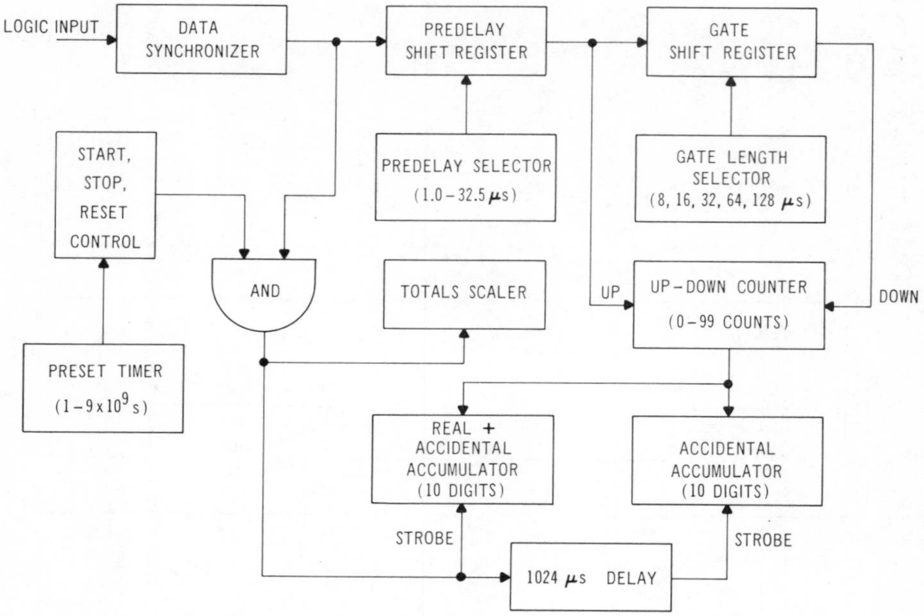

FIG.3. HLNCC shift register coincidence logic block diagram.

include high and low voltage power supplies, six channels of amplification and discrimination, shift register coincidence circuitry, and control, display and readout circuitry. An HP-97 programmable calculator is interfaced to the system for automatic data transfer and calculation. A standard RS-232C data communications interface is also provided for the automatic transfer of data to a terminal, computer, or tape recorder. The preamplifier input signals are processed by six linear pulse-shaping amplifiers (with 0.5 μs time constants) and six discriminators. The six discriminator outputs are ORed to form a single standard logic signal for the shift register (SR) coincidence logic.

A simplified block diagram of the shift register coincidence counter logic is shown in Fig. 3. The circuitry is based on a concept of Böhnel [1] as developed by Stephens, Swansen, and East [2] and improved by Swansen, et al. [3] for the present instrument. The predelay and gate are shift registers operating with a 2 MHz clock; e.g., a 32 μs gate uses 64 shift register stages with 0.5 μs/stage. The predelay SR has a selectable length from 1.0 to 32.5 μs in 0.5 μs steps. The predelay must be at least as long as the recovery time of the detector and amplifier stages; otherwise, the system dead time affects the R+A gate more than the A gate and introduces a bias on the net real count R. The gate SR has a selectable length of 8, 16, 32, 64, or 128 μs.

FIG.4. HLNCC and a germanium detector assaying a ZPPR fuel drawer mockup.

The standardized logic input pulses are first synchronized with the 2 MHz SR clock and then sent through the predelay SR and gate SR. The number of pulses in the gate at any time is determined by an up-down counter. A pulse entering the gate increments the counter and a pulse leaving the gate decrements the counter. Thus the number of pulses in the gate is just the count in the up-down counter. The contents of the up-down counter are added to the R+A and A accumulators for each input pulse as indicated by the strobe arrows in Fig. 3; the strobe to the R+A accumulator occurs immediately as the pulse leaves the synchronizer, whereas the strobe to the A accumulator is delayed by 1024 μs. Because a neutron pulse which enters the predelay shift register is produced at a later time than those neutron pulses already in the gate shift register, the R+A and A accumulators actually tally events which precede the neutron pulses which strobe the accumulators; this is functionally equivalent to the conceptual timing diagram in Fig. 2.

The principal advantage of the shift register coincidence circuitry is that each input pulse generates its own R+A and A gates; it is not necessary for one gate to close before the next can open. This feature greatly reduces the system dead time and allows operation of the HLNCC to counting rates of several hundred kHz.

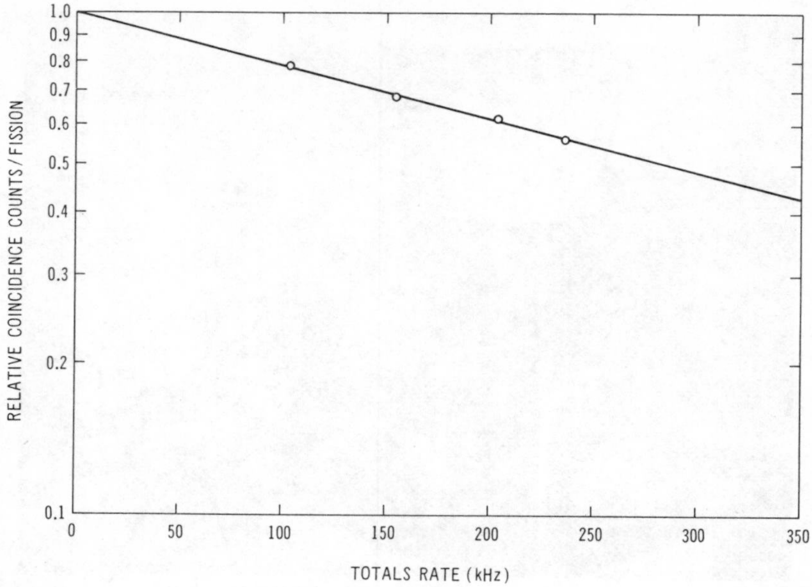

FIG.5. Relative coincidence counts/fission versus the totals count rate.

A preset timer allows measurements to be made from 1 s to 9 x 10^9 s. The timer in conjunction with the stop, start, and reset controls opens an AND gate to allow the collection of data. Each pulse passing through the AND gate is counted in the totals scaler. The data from the SR logic circuit (measurement time, total counts, real-plus-accidental coincidence counts, and accidental coincidence counts) are displayed simultaneously on the front panel.

The instrument derives all dc power supply voltages from dc-dc converters operating off of a single dc supply voltage, which is obtained either from a 100/115/230 Vac, 50/60 Hz line or a battery (10-30 Vdc). Power consumption is approximately 18 W.

5. Applications

5.1 Introduction

The HLNCC is a versatile coincidence counter that can be used to assay a large variety of plutonium sample types including oxide, mixed oxide, carbide, metal, fuel rods, fast critical assembly plates, solutions, scrap and waste. The instrument determines the effective ^{240}Pu content of the sample, after corrections are made for geometric, matrix, dead time, and multiplication effects. Total plutonium content is determined from the isotopic composition, which is either assumed known or is estimated from gamma-ray measurements made with a germanium detector. Figure 4 shows the HLNCC being used with a germanium detector for the assay of fast critical assembly fuel drawer mockups [4].

5.2 Calibration Source

A ^{252}Cf source is shipped with the HLNCC in the top end plug. The source can be removed from the end plug and mounted on a bracket near the center of the detector well for calibration of the system. Assay results in the field can be normalized to the ^{252}Cf response, which in turn is normalized to the response of plutonium standards at a calibration facility. The ^{252}Cf source strength is $\sim 10^4$ neutrons/s, which provides a detector count rate of $\sim 10^3$ counts/s.

5.3 Dead Time

Figure 5 shows the relative coincidence counts per fission for the HLNCC vs. the totals counting rate. Dead time corrections can be made easily for totals counting rates <250 kHz. The correction factor is $\exp(\delta T)$, where T is the totals counting rate and δ is the effective coincidence dead time (~ 2.4 μs).

5.4 Geometric Effects

The response of the HLNCC depends on the position of the sample and the presence or absence of the end plugs and other scattering material. The coincidence rate falls off more rapidly than the totals rate near the ends of the detector well because the coincidence rate varies as the square of the absolute detector efficiency. With the end plugs installed, the totals rate for a point source falls off $\sim 7\%$ from the center of the detector to the end of the active region; the coincidence rate falls off $\sim 14\%$. Axial and radial response curves will be presented in ref. 8.

5.5 Matrix Effects

When neutron coincidence counters are used to measure bulk plutonium samples that are combined with large amounts of nonfissionable matrix materials, it is important to understand the influence of the matrix materials on the assay results so one can minimize the sensitivity of coincident-neutron assay techniques to the physical form and spatial configuration of such matrix materials.

A series of measurements was made by placing a ^{252}Cf source at the center of cans containing sand, ash, MgO and various amounts of polyethylene in the form of chips and annuli, and by counting the cans at the center of the HLNCC detector well. An empty can with the ^{252}Cf source at its center served as the standard. The coincidence response was enhanced by 1-6% relative to the standard for the nonhydrogenous materials and the polyethylene chips. The polyethylene annuli (wall thickness = 0.5 to 4.3 cm) changed the response by ±17%.

Matrix effects are thus shown not to be large except for samples with very high hydrogen content. Matrix corrections can be determined directly with a source-addition technique [5].

5.6 Multiplication Effects

5.6.1 Introduction

Neutron coincidence counting of large plutonium metal and plutonium oxide samples is strongly influenced by neutron multiplication, which is due to spontaneous fission neutrons or (α,n) neutrons inducing fissions in the plutonium samples. This increases the total neutron production rate and the average multiplicity of the neutron-producing events. Because the coincidence counting rate is much more sensitive to multiplication than is the total neutron counting rate, the ratio of coincidence to totals counting rates can be used to correct for the multiplication effects in some sample categories.

5.6.2 Multiplication in metal plutonium samples

For plutonium metal samples, where (α,n) reactions are negligible, a multiplication correction due to Ensslin [6] can be performed. Let M represent the neutron multiplication, $\overline{\nu}$ the average neutron multiplicity for spontaneous fission in the sample, ρ the ratio of the coincidence to totals count rates for a small nonmultiplying sample (corrected for background), and ρ_M the same ratio for a sample with multiplication M. Then the multiplication is

$$M \cong \frac{1}{\overline{\nu}}\left[1 + (\overline{\nu}-1)\ \frac{\rho_M}{\rho}\right]$$

and the corrected net real coincidence count rate R_c is

$$R_c \cong \frac{R_m}{M}\ \frac{\rho}{\rho_M}\ , \qquad (1)$$

where R_m is the measured net real coincidence count rate.

Two groups of ZPPR [4] plutonium fuel plates were counted simultaneously in a thermal neutron well counter with various geometric couplings. The uncorrected net real coincidence response per second is plotted vs. the geometric separation of the groups as the upper curve in Fig. 6. The ratio ρ was determined by measuring one small ZPPR plate. The corrected response is shown as the lower curve in Fig. 6 and is constant within statistical errors.

These measurements indicate that this procedure can correct the assay of plutonium metal in any geometry for multiplication effects, whereas calibration curves or computed parametric response functions cannot correct for an arbitrary geometry.

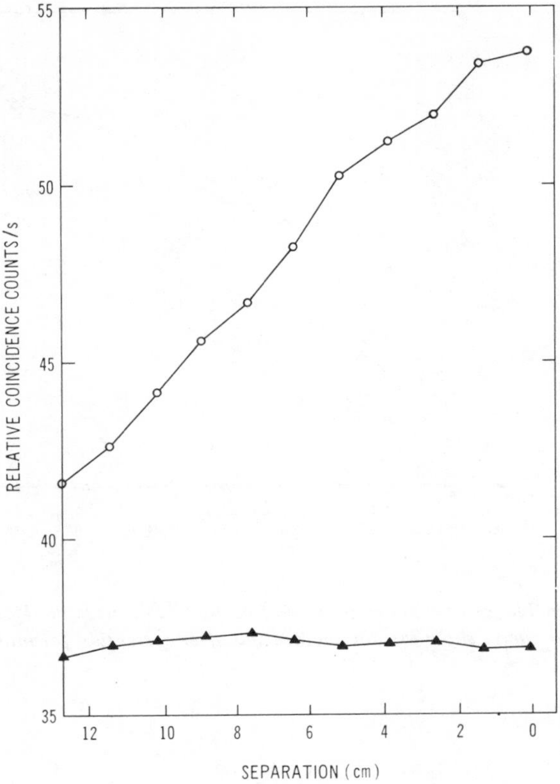

FIG.6. Relative coincidence counts per second vs. the separation between two groups of ZPPR fuel plates. The upper and lower curves show the data before and after correction for multiplication, respectively.

5.6.3 Multiplication in plutonium oxide samples

If the ratio of (α,n) to spontaneous fission neutrons (α) is known or can be measured, a multiplication correction due to Ensslin [6] can be applied to plutonium oxide samples. The multiplication M is

$$M = \frac{1}{\bar{\nu}}\left[1 + \frac{(\bar{\nu}-1)}{1+\beta} \frac{\rho_M}{\rho}\right] , \qquad (2)$$

where β is a function of the multiplication, the ratio of (α,n) to spontaneous fission neutrons, and the average neutron multiplicity for induced fission, and where the other parameters

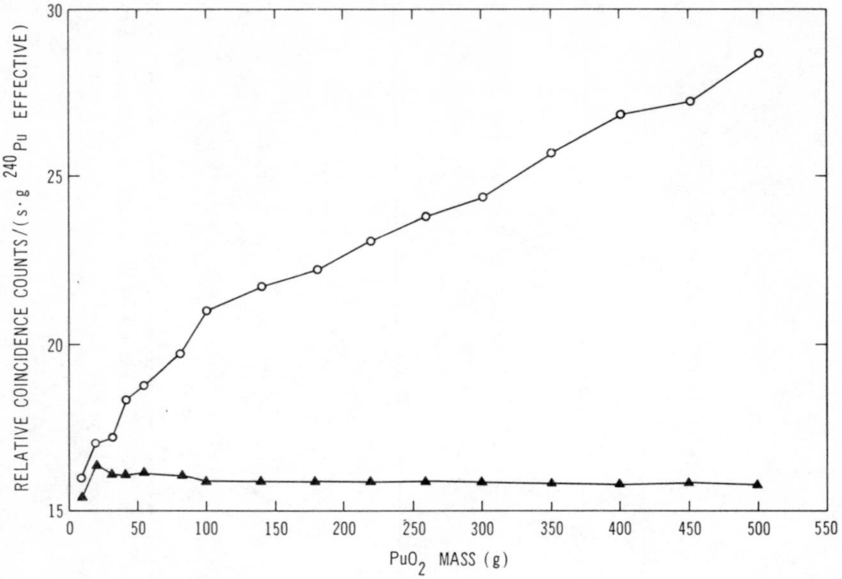

FIG. 7. Relative coincidence counts per gram of effective ^{240}Pu versus the PuO_2 sample mass. The upper and lower curves show the data before and after correction for multiplication, respectively.

have the same meaning as in section 5.6.2. The multiplication M is determined from Eq. (2); the corrected net real coincidence response is then given by Eq. (1). Multiplication corrections for dense PuO_2 were studied by observing the totals and coincidence responses from a thermal neutron well counter as a sample container was filled with PuO_2 powder. Uncorrected and corrected coincidence responses per gram of effective ^{240}Pu are shown in Fig. 7. Values for α and ρ were determined from the measurement of the smallest sample. The corrected response curve is flat within $\pm 3\%$. Also, fluctuations in the uncorrected response due to slight variations in powder geometry have been smoothed out.

6. Conclusion

The HLNCC is a portable, thermal neutron coincidence counter designed primarily for the assay of samples with large plutonium content. The instrument determines the effective ^{240}Pu content of the sample, after corrections are applied for dead time, geometric, matrix, and multiplication effects. Total plutonium content is determined from the isotopic composition.

The HLNCC is designed to have good detection efficiency for coincidence counting and moderate mass for portability. The instrument is useful for the assay of many sample categories, such as cans of PuO_2 powder, fast critical assembly fuel drawers, and fuel rod assemblies.

The use of shift-register coincidence counting logic and six parallel analog signal processing channels permits neutron counting rates greater than 250 kHz with moderate dead time corrections. Data can be transferred automatically from the electronics package to a Hewlett-Packard HP-97 programmable calculator for on-line analysis; a standard data communications interface is also included.

Engineering details on the electronics package are available [7]; detailed information on the HLNCC system, its operating procedures and characteristics, and coincidence counting applications will be available in the near future [8].

REFERENCES

[1] BÖHNEL, K., Die Plutoniumbestimmung in Kernbrennstoffen mit der Neutronen Koinzidenzmethode, Karlsruhe Rep. KFK2203 (1975); also AWRE translation No.70 (54/4252) (1978).

[2] STEPHENS, M.M., SWANSEN, J.E., EAST, L.V., Shift Register Neutron Coincidence Module, Los Alamos Scientific Laboratory Rep. LA-6121-MS (1975).

[3] SWANSEN, J.E., ENSSLIN, N., KRICK, M.S., MENLOVE, H.O., New Shift Register for High Count Rate Coincidence Applications, Los Alamos Scientific Laboratory Rep. LA-6788-PR (1976) 4.

[4] SAPIR, J.L., MENLOVE, H.O., TALBERT, W.L., DE CAROLIS, M., BEYER, N., Calibration of the IAEA High-Level Neutron Coincidence Counter for Plutonium Plates in ZPPR Drawers, Los Alamos Scientific Laboratory Rep. LA-6849-PR (1977) 4.

[5] MENLOVE, H.O., WALTON, R.B., 4π Coincidence Unit for One-Gallon Cans and Smaller Samples, Los Alamos Scientific Laboratory Rep. LA-4457-MS (1970) 27.

[6] ENSSLIN, N., (to be published).

[7] ALTMAN, D., "High Level Neutron Coincidence Counter (HLNCC): Maintenance Manual", EG&G Rep. EGG-1183 **5091** (1978).

[8] KRICK, M.S., et al., Los Alamos Scientific Laboratory Rep. (to be published).

DISCUSSION

H.P. FILSS: How would higher isotopes (americium or even uranium) affect your neutron coincidence result?

M.S. KRICK: The effect of uranium on the coincidence counter response is to increase or decrease the induced multiplication, depending on the nature of the sample. The effect of americium on the response is induced multiplication of the

(α,n) neutrons produced by americium alpha particles on light elements in the sample.

E.W. LEES: On this same point perhaps I can add that the HLNCC is sensitive to any isotope which spontaneously fissions; ^{241}Am has a negligible spontaneous fission rate, but the Cm isotopes could prove a problem.

NON-DESTRUCTIVE ASSAY OF LARGE QUANTITIES OF PLUTONIUM

M. DE CAROLIS, W. ALSTON, N. BEYER,
A. RAMALHO, D.E. RUNDQUIST
Department of Safeguards,
International Atomic Energy Agency, Vienna

Abstract

NON-DESTRUCTIVE ASSAY OF LARGE QUANTITIES OF PLUTONIUM.

A combination of passive neutron coincidence counting and high-resolution gamma-spectrometry measurements is used in nuclear facilities by the IAEA to verify the operator's statement of Pu content of items having different geometry, different composition and different quantities of Pu. A compact and readily transportable High-Level Coincidence Counter (HLNCC) developed by the Los Alamos Scientific Laboratory is used for the passive neutron coincidence measurements. Gamma-spectrometry measurements are performed using a high purity intrinsic germanium detector with a portable multichannel analyser SILENA – System 27. These measurements provide verification of the Pu isotopic composition which is then used in conjunction with the neutron coincidence measurements to determine the total Pu content in a sample. The two techniques are complementary and necessary for inspection purposes. In a typical field situation the inspector is faced with a number of measurement problems and problems relating to the availability of Pu standards. In one situation standards were developed from production material by sampling and subsequent mass-spectrometric analysis. The gamma-ray spectrometer calibration was established from an intrinsic efficiency calibration based on one of the Pu isotopes. A description is given of the instruments and techniques used, with particular emphasis on the field applications and measurement results obtained with bulk PuO_2 and Pu mixed-oxide powder. The measured quantities ranged from approximately one hundred grams to about two kilograms.

1. INTRODUCTION

The quantitative determination of plutonium in bulk material is one of the more important safeguards measurement problems faced by the IAEA. A major step forward in the achievement of technically credible safeguards was the development and application of techniques for the in-field verification of an operator's statement of large quantities of Pu material. The technique employs passive neutron coincidence counting for the determination of the ^{240}Pu effective content and high-resolution gamma-spectrometry measurements for the Pu isotopic composition. The measurements are to verify the Pu content of items having varying geometric configurations, chemical compositions and quantities which range from approximately 100 g to over 2 kg Pu.

A number of constraints are placed on measurements performed in the field that lead to problems not normally encountered when similar measurements are taken in a laboratory environment. Generally the inspector is faced with the prospect of large measurement uncertainties caused by the relatively short analysis times that are dictated by the operating regime of the facility, a high radiation background, the presence of electrical interference and problems in optimizing the geometrical layout of the equipment. A major problem is that of calibration, particularly for plutonium, since in some facilities calibration standards are not available. It is also not always possible to ship samples for analysis with the result that calibrations may have to be performed at remote facilities.

This paper summarizes the current approach by the IAEA for plutonium verification and reviews some of the experience gained to date in such measurements.

2. DESCRIPTION OF THE INSTRUMENTS

For the passive neutron measurements, a High Level Neutron Coincidence Counter (HLNCC) (Fig.1) developed by the Los Alamos Scientific Laboratory is used. The well neutron detector consists of 18 ^{3}He proportional counters embedded in a hexagonal array of six polyethylene slabs. It is combined with an electronic measuring unit based on a coincidence logic which separates the time-correlated spontaneous fission neutrons from the random neutrons (those originating from α,n reactions in the sample or from external sources) and it is used to assay the ^{240}Pu effective content of plutonium-bearing samples. The system, which is compact and suitable for inspectors to transport, is described more fully in the joint paper from LASL by Krick et al. [1].

Gamma-ray spectrometric measurements are performed using a large-volume high-purity intrinsic germanium detector (with approximately 14% efficiency and 1.9 keV FWHM resolution at 1.332 MeV of ^{60}Co), combined with a portable multichannel analyser Silena System 27 (Fig.2).

The main characteristics of the MCA System 27 are as follows:

MCA memory	1024 channels
Maximum count per channel	10^6
Analog to digital converter:	
Number of channels	2048
Clock rate	50 MHz
Spectrometric amplifier	
Data processor	
Data display on the cathode-ray tube	
HV power supply	
Digital cassette magnetic tape recorder.	

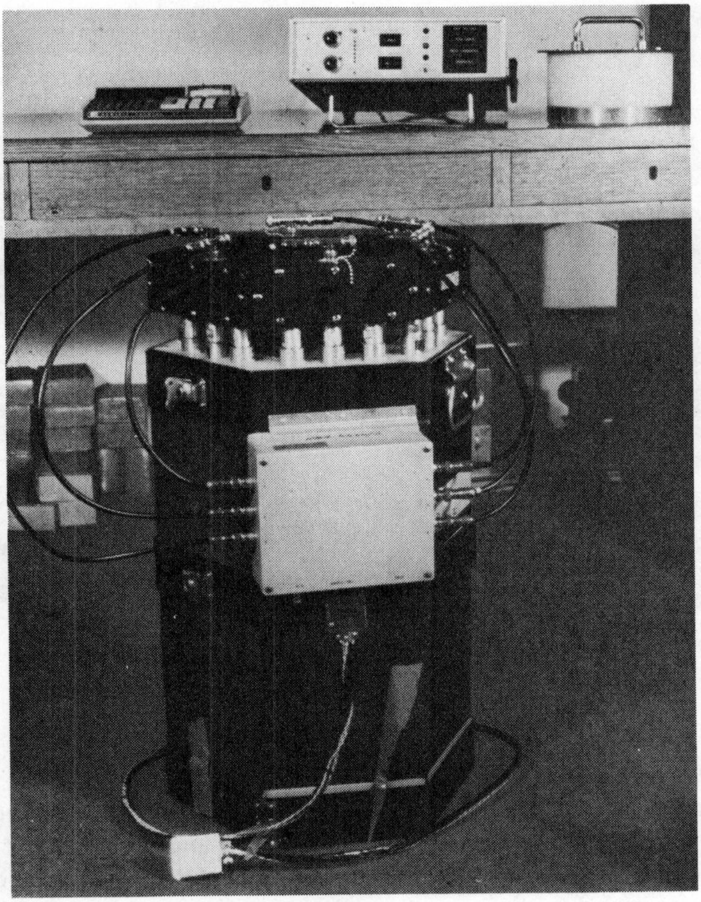

FIG.1. HLNCC system.

3. CALIBRATION OF THE EQUIPMENT

To achieve on site the verification of special nuclear material quantities, it is necessary to have either representative standards available at the time of measurement or a valid calibration previously performed. Physical standards for NDA measurements of nuclear material were considered in detail by an Agency Advisory Group [2].

It is often not possible to use materials from the facility under safeguards for the preparation of calibration standards, for reasons of accessibility for sampling, shipment problems of Pu-bearing samples, etc. In these cases the calibration of the

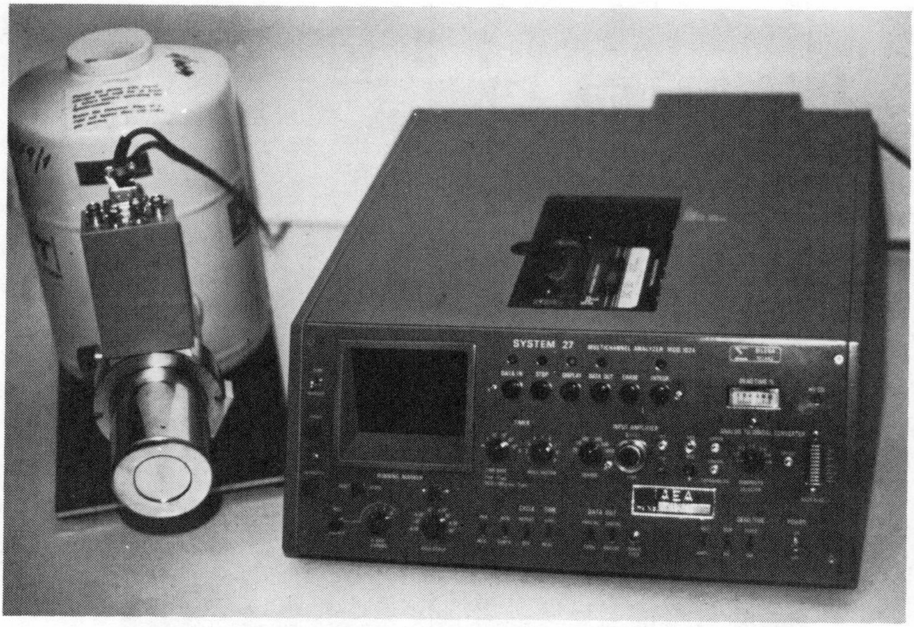

FIG.2. High-resolution gamma-spectrometry system.

equipment is performed at another facility, often in another country. A "normalization" coincidence count-rate measurement, using a low-activity (~ 10^4 n/s) ^{252}Cf source, which is a spontaneously fissioning isotope, is carried out at the same time. This normalization source forms an integral part of the HLNCC and, at the facility where the inspection takes place, it is used for a second normalization and removed prior to the commencement of measurements. The readings obtained are then adjusted by a function of the ratio of the two normalization measurements to correct for differences in the detector responses in the different measurements, taking account of the decay of ^{252}Cf in the interim period.

3.1. Calibration of the HLNCC

For a recent inventory involving bulk plutonium oxides, standards were developed from production material by sampling and accurately weighing a set of containers covering the range of concentration and isotopic content of Pu material to be assayed. Three sets of Pu calibration standards were prepared, for PuO_2 powder, $PuO_2 + UO_2$ mixed oxide and PuO_2 mixed with natural uranium oxide. The results of the mass spectrometric analyses for the different Pu samples selected for the calibration are shown in Table I.

The calibration of the HLNCC was performed by taking a set of four readings for each calibration standard, each for sufficient counting time to ensure counting statistics of better than 1%.

Table I shows the essential data for the calibration of two HLNCC systems. For each set of Pu material selected to be considered as calibration standard, the total Pu, ^{240}Pu effective (from the destructive analysis), the number of coincidences registered with both systems, and the difference (in percent) from the calibration curve of best fit, are reported.

It is clear that a good fit is obtained using a parabolic calibration curve. However, within the limits of error of the measurements, the range of materials measured and the type of electronics used, a linear calibration curve would also be sufficient. In this particular application, the effects of multiplication were not serious and were taken into account by the calibration curve.

However, a simple procedure for self-multiplication correction has to be studied for cases when large quantities of Pu have to be measured and the measuring conditions are not particularly optimal, as for the present application, where representative working standards were available.

3.2. Calibration of the gamma-spectroscopy system

While Pu standards are required in order to calibrate the HLNCC, it is necessary to know the isotope constituents of the plutonium to obtain total plutonium. The method for isotopic gamma-spectrometric analysis makes use of an intrinsic calibration independent of parameters such as the type of material, self-absorption, container and geometrical conditions. Further details of this approach are given in Section 4.

4. Pu ISOTOPIC MEASUREMENTS (^{241}Pu/^{239}Pu)

To correlate the quantity of ^{240}Pu effective measured with the high-level neutron coincidence counter, non-destructive high-resolution gamma-spectrometry measurements are carried out to determine the plutonium isotopic composition.

Considering the large number of samples that normally have to be measured, the analysis time for gamma spectrometry is in general limited to the same time period as for the HLNCC, i.e. about 5–10 minutes. The measurement time is often not sufficient to accumulate statistically acceptable intensities for ^{240}Pu and ^{238}Pu gamma lines and, for this application, it is therefore assumed that verification of ^{239}Pu to ^{241}Pu ratio information is adequate for acceptance of the operator statement on the Pu isotopic composition.

The NDA method used for isotopic composition determination does not give the absolute value of the single Pu isotopes, but the relative value, i.e. the ratio of

TABLE I. CALIBRATION DATA FOR TWO HLNCC SYSTEMS

Composition	Plutonium content[a]		No. of coincidences		Parabolic calibration curve best fitting		% difference from calibration curve	
	Total gr	^{240}Pu g eff.	HLNCC No. 2	HLNCC No. 3	^{240}Pu No. 2 g eff.	^{240}Pu No. 3 g eff.	HLNCC No. 2	HLNCC No. 3
PuO_2 powder	911.0	270.28	2836.52	2807.2	275.53	269.29	−1.94	0.37
	1971.6	578.53	6821.40	7053.3	573.91	579.42	0.80	−0.15
	2442.4	718.19	9123.79	9230.0	720.29	717.76	−0.29	0.06
	385.7	116.67	945.07	953.58	113.93	117.21	2.35	−0.46
$PuO_2 + UO_2$ (HEU) Mixed oxide	199.8	35.76	313.8	315.0	34.16	34.72	4.47	2.91
	669.3	120.24	1552.8	1536.4	117.58	117.33	2.21	2.42
	628.9	112.76	1500.8	1498.7	115.56	115.70	−2.48	−2.61
	128.9	23.08	222.0	211.98	25.06	24.94	−8.58	−8.06
	229.1	41.30	379.3	371.2	40.41	39.87	2.15	3.46
	460.4	82.69	903.9	919.1	83.05	83.28	−0.44	−0.70
$PuO_2 + UO_2$ (nat) Mixed oxide	62.93	13.29	95.1	97.3	12.43	12.60	6.47	5.19
	78.1	16.99	135.5	134.9	18.52	18.16	−9.01	−6.89
	134.6	29.25	209.9	215.1	28.51	28.70	2.53	1.88
	215.0	46.84	402.4	415.8	46.90	47.07	−0.13	−0.49
	221.4	47.95	418.8	427.5	47.97	47.79	−0.04	0.33

[a] Plutonium-content values obtained after sampling, weighing and chemical and mass-spectrometric analysis performed at the IAEA Safeguard Laboratory, Seibersdorf.

the isotopes of $^{241}Pu/^{239}Pu$. The advantage of this method is that it does not require Pu standards for calibration. The determination by non-destructive measurements of the plutonium isotopic ratios using the gamma-spectrometric technique has been considered by several authors [3–7].

To determine the isotopic ratio $^{241}Pu/^{239}Pu$ by gamma spectrometry it is necessary to measure the intensities of the gamma rays emitted by the nuclei of the corresponding isotope. To reach this condition it is necessary

(i) To perform a correct evaluation of peak areas, even if they are not well resolved;
(ii) To determine the overall relative efficiency curve of the complete measuring system including the sample attenuation.

To calculate the nuclear decay and the absolute branching intensities values, the data given by Gunnink [8] is used. However, differences between the values reported by Gunnink and those reported by the ANSI standard N 15–23 [8–9] indicate an uncertainty in the measurements of isotope ratios.

To perform the $^{241}Pu/^{239}Pu$ isotopic ratio, different gamma peak characteristics of the two isotopes were considered. For ^{241}Pu, the peak at 370.934 keV from the ^{237}U daughter of ^{241}Pu was chosen, assuming that a decay equilibrium has been established, because of the negligible contamination from the ^{241}Am and because it reduces the energy range used for the isotopic analysis. For ^{239}Pu, the peak at 413.712 was used for the isotopic ratio although the peaks at 345.014, 393.0 and 451.474 also gave good results.

To measure the isotopic ratio $^{241}Pu/^{239}Pu$ the overall relative efficiency calibration (ORE) of the measuring system, including the Pu sample, was determined. The overall relative efficiency calibration takes into account the sample attenuation, geometry, detector efficiency and electronic set-up.

To make the overall relative efficiency calibration for a given energy interval, a few well-known and resolved energy lines of ^{239}Pu were selected as follows:

$$E_i = 129.294;\ 345.014;\ \left.\begin{matrix}392.53\\393.14\end{matrix}\right\};\ 413.712;\ 451.474\ \text{keV}$$

The net peak areas for the above energy lines and for the ones at 370.934 keV, were obtained using a Gauss 6 programme.

Table II shows the results of the gamma-spectrometric measurements performed on the PuO_2 standards. The errors of the NDA gamma spectrometric analysis are of the same order as the statistical error in the peak areas selected for the overall relative efficiency calibration.

To utilize the NDA techniques more effectively it would be very useful to obtain a rough evaluation on site of the nuclear material quantities as well as

TABLE II. $^{241}Pu/^{239}Pu$ ISOTOPIC RATIO MEASUREMENTS ON PuO_2 POWDER

Sample	Analysing time (s)	$^{241}Pu/^{239}Pu$ ratio by destructive analysis	$^{241}Pu/^{239}Pu$ ratio by gamma spectrometry	Difference from analysis value (%)
1	600	0.117	0.108	7.7
2	900	0.117	0.114	2.6
3	1000	0.115	0.120	−4.3
4	1000	0.128	0.130	−1.3
5	600	0.114	0.108	5.3

ensuring reasonable statistics and resolution. To do this it is necessary to develop a PHA containing a microprocessor for use in the field. This will help optimize measurement times within the constraints at the facility and provide clean data for subsequent analysis at IAEA Headquarters.

5. $^{240}Pu/^{239}Pu$ ISOTOPIC RATIO

Investigations of the ^{240}Pu gamma rays at different energy values have been performed in order to find the most suitable energy line to be used for $^{240}Pu/^{239}Pu$ isotopic ratio if an experimental confirmation is necessary.

The ^{240}Pu gamma ray at 642 keV is not the most suitable one because the analysis is complicated by ^{241}Am gamma-ray interference, particularly at 641.4 keV and at other energy values in this energy region. Moreover, owing to the long counting time required to accumulate sufficient counts for good statistics, the ^{240}Pu gamma-ray peak at 642 keV is not interesting for safeguards purposes, especially when in-field measurements are considered.

Although the ^{240}Pu gamma ray at 160.3 keV is affected by the interference of other isotopes, it is considered to be the most suitable for the ratio of this isotope with ^{239}Pu. Corrections for the contribution of the isotopes ^{239}Pu, ^{241}Pu and ^{241}Am were applied. For these corrections, a well-resolved gamma ray of the same interfering isotope was selected and its intensity value was converted to the one at 160.3 keV using the known branching intensity and the overall relative efficiency values for the two energies. To make the overall relative efficiency calibration, the ^{239}Pu energy line at 129.294 keV was used in addition to the lines reported in Section 4.

6. APPLICATIONS IN THE FIELD

The IAEA's combined HLNCC/gamma-spectroscopy systems have been used in facilities containing plutonium in a number of countries in the world. The application has so far centred mainly on verification measurements of mixed oxides (fuel rods, fuel assemblies and scrap), plutonium powder and plutonium coupons.

The wide range of locations of these facilities around the world leads to particular problems for transport. The equipment is shipped by air packed in containers specially designed to minimize the risk of damage during transport. The transport containers can be easily and quickly handled, unpacked and repacked. The equipment itself is designed to be compact, simple to operate and readily assembled in a short time. In future, for a nuclear facility where the frequency activity verification is rather high, it will be worthwhile keeping the measuring NDA equipment fixed at the facility.

The types of measurement performed routinely are those consistent with the IAEA's verification activities for flow and inventory of nuclear materials at strategic points [10] in facilities. The frequency of these activities depends on the throughput of the facility and the nature of the safeguards agreement and can range from twice annually (for physical inventory verification) to regular weekly or even daily visits for the verification of flow or for short-term detection of diversion.

Although the HLNCC is designed to operate as a free standing unit, in many cases it is necessary to have auxiliary equipment specially designed. Such equipment includes a special mechanical hardware for the measurement of very long fuel assemblies installed permanently at a mixed-oxide fuel fabrication facility and a specially designed support for trays containing plutonium-bearing coupons. For the high-resolution gamma-spectrometry system, special collimators and support had to be constructed to ensure a geometry for optimum measurement conditions.

To minimize the counting time, the equipment is set up in such a way that counting on the item to be verified can be carried out simultaneously by the HLNCC and the gamma spectrometer. For items which fit completely into the HLNCC, the gamma spectrometer is lined up to view the nuclear material through the body of the HLNCC, while for other items (e.g. fuel assemblies) the spectrometer is located to view the nuclear sample directly.

Measurement times are determined not only by the composition, quantity and geometry of the item to be verified, but also by the IAEA's verification goals, the requirement to minimize the IAEA presence and possible interference in the operation of the plant and the type of sampling plan used ("attributes" or "variables"). To obtain good quantitative results with the gamma-spectroscopy system it is necessary to count for relatively long periods (typically 30–60 min) although good statistics can be obtained with the HLNCC with typical counting times of 5–10 min. Both instruments are combined to give quick "attributes" (test for the presence of plutonium) results with typical counting times of up to 5 min.

Experience has been gained in measuring the following classes of plutonium-bearing material using the techniques similar to those outlined above:

Plutonium dioxide powder
Mixed-oxide scrap (heterogeneous and homogeneous)
Plutonium alloy coupons
Mixed-oxide fuel assemblies and rods.

For these materials the plutonium content varied typically from 50 g to about 2 kg, although some measurements have been performed on smaller quantities. Errors in the measurement of the plutonium content using these techniques in the field are estimated to range from 3 to 10%, compared with the calibration based on chemical and mass-spectrometric analysis. Under more favourable conditions the error may even be smaller.

The HLNCC operates well in relatively high neutron backgrounds found in or near plutonium powder stores and mixed-oxide assembly stores and in the majority of cases no correction for background has been found to be necessary. More care has to be exercized in choosing the orientation of the gamma spectrometer although suitably placed shielding is used to minimize the effects of the background radiation.

Good consistency has been reported for measurements of material in all these categories, particularly where it has been possible to calibrate the systems with materials from the facility where the measurements were made, since in these cases the effects brought about by differences of composition, physical structure and measurement geometry were minimized.

Problems have been experienced in cases where the calibration had to be performed on material available in other facilities. In these instances higher systematic errors resulting from differences in geometry and composition between the measurement sample and the calibration sample were recorded.

The ^{252}Cf calibration source was found to be useful for checks in the field on the operation and calibration of the equipment, and particularly to correct such problems as the loss of one detector from the assay measurements system.

7. CONCLUSIONS AND RECOMMENDATIONS

The HLNCC coupled with the gamma spectrometer described in this paper provides a fast and reliable method for determining plutonium content of materials in quantities, physical form and compositions found in the nuclear industry today. The system has proved to be of considerable use in safeguards applications and has significantly improved the IAEA's capability for the measurement of plutonium-bearing materials.

The optimum use of the HLNCC by the IAEA inspectorate is being improved as more experience is gained. The following is a summary list of some requirements and possible improvements that we think will facilitate more effective and efficient use of the technique:

(i) The development of appropriate standards covering the range of the material quantities, characteristics and types and sizes of the containers;

(ii) To achieve on site the verification of the material quantity, representative standards or a valid calibration are required at the time of measurement;

(iii) A single program capability should exist to provide at least a rough evaluation of the special nuclear material and to indicate a gross difference for each sample between the operator and inspector data at the time the measurement is made;

(iv) If possible, a simple program should be added, in the neutron measuring system, that automatically checks the reliability of the data, e.g. by evaluating the compatibility of the gross, accidental and coincidence counts;

(v) To provide better statistical precision for the gamma-peak areas considered for the different plutonium isotopes without increasing the analysing time too much, a large-volume intrinsic Ge detector connected with a fast electronic preamplifier, amplifier and analog-to-digital converter circuit should be used together with a multichannel analyser memory of at least 2000 channels.

REFERENCES

[1] KRICK, M.S., et al.; "A portable neutron coincidence counter for assaying large plutonium samples", IAEA-SM-231/50, these Proceedings, Vol. II.

[2] The Use of Physical Standards in Inspection and Measurements of Nuclear Materials by Non-Destructive Techniques, Advisory Group Meeting 22–26 August 1977, IAEA AG-112 (1977).

[3] KRANER, H.W., On the use of gamma-ray spectroscopy to determine Pu isotopic abundances in plutonium sources, BNL 50237 (T-573) (May 1970).

[4] LEMMING, J.F., et al., Gamma-ray isotopic ratio measurements for the plutonium inventory verification programme, Monsanto Research Corp., MLM-2312, TID-4500 (Aug. 1976).

[5] DRAGNEV, T.N., Intrinsic self-calibration of non-destructive gamma spectrometric measurements, IAEA/STR-60, August 1976.

[6] GUNNINK, R., EVANS, J.E., In-line measurement of total and isotopic plutonium concentrations by gamma-ray spectrometry, UCRL-52220 (Feb. 1977).

[7] GUNNINK, R., Plutonium Isotopic Measurements by Gamma-Ray Spectrometry, MLM-2177 (Oct. 1974).

[8] GUNNINK, R., EVANS, J.E., PRINDLE, A.S., A re-evaluation of the gamma-ray energies and absolute branching intensities of U-237, 238, 239, 240, Pu-240 and Am-241, Lawrence Livermore Laboratory, UCRL-52139 (Oct. 1976).

[9] American National Standard N15-22-1975.

[10] ALSTON, W., BAHM, W., FRITTUM, H., SHEA, T., TOLCHENKOV, D., "Model for the application of IAEA safeguards at mixed-oxide fuel fabrication facilities", IAEA-SM-231/111, these Proceedings, Vol. I.

DISCUSSION

V.M. GRYAZEV (*Chairman*): The method you describe is very interesting but I think rather complicated for practical use. I have two questions. First, how do you take account of impurities of light elements and their effect on the total neutron radiation?

M. DE CAROLIS: The effect of impurities on neutron multiplication still has to be investigated.

V.M. GRYAZEV (*Chairman*): If calibration has to be performed at a remote facility, will you not have problems modelling all the possible solutions or mixtures?

M. DE CAROLIS: When calibrating the measuring equipment at a remote facility where representative working standards are not available there are a few parameters such as sample geometry, matrix effects, different isotopic composition, etc. which have to be considered for the construction of a calibration curve.

A.G. HAMLIN: The last two columns of Table I contain figures on precision of measurement. These figures appear to show a deterioration in precision from plutonium oxide to plutonium plus uranium oxide (HE) and then to plutonium plus uranium oxide (LE). Since calibration was made against like standards one wonders what the reason could be. A possible explanation is the decreasing amount of plutonium in the three groups of samples. Could you say whether this is the explanation or whether some other effect is involved?

M. DE CAROLIS: The working standards were made available on the basis of mass spectrometry which provides information on plutonium only. There may be a certain additional error using NDA equipment. Moreover, the effect of multiplication – although sensitive to the amount of Pu in a container – is stronger in the case of Pu plus uranium oxide owing to the (α,n) reaction and the matrix composition.

It is believed that the efficiency of the measuring system can be influenced by the presence of uranium in the sample.

EXPERIMENTAL AND THEORETICAL OBSERVATIONS ON THE USE OF THE EURATOM VARIABLE DEAD-TIME NEUTRON COUNTER FOR THE PASSIVE ASSAY OF PLUTONIUM

E.W. LEES
Nuclear Physics Division,
AERE Harwell,

F.J.G. ROGERS
Nuclear Materials Accounting Control Team, Harwell,
United Kingdom

Abstract

EXPERIMENTAL AND THEORETICAL OBSERVATIONS ON THE USE OF THE EURATOM VARIABLE DEAD-TIME NEUTRON COUNTER FOR THE PASSIVE ASSAY OF PLUTONIUM.

Careful experimentation with the Euratom Variable Dead-Time Counter (VDC) has highlighted several factors which must be heeded when using the instrument to assay Pu samples. (i) An accurate determination of the dead time of the scalers is essential to avoid unphysical calibration curves; (ii) care must be taken in extrapolating calibration curves taken over a small range of ^{240}Pu masses to large mass values; and (iii) the calibration curves differ for PuO_2 samples and for mixed PuO_2/UO_2 samples. An outline of how to calculate the theoretical response of a VDC to various samples is given. Neutron multiplication effects are shown to modify the linear response of the instrument with respect to ^{240}Pu mass. The factors affecting the extent of the neutron multiplication, such as sample mass, density, geometrical arrangement, chemical and isotopic composition, are investigated. Finally, limited comparisons between the experimental data and the theoretical calculations indicate provisional validity of the calculations.

1. INTRODUCTION

In the first part of this paper, the experimental experience gained with the Euratom Variable Dead-time Counter (VDC) is discussed in detail and some possible problems are outlined. The second half of the paper contains theoretical calculations of the response of the VDC and also attempts to explain some of the experimental results observed in the earlier pages.

During 1976, Euratom safeguards inspectors used a VDC for measuring the spontaneous fission (S.F.) neutrons from plutonium samples during their inspections of three UKAEA sites.

These results, which were made available to UKAEA staff indicated that the VDC had potential for the quantitative determination of plutonium stocks. We were allowed to borrow the instrument for a period of a few weeks, during which time we made calibration measurements and did additional work with stores samples containing up to 1.85 kg plutonium.

2. THE VARIABLE DEAD TIME PASSIVE NEUTRON COUNTER (VDC)

2.1 The Detector Assembly

This consists of a hexagonal array of six polyethylene slabs, each containing three helium-3 tubes, and has top and bottom slabs of polyethylene. A fuller description will be found in references [1] and [2] .

2.2. The Electronics

The system consists of pulse amplification and discriminator circuits for each of the six slabs, a pulse mixing circuit, and one fast plus four slow pulse counters with nominal dead times of 16, 32, 64, and 128 microseconds.

The observed counts in the five scalers are used to calculate the S.F. neutron count for the sample.

According to Birkhoff [1], the S.F. neutron count X_i may be derived from the equation

$$X_i = c_o - \frac{c_i}{1 - c_i t_i} \qquad (1)$$

where c_o = count in zero dead time scaler
and c_i = count in scaler with dead time t_i

In practice the equation does not give a linear correlation between X_i and the equivalent mass of ^{240}Pu (M_{240} e) in the sample. The equation may be modified to

$$X_{io} = X_i\ (1+b\,C_i) \qquad (2)$$

where b is a linear correlation factor derived from the experimental results from a set of standards. Alternatively the non-linearity can be accepted and the X_i values fitted to a polynomial of form

$$M(240\ e) = a + b\,x_i + c\,(x_i)^2 + \ldots\ldots \qquad (3)$$

Stanners [3] proposes a refinement of the Birkhoff equation and his paper explains the derivation of his modified version:

$$Y_i = \frac{X_i}{1-E_1 - \frac{c_i t_i}{1-c_i t_i} E_2} \qquad (4)$$

where $E_1 = e^{-t_i/q} .\ E$

$E_2 = q/t_i\ (1-e^{-t_i/q}).\ E$

and $E = 1 - q\ \ c_i(1 - e^{-t_i/q})$.

q is a constant related to the mean time lapse between the arrival of the first and subsequent neutrons from one fission event.

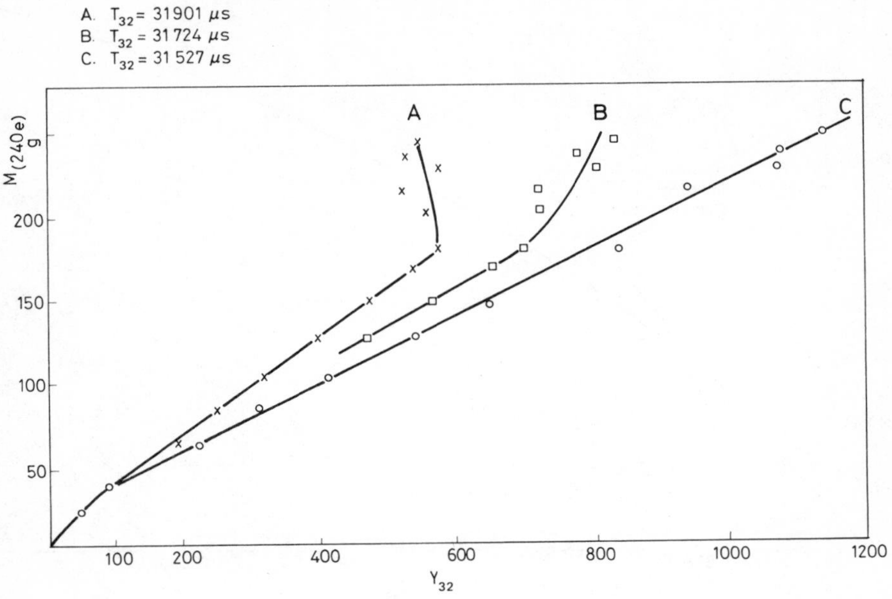

FIG.1. Comparison of Y_{32} values from the same data set calculated with different dead times.

3. INSTRUMENT CALIBRATION

3.1. Determination of Actual Scaler Dead Times

For random neutrons alone, $X_i = 0$; therefore:-

$$t_i = \frac{1}{c_i} - \frac{1}{c_o} \qquad (5)$$

A series of measurements at count rates between 3,500 cps and 45,000 cps were made with an americium-beryllium source. The t_i values were calculated for each result. No variations in t_i with count rate were noted, and the standard error of each dead time was < 0.15% of the mean value. These values were used for all our subsequent calculations. The effect of using an incorrect dead time to calculate the Y_{32} values for a set of plutonium/uranium mixed oxide sources are shown in Figure 1. Set C is calculated with our t_{32} = 31.527 µs. Set B used 31.724 µs, and set A 31.901 µs. We do not know whether the dead times for sets A and B were determined with insufficient accuracy, or whether these constants alter over a time scale of months. Clearly, a periodic recheck is desirable.

3.2 The Effect of Source Position

A set of plutonium oxide standards were grouped to simulate a 'can' 23 cm high and 10 cm in diameter. This can was counted at varying heights above the base of the detector assembly, and

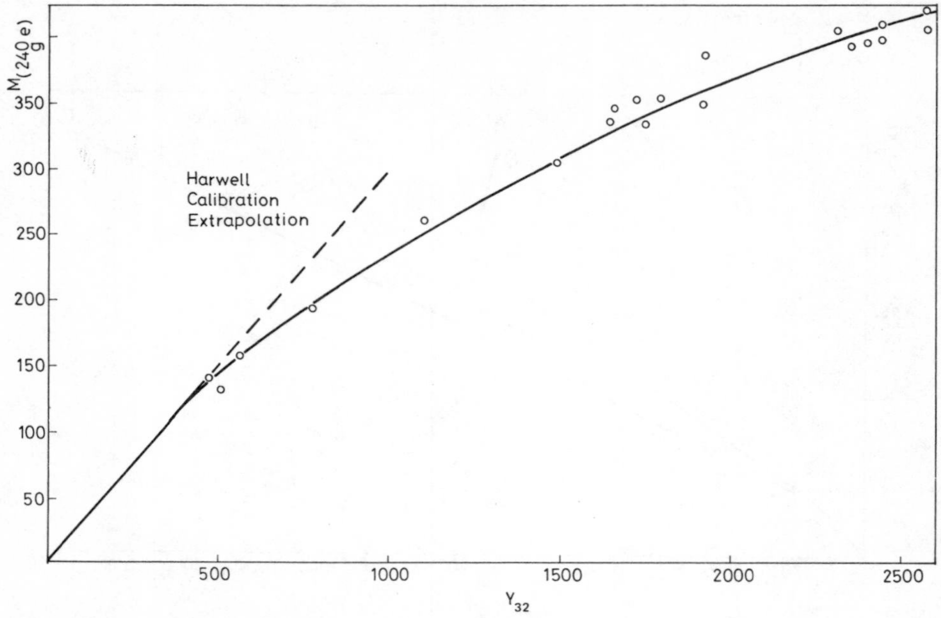

FIG.2. Plutonium oxide samples plotted to compare with the Harwell calibration.

the position of the maximum count-rate determined. A movement within ± 5 cm of this optimum position produced a diminution of count by about 2%. A 10 cm variation dropped the count by 10%, and a 20 cm variation lowered the count by 40%. Clearly, correct positioning of a sample is most important.

The same can was stationed at the optimum height, and displaced laterally. Up to 5 cm displacement showed no significant count rate variation. At 10 cm there was an enhancement of 8%.

3.3 Calibration Against Plutonium Oxide Standards

These standards consisted of plutonium oxide pellets in steel cans 1.27 cm in diameter and with lengths from 12.7 cm to 23 cm depending on the content. Used singly or together they covered a plutonium oxide weight from 50 to 650 g, with $M_{(240\ e)}$ from 9 to 118 grams. 3 consecutive counts were made for each plutonium weight, and the mean of each set used to calculate all 8 X_i and Y_i values.

Comparisons were made between the different values to identify (a) the relative accuracy of X and Y, and (b) the effect of the 4 alternative time constant scalers.

TABLE I. Y_{32} VALUES FOR A SET OF SIMILAR PuO_2 CANS

Can No.	Pu wt (kg)	$M_{(240\ equiv.)}$ (g)	Y_{32}	$Y_{32/g\,240}$
1	1.827	374.9	1753.6	4.678
2	1.816	372.6	1754.9	4.710
3	1.815	372.4	1776.9	4.771
4	1.817	372.8	1800.7	4.830
5	1.806	370.6	1617.1	4.363
6	1.804	370.2	1732.2	4.679
7	1.807	370.8	1778.8	4.797
8	1.820	373.5	1716.2	4.595
9	1.806	370.6	1893.8	5.110
10	1.815	372.4	1797.9	4.828

Mean $Y_{32/g240} = 4.736 \pm 0.191$ (4.0% 1 σ)

A linear regression fit to the X_{32} values gives

$M_{(240\ e)} = 0.46899\ X_{32} - 3.055$

The regression coefficient $r = \dfrac{S_{XM}}{S_{X}.S_{M}} = 0.99597$ and the standard error of estimate of $M_{(240\ e)} = 3.045$g (1 σ).

Similar treatment of the Y_{32} values gives

$M_{(240\ e)} = 0.28992\ Y_{32} + 0.8653$

$r = 0.99920$, and the standard error of estimate of $M_{(240\ e)}$ is 1.358g (1 σ).

Similar comparisons showed that the choice of time constants was not critical. The Y_{32} and Y_{64} values gave slightly better fits than the two others, and we chose Y_{32} for future calculations.

3.4 Effect of Counting Plutonium in the Presence of Random Neutrons

Three sets of plutonium standards, with $M_{(240\ e)} = 36.2$, 72.5 and 117.8 g, were counted with and without added random neutrons from an americium beryllium source, up to a maximum gross neutron count of 27,000 cps. All results with a count below 25,000 cps, irrespective of the S.F. : Random Neutron ratio gave consistent results. Above this count rate the Y_{32} values appeared to increase by about 5% above the normal value for that plutonium weight.

TABLE II. Y_{32} VALUES FOR A SERIES OF PLUTONIUM/URANIUM MIXED-OXIDE SAMPLES

Can No.	Gross wt of can (s) (kg)	Pu wt of can (s) (g)	Y_{32}	$M_{(240\ equiv.)}$ declared (g)	$M_{(240\ equiv.)}$ calc (g)	Difference calc/declared (%)
1	–	100.3	48.5	20.6	23.4	(13.7)
2	1.232	183.6	93.7	37.7	35.1	(6.8)
3	1.874	238.0	156.2	48.8	50.3	(–3.1)
4	1.970	289.6	190.6	59.4	58.2	(1.9)
5	2.314	337.8	254.5	69.3	72.3	(–4.3)
6	1.920	410.9	289.4	84.3	79.6	(5.6)
7	2.157	511.2	396.7	104.9	100.8	(3.9)
8	2.550	545.4	455.4	111.9	111.8	(0.1)
9	2.224	643.6	591.1	132.0	136.2	(–3.2)
10	2.406	701.6	651.1	143.9	146.9	(–2.1)
11	2.819	815.8	778.7	167.4	170.1	(–1.6)
12	2.974	860.7	841.1	176.6	182.0	(–3.0)
12+1	–	961.0	908.4	197.2	195.3	(1.0)
12+2	4.206	1044.3	943.8	214.3	202.7	(5.4)
12+3	4.848	1098.7	1074.9	225.4	232.1	(–3.0)
12+4	4.944	1150.3	1064.0	236.0	229.5	(22.8)
12+5	5.288	1198.5	1143.9	245.9	249.3	(–1.4)

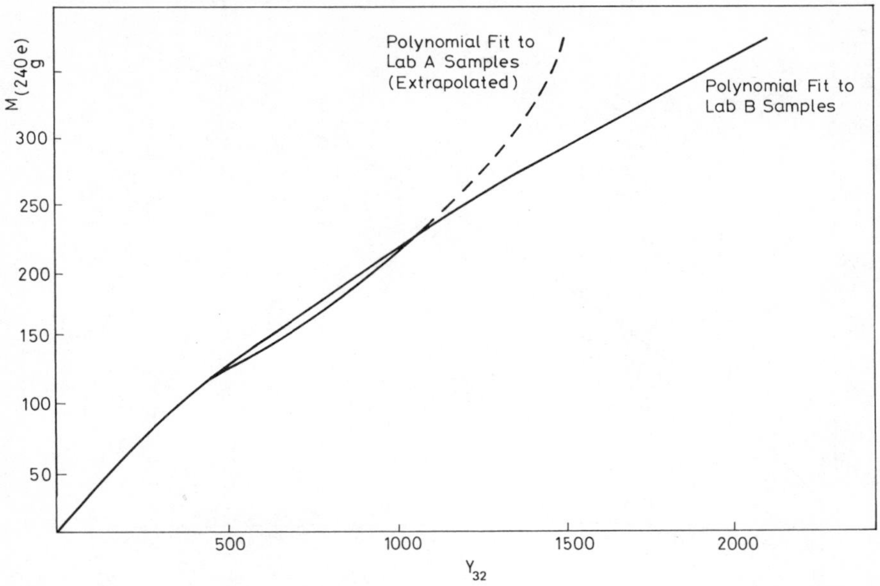

FIG.3. Two calibration curves for $PuUO_2$ mixed-oxide samples. (Polynomial fitting).

4. RESULTS FROM PLUTONIUM SAMPLES IN STORES AND LABORATORIES

4.1 Plutonium Oxide Samples

Figure 2 shows the results for a random selection of crude samples, with $M_{(240\ e)} \leqslant 250g$, plotted against an extrapolation of the standard calibration (Para. 3.3. et seq.) Clearly the linear regression does not hold above the level of the calibration sources ($M_{(240\ e)} \leqslant 118g$).

Consistency checks for repeat counts on one sample at $M_{(240\ e)} \leqslant 118g$ had shown a 1 standard deviation of $\pm$ 5% or better. Table I shows the consistency of results on a set of 10 plutonium oxide cans, all of similar weight and dimensions. The can to can variations are of similar order to those experienced in repeat scans on one sample.

4.2 Plutonium/Uranium Mixed Oxide Samples

The majority of samples found in the UKAEA stores and laboratories are in this form. The plutonium content is usually ~ 30% relative to the total, and the uranium varies from depleted to 93% enrichment.

The maximum $M_{(240\ e)}$ per can did not exceed 180g. To match the range of plutonium oxide samples, results for $180g < M_{(240\ e)} < 250g$, two cans were placed side by side. As for the pure oxide samples, the results for the mixed oxide samples do not lie on a straight line.

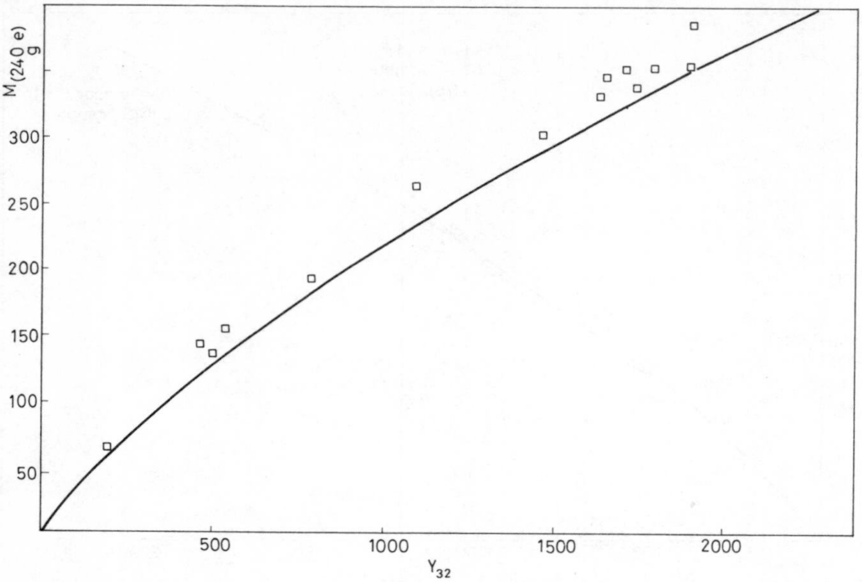

FIG.4. Plutonium-oxide sample results compared with the mixed-oxide calibration.

A three degree polynomial regression was derived and gives rise to the differences shown in Table II.

The derived polynomial was then used to calculate $M_{(240\ e)}$ values for a further selection of 31 mixed moxide cans. One result , for a can with $M_{(240\ e)}$ nominally of 28.5g, was 45% in excess of the declared value, the other 30 all fall within $\pm$ 10%. The anomalous result will be re-investigated when time permits.

The same polynomial was used to determine the $M_{(240\ e)}$ of a set of 46 cans at another site. The 39 cans with $M_{(240\ e)} \leqslant$ 250g gave results comparable with those in the last paragraph. Again the odd low can gave a very high result.

For the seven cans with $M_{(240\ e)}$ declared at 300 grams the results were far too high. Again the extrapolation of the calibration curve calculated for a limited range of $M_{(240\ e)}$ failed when applied to higher samples. A polynomial regression to the 46 results from this set does not differ greatly in the range $M_{(240\ e)} \leqslant$ 250g, but is of markedly different shape above this value. Figure 3 compares the two regression curves.

4.3 Intercomparison between mixed oxide and pure plutonium oxide samples

Figure 4 shows the individual points for plutonium oxide samples plotted against the regression curve from the plutonium/ uranium oxide samples.

Invariably the mixed oxide results give higher Y_{32} for equivalent $M_{(240\,e)}$ than the pure plutonium oxide.

5. CONCLUSIONS FROM THE EXPERIMENTAL WORK, AND FUTURE PLANS

5.1 The UKAEA V.D.C.

At the time of writing the Nuclear Materials Accounting Control Team (NMACT) of the UKAEA are commissioning a VDC. The counter assembly is of Euratom design and manufacture, the electronics of AERE Harwell origin. The only major difference between the two systems is that the NMACT instrument will have Böhnel [4] type shift register circuits as well as the VDC type.

Preliminary calibrations will be carried out as for the Euratom instrument, but with more time available it will be possible to investigate the real improvement in accuracy obtained by longer counts or more repeat counts per sample.

We also intend to investigate the practical effects of the density and mode of packing samples, to form a greater range of calibration curves for different materials, and above all to extend the UKAEA experience already reported [5] on the merits of shift register versus VDC circuits for coping with large samples.

NMACT also intends to co-operate with Nuclear Physics Division, AERE, in the investigation of the performance of the detector assembly.

Once the preliminary calibration work is completed, NMACT looks forward to a joint experiment with the IAEA, making a direct comparison with the Los Alamos designed Portable High Level Neutron Coincidence Counter (HLNCC). [8]

6. THEORETICAL RESPONSE OF VDC

It is important to understand completely the response of any equipment used in Safeguards work in order to assess its limitations and also to have confidence regarding the deployment of the device in the 'field'. For these reasons, the calculation of the theoretical response of coincidence neutron counters has been undertaken at Harwell : the Euratom VDC system of analysis will be described specifically although the method can be extended to the shift register system.

6.1 Simplified approach

It is the fact that spontaneous fission(SF) usually results in 2 or more neutrons being released simultaneously that allows us to use the VDC. Although the VDC moderator slows down the S.F. neutrons and spaces their detection probabilities in time, they still remain time correlated over a time interval commensurate with the dieaway time of the detector which for the Euratom VDC is 30 μs [1]. Thus one is in effect measuring a property which depends on the number of neutrons emitted during the S F event - this is called the neutron multiplicity. In fact, the quantity Y is defined by Stanners [3] as

$$Y = \sum_{i=1}^{m} (n_i - 1) \qquad (6)$$

where m is the number of detected pulse groups per second with n_i ($\geqslant 2$) pulses in the i th group, i.e. Y is the number of neutrons detected in the VDC assembly per SF less one multiplied by the fission rate.

Now the multiplicity of neutrons emitted in fission is well known; Terrell [6] has shown that the emission probabilities can be described by a Gaussian centred around $\bar{\nu}$ (the average number of neutrons released per fission) with a distribution width parameter σ. The quantity of interest is not the actual multiplicities of the fission process, but those measured with a detector of finite efficiency ε. The probability of observing i neutrons from a fission event characterised by an average release of $\bar{\nu}$ and width σ is

$$\text{PROB}\ (i, \varepsilon, \bar{\nu}) = \varepsilon^i \sum_{j=i}^{N} \frac{j!}{(j-i)!\,i!} (1-\varepsilon)^{j-i} P_j \quad (7)$$

where N is the maximum multiplicity considered ($\leqslant 9$) and P_j is the probability of the fission event releasing j neutrons.

For the SF of ^{240}Pu, $\bar{\nu}_{SF} = 2.15$, $\sigma = 1.1$ and for the Euratom VDC $\varepsilon = 0.062$ for a fission spectrum [1]. Comparison with equation (6) shows that we may define the theoretical value of Y as

$$Y = F . \sum_{i=2}^{N} (i-1)\ \text{PROB}\ (i, \varepsilon, \bar{\nu}_{SF}) \quad (8)$$

where F is the SF rate. This clearly shows the linear relationship between Y and ^{240}Pu mass in the absence of neutron multiplication.

If one of the neutrons from the SF of ^{240}Pu induces a 'fast' fission (FF) in the surrounding fissile or fertile material, this is essentially co-incident in time with the original SF. Thus this type of event effectively appears to the VDC as a single fission event with a much larger average number of neutrons being emitted than for SF i.e.

$$\bar{\nu}_{eff} = \bar{\nu}_{SF} + \bar{\nu}_{FF} - 1 \quad (9)$$

and since the individual $\bar{\nu}$ distributions are Gaussian,

$$\sigma_{eff} = \left[\sigma_{SF}^{\ 2} + \sigma_{FF}^{\ 2}\right]^{\frac{1}{2}} \quad (10)$$

To take account of neutron multiplication effects from FF induced by neutrons from (α, n) reactions in actinide oxides, we must include a term with an effective average number of neutrons equal to $\bar{\nu}_{FF}$. Thus defining

$$S(\bar{\nu}) = \sum_{i=2}^{N} (i-1)\ \text{PROB}\ (i, \varepsilon, \bar{\nu}) \quad (11)$$

we may write for the case when neutron multiplication effects are considered that

$$Y = (F - F_{FF})\ S(\bar{\nu}_{SF}) + F_{FF}\ S(\bar{\nu}_{eff}) + F_{\alpha n}\ S(\bar{\nu}_{FF}) \quad (12)$$

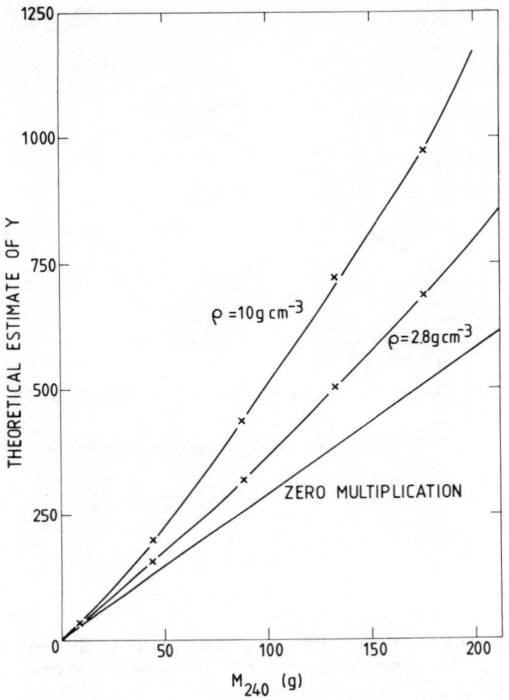

FIG.5. PuO_2 spheres. (239/240/241 is 0.7687/0.1995/0.0318).

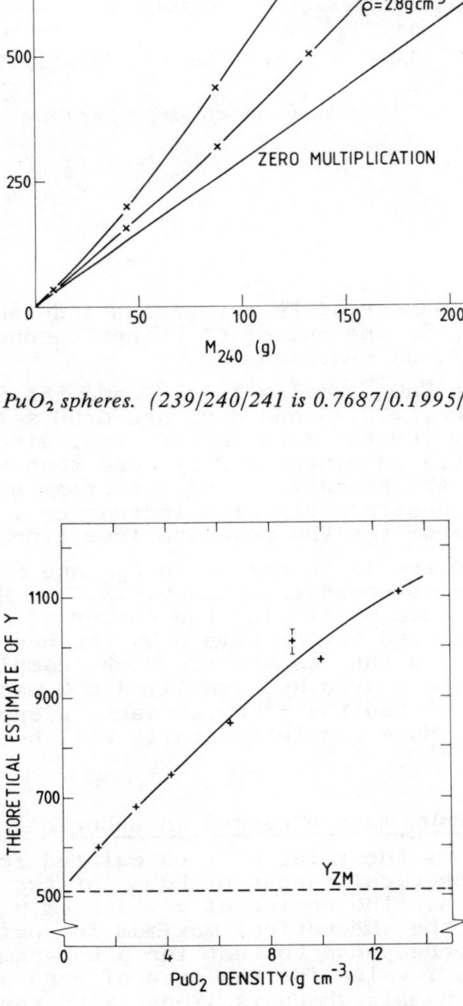

FIG.6. 1 kg PuO_2. Variation of Y with PuO_2 density (239/240/241 is 0.7687/0.1995/0.0318).

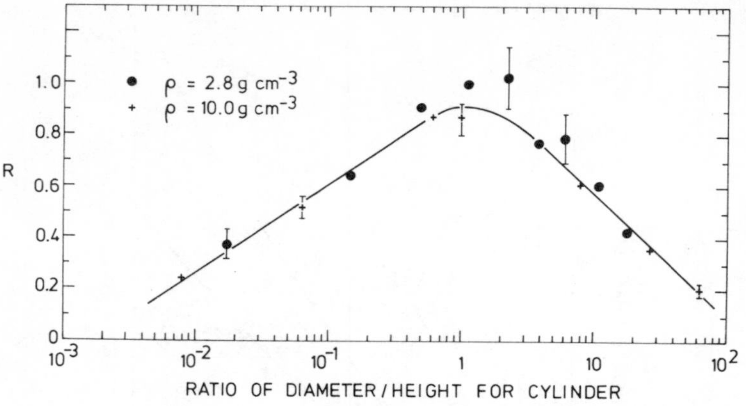

FIG. 7. Multiplication effects for 1 kg PuO_2.

where F_{FF} is the number of FF per second induced by the SF neutrons and $F_{\alpha n}$ is the number of FF per second induced by neutrons from (α, n) reactions.

Now $\bar{\nu}_{eff}$ for PuO_2 samples is 4.27 and the S values in equation (11) for $\bar{\nu}$ = 2.15 and 4.27 are 6.96 x 10^{-3} and 2.96 x 10^{-2} respectively for the Euratom VDC, i.e. although $\bar{\nu}$ has not quite doubled, S($\bar{\nu}$) has increased by more than a factor 4. It is the fact that the probability of detecting higher multiplicity events increases rapidly with increasing $\bar{\nu}$, which accounts for the departure of the VDC response from linearity.

All that remains is to calculate F_{FF} and $F_{\alpha n}$ in equation (12) and they are obtained from the U.K.A.E.A. Monte Carlo code MONK which is capable of listing the number of FF in each nuclide type contained in the sample by the neutrons from the original ^{240}Pu SF as they escape from the sample volume. A volume source was employed and usually 4,000 neutrons were tracked. Unless stated the ^{240}Pu content is approximately 20% of the total Pu. More complete details will be found in [7].

7. RESULTS

7.1 Increasing PuO_2 mass arranged in spherical geometry

Figure 5 shows the value of Y calculated from equation (7) for increasing spherical masses of PuO_2 for two different material densities. The choice of ρ = 10g cm^{-3} corresponds to a value close to the theoretical maximum for oxide fuel and ρ = 2.8g cm^{-3} corresponds to that for a non-compressed powder. Also shown is the Y value for the case of zero neutron multiplication in the sample which is linear with respect to ^{240}Pu mass.

TABLE III. EFFECT OF VARIOUS MATRIX COMPOSITIONS ON SIGNAL FROM A FIXED MASS OF ^{240}Pu (176.1 g).
Y_{ZM} = 512 and 4000 neutrons tracked.

	Composition[a]	Total Mass (kg)	FF	Y_{TOT}
DENSITY = 2.8 g cm^{-3}	PuO_2	1	157	681
	PuO_2 - 5% is 240	3.6	260	848
	30% PuO_2 / 70% UO_2 235/238 is 99/1	3.3	216	729
	30% PuO_2 / 70% UO_2 235/238 is 1/99	3.3	117	630
DENSITY = 10 g cm^{-3}	PuO_2	1	430	975
	PuO_2 - 5% is 240	3.6	953	1741
	30% PuO_2 / 70% UO_2 235/238 is 99/1	3.3	703	1134
	30% PuO_2 / 70% UO_2 235/238 is 1/99	3.3	302	815
	Pu metal Density = 19 g cm^{-3}	0.9	909	1327

[a] Unless stated Pu is 239/240/241 = 0.7687/.1995/.0318

7.2 Variation of density for a fixed spherical mass

Figure 6 shows the value of Y as a function of PuO_2 density for a fixed mass of ^{240}Pu (176.1g) arranged in spherical geometry; the zero multiplication signal is illustrated by the dashed curve. Obviously the neutron multiplication is sensitive to the material density, and ~40% difference in VDC response may be expected for 1 kg PuO_2 in close packed fuel pellet form as compared to loose powder.

7.3 Effect of cylindrical geometry

Normally the samples measured in Britain are doubly sealed in cylindrical cans. The neutron multiplication in a cylindrical geometry will be less than that for spherical geometry and will

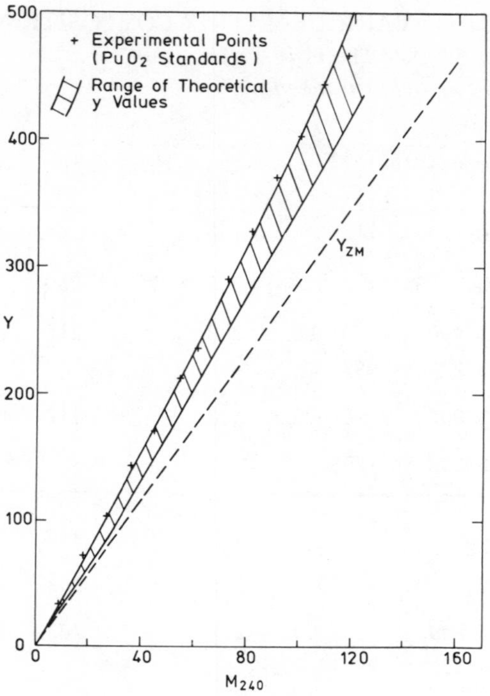

FIG.8. Attempt to compare experimental measurements and theoretical predictions of Y for PuO_2.

depend on the ratio of the diameter to the height of the cylinder. This ratio will vary from sample to sample and an attempt was made to relate multiplication in a cylindrical geometry to that for a sphere of the same mass and density by defining

$$R = \frac{Y_{CYL} - Y_{ZM}}{Y_{SPH} - Y_{ZM}} \quad (13)$$

where Y_{CYL} and Y_{SPH} refer to the Y values calculated allowing for neutron multiplication in a cylinder and a sphere respectively and Y_{ZM} is the value for zero multiplication.

Thus R essentially defines the fraction of the increase in VDC signal for cylindrical geometry to be expected relative to a sphere of the same mass and density. Figure 7 illustrates the behaviour of R as a function of cylinder diameter/height for 1kg of PuO_2 and R, as expected, is independent of density. For the diameter/height ratios frequently used in British stores, the multiplication effect is 80 to 90% of that for a sphere.

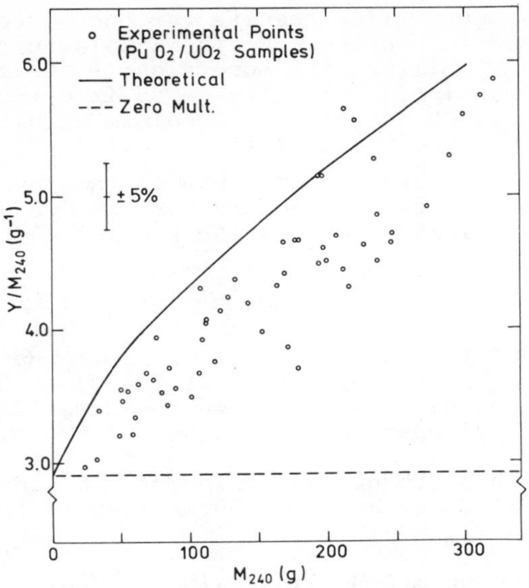

FIG. 9. Results similar to those for the mixed PuO_2/UO_2 samples discussed in Section 4.2.

7.4 Effects of Matrix Material

Table III lists the results for neutron multiplication for the same mass of ^{240}Pu (176.1g) in various matrix compositions arranged is spherical geometry. For the same material density, the PuO_2 sample with the lower content of ^{240}Pu(5.6%) would give the largest response and the UO_2/PuO_2 mixed oxide with natural U would give a lower response than the PuO_2 sample with 20% ^{240}Pu. This appears to be in contrast to the experimental observations outlined in section 4.3 earlier, but the majority of the mixed oxide samples were in fuel pellet form (i.e. high density material) whereas the PuO_2 samples of similar ^{240}Pu mass were in powder form.

8. DIRECT COMPARISON WITH EXPERIMENT

From the above, it is clear that the geometry, material density and matrix composition will critically affect the response of the VDC. Unfortunately the geometry and material density are not known for most of the samples measured to date. However, an attempt has been made in figure 8 to compare experimental measurements and theoretical predictions of Y for PuO_2 samples discussed in section 3.3 by making assumptions about the sample geometry. The experimental points appear to lie on the upper range of the calculated response and clearly indicate that multiplication is taking place even for these low mass values.

Figure 9 shows similar results for the mixed PuO_2/UO_2 samples discussed in section 4.2; the Y value per gram of ^{240}Pu has been plotted to illustrate more clearly the effects of neutron multiplication. The solid curve is a theoretical estimate and although too high in absolute magnitude, it does reproduce the observed trend of the results.

It is hoped to make measurements on samples whose geometry and density are well defined in the near future and that these will be a more stringent test of the present calculations.

ACKNOWLEDGEMENTS

Our grateful thanks are due to Dr U. Miranda (Euratom, Luxembourg) for the loan of the Euratom VDC; to Dr W. Stanners and his colleagues of Euratom Safeguards Directorate for their co-operation and discussions on the use of the V.D.C.

We also wish to thank the following U.K.A.E.A. staff for helpful discussions and active co-operation in the work described here:-

Mr A.S. Adamson and Mr A.G. Wain of NMACT
Dr B.W. Hooton, Dr J.W. Leake and Dr K.P. Lambert of A.E.R.E. Harwell.

REFERENCES

[1] BERG, R., SWENNEN, R., BIRKHOFF, G., BONDAR, L., LEY, J., and BUSCA, B. On the determination of the Pu-240 in solid waste containers by spontaneous fission neutron measurements. Euratom Rep EUR 5158 (1974).

[2] BIRKHOFF, G., BONDAR, L., and COPPO, N. Variable dead time neutron counter for tamper resistant neutron measurements of spontaneous fission neutrons. Euratom Rep EUR 4801 (1972).

[3] STANNERS, W. A brief note on the analysis of VDC results. Euratom, Luxembourg (1974) Priv. Comm.

[4] BOHNEL, K. Die Plutoniumbestimmung in Kernbrennstoffen mit der Neutronen Koinzenmethode. KFK Karlsruhe report KFK 2203 (1975).

[5] LAMBERT, K.P., and LEAKE, J.W. A comparison of the VDC and shift register neutron coincidence systems for 240-Pu assay. AERE Harwell Report R-8303 (Revised). In Preparation (1978).

[6] TERREL, J. Physics and Chemistry of Fission. IAEA Vol.II, Salzburg (1965) pp.3-24.

[7] LEES, E.W., and HOOTON, B.W., to be published as AERE Report R-9168.

[8] MENLOVE, H.O., ENSSLIN, N., EVANS, M.L., HATCHER, C.R., KRICK, M.S., MEDINA, E., and SWANSEN, J.E. Instruction Manual for the Portable High Level Neutron Coincidence Counter, Los Alamos, (1977).

DISCUSSION

L.A. STANCHI: I am pleased to see the good shape of your curves, but I disagree with the approach used. You can find imperfections in the analog circuits, but what you are calling "dead time" is produced in the digital part. The maximum variation in this dead time is ± half the period of the quartz stablilized clock, or about 30 ns. Your correction of 32 μs to 31.5 μs is completely arbitrary.

Equation (1) of your paper needs a correction for the dead time resulting from the detectors and analog circuits. What you are measuring as "total count" is already depleted as a result of this dead time, and C_0 in equation (1) should be divided by $1 - C_0\tau$, where τ is the dead time of the detectors and analog circuits. Equating X_i to zero when using a pure α,n source can be a means of obtaining τ, keeping t_i at its nominal value and neglecting the effect of cascaded dead times.

F.J.G. ROGERS: We know that the zero dead-time scaler does in fact have a finite dead time (equal to 0.80 μs for our own instrument). Comparisons for a set of standards, with the calculations corrected to allow for this, did not differ appreciably from those carried out using "arbitrary" values as determined in the paper.

Clearly, the assigned dead times giving the reverse curve effect seen in Fig.1 were wrong. We believe this is because they were determined by choosing values for t_i which gave a linear regression for the ^{240}Pu equivalent mass against the Y_i values. Owing to neutron multiplication, this method is incorrect and can produce the misleading results shown.

W. STANNERS: Our experience with the VDC counter on various types of plutonium-bearing material suggests that a very large calibration campaign would be required if this method is to be more than a consistency check in the field for samples of similar content and geometry. On the basis of your experience with the Monte Carlo method, do you think that Monte Carlo calculations could be used to reduce the amount of empirical calibration required?

E.W. LEES: Not in the foreseeable future, I fear, unless sample geometry and densities could be specified more accurately.

R. SHER: Rather than calculating multiplication effects on the basis of the size, type, or configuration of the sample, as you have done, can you use methods such as Dowdy's or Ensslin's in which multiplication is calculated on the basis of various inherent properties of the counts themselves?

E.W. LEES: We have not so far tried this approach.

NON-DESTRUCTIVE ASSAY TECHNIQUES FOR IRRADIATED FISSILE MATERIAL IN EXTENDED CONFIGURATIONS

P. FILSS
Institute of Chemical Technology,
Kernforschungsanlage Jülich GmbH,
Jülich, Federal Republic of Germany

Abstract

NON-DESTRUCTIVE ASSAY TECHNIQUES FOR IRRADIATED FISSILE MATERIAL IN EXTENDED CONFIGURATIONS.

The paper presents a non-destructive assay (NDA) system for the direct determination of irradiated fissile material in extended waste boxes and fuel elements. It is based on active neutron interrogation with an Sb-Be neutron source and specific attenuation of the source neutrons relative to the fission neutrons before registration. Based on previous experience with a small sample system, construction and performance of this assay system are reported. The system operates in a hot cell in the presence of some 100 Ci of fission products. The count-rate, obtained from source neutrons, was finally equivalent to 60 mg ^{235}U. This value indicates the lower detection limit of the system. One part of the system (I) is intended for small samples mainly for calibrating purposes. The samples stand there fixed during the measurement. In the other part of the system (II) the samples continuously move during the measuring turn. For waste boxes of 16.5 cm diameter and 25 cm high, the relative counting efficiency in system II is 50% compared with system I. Additional parameters were determined, which influence the measuring accuracy of the waste box. Different packing positions change the result by 5%, the addition of 500 g metal wool by 2% and the measurement of eight sub-samples as a whole by 4%. Performance is demonstrated with irradiated fuel elements of the AVR reactor at burnup values between 0 and 170 000 MW·d/t. Many types of fuel element from water reactors will fit within the dimensions of this measuring system II.

INTRODUCTION

An important research activity of the Kernforschungsanlage Jülich (KFA), Federal Republic of Germany, is the high-temperature reactor in combination with the U-Th fuel cycle. Reprocessing of this type of fuel is being investigated at the Institute of Chemical Technology. The fuel elements of the AVR test reactor (15 MW(e)) can be used for experimental reprocessing studies. They are graphite balls weighing 200 g, originally containing 1 g ^{235}U (93% enrichment) and 5 g thorium as mixed-oxide particles. They reach high burnup values and consequently contain a high content of fission-product radioactivity. Typical values are as much as 100 Ci of fission products per sample (fuel element) and dose-rate values of up to some 1000 R/h at 10 cm distance.

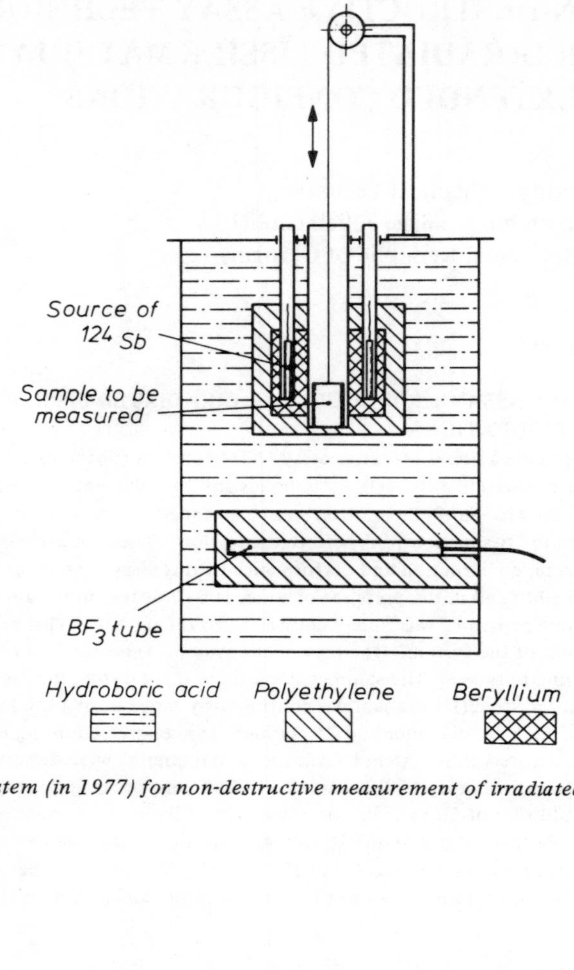

FIG.1. Assay system (in 1977) for non-destructive measurement of irradiated fissile material in small samples.

Reprocessing of this type of fuel begins with burning the graphite, which is followed by dissolving the remaining ash. Liquid solutions, which give access to chemical analysis, are obtained relatively late. The non-destructive analysis of solid material samples is therefore of great significance in the head-end of the reprocessing. Under the radioactivity and dose-rate conditions mentioned no conventional assay system worked. Therefore, a new measuring method and assay system was developed for this kind of radioactive sample, which operates at dose-rate values of 100 – 1000 R/h and under hot-cell conditions. An assay system for small samples [1] was first developed and tested successfully. This active neutron interrogation system makes use of the ^{124}Sb-Be photo-neutron source with a low neutron energy far below any fast fission threshold. The photo-neutron source needs hot-cell shieldings because of the strong ^{124}Sb γ-radiation.

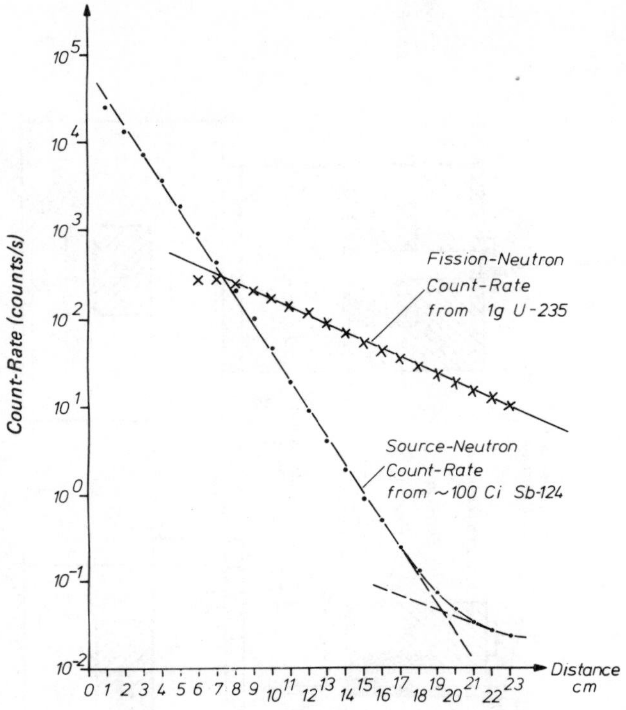

FIG.2. Variation of the source neutron count-rate (from Sb-Be) and the fission neutron count-rate (from fissile material in the sample) with varying distance between source and counter (thickness of hydroboric acid layer in Fig.1).

Since hot-cell shieldings must anyway be provided for handling reprocessing samples, the necessity of shieldings is no obstacle. The ^{124}Sb radioactivity may reach 100 Ci, which is of the same order of magnitude as the fission-product radioactivity expected.

Figure 1 shows the assay system previously developed for small samples. The photo-neutrons are generated in the core of metallic beryllium surrounding the small sample. Fission neutrons (fast, prompt) are generated by the source neutrons in any sample containing fissile material. According to their different energy, both kinds of neutron are differently attenuated on their way to the neutron counter at the bottom, which is a moderated BF_3-tube. Figure 2 from Ref.[2] shows the attenuation in concentrated boric acid, which is discussed in more detail in Ref.[3]. By this specific attenuation the contribution of the source

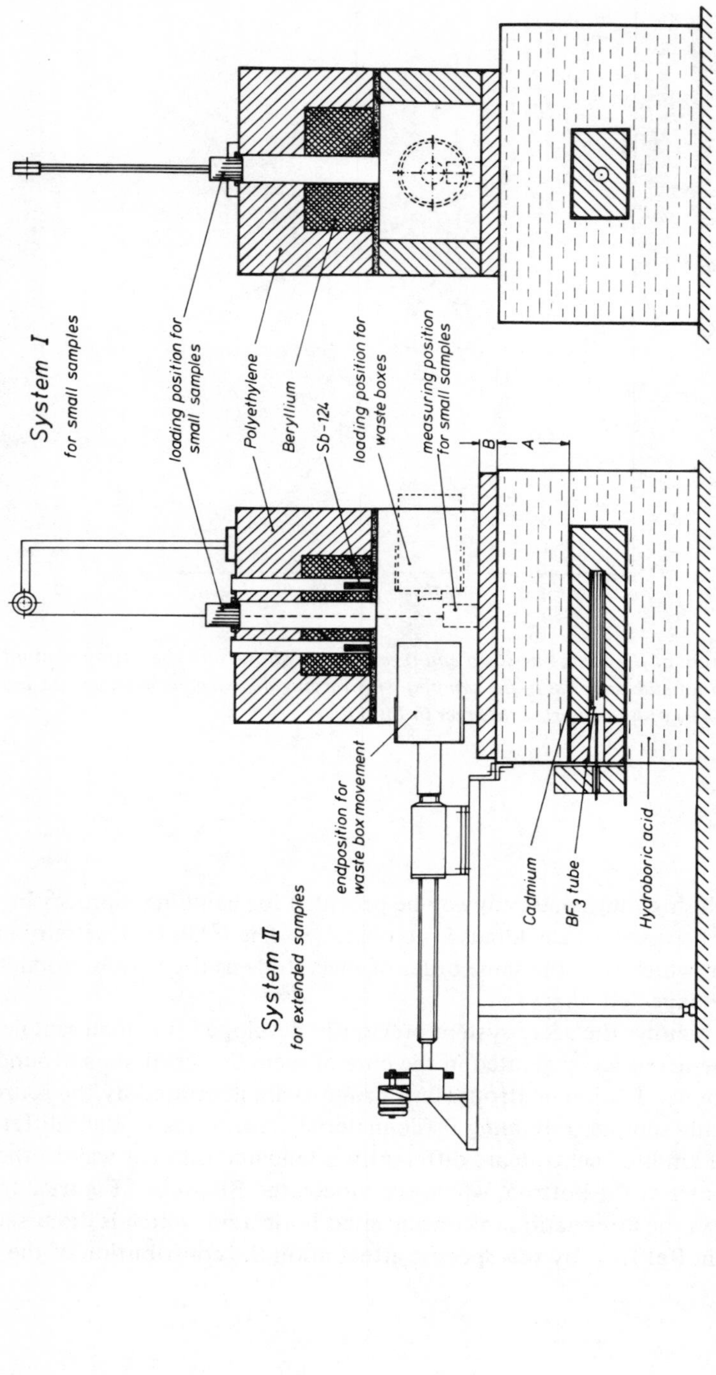

FIG.3. Assay system for extended samples containing mixtures of fissile and radioactive material.

neutrons to the total neutron count-rate was effectively reduced to a very low background level. The neutron count-rate from fissile material was therefore determined by subtraction of the source neutron count-rate from the total count-rate. The BF_3 tube can operate at the mentioned dose-rate in a hot-cell installation.

1. THE NECESSITY FOR ANALYSING EXTENDED SAMPLES

The assay system seen in Fig.1 worked reliably and had a very low background count-rate, equivalent to approximately 5 mg ^{235}U. But the sample size in this set-up was limited to a diameter of approximately 6 cm. Larger samples must be expected in the course of reprocessing experiments, and as fuel elements from water reactors. This kind of sample cannot simply be subdivided and put into small sample containers. For example, to handle and pack extended waste samples 5-litre boxes 165 mm diameter and 250 mm high are used at KFA Jülich. They can be filled with filter packages or other radioactive configurations of fissile material. Possible direct control of these radioactive boxes by NDA techniques will avoid excessive sampling and preparation work with radioactive material. Samples may be measured directly, which otherwise can only be controlled by book inventory.

The purpose of this work is to modify and extend the assay system for small samples (Fig.1) in order to handle larger samples of the above-mentiond size and similar configurations of fissile material.

2. GENERAL CONSIDERATIONS FOR LAYOUT AND CONSTRUCTION

As mentioned the sample size is no longer small in comparison with typical neutron transport properties and the dimension of the neutron counter. Therefore an attempt will be made to move the extended sample continuously through an open geometry of the low-energy Sb-Be neutron source. To re-measure fission neutrons selectively, the same type of moderated BF_3 tube will be used as counter, preceded by an equivalent attenuation layer. This kind of moving and averaging has been successfully used in other systems, mainly by the Los Alamos group [4]. The attenuation curves shown in Fig.2 can be understood from scattering and diffusion theory as general relations for the one-dimensional case [3]. Similar attenuation properties are expected for equivalent layers of hydrogenous material combined with thermal neutron absorbers in a one-dimensional geometry. More detailed calculations, such as those for a sigma pile, cannot be readily performed because the assay system must provide considerable open space in order to bring extended samples to the measuring position. From Fig.2 the optimum attenuation layer A was expected to be around 18 cm. After the construction of a preliminary system it is necessary to reduce the background count-rate as much as possible. No quantity of fissile material can be determined under hot-cell conditions if it gave rise to only a small fraction of the background count-rate. The frequent use of calibrating samples must be anticipated. It would be advantageous if the same spherical fuel element can be used as standard as in the earlier system. The space-integrated count-rate should not be too dependent on the packing position of the fissile material in the sample box. This is important for waste box contents that are not exactly known.

3. FINAL LAYOUT OF AN ASSAY SYSTEM FOR EXTENDED SAMPLES

The assay system shown in Fig.3 was chosen at the end of an extensive development series. It incorporates two independent measuring systems – system I for small samples and system II for extended samples. All types of sample that could be handled in Fig.1 can be measured in system I. The small samples are loaded at the top and transported by a lift to the measuring position indicated in Fig.3. This measuring position is situated outside the beryllium core in the space between the neutron source and the attenuation layer (A + B) of the counting region. A + B is composed of 18 cm borated water + 5 cm polyethylene. This measuring position for small samples is not completely surrounded by reflector material as was the case in Fig.1. The source neutron spectrum turned out to be somewhat more energetic, which is advantageous for the assay of concentrated fissile material.

Between the Sb-Be source at the top and the attenuation layer A + B of the counter an open space remains. It extends in a horizontal direction with a cross-sectional area of 21×28 cm^2. This open throughput renders the measurement of extended samples possible in system II.

Long samples such as water reactor fuel elements can be moved through it in a horizontal direction for measurement. The neutron flux density and the neutron counting efficiency for fissile material have their maximum in the centre of the system as can be seen from Fig.5. The most effective position for neutron interrogation and measuring is practically the same region, which is used by the lift of system I. By contrast with system I the sample does not remain fixed in its position during a measuring turn but moves horizontally across the centre of the system. A screw-like driving mechanism permits waste boxes of 16.5 cm diameter and 25 cm length to be moved and rotated with constant speed. This type of waste box is loaded at the right side, as indicated in Fig.3, and moved over a 35-cm length in 1000 s, which is exactly the measuring time. By the combined movement and rotation, an average is obtained over the different parts of the box. The measuring site is surrounded by polyethylene and

TABLE I. COUNT-RATES (counts/s) FROM SOURCE NEUTRONS AND FISSION NEUTRONS IN AN ASSAY SYSTEM SIMILAR TO FIG.3, SHOWING THE NECESSITY OF THERMAL NEUTRON ABSORBERS

	Without Cd layer between attenuation layer A and counter	With Cd layer between attenuation layer A and counter
Source neutron count-rate	65.7	0.8
Additional neutron count rate from 1 g ^{235}U	4.9	2.6

beryllium, both of which are good reflector materials. It must be pointed out that this geometry is not safe from the aspects of critiçality, not even for ^{235}U. This can be seen from Fig.1.A.3 in Ref.[5]. It is strictly necessary to avoid measuring any sample that is entirely unknown. Only if the range of fissile material is approximately known, e.g. below 100 g ^{235}U, will no problem arise from criticality.

Otherwise, *by bringing slightly undercritical masses into the good reflecting geometry, critical accidents may occur.* Figure 3 shows the assay system in its final state. The most effective modifications, which led to this state and to the acceptable source neutron count-rate, are mentioned in the following.

(a) The selective neutron attenuation in poisoned hydrogenous material could be demonstrated in a preliminary system without the lateral polyethylene bricks where pure water was used instead of boric acid. In that assay system the neutron count-rate was registered with and without the cadmium sheet between the attenuation layer (A + B) and the moderated BF_3 tube counter. The measured count-rate reported in Table I shows that the cadmium layer reduces the source neutron count-rate by a factor of 80. This result indicates that the selective attenuation of different neutron streams even works in this modified geometry. The elimination of thermal neutron diffusion is effective for source neutrons as shown in Ref.[3]. The quantitative agreement with attenuation curves from optimal shielded systems as in Fig.2 is not as good as was expected. Therefore, the first choice for the attenuation layer thickness of 18 cm was finally changed to 23 cm. The source neutron count-rate was still relatively large in the system described, i.e. equivalent to 300 mg ^{235}U. It was further diminished by subsequent modifications.

TABLE II. NEUTRON COUNT-RATES (counts/s) IN THE ASSAY SYSTEM (ACCORDING TO FIG.3) INFLUENCED BY ADDITIONAL POLYETHYLENE MATERIAL IN FRONT OF THE COUNTER REGION

	Neutron count-rate: Without	additional polyethylene block in front of the counter region: With
Source neutron count-rate	3.4	1.1
Fission neutron count-rate from 1 g ^{235}U	38.0	38.0

(b) Originally the irradiation section necessary for system II was open in all lateral directions. The neutron source section was only supported by four metal pillars. It was then closed at the front and at the back by 10-cm-thick lateral polyethylene bricks, which diminished the neutron escape from the source, and the formation of an intense scattered neutron background. The source neutron count-rate was reduced by another factor of 3.

(c) Additional shielding was very effective in the front section of the BF_3 tube moderator. A 10-cm-thick polyethylene block, slightly wider than the BF_3 tube moderator, reduced the source neutron background by another factor of 3 as shown in Table II.

The right column of Table II includes typical count-rate values, which were obtained in the final state of the system and at a typical intensity. The source neutron count-rate is equivalent to 30 mg ^{235}U in the measuring geometry of system I.

4. PERFORMANCE IN FISSILE MATERIAL DETERMINATION WITH UNIRRADIATED AND IRRADIATED MATERIAL

The strength of the Sb-Be neutron source decreases with the half-life of ^{124}Sb of 60.3 days. This changing intensity requires special means to compare the measured count-rates of different days. This is accomplished by comparison of the measured count-rate with that of a standard sample on the same day. An

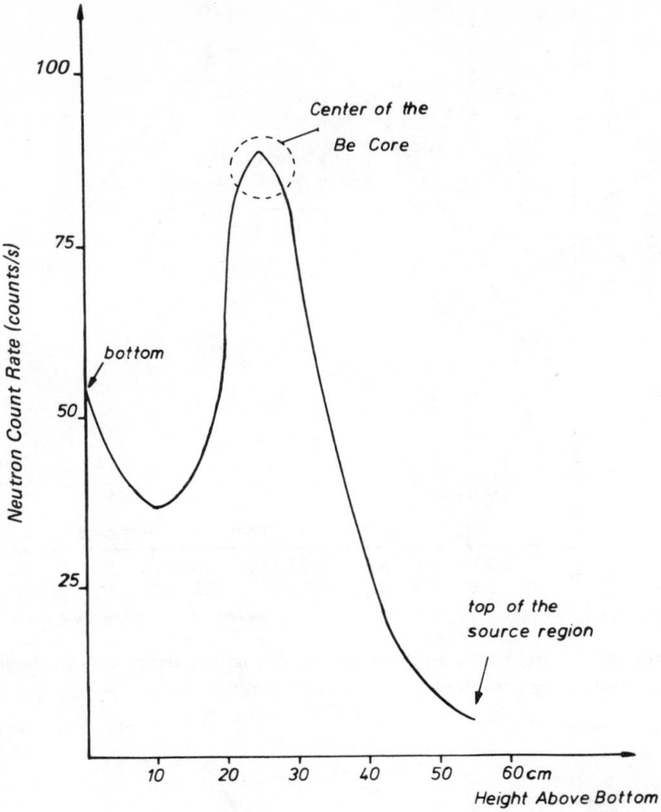

FIG.4. Neutron count-rate from a 1-g ^{235}U sample, changing with different positions of the lift in system I of Fig.3.

unirradiated AVR fuel element containing 1 g ^{235}U was used as standard for all types of sample. The background count-rate C_0 was measured without any fissile material in the system. The count-rate of the standard C_{st} was measured with the 1 g standard in the measuring position of system I. Sample count-rate C was measured with an unknown sample in system I or II. All count-rates were taken in counts per seconds. From these measured count-rates, the "standard count-rate" S is calculated by the following formula

$$S = \frac{C - C_0}{C_{st} - C_0}$$

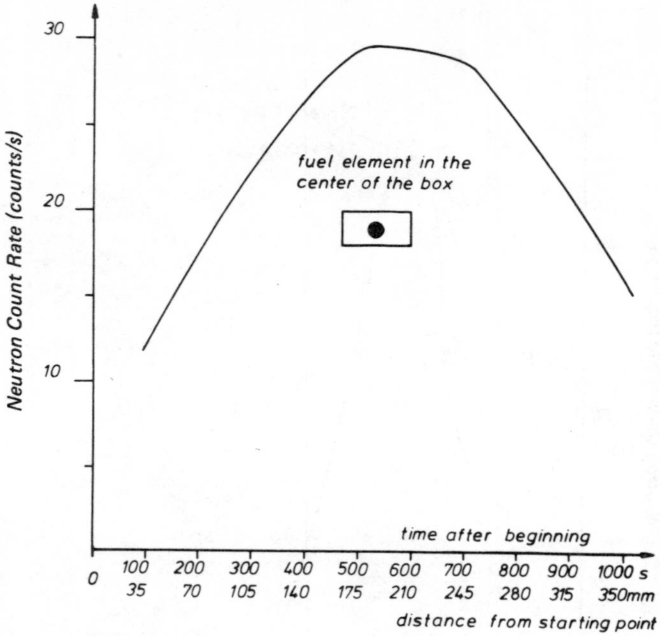

FIG.5. Neutron count-rate from a 1-g ^{235}U sample, changing during the movement of the extended waste box through the centre of system II in Fig.3.

The standard count-rate is that relative to this special 1-g standard. It is independent of the intensity of the source and its occasional replacement, which equally influences all count-rates measured. It is dependent on the fissile content of the sample, on the use of system I or system II and, to a minor degree, on the composition of the sample.

Small samples are measured in system I, which is loaded on top. Figure 4 shows the dependence of the neutron count-rate on the position of the lift with the standard. The maximum count-rate is obtained in the centre of the Be core. A similar neutron field is used as the measuring position in Fig.1. Because of better reproducibility and comparison with system II, the bottom position is chosen as the standard measuring position (0 cm) in this system. During measurement, the sample remains at its fixed position in system I. As reported in Ref.[6]. the calibration curves for differently loaded AVR fuel elements deviate slightly from the old system (Fig.1). The relation between the content of ^{235}U and the standard count-rate can be described as proportionality in the range between 0 and 1 g ^{235}U. So, the standard count-rate for AVR fuel elements is equal to their content of fissile uranium measured in grams. In other cases,

TABLE III. NEUTRON COUNT-RATES (counts/s) FROM FISSILE MATERIAL (AVR FUEL ELEMENT, CONTAINING 1 g ^{235}U AT DIFFERENT POSITIONS AND WITH DIFFERENT ADDITIVE MATERIAL IN AN EXTENDED WASTE BOX

Position of the sample in the waste box	Standard count-rate:			
	Without additional matrix	With additional matrix	material consisting of	
1. 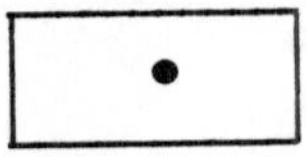	0.550	0.539	700 g	copper wool
2. 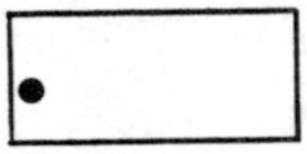	0.516	0.526	550 g	brass wool
3. 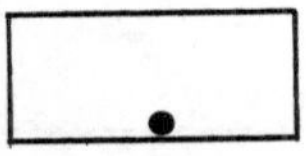	0.533	0.536	550 g	iron wool
4. 	0.502	0.522	470 g	paper
ϕ	0.525	0.531		

the preparation of similar standards is necessary to obtain the relation between count-rate and fissile content.

Extended samples are measured in system II. Waste boxes are loaded at the right side and then moved and rotated horizontally through the centre of the system. During the measuring time of 1000 s, the box is moved through 35 cm to the left. During this measuring turn, different parts of the box get into the most intense part of the neutron field. It is practically the same region where,

TABLE IV. COMPARATIVE MEASUREMENTS OF 8 FUEL ELEMENTS, WITH BURNUP VALUES BETWEEN 0% fifa AND 110% fifa AS INDIVIDUAL SAMPLE IN SYSTEM I AND, AS A WHOLE, CONTAINED IN A WASTE BOX IN SYSTEM II OF FIG.3
(The dose-rate of the 8 fuel elements was 3000 R/h at 10 cm distance)

No.	Sample name	Burnup [fifa] (%)	Standard count-rate (counts/s)	Relative content of ^{137}Cs (γ-ray intensity at 662 keV)
1	UCC	0	1.000	0
2	BE-A	34	0.664	19
3	BE-B	64	0.382	31
4	50/24	~ 110	0.143	46
5	50/17	~ 90	0.170	51
6	50/20	~ 70	0.314	38
7	50/14	~ 50	0.527	28
8	50/12	~ 50	0.592	28
System I: Sum of 1 – 8			3.792	
System II: Waste-box, filled with above 8 samples			1.890	

alternatively, the lift moves in system I. This can be seen from Fig.5, which indicates the count-rates at different times of the movement. Usually only the averaged count-rate is measured during 1000 s. This is the usual counting time in system II. It is needed for the movement of the waste box. To evaluate the influence of different packing positions, the same AVR fuel element was packed at the positions indicated in Table III, the first column of which shows the obtained standard count-rates. They deviate from the average by at most 5%. The mean value indicates that the same sample, an AVR fuel element, exhibits different count-rates in systems I and II. On being packed into a waste box and moved during the measurement its count-rate is only 52.4% of that in system I. Extended waste boxes will usually contain some additional material. Therefore, metal wool or paper was added to the differently packed waste boxes. The second column of Table III shows that the additional material indicated does not influence the measured count-rate by more than 2%. This result is very important to the measurement of waste boxes with changing composition. The result will probably not apply to large amounts of water shifting the whole neutron spectrum. But water can be excluded in waste boxes, where only solid material is permitted. So the count-rate from waste boxes is mainly dependent on the fissile content and only to a minor degree on packing and additional material.

Another comparison between systems I and II of Fig.3 was obtained by the measurement of irradiated fuel elements of the AVR test reactor. As reported in Table IV, eight irradiated fuel elements were first measured individually in the measuring position of system I. Then they were packed, together with some steel wool, into an extended waste box and measured as a whole in system II. The caesium content, indicated in the last column of Table IV, is the result of different burnup values. It has no influence on the measurement of the fissile content.

Plutonium can be excluded in these samples since the fresh fuel elements were loaded with highly enriched uranium. The fissile content in this case is simply related to the sum of $^{233}U + ^{235}U$ as shown in Ref.[3]. For AVR fuel elements, the standard count-rate from system I is equal to the fissile content in grams, as mentioned above. The total content of fissile uranium U_I of the eight samples is therefore the sum of the standard count-rates.

$$U_I = (1.00 + 0.664 + 0.382 + 0.143 + 0.170 + 0.314 + 0.527 + 0.592)\ g\ U_{fissile} = 3.79\ g\ U_{fissile}$$

When the waste box, filled with the total of 8 fuel elements, was measured in system II, a standard count-rate N = 1.89 counts/s was obtained. According to Table III, the averaged efficiency in system II is 0.53 as compared with system I. With this efficiency, a content, U_{II}, is calculated.

$$U_{II} = \frac{1.89}{0.53}\ g\ U_{fissile} = 3.57\ g\ U_{fissile}$$

Both determinations of the total content of fissile uranium U_I and U_{II} are in close agreement with a 6% deviation between each other. They would have been identical if the relative efficiency for system II had been taken to be 50%. Self-shielding and flux depression in these more densely packed boxes make a slightly diminished efficiency reasonable. This type of packing is typical for waste boxes.

A relative efficiency of 50% as compared with system I is therefore recommended for the measurement of waste boxes in system II. If a high precision is needed, samples with a different composition need calibration with standard samples of similar composition. Samples of other geometrical shapes are acceptable as long as they fit into the waste boxes used. Re-calibration of system II relative to system I becomes necessary if another driving system is used for differently shaped samples. A new determination of the source neutron count-rate must be included in this re-calibration if the cross-sectional area of the tunnel is modified. This can be achieved by changing the lateral polyethylene bricks. A wider tunnel section may be necessary for some types of fuel element from water reactors. The assay system itself is equally well suited for the measurement of non-radioactive and radioactive fissile material in extended samples. This was demonstrated with fuel elements of the AVR test reactor at burnup values between 0 and 170 000 MW · d/t.

ACKNOWLEDGEMENT

This research was kindly supported by the IAEA under Research Contract No.1931/RB. It was performed at the Institute for Chemical Technology (ICT) of the KFA Jülich, Federal Republic of Germany. Thanks are due to Professor Dr. E. Merz for his stimulating interest in NDA techniques for reprocessing problems, and to Mr Hausmann, Mrs Hrastnik and Mr Moersch for their assistance in construction and measurement.

REFERENCES

[1] FILSS, P., "Non-destructive control of fissile material in solid and liquid samples arising from a reactor and fuel reprocessing plant", Safeguarding Nuclear Materials (Proc. Symp. Vienna, 1975) 2, IAEA, Vienna (1976) 471.

[2] FILSS, P., Fissile isotope determination in irradiated fuel elements and waste samples using Sb-Be neutrons, Trans. Am. Nucl. Soc. **20** (1975) 341.

[3] FILSS, P., "Fissile isotope determination in irradiated fuel elements and waste samples using Sb-Be neutrons, Nuclear Energy Maturity, Progress in Nucl. Energy Series, Pergamon Press (1976) 66–80.

[4] ATWELL, T.L., MARTIN, E.R., MENLOVE, H.O., "In-plant non-destructive assay of HTGR fuel materials", Safeguarding Nuclear Materials (Proc. Symp. Vienna, 1975) 2, IAEA, Vienna (1976) 501.

[5] UKAEA, Handbook of Criticality Data **1**, UKAEA, Risley, Warrington, Lancashire (1965).

[6] STEIN, G., et al., IAEA-SM-231/42, these Proceedings, Vol.II.

DISCUSSION

E. DERMENDJIEV: During the neutron interrogation of irradiated AVR pebbles in your system you should expect a certain number of fissions of the

^{233}U contained in spent U-Th fuel. The average number of neutrons per thermal fission in ^{233}U is larger than the $\langle\nu_{th}^5\rangle$ of ^{235}U ($\langle\nu_{th}^3\rangle/\langle\nu_{th}^5\rangle \sim 1.025$). On the other hand, you calibrate the system with fresh pebbles (^{235}U only). Did you estimate this effect?

P. FILSS: Your question relates in general to the behaviour of different fissile nuclides in this active neutron interrogation system. This matter has not yet been dealt with experimentally. For a numerical estimation it may be assumed that samples with low concentrations of fissile material will not disturb the interrogating neutron flux ϕ. Then the average number of fission neutrons generated may be estimated to be proportional to

$$\phi \times \sigma \times \bar{\nu} \times N$$

where σ = fission cross-section
$\bar{\nu}$ = as defined by you and
N= number of special nuclide atoms in the sample.

The product of $\sigma \times \bar{\nu}$ is thus the specific valuation factor for different fissile nuclides. In the case of ^{233}U, $\sigma \times \bar{\nu}$ is only 94.5% of the value for ^{235}U.

Without this estimated numerical correction, calibration with ^{235}U followed by subsequent use for ^{233}U determination may lead to 5.5% underestimation of ^{233}U. The actual discrepancy with AVR fuel elements must have been much less, since the bred content of ^{233}U was always below the residual content of ^{235}U. The unmodified calibration curve for ^{235}U was therefore applied.

J. BOUCHARD: The neutron interrogation system described in your paper has apparently been used to date for assaying only highly enriched uranium fuels or thorium fuels.. Have you considered the possible limitations of this system when used for irradiated fuels from the uranium-plutonium cycle in view of the strong neutron emission resulting from, in particular, the large amounts of curium contained in such fuels?

P. FILSS: Our practical experience is so far restricted to irradiated and unirradiated fuel from the thorium cycle. Irradiated fuel of this type was once analysed to determine the irregular contribution from other sources by removing the ^{124}Sb and counting the remaining neutrons. The number of irregular neutrons was found to be 3% in the worst cases (burnup above 100 000 MW · d/t and cooling time ~ 1 month). Comparison with similar work done at other laboratories suggests that spontaneous neutron emitters, such as are present in Pu samples, will not lead to more irregular neutrons as long as the typical parameters of operation are not drastically changed.

NON-DESTRUCTIVE ACTIVE ASSAY OF URANIUM-235 AND PLUTONIUM-239 IN NUCLEAR REACTOR FUEL

Y. MATSUDA, T. TAMURA, T. MURATA
Nippon Atomic Industry Group Co., Ltd,
NAIG Nuclear Research Laboratory,
Ukishima, Kawasaki-shi, Japan

Abstract

NON-DESTRUCTIVE ACTIVE ASSAY OF URANIUM-235 AND PLUTONIUM-239 IN NUCLEAR REACTOR FUEL.

The feasibility of a method for determining the ^{235}U and ^{239}Pu contents in Pu-U mixed fuel by measuring prompt and delayed fission neutrons in neutron-induced fissions has been experimentally investigated. Mixed (Pu-U) oxide fuel rods of six different types, and a LWR-type fuel pellet irradiated to about 6000 MW · d/t were used as fuel samples. Neutrons irradiating the samples were generated by bombarding a thick Li target with protons from the NAIG 4-MV Van de Graaff accelerator. In the case of the sample fuel rods the irradiation was made with fast neutrons with a maximum energy of 350 keV, and an average energy of about 150 keV. It is found that the measured mass of ^{235}U and ^{239}Pu agree well with the nominal mass of each isotope in the fuel rods and that the accurate non-destructive assay of Pu-U mixed fuel rods will be made by the present technique with good standard samples. The fuel pellet was irradiated with thermalized neutrons. To shield the high gamma radiation of the pellet, it was contained in a $14 \times 14 \times 22$ cm^3 lead cask, which was surrounded by polyethylene bricks. Detecting efficiencies of prompt and delayed neutrons of the measuring system were obtained with UO_2 small disks as standard samples. The fission ratio of ^{239}Pu to ^{235}U for the measuring system was estimated by using a multigroup neutron diffusion calculation code. The contents of ^{235}U and ^{239}Pu in the fuel pellet were obtained as mass of ^{235}U: 109 ± 3 mg, and mass of ^{239}Pu: 13 ± 2 mg. The result for ^{235}U is consistent with the result of the mass-spectrometric analysis of uranium isotopic content ratio. No such data were available for ^{239}Pu.

INTRODUCTION

The amounts of Pu-U mixed nuclear materials such as FBR fuels, Pu-recycle LWR fuels and spent fuels are increasing rapidly with the expansion of the nuclear industry. There is a currently strong demand, within the framework of safeguards techniques, to develop non-destructive techniques for determining the fissile materials in Pu-U mixed fuels.

The wide variety of Pu and U enrichment, and the strong gamma-ray field of the spent fuels make it difficult to determine Pu and U contents in Pu-U mixed fuels by the ordinary methods of passive or active assay techniques.

TABLE I. TOTAL YIELDS AND GROUP CONSTANTS OF DELAYED NEUTRONS OF ^{235}U AND ^{239}Pu

Element	U-235		Pu-239	
Energy	En $\leqq$ 4.0 MeV		En $\leqq$ 4.0 MeV	
Total yield	0.01668±0.00070 n/s		0.00645±0.00040 n/s	
Group	$T_{\frac{1}{2}}$ (s)	Relative yield (%)	$T_{\frac{1}{2}}$ (s)	Relative yield (%)
1	54.51	0.038	53.75	0.038
2	21.84	0.213	22.29	0.280
3	6.00	0.188	5.19	0.216
4	2.23	0.407	2.09	0.328
5	0.496	0.128	0.549	0.103
6	0.179	0.026	0.216	0.035

Several methods for measuring Pu and U contents in Pu-U mixed fuels with neutron active assay have recently been proposed and feasibility studies have been performed [1–3].

In this work, the feasibility of a method of determining the ^{235}U and ^{239}Pu contents in the samples of FBR and ATR types, and fresh and spent LWR-type fuels, by measuring prompt and delayed neutrons in neutron-induced fissions, has been experimentally investigated, and the contents of ^{235}U and ^{239}Pu in the fuel samples have been determined non-destructively.

1. ADOPTED METHOD

The delayed neutron yield of ^{239}Pu is much smaller than that of ^{235}U, while prompt neutron yields of both nuclides do not differ as much as those of delayed neutrons. So, it is possible to determine the contents of ^{235}U and ^{239}Pu in the Pu-U mixed fuel by measuring prompt and delayed neutrons in fission induced by the neutrons which have energy below the fission threshold of ^{238}U. The delayed neutron data of ^{235}U and ^{239}Pu are shown in Table I [4].

Other fissionable nuclides contained in the fuels are ^{240}Pu, ^{241}Pu and ^{242}Pu. For ^{240}Pu and ^{242}Pu, the average fission cross-section for sub-threshold neutrons is so small, and for ^{241}Pu the content is so low, that these nuclides have little effect on the present assay.

Taking into account the fissions of ^{235}U and ^{239}Pu, and neglecting the self-shielding effect of fuel samples, the prompt neutron counts P and the delayed neutron counts D are expressed as

$$P = a_1 \cdot N_5 + a_2 \cdot N_9$$

$$D = b_1 \cdot N_5 + b_2 \cdot N_9 \qquad (1)$$

where N_5 and N_9 are the mass of ^{235}U and ^{239}Pu in the effective region of the fuel sample, respectively. The coefficients a_1 and a_2 are the prompt neutron counting response to the unit mass of ^{235}U and ^{239}Pu, respectively. The coefficients b_1 and b_2 are the response for the delayed neutrons. A dimensionless parameter, known as the discrimination ratio, is defined by $(a_1 \cdot b_2 / a_2 \cdot b_1)$, and to obtain a better discrimination between ^{235}U and ^{239}Pu, the parameter must be notably smaller than unity.

The prompt neutron counting response (a_1 and a_2) is proportional to the number of neutrons per fission and to the fission cross-section of each isotope. The response of the delayed neutron counts (b_1 and b_2) is given by

$$b = Y \cdot F \qquad (2)$$

where F is the fission rate for unit mass of fissile material during irradiation and Y is the delayed neutron response for unit fission rate. The factor Y depends not only on delayed neutron group yield, but also on the irradiation and counting conditions. For n cycles of repetitive measurement with cycle time T; t_1 seconds irradiation, t_2 seconds cooling, t_3 seconds counting and t_4 seconds waiting prior to next irradiation, Y is given by

$$Y = \sum_{i=1}^{6} \left[\left\{ \sum_{j=1}^{n} (n-j+1) \cdot e^{-(j-1)\lambda_i T} \right\} \frac{\epsilon_i \eta_i}{\lambda_i} (1 - e^{-\lambda_i t_1}) \times (1 - e^{-\lambda_i t_3})\, e^{-\lambda_i t_2} \right] \qquad (3)$$

where i : delayed neutron group index

λ_i : decay constant of i-th delayed neutron group

η_i : absolute yield of i-th delayed neutron group per fission

ϵ_i : detecting efficiency for i-th delayed neutron group.

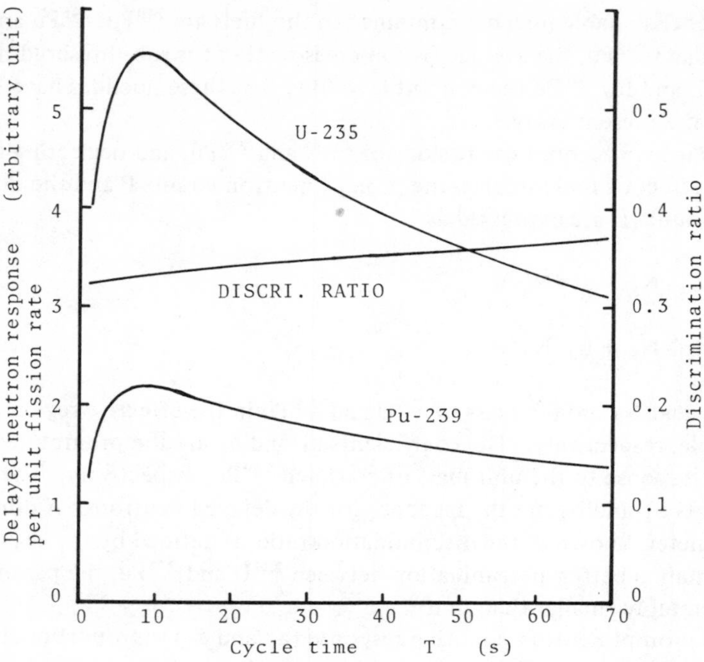

FIG.1. Delayed neutron response for ^{235}U and ^{239}Pu, and discrimination ratio of ^{239}Pu to ^{235}U, as a function of cycle-time.

Neutrons irradiating the fuel samples were generated by bombarding a thick Li metal target with protons from the NAIG Van de Graaff accelerator. The accelerated proton beam from the accelerator was deflected during the delayed neutron counting and, in addition to the deflection, to reduce the background neutron a mechanical beam shutter was adopted to stop the stray proton beam. It took about 0.3 s to drive the shutter.

The dependence of Y on the measurement cycle time T for ^{235}U and ^{239}Pu is shown in Fig.1. In this calculation it was assumed that the total measurement time $n \cdot T = 1980$ s, cooling and waiting time $t_2 = t_4 = 0.3$ s and the efficiencies ϵ_i did not depend on delayed neutron group i. Figure 1 also shows the discrimination ratio $(Y_9 \cdot \nu_5 / Y_5 \cdot \nu_9)$ which corresponds to the ratio $(a_1 \cdot b_2 / a_2 \cdot b_1)$, where ν_5 and ν_9 are the number of neutrons per fission of ^{235}U and that of ^{239}Pu, respectively. As is seen in Fig.1, the maximum delayed neutron response is obtained at about $T = 7 \sim 9$ s. The lower the discrimination ratio the better the discrimination obtained between N_5 and N_9. So, the optimum cycle time was chosen as $T = 6.6$ s; irradiation time 3 s, cooling time 0.3 s, counting time 3 s, and waiting time 0.3 s. Each fuel sample was measured in 300 cycle times.

2. ASSAY OF NON-IRRADIATED PuO_2-UO_2 MIXED-OXIDE FUEL RODS

Mixed-oxide fuel rods of six different types were used as the fuel sample – three different FBR fuel rods with the same diameter, an ATR fuel rod, a Pu-recycle LWR fuel rod, and a LWR fuel rod not containing Pu. The characteristics of these fuel rods are given in Table II, which also lists the sample numbers.

Irradiating fast neutrons, with a maximum energy of 350 keV and an average energy of about 150 keV, were produced by bombarding a thick Li metal target with 2.1-MeV proton from the accelerator. The total neutron yield was about 3×10^8 n/s. The neutrons are energetic enough to penetrate the samples but not too energetic to cause fissions in ^{238}U.

The experimental arrangement is shown schematically in Fig.2. Prompt neutrons were detected with a ^4He recoil proportional counter (1 in. dia., 12 in. long, 5 atm), and delayed neutrons with three ^3He proportional counters (1 in. dia., 8 in. long, 4 atm) in Cd-covered polyethylene moderator brick of $10 \times 10 \times 20$ cm^3. Between the fuel sample and the delayed neutron detector a tantalum sheet 0.5 mm thick and a polyethylene sheet 6 mm thick containing B_4C (50%) were placed to reduce the epicadmium and resonance region neutrons scattered from the detector. Prompt neutrons were counted during "on" irradiation and delayed neutrons were counted during "off" irradiation.

Using the nominal contents of ^{235}U and ^{239}Pu in unit length of each fuel rod, the coefficients a_1, a_2, b_1 and b_2 in Eq.(1) were obtained with the least-squares method to the measured counts of prompt and delayed neutrons for six fuel samples. The discrimination ratio ($a_1 b_2 / a_2 b_1$) was found to be 0.354. The contents of ^{235}U and ^{239}Pu in unit length of a fuel rod are determined by using these coefficients and the measured counts of prompt and delayed neutrons for the fuel sample. The comparison between the measured mass (the ordinate) and the nominal mass (the abscissa) for ^{235}U and for ^{239}Pu is shown in Fig.3. The error indicated in the figure includes the statistical error only. Good agreements are obtained as shown in the figure between the measured and the nominal mass. For the fuel sample of high ($^{239}Pu/^{235}U$) ratio, the background neutrons from the spontaneous fission and the (α, n) reaction make the error especially large. For example, in the case of No.4 sample the S/N ratio was 0.18 for the delayed neutron measurement, and the statistical error of the measurement $\pm 7.7\%$, though that of the prompt neutron measurement was $\pm 1.4\%$. To achieve a statistical error better than 1% for delayed neutron counting with the present measuring system, it is necessary to irradiate the fuel sample with a neutron intensity of around 5×10^9 n/s.

Examining Fig.3 in detail, it is found that the fuel rods of different diameter show a somewhat large discrepancy, more than the statistical error. To

TABLE II. CHARACTERISTICS OF (Pu-U) MIXED-OXIDE SAMPLES FOR THE EXPERIMENT

Sample No.		1	2	3	4	5	6
Reactor type		FBR	FBR	FBR	LWR Pu recycle	ATR	LWR
Rod diameter	(mm)	6.3	6.3	6.3	9.98	16.85	11.8
Pellet diameter	(mm)	5.5	5.5	5.5	8.7	15.0	10.0
PuO_2 / $(U,Pu)O_2$	(%)	40.06	20.10	17.97	3.95	0.49	—
Pu-238	(%)	—	—	0.038	0.045	0.039	—
Pu-239	(%)	91.337	91.337	90.184	90.279	90.319	—
Pu-240	(%)	7.854	7.854	8.501	8.449	8.490	—
Pu-241	(%)	0.774	0.774	1.179	1.142	1.074	—
Pu-242	(%)	0.034	0.034	0.097	0.085	0.077	—
U-235 enrichment	(%)	19.91	89.89	60.11	0.72	0.72	2.99
U-235 / 1 cm rod length	(g)	0.251	1.388	1.078	0.0339	0.1133	0.215
Pu-239 / 1 cm rod length	(g)	0.771	0.355	0.355	0.1801	0.0721	—

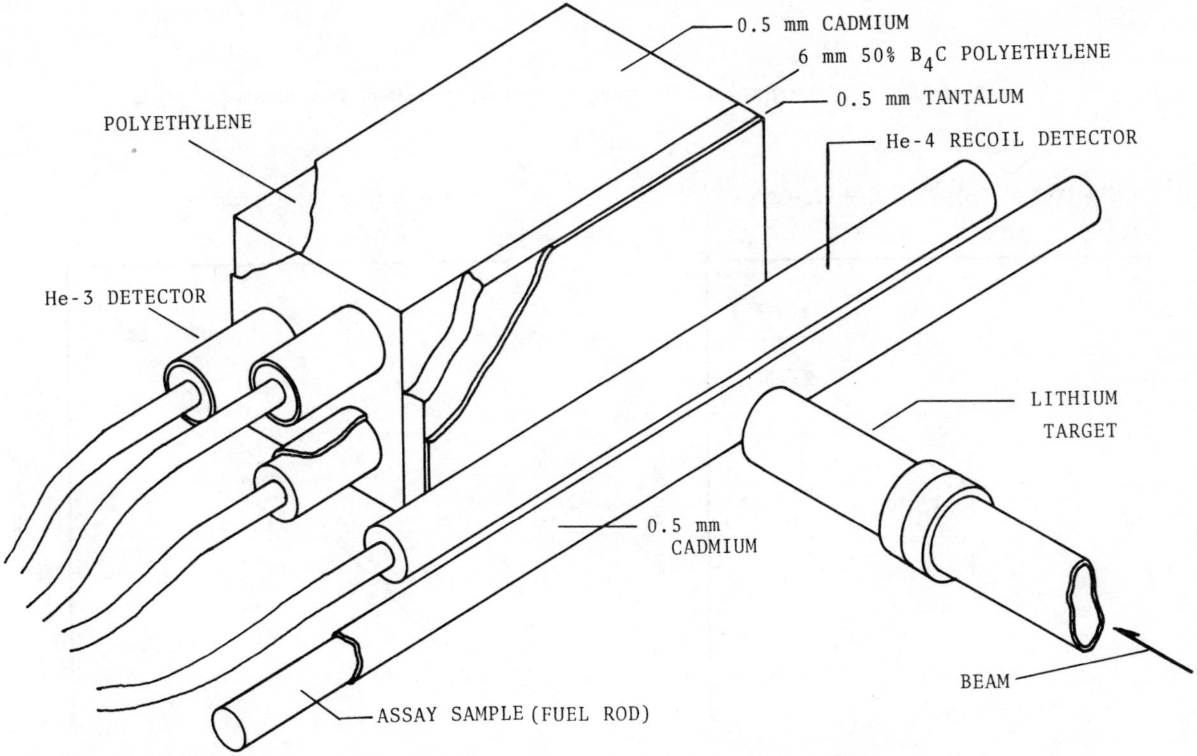

FIG.2. Arrangement for Pu-U mixed-oxide fuel rod assay by measuring prompt and delayed neutrons.

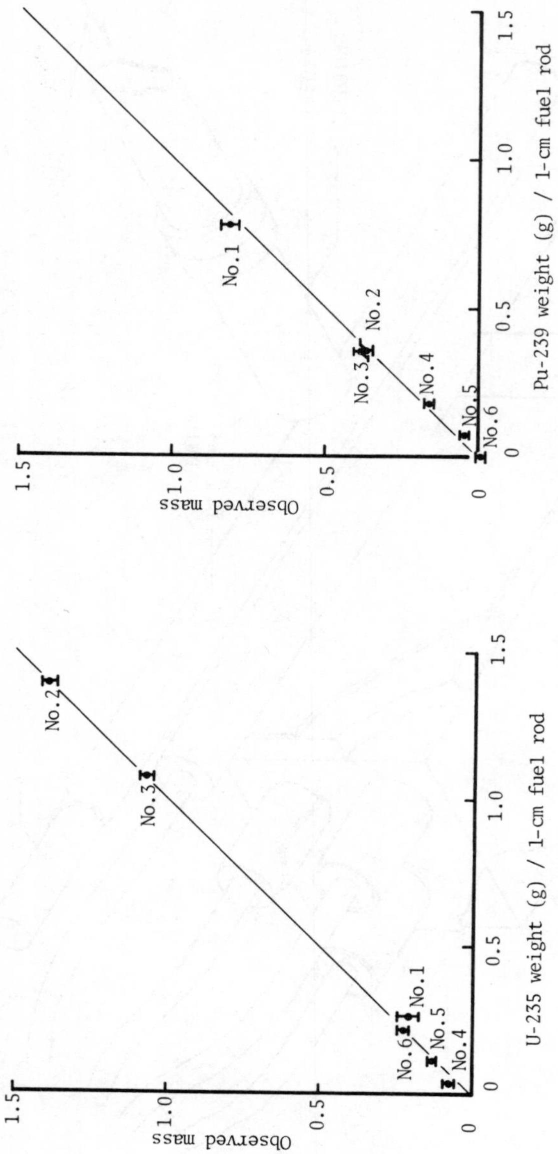

FIG.3. Comparison of the present result and nominal weight for Pu-U mixed-oxide fuel rods.

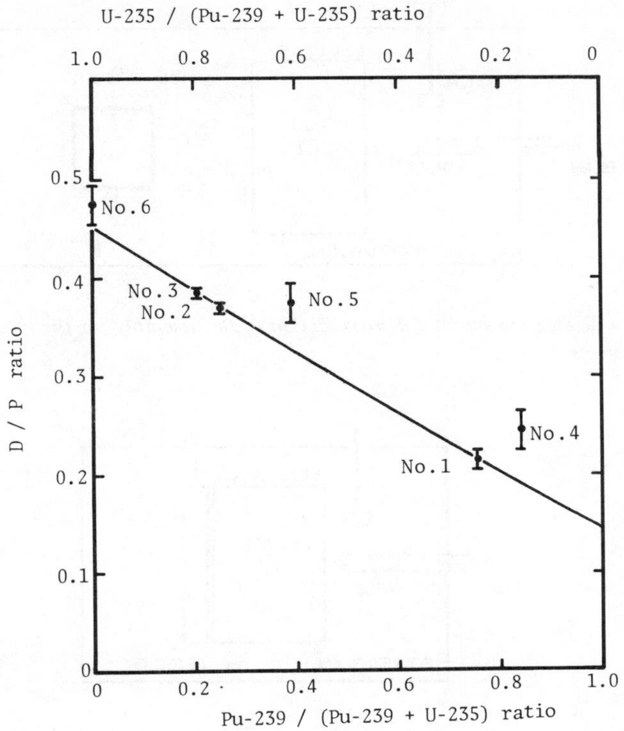

FIG.4. Correlation between D/P ratios and $^{239}Pu/(^{239}Pu + ^{235}U)$ ratios.

investigate this tendency more clearly, the measured ratios, D/P, of delayed neutron to prompt neutron counts were plotted against the nominal mass ratios; $^{239}Pu/(^{235}U + ^{239}Pu)$. The result is shown in Fig.4, where the solid line was calculated according to Eq.(1) with four coefficients, which were determined with the least-squares method for the three FBR-type rods. The data of the D/P ratio for LWR- and ATR-type fuel rods, which have a larger diameter, do not seem to be consistent with the calculated curve.

Two main causes may be considered for the discrepancy. First, the geometrical factor for the neutron-detecting efficiencies would not be identical for different sizes of fuel rods. Second, fission of ^{238}U induced by fission neutrons may have a different effect for the different sizes of fuel rods and for different content ratios of ^{238}U.

In conclusion, it can be said that, if the measuring system is calibrated with standard samples with the same rod diameter and with fissile content ratios not so different from the fuel samples to be measured, an accurate assay will be achieved.

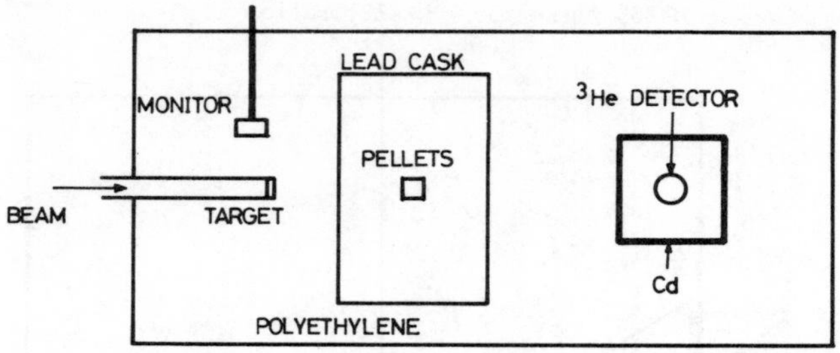

FIG.5. Schematic diagram showing a cross-section of the assembly for the measurement of prompt neutrons.

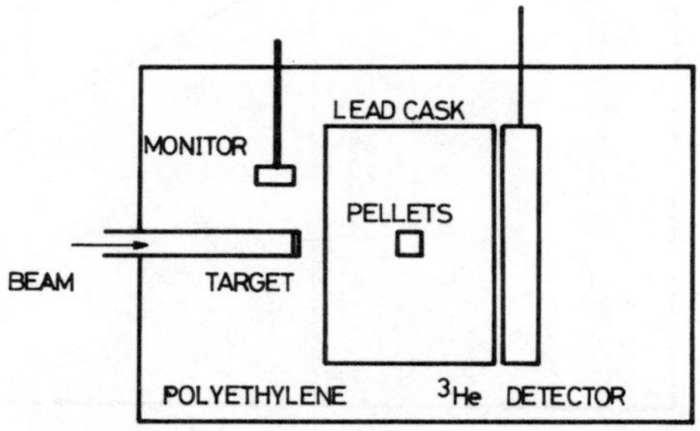

FIG.6. Schematic diagram showing a cross-section of the assembly for the measurement of delayed neutrons.

3. ASSAY OF AN IRRADIATED LWR FUEL PELLET

A LWR fuel pellet (2.6% ^{235}U enriched UO_2) irradiated to about 6×10^3 MW · d/t was used as a fuel sample. The size of the sample is 12.6 mm dia. and 6 mm long; and its weight is 6.6 g.

To shield the high-intensity gamma ray from the fuel sample, it was contained in a lead cask ($14 \times 14 \times 22$ cm^3), which was surrounded by polyethylene bricks, and irradiated with neutrons from outside the cask. To obtain good detecting efficiency and a good S/N ratio, the measurement assemblies for prompt and delayed neutrons were arranged independently. The experimental arrangement is shown schematically in Fig.5 for measuring prompt neutrons, and in Fig.6 for measuring delayed neutrons. The neutrons irradiating the fuel sample were produced by the $^7Li(p,n)^7Be$ reaction and moderated with polyethylene bricks surrounding the cask.

For prompt neutron measurement, the maximum energy of the neutrons for irradiation was 230 keV. Such low-energy neutrons were moderated and attenuated considerably on the way to the ^{3}He neutron detector, which was placed far from the neutron-generating target, and mostly absorbed by the Cd sheet around the detector. On the other hand, high-energy fission neutrons originating from the fuel sample were not so strongly attenuated and, after penetrating through the Cd sheet, were moderated by polyethylene and then detected with the detector.

For delayed neutron measurement, to obtain good detecting efficiency, the intensity of the neutrons irradiating the fuel sample was raised about ten times more than that of the prompt neutron measurement by increasing the energy of protons bombarding the Li target. Delayed neutrons were detected by a ^{3}He detector placed near the cask. The cycle time of irradiation and counting was the same as that of the measurement described in Section 3.

The neutron counting responses of the measuring systems for ^{235}U (a_1 and b_1) were determined experimentally by using ten pieces of non-irradiated UO_2 disks (2.99% enriched, 10 mm dia. × 1 mm length) as standard samples. No suitable Pu sample was available, and the responses for ^{235}U were converted into the responses for ^{239}Pu using a theoretical calculation. The conversion factor, a_1/a_2, for the prompt neutron measurement is given by

$$\frac{a_2}{a_1} = \frac{\nu_9 \cdot F_9}{\nu_5 \cdot F_5} \tag{4}$$

where F_9 and F_5 are the fission rate for unit mass of ^{239}Pu and ^{235}U, respectively. The ratio of fission rate, F_9/F_5, was calculated with a one-dimensional 23-thermal-group neutron-diffusion calculation code. The conversion factor, b_2/b_1, for the delayed neutron measurement is given by

$$\frac{b_2}{b_1} = \frac{Y_9 \cdot F_9}{Y_5 \cdot F_5} \tag{5}$$

where Y is the delayed neutron response for unit fission rate defined in Eq.(3).

Using the coefficients, a_1, a_2, b_1 and b_2, the contents of ^{235}U and ^{239}Pu in the fuel pellet were obtained as

mass of ^{235}U : 109 ± 3 mg

mass of ^{239}Pu : 13 ± 2 mg

The present result for ^{235}U is consistent with the result of mass-spectrometric analysis of U isotopic content ratio. No such data were available for ^{239}Pu.

4. CONCLUSION

The method of determining the ^{235}U and ^{239}Pu contents in Pu-U mixed fuel by measuring prompt and delayed fission neutrons in neutron-induced fissions has been verified experimentally.

An accurate non-destructive active assay of Pu-U mixed fuels can be performed by calibrating the measuring system with standard samples of the same size.

For Pu-U mixed fuel containing Pu recovered from spent LWR fuel and for high burnup LWR fuels, the fission contribution of ^{241}Pu cannot be ignored, and to determine the ^{241}Pu content it is necessary to make other types of measurement in addition to the present method.

Spent fuels generally have not only high gamma-ray emission, but also relatively high neutron emission, which depend on irradiation history. We have already measured the neutron emission rate [5] from spent BWR-type fuels, and intend to develop a practical technique of the method for the assay of fissile contents in spent fuels.

ACKNOWLEDGEMENTS

The authors would like to express their appreciation to Power Reactor & Nuclear Fuel Development Corporation for lending them the mixed-oxide fuel rods. They are grateful to Director Dr. N. Kawai for his encouragement and his valuable discussions, and Messrs T. Nakakita and T. Yokota for their assistances in the course of work.

REFERENCES

[1] BIRKHOFF, G., BONDAR, L., LEY, J., Determination of the ^{235}U, ^{239}Pu and ^{240}Pu Contents in Mixed Fissile Materials by Means of Active and Passive Neutron Techniques, EUR 4778e (1972).

[2] GOZANI, T., Active Direct Measurement of Residual Fissile Content in Spent Fuel Assemblies, Electric Power Research Institute Rep. EPRI278-1 (July 1975).

[3] SAWAN, M., CONN, R.W., Neutron pulses slowing down in heavy media analysis with applications of the lead spectrometer, Nucl. Sci. Eng. **54** (1974) 127.

[4] COX, S.A., Delayed neutron data-review and evaluation, ANL/ND-5 (April 1974).

[5] TAMURA, T., et al., Autumn Meeting of the Atomic Energy Society of Japan (1976) to be published.

NEW NEUTRON CORRELATION MEASUREMENT TECHNIQUES FOR SPECIAL NUCLEAR MATERIAL ASSAY AND ACCOUNTABILITY

E.J. DOWDY, C.N. HENRY, A.A. ROBBA, J.C. PRATT
Los Alamos Scientific Laboratory,
Los Alamos, New Mexico,
United States of America

Abstract

NEW NEUTRON CORRELATION MEASUREMENT TECHNIQUES FOR SPECIAL NUCLEAR MATERIAL ASSAY AND ACCOUNTABILITY.

Assay of plutonium samples using the moments of the neutron count distribution from a well counter has been demonstrated. Samples of various composition, and ranging in mass from a few tens of grams to several kilograms have been used in this demonstration. Several advantages over conventional coincidence counting methods have been realized. Among these are the ability to account for sample multiplication and the ability to detect and correct for detection system malfunction without the loss of even a single analysis cycle.

INTRODUCTION

Nuclear reactor "noise" (fluctuations in the pulse trains from neutron detection systems) has been analyzed[1] for many years to extract the contained information about core physics parameters. In the time domain, the major categories of such methods are: the time correlation analyses among neutron detection pulses, with the Rossi-α method being perhaps the best known example; and the analyses of the moments of the neutron count distribution, with the Feynman variance technique being the earliest version. There is an analogy between the former category and the various coincidence counting techniques using well counters for the assay of neutron emitting samples of special nuclear materials (SNM): coincidence counting is essentially an integration of the neutron dieaway curve which is the usual result of the Rossi-α or similar methods in reactor work. Until now there has been no SNM assay analogue of the latter category of reactor noise measurements, i.e., the analysis of the moments of the neutron count distribution. The extractable information in this data analysis technique is expected to be sufficient for assay purposes so we have begun a study of its potential for this application.

This report contains the results of the initial phases of the study in which neutron emitting samples were counted in a typical well counter. We have confirmed the utility of the analysis of the moments of the neutron detection distribution for assay of various chemical composition plutonium samples ranging in mass from a few tens of grams to several kilograms. For large metal samples, and presumably for large, dense compound samples, there is a significant neutron multiplication effect, precluding assay by the conventional coincidence counting technique. However, the moments of the count distribution method provides information that permits the determination of the mass of contained spontaneously fissioning isotopes even if there is a significant sample multiplication.

The following sections contain a brief outline of the basic theory of the moments analysis technique, a description of the experimental procedures, and a presentation of the results of the measurements that have been made to demonstrate the validity of the technique.

THEORY

Multiplication

The formalism for the moments method[2] applied to the assay of samples with only fission sources of neutrons is developed using the following symbols and definitions:

ε ≡ fission detection efficiency (isotope average) of the neutron detector.

$s(t_{ij})$ ≡ probability per unit time of survival, to time t_j, of a neutron injected into the detector at time t_i. Since we use a moderated thermal neutron detector, this is the capture time distribution of the detector.

ν ≡ number of neutrons emitted in a single fission event (variable from event to event).

$<\nu>$ ≡ average number of neutrons emitted in a single fission event.

M ≡ multiplication of the contained fissionable material <u>in the detector</u>.

F ≡ average fission rate of the contained material.

$P(t_1/t_o)$ ≡ probability per unit time of a detection event at time t_1 following a fission event at time t_o.

$P(t_2/t_o)$ ≡ probability per unit time of a detection event at time t_2 following a fission at time t_o and a detection event at time t_1 - both neutrons from the same fission chain.

$P(t_1,t_2/t_o)$ ≡ joint probability, per unit time squared, of detection events t_1 and t_2 - both neutrons from the same fission chain starting at t_o.

$\Pi(t_1,t_2/t_o)$ ≡ joint probability, per unit time squared, of detection events at t_1 and t_2 with both neutrons from the same fission chain starting at t_o, for all times t_o preceding t_1.

For a fission event occurring at time $t = t_o$ in which ν neutrons were emitted, resulting ultimately in $M\nu$ neutrons in the saturated fission chain, the probability of one of these being detected in dt_1 at t_1 is

$$P(t_1/t_o)dt_1 = M\nu \frac{\varepsilon}{<\nu>} s(t_{1o})dt_1$$

The probability of a count in dt_2 at t_2 resulting from the same fission event, assuming one of the neutrons was detected in dt_1 at t_1, is:

$$P(t_2/t_o)dt_2 = (M\nu-1)\frac{\varepsilon}{\langle\nu\rangle} s(t_{2o})dt_2$$

$$\therefore P(t_1,t_2/t_o)dt_1dt_2 = \frac{M\nu(M\nu-1)}{\langle\nu\rangle^2}\varepsilon^2 s(t_{1o})\, s(t_{2o})\, dt_1dt_2$$

The number of fissions occurring in dt_o at t_o is $F\,dt_o$. Thus, the joint probability of a count in dt_1 at t_1 and a count in dt_2 at t_2 from a common ancestor fission event is obtained by integrating the product of the probability above and the fission rate over all preceding t :

$$\begin{aligned}\Pi(t_1,t_2/t_o) &= \int_{t_o=-\infty}^{t_1} P(t_1,t_2/t_o)dt_1dt_2\, F\, dt_o\\ &= \varepsilon^2 F\frac{\langle M\nu(M\nu-1)\rangle}{\langle\nu\rangle^2}\int_{t_o=-\infty}^{t_1} s(t_{1o})\, s(t_{2o})\, dt_1dt_2dt_o\\ &= \varepsilon^2 F\frac{\langle M\nu(M\nu-1)\rangle}{\langle\nu\rangle^2} S'(t_{12})\end{aligned}$$

$$\text{where } S'(t_{12}) \equiv \int_{t_o=-\infty}^{t_1} s(t_{1o})\, s(t_{2o})\, dt_1dt_2dt_o$$

In addition to counts occurring in dt_1 and dt_2 due to common ancestor fission events, there will be random counts occurring due to fission events belonging to unrelated fission chains. For a stationary system, the fission rate is constant and the probability of a pair of random counts, one in dt_1, the other in dt_2, is:

$$(\varepsilon F\, dt_1)\,(\varepsilon F\, dt_2) = \varepsilon^2 F^2\, dt_1\, dt_2$$

The total probability of a pair of counts, one in dt_1, the other in dt_2, is the following sum:

$$F^2\varepsilon^2\, dt_1\, dt_2 + F\varepsilon^2\frac{\langle M\nu(M\nu-1)\rangle}{\langle\nu\rangle^2} S'(t_{12})$$

Consider a counting time interval, T, in which the number of counts is C. The expected number of pairs of counts for a large number of intervals, N, is given by:

$$\frac{\langle C(C-1)\rangle}{2} = \frac{1}{N}\sum^{N}\int_{t_2=0}^{T}\int_{t_1=0}^{t_2}\left\{F^2\varepsilon^2\, dt_1dt_2 + F\varepsilon^2\frac{\langle M\nu(M\nu-1)\rangle}{\langle\nu\rangle^2} S'(t_{12})\right\}$$

i.e.,

$$\frac{\langle C(C-1)\rangle}{2} = \left\langle\frac{F^2\varepsilon^2 T^2}{2}\right\rangle + \left\langle F\varepsilon^2\frac{\langle M\nu(M\nu-1)\rangle}{\langle\nu\rangle^2}\int_{t=0}^{T}\int_{t=0}^{t} S'(t_{12})\right\rangle$$

But, $\langle F\varepsilon T\rangle = \langle C\rangle$, so that

$$\frac{\langle C^2\rangle - \langle C\rangle}{2} = \frac{\langle C\rangle^2}{2} + \frac{\langle C\rangle}{T}\varepsilon\frac{\langle M\nu(M\nu-1)\rangle}{\langle\nu\rangle^2}\int_{t_2=0}^{T}\int_{t_1=0}^{t_2} S'(t_{12})$$

Rearranging and dividing by $\frac{<C>}{2}$:

$$\frac{<C^2> - <C>^2}{<C>} = 1 + \frac{2\varepsilon}{T} \frac{<M\nu(M\nu-1)>}{<\nu>^2} S(T)$$

$$\text{where } S(T) \equiv \int^{T} \int^{t_2} S'(t_{12})$$

The left-hand side is the ratio of the variance of the counts in an interval of width T to the mean of the counts in the interval. For counts distributed according to a Poisson distribution, the variance-to-mean ratio is unity. Thus, the equation informs us that for a fissioning system, the variance-to-mean ratio exceeds unity by an amount

$$Y \equiv \frac{2\varepsilon}{T} \frac{<M\nu(M\nu-1)>}{<\nu>^2} S(T)$$

which is dependent on the multiplication of the sample and independent of the count rate. So, the first and second moments of the neutron count distribution can be used to obtain information about sample multiplication. However, there is no direct information about sample mass in Y.

Mass

Information about the mass of the sample can be obtained from the first two moments of the count distribution, even in the presence of high backgrounds or large indigenous (α,n) sources. An approach, which is similar to conventional coincidence counting is taken, viz., separation of correlated and uncorrelated counts. The incoming pulse train consists of a sum of random [background and (α,n)] neutron and correlated [spontaneous and induced fission] neutron detection pulses. It can be shown[3] that the following combination of the moments

$$Q \equiv \overline{C^2} - \overline{C}^2 - \overline{C}$$

contains only those pulses arising from spontaneous fission neutrons or neutrons from fissions induced either by spontaneous fission or (α,n) neutrons. In the absence of multiplication, Q is determined solely by the spontaneous fission neutron source strength, which can be related to the mass of the even isotopes. Assuming an isotopic analysis is available or has been measured, the assay can be made.

EXPERIMENTAL METHOD

A large set of plutonium samples, described in Table I, has been used in a test of the method for neutron emitting samples. Each was placed in a typical well counter used for SNM assay.[4] The counter was operated in the low efficiency mode, with an efficiency for ^{252}Cf spontaneous fission neutrons of 3.25%. The logic pulses from the detection circuitry were interfaced with a μNOVA computer via a timer/scaler interface board, which counted the pulses for the time interval specified by the software. The detector deadtime, τ, (required for corrections to the data), the width of the time intervals, T, and the number of time intervals, N, for the analysis, were entered using a keyboard terminal. The incoming pulses were scaled to obtain C, squared to obtain C^2 and running sums ΣC and ΣC^2 kept to the end of the analysis time. Averages $<C^2> = \frac{1}{N} \sum^{N} C^2$ and $<C> = \frac{1}{N} \sum^{N} C$ were calculated at the end of the

TABLE I. SNM ASSAY STANDARDS

Sample ID	Description	Pu Mass (g)	"Effective" ^{240}Pu Mass (g)$^{+}$
STD-2	Ash, PuO_2, DE*	21	1.22
STD-5	PuO_2, MgO	60	3.838
STD-7	PuO_2, MgO	120	7.672
STD-8	PuO_2, MgO	240.1	15.35
STD-9	PuO_2, MgO	480.35	37.39
STD-10	PuO_2	60	4.599
STD-11	PuO_2, NaCℓ	60	4.599
STD-12	PuO_2, $Aℓ_2O_3$	60	4.599
STD-13	PuO_2, LiCℓ	60	4.599
STD-14	PuO_2, NaCℓ	120	9.196
STD-15	PuO_2, $Aℓ_2O_3$	120	9.197
STD-16	PuO_2, LiCℓ	98	7.515
PEO-381	PuO_2	615	64.97
PEO-382	PuO_2	556	54.93
PEO-385	PuO_2 (impure)	459	43.38
PEO-447	PuO_2	777	81.36
ZPRN-7-052	Pu metal	220	25.96†
ZPRD-1-122	Pu metal	26	2.99†
ZPRD-1-127	Pu metal	26	2.99†
ZPRH-2-022	Pu metal	71	15.22†
ZPRH-2-020	Pu metal	71	15.22†
ZPRN-7-731	Pu metal	220	25.30†
ZPRS-7-194	Pu metal	142	12.35†
ZPRS-7-240	Pu metal	142	12.35†
ZPRC-2-379	Pu metal	69	3.45†
ZPRC-2-494	Pu metal	69	3.45†
643-001	Pu metal	1678	98.33
643-002	Pu metal	2200	130.80
643-003	Pu metal	1729	101.67
643-004	Pu metal	585	34.46
643-005	Pu metal	1377	81.66
621-000	Pu metal	1006	57.66
620-000	Pu metal	2001	115.31
JX 4268	Pu metal	2962	142.77†
JX 4269	Pu metal	2962	141.29†
JX 4570	Pu metal	4086	211.25†
JX 4572	Pu metal	3193	160.29†
JU 125	Pu metal	2174	109.57†
NHP-1	Pu metal	1968	117.10†
COMB-1	Pu metal	2684	155.99
COMB-2	Pu metal	3269	190.45
COMB-3	Pu metal	2684	155.99
COMB-4	Pu metal	3269	190.45
COMB-5	Pu metal	3957	269.57
COMB-6	Pu metal	3007	172.97
COMB-7	Pu metal	3592	207.43
COMB-8	Pu metal	3007	172.97

+ Even isotope masses weighted by spontaneous fission decay constant and neutron multiplicity values relative to those of ^{240}Pu.

* Diatomaceous earth.

† ^{240}Pu mass only.

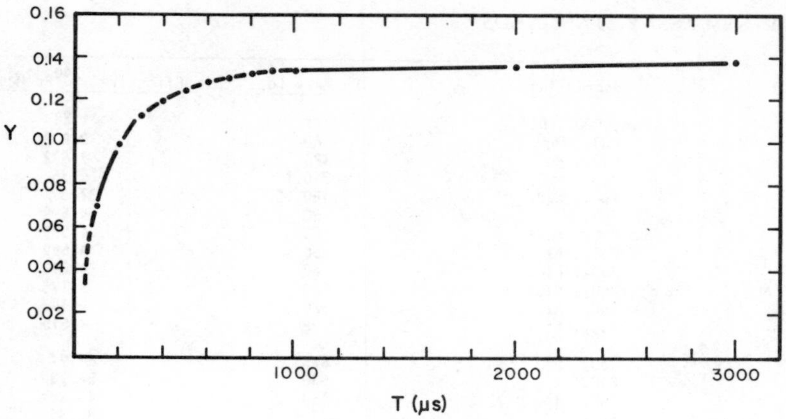

FIG.1. Y versus the time interval width, T, for ^{252}Cf spontaneous fission. The "saturated" value is attained in approximately 1000 μs. Addition of random neutrons does not affect the shape of the curve. Uncertainties in data points are of the same size as the symbol.

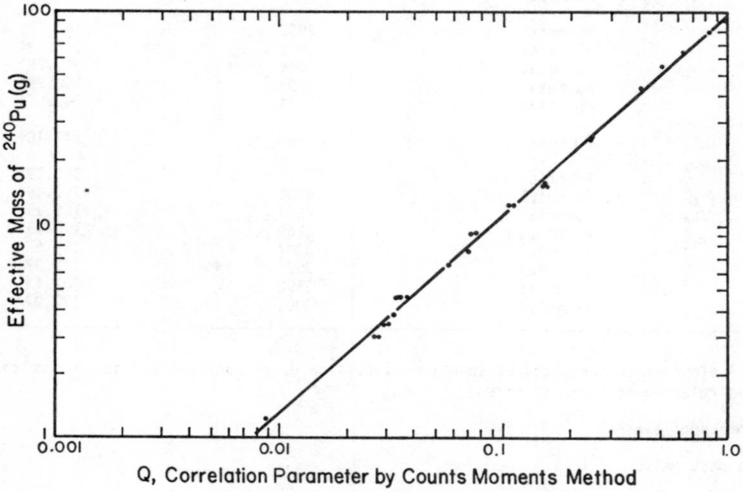

FIG.2. Q versus ^{240}Pu effective mass, for oxide, oxide plus additive, and metal samples less than 220 g listed in Table I. Measurement time was 2000 s for each sample. The correlation coefficient is 0.998.

cycle and the deadtime corrected values for Y and Q output via the terminal. The time interval count distribution for $C \leq 150$ was also printed as a diagnostic aid. This allows the operator to ascertain if electronic transients occurred during the measurement period, a capability that is not found in other assay equipment such as variable deadtime counters or coincidence counters. Software calculation times were negligible compared with data collection times.

Two parameters of the measurements needed to be predetermined, the "optimum" interval width, T, and the system deadtime, τ. The criterion used for "optimum" T is that the value of Y be "saturated". To determine how this criterion would be satisfied, a measurement of the value of Y for ^{252}Cf spontaneous fission neutrons as a function of T was made. The results are shown in Figure 1. We chose the value T = 2000 μs for all subsequent measurements based on these results. It is worth noting that the moments method does not require narrow count gates as do other assay methods, and it is also worth noting that unlike other assay methods, the optimum T is not affected by increasing (α,n) contamination. This latter statement has been verified by measurements of Y versus T for a ^{252}Cf source and an (α,n) source with a strength 10 times that of the ^{252}Cf source. The resultant curve shape is indistinguishable from that of Figure 1.

If system deadtime is explicitly included in the development of the expression for Y, then for a random neutron source, $Y = -2\overline{C}\,\frac{\tau}{T}$. A measurement of Y for such a source permits the determination of τ. We made such measurements for an AmLiF source with the result that $\tau = 1.795 \pm 0.019$ μs. All data were corrected for deadtime effects. The deadtime value was redetermined daily as an equipment check, with no significant variation observed. The same μNOVA code was used for both sample and deadtime measurements, with the operator indicating the required measurement.

RESULTS

Figure 2 is a plot of the measured values of Q versus the effective mass of ^{240}Pu for all of the oxide, oxide plus additives and the metal samples (with mass $\leq$ 220 g) described in Table I. As is obvious from the good correlation coefficient, reliable estimates of sample mass can be obtained using this technique. As an estimate of the precision, we have made a comparison with the new LASL shift register coincidence module[5] results. The logic pulses from the detection circuitry were processed simultaneously in the μNOVA computer to give the corrected values for Q, and in the shift register coincidence module to give the values for $\mathring{R}$, the coincidence count rate. The $\mathring{R}$ values were also corrected for deadtime effects.[6] The results of the comparison are given in Table II. The cumulative sum mass errors are seen to be comparable for the two techniques, leading us to conclude that the measurement precisions are comparable for these samples. A comparison of Q and $\mathring{R}$ for all the samples of Table I is provided by Figure 3. This test was not meant to reveal the ultimate precision in either technique, but only to compare our new technique with an established one.

For the larger metal samples, in which there is a nonnegligible multiplication effect, sample assay can't be done by conventional coincidence counting methods. The moments method can still provide mass estimates for such samples, if a combination of the parameters Q and Y are used. Figure 4 is a plot of the results of simultaneous measurements of Y and Q/^{240}Pu mass for the large (> 500 g) metal standard samples described in Table I. Since

TABLE II. COMPARISON OF MOMENTS METHOD AND SHIFT REGISTER COINCIDENCE METHOD FOR SNM ASSAY

Sample ID	Accepted Value of Effective ^{240}Pu Mass (g)	$Q / {}^{240}Pu$ Mass ($\times 10^{-3}$)	Estimated Mass from Q Correlation a)	$\dot{R} / {}^{240}Pu$ Mass	Estimated Mass from $\dot{R}$ Correlation b)	Δ_Q (g) c)	$\Delta_{\dot{R}}$ (g) d)
STD-2	1.22	7.21	1.00	1.18	1.01	+ 0.22	+ 0.21
STD-5	3.838	8.41	3.66	1.34	3.62	+ 0.18	+ 0.22
STD-7	7.672	9.11	7.92	1.55	8.39	- 0.25	- 0.72
STD-8	15.35	9.91	17.2	1.57	17.0	- 1.8	- 1.6
STD-9	37.39	9.68	41.0	1.55	40.7	- 3.6	- 3.3
STD-10	4.599	8.11	4.22	1.26	4.07	+ 0.38	+ 0.53
STD-11	4.599	7.28	3.79	1.18	3.82	+ 0.81	+ 0.78
STD-12	4.599	7.57	3.94	1.35	4.39	+ 0.66	+ 0.21
STD-13	4.599	7.96	4.14	1.22	3.96	+ 0.46	+ 0.64
STD-14	9.196	7.82	8.14	1.29	8.37	+ 1.06	+ 0.83
STD-15	9.197	8.18	8.52	1.29	8.33	+ 0.68	+ 0.87
STD-16	7.515	7.62	6.49	1.28	6.77	+ 1.02	+ 0.74
PEO-381	64.97	9.81	72.2	1.53	70.1	- 7.2	- 5.1
PEO-382	54.93	9.16	57.0	1.51	58.4	- 2.1	- 3.5
PEO-385	43.38	9.45	46.4	1.50	45.7	- 3.0	- 2.3
PEO-447	81.36	10.2	94.0	1.66	94.9	-12.6	-13.5
ZPRN-7-052	25.96	9.58	28.2	1.61	29.4	- 2.2	- 3.4
ZPRD-1-122	2.99	9.03	3.06	1.38	2.92	- 0.07	+ 0.07
ZPRD-1-127	2.99	8.83	2.99	1.34	2.83	- 0 -	+ 0.16
ZPRH-2-022	15.22	10.2	17.5	1.62	17.4	- 2.3	- 2.2
ZPRH-2-020	15.22	9.86	17.0	1.59	17.0	- 1.8	- 1.8
ZPRN-7-731	25.30	9.78	28.0	1.56	27.8	- 2.7	- 2.5
ZPRS-7-194	12.35	8.61	12.0	1.39	12.1	+ 0.35	+ 0.2
ZPRS-7-240	12.35	8.88	12.4	1.42	12.4	- 0 -	- 0 -
ZPRC-2-379	3.45	8.81	3.44	1.41	3.44	+ 0.01	+ 0.01
ZPRC-2-494	3.45	8.58	3.35	1.37	3.32	+ 0.10	+ 0.13
	Σ = 473.7	< 8.83×10^{-3} >	Σ = 507.6	< 1.42 >	Σ = 508.2	Σ = -33.8	Σ = -34.3

a) Using measured Q values and the average of Column 3.

b) Using measured $\dot{R}$ values and the average of Column 5.

c) Column 2 - Column 4

d) Column 2 - Column 6

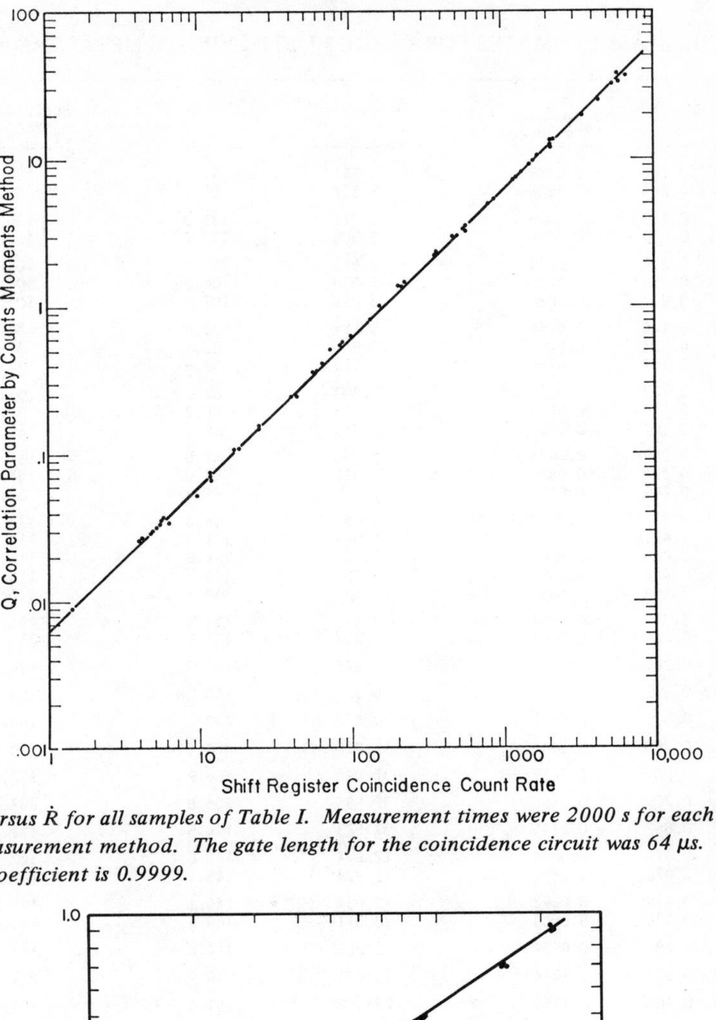

FIG. 3. Q versus Ṙ for all samples of Table I. Measurement times were 2000 s for each sample and each measurement method. The gate length for the coincidence circuit was 64 μs. The correlation coefficient is 0.9999.

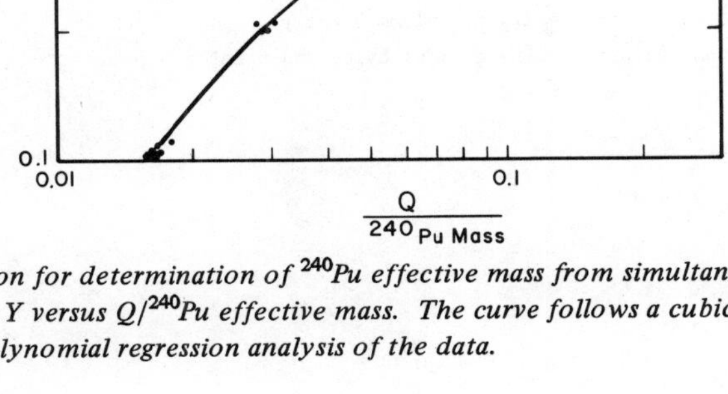

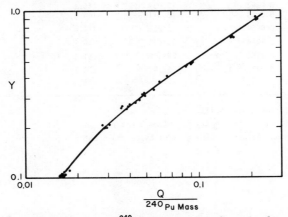

FIG. 4. Calibration for determination of ^{240}Pu effective mass from simultaneous Y and Q measurement: Y versus Q/^{240}Pu effective mass. The curve follows a cubic equation suggested by a polynomial regression analysis of the data.

TABLE III. MASS ESTIMATES FOR LARGE MULTIPLYING Pu METAL SAMPLES

Sample #	Y	$\frac{Q}{^{240}\text{Pu Mass}}$ [a]	Q(measured)	^{240}Pu Mass[b]	Accepted Value of ^{240}Pu Mass	Δ^c(%)
643-001	0.266	0.0389	3.551	91.3	98.33	- 7.2
	0.264	0.0386	3.534	91.6	98.33	- 6.9
643-002	0.259	0.0376	5.012	133.3	130.80	+ 1.9
	0.259	0.0376	5.008	133.2	130.80	+ 1.8
643-003	0.200	0.0276	2.972	107.7	101.67	+ 5.9
	0.198	0.0274	2.928	106.9	101.67	+ 5.1
	0.207	0.0287	3.092	107.7	101.67	+ 6.0
643-004	0.106	0.0166	0.554	33.4	34.46	- 3.1
	0.104	0.0165	0.546	33.1	34.46	- 3.9
	0.107	0.0168	0.570	33.9	34.46	- 1.6
643-005	0.103	0.0165	1.357	82.2	81.66	+ 0.7
	0.110	0.0169	1.456	86.2	81.66	+ 5.5
	0.101	0.0164	1.359	82.9	81.66	+ 1.5
621-000	0.268	0.0393	2.263	57.6	57.66	- 0.1
	0.274	0.0406	2.315	57.0	57.66	- 1.1
	0.268	0.0394	2.262	57.4	57.66	- 0.4
620-000	0.471	0.0870	9.956	114.4	115.31	- 0.8
	0.471	0.0871	9.954	114.3	115.31	- 0.9
	0.468	0.0863	9.907	114.8	115.31	- 0.4
	0.470	0.0868	9.956	114.7	115.31	- 0.5
	0.469	0.0864	9.922	114.8	115.31	- 0.4
COMB-1	0.317	0.0494	7.475	151.3	155.9	- 2.9
	0.321	0.0502	7.606	151.5	155.9	- 2.8
COMB-2	0.281	0.0418	8.245	197.2	190.45	+ 3.4
COMB-3	0.372	0.0618	9.322	150.8	155.9	+ 3.2
COMB-4	0.339	0.0541	10.477	193.7	190.45	+ 1.7
COMB-5	0.295	0.0447	12.270	274.5	270.60	+ 1.4
COMB-6	0.906	0.2225	37.098	166.7	172.97	- 3.6
COMB-7	0.709	0.1582	32.595	206.0	207.43	- 0.7
COMB-8	0.882	0.2146	37.523	174.8	172.97	+ 1.1
JX4268	0.476	0.0884	12.874	145.6	142.77	+ 2.0
	0.478	0.0891	12.924	145.1	142.77	+ 1.6
JX4269	0.479	0.0892	13.034	146.1	141.29	+ 3.4
	0.478	0.0889	13.011	146.4	141.29	+ 3.6
NHP-1	0.205	0.0284	3.254	114.6	117.10	- 2.2
JX4570	0.697	0.1544	33.494	216.9	211.25	+ 2.7
JX4572	0.703	0.1563	25.206	161.3	160.29	+ 0.6
JU125	0.310	0.0477	5.397	113.2	109.57	+ 3.3
					Average =	2.5%

[a] from polynominal least squares fit.

[b] obtained by dividing column 4 entry by column 3 entry.

[c] column 5 entry minus column 6 entry divided by column 6 entry

$Q/^{240}Pu$ mass is a constant for nonmultiplying samples, the fact that this ratio is not constant for these samples indicates multiplication. Indeed, the largest sample, #JX4570, would have a free field multiplication of approximately M=2, and an even larger M inside the well counter. Even so, a reliable estimate of its mass can be obtained by the simultaneous measure- use of Y and Q. A demonstration of the procedure is provided in Table III. A polynominal fit to the data provided the expected value of $\frac{Q}{^{240}Pu_{mass}}$ (column 3) for a given Y value (column 2). The mass estimates were obtained by using the measured Q value and the calculated $\frac{Q}{^{240}Pu_{mass}}$ values obtained from the fit. These ^{240}Pu mass estimates are on the average within 2.5% of the declared masses, verifying the potential of the technique for accurate assay of even large multiplying masses. A very important feature of the moments technique is also revealed in the data of Table III. An examination of column 6 shows that COMB-1 and COMB-3 have the same mass, COMB-2 and COMB-4 have the same mass, and COMB-6 and COMB-8 have the same mass. These sets have the same masses because they are made up from the same pieces of plutonium. The geometries of the combinations were intentionally altered to change the multiplication of the pieces. Accurate assay was still possible, verifying that the effects of multiplication are separately accounted for. In the same way, the simultaneous measurement of Q and Y can account for changes in the fission detection efficiency. As an example, the three separate measurements of sample #643-003 vary by approximately 5% in the individual values of Q and Y but less than 1% in the assay value of the ^{240}Pu mass.

DISCUSSIONS AND CONCLUSIONS

This initial study has confirmed the utility of the moments method for SNM assay applications. We have assayed samples ranging from a few tens of grams to those as large as criticality safety limits allow. The measurement precision has been demonstrated to be as good as the measurement precision of the coincidence counting technique. The advantages provided by the moments method include: the simplicity of the associated circuitry (we have built hard-wired versions in four-wide NIM modules); the fixed counting period, which is determined only by the counter characteristics and not by random neutron source strength contamination; and, perhaps most importantly, the moments method provides a diagnostic capability unavailable with the other techniques. Equipment malfunction can be immediately signaled by an inordinately skewed count probability distribution output <u>for each sample</u>. In our newest software versions, inordinately large single interval counts are automatically eliminated so that even a single measurement cycle is not lost. This is considered an extremely important characteristic of counting stations in DYMAC-type assay applications so that processes can be operated without interruptions except for discovery of real material differences.

We are in the process of extending the moments method to the assay of ^{235}U using an active well counter.[7] Additional algorithms and detection schemes are being examined to attempt extraction of additional information from the moments of the neutron count distribution.

REFERENCES

[1] PACILIO, N., "Reactor Noise Analysis in the Time Domain", AEC Critical Review Series TID 24512 (1969).

[2] DeHOFFMANN, F., "The Statisitcal Aspects of Pile Theory", in the Science and Engineering of Nuclear Power, Addison-Wesley Press, Inc. (1949).

[3] HANSEN, G. E., private communication, 1978.

[4] Many detectors of this type have been constructed at various laboratories. Los Alamos experience began with a detector described in Los Alamos Scientific Laboratory report, LA-3921-MS (1968).

[5] Nuclear Safeguards Research Program Status report, LA-6788-PR, compiled by Joseph L. Sapir (June, 1977).

[6] AUGUSTSON, R. H., Appendix of "DYMAC" Demonstration Program: Phase I Experience" (February, 1978).

[7] Nuclear Safeguards Research and Development Program Status report, LA-7030-PR, compiled by Joseph L. Sapir (March, 1978).

CALORIMETRIC SYSTEMS DESIGNED FOR IN-FIELD NON-DESTRUCTIVE ASSAY OF PLUTONIUM-BEARING MATERIALS

C.T. ROCHE, R.B. PERRY, R.N. LEWIS,
E.A. JUNG, J.R. HAUMANN
Argonne National Laboratory,
Argonne, Illinois,
United States of America

Abstract

CALORIMETRIC SYSTEMS DESIGNED FOR IN-FIELD NON-DESTRUCTIVE ASSAY OF PLUTONIUM-BEARING MATERIALS.

Calorimetric assay provides a precise non-destructive analytical (NDA) method for determining sample plutonium content based on the heat emitted by decaying radionuclides. Calorimetry has a plutonium-detection sensitivity of 20 ppm, and power measurement precision better than 0.1% is obtainable. It is insensitive to the chemical form of the plutonium and is independent of measurement-bias problems due to sample geometric configuration and sample matrix composition. Also, the ability of an operator to calibrate a calorimeter using electrical-heat standards eliminates the necessity of transporting plutonium calibration sources. These considerations make calorimetry an important assay tool which can be used by itself or for calibrating more rapid, but less accurate, NDA techniques. The total plutonium mass of the sample may be obtained non-destructively by combining the calorimetric power measurement with a gamma-spectrometer analysis of the sample isotopic content. Conventional calorimetric design measures the temperature rise of a plutonium-containing sample chamber in contact with a large water-bath heat sink. This design lacks the mobility needed by inspection personnel. The Argonne National Laboratory (ANL) air-chamber isothermal calorimeters are low-thermal capacitance devices which eliminate the need for large, constant-temperature heat sinks. A bulk calorimeter designed to measure sealed containers holding up to 3 kg Pu, and a small-sample calorimeter designed to measure mixed-oxide fuel pellets and powders are discussed. The operational characteristics of these instruments are described, and the results of sample assays presented.

I. INTRODUCTION

Increased emphasis on on-site verification of nuclear material inventories has magnified the need for techniques capable of in-field nondestructive analysis (NDA) of plutonium. Because of the large variations encountered in the chemical form of plutonium, the nonstandard measurement geometries, and the presence of radiation absorbers, there are often large uncertainties associated with on-site assays. Calorimetry has provided a precise NDA method for the determination of plutonium content in the laboratory.[1] It has also been applied routinely to help satisfy plant-shipper/receiver accountability requirements.[2] The lower limit of sensitivity of this technique is approximately 0.1 g Pu-239 in a one-liter container.[3] However, it is for the analysis of dense, heterogeneous mixtures of relatively high Pu content that the method is best suited. Linear sample power-mass relations have been obtained for items containing plutonium in the kilogram region. A measurement precision of better than 0.1% is readily attainable.[4,5] This is not true of gamma-ray and neutron techniques where photon absorption and neutron multiplication cause considerable nonlinearity in assaying high mass samples.[6,7] Among the advantages of calorimetric assay are its insensitivity to the chemical form of the plutonium and its independence from measurement-bias problems due to sample geometric configuration and sample matrix composition. Also, the ability of an operator to calibrate a calorimeter with electrical heat eliminates the necessity of transporting plutonium calibration sources. These considerations suggest calorimetry as an important in-field assay tool which could be used by itself or as a selective cross-calibration of more rapid, but less accurate, NDA techniques. Argonne isothermal calorimeters have been constructed as low-thermal-capacitance devices in which the water-bath heat sinks have been eliminated. This design has shown promise of reducing the assay time and of increasing the mobility of the system.[8] This report will discuss work in progress on a set of calorimeters designed to assay the types of materials normally encountered by IAEA personnel. A bulk (storage container) calorimeter, designed to measure sealed cans holding up to 3 kg Pu, and a small-sample calorimeter, designed to measure mixed-oxide fuel pellets and powders, will be reviewed.[5,9] The operational characteristics of the instruments will be described, and the results of sample assays will be presented.

II. DISCUSSION OF CALORIMETRIC TECHNIQUE

Calorimetric analysis of plutonium relies on the ability of the assay device to measure thermal power with high precision and accuracy. Thermal energy is generated by the absorption and degradation of the radiation released in the decay of the Pu isotopes. The principal decay modes of the isotopes of interest produce alpha particles and low-energy beta and gamma rays, which are easily converted to heat energy within the measurement chamber. In unirradiated fuel, only the isotopes Pu-238–242 and Am-241 have high enough specific activities to contribute to the total heat. Thus we may relate thermal power to grams of plutonium by using the proper conversion factor, known as the effective specific power (P_{eff}). P_{eff} has been defined as the weighted-average isotopic power per gram of sample.[5]

$$P_{eff} = P_S/M_t = \sum^{n} M_i P_i/M_t = \sum (KRQ\lambda)_i$$

where

P_S	=	sample power
M_t	=	sample plutonium mass = Σ mass of Pu isotopes
P_i	=	isotopic thermal power/gram
R_i	=	M_i/M_t
Q_i	=	isotopic total decay energy
λ_i	=	isotopic decay constant
K_i	=	isotopic normalized constant

TABLE 1. Pu ANALYSIS PERFORMED BY THE ANL BULK CALORIMETER

A: Assay of ZPPR Mixed-Oxide Fuel Rods

Sample Composition (Pu,U)O_2 Pu - 13.25%
Nominal Isotopic[a] ^{238}Pu = 0.05% ^{239}Pu = 86.6%
^{240}Pu = 11.5% ^{241}Pu = 1.7%
^{242}Pu = 0.02% ^{241}Am = 0.2%

Effective Specific Power[b] : 3.54 mW/g

No. of Rods	Sample Power(mW)	Sample Mass(g)[d]	Reported Book value
6	233 ± 1	65.8 ± 0.3	70.9
13	528 ± 2	149.2 ± 0.6	154.0
25	1006 ± 1	284.2 ± 0.3	296.3
32[c]	1286 ± 1	363.3 ± 0.3	379.3
32	1286 ± 1	363.3 ± 0.3	379.3
45	1805 ± 2	509.9 ± 0.6	533.3

[a] Batch Isotopic analysis reported 8/11/70

[b] Effective Specific Power calculated from Batch Isotopic Data 7/1/78

[c] The first assay of 32 rods was performed with the rods in a close packed geometry, the second with maximum spatial separation.

[d] Uncertainties in the Isotopic Specific Power were not included in this analysis

B: Assay of PuO_2 Scrap Samples

Sample Composition PuO_2, Al_2O_3, Pu - 1%
Nominal Isotopic[a] ^{238}Pu = 0.02% ^{239}Pu = 90.7%
^{240}Pu = 8.4% ^{241}Pu = 0.9%
^{242}Pu = 0.05%

Effective Specific Power[b] - 2.93 ± 0.02 mW/g

Sample Power(mW)	Sample Mass(g)	Reported Book Value
208 ± 1	71.0 ± 0.6	70.62
107 ± 1	36.5 ± 0.4	35.62
62 ± 2	21.2 ± 0.7	20.62
10 ± 3	3.4 ± 1.1	5.62

[a] Batch Isotopic Reported 10/30/67

[b] Effective Specific Power determined by Gamma-Spectrometry on 5/12/78

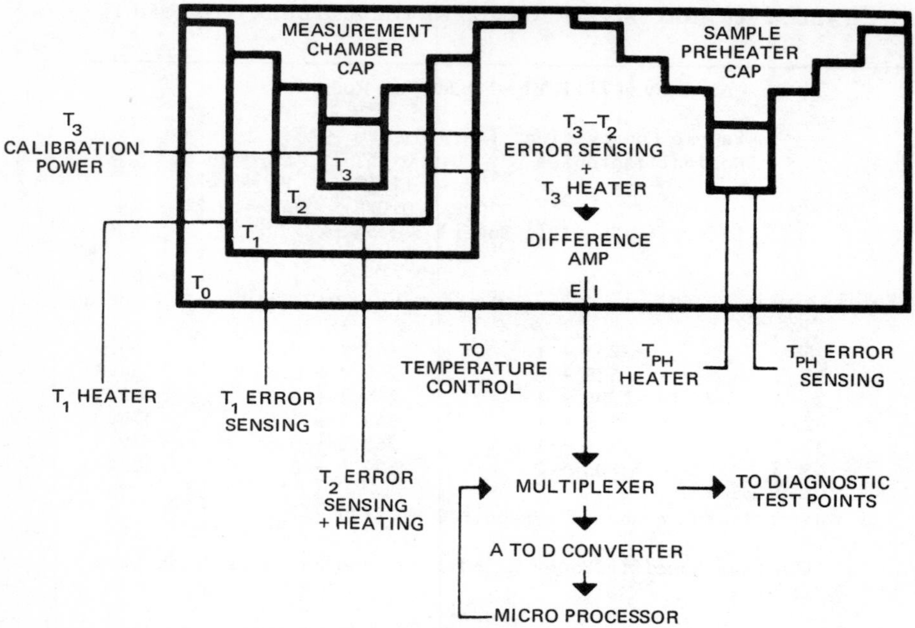

FIG.1. Schematic representation of ANL isothermal calorimeter including measurement and control components.

The ANSI standard on calorimetry suggests two methods for determining P_{eff}.[3] The methods are distinguished by whether an isotopic analysis is conducted. The empirical method requires that a combined chemical and calorimetric analysis be performed on a set of representative samples to determine the total Pu content per watt of measured power. Samples must be reassayed at a later time to account for changes due to radioactive decay. The second method requires that an isotopic analysis be performed to determine the content of Pu-238-242 and Am-241. The P_{eff} may be calculated at any future time by applying the known decay laws.[3]

The determination of P_{eff} often accounts for the largest portion of the uncertainty associated with calorimetric assay. In certain circumstances this may be avoided by accepting the agreement between heat output as determined by the shipper and that determined by the receiver. While this method of operation may not be desirable for material in long term storage, the change in power output resulting during normal shipment between facilities would not be significant. During a one-month period, ZPPR-fuel power output will increase less than 0.15% (see Table 1).

III. INSTRUMENT DESIGN

In classical heat-flow calorimeters, the temperature difference developed across a thermal resistance linking the sample chamber and a heat sink is proportional to the power produced by the samples. ANL calorimeters

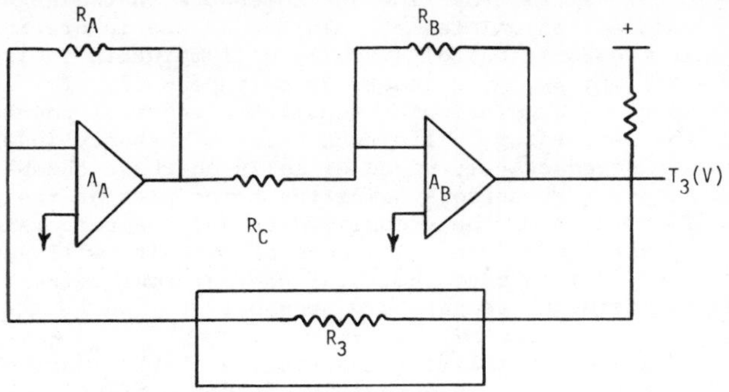

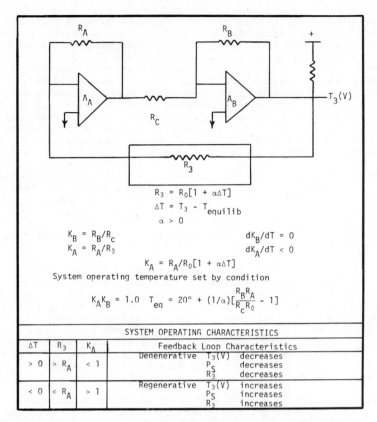

$$R_3 = R_0[1 + \alpha\Delta T]$$
$$\Delta T = T_3 - T_{equilib}$$
$$\alpha > 0$$

$K_B = R_B/R_C$ $\quad dK_B/dT = 0$

$K_A = R_A/R_3$ $\quad dK_A/dT < 0$

$$K_A = R_A/R_0[1 + \alpha\Delta T]$$

System operating temperature set by condition

$$K_A K_B = 1.0 \quad T_{eq} = 20° + (1/\alpha)[\frac{R_B R_A}{R_C R_0} - 1]$$

SYSTEM OPERATING CHARACTERISTICS

ΔT	R_3	K_A	Feedback Loop Characteristics
> 0	> R_A	< 1	Degenerative $T_3(V)$ decreases P_S decreases R_3 decreases
< 0	< R_A	> 1	Regenerative $T_3(V)$ increases P_S increases R_3 increases

FIG. 2. Resistance thermometry and feedback control circuitry in ANL isothermal calorimeters.

are designed as servo-controlled devices in which electrical power is provided to maintain the unit at a constant temperature above ambient. No temperature gradients are developed during an analysis. When a heat-producing source (P_S) is assayed, the electrical power necessary to maintain the steady state condition (P_C) is reduced from the empty chamber baseline power (P_0). Thus the sample-produced power may be determined from the differences between the calorimeter-applied powers with and without a sample.

$$P_S = P_0 - P_C$$

In the Argonne design, the calorimeter is most simply viewed as a constant temperature oven composed of a series of concentric chambers (Fig. 1). Each of these chambers is constructed from an aluminum cylinder upon which resistive heating coils and heat sensors are mounted. The cylinders are separated from one another by high thermal-resistance material. The ends of the cylinders are similarly protected by nonconducting plugs and by pancake-type heater coils. Alternating zones of high and low thermal conductivity in this manner tends to minimize the effects of localized external temperature changes. A temperature profile is established within the calorimeter to eliminate axial heat flow and to ensure that a negative radial temperature gradient is maintained. In the schematic drawing in Fig. 1, the electronic feedback control circuits will maintain the relation $T_3 > T_2 > T_1 > T_0 > T_{ambient}$. The inner two cylinders (T_3, T_2) act as the measurement chamber. The calorimeter-supplied electrical power is adjusted to maintain the temperature difference between these cylinders to ± 20 microdegrees. Noninductively wound Ni coils on these chambers act as both heaters and temperature sensors using the principles of resistance thermometry. The T_3 Ni coil may be visualized as being a temperature-sensitive element in a resistive bridge. The control circuits will supply sufficient electrical power (P_C) so that the total thermal source ($P_S + P_C$) will be adequate to maintain the temperature necessary to "null" the bridge.

Figure 2 gives a simplified description of the inner-chamber control circuit. The system works on a negative feedback principle. This causes the electrical power applied to R_3 to be increased or decreased in response to the chamber temperature. The outer shells act as protective buffers for the inner measurement chambers. These are adjusted by a series of YSI thermistors and copper heating coils so that the measurement chamber is unaffected by changes in the ambient temperature.

Data manipulation for both the small-sample and the bulk calorimeters is performed by a microprocessor-based data acquisition system (DAS). The hardware for this module was designed to be common to all ANL calorimeters, thus facilitating operator training and use. The DAS hardware consists of a microprocessor, a 12-bit analog-to-digital converter (ADC), 8K bytes of nonvolatile memory, and 1K bytes of scratch-pad memory. A number of input-output (I/O) peripherals, including a printer and a keyboard, are also included. The entire unit is housed in an attache case and weighs approximately 5 kg. (Detailed circuit diagrams and program listings are supplied in the instrument design manual [9]).

The data-analysis program resides on the nonvolatile memory. While the code will differ slightly with each calorimeter, the basic data-acquisition and -handling routines have been structured to appear identical to the user. Among the routines included are a set of programs to calculate P_{eff} from the sample isotopic data. The code to correct the isotopic data for radioactive decay is also available. Standard assay operation proceeds as follows: The DAS monitors the electric power applied to the measurement chamber. The analog signal is digitized, and the average and the standard deviation of a predetermined number of power measurements is determined. The number of measurements needed to properly describe the behavior of the calorimeter at equilibrium depends upon the thermal time constant of the device and the sample being assayed. The power applied during an actual assay will be compared to either an electrical-heat calibration curve or to a zero-source power baseline reading. The sample power and its associated uncertainty are then calculated.

The small-sample calorimeter was designed as a portable, rapid assay device for analyzing small quantities of Pu, such as fuel pellets and mixed-oxide powders. The maximum sample power output which can be assayed is 32 mW (approximately 10 g of fast reactor Pu). The system is contained in two briefcase-sixed packages and weighs a total of 18 kg. It consists of the calorimeter, the sample preheater, and a microprocessor-controlled data analysis system. The unit has been designed for in-field operation and is capable of operating under a sizable range of voltage and temperature conditions. It has a measurement cycle of 20 min with a measurement precision of 0.1%.[5] (A detailed operating description for this device, including circuit diagrams, is available in the manual *ANL Small-Sample Calorimeter: System Design and Operation*, C. T. Roche, et al.[9])

The bulk calorimeter has been designed to assay large canisters of Pu-containing material normally found in fuel-processing facilities. The chemical form of this material may vary from Pu-metal alloys to mixed-oxide (MOX) powders. The material may be in any physical configuration from finished product to scrap. In addition, since it is often highly desirable that these canisters not be opened, it is possible that the exact constitutents and their geometric arrangement will be unknown. In developing a device which addresses these problems, we attempted to construct a system which would be capable of accurately assaying large concentrations of Pu, be relatively insensitive to ambient temperature fluctuations, and be transportable between facilities. The bulk calorimeter consists of 5 servo-controlled cylinders separated from each other by heat-conducting epoxy. The system power supplies and the control circuitry are located in a NIM standard power bin. The unit also includes a sample preheater and a microprocessor-controlled DAS. The calorimeter will assay cans up to 11.1 cm in diameter by 33 cm in length. The total sample power output may be as high as 26 W, which is equivalent to approximately 3 kg of high burn-up recycle Pu. Preliminary estimates of the sample-power measurement precision are less than 0.1%. The unit has been operated successfully in areas undergoing large temperature fluctuation. The temperature instability contribution to the total system uncertainty is less than 0.02%/°C. Unlike the small-sample calorimeter in which sample encapsulation is controlled by the analyst, the bulk calorimeter will often be assaying items with poor heat=transfer properties. This includes scrap containers in which the heat-emitting material is doubly wrapped in polyethylene within the storage canister. Under these conditions, the heat-transfer properties of the sample will provide the limiting time constant governing the equilibration period. An equilibration period of 5 hr was necessary for the material assayed in Table 1. Equilibration prediction techniques and more accurate sample preheating procedures are being examined as possible ways to reduce the assay time.

IV. EXPERIMENTAL EVALUATION

A series of experiments were performed as part of the program to evaluate the measurement reliability of these calorimeters. The samples assayed during these experiments were representative of the forms of plutonium encountered in the nuclear fuel cycle. The experimental procedure adopted required that the effective specific power and the thermal power output be determined for each sample. All P_{eff} were determined with the computational method (see Section II). With the exception of the ZPPR mixed-oxide fuel rods (Table 1), the isotopic composition of the samples was determined prior to calorimetric assay with a gamma-ray spectrometric technique.[10] The ZPPR rods had been well-characterized by previous chemical and mass-spectrometric analysis.[11] The gamma-spectrometric measurement used a low-energy photon spectrometer (LEPS) to analyze the 90-110 keV region of the spectra. When the analysis is limited to this small energy range, the effects of gamma-ray absorption and sample nonuniformity can be ignored. Since gamma-assay of Pu samples does not determine Pu-242 content, the isotopic ratios are slightly biased. However, only low burn-up material is considered in these experiments, and the added uncertainty in P_{eff} is less than 0.2%. Typical precisions (1 σ) obtained in the determination of P_{eff}

for Pu are 0.5%, while precisions of 2.5% are obtained from analysis of mixed-oxide fuels (Pu, U). The increased uncertainty results from the presence of U X-rays in the 90-110 keV region.

The calorimeter must be electrically calibrated before assaying samples. In this procedure, a precision voltage source supplies measured voltage increments to a Ni coil wound around the inner measurement chamber. The feedback control circuitry will correspondingly decrease the power to the heater-sensor coil to maintain the proper temperature balance. This simulates a series of radioactive heat standards being placed in the measurement chamber. The entire procedure is controlled automatically by the DAS. The microprocessor outputs the input calibration power and the calorimeter-supplied power. A linear least-squares fit to the data indicates a slope of -1.001 ± 0.003 with an intercept of 31.434 ± 0.003 mW for the small-sample calorimeter. A slope of -1.001 ± 0.001 and an intercept of 26.499 ± 0.008 W are found for the bulk calorimeter. Electrical calibration has the advantage of being traceable to high-precision electrical standards through organizations like the U. S. National Bureau of Standards. It also eliminates the necessity of transporting Pu heat standards.

Table 1 shows the results of assays performed by the bulk calorimeter. The samples in the first experiment consisted of stainless steel-encapsulated MOX fuel rods in a storage container 11 cm in diameter and 18 cm long. The fuel rods were placed in the canister, assayed, removed and replaced in a different arrangement, and then reassayed. Within the limit of measurement error, the assay was unaffected by the geometric arrangement of the rods. In addition, by changing the number of fuel rods assayed, we obtained a linear variation of measured sampled mass to reported mass, with a slope of 0.962 ± 0.001. This behavior differs from neutron and gamma-ray assay of large quantities of Pu where the geometric arrangement of the material may result in errors due to neutron multiplication or photon absorption.[6,7]

The second experiment was performed on a set of realistic scrap samples. The samples were constructed from a mixture of plutonium- and aluminum-oxide powders. The cans were agitated between measurements to test the system response to nonuniform changeable matrices. This rearrangement of the sample-matrix distribution had no effect on the assay. The calorimeter gives reasonable results for samples emitting as little as 60 mW, which is less than 0.25% of the full power of the system. The failure of the calorimeter to accurately assay the 5-g scrap sample sets a lower limit on the sensitivity of the device. This limit is a function of the calorimeter full-power setting and not an inherent limit of the design. A similar design with a full-scale power setting of 10 W was able to assay the sample accurately within the uncertainty of the system.

The results of the set of assays using the small-sample calorimeter are shown in Table 2. In these experiments, both mixed-oxide fuel and Pu-metal alloy fuel were analyzed.[5] The material was double-encapsulated in Al sample holders prior to assay. This encapsulation system was designed to maximize the rate of heat transfer, thus minimizing the sample assay time. (The capsules containing MOX pellets were sealed by W. Ulbricht of New Brunswick Laboratory.) The sample power was determined in a 4-min measurement following a 15-min equilibration period. The 1 σ errors given in the power measurements include contributions from the system temperature stability, the sample heat distribution uncertainty, and the effect of multiple analyses of individual samples. The principal error contribution to the final conversion to grams of Pu is the uncertainty in the gamma-ray-

TABLE II. Pu ANALYSIS PERFORMED BY THE ANL SMALL SAMPLE CALORIMETER

A: Assay of Mixed-Oxide Fuel Pellets

Sample Composition - Pu = 11.5%, U = 76.5%, O = 12%
Nominal Isotopic - ^{238}Pu = 0.1%, ^{239}Pu = 86.5%,
^{240}Pu = 11.8%, ^{241}Pu = 1.5%,
^{242}Pu = 0.2%, ^{241}Am = 0.7%

Peff (mW/g)	Sample Power (ms)	Sample Mass (g)	Reported Book Value
3.63 ± 0.08	11.13 ± 0.05	3.07 ± 0.07	2.91
8.87 ± 0.11	6.33 ± 0.02	1.64 ± 0.05	1.69

B: Assay of ZPR-3 Alloy [5]

Sample Composition - Pu = 98.79%, Al = 1.17%
Nominal Isotopic - ^{238}Pu = 0.01%, ^{239}Pu = 95.2%
^{240}Pu = 4.5%, ^{241}Pu = 0.2%
^{242}Pu < 0.2%, ^{241}Am = 0.2%

Peff (mW/g)	Sample Power (ms)	Sample Mass (g)	Reported Book Value
2.48 ± 0.01	4.261 ± 0.005	1.72 ± 0.02	1.72
2.49 ± 0.01	3.733 ± 0.004	1.50 ± 0.01	1.50
2.48 ± 0.01	4.409 ± 0.005	1.78 ± 0.01	1.76
2.51 ± 0.02	4.263 ± 0.005	1.70 ± 0.02	1.71

determined value of P_{eff}. However, in all cases the calorimetrically determined Pu content came well within two standard deviations of the reported book value.

REFERENCES

1. S. R. Gunn, *Radiometric Calorimetry: A Review,* Nuclear Instruments and Methods, 29(1) (1967).

2. W. W. Rodenburg, *An Evaluation of the use of Calorimetry for Shipper-Receiver Measurements of Plutonium,* MLM - 2518.

3. ANSI N15. 22-1975, *American National Standard Calibration Techniques for the Calorimetric Assay of Pu-Bearing Solids Applied to Nuclear Materials Control.*

4. D. M. Bishop and I. M. Taylor, *Calorimetry of Plutonium Recycle Fuels: Applications, Incentives, and Needs,* Proceedings of the Symposium on the Calorimetric Assay of Plutonium, p 75 (1973).

5. C. T. Roche, R. B. Perry, R. N. Lewis, E. A. Jung and J. R. Haumann, *A Portable Calorimeter System for Nondestructive Assay of Mixed-Oxide Fuels*, Nondestructive and Analytical Chemical Techniques in Nuclear Safeguards Analysis, ACS Symposium Series, (1978).

6. R. B. Perry, R. W. Brandenburg, N. S. Beyer , R. C. Gehrke and D. J. Henderson, *Effect of Induced Fission on Plutonium Assay with a Neutron Well Coincidence Counter*, ANL-8029.

7. J. E. Cline, *A Relatively Simple and Precise Technique for the Assay of Plutonium Waste*, ANCR-1055.

8. C. W. Cox, C. J. Renken, R. W. Brandenburg, R. B. Perry and G. A. Youngdahl, *Electric Heat Balance Calorimeter*, ANL-78-32.

9. C. T. Roche, R. B. Perry, R. N. Lewis, E. A. Jung and J. R. Haumann, *ANL Small-Sample Calorimeter: System Design and Operation*, ISPO-13.

10. R. Gunnink, *A System for Plutonium Analysis by Gamma-Spectrometry*, UCRL-51577 (1974).

11. N. S. Beyer, R. N. Lewis and R. B. Perry, *Small, Fast-Response Calorimeters Developed at ANL for the Assay of Plutonium Fuel Rods*, Proceedings of the Symposium on the Calorimetric Assay of Plutonium, p 99 (1973).

AUTORADIOGRAPHIC INVENTORY METHODS

S.B. BRUMBACH, R.B. PERRY
Non-destructive Assay Section,
Argonne National Laboratory,
Argonne, Illinois,
United States of America

Abstract

AUTORADIOGRAPHIC INVENTORY METHODS.

Described are autoradiographic techniques which can verify the number and SNM content of plutonium- and uranium-containing fuel elements. These techniques are applied to fast critical assembly fuel elements and to low-enriched uranium in LWR fuel assemblies. Autoradiographic images are formed by the spontaneously emitted X- and gamma-rays from the fuel elements striking X-ray film in contact with the fuel elements or their containers. Autoradiography allows a large number of items to be examined in a minimum inspection time and with minimum effect on the facility. Results are presented for fast critical assembly fuel in a variety of storage modes as well as in fast critical assemblies themselves. Results are also presented for low-enriched uranium rods in unirradiated LWR fuel assemblies. In all cases, missing fuel elements or substitution of elements containing inert material or depleted uranium were detected.

I. INTRODUCTION

Effective safeguards require frequent verification of special nuclear material (SNM) inventories. Inventory techniques are needed which can verify the piece count and can provide an SNM attribute check for large numbers of SNM-containing items while minimizing inventory time, personnel radiation exposure, and interference with normal operations of the facility under inspection. Such inventory techniques are particularly important at power reactors, fuel-fabrication facilities, and fast critical assemblies. One method which addresses these inventory requirements is autoradiography. In this method, spontaneously emitted radiation from the various isotopes in urnaium- and plutonium-containing fuels exposes film in contact with fuel elements or their containers. With favorable source-film geometry, clear images of individual SNM-containing elements can be obtained *in situ*, without handling fuel elements or opening containers. Counting the images on the film provides a piece count of fuel elements in containers or arrays. Visual inspection of the image provides an attribute check that the fuel element contains SNM. Autoradiography is a qualitative, rather than a quantitative technique, and does not provide a direct measure of the mass of fissil material contained in a fuel element. Autoradiography does provide an indication that a fuel element contains the expected concentration of radioactive material and that the emitted radiation qualitatively duplicates that from SNM-containing fuel elements. Recently, autoradiographic methods have been applied to inventory verifications of plutonium-containing fast critical assembly fuels.[1] The technique has also been applied to uranium in LWR fuel assemblies.[2,3]

FIG.1. Autoradiograph of thirteen plutonium-containing plates and one solid stainless-steel plate in an Argonne storage canister.

II. APPLICATIONS TO PLUTONIUM

A. General

The photon radiation from the various isotopes in plutonium fuels can efficiently expose X-ray film. The photon energy spectrum from most plutonium fuels is dominated by the 60 keV gamma-ray from Am-241. Because the principal fissil isotope, Pu-239, contributes only a minor part of the total photon flux, autoradiographic verifications of fissile content are indirect.

B. Fast Critical Assembly Fuel in Storage

The autoradiographic method has been sucessfully applied to inventories of plutonium-containing fuel plates at fast critical assemblies operated by the Argonne National Laboratory. Most of the fuel plates in the Argonne inventory contained an alloy of plutonium, depleted-uranium, and molybdenum. The plates were typically 5.08-cm wide, 0.64-cm thick, and from 2.54- to 20.32-cm long; they were clad in stainless steel, and were stored in two types of aluminum canisters. Fuel plates at the Argonne-Idaho site were stored on their edges in flat, rectangular canisters. Plates were separated by aluminum posts. Autoradiographs of the contents of a sealed canister were obtained by simply placing a cassette containing X-ray film under the canister. The film was exposed, removed and finally processed. Since the film can be quickly placed and removed, many autoradiographs can be obtained simultaneously. For Kodak Industrex-AA film and plutonium containing 11.5% Pu-240, exposures of 45 minutes were adequate. Shorter exposures were used for material containing larger fractions of Pu-240. Fuel plates in the canister were easily counted by the counting of their images on film. Missing plates or the substitution of plates made of steel, lead, or uranium were easily detected. Figure 1 shows an autoradiograph of a canister containing 14 plates, each 10.1-cm long.

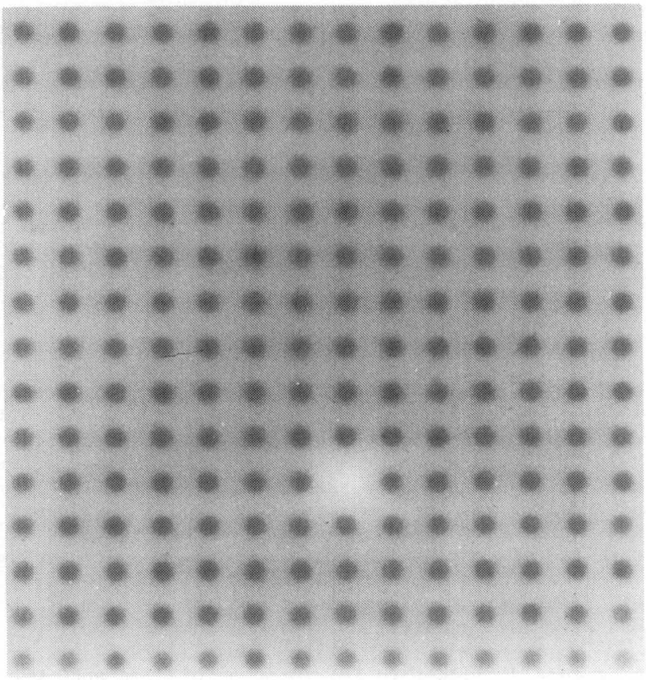

FIG.2. Autoradiograph of plutonium-containing fuel rods in a storage drawer at the Reactor Research Institute, Switzerland.

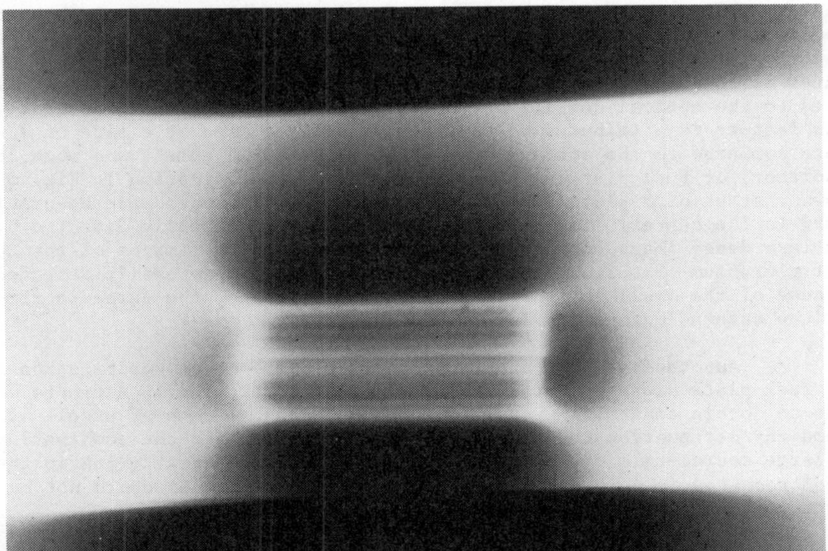

FIG.3. Autoradiograph of a stack of eight plutonium-containing plates with a depleted uranium plate in the centre.

Thirteen plates contained plutonium, while the plate at the center right contained only stainless steel. An image of the stainless steel plate was formed by scattered radiation from adjacent plates, but it is readily distinguished from the plutonium plates. Autoradiographic images were also used to verify plate length. A missing element in an array of plates was easily detected.

The contents of sealed fuel-plate storage canisters at the Argonne-Illinois site were also verified by autoradiography. The Illinois canisters had cylindrical geometry, and autoradiographs were obtained by wrapping X-ray film cassettes around the outer circumference of the canisters. Plates in the canisters could be counted, and missing plates, or substituted inert material or uranium could be readily detected.

The Proteus critical assembly at the Reactor Research Institute in Switzerland uses mixed-plutonium- and depleted-uranium-oxide fuel rods in its fast core. These rods are 0.67 cm in diameter, 5.6-cm long, and are clad in 0.01-cm thick aluminum. Aluminum block drawers, drilled with a square array of holes to accommodate the rods, are used for storage. A sheet of X-ray film placed under a storage drawer showed autoradiographic images of the individual fuel rods, as seen in Fig. 2. The film used was Kodak AA, and exposure was for 7 minutes for the 17% Pu-240 material. The pale image in the fifth row from the bottom of Fig. 2 is from a rod containing depleted uranium, which has a much lower photon emission rate than does a rod containing plutonium. A void rod position was characterized by a very faint circular image due to radiation from neighboring rods. Void positions were easily detected by visual inspection.

It is more difficult to interpret autoradiographs when the fuel elements are stored very close together, as is done at some facilities. An autoradiograph of a stack of plates is shown in Fig. 3. The plates contained a 99% Pu, 1% Al alloy and were clad in stainless steel. They were 5.08-cm wide, 7.62-cm long, and 0.32-cm thick. An 0.1-cm thick sheet of stainless steel between the plates and the film simulated a storage container wall. One feature of Fig. 3 common to all autoradiographs of thin plates is the greater optical density of the film around the plates compared to the optical density of the film directly under the plate edge. This feature is attributed to the greater surface area of a side of a plate compared to the surface area of an edge. This effect was much less important for fuel plates 0.64-cm thick. The autoradiograph in Fig. 3 shows a stack of 8 plutonium-containing plates, with one depleted-uranium plate in the center. The depleted-uranium plate was easily detected by its less dense image and by the reduced density of the images of the adjacent plutonium plates. Voids in plate stacks were also easily detected. Because of the small surface area of the plate edges, the exposure time for the autoradiograph in Fig. 3 was 4 hours.

Autoradiography cannot be used for piece-count verification if the fuel-plate storage containers are made of thick steel. Attempts were made to obtain images of a stack of plates through 1.2 cm of steel. The gamma-ray attenuation resulted in long exposures, while the combination of large source-film distance and the large amount of scattering in the steel resulted in very diffuse images. Individual plates could not be distinguished.

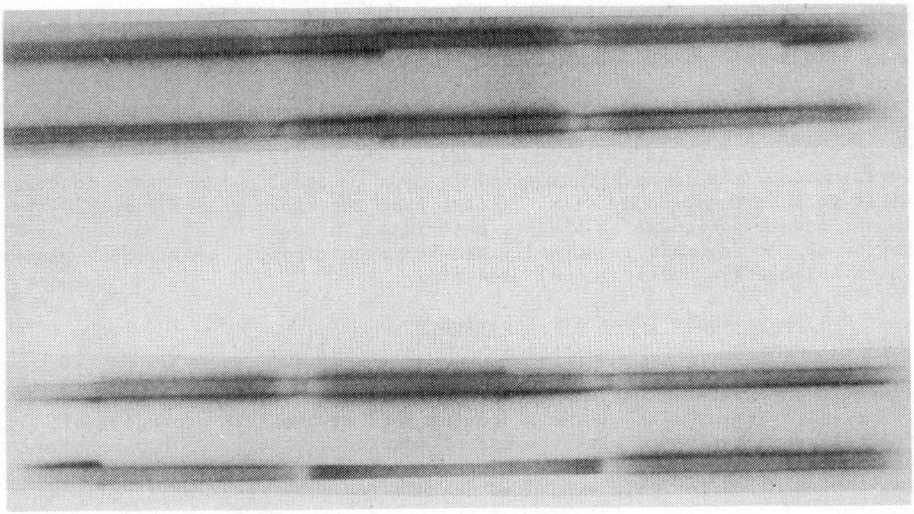

FIG.4. In-situ autoradiographs of plutonium-containing plates in a reactor drawer.
Above: Six plutonium-containing plates. Below: Five plutonium-containing plates and one void.

C. *In Situ* Verification of Fuel in Assembly Cores

A large fraction of the SNM inventory at a fast critical assembly is normally contained in the assembly core. Verification of this in-core material is an important part of fast critical assembly safeguards. Autoradiography has been successfully applied to in-core verification at the split-table type assemblies operated by Argonne National Laboratory.

The Argonne assemblies consist of pairs of arrays of fuel-containing drawers held in place by a steel matrix. Fuel plates are loaded into a drawer on their edges, normally forming one or two columns. The remaining space in the drawers is filled with such diluent materials as depleted uranium, sodium, graphite, stainless steel, iron oxide, or aluminum. Autoradiographs of the contents of an assembly drawer were obtained by slipping an X-ray film packet between the top of the contents of a drawer and the steel matrix. Exposures of 15 minutes were adequate with Kodak-AA film for 0.64-cm plates and plutonium containing 11.5% Pu-240.

Examples of autoradiographs of fuel plates in reactor drawers are shown in Fig. 4. The autoradiograph on the top shows six plutonium-containing plates, each 7.62-cm long and 0.32-cm wide, arranged in two rows. Kodak-AA film and approximately one-hour exposures were used. The plate images are characterized by well-defined outlines of the plates resulting from scattering by the diluent material adjacent to the plutonium-containing plates. When a plate made of uranium or inert material was substituted for a plutonium plate, the dummy plate formed virtually no image and was immediately obvious. Less obvious, but still detectable, was a missing plate. An image of a void is shown in the center of the lower row of the autoradiograph on the bottom of Fig. 4. The void does not have the sharp outline characteristic of plutonium plates. An image is formed because, it it believed, the void acts like a collimator for radiation from the neighboring plates.

In addition to split-table type fast critical assemblies, some vertical assemblies are also in use. In contrast to the split-table assemblies in which fuel in the core is readily accessible for autoradiographic verification, the in-core fuel elements of a vertical assembly are inaccessible to interrogation by film. Access from the sides of the assembly is not possible because of blanket materials. Access through the top or bottom of the assembly is normally blocked by horizontal positioning plates which support the vertical fuel elements.

D. Large-Scale Inventory Experience

Autoradiographic methods have been applied to several large-scale inventory verifications at the Argonne fast critical assembly facilities. Two inventories of the plutonium fuel at the ZPPR storage vault in Idaho were conducted with the use of autoradiography. On the more recent occasion, the contents of 800 storage canisters were verified. One day was required for film exposure, one day for film processing, and one day for film analysis. The maximum radiation exposure to any person participating in the verification was 50 mR. Two autoradiographic inventory verifications have also been conducted at the Argonne-Illinois site at the storage vault for the ZPR-6 and ZPR-9 assemblies. On one occasion, 172 fuel-plate canisters were verified with a total of four man-days of effort. Radiation exposure was 160 mR for the one individual who obtained the autoradiographs. A full-scale, *in situ* inventory verification was also performed for the core of the ZPR-9 assembly. Approximately 400 fuel drawers were examined, requiring two people working in the reactor cell for one shift. Radiation dose to each of those people was about 80 mR. Another man-day was required to process the film. In all cases, these autoradiographic inventories resulted in significant reductions in facility interruption, in total effort, and in radiation exposure when compared to previous inventory methods.

E. Discussion

Autoradiography provides a simultaneous piece count and attribute check for plutonium-containing fuel elements. The technique is not quantitative and does not provide a direct measure of the mass of fissile plutonium. It does provide assurance that an item of the approriate size, shape, and location contains radioactive material with the specific activity of some reference material.

Autoradiography can also provide a verification that the energy distribution of the radiation forming the image is qualitatively like that of the reference material. This energy-distribution verification can be improved by the use of thin absorber foils placed between the film and the radiation source. The ratio of the image density with and without the foil depends on the photon energy profile of the sample radiation. As an example, by using 0.05-cm thick copper foil, 0.05-cm thick lead foil, and 0.025-cm thick cadmium foil, it was very easy to distinguish between images formed by plutonium-containing fuel elements and images formed by the higher-energy gamma radiation from Co-60 or Ir-192 sources. In addition, the image-density ratio method was able to distinguish between fuel elements containing 11.5% Pu-240 and elements containing 27% Pu-240.

The specific activity of the radiation source can be assured by using an exposure calibration source when making inventory-verification exposures. Such a calibration requires that a small source of known activity, which produces a known image density for a specific exposure time, be attached to a piece of film used in an inventory. After film processing, this reference image can give assurance that film sensitivity and film-processing conditions were correct, and that the fuel elements under investigation had the expected activity.

Because of the qualitative nature of autoradiographic methods, autoradiography should be supported by quantitative, nondestructive assay (NDA) measurements in any inventory system. One appropriate companion technique to autoradiography is high-resolution gamma-ray spectroscopy to assure the presence of fissile plutonium in the expected concentrations, at least in the surface region of the fuel element.[4] Because of self-absorption, all gamma-ray techniques, including autoradiography, suffer from the problem of sensitivity only to plutonium in the surface region. Active-neutron techniques can provide added assurance that fissile material is distributed throughout the fuel element.[4] A potentially powerful companion technique to autoradiography for *in situ* assembly-core verification is a measurement of the excess reactivity for that core.[4]

The appropriate role of autoradiography, and its coordination with other NDA techniques, must be determined by the credible diversion paths and by the specific safeguards objectives applicable to any safeguards system. Inventory verifications performed by international organizations must consider diversion strategies where wide-scale institutional efforts can be made to conceal SNM diversions. Such a diversion strategy could include the fabrication of fuel elements which contain little or no fissile material but which closely duplicate the response of genuine fuel elements when examined by autoradiography or NDA techniques. The effort to detect such substitution would obviously require the application of several techniques. In other cases where diversion might be by subnational groups, the fabrication of sophisticated replacement fuel elements would be less likely. In this instance, autoradiography would require less support by other measurement methods. In any case, the value of the autoradiographic method is in its ability to provide qualitative assurance that a large number of items are present in the quantity and location expected and have the specific radiation activity appropriate for those items. This assurance can be provided with a minimum of inspector effort and of facility impact.

III. APPLICATIONS TO URANIUM

A. General

Much of the SNM in world commerce is in the form of low-enriched uranium in fuel assemblies for light water reactors (LWR). Effective safeguards of the uranium in unirradiated assemblies requires a verification of the U-235 content of all the rods in an assembly. At present, several NDA techniques can be used to verify the content of isolated, individual LWR fuel rods or the rods on the periphery of fuel assemblies. The only methods which can examine fuel rods in the interior of an assembly are autoradiography and an active-neutron technique. [5] There are some important differences between autoradiography as applied to uranium and as applied to plutonium. One difference is the longer exposure time caused by the lower gamma-ray emission rates for uranium isotopes. If the methods used to obtain images of plutonium-containing fuel elements were applied to uranium-containing fuel elements of equal fissile content, exposures of several days would be required to obtain equal image density. This exposure-time problem can be partially overcome by using intensifying screens with high-speed films. A survey of films and intensifying screens indicated that Kodak Lanex screens (coated with gadolinium and lanthanum oxides) and Kodak-XR film minimized the exposure time for uranium-containing fuels.[3] The slightly slower Kodak Ortho-G film was normally used because of its compatibility with automatic industrial processing equipment. Exposure times in fuel assemblies were typically 3-4 hours with Ortho-G film and 2 hours with XR film. Polaroid film was also evaluated.[3] Although easier to process, image quality was not as good as conventional X-ray films.

A second problem encountered in the autoradiography of U-235 in LWR fuel is the contribution to image density from U-238. The most intense gamma ray from low-enriched LWR fuel is the 185 keV gamma ray from U-235. Decay daughters dominate the gamma-ray spectrum from U-238. The largest contribution to the total photon flux from both U-235 and U-238 are X-rays resulting from alpha decay. These X-rays are more intense from U-235 because of its shorter half-life. However, measurements made with a multichannel analyzer and an NaI(Tl) detector showed that, in the 66-283 keV photon-energy range, the total photon flux from a rod containing depleted uranium was about 65% of the photon flux from a rod containing 3%-enriched uranium.[3]

In addition to gamma radiation, the Pa-234m daughter of U-238 also emits 2.29 MeV beta radiation and associated bremsstrahlung photons which penetrate common fuel-element cladding and which efficiently expose film. These contributions from Pa-234m radiations were significantly reduced by putting an additional 0.046-cm thick stainless steel absorber between the film packet and the fuel rod. The resulting U-235 sensitivity with Ortho-G film, Lanex screens, and a total of 0.076 cm stainless steel absorber is shown in Fig. 5, and was the greatest sensitivity to U-235 enrichment achieved.

B. Low-Enriched Uranium: LWR Assemblies

A model boiling water reactor (BWR) fuel assembly containing 3%-enriched uranium fuel rods was constructed to evaluate autoradiographic methods for use in actual LWR assemblies. Autoradiographs were obtained by simply inserting a film cassette between two rows of rods. Autoradiographs obtained in this model assembly indicated an additional problem: background contributions from other rods in the assembly. The density of an autoradiographic image of any single rod in an assembly has a major contribution from all other rods in that assembly. Lead shielding can be used to reduce this effect. The shielding configuration used in this study incorporated approximately 0.076-cm thick lead foil in the film packet between the film and the row of rods not under investigation. This reduced background contributions from rods on the "back" side of the film. A second lead foil was placed parallel to the film, one row removed on the "front" side, to further reduce the background.

The cassettes used to obtain these autoradiographs normally contained Kodak Ortho-G film at the center. On either side of the film was a Kodak Lanex intensifying screen. On the "back" side of the screen was lead foil for background reduction. On the "front" side was stainless steel beta absorber. All these items were enclosed in 3.5- or 7-cm wide

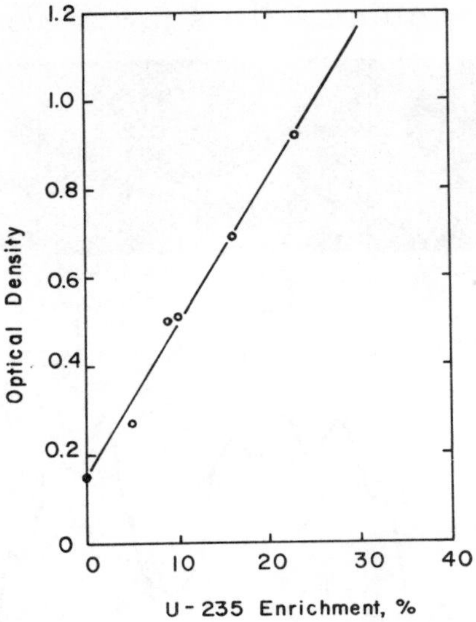

FIG. 5. Single-rod image density as a function of ^{235}U enrichment.

lighttight Kodak ready-pack film cassettes. These cassettes and the lead shielding for the "front" side of the film were then enclosed in polyethylene envelopes to prevent contamination of the fuel rods. Depending on the thickness of beta and gamma absorbers, the total cassette thickness varied between 0.25 and 0.36 cm. This provided a loose fit between rows of rods, and there was no rod displacement.

Even with the lead shielding, the contribution to image density of a particular rod position by other rods in the assembly was large. This contribution was seen in an autoradiograph of a row of 7 rods with the center rod position vacant. The densities of the rod images on either side of the void were 0.77 optical-density units (3.5-hour exposure). The density at the center of the void was 0.60.

Since 65% of the photon flux from a 3%-enriched uranium rod was attributed to U-238, it was anticipated that 65% of the image density would also be contributed from the U-238. When a rod containing depleted uranium occupied the central position of a row in the model assembly (3.5-hour exposure), its image density was 0.71. This is 65% of the difference between the density of the adjacent 3%-enriched rods and the density of the vacant position. Thus, the density difference between a 3%-enriched rod and a depleted rod was only 0.06 optical density units. Even though an enrichment change from 0.2% to 3% U-235 represents a very large relative change in U-235 content, it is associated with a small relative change in image density.

The uncertainty in making an autoradiographic U-235 enrichment measurement depends on the uncertainties in reproducing image densities for equal enrichment and on the reproducibility of the density measurement.

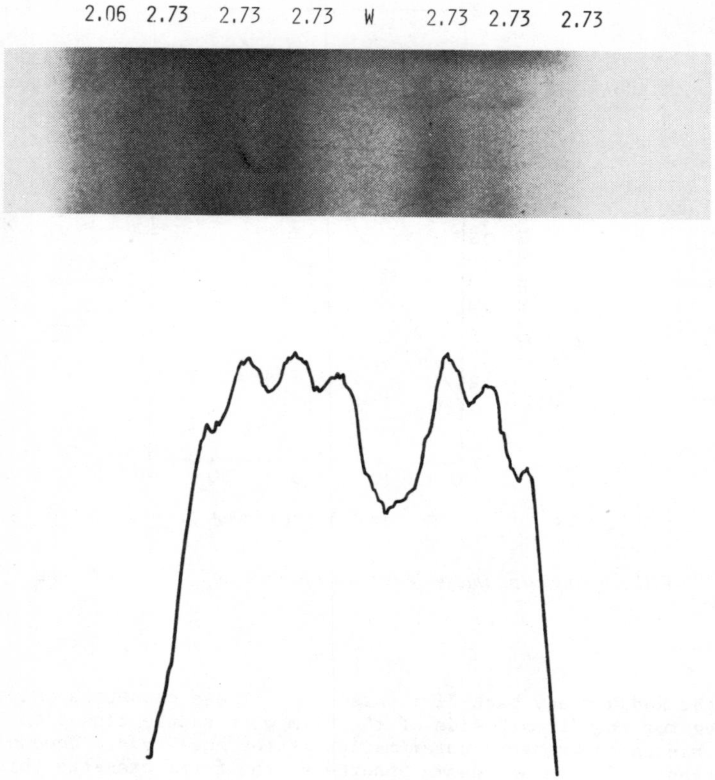

FIG.6. Autoradiography and scanning densitometer trace of a row of rods in a BWR fuel assembly. Rod W contains no uranium.

Typical optical densitometers for X-ray film have repeatability uncertainties of ± 0.01 optical-density units. The standard deviation of image densities for 8 autoradiographs of the same row of rods in the model assembly was 0.012. At a confidence level of 95% (2 standard deviations), this represents 40% of the density difference between a depleted-uranium rod and a 3%-enriched rod. This means that for a 3%-enriched rod, an autoradiographic enrichment measurement would be 3 ± 1.2% U-235. A rod enriched to 2% would have overlapping confidence intervals with a 3%-enriched rod, and this enrichment difference could not be reliably detected.

In some cases, the image density difference for rods of equal enrichment can be greater than 0.012 optical-density units. These differences have been attributed to nonuniform contact between the X-ray film and the intensifying screens. Since reactor operators and fuel fabricators do not allow fuel rods to be put under stress, the film cassettes must fit rather loosely between the rows of an assembly. Some other mechanism must be found to provide good screen-film contact in order to reduce this source of uncertainty.

Several autoradiographs were obtained in the model assembly in which one or more 3%-enriched uranium rods were replaced by rods containing depleted uranium, or in which rod positions were simply left vacant. In all cases, thse simple defects were detected by visual inspection of the autoradiographs. In all cases, an accurate count of enriched fuel rods could be made.

The autoradiographic technique was also field tested at a boiling water reactor (BWR) site and at a pressurized water reactor (PWR) site. The technique was approved by the reactor operators for use with their fuel assemblies. An autoradiograph of a row of rods in a BWR assembly and its scanning-densitometer trace are shown in Fig. 6. The numbers above the rod images refer to the percent U-235 enrichment in each rod. "W" refers to a water rod which contains no uranium. Figure 6 is very typical of autoradiographs of rows of low-enriched uranium rods. The images of the end rods have less density and are less well-defined than the images of interior rods. The total image density increases towards the center of the assembly. This was attributed to higher background at the center. The 2.06%-enriched rod at the left appears to have a greater image density than the 2.73%-enriched rod on the right. The rods adjacent to the end rods, both 2.73% enriched, do not have equal image densities.

Autoradiographs were also obtained for PWR assemblies. A PWR assembly presents a problem because of the presence of control-rod thimbles distributed throughout the fuel-rod matrix. These control-rod thimbles have larger diameters than the fuel rods. Consequently, it is difficult to fit a film cassette between rows of rods and still maintain uniform film-screen contact. In the PWR autoradiographs which were obtained, all fuel rods were visible and could be distinguished from the control-rod thimbles. However, rods of equal enrichment did not always produce images of equal density. This was attributed to the nonuniform film-screen contact.

C. Highly Enriched Uranium

Tests have recently been initiated to evaluate the applicability of autoradiography to highly enriched uranium. Preliminary results indicate that autoradiography will be very useful for verifying highly enriched uranium in the core of split-table type fast critical assemblies. Images of highly enriched uranium plates were obtained on film which was in reactor drawers during and shortly after a period of reactor operation. Some of the image density was caused by radiation from fission products. The contribution to image density from fission radiation during reactor operation has not yet been determined. This in-core uranium technique has the advantage over the in-core plutonium technique of providing a more direct indication of the presence of fissile material in the fuel elements.

D. Discussion

Autoradiography provided a means of verifying that rods in the interior of an unirradiated fuel assembly did contain U-235. This indication of U-235 content was only relative and was not an absolute-enrichment measurement. Autoradiographs need to be supported by quantitative, absolute-enrichment measurements on the exterior rods of the assemblies. Once enrichment of exterior rods is established, autoradiographs can verify enrichment consistency throughout the assembly.

The uncertainty in even relative enrichment determinations of 3% U-235 fuel rods was typically ± 1% U-235. These uncertainties precluded the reliable detection of the small enrichment variations normally found in rows of BWR assemblies. These uncertainties were due to poor film-screen contact, to the large radiation background present in the interior of fuel assemblies, and to the large contribution to image density from U-238.

Despite its lack of sensitivity to small enrichment changes, the autoradiographic technique was consistently able to detect dummy rods, missing rods, and rods containing depleted uranium. These defects were detected by visual inspection of processed film. The technique has the advantage of being simple to apply and of requiring simple, portable equipment. The technique was acceptable to the utility operating the reactors at which field tests were performed. Since interior rods in assemblies are not subject to verification under present international safeguards procedures, the autoradiographic technique may prove useful in providing more effective safeguards of the U-235 contained in unirradiated fuel assemblies deployed internationally.

ACKNOWLEDGEMENT

We thank the staffs of the Reactor Research Institute, Würenlingen, Switzerland, the Nuclear Research Center, Karlsruhe, Germany and the Commonwealth Edison Company for their cooperation and assistance in obtaining some of the data presented here.

REFERENCES

1. S. B. Brumbach and R. B. Perry, *Autoradiographic Technique for Rapid Inventory of Plutonium-Containing Fast Critical Assembly Fuel*, ANL-77-67 (1977).

2. H. Aiginger, C. M. Fleck, W. Pochmann, E. Unfried, and P. Wobrauschek, *Investigation of the Distribution of Fissile Material in Subassemblies by Means of X-ray Film Technology*, IAEA-76-CL 7325 (1976).

3. S. B. Brumbach and R. B. Perry, *Autoradiography as a Safeguards Inspection Technique for Unirradiated LWR Fuel Assemblies*, ANL-78-27 (1978).

4. D. D. Cobb and J. L. Sapir, *Preliminary Concepts for Materials Measurement and Accounting in Critical Facilities*, LA-7028-MS (1978).

5. J. D. Brandenberger, E. Medina, and H. O. Menlove, *Portable Neutron Assay System for LWR Fuel Assemblies*, LA-6849-PR (1977).

DISCUSSION

K.J. QUEALY: Autoradiography seems to offer the potential for simple solutions to difficult safeguards problems. Have you any plans to extend and refine your work in this field?

R.B. PERRY: Work on the application of autoradiography for inventory verification of fast critical assembly fuel is continuing and will be extended to high-enriched uranium fuel. The IAEA has requested additional work to improve sensitivity for LWR fuel but we are not optimistic that significant improvement can be obtained because of background radiation from ^{238}U and background radiation from other fuel rods in the assembly.

A CORRECTION FOR VARIABLE MODERATION AND MULTIPLICATION EFFECTS ASSOCIATED WITH THERMAL NEUTRON COINCIDENCE COUNTING*

N. BARON
Los Alamos Laboratory,
Los Alamos, New Mexico,
United States of America

Abstract

A CORRECTION FOR VARIABLE MODERATION AND MULTIPLICATION EFFECTS ASSOCIATED WITH THERMAL NEUTRON COINCIDENCE COUNTING.

A correction is described for multiplication and moderation when doing passive thermal-neutron coincidence counting non-destructive assay measurements on powder samples of PuO_2 mixed arbitrarily with MgO, SiO_2, and moderating material. The multiplication correction expression is shown to be approximately separable into the product of two independent terms; F_{Pu}, which depends on the mass of ^{240}Pu, and $F_{\alpha n}$, which depends on properties of the matrix material. Necessary assumptions for separability are (1) isotopic abundances are constant, and (2) fission cross-sections are independent of incident neutron energy: both are reasonable for the 8% ^{240}Pu powder samples considered here. Furthermore, since all prompt fission neutrons are expected to have nearly the same energy distributions, variations among different samples can be due only to the moderating properties of the samples. Relative energy distributions are provided by a thermal-neutron well counter having two concentric rings of ^{3}He proportional counters placed symmetrically about the well. Measured outer-to-inner ring ratios raised to an empirically determined power for coincidences, $(\dot{N}^I/\dot{N}^O)^Z$, and singles, $(\dot{T}^O/\dot{T}^I)^\delta$, provide corrections for moderation and $F_{\alpha n}$, respectively, and F_{Pu} is approximated by M_{240}^X/M_{240}. The exponents are calibration constants determined by a least-squares fitting procedure using standards data. System calibration is greatly simplified using the separability principle. Once appropriate models are established for F_{Pu} and $F_{\alpha n}$, only a few standards are necessary to determine the calibration constants associated with these terms. Since F_{Pu} is expressed as a function of M_{240}, correction for multiplication in a subsequent assay demands only a measurement of $F_{\alpha n}$. The effective mass of ^{240}Pu corrected for multiplication and moderation for these kinds of powder samples using this detection system is

$$M_{240} = \exp\left\{\frac{\ln\left[\dfrac{\dot{N}^I + \dot{N}^O + 2\cdot(\dot{N}^I\cdot\dot{N}^O)^{1/2}}{(\dot{T}^O/\dot{T}^I)^{3.295}\cdot(\dot{N}^I/\dot{N}^O)^{2.255}\cdot 31.45}\right]}{1.065}\right\}$$

---

* Work performed under the auspices of the United States Department of Energy.

INTRODUCTION

At the Los Alamos Scientific Laboratory, passive thermal-neutron coincidence counting is one of several techniques used to nondestructively assay spontaneously fissioning material. Each spontaneous fission (SF) of a fertile nucleus emits an average of several prompt neutrons. Therefore the coincident-neutron count rate measured in the presence of fertile material is proportional to the material's mass. The samples considered here are assumed to consist of only isotopes of ^{239}Pu and ^{240}Pu. If other fertile isotopes are present in addition to ^{240}Pu, then the quantity measured is the ^{240}Pu effective mass M_{240} which is defined to be the ^{240}Pu mass which will produce the same number of neutrons per second from spontaneous fissions as actually produced by all fertile isotopes in the mixed sample.

The proportionality constant is referred to here as the specific response rate SRR and defined as

$$SRR \equiv \frac{\dot{N}_{SF}}{eff \cdot M_{240}} \tag{1}$$

where $\dot{N}_{SF}$ is the count rate of coincident-neutron pairs resulting only from spontaneous fissions, and eff is the relative efficiency with which the detector counts such pairs. However, the observed coincidence count rate $\dot{N}$ is the sum of the count rates of coincident-neutron pairs due to both SF- and induced fission-events. In fact fissions always will be induced by SF-neutrons and their progeny. We refer to the contribution to $\dot{N}$ by these induced fissions as the self-multiplication coincidence count rate $\dot{N}_{SM}$. Furthermore, nondestructive assay (NDA) measurements often are made on samples of PuO_2 powder that are contaminated with oxides or salts of light weight elements having large (α,n) reaction cross sections. Since both ^{239}Pu and ^{240}Pu are sources of radioactive decay α-particles, the presence of such contaminants results in a significant number of (α,n)-neutrons which can induce fissions. The contribution of these induced fissions to $\dot{N}$ is referred to as the induced multiplication coincident count rate $\dot{N}_{IM}$.

The measured neutron coincidence count rate can be written now as

$$\dot{N} = \dot{N}_{SF} + \dot{N}_{SM} + \dot{N}_{IM} \tag{2}$$

Using the above relation, the quantity $\dot{N}_{SF}$ that must be extracted from the measured rate in order to calculate the constant of the system SRR is given by

$$\dot{N}_{SF} = \dot{N} \cdot \left[\frac{\dot{N}_{SF}}{\dot{N}_{SF} + \dot{N}_{SM} + \dot{N}_{IM}} \right] \tag{3}$$

Defining the bracketed term in Eq. 3 to be the inverse of the correction factor IF for induced fissions where

$$IF \equiv 1 + \frac{\dot{N}_{SM}}{\dot{N}_{SF}} + \frac{\dot{N}_{IM}}{\dot{N}_{SF}} \tag{4}$$

allows Eq. 3 to be rewritten as

$$\dot{N}_{SF} = \frac{\dot{N}}{IF} \tag{5}$$

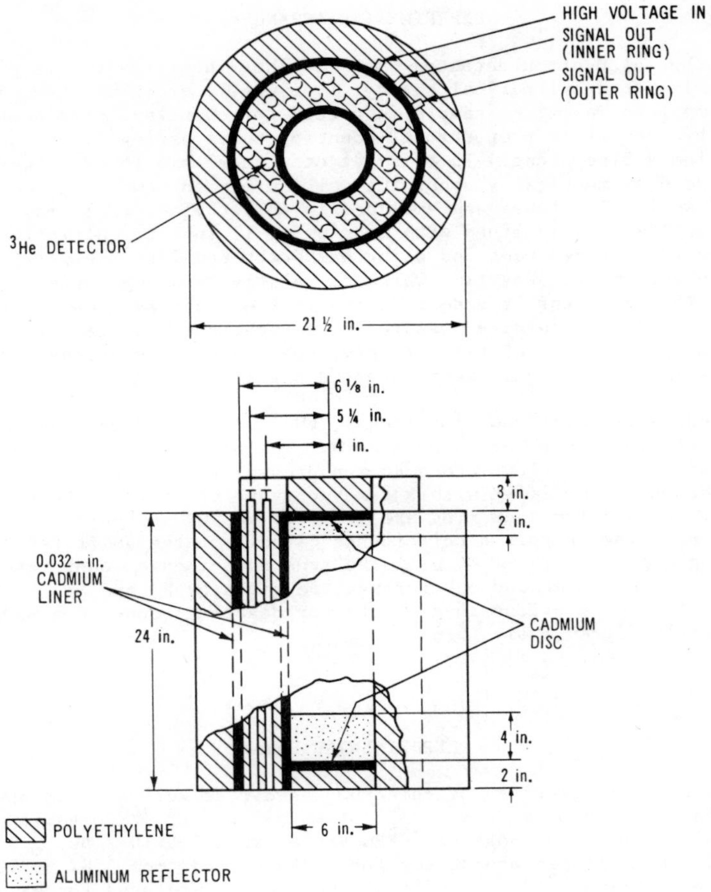

FIG.1. A two-ring thermal-neutron well counter.

Substituting Eq. 5 into Eq. 1 gives

$$SRR = \frac{\dot{N}}{IF \cdot eff \cdot M_{240}} \tag{6}$$

The object here is to investigate 1) the multiplication correction factor IF which accounts for fissions induced by SF- and (α,n)-neutrons and their progeny in powder samples of PuO_2 mixed arbitrarily with matrix materials of SiO_2 and MgO, and 2) variations of the detection efficiency eff of coincident-neutron pairs, where eff is a function only of the fission neutron's energy distribution. Since all fission neutrons are born with nearly the same energy distribution (1), any variation of eff among samples can only be due to differences in their moderating properties. Thus a study of eff implies a study of the relative moderating property of the sample.

EXPERIMENTAL ARRANGEMENT

The thermal neutron detector used for this investigation is pictured in Fig. 1. It has a cylindrical cavity, i.e., well, of about 15-cm diam. x 51-cm long into which the sample is placed for counting. The neutrons are counted by ^3He filled proportional counting tubes having dimensions of 2.5-cm diam x 51-cm long. In this detector there are 16 such counting tubes placed symmetrically about the well in each of two rings concentric with the well. The tubes are embedded in polyethylene (CH_2) moderating material. The well is lined with cadmium to 1) inhibit multiplication caused by thermalized neutrons being backscattered into the well, and 2) provide a system die-away time which is relatively independent of the sample. This detector is undermoderated. The die-away times of the inner and outer rings are readily measured by a technique reported previously (2) to be about 30- and 45-μS respectively. Design considerations for thermal neutron well counters have been reported previously (3).

To reduce deadtime counting losses, the counting tubes in each ring are divided into two groups of 8 tubes each. The outputs of the 8 tubes in each group are paralleled into a common preamplifier whose output signals are subsequently amplified with a time constant of .5 μS. The two amplifier outputs for each ring are fed into an "OR" circuit and the logic output signal then enters a coincidence circuit of the shift register type (4) having a gate width of 64 μS. The true coincidence count rates measured for the inner and outer rings are given by $\dot{N}^I$ and $\dot{N}^O$ respectively. Corrections to the data for deadtime losses are made by the method reported previously (3,5).

EXPERIMENTAL DATA

Standards were prepared of pure PuO_2 powder as well as PuO_2 powder mixed with either MgO or SiO_2. The atoms per cm^3 of ^{240}Pu varied from $1x10^{19}$ to $3.4x10^{20}$ and those of ^{239}Pu varied from $1.5x10^{20}$ to $3.0x10^{21}$. The isotopic abundances for ^{240}Pu varied from 5.8% to 10.6%, the remainder being ^{239}Pu. Identical containers were used for each standard and, except for samples of pure PuO_2, the fill heights were approximately constant. The singles and coincidence neutron count rates for each sample were measured using the detector pictured in Fig. 1. The results of these measurements are listed in Table I. Another data set is listed in Table II for the same standards surrounded by 559 g of CH_2.

In Fig. 2 is shown a plot of $\dot{N}/M_{240}$ vs. M_{240} for the data of the detector's inner ring listed in Table I. The results for the outer ring are qualitatively similar. In all cases the values of $\dot{N}/M_{240}$ are observed to vary monotonically with the values of M_{240} mixed with a particular matrix material. This clearly demonstrates the relation of self-multiplication and M_{240}. Furthermore the responses in Fig. 2 are grouped according to the matrix material in each sample. This suggests the possibility that the dependence of multiplication on the mass of plutonium, to some approximation, is independent of the manner by which the matrix material affects the multiplication. If this were so, the multiplication correction factor IF would be separable into the product of two independent functions; one dependent on the mass of plutonium and the other on the properties of the matrix material. This idea is pursued further in the next section.

ANALYSIS

Multiplication Correction IF

To investigate the dependence of IF on the plutonium and matrix material expressions must be written for $\dot{N}_{SF}$, $\dot{N}_{SM}$, and $\dot{N}_{IM}$. In the following schematic treatment, integration over energy and space is implied. The probability that any two neutrons born by fission process q leak out of the sample and into the detector and are detected subsequently is defined as

$$P_{\dot{N}_q} \equiv [\nu_q \cdot S(E_1) \cdot L_1 \cdot D_1] \cdot [(\nu_q - 1) \cdot S(E_2) \cdot L_2 \cdot D_2] \qquad (7)$$

where

ν_q is the number of prompt neutrons emitted by fission process q

$S(E_1)$ and $S(E_2)$ are the probabilities that at birth one of the two prompt neutrons will have an energy between E_1 and $E_1 + dE_1$ and the other an energy between E_2 and $E_2 + dE_2$

L_1 and L_2 are the probabilities that the two fission neutrons leak into the detector

D_1 and D_2 are the probabilities that the two fission neutrons that entered the detector will be counted.

Since all prompt fission neutrons have very nearly the same energy distribution (1), the probability distributions S, L, and D are the same for all fissions processes. Thus we define for each fission neutron

$$R \equiv S \cdot L \cdot D \qquad (8)$$

Substituting Eq. 8 into Eq. 7 gives

$$P_{\dot{N}_q} = \nu_q \cdot (\nu_q - 1) \cdot R_1 R_2 \qquad (9)$$

With the aid of Eq. 9, the count rate of coincident-neutron pairs generated by fission process q can be written

$$\dot{N}_q = W_q \cdot P_{\dot{N}_q} \qquad (10)$$

where W_q is the number of q-type fissions/S. Therefore

$$\dot{N}_{SF} = \eta_{SF} \cdot \nu_{SF} \cdot (\nu_{SF} - 1) \cdot R_1 \cdot R_2 \qquad (11\text{-a})$$

$$\dot{N}_{SM} = \eta_{SF} \cdot \nu_{SF} \cdot S(E) \cdot P_{SM} \cdot \nu_{SM} \cdot (\nu_{SM} - 1) \cdot R_1 \cdot R_2 \qquad (11\text{-b})$$

$$\dot{N}_{IM} = \eta_{\alpha} \cdot S_{\alpha}(E_{\alpha}) \cdot P_{\alpha n} \cdot S_{\alpha n}(E_n) \cdot P_{IM} \cdot \nu_{IM} \cdot (\nu_{IM} - 1) \cdot R_1 \cdot R_2 \qquad (11\text{-c})$$

where

η_{SF} is the number of SF-events/S

P_{SM} is the probability a SF-neutron will induce a fission and is a function of the areal densities of ^{239}Pu and ^{240}Pu

η_{α} is the number of α-particles produced by the radioactive decay of plutonium nuclei per second

$S_{\alpha}(E_{\alpha})$ is the probability an alpha particle will have an energy between E_{α} and $E_{\alpha} + dE_{\alpha}$

TABLE I. COINCIDENCE COUNTING MEASUREMENT RESULTS *(No CH_2 placed about samples)*

Sample ID	Dominant Matrix Material Mixed With PuO_2	Mass of Pu in PuO_2 (grams)	Effective Mass of ^{240}Pu (grams)	True Coincidence Count Rates Corrected for Bkgd & Deadtime		"Singles" Count Ratios, $\dot{T}^O/\dot{T}^I$	Specific Response Rates, Uncorrected for Matrix Effect		Specific Response Rates Uncorrected for Relative Coincidence Detection Efficiency			Specific Response Rate Corrected for Efficiency, i.e., Moderation,[d] SRR^B
				Outer Ring, $\dot{N}^O$ (counts/S)	Inner Ring, $\dot{N}^I$ (counts/S)		Outer[a] Ring, $SRR^O(Pu)$	Inner[a] Ring, $SRR^I(Pu)$	Outer[b] Ring, $SRR^O \cdot eff$	Inner[b] Ring, $SRR^I \cdot eff$	Both[c] Rings, $SRR^B \cdot eff$	
705-004	SiO_2	20.98	1.22	10.63	16.42	.8106	8.60	13.29	17.18	26.54	86.43	32.42
3	SiO_2	60.29	4.693	43.28	67.92	.8098	8.34	13.09	16.71	26.23	84.81	30.70
4	SiO_2	60.02	3.837	35.38	55.01	.8088	8.45	13.14	17.00	26.43	85.82	31.72
6	SiO_2	120.0	7.672	74.98	115.14	.8139	8.56	13.15	16.87	25.91	84.59	32.16
5	MgO	60.03	3.838	39.25	61.19	.8428	9.37	14.61	16.46	25.67	83.24	30.58
7	MgO	120.0	7.672	82.49	128.58	.8412	9.42	14.68	16.65	25.95	84.17	30.94
8	MgO	240.1	15.35	180.83	280.86	.8371	9.86	15.32	17.72	27.53	89.42	33.13
10	(none)	60.03	4.599	43.83	68.04	.8201	8.63	13.40	16.59	25.75	83.68	31.04
385	(none)	459.0	43.39	485.23	760.95	.8244	8.75	13.73	16.54	25.93	83.89	30.41
382	(none)	556.0	54.93	634.18	987.13	.8261	8.90	13.85	16.70	25.99	84.36	31.11
381	(none)	615.0	64.97	758.64	1186.93	.8263	8.90	13.93	16.69	26.12	84.57	30.82
447	(none)	779.0	81.36	1000.91	1575.18	.8302	9.24	14.55	17.06	26.86	86.73	31.20

[a] Calculated using Eqs. 23, 22-b.

[b] Calculated using Eqs. 21, 22-a, b.

[c] Calculated using Eqs. 29, 21, 22-a, b.

[d] Calculated using Eqs. 31, 32.

TABLE II. COINCIDENCE COUNTING MEASUREMENT RESULTS *(559 g CH_2 placed about samples)*

Sample ID	Dominant Matrix Material Mixed With PuO_2	Mass of Pu in PuO_2 (grams)	Effective Mass of ^{240}Pu (grams)	True Coincidence Count Rates Corrected for Bkgd & Deadtime		"Singles" Count Ratios, $\dot{T}^O/\dot{T}^I$	Specific Response Rates, Uncorrected for Matrix Effect		Specific Response Rates Uncorrected for Relative Coincidence Detection Efficiency			Specific Response Rate Corrected for Efficiency, i.e., Moderation,[d] SRR^B
				Outer Ring, $\dot{N}^O$ (counts/S)	Inner Ring, $\dot{N}^I$ (counts/S)		Outer[a] Ring, $SRR^O(Pu)$	Inner[a] Ring, $SRR^I(Pu)$	Outer[b] Ring, $SRR^O\cdot eff$	Inner[b] Ring, $SRR^I\cdot eff$	Both[c] Rings, $SRR^B\cdot eff$	
705-004	SiO_2	20.98	1.22	10.62	19.85	.7519	8.59	16.06	21.99	41.10	123.22	30.07
3	SiO_2	60.29	4.693	44.23	80.37	.7492	8.52	15.49	22.07	40.10	121.67	31.65
4	SiO_2	60.02	3.837	36.24	64.87	.7521	8.65	15.49	22.13	39.61	120.95	32.54
6	SiO_2	120.0	7.672	76.24	136.51	.7554	8.70	15.59	21.94	39.28	119.93	32.23
5	MgO	60.03	3.838	42.05	76.93	.7856	10.04	18.37	22.24	40.68	123.08	31.54
7	MgO	120.0	7.672	87.87	164.22	.7839	10.03	18.75	22.38	41.82	125.39	30.60
8	MgO	240.1	15.35	187.57	343.86	.7799	10.23	18.76	23.21	42.55	128.61	32.80
10	(none)	60.03	4.599	45.20	80.75	.7684	8.90	15.90	21.20	37.87	115.74	31.26
385	(none)	459.0	43.39	499.65	913.27	.7642	9.01	16.47	21.86	39.95	120.91	31.02
382	(none)	556.0	54.93	654.38	1202.91	.7653	9.18	16.88	22.17	40.76	123.05	31.19
381	(none)	615.0	64.97	781.80	1449.98	.7655	9.17	17.01	22.12	41.03	123.40	30.63
447	(none)	779.0	81.36	1051.32	1915.94	.7702	9.71	17.69	22.95	41.82	126.73	32.76

[a] Calculated using Eqs. 23, 22-b.

[b] Calculated using Eqs. 21, 22-a, b.

[c] Calculated using Eqs. 29, 21, 22-a, b.

[d] Calculated using Eqs. 31, 32.

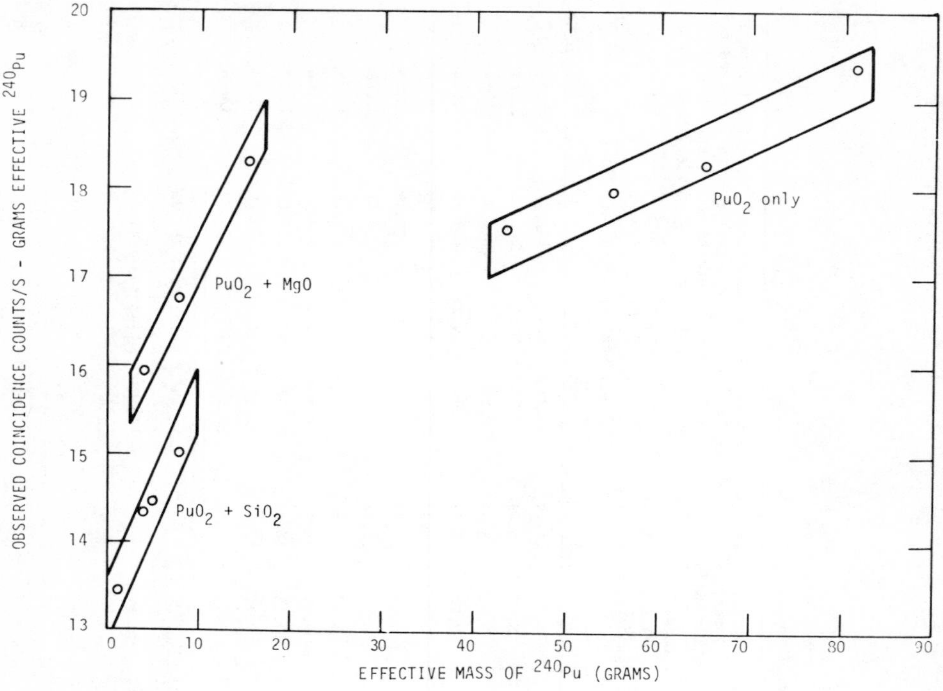

FIG.2. Observed coincidence count-rate/gram effective ^{240}Pu vs grams effective ^{240}Pu.

$P_{\alpha n}$ is the probability a α-particle will induce a (α,n) reaction and is a function of the reaction's cross section and Q value as well as the intimacy of the mixture of plutonium and matrix material.

$S_{\alpha n}(E_n)$ is the probability the (α,n)-neutron will have an energy between E_n and $E_n + dE_n$

P_{IM} is the probability a (α,n)-neutron of energy E_n will induce a fission and is a function of the areal densities of ^{239}Pu and ^{240}Pu

P_{IM} and P_{SM} differ only because of the relative energies of the (α,n)- and fission-neutrons. This relationship is expressed as

$$P_{IM} = P_{SM} \cdot B(E_n) \tag{12}$$

where $B(E_n)$ is a function of the (α,n)-neutron's energy and is a unique property of the matrix material and its intimacy of mixture with the plutonium. In addition η_α is linearly proportional to both the masses of ^{240}Pu and ^{239}Pu. Invoking the constraint that the isotopic abundances of ^{240}Pu and ^{239}Pu are constant for all samples, P_{SM} and η_α can be considered functions of only ^{240}Pu and the ratio η_α/η_{SF} can be written

$$\frac{\eta_\alpha}{\eta_{SF}} = K \tag{13}$$

where K is a constant.

Substituting Eqs. 11-a, b, c, 12, and 13 into Eq. 4 gives

$$IF = 1 + P_{SM} \cdot \left[\frac{S(E)\cdot\nu_{SM}\cdot(\nu_{SM}-1)\cdot R_1\cdot R_2}{(\nu_{SF}-1)\cdot R_1\cdot R_2} + \frac{K\cdot S_\alpha(E_\alpha)\cdot P_{\alpha n}\cdot S_{\alpha n}(E_n)\cdot B(E_n)\cdot\nu_{IM}\cdot(\nu_{IM}-1)\cdot R_1\cdot R_2}{\nu_{SF}\cdot(\nu_{SF}-1)\cdot R_1\cdot R_2} \right] \quad (14)$$

where we have factored the quantity P_{SM} from the bracketed term by assuming its value is not strongly dependent on the incident neutron's energy. Since P_{SM} is the only quantity in Eq. 14 that can vary among samples because of changes in the characteristics of the plutonium, namely its areal density, we define

$$f_{Pu} \equiv P_{SM} \quad (15)$$

The bracketed term in Eq. 14 is designated $f_{\alpha n}$ to indicate its value can vary among samples only by varying the matrix material which controls the quantities $P_{\alpha n}$, $S_{\alpha n}(E_n)$, $B(E_n)$, and $\nu_{IM}(E_n)$ where $\nu_{IM}(E_n)$ is linearly proportional to E_n (6). The expression for IF can now be written

$$IF = 1 + f_{Pu} \cdot f_{\alpha n} \quad (16)$$

which implies that the plutonium and matrix material affect the multiplication independently of one another. We now define the functions

$$F_{Pu} \equiv f(P_{SM})$$

and

$$F_{\alpha n} \equiv f(P_{\alpha n},\ S_{\alpha n},\ B,\ \nu_{IM})$$

so that Eq. 16 is rewritten to some approximation as

$$IF \equiv F_{Pu} \cdot F_{\alpha n} \quad (17)$$

The functions F_{Pu} and $F_{\alpha n}$ will be determined empirically.

The factor F_{Pu} is proportional to the multiplication caused by SF- and (α,n)-neutrons in the presence of ^{240}Pu and ^{239}Pu areal densities or, equivalently, M_{240}. F_{Pu} is referred to as the plutonium multiplication correction and is a function of only M_{240}.

The factor $F_{\alpha n}$ is that part of IF which accounts for variations in the induced fission coincidence count rate due to different matrix materials. Its value is proportional to 1) the probability that a random neutron of energy E_n will be generated by the (α,n) reaction on the matrix material, 2) the function $B(E_n)$ which accounts for the difference between the probabilities of inducing a fission by SF- and (α,n)-neutrons, and 3) the multiplicity $\nu_{IM}(E_n)$ which determines the relative detection efficiency of coincident-neutron pairs generated by (α,n)-neutrons. $F_{\alpha n}$ is referred to as the matrix multiplication correction. Even among samples having the same neutron moderation, the matrix correction is not expected to be constant for a particular matrix material. $P_{\alpha n}$ and E_n, and therefore $F_{\alpha n}$, are dependent upon the intimacy of mixture of PuO_2 and matrix material. Thus perturbations about some average $F_{\alpha n}$ are expected due to variations in the intimacy of the mixture.

Plutonium Correction F_{Pu}

Under the constraint of constant isotopic abundance, an increase in the mass of plutonium in a sample results in a linear increase in the number of plutonium target nuclei available for the (n,f) induced fission reaction. Since the average number of prompt neutrons associated with SF-events is greater than unity, the source neutrons available to induce the (n,f) reaction increase even more rapidly. Consequently it is expected that induced fissions will account for a larger percentage of the observed coincidence counts as the mass of plutonium in the sample increases. These anticipated results are observed empirically. This suggests that F_{Pu} can be defined as

$$F_{Pu} \equiv \frac{M_{240}^{X}}{M_{240}} \tag{18}$$

where the exponent X is an empirically determined constant.

Matrix Correction $F_{\alpha n}$

The principal problem in this scheme to assay M_{240} is the identification and measurement of the function $F_{\alpha n}$. This quantity must vary monotonically with E_n and $P_{\alpha n}$ as does IF. Fortunately a quantity that also varies monotonically with E_n and $P_{\alpha n}$ can be defined and measured using a detector like that pictured in Fig. 1. A two-ring detector can provide information on the energy distribution of neutrons which enter the detector. This information is obtained by measuring the ratio of the neutron singles count rates in the outer and inner rings, $\dot{T}^O/\dot{T}^I$, where superscripts O and I refer to the outer and inner rings respectively. The singles count rate in each ring is composed of the sum of the fission neutrons, i.e., SF- , SM- , and IM-neutrons, as well as the (α,n)-neutrons. It was previously noted that the energy distributions of prompt fission neutrons do not vary among samples having similar moderation properties. However, the larger the average energy E_n of the (α,n)-neutrons the harder will be the singles' spectrum thereby increasing the ratio $\dot{T}^O/\dot{T}^I$. Furthermore an increased value of $P_{\alpha n}$ causes equal percentage increases in both the outer and inner ring partial single count rates due to (α,n) neutrons. Consequently the ratio $\dot{T}^O/\dot{T}^I$ is increased provided the average (α,n)-neutron energy $\overline{E_n}$ is greater than the average fission neutron energy $\overline{E_f}$,

$$\overline{E_n} > \overline{E_f} \tag{19}$$

Therefore both IF and $\dot{T}^O/\dot{T}^I$ have the same qualitative dependence on E_n and $P_{\alpha n}$.

Prompted by this qualitative similarity a function of $(\dot{T}^O/\dot{T}^I)$ is defined where

$$F_{\alpha n} \equiv (\dot{T}^O/\dot{T}^I)^{\delta} \tag{20}$$

for samples containing matrix materials that meet the criterion of Eq. 19 where δ is an empirically determined constant.

Calibration Constants δ and X

Substituting Eqs. 18 and 20 into Eq. 17 and using the result in Eq. 6 allows the specific response rates for the outer and inner rings to be written

$$SRR^{O(I)} = \frac{\dot{N}^{O(I)}}{eff \cdot (\dot{T}^O/\dot{T}^I)^\delta \cdot M_{240}^X} \qquad (21)$$

Since the standards used for the data listed in Table I were fired at the same temperature during their preparation, they are expected to have the same moderation properties and therefore a constant value of eff. Consequently the quantities $[SRR^{O(I)} \cdot eff]$ calculated using Eq. 21 and the data listed in Table I must be constants if the separability concept for IF (Eq. 17) and the assumed forms for F_{Pu} (Eq. 18) and $F_{\alpha n}$ (Eq. 20) are valid for these powder samples. Applying a least squares fitting procedure using the data of Table I in the expressions for $[SRR^{O(I)} \cdot eff]$ obtained from Eq. 21, the calibration constants δ and X are calculated to be the same for the inner and outer rings where

$$\delta = 3.295 \qquad (22\text{-a})$$

$$X = 1.065 \qquad (22\text{-b})$$

Using Eqs. 22-a and b, values of $[SRR^{O(I)} \cdot eff]$ are calculated and listed in Table I for each sample and the mean values are

$$[SRR^O \cdot eff]_{\substack{Av \\ \text{Table I}}} = 16.87 \pm 2.15\% \qquad (22\text{-c})$$

$$[SRR^I \cdot eff]_{\substack{Av \\ \text{Table I}}} = 26.28 \pm 2.16\% \qquad (22\text{-d})$$

Similar calculations are made for the data in Table II using values of δ and X listed in Eqs. 22-a and b. The results are listed in Table II and the mean values are

$$[SRR^O \cdot eff]_{\substack{Av \\ \text{Table II}}} = 22.20 \pm 2.30\% \qquad (22\text{-e})$$

$$[SRR^I \cdot eff]_{\substack{Av \\ \text{Table II}}} = 40.55 \pm 3.16\% \qquad (22\text{-f})$$

The constant values of the quantities $[SRR^{O(I)} \cdot eff]$ given by Eqs. 22-c, d, e, f, lend credence to the separability approximation for IF described by Eq. 17 and the assumed form for F_{Pu} and $F_{\alpha n}$. The calibration constants δ and X could be determined also using the data of Table II. Although the values of δ are different, the value of the assay subsequently will be shown to be unchanged.

Modified specific response rates SRR(Pu) that are uncorrected for the matrix multiplication are defined as

$$SRR^{O(I)}(Pu) \equiv \frac{\dot{N}^{O(I)}}{M_{240}^X} \qquad (23)$$

Using this definition in Eq. 21 gives

$$SRR^{O(I)}(Pu) = [SRR^{O(I)} \cdot eff] \cdot (\dot{T}^O/\dot{T}^I)^{3.295} \qquad (24)$$

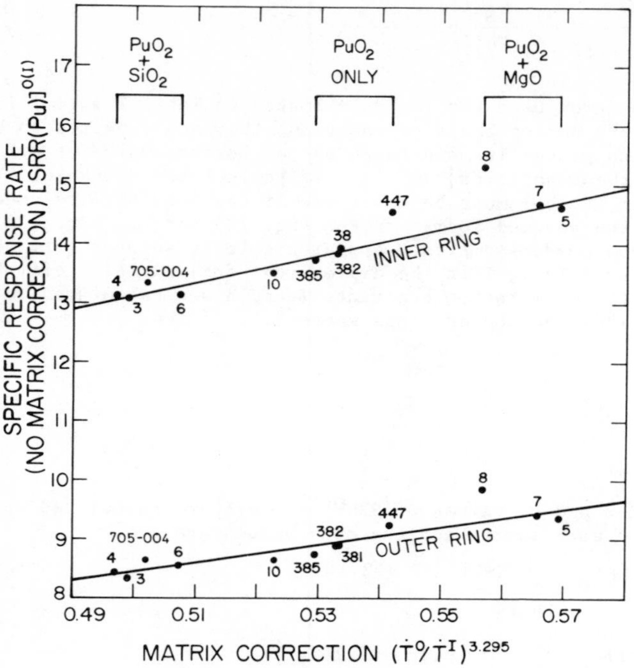

FIG.3. The specific response rate uncorrected for matrix effect vs matrix correction.

Also listed in Table I are values of $SRR^{O(I)}(Pu)$ calculated using Eqs. 23 and 22-b. Pictured in Fig. 3 are plots of these values as a function of the matrix correction $(\dot{T}^O/\dot{T}^I)^{3.295}$. The linear statement of Eq. 24 is seen to be obeyed reasonably well. This implies that the matrix correction defined by Eq. 20 can be determined in the presence of an arbitrary combination of these matrices. As shown in Fig. 3, the clustering of the values $(\dot{T}^O/\dot{T}^I)^{3.295}$ for a particular matrix material indicates that they are indeed determined primarily by the matrix material and are independent of M_{240}.

Response of "Outer + Inner Ring"

Counting statistics can be improved and the measurement uncertainty reduced by calculating a single response obtained by combining the counting data of both rings of a two-ring detector. The count rate expected by paralleling the amplified signals from both rings into the same coincidence circuit can be obtained from the count rates of each individual ring by the following analysis.

The average coincidence detection efficiency of SF-neutrons in the outer (inner) ring is defined as

$$\varepsilon^{O(I)} \equiv \frac{SRR^{O(I)}}{a} \qquad (25)$$

TABLE III. MEAN VALUES OF SPECIFIC RESPONSE RATES

	No polyethylene about sample		559 grams polyethylene about sample	
	$SRR^B \cdot eff$ (a)	SRR^B (b)	$SRR^B \cdot eff$ (a)	SRR^B (b)
$\begin{pmatrix}\text{Specific}\\\text{Response}\\\text{Rate}\end{pmatrix}_{Av}$	85.14	31.35	122.73	31.52
$\sigma\begin{pmatrix}\text{Specific}\\\text{Response}\\\text{Rate}\end{pmatrix}_{Av}$	±2.02%	±2.66%	±2.73%	±2.85%

a) Calculated using Eqs. 29, 21, 22-a, b.

b) Calculated using Eqs. 31, 32.

where a is the number of spontaneous fissions per S per gram of effective ^{240}Pu. Consequently the average probability that one member of a pair of coincident neutrons will be detected by a counter in the outer (inner) ring can be written as

$$p^{O(I)} = (\varepsilon^{O(I)})^{1/2} \tag{26}$$

Analogous to Eq. 25, the average efficiency of counting a pair of coincident neutrons by both rings of a two-ring detector is defined to be

$$\varepsilon^B \equiv \frac{SRR^B}{a} \tag{27}$$

where

$$\varepsilon^B = (p^I)^2 + (p^O)^2 + 2 \cdot p^I \cdot p^O \tag{28}$$

With the aid of Eqs. 25-28, we obtain

$$SRR^B = SRR^I + SRR^O + 2 \cdot (SRR^I \cdot SRR^O)^{1/2} \tag{29}$$

where $SRR^{O(I)}$ are defined by Eq. 21. Using the values of $[SRR^{O(I)} \cdot eff]$ listed in Tables I and II, values of $[SRR^B \cdot eff]$ are calculated and also listed in these tables. Their mean values are tabulated in Table III.

Correction for Moderation

Information about the fission neutron energy distribution is provided by the neutron coincidence data of a two-ring detector. A greater neutron

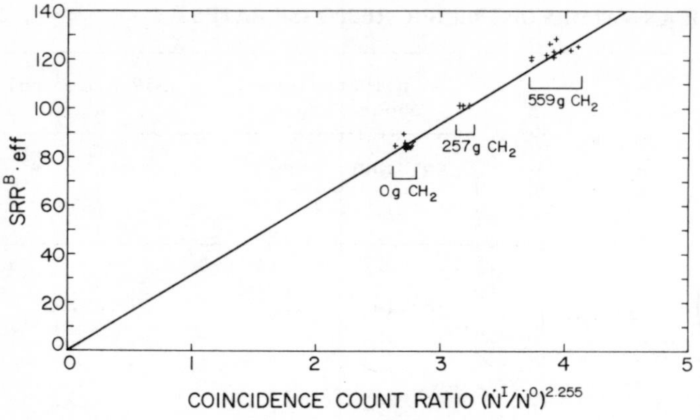

FIG.4. The specific response rate due to both rings, uncorrected for relative coincidence detection efficiency, vs coincidence count ring ratio.

moderation results in a softer fission neutron spectrum thereby increasing the coincidence count ratio $\dot{N}^I/\dot{N}^O$. A function of this ratio is defined which relates a sample's moderating property and the relative detection efficiency eff of the coincidence, i.e., fission, events where

$$\text{eff} \equiv (\dot{N}^I/\dot{N}^O)^Z \tag{30}$$

and Z is an empirically determined constant. Substituting Eqs. 30, 22-a, b into Eq. 21 and using the result in Eq. 29 gives

$$\text{SRR}^B = \frac{\dot{N}^I + \dot{N}^O + 2 \cdot (\dot{N}^I \cdot \dot{N}^O)^{1/2}}{(\dot{N}^I/\dot{N}^O)^Z \cdot (\dot{T}^O/\dot{T}^I)^{3.295} \cdot M_{240}^{1.065}} \tag{31}$$

Using the data of Tables I and II in the above expression, a least squares fitting procedure gives

$$Z = 2.255 \tag{32}$$

and

$$(\text{SRR}^B)_{Av} = 31.45 \pm 2.76\% \tag{33}$$

The values of $[\text{SRR}^B \cdot \text{eff}]$ listed in Tables I and II are pictured as a function of $(\dot{N}^I/\dot{N}^O)^{2.255}$ in Fig. 4. The slope is SRR^B. Also pictured there are the results of calculations using data taken for standards 3, 4, and 6 surrounded by only 257 g of CH_2. They are observed to be consistent with those of Tables I and II having 0- and 559-g of CH_2 respectively about the samples. This indicates the efficiency correction eff given by Eqs. 30 and 32 is valid for this detector over very large ranges of neutron energy moderation. One should note here that changes in sample moderation are approximated using discrete moderating material. The extent of agreement between intimate and discrete cases cannot be stated until standards having intimately mixed moderating material are constructed in order to measure this effect. The moderation correction is expected to be independent of whether the moderating medium is discrete or intimately mixed with the plutonium powder if the fission cross section is independent of the incident neutron's energy; an assumption made to establish the principle of separability. This is true for neutron energies as low as about 10 eV in the presence of ^{239}Pu. It is therefore assumed the two cases agree for a reasonable level of moderation.

Listed in Table III are values of $[SRR^B \cdot eff]_{Av}$ and $[SRR^B]_{Av}$ calculated for 0- and 559-g of CH_2 placed about the sample. The ratio of mean responses it no moderation corrections are made is given as

$$\frac{[SRR^B \cdot eff\ (559\ g\ of\ CH_2)]_{Av}}{[SRR^B \cdot eff\ (0\ g\ of\ CH_2)]_{Av}} = 1.45 \pm 3.08\%$$

The ratio of the mean responses upon correcting for moderation is given as

$$\frac{[SRR^B (559\ g\ of\ CH_2)]_{Av}}{[SRR^B (0\ g\ of\ CH_2)]_{Av}} = 1.01 \pm 3.78\%$$

The 45% increase in specific response rate caused by 559 g of CH_2 moderating material is eliminated. This correction for such grossly different moderating properties indicates the validity of correcting for moderation effects by using a function of the coincidence count ratio $\dot{N}^I/\dot{N}^O$.

Finally, if calibration constants δ and X defined by Eq. 21 were determined using the data of Table II, the only effect is to increase the value of Z in Eq. 30 and hence the relative coincidence detection efficiency. However, the value of SRR^B given by Eq. 31 is unchanged. This implies that the assayed value of M_{240} calculated with the aid of Eq. 31 after determining the system's calibration constants δ, X, SRR^B_{Av}, and Z is independent of the degree of moderation in the several standards used for calibration. It is important only that standards of similar moderation be used to obtain the calibration constants.

Effective Mass of ^{240}Pu

The effective mass of ^{240}Pu in a sample of PuO_2 powder having arbitrary neutron moderation properties and combined arbitrarily with MgO and SiO_2 for which $\dot{N}^{O(I)}$ and $\dot{T}^{O(I)}$ are measured can be calculated using Eqs. 31, 32, and 33. The final expression for the detector pictured in Fig. 1 is

$$M_{240} = \exp\left\{\frac{\ln\left[\dfrac{\dot{N}^I + \dot{N}^O + 2 \cdot (\dot{N}^I \cdot \dot{N}^O)^{1/2}}{(\dot{T}^O/\dot{T}^I)^{3.295} \cdot (\dot{N}^I/\dot{N}^O)^{2.255} \cdot 31.45}\right]}{1.065}\right\} \qquad (34)$$

EXPERIMENTAL RESULTS

Using the two-ring well counter pictured in Fig. 1 and the techniques described above to correct for moderation and multiplication effects, a standard deviation of less than 3% is obtained for the system's specific response rate. This compares with deviations of 16% and 12% for the inner and outer rings respectively in the absence of multiplication and moderation corrections. These figures reflect the expected accuracy for an assay of effective mass of ^{240}Pu in a powder sample of PuO_2 mixed arbitrarily with MgO, SiO_2, and moderating material. These results are based on the analyses of 24 samples among which 1) masses of total plutonium varied from 21- to 779-grams, 2) plutonium densities varied by a factor of 19, 3) a variable moderation effect changed the specific response rate by 45%, 4) the total multiplication varied by 33%, and 5) the multiplication due to different matrix materials varied by 15%.

DISCUSSION AND CONCLUSIONS

To perform accurate nondestructive assays of fertile PuO_2 powder material, it is important to correct for multiplication and moderation effects. Two important principles are discussed which guide the development of a model to correct for these phenomena. Generally, attempts to develop techniques to account for multiplication and moderation will be more fruitful if these principles are recognized.

The first of these states that the ratio of the induced fission--to spontaneous fission--coincident count rate, $\frac{\dot{N}_{SM} + \dot{N}_{IM}}{\dot{N}_{SF}}$, is separable into two independent terms. One term, designated f_{Pu}, is controlled by the areal density of plutonium in the sample. The other term $f_{\alpha n}$ is controlled by the characteristics of the matrix material including, primarily, the probability of generating a random neutron of energy E_n via an (α,n)-reaction. This principle is established by invoking the assumptions that 1) the probability of inducing a fission is independent of the incident neutron's energy and 2) the plutonium isotopic abundances are constant for all samples. Let us now investigate the degree of validity of these assumptions.

Since 89.4% to 94.2% of the plutonium in the samples studied consist of the fissile isotope ^{239}Pu, to a good approximation the induced fission probability is energy independent. Furthermore the only quantities in IF that can depend on the ^{240}Pu isotopic abundance IA_{240} are F_{Pu}, defined by Eqs. 18 and 22-b, and the ratio η_α/η_{SF}, which appears in the expression for $f_{\alpha n}$ given by the bracketed term in Eq. 14 and Eq. 13. Using a system that was calibrated with standards having IA_{240} equal to 8%, assume a NDA measurement of a 10% sample; i.e., an increase in IA_{240} by a factor of 1.25. The value of M_{240} in the 10% sample is accounted for in the definition of F_{Pu} given by Eq. 18. Associated with this value of M_{240} is an implied mass of ^{239}Pu, M_{239}, governed by the system's calibration constants obtained using 8% standards. However, the value of M_{239} associated with the 10% sample is 2% less. Since the number of induced fissions is indicated by Monte Carlo calculations to be about 25% of the observed coincidence count rate $\dot{N}$, the value of $\dot{N}$ is expected to decrease by .5% which is the implied error in F_{Pu}. In addition, a straightforward calculation shows η_α/η_{SF} decreases by about 15% for this change in IA_{240}. However η_α/η_{SF} is proportional to $\dot{N}_{IM}/\dot{N}_{SF}$ which Monte Carlo calculations indicate is about 5% for these types of samples where $\dot{N}_{SF}/\dot{N}$ is about 75%. Thus the value of the observed coincidence counts is expected to decrease by .6% which is the implied error in $F_{\alpha n}$. The total error in IF is expected to be the sum of the errors in F_{Pu} and $F_{\alpha n}$ or only about $\pm$ 1.1% for variations of IA_{240} of $\pm$ .02 about a mean of .08. Therefore both assumptions used to prove separability have a high degree of validity for these powder samples.

System calibration is greatly simplified using the separability principle. Once appropriate models are established to represent $F_{\alpha n}$ and F_{Pu}, only a few standards should be necessary to establish the calibration constants associated with these terms. If F_{Pu} can be expressed as a function of M_{240}, then the correction for multiplication in a subsequent assay demands only a measurement of $F_{\alpha n}$ to establish the effect of the matrix material.

The second important principle is the recognition that all fission neutrons in passive systems have nearly the same energy distribution. This implies that relative spectroscopic information measured among samples should allow a correction for variation in neutron detection efficiency eff due to varying neutron energy moderation.

The model described here corrects for multiplication and moderation encountered in samples of PuO_2 powder mixed arbitrarily with MgO, SiO_2, and moderating material. A two-ring detector like that pictured in Fig. 1 is used to provide the spectroscopic information to correct for eff and $F_{\alpha n}$, and F_{Pu} is expressed as a function of the mass of ^{240}Pu.

Finally the magnitude of the singles count ring-ratio $(\dot{T}^O/\dot{T}^I)^\delta$ for the types of powder samples studied here serves to identify the contaminant that generates the (α,n)-neutrons. Such information is useful in process control.

REFERENCES

(1) MARION and FOWLER, Fast Neutron Physics, Part II, Interscience Publishers (1963) 2036.

(2) BARON, N., Los Alamos Scientific Laboratory Report LA-6849-PR (1977) 40.

(3) ENSSLIN, N., EVANS, M. L., MENLOVE, H. O., and SWANSEN, J. E., INMM, Summer 1978.

(4) SWANSEN, J. E., ENSSLIN, N., KRICK, M. S., and MENLOVE, H. O., LA-6788-PR (1976) 4.

(5) BARON, N., Los Alamos Scientific Laboratory Report LA-7030-PR (1977) 76

(6) KEEPIN, G. R., Physics of Nuclear Kinetics (1965) 54

DISCUSSION

H.J. WEBER: Does the formalism for correction of multiplication effects apply only to optimally moderated neutron well counters?

N. BARON: The correction factor F_{Pu} is independent of the die-away time of the detector. However, this is not true of $F_{\alpha n}$. I should imagine a system could be designed so that the sensitivity of $F_{\alpha n}$ to changes in (α, n)-neutron energies is maximized.

DEVELOPMENT OF A NEW SURVEY ASSAY METER TO MEET IAEA AND UNITED STATES DEPARTMENT OF ENERGY REQUIREMENTS

M.S. ZUCKER, R.L. CHASE, D. STEPHANI,
S. FIARMAN, J.L. ALBERI
Brookhaven National Laboratory,
Upton, New York,
United States of America

Abstract

DEVELOPMENT OF A NEW SURVEY ASSAY METER TO MEET IAEA AND UNITED STATES DEPARTMENT OF ENERGY REQUIREMENTS.

An instrument has been developed to supplant the Eberline Stabilized Assay Meter (ESAM) which has played an important role in safeguards, but has certain deficiencies and faults which gradually became apparent during its widespread usage. The result of this experience was that knowledgeable users felt an improved version of this instrument was needed which incorporated more recent electronic technology. An effort was made by the IAEA and the US Department of Energy to implement this. The result is the BNL Survey Assay Meter (BSAM) developed by Brookhaven National Laboratory's Technical Support Organization, and Instrumentation Division. The BSAM was designed after detailed analyses of the ESAM circuitry, operating experience with the ESAM by BNL, LASL, the IAEA, and other groups, and a review of the features considered desirable in a future instrument. Thus, the BSAM incorporates such operational features of the ESAM with improvements, together with several new features, which materially decrease the effort required to make measurements and reduce data to meaningful quantities. As in any engineering endeavour, compromises had to be made while carrying out the design goals. A proper balance between operation simplicity and the sophistication of the device was necessary. Moreover, desirable features such as portability, ruggedness, low current consumption, facility for data manipulation and reduction, capability for operation independently of line power, effective utilization of low noise, high resolution, or small detectors in addition to NaI-photomultipliers, often conflict to a greater or lesser degree with one another. One of the BSAM's more innovative features is the interfacing of a built-in pocket calculator to the nuclear electronics comprising the basic unit. The calculator serves as a digital readout and automatically performs calculations connected with the instrument's operation and it has statistical functions such as linear regression, standard deviation, etc., and can be used independently for other calculations. It also simplifies the establishment of a proper normalization condition for one single-channel condition relative to the other for Compton or environmental background subtraction. The time formerly required for this latter procedure often took up to one or two hours, but has now been reduced by one or two orders of magnitude. Another new feature is a set of internal preset conditions for the more commonly met assay situations, ^{235}U (via the 185.6-keV peak), Pu (375 ± 75 keV complex) and Pu (375–450 keV, to minimize ^{241}Am contributions). This feature frees the operator from making adjustments in the field in these common situations. However, externally available controls also exist that can be field-adjusted to meet unusual requirements, non-standard detectors, etc. The BSAM design accommodates germanium and cadmium telluride diode detectors and proportional counters besides photomultiplier-based detectors, without significant loss in resolution or signal-to-noise ratio. The instrument is packaged so that its outer fiberglass case is the normal carrying container. It can be slung over the operator's shoulder freeing hands for the detector, calculator, etc., or it can be used on a table. The internal rechargeable cells make the unit independent of power lines for a normal work day.

INTRODUCTION

A combination of nuclear instrumentation components consisting essentially of a pulse amplifier, two single-channel analyzers, a scaler, a detector bias supply, and a detector (usually a NaI-p.m. tube) has proven to be a particularly effective tool for non-destructive assay of nuclear material.

This combination has to be viewed in the context of a whole hierarchy of instrumental set-ups primarily for detecting gamma rays. The simplest of these, consisting of a gamma detector, its bias supply, an amplifier and a simple discriminator and scaler provides minimal information, but is easy to operate. The most complicated would have a high resolution detector, bias supply, amplifier and a multichannel analyzer based on a computer, capable of extracting every last bit of information from the gamma spectrum. It would require a high degree of training for the operator, and not be very portable. A system consisting of detector, bias supply, and two single-channel analyzers, occupies a middle ground as to its complexity, the demands it makes on the operator, and the amount of information which it can furnish. This last has proven to be a rather fortuitous combination.

With the components (except for the detector) packaged into one unit and commercially sold by the Eberline Instrument Company, this combination has become the single most widely used instrument in the safeguards field. It has the virtues of being compact, easily transportable, widely applicable, and relatively easy to take data with and interpret the results thereof. Thus, it allows use by personnel who are either relatively untrained or are simply too involved with other details to take advantage of more complex equipment. In a field which propective instrument manufactures shy away from because of low volume, the Eberline Stabilized Assay Meters, or "SAM" in its various models, have proved an unusual commercial success. Well over 300 units have been sold over the decade of its existence, although perhaps the majority of these are used in health physical applications.

While technical and commercial success imply a certain perfection in the original conception and construction, the importance of the instrument in the safeguards field prompted a fresh look at how it might be improved, particularly so as to take advantage of recent developments in electronics and nuclear instrumentation.

Our effort at Brookhaven National Laboratory at updating the Eberline SAM (ESAM) commenced less than two years ago, following a much longer period of exposure to its virtues and flaws, and has culminated in what we think will be a worthy successor. The instrument we developed, referred to as the Brookhaven Survey Assay Meter (BSAM), will do everything that the ESAM was capable of, though presumably better, and has taken advantage of newer technology and ideas to add many significant facilities and qualities that the ESAM does not have. In the following, we endeavor to discuss these improvements and the design philosophy behind them.

DETECTORS

The BSAM unit was designed from the start with the intent of using it with all the types of detectors of safeguards interest, namely, inorganic (e.g., NaI) or organic crystal-photomultiplier tube combinations, proportional counters, high resolution ("intrinsic" Ge), and moderate resolution (CdTe) solid state detectors.

At present, and in the near future, the detector most used will be the NaI-p.m. tube. Despite its resolution and drift problems, its efficiency, sensitivity, and inherent gain (which confers on it good signal-to-noise ratio) make it still the most cost-effective detector of gamma rays in most situations.

A survey of the present usage of the ESAM NaI p.m. tube detector indicated that it most often is used with its 5 cm. D frontal area severely stopped down (collimated). Taking this together with the fact that the NaI thickness ($\sim$1 cm) was optimum for ^{235}U detection, but not for Pu which has become much more important since the ESAM was first developed, and that a significantly smaller p.m. tube would vastly reduce the weight of the shielding required, led to the choice of a 19mmDX19mmH detector mounted on a 19mm D photomultiplier tube. Such small p.m. tubes are usually not as good in resolution as the larger ones, but in this case, the chosen vendor, Bicron, is able to claim significantly better resolution than that of the ESAM detectors.

To go with the new smaller detector, two special shield collimators were designed. Both collimators have a snap-on front piece giving an angular resolution of about 50 degrees. Without the front piece, the front surface of the detector is nearly flush with the front of the shield. In both, the shielding has been graded so that most of the shielding weight is where it does the most good, the design criterion being <1% transmission through the collimator at 186 keV in one case and <10% at 450 keV in the other.

As is well known, p.m. tube gain changes with temperature, and exhibits fatigue and hysteresis effects, even permanent loss of gain, depending on the radiation exposure. Moreover the intensity, wavelengths, and time characteristics of the scintillations produced in the crystal change with temperature. For this reason, any such detector used in a varying environment must have its gain stabilized in some manner.

At the present time the best stabilization method available for a portable instrument is to fix a small ($\sim$1000 dps) source of alpha particles, ^{241}Am, in the neighborhood of the crystal so that they produce a steady average count rate of good resolution ($\sim$4-8%) pulses in the p.m. tube.

The Bicron tube uses an ^{241}Am source deposited on a foil located so as to bombard the detector crystal directly with alpha particles as opposed to the ESAM detectors made by Harshaw in which a specially doped NaI seed crystal with ^{241}Am added to it is used as a source of light pulses which shine through the crystal to the p.m. tube. Both systems seem to work equally well although each has its own advantages. In either case, these alpha produced pulses are then used as a reference for a feedback system which adjusts the gain of the photomultiplier tube (by varying the H.V. supplied to it) so as to keep the alpha pulses at constant amplitude, and thus supposedly also control the pulse height of gamma produced scintillations.

This is the method used in the ESAM, and is basically the method used in the present BSAM. However, as has been known for several years now, this stabilization scheme suffers from a basic flaw. Unfortunately, the temperature dependence of the pulse height of alpha particles is not the same as that of gamma rays. Without going into details here, let it suffice to say that the light emission process depends upon the density of ionization, as are the various non-radiative recombination processes competing with it, and this density is quite different for the alpha particles than for the electrons produced by gamma interaction with the scintillator. The magnitude of the difference in temperature coefficient is a function of variables involved in the production of scintillators which are not yet well understood. It also depends on the band width of the pulse amplifier, this latter stemming from the fact that several different light producing mechanisms are involved, each with different time dependence, each affected differently by temperature. Typically, increasing temperature decreases the gain with the change being most rapid at low temperatures, the temperature coefficient for the alpha pulses being of the order of 1.5 to 2 times that for gamma ray produced pulses. The details are strongly dependent on manufacturing procedures probably not under good control by the manufacturer. The above discussion emphasizes the scintillator, but the p.m. tube itself, of course, changes gain and spectral response with temperature as well, and the net result depends on both.

With the above situation, stabilizing on the ^{241}Am alpha peak will keep that peak constant but allow what is really of interest, the gamma peak, to shift, though at a reduced rate and in the opposite sense than if it were not controlled at all.

To improve on this situation, a thermistor was embedded in the crystal in the same place as the alpha source. This senses the crystal-p.m. tube temperature and produces a signal of the right sign and approximate magnitude to compensate for the difference in alpha peak and gamma peak behaviour with temperature. This compensation can only be approximate over the (allowed) temperature range of the detector, which is 5°C to 45°C, because the actual curvilinear temperature characteristic is being linearly approximated and because there is some variation in these characteristics between nominally identical detectors. However, this alone should produce significant improvement in stability for the BSAM compared to the ESAM.

All NaI-p.m. tube detector assemblies have two basic parameters, gain and resolution. Those NaI-p.m. tubes which use a ^{241}Am source implant for stabilization have two more, the gamma equivalent energy of the alpha particle pulse and the alpha particle count rate. The ^{241}Am alpha has approximately 5.5 MeV but the pulse height it ultimately produces may be equivalent to a gamma ray of appreciably less energy. The manufacturer has some control of this (by the amount of doping in the case of the Harshaw implanted seed, and by the use of absorbers in the Bicron process). A gamma equivalent energy is specified which is in excess of the highest energy gamma of interest so that for the precision with which the gamma equivalent can be made ($\sim$100-200keV) and the alpha peak resolution (4-8%) there will be no interference. The highest energy of general interest in safeguards is the ^{208}Tl gamma at 2.65 MeV, hence the nominal value 3MeV has been chosen. In order to accommodate tubes differing from this, or having alpha count rates markedly different from the typical 800-1000 cps, adjustments for each of these has been built in. These will allow the BSAM to be adjusted for optimum performance of the stabilization circuit with any of the so-called "stabilized" NaI p.m. tube detectors.

The servo for controlling the p.m. tube gain has appreciably more amplification, a greater capture range, and a visual panel indication (LED) indicating in which direction the detector HV bias control should be changed in order to lock the circuit into stabilization. With the HV bias control left at any reasonable previous value such as the last used operating point, the circuit will stabilize as soon as the BSAM is turned on, and there will be no need to raise the bias voltage from zero. The bias polarity can be changed via a switch on the bias supply card.

In the event a non-stabilized tube, or another type of detector is used, the HV bias becomes an adjustable regulated high voltage supply, suitable for use with proportional counters, germanium diode detectors, etc.

It is known that in order for the ^{241}Am scheme to work best the amplifier must have good low frequency response, i.e., large amplifier time constants, because the slower components of the light output seem to show better agreement between alphas and electrons. This argues for large amplifier time constants, but this would limit the count rate capability. A compromise was reached in making the amplifier pulses to be approximately double differentiated with a 4-5 μsec zero crossing after initial rise.

Given the complexity of the HV stabilization scheme based on ^{241}Am, the question arises as to whether any scheme other than this classical approach should have been considered. Certain methods can be dismissed easily. A built-in high energy gamma source would contribute too many Compton gammas to the lower lying regions where the gammas of interest are. A low-energy gamma emitter on the other hand would ride on a backgound of Compton gammas from the gammas being examined and since this background can vary in shape, the operating point would tend to be shifted accordingly. A method which works in situations where a particular gamma emitting sample is assayed for an appreciable time uses the full energy "photo" peak of interest for the stabilization. This is tempting to consider but probably not practical in an instrument which is used for relatively short runs on samples which may have differently shaped spectra, and for series of measurements that may have appreciable waiting times between them during which the servo would have no reference. (Of course, it perhaps could be given a short-term memory, etc.)

A method which was tried, in fact had considerable effort put into it, was to use a pulsed light-emitting diode implanted in the crystal as a source. This scheme was devised by Bicron, and used successfully, but only in temperature controlled environments, i.e., it would compensate for fatigue, count rate effects, etc. This method can be criticized immediately in that it bypasses the crystal and thus tests only the p.m. tube and amplifier. However, even disregarding this, there were problems in that the LED spectrum and intensity changed with temperature, and doubt arose as to whether diodes or diode-p.m. tube combinations would have uniform enough specifications for a device that was to be produced in quantity. It was finally decided that compensating for temperature and device non-uniformity could be done, but this would involve perhaps too many necessary adjustments for an instrument destined for commercial production. There are some of us who think that this method was discarded too soon. In any case the ^{241}Am stabilization scheme, used in the ESAM, but augmented with a thermistor for a higher order of compensation, was decided upon for the BSAM.

In contrast, making the BSAM work with other detectors was much easier. There was some problem in reducing noise levels due to interaction of the digital with the analogue (amplifier, etc.) circuitry in the close-packed confines of the instrument, but this was largely overcome and as a result only about 100 eV of broadening is added in the region of ^{57}Co (122, 136 keV). A high resolution intrinsic germanium detector (∿640eV at ^{57}Co as measured with a state-of-the-art amplifier) had an apparent resolution of 750eV when used with the BSAM. (The resolution of the channel controls is fine enough to make use of this high a resolution.) Another problem which had to be solved was providing preamp power for high resolution solid state detectors directly from the BSAM. Since the BSAM was specifically designed for portable application free of power lines, the amount of power was strictly limited. It proved possible, however, in conjunction with ORTEC and PGT to develop appropriate preamps. These give state-of-the-art line widths, yet consume in toto less current (∿9-10mA) than is usually supplied to a front-end cooled FET alone, and require only a single +10V supply. There is only a small compromise in preamp dynamic range, i.e., count rate capabiltiy.

All detectors used with the BSAM need only be supplied with an appropriate preamplifier and power cord which plugs into the unit and provides an appropriate polarity-adjustable bias supply, preamp power, signal lead, and in the case of the NaI-p.m. tube newly designed for the BSAM, thermistor leads. To allow use of the ESAM p.m. tubes, which may have an advantage with their greater area, an appropriate preamp adapter plug is supplied with each BSAM.

Figure 1 shows the BSAM surrounded by some of the various detectors it can be used with. The smallest detector is a CdTe diode, still under development. In the lower left is the ESAM detector; attached to the BSAM is the specially developed NaI-p.m. tube without its shield-collimator. In the upper right is a special hand-held high resolution intrinsic germanium detector (∿540eV at ^{57}Co) which can contain enough liquid N_2 for ∿8 hrs. of operation, and which will suffer only negligible loss in resolution when used with the BSAM. This unit and a similar one by PGT were developed for the IAEA to be used with the BSAM or other instrumentation.

Figure 2 illustrates the self-contained nature of the BSAM; with the cover off and the side panel in place the unit is ready for travelling.

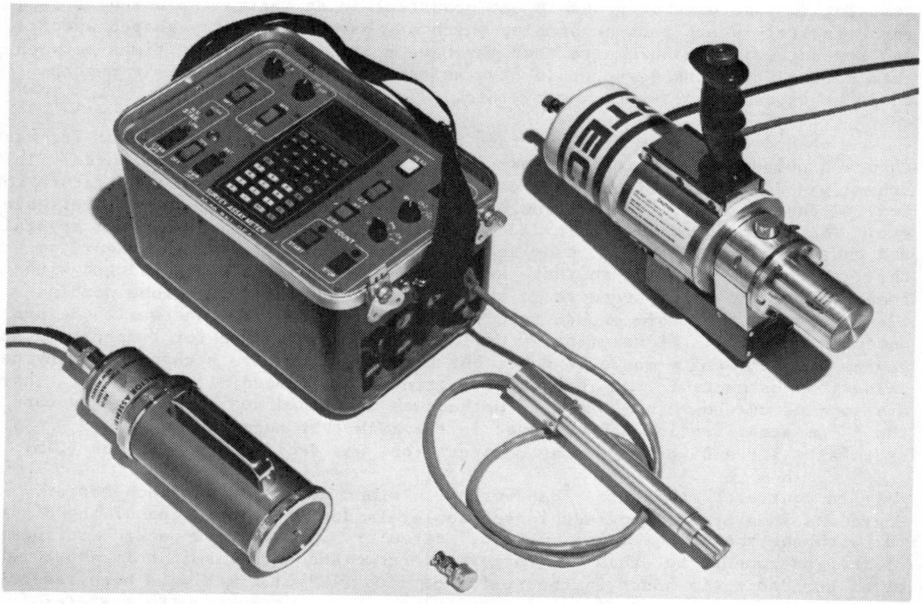

FIG.1. The BSAM and some of the wide variety of detectors it is designed to accommodate. From left to right, the ESAM RD19 (5 cm D NaI), a CdTe detector, a small-diameter NaI-photomultiplier tube detector designed especially for the BSAM (without collimators), and a newly developed hand-held liquid N_2 cooled "intrinsic" germanium detector. Proportional counters are also usable with the BSAM but have not been shown.

INTERFACED CALCULATOR

A novel feature of the BSAM is a pocket calculator physically and electronically integrated with the nuclear detection circuitry. This calculator serves three functions: (i) The light emitting diode display serves as a data readout for the instrument. (ii) Certain calculations are performed

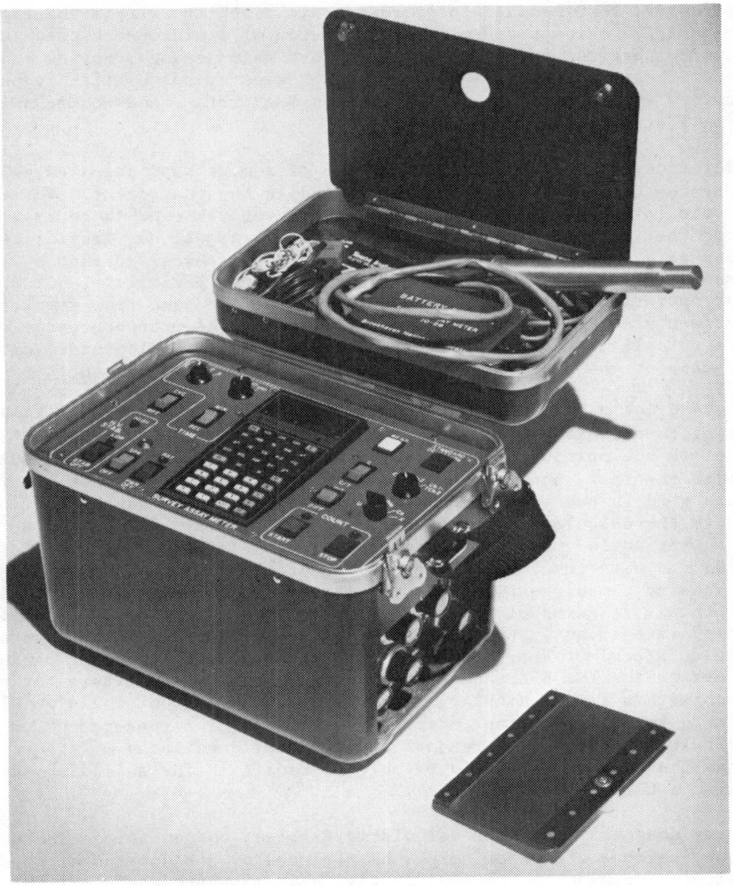

FIG.2. The BSAM outer containment is its own travelling case. The lid stores the basic (NaI-photomultiplier tube) detector. The side-control panel has a locking cover.

automatically or semi-automatically which greatly facilitate the basic operation of the BSAM, such as setting up and automatically performing background subtraction, calculating the standard deviation of counts, allowing a simple procedure for calibrating the instrument for enrichment measurement, and enabling a semi-automatic procedure for calculating enrichment from assay results. (iii) The calculator can also be used independently of the BSAM, provided, of course, memory locations being used by the BSAM are not overwritten; the BSAM also has priority in making use of the calculator. The calculator used in the present versions of the BSAM is a Texas Instrument 51A, chosen because it has every function conceivably of use to an inspector, e.g., logarithms and exponentials for attenuation calculations, various statistical functions, such as standard deviation and linear regression, etc., and was easy to interface to the rest of the BSAM by virtue of the logic levels, etc.

The control for the calculator resides in circuitry within the BSAM made up of discrete logic chips performing the function of a microprocessor, in effect making the connections in proper sequence just as a person pressing the calculator keyboard buttons would. Which program or type of calculation is performed is determined by the position of the various BSAM rotary and rocker switch and pushbutton controls.

Under study at present is the question of a more sophisticated calculator or microprocessor equivalent to be interfaced to the instrument. While the present calculator is sufficient for needs presently known, there is a likelihood that as experience with the instrument grows, new facilities may prove desirable. Perhaps more important, it is now perceived that a programmable calculator would not only give greater flexibility, but would also make many operations more fully automatic, and at the same time simplify some of the interface electronics. This latter is because the microprocessor would now only have to call a stored program or subroutine in the calculator, rather than directly causing the program to be executed as is done at present.

The case for a completely dedicated microprocessor that would take over the calculating function as well as program execution comes about somewhat obliquely and is not in our opinion as good as the case for a programmable calculator. It starts with the fact that pocket calculator displays, the present one included, are not as good as one would wish. While they have the advantage of coming "free" with the calculator, they suffer from being small and having a restricted visual viewing angle. One reason for this is the manufacturer's need to balance off visibility with the current consumption of the light emitting diode display, this item being the biggest source of power drain in the typical calculator. Thus one thinks in terms of providing a separate display for the BSAM of a type more appropriate to an instrument for field use, discarding one advantage of the pocket calculator. Of course, the need to conserve display power consumption is also important for the BSAM, however, there are options available for saving power even with a larger display. For example, the readout could be blanked except for a fixed short time interval, at the end of a measurement or calculation, just long enough for a readout under normal conditions. In case the operator misses the readout, it would be able to recall it for an additional moment by pressing a button "readout recall".

Having done away with the calculator display, one questions the need for a calculator, which could in principle be replaced by a dedicated microprocessor. Admittedly it would be impractical to give such microprocessor all the facility of a good pocket calculator, but then, the inspector nowadays presumably has his own. We feel however that particularly in a portable as opposed to a work-table type of operation, having a calculator located on the instrument itself is an obvious advantage. It is certainly more readily accessible and requires no free hand other than that occupied with operating the BSAM.

A dedicated microprocessor has its difficulties. Aside from requiring considerable development engineering time, to make them very general and flexible requires an impractical number of push-button controls, and it is difficult to make changes in the program compared to a programmable calculator.

To put matters in perspective, one should realize that, after all, a pocket calculator is itself a microprocessor of a very general type, fully engineered with all the details already worked out by the manufacturer and ready to be used. If the display is not as appropriate as desirable, it might be preferable to accept that as an engineering compromise, or perhaps provide a separate display, as the most cost-effective solution.

PRECALIBRATION FEATURES, SINGLE-CHANNEL CONDITIONS AND DETECTOR BIAS VOLTAGE

Another novel feature of the BSAM are precalibrated (i.e., factory or laboratory set) baseline and window widths, controlled by internal adjustments which automatically set the instrument up for certain common measurement situations. These are selected by a rotary switch and mean that the operator no longer has to adjust the instrument in the field prior to use for these common situations. The switch simply selects internally derived biases for application to the single-channel baseline and window width controls. One of these is measurement of the ^{235}U 186keV in the presence of a ^{238}U Compton background. Others are the whole (375±75)keV Pu complex, the complex from (>375 to 450)keV to minimize ^{241}Am contributions to the Pu assay, and the 662keV ^{137}Cs peak which can be used in conjunction with a ^{137}Cs check source in a standard geometry to test the functioning of the instrument. The BSAM also has externally available multi-turn baseline and window width controls which enable completely arbitrary adjustment. For each precalibrated condition, the BSAM amplifier gain is set at an appropriate value. While under external control the amplifier gain is independently variable.

The precalibrated single-channel conditions of course depend upon the fixed relation between the gamma equivalent energy of the alpha particle peak and the various gammas of interest. It is for this reason that adjustment is available in the BSAM so that "stabilized" detectors having different gamma equivalent energies can be accommodated. The stability of the electronics and the gain of the stabilizer circuit has been made high enough for this relation between gamma and alpha equivalent gamma energy to be maintained accurately over a wide range of temperatures.

As mentioned above, the H.V. need not be adjusted at the start of each series of measurements, but will be close to optimum value if it was set approximately correctly on the previous occasion of use.

HUMAN ENGINEERING FACTORS AND OTHER FEATURES

We started with the premise that the operator of the instrument, either in an inspection or some other measurement situation, generally is involved with enough distractions so that features which saves time and effort are not merely worthwhile but necessary. Separating the controls, etc., into two groups is one of the ways this premise has been implemented. The basic operating controls, including those pertaining to the preset conditions, and the readout are located on a main panel. In most situations the features on this main panel will be the only ones used, and the other adjustments which will be of interest in only special situations, such as changing detectors or detector types, or to accommodate special experimental or measurement requirements, are located in a subsidiary side panel with which the operator usually will not have to bother with. The side panel, accessible through a removable door, has the controls pertaining to operator controlled variation of both single channel baselines and window widths (four multi-turn potentiometers with digital readout), a detector bias adjustment (also a digital readout multi-turn potentiometer), inspection points for the amplifier and both single channel outputs, a recessed screwdriver operated multi-turn potentiometer for alpha gamma-equivalent energy adjustment, and connector sockets for power input and for preamplifier signal, power, and detector bias, the latter all being in a single connector.

The five multi-turn potentiometers for adjusting the single channel conditions and detector bias all have highly visible scale-of-1000 numerical readouts.

A problem with an instrument to be used in areas where line power may be either 220V or 110V is preventing accidental connection to the wrong voltage. The approach taken was to provide two separate line cords. One is terminated at the line end with the appropriate male line plug for 220V, the other with the standard 110V plug. The other end of each cord is terminated with a plug with internal wiring such that the line power is applied to the two primary windings of a power transformer being either in series or parallel, as appropriate.

In order that the operator not have to keep watching the instrument for a measurement to end, a brief audible signal is emitted by a small loudspeaker when the assay is over. The same loudspeaker is also used to emit a sound pulse every time an event is registered by the single channel conditions so that the BSAM can be employed in the manner of the classical Geiger counter to search for sources of radioactivity. Finally, in case the internal scalers overflow their capacity of 10^6-1 counts, a continuous signal is emitted as a warning until the instrument is reset. For extremely noisy environments, an earphone and chassis jack are provided; the audible feature can be switched off when it is not desired.

The BSAM chassis is of aluminum enclosed in an outer case of fiberglass. With the fiberglass cover (which incidently holds the detector and collimator-shield, the battery charger, line cords, and other accessories) in place the unit is fully protected and self contained, and can be treated as hand carried luggage without the need for any further containment. Figure 2 illustrates how the instrument packs for transportation, with the cover being used to carry the detector and various accessories, and the side panel being closed off with its own locking panel. For shipment as freight, a fiberglass container is furnished for additional protection. Even with this, the resulting package is smaller than the suitcase required for transportation for the ESAM.

The BSAM, which with its internal rechargeable battery weighs about as much as the ESAM chassis, can be carried in a portable operation by a sling over the operator's shoulder so as to leave one arm free for manipulating the controls while the other arm holds the detector.

The most convenient features for the operator however have to do with certain of the facilities which the interfaced calculator allows. With the ESAM, adjustment of the width of the channel used to monitor ^{238}U Comptons for subtraction from the channel used for the ^{235}U 186keV peak plus ^{238}U Compton background had to be done laboriously by trial and error. With the BSAM the window width of the channel used in the subtraction can be set arbitrarily. A single measurement with a depleted sample, taken over a long enough period to give good statistics, and a few manipulations on the calculator, and the factor f with which the ^{238}U sensing channel count has to be multiplied by before being subtracted from the channel bracketing the 186 keV peak will be stored in the calculator memory. From that point on, the net quantity will automatically be calculated at the end of each measurement period and the result displayed. The BSAM procedure for setting up the instrument for background subtraction has reduced the time formerly required by one or two orders of magnitude.

Measurements of enrichment and in fact the initial calibration of the instrument are similarly made much easier than they were with the ESAM. The slope and intercept of the linear calibration for enrichment are obtained from two standards with known different enrichments and are stored in memory. The net count, (Ch1-fCh2), is multiplied by the slope and added to the intercept to yield the enrichment directly. In the most recent version of the BSAM this can be done with a single substitution operation.

The calculator also serves to automatically calculate the count per unit time, and as mentioned above, the standard deviation, of either channel, and of the difference between one channel and a constant factor times the other, requires only a single button be pushed.

Another feature which is useful is a control which when activated causes the instrument to reset itself and repeat the measurement with the previous result remaining in display until the new result displaces it. This "digital ratemeter" mode is useful for example when the instrument is used in area surveys, where the count summation over the time interval gives the operator a more objective answer than would an analogue meter subject to continuous variation because of the statistical fluctuations in count rate.

The instrument will automatically readout in counts or counts per unit time (either minutes or seconds), as desired.

Serviceability of the instrument has been enhanced by the method of construction. After the inner chassis has been removed from its fiberglass outer container, the front panel can be swung away from the rest of the instrument to reveal the back plane of card sockets into which the seven printed circuit cards comprising most of the electronics are inserted. This facilitates signal and voltage measurements at the pin outlets of each card. In addition, a printed circuit board extender is furnished allowing in situ measurement on the boards themselves. Finally, an inventory of boards equal to about 50% of the BSAM production to date has been established to be divided between the IAEA in Europe and BNL in the U.S. so as to enable substitution in case of suspected failure. The built in precalibrated single channel analyzer conditions and amplifier and SCA outputs also facilitate diagnosis.

The BSAM has been produced in two versions as regards power supply. In one, the instrument is powered in a manner similar to that of most pocket calculators, that is to say, from an internal rechargeable NiCd storage battery, either completely free of the power line, or attached to it via a small charger which can be packed in the storage lid of the BSAM when not in use. In the second version, power is supplied via a regulated potted power supply which occupies, in those BSAM units in which it is installed, the volume normally reserved for the NiCd battery. The impetus for the development of this alternate power supply was the suggestion that is would result in less weight for the BSAM. It turned out that while some weight was saved, the difference is not significant and not worth the independence of line power and freedom from line noise and transients that the storage battery version confers. The two types of units are readily converted one to another. If the eight hour life of the BSAM battery is insufficient, it is possible to replace the discharged set of batteries with one which is charged, or after a small delay, the unit can be run off the battery charger. The BSAM battery can be charged overnight. Of course, with line power available and using the charger the BSAM will run indefinitely.

In contrast to the full work day battery capacity of the BSAM, the ESAM battery for portable application, which is fairly heavy-about equal in weight to the ESAM chassis itself and has to be carried separately, would power the ESAM for only about six hours, and then only if stringent measures to economize the power drain (such as shutting off the display when not being read) were taken.

It is recognized that perhaps the majority of applications of the BSAM will be of a sedentary nature in which the instrument is used on a table top with AC power readily available. However, in order for the unit to be useful in those cases where power lines are unavailable or too noisy and unreliable, the BSAM had to be designed to work off the self-contained battery, and for a reasonable length of time, say one work day. This required very careful design of the components of the system in order that they function at a level expected of modern nuclear electronics and yet be economical of power. Several novel design techniques were used, e.g., current mirrors as loads in the amplifier and voltage regulators requiring minimal forward drops to operate. The design goal was met with only unimportant compromises with performance. Moreover, only standard, readily available electronic components were used.

With the ESAM one could accumulate data in a scaler that could hold either one channel, or the other, or the simple difference, exclusively. The BSAM however, uses a separate scaler to record each channel and then uses these to calculate on demand the difference, or any other quantity of interest (standard deviation, counts per unit time, etc). In this manner no information pertaining to a measurement is lost from the BSAM until the operator resets the instrument for another measurement.

ENVIRONMENTAL AND FIELD TESTING

As production of some eleven units together with spare printed circuit cards drew to a close, a program of testing the units began which is still underway.

The behaviour with temperature change is of course of prime importance. Initial work was done with a makeshift enclosure cooled with solid CO_2 and heated by electric resistor elements. The temperature limits of the range studied were those specified by Bicron as those for which their guarantees would hold, 5°C to 45°C, though it was soon decided that to avoid accidentally destroying the expensive photomultiplier detectors these limits had to be approached very carefully. An electronic thermometer noted the temperature and whether equilibrium had been reached while the pulse height spectrum of a ^{137}Cs source was monitored at the BSAM amplifier output with a multichannel analyzer. This allowed measurements of the shift of the ^{137}Cs and the ^{241}Am peaks as a function of temperature for a variety of conditions (servo on or off, different p.m. tubes, etc.). At present this type of testing is being continued using an environmental chamber which allows simultaneous control of the relative humidity.

Preliminary work has begun on vibration testing. It is planned to subject the BSAM to repetitive accelerations of up to 4 g along (in turn) each of three mutually perpendicular axes.

Shock testing has been approached gingerly, the goal being a 1 m. drop onto wood planks over a concrete floor with arbitrary orientation of the unit.

Underwriter or equivalent certification for the BSAM is being looked into. There are problems of cost, time, and the fact that standards are not uniformly accepted on a worldwide basis, and even are contradictory.

Finally, and most important are field trials under actual working conditions, which have already begun at two U.S. DOE sponsored nuclear energy research centers. At one, the unit has been used in a laboratory environment up till now but is slated to be used in hold-up measurement in an area without AC power lines. At the other, which is a major U.S. reprocessor of spent fuel and receiver of recoverable scrap, the BSAM is being used in conjunction with a high resolution coax intrinsic germanium detector for assaying ^{235}U in recycled uranium in which interferences from inbred nuclides rule out the use of an NaI detector.

MANUAL

A manual on the BSAM, some 80 pages long, fully illustrated with photographs, drawings, and circuit schematics, designed to be a self-teaching text which details internal as well as external adjustments and typical operating procedures, has been produced and is undergoing revision in the light of operating experience. Sample copies are available.

COMMERCIAL PRODUCTION

Some four well known U.S. and one Canadian firm have expressed stong interest in commercial production of future BSAM units. Unit prices have been estimated to be in the $3,000 to $4,000 range, with units becoming available as soon as three or four months after plans are released.

ACKNOWLEDGEMENT AND CONCLUSIONS

We would like to take this opportunity to thank the many individuals, too numerous to mention here, and various groups, in particular the U.S. DOE OSS, the IAEA Division of Development, groups at Argonne N.L. and L.A.S.L., and in the U.S. NRC, who have had a direct input in this effort in setting guidelines, establishing specifications, and sharing experiences.

We think the result, the BSAM, will prove a worthy successor to the ESAM, superior in those features common to both, and with many new features that make it easier and more reliable to use. We realize that perfection is a goal that can only be approached, and that, only through further development work. Continued work is also necessary to prevent obsolescence and to be responsive to the needs of the safeguards community.

DISCUSSION

W.C. BARTELS: I understand that the new survey assay meter has better temperature stabilization than the previous model. Could you comment on the improvements accomplished by your two-year effort.

M.S. ZUCKER: The new stabilization circuitry has a much greater open loop gain and a bigger capture window. In addition, since the temperature behaviour of the alpha particle pulses used for stabilization is somewhat different than that of the gamma pulses whose amplitude we wish to control, a thermistor embedded in the NaI crystal is used to furnish an approximate correction signal compensating for that difference.

Some other advantages of the new stabilizer circuitry are: (i) if the circuit is out of the proper stabilization range, there is an (LED) indication as to how the bias should be changed to bring the system under control; (ii) if the high-voltage adjustment has been left near the right value, the servo will lock in when the power is turned on, i.e. the HV need not be raised from zero; (iii) there are adjustments available to compensate for the gamma equivalent energy of the alpha particle and the alpha count rate, so that a wide range of stabilized NaI-photomultiplier tubes can be used.

RECENT EXPERIENCE WITH A MOBILE SAFEGUARDS NON-DESTRUCTIVE ASSAY SYSTEM*

R.J. SORENSON, J.E. FAGER, F.P. BRAUER
Battelle, Pacific Northwest Laboratories,
Richland, Washington,
United States of America

Abstract

RECENT EXPERIENCE WITH A MOBILE SAFEGUARDS NON-DESTRUCTIVE ASSAY SYSTEM.

A mobile, real-time, non-destructive assay system for nuclear material safeguards applications has been designed, assembled, and evaluated. The system is designed to be used by either an independent agency for verification of prior measurements or by plant personnel for various sample measurements. The system consists of electronic and computer-support equipment mounted in a specially constructed vehicle. This vehicle also carries passive and active neutron and gamma-ray measurement equipment that is operated outside the vehicle. Currently, the analysis capabilities include gross sample weight, neutron counting, spontaneous fission neutron counting, gamma-ray spectrometry and fissile material detection by fissions induced with a neutron source. The minicomputer mounted in the vehicle is used for measurement control, data acquisition, and data analysis. Recent field experience with the system includes handling and measuring plutonium metal, plutonium oxide, and plutonium nitrate. A variety of fuel research materials has also been measured, including ^{233}U, ^{235}U, plutonium, and thorium in various matrices. The system has also been used to measure amounts of material received, stored, or shipped. Field measurements are now under way on a variety of fuel-cycle waste materials, such as low-enriched ^{235}U, high-enriched ^{235}U, and plutonium in heterogeneous matrices. During field use, a number of practical problems were encountered that are as important as the technical considerations in achieving results with the system. The question of calibration standards and attempts to operate without such standards are also discussed.

INTRODUCTION

In answer to the need for a real-time, nondestructive assay (NDA) system that can provide timely and accurate measurements in the field, a mobile verification and measurement system has been designed, developed, and evaluated for nuclear material safeguards applications. This system is designed to be used by either an independent agency for verifying prior measurements or by plant personnel for sample measurements when material is present at a number of locations and in a variety of forms. In a facility such as a research complex, it is not practical to have measurement equipment located wherever nuclear safeguards, radiation exposure and logistics considerations limit

---

* Work performed under US Department of Energy Contract No. EY-76-C-06-1830.

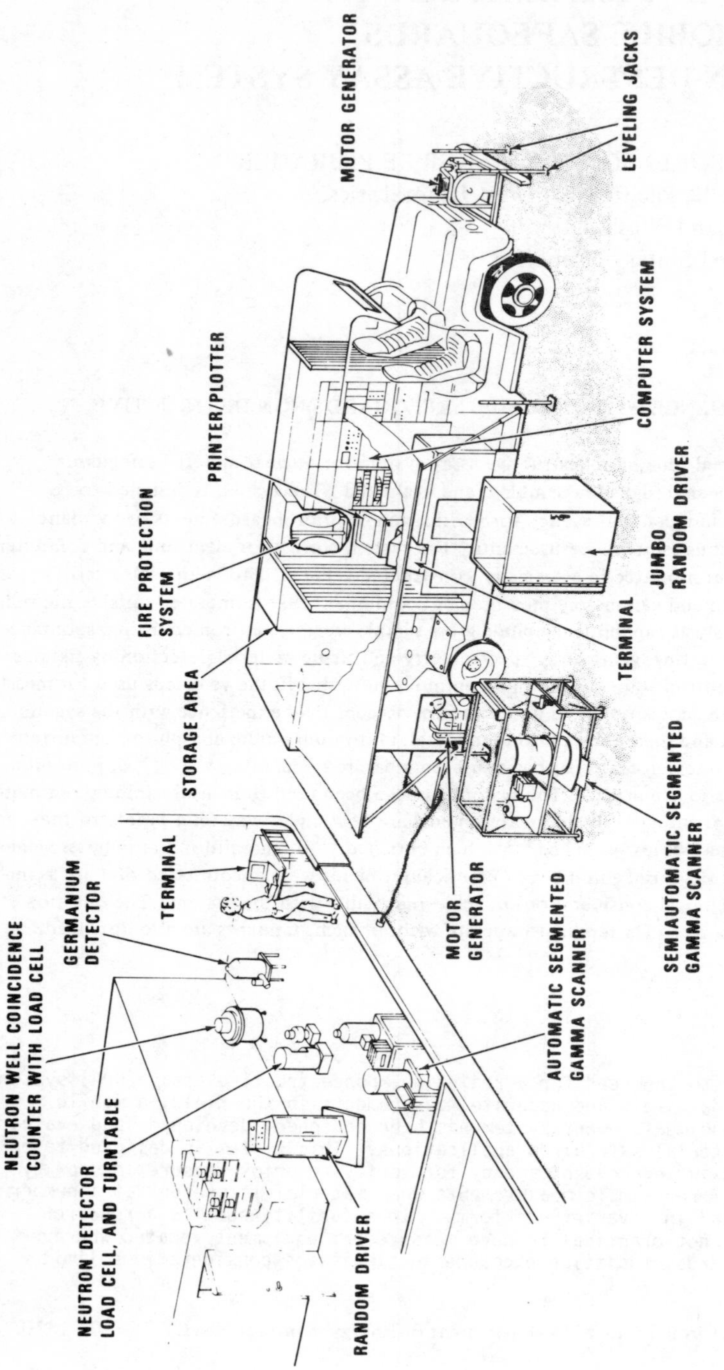

FIG.1. Schematic view of the mobile non-destructive assay verification and measurement system.

FIG.2. Mobile safeguards vehicle.

the ease with which nuclear material can be moved. Prior papers have described the need for this mobile NDA system and reported on early work in this field [1-5]

The system requirements were: mobility, ease of set-up, operational simplicity and flexibility, variety of measurement capabilities, and real-time control and data analysis. Earlier work demonstrated that the limitations of manually-operated NDA equipment reduced the effectiveness of inspection measurements. With the manually-operated equipment, only a few verification measurements could be performed in the field. This diminished the safeguards value or impact of the overall inspection effort. In addition, with the sophistication of newer instrumentation, automation of data handling becomes more important for control, acquisition, storage, display, processing and reporting of the data obtained during verifications.[1]

As mentioned above, the system described here has been designed to be used as a primary measurement system where material is present at a number of locations and in a variety of forms. In this mode of operation, high sample analysis rate, short assay time, and other inspector-related requirements are not needed. High precision, versatility, computer routines, and other primary measurement requirements are available with this same instrumentation.

SYSTEM DESCRIPTION

The mobile safeguards system, shown in Figure 1, consists of a specially constructed vehicle that contains electronic signal and data processing equipment and carries portable radiation measurement equipment. Before measurements are made, the necessary detection equipment and a minicomputer terminal are moved from the vehicle to the measurement station, which may be up to 70 m

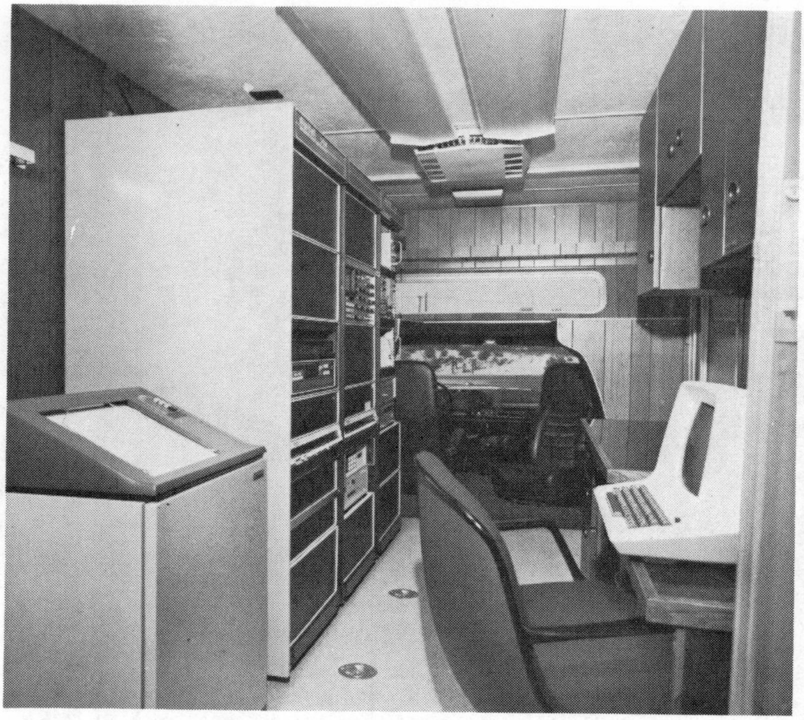

FIG.3. Electronic and computer equipment inside the mobile safeguards vehicle.

from the vehicle. The detection equipment is connected via instrument cables to the equipment mounted in the vehicle. The size and amount of equipment that must be moved from the vehicle to the measurement station has been minimized for each desired analysis.

The vehicle, pictured in Figures 2 and 3, was constructed on an extensively modified 2-ton truck chassis (8800 kg gross vehicle weight). An enclosure on the truck, designed to our specifications by a recreational vehicle manufacturer, houses most of the electronic instrumentation in a fixed position and carries the detection equipment along with the necessary cables. The enclosure has reinfored steel frames and is open to the cab of the truck. The interior can be separated by dust-proof, accordion type doors to form three compartments: 1) a cab area, 2) a fixed electronic instrumentation and office area, and 3) a storage area for equipment and detectors used outside the vehicle. Built-in equipment includes shock mounted instrument racks, storage cabinets, and a desk. The vehicle is equipped with both mechanical and hydraulic stabilizing and leveling jacks. Instrument power, air conditioning, heating, and lighting can be supplied from two motor generators mounted on the vehicle or from external 110-V or 220-V AC power. A hydraulically-operated platform outside the large rear access doors to the vehicle moves equipment to and from the ground.

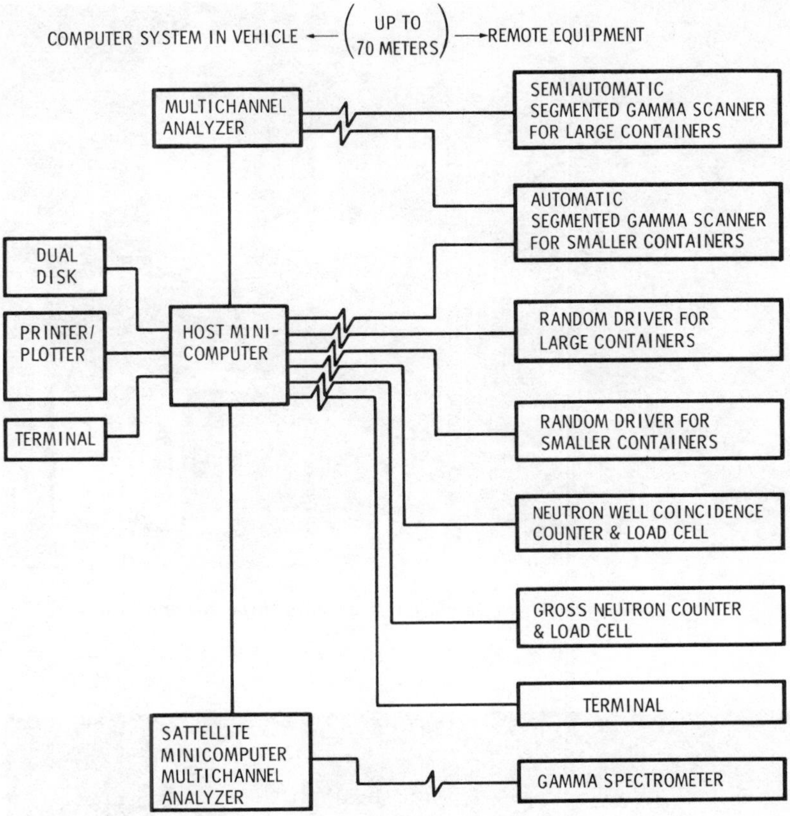

FIG.4. Block diagram of mobile non-destructive assay system.

NONDESTRUCTIVE ASSAY

All the equipment used in the mobile system is "state-of-the-art" and commercially available.[5] The equipment was modified at Pacific Northwest Laboratory to minimize the instrumentation to be moved from the vehicle to the measurement area and to provide for remote operation with up to 70 m of cable between the vehicle and the measurement area. Figure 4 is a block diagram of the NDA system. Various other configurations are possible to suit specific measurement needs.

The following is a brief description of each of the NDA systems. These systems have been described previously in greater detail.[5-7]

- Passive Neutron Measurements. Two passive neutron systems are included: a gross neutron counter with load cell and turntable, and a neutron well coincidence counter with load cell. These detectors are used for passive assay with total neutron and fission netron measurements. The load cells are used to automatically record gross sample weight simultaneously with the passive neutron measurement. Figure 5 shows the gross neutron counter load cell, and a gamma-ray spectrometric system in simultaneous operation.

FIG.5. *Non-destructive assay equipment configured for inspector application.*

FIG.6. *Segmented gamma-scan assay system.*

- Random Drivers. Two commercial random drivers were selected for active and passive assay of fissile materials. A standard-size random driver was selected for assay of samples in the 250-ml to 19-liter size range. A larger unit was selected for assay of material in containers with volumes of up to 200 liters. Turntables were mounted in the bottom of the random drivers to permit sample rotations during analysis. Matrix corrections may be derived from the thermal neutron monitoring ^{3}He tubes that were also installed in the detection chamber. The random drivers can be operated remotely under computer control.

- Gamma-Ray Detectors. Four intrinsic germanium (Ge) detectors were obtained for gamma-ray spectrometric measurements. The detectors are normally operated with the detector bias supply and amplifier close to the detector and with up to 70 m of signal cable between the detector and the analogue-to-digital convertor (ADC), which is located in the vehicle. One of the Ge detectors used for plutonium isotopic verification is included in Figure 5. An initial concern was the effect of having long signal cables between the high-resolution Ge detectors and the ADC, located in the vehicle. Measurements that were made with several different cable lengths have shown that only a slight resolution loss occurs when longer signal cables are used.[5-7]

- Gamma-Ray Spectrometric Systems. The vehicle contains two multichannel analyzers. One of the analyzers is a commercial 4096 channel hard-wired analyzer with cathode ray tube (CRT) display. Either the total spectrum or selected regions of the spectrum can be transferred to a PDP-11 host computer. The analyzer can be operated under manual or computer control. The second multichannel analyzer consists of a satellite PDP-11 computer with a 200-MHz, gain-stabilized ADC connected to the computer through a direct-memory-access (DMA) interface. The satellite and host computer are connected by an asynchronous serial line.

- Segmented Gamma Scan Assay System. Two of the Ge detectors and the hard-wired analyzer can be used as part of the segmented gamma scan assay system. A commercial system was acquired and modified to increase its portability so that it could be operated remotely on 70 m of cable with the analyzer and computer located at the measurement site as shown in Figure 6. Transmission sources are used for absorption corrections. A larger, portable semiautomatic, segmented gamma scan system has recently been designed and built for field applications requiring measurements on 200-liter waste drums and large filler boxes.

COMPUTER SYSTEM

A minicomputer system in the vehicle is used to control data acquisition, accumulate measurement results, perform real-time data analysis, store results, and prepare reports. The system consists of two PDP-11/05 minicomputers, two 1.2 million-word disk units, CRT terminals, a graphic CRT terminal, a hard-copy data terminal, a high-speed printer-plotter and interfaces for measurement equipment. Any of the terminals can be operated remotely with 70 m of cable. The computer and associated instrumentation is shown in Figure 3.

SOFTWARE DESCRIPTION

The operating software system is a RK-05 disk based RT-11 system using FORTRAN and BASIC. Software packages have been assembled for a variety of uses. These include specific programs for calibration and measurements with each detector system, programs for integrated verification and inspection

```
INSPECTOR NUMBER?(7)
INSPECTOR NUMBER    FACILITY NUMBER     VAULT LOCATION      INSPECTION MODE
      7.              CAN S-73          303C TUBE B-17        1002 GROSS

PROCEED WITH ANALYSIS OF        7. ? (YES)
PLACE SAMPLE CAN S-73           ON TURN TABLE
PAUSE -- WAITING FOR SAMPLE TYPE RETURN WHEN READY


MCA STARTED AT   14:57:48 FOR  100. SECONDS


ISOTOPIC ANALYSIS RESULTS FOR     6.  CAN 22-7-122            DATE 28-MAY-77
                FACILITY               NDA INSPECTION
          WEIGHT                     WEIGHT                  DIFFERENCE
         PER CENT     ERROR        PER CENT     ERROR         PER CENT
PU-238     0.000      0.000          0.000      0.000           0.0
PU-239    91.640      0.020         92.826      2.780          -1.3
PU-240     7.930      0.020          6.785      0.977          14.4
PU-241     0.384      0.005          0.389      0.012          -1.3
                  FACILITY ANALYSIS VERIFIED


NEUTRON COUNTING RESULTS FOR        7.  CAN S-73
                  FACILITY               INSPECTOR              DIFFERENCE
               VALUE   ERROR          VALUE   ERROR        ABSOLUTE   PERCENT
GROSS WEIGHT   1285.0   1.0           1285.0  15.0            0.0       0.0
PU WEIGHT        70.9   1.0             45.3   5.0          -25.6      36.1
                ***********************************************
                ****FACILITY AND NDA ANALYSIS DO NOT AGREE*****
                ***********************************************

INSPECTOR NUMBER?
```

FIG.7. Example of inspector verification report.

operation, and data storage and retrieval programs. Measurement programs include:

- gross neutron counter and load cell operations, data acquisition, and data analysis
- neutron well coincidence counter and load cell operation, data acquisition, and data analysis
- gamma-ray spectrometric analysis, including automatic peak search and peak area calculations
- automatic and semiautomatic segmented gamma-scan assay of low-density materials for uranium or plutonium
- plutonium isotopic analysis by gamma-ray or x-ray spectrometry
- automatic random driver assay of low-density, heterogenous matrix waste for uranium or plutonium.

An integrated plutonium verification and inspection program has been developed. The minicomputer system mounted in the vehicle is used for measurement control, data acquisition, and data analysis. In the inspection mode, information concerning containers to be verified can be entered into the minicomputer system before the measurements are taken by the inspection personnel. The computer then instructs the inspector via the terminal on the analysis operation required at the measurement site. The computer also 1) controls the weight, neutron, and gamma-ray measurement equipment, 2) acquires the measurement

TABLE I. MEASUREMENT CAPABILITIES OF THE MOBILE NON-DESTRUCTIVE ASSAY SYSTEM

Purpose	Equipment	Measurement
Pu verification or measurement	Random driver	$^{239,\,240}Pu$
	Gross neutron counter	^{240}Pu
	Neutron well coincidence counter	^{240}Pu
	Gamma spectrometer	$^{238,\,239,\,240,\,241}Pu$
	Load cells	Total weight
U verification or measurement	Random driver	$^{233,\,235,\,238}U$
	Neutron well coincidence counter	^{238}U
	Gamma spectrometer	$^{233,\,235,\,238}U$
	Load cells	Total weight
Th measurements	Gamma spectrometer	^{232}Th
Scrap measurements	Random driver	$^{233,\,235}U$, ^{239}Pu
Waste and low-density scrap measurements	Segmented gamma scan	^{235}U, ^{239}Pu, other
Material identification	Gamma spectrometer	Any gamma emitter

data, 3) reduces the data to concentrations and isotopic ratios, 4) compares the measurement results with the values previously stored in the computer, 5) notifies the inspector via the terminal of the results immediately after the measurement, 6) stores the results, and 7) prepares data reports.

The computer data analysis response times are essentially instantaneous. Figure 7 shows an example of inspection measurements observed on the CRT. Analyses can be performed with counting times of 50 to 200 seconds. Inspection responses are circled in Figure 7; all other information is supplied by the computer.

Data storage and retrieval programs have been developed to store both raw data and assay results on disk files for future reference and retrieval so that measurements can be compared with the past data. The measurement parameters associated with the system have been evaluated in order to reduce measurement time and the amount of equipment required to be moved from the vehicle to the measurement site. These parameters include the effect of counting time, the length of cables between different parts of the system, the selection of components, and the method of data analysis.

MEASUREMENT CAPABILITIES

The measurement capabilities available in the mobile system are suitable for a large variety of nuclear material safeguards situations. The NDA capabilities of the system include receipts, shipments, scrap, and waste measurement and are summarized in Table I.

Since different, independent methods can be used to measure some materials, data required for estimating systematic errors can be obtained. Thus both active and passive plutonium measurements can be compared.

The real-time response of the system simplifies error or discrepancy detection while the data storage features provide needed records for detailed evaluations. The computer equipment is sufficient for incorporation of a controlled materials information system.

FIELD EXPERIENCE

Field experience with the mobile system includes a variety of inventory verification and operator measurement activities at several locations. Since early 1977, the system has been used for performing inventory verifications for the various contractors at the U.S. Department of Energy's Hanford complex.

Calibration standards have not been available for the plutonium measurements. Using the material custodian's prior measurement data provides one method of alleviating the problem of no standards. To do this, randomly selected items from the same strata and with prior mass and isotopic characteristics are measured by one or more of the NDA methods available with the mobile system. These data are used as a temporary calibration reference curve; additional items are measured and compared to the temporary calibration points. As more items are measured, the calibration curve is periodically refined to include all of the data from the same strata. Any difference between the prior measurement and the verification measurement, based upon the best calibration available, is reported to the inspector by the computer system at the completion of each measurement. Various statistical tests can be used to determine whether the differences are significant and whether the verification measurement fits the current calibration curve. Thus, the inspector can use the information to question the material custodian regarding the prior measurement. An example of such a calibration curve plotted by the mobile NDA computer system is shown in Figure 8.

The system also has been used to measure the amounts of material received, stored, or shipped. The required precision for these types of determinations can be achieved by lengthening the measurement period. The measurement data are also saved on disk files by the computer system. If the verification measurements are repeated at a later date, the results can be compared with prior measurements.

Miscellaneous laboratory materials have also been measured. These items usually contain small quantities of nuclear material in a wide variety of containers and sample matrices. The exact concentration or composition of the material may not be well characterized. It is economically impractical to prepare standards for each small lot of material. The mobile system is also used to obtain reference measurements for the laboratory material.

Field experience measuring low-enriched ^{235}U, high-enriched ^{235}U, and plutonium in large waste drums in a heterogeneous matrix will be obtained during the summer and fall of 1978. This will provide an important evaluation of the mobil NDA system ability to measure uranium in low-level waste material.

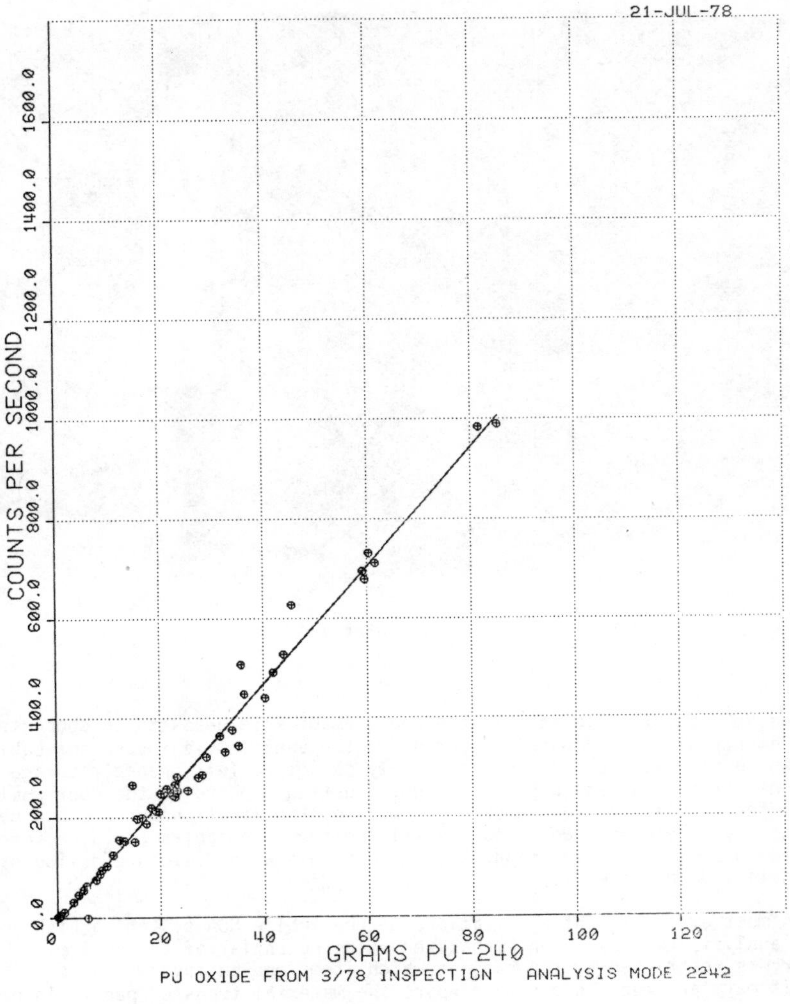

FIG.8. Calibration curve from a recent inspection.

PRACTICAL PROBLEMS

A number of nontechnical, practical problems were encountered during field use. A reliable supply of power was not always available for continuous operation of the NDA equipment. Two self-contained 5-kW motor generator units were added to the vehicle to alleviate this problem. They have provided a very reliable supply of power, but are limited in operating life by the amount of gasoline carried in the vehicle. The 200-liter gasoline tanks provide about 20 hours of motor generator operation. Use of the equipment for long periods of time is best performed where external power is available. Also, there is a trade-off between the option of bringing gasoline to the vehicle and disconnecting the NDA equipment and driving the vehicle to a source of gasoline.

FIG.9. Instrument cables.

The NDA equipment located at the measurement site needs to be connected to the electronic and computer equipment in the vehicle via instrument cables, as shown in Figures 1 and 9. Doors usually cannot be left open at storage locations for the cables to pass through. Posting a guard at the door would fulfill the regulatory requirement. At most of the locations, there has been adequate space between a door and its sill so that the cables could be forced under the door. In one location, a permanent hole with a closure device has been installed in the wall.

In most of the recent applications of the mobile NDA system, a high sample analysis rate has been emphasized. The potential of the system, however, is restricted by the rate at which items can be moved. In some cases it may take much longer to prepare the material transfer papers in order to move the nuclear material than it does to make the measurement.

The long signal cables between the high-resolution Ge detectors and the analog-to-digital converter located in the vehicle have not been a problem. No serious problem has been encountered with the computer or electronic equipment as a result of the vehicle design. Proper preparations to protect the disk heads are necessary before moving the vehicle. The computer system can normally be operational in less than an hour after the vehicle is moved.

ACKNOWLEDGEMENTS

The continuing support of A. C. Walker and H. E. Ransom of the Department of Energy, Richland Operations Office and H. L. Henry and C. O. Harvery of Battelle, Pacific Northwest Laboratories is greatefully acknowledged.
L. M. Browne was primarily responsible for the vehicle design and acquisition.

The mechanical and electronic design, acquisition and installation assistance supplied by J. H. Kaye, H. G. Rieck, Jr., and W. A. Mitzlaff was necessary for success of this project.

REFERENCES

[1] SORENSON, R. J. KAYE, J. H., "Automated Data Acquisition and Analysis System for Inventory Verification," BNWL-B-474, Battelle, Pacific Northwest Laboratories, Richland, Washington (1974).

[2] WALKER, A. C., STEWART, K. B., SCHNEIDER, R. A., SORENSON, R. J., "Verification of Plutonium Inventories," Safeguarding Nuclear Materials Vol. I, IAEA-SM-201/29, International Atomic Energy Agency, Vienna, pp 517-533 (1976) and BNWL-SA-5342, Battelle, Pacific Northwest Laboratories, Richland, Washington (1975).

[3] COHEN, I., GUNDERSEN, G., "Field Experience with the AEC Measurement Van," CONF 710617-3, Proceedings Institute of Nuclear Materials Management (1971).

[4] SCHNEIDER, R. A. SORENSON, R. J., "Safeguards Inspection Activities and Strategies Using the Hanford Complex as a Model," BNWL-1805, Battelle, Pacific Northwest Laboratories, Richland, Washington (1974).

[5] BRAUER, F. P., FAGER, J. E., KAYE, J. H., SORENSON, R. J., "A Mobile Nondestructive Assay Verification and Measurement System," Journal of the Institute of Nuclear Materials Management, VI II (1977) 680-868, Washington, DC

[6] BRAUER, F. P., FAGER, J. E., KAYE, J. H., SORENSON, R. J., "A Mobile Computerized Gamma-Ray Spectrometric Analysis and Data Processing System," PNL-SA-6571, Battelle, Pacific Northwest Laboratories, Richland, Washington (1978), and Trans. American Nuclear Society, 28 (Suppl. 1), May 1-4, 1978, p. 71.

[7] FAGER, J. E., BRAUER, F. P., "Rapid Nondestructive Plutonium Isotopic Analysis," PNL-SA-6601, Battelle, Pacific Northwest Laboratories, Richland, Washington (1978).

DISCUSSION

A.G. HAMLIN: Can you give us an idea of the capital and operating costs for the laboratory described and of the average number of verifications that can be handled per day?

R.J. SORENSON: The mobile system was put together over a period of time using existing equipment plus certain equipment which was specially purchased. However, I would estimate the capital cost to be in excess of US $250 000. We have not calculated operating costs, but the system can be operated by one or two persons.

The number of verification samples that can be measured per day depends on both the operator and the type of material. Assuming uniform sample types and an operator who will promptly move the items, we have measured samples at the rate of 5 minutes/sample for 6 hours a day, or about 75 samples per day. The instrument counting time, however, was only 30 to 60 s/sample.

NON-DESTRUCTIVE MEASUREMENTS OF URANIUM AND THORIUM CONCENTRATIONS AND QUANTITIES

T.N. DRAGNEV, B.P. DAMJANOV, J.S. KARAMANOVA
Institute of Nuclear Research and Nuclear Energy,
Bulgarian Academy of Sciences,
Sofia, Bulgaria

Abstract

NON-DESTRUCTIVE MEASUREMENTS OF URANIUM AND THORIUM CONCENTRATIONS AND QUANTITIES.

The passive X-ray fluorescent-gamma spectrometry method and technique for uranium concentration measurements was developed and tested. It is based on the measurement of the intensity ratios of self-excited K_α X-rays of uranium to the intensity of the combined peak with 92.8 keV average energy. The last peak has 92.367 and 92.792 keV gamma rays of ^{234}Th, representing the activities of ^{238}U and its daughter isotopes, and 93.35 keV Th K_α X-rays representing the activities of ^{235}U and its daughters. The results of the measurements do not depend on the size and the shape of the measurements. The procedure is developed to take automatically into account the presence of any absorber or cladding between the measured sample and the detector. The attainable precision of the measurements (at 95% confidence level) is 0.2 – 0.3%. If combined with enrichment measurements, and after suitable empirical calibration, the method can be used without standards. Gamma-spectrometric measurements of ^{238}U and ^{232}Th are based on the daughter isotopes' gamma activities. However, this is correct only when there is a corresponding equilibrium between ^{238}U and ^{232}Th and the daughter isotopes' activities. Where such equilibrium is not reached the status of the daughter products' activities regarding equilibrium, has to be taken into account. Two methods of quantitative corrections are proposed: (i) The use of an absolute determination of the ^{228}Ac/^{224}Ra activity ratio through self-calibrated measurements and individual activities and their correlation with the equilibrium activities. (ii) Use of two of the same sample measurements at two different moments during the unrestored equilibrium and the correlation of the measurement results with the ^{232}Th activity. This method can be generally applied.

1. PASSIVE X-RAY FLUORESCENT-GAMMA SPECTROMETRIC MEASUREMENTS OF URANIUM CONCENTRATION

Measurements of ^{235}U quantities in different U-containing items include the determination of both the ^{235}U enrichment (isotopic composition) and the U concentration.

There are publications, methods and techniques measuring the U concentration [1, 2] and the ^{235}U enrichment [3–5]. However, up to now there appear to be no passive methods of measuring U concentration.

The main drawbacks of existing active non-destructive methods for measuring U concentration are:

(1) It is necessary to have an additional rather intense radioactive source for the excitation of X-ray fluorescence radiation of uranium, which is difficult and inconvenient, particularly for inspectors travelling in different countries.

(2) All non-destructive X-ray fluorescent techniques are only applicable for measuring the so-called "bare" materials, i.e. materials unscreened by other materials. So even, from the safeguards aspect, important items such as clad fuel elements and assemblies are not measurable by these methods.

The main advantage of these methods is that they are fast. This paper describes a passive non-destructive method for U concentration measurements that has been tested.

The basic idea is to use the intensity of the self-excited (excited by the α-, β-, and γ-ray radiations of the measured sample's nuclei themselves) UK_αX-rays as a measure of the sample's U concentration. As is known, the intensity of the excited X-rays depends both on the intensity of the radiation that has excited them, and on the U concentration of the measured sample. The same should be correct for self-excited X-rays, to even a greater degree because the exciting radiation is more homogeneous. In this case the intensities of the inner radiations are correlated with the intensity of the unresolved ~92.8 keV peak, which includes 92.367 and 92.792 keV gamma-rays representing ^{238}U and its daughter activities and 93.35 keV $ThK_{\alpha 1}$, X-rays representing ^{235}U and its daughter activities. So it is proposed that the intensity ratio of 94.460-keV UK_{α_2} X-ray line to the unresolved 92.8-keV peak be used as a concentration measure. This ratio has the following important advantages:

(i) Energy-wise the lines are very close to each other, so the corrections for relative changes of the overall detecting efficiencies are not large;

(ii) The compound peak with an average energy of ~92.8 keV (including 92.367, 92.792 and 93.36 keV lines) represents all possible sources of inner exciting radiations.

To correct the different overall detecting efficiencies (ODE), the intensities of the two UK_α X-ray lines are used. A linear dependence of the overall relative efficiency from the energy in this narrow range is presumed.

Powder and pellet samples of different size, shape, concentrations and enrichments were measured many times. The results of some of the measurements are given in Table I. It was established that:

TABLE I. COMPARISON OF THE CHEMICAL AND X-RAY FLUORESCENT-GAMMA-SPECTROMETRIC DETERMINED CONCENTRATIONS OF SOME PELLET SAMPLES

Sample	Chemical analysis		Our method	
	C (%)	±ΔC	C (%)	±ΔC
A	87.97	0.05	87.97	0.13
B	87.84	0.08	88.01	0.13
C	87.54	0.06	87.45	0.08
D	87.78	0.08	87.51	0.16
E	87.80	0.09	87.69	0.08
F	87.74	0.08	87.80	0.11
G	84.65	0.09	84.68	0.13

(i) The results (ratios) do not depend on the size and the shape of the measured samples, which make the calibration very convenient;

(ii) The precision of the measurements depends mainly on the measurement time, and the attainable precision is in the range of 0.2–0.3% at 2 σ level;

(iii) The sensitivity of the concentration measurements – the relative changes of the intensity ratios (ΔR/R) to the relative changes of the corresponding concentrations (ΔC/C) – is quite high. For the measured range of concentrations it was roughly a 4% change of intensity ratio at a 1% change of concentration.

(iv) The intensity ratios depend on the enrichment of the measured sample. The relative changes of the intensity ratios (ΔR/R) to the corresponding relative enrichment changes (ΔEN/EN) are only about 0.3% changes of the intensity ratios at 1% change of the enrichment – Table II. So the eventual errors in the enrichment of the measured sample will not introduce significant errors in its concentration determination. However, to measure the concentration correctly through this intensity ratio one has to know or measure the enrichment of the measured sample. Probably the optimal manner of measurements, when the enrichment is known, are the combined measurements – enrichment as it is proposed in IAEA-SM-231/130[1] – and the concentration using the intensity ratios. Finally, if the correct calibration is made both determinations can be made without the use of any standards. This is important in many cases and particularly for IAEA Nuclear Materials Safeguard applications.

[1] DRAGNEV, T.N., DAMJANOV, B.P., these Proceedings, Vol.I.

TABLE II. RESULTS OF INTENSITY RATIO MEASUREMENTS FOR CONCENTRATION DETERMINATIONS OF SAMPLES WITH DIFFERENT ENRICHMENTS

Sample	Powders				Pellets		
Enrichment (%)	1.80	2.38	3.19	3.96	2.27	2.55	3.04
$R = \frac{I_{K_{\alpha 1}U}}{I_{92.8}}$	1.112	1.209	1.376	1.469	1.313	1.362	1.440
$(\pm)\Delta R$	0.002	0.002	0.003	0.004	0.003	0.002	0.003

(v) The studies of the particle-size effects on these measurements are not yet finished (suitable samples were not available), but the results on several samples with different particle sizes have demonstrated that, for the proposed passive method, these effects are weaker than those for active measurements [7].

1.2. Effects of different absorbers between the measured sample and the detector

As the attenuation factors for gamma- and X-rays with different energies in each absorber are different, the intensity ratios depend upon the material and the thickness of absorbers (e.g. cladding) between the measured sample and the detector. To investigate this dependence, special measurements were carried out. One of the samples was measured many times while different absorbers were inserted between the sample and the detector. Some of the measurements results are given in Table III. Two intensity ratios were determined for each absorber. The intensity ratio of the compound U peak with an average energy of about 92.8 keV and a 94.66 keV peak of $UK_{\alpha 2}$ line (a concentration measure) – R_c and corresponding intensity ratios of two UK_α lines (R_α – 98.441 keV, $K_{\alpha 1}$ line and 94.660 keV, $K_{\alpha 2}$ line. Using the least-squares method the empirical dependance between the absorber-correcting factor (F) and the K_α intensity ratios, R, was established. For the absorber-correcting factor (F) the ratios of R_{c_i} with different absorbers to the R_{c_0} of a "bare" (unscreened) sample were used. Using this dependance, the correcting factor F can be determined each time from the measured R_α ratio. Then the measured concentration ratio $(R_c)_m$ is multiplied by factor F, and the corrected $(R_c)_c$ is obtained and used to determine the U concentration. As the values of the energies of the two peaks used for concentration measurements were very near each other, this procedure worked very well, as can be seen from Table III. So, using this procedure, it is not necessary to know the

TABLE III. RESULTS OF THE CONCENTRATION MEASUREMENTS OF A SAMPLE WHILE DIFFERENT ABSORBERS ARE INSERTED BETWEEN THE DETECTOR AND THE SAMPLE

Thickness × 0.65 mm	$\frac{I_K}{I_K}$	$R_c = \frac{I_K}{I_{92.8}}$	$\frac{R_{c_0}}{R_{c_i}}$	$\left(\frac{R_{c_0}}{R_{c_i}}\right)_{(calc.)}$	$R_c \cdot \left(\frac{R_{c_0}}{R_{c_i}}\right)_{(calc.)}$
0	0.597	0.872	1.000	1.005	0.872
1	0.581	0.878	0.993	0.978	0.859
2	0.573	0.912	0.957	0.964	0.879
3	0.558	0.924	0.944	0.944	0.872
4	0.550	0.936	0.932	0.934	0.875
6	0.521	0.960	0.909	0.906	0.870
7	0.517	0.967	0.902	0.903	0.873
					Mean 0.871

Stand. dev. ±0.006(0.71%).
Stand. error ±0.002(0.27%).

cladding material or its thickness. The procedure automatically takes into account the absorber influence.

A method based on similar ideas is being developed for Th concentration measurements.

2. GAMMA-SPECTROMETRIC METHODS FOR MEASURING ^{238}U and ^{232}Th QUANTITIES

Gamma-spectrometric measurements of the activity of special nuclear material (SNM) is a very convenient method for determining their isotopic concentration quantities and their isotopic composition.

Unfortunately, however, some SNM isotopes do not emit gamma rays suitable for non-destructive measurements. This is true even for such widespread isotopes as ^{238}U and ^{232}Th, which hardly emit gamma rays. In such cases the activity of their daughter products can be used when there is equilibrium between the activity of the measured isotopes and its daughter product. This cannot be done directly when the equilibrium between these activities is broken. It is known, for example, that chemical treatments of U and Th connected with their purifications and enrichment disturb the radioactive equilibrium between ^{238}U and ^{232}Th and their daughter products. The equilibrium for ^{238}U-$^{234}Pa^m$ will be restored

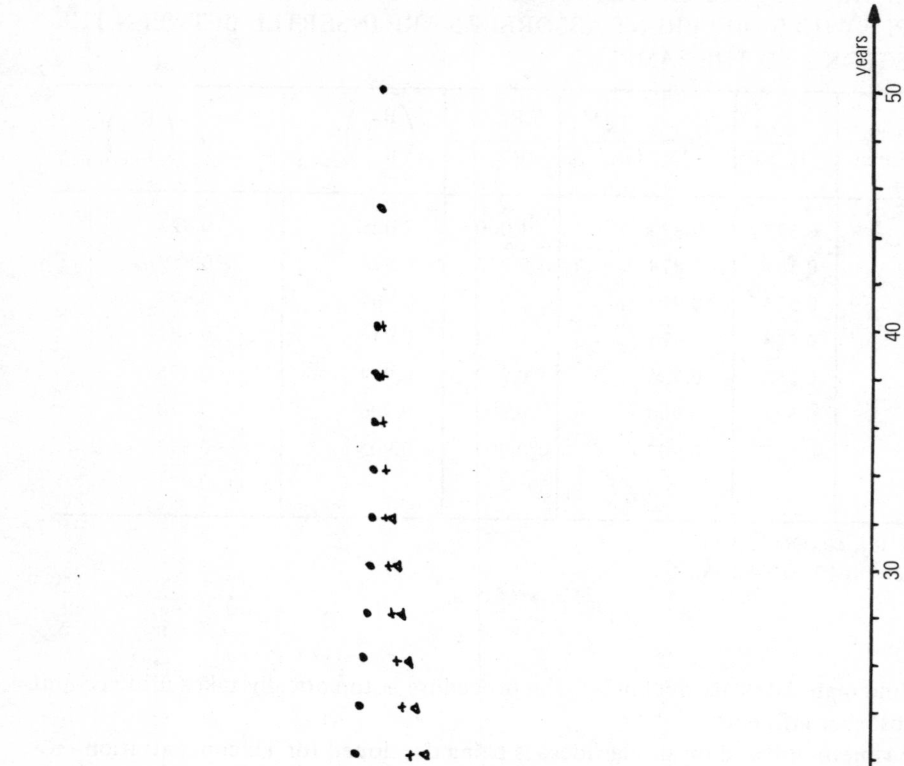

FIG.1. ^{228}Ac and ^{228}Th activity and activity ratios in the time after removing ^{228}Ra from a Th sample.

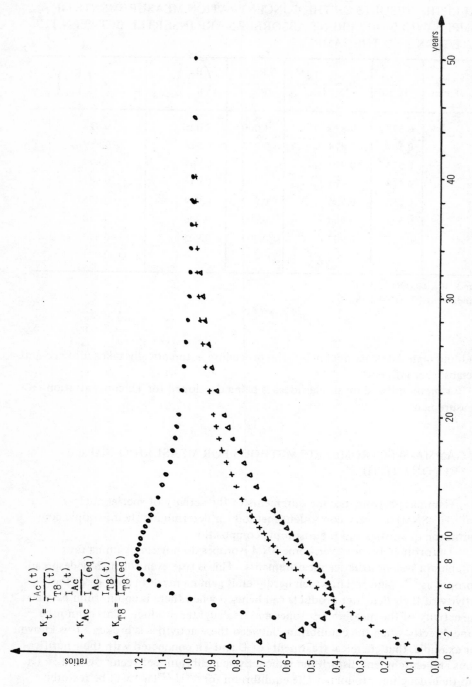

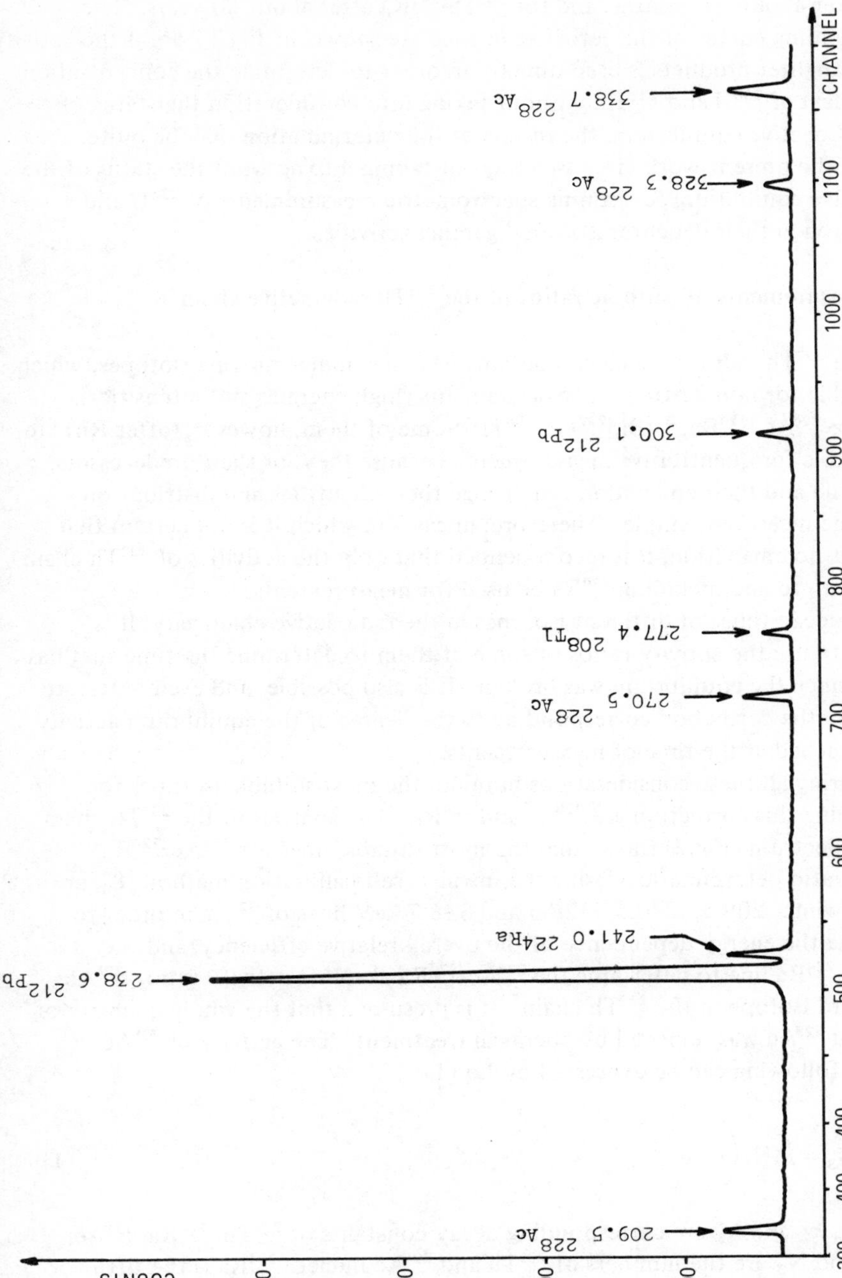

FIG.2. ^{232}Th radioactive chain gamma spectrum.

again after about six months, and for ^{232}Th-^{241}Ra after about 60 years. The corresponding curves of the activities in time are shown in Fig.1. So, if the activity of the daughter products is used directly in order to determine the concentration (quantities) of ^{238}U and ^{232}Th, without taking into consideration the status of their radioactive equilibrium, the results of the determination may be quite wrong. The present work gives two ways of taking into account the status of the radioactive equilibrium for gamma-spectrometric measurements of ^{238}U and ^{232}Th through their daughter isotopes' gamma activities.

2.1. Measurements of isotopic ratios of the ^{232}Th radioactive chain

The ^{232}Th radioactive chain contains several gamma-emitting isotopes, which are suitable for non-destructive measurements (high energies and intensities). These are ^{228}Ac, ^{224}Ra, ^{212}Bi, ^{212}Pb, ^{208}Tl. Some of them, however, (after Rn) are not suitable for quantitative measurements because they, or their predecessors, are volatile and their emanation can change their quantities and distribution within the measured sample. Therefore, in cases in which it is not certain that there was no emanation, it is recommended that only the activities of ^{232}Th chain isotopes up to and including ^{224}Ra be used for measurements.

As decay times of different isotopes in the radioactive chain vary, it is possible to use the activity ratios of some of them to determine the time that has elapsed since the equilibrium was broken. It is also possible, and even better, to determine the correction corresponding to the degree of the equilibrium activity ratios, reached at the time of measurements.

Bearing all these considerations in mind, the most suitable isotopes for determining this correction are ^{228}Ac and ^{224}Ra. The analysis of the ^{232}Th chain gamma spectrum (Fig.2) shows that the most suitable lines for $^{228}Ac/^{224}Ra$ isotopic ratio determination, using the intrinsic self-calibration method [8], are the following: 209.5, 270.5, 328.3 and 338.7 keV lines of ^{228}Ac in order to determine the energy dependence of the overall relative efficiency, and the 241-keV ^{224}Ra line to determine the $^{228}Ac/^{224}Ra$ absolute activity ratio. ^{228}Ac is the third isotope in the ^{232}Th chain. It is presumed that the whole quantity of ^{228}Ra and ^{228}Ac was removed by chemical treatment. The activity of ^{228}Ac in the time following can be expressed by Eq.(1):

$$\lambda_3 N_3 = \lambda_1 N_1 \left(1 - e^{\lambda_2 t}\right) \tag{1}$$

where λ_1, λ_2 and λ_3 are corresponding decay constants of ^{232}Th, ^{228}Ra, ^{228}Ac; and N_1 and N_3 are the numbers of ^{232}Th and ^{228}Ac nuclei. ^{224}Ra is the fifth isotope in the ^{232}Th chain. It comes immediately after ^{228}Th. As ^{228}Th chemical behaviour is the same as that of ^{232}Th, there will be no change in their isotopic ratio by chemical treatment. However, after treatments which change the isotopic abundances of its predecessors, there will be a change of ^{224}Ra abundance and also activity, because the equilibrium between the rates of its formation and decay will be disturbed. The ^{224}Ra decay time is much shorter than that of ^{228}Th and, as a result, there will soon be a dynamic equilibrium between them. The corresponding equation is:

$$N_4 = \frac{\lambda_5}{\lambda_4 - \lambda_5} N_5 \qquad (2)$$

or with sufficient accuracy it can be written:

$$\lambda_5 N_5 = \lambda_1 N_1 \left[1 + \frac{\lambda_4}{\lambda_4 - \lambda_2}\left(e^{-\lambda_4 t} - e^{-\lambda_2 t}\right)\right] \qquad (3)$$

where λ_2, λ_4 and λ_5 are corresponding decay constants of ^{228}Ra, ^{228}Th and ^{224}Ra; and N_1 and N_5 are numbers of ^{232}Th and ^{224}Ra nuclei. So, the ratio of these two activities can be expressed by Eq.(4):

$$\frac{\lambda_3 N_3}{\lambda_5 N_5} = \frac{1 - e^{-\lambda_2 t}}{1 + \frac{\lambda_4}{\lambda_4 - \lambda_2}\left(e^{-\lambda_4 t} - e^{-\lambda_2 t}\right)} \qquad (4)$$

The relative changes of these two activities and their ratio in time are shown in Fig.1 and Table IV.

Using these activities and ratio it is possible at the time of the measurements to determine the ratio of ^{228}Ac/^{224}Ra to their equilibrium activity ratio, and finally the ^{232}Th activity and concentration or quantity in the given sample by passive gamma-spectrometric measurements.

It is important to indicate, however, that this method of determining the equilibrium activity of ^{232}Th daughter isotopes, which is equal to its own activity, is correct only once the equilibrium has been destroyed. If, in time intervals longer than several days, treatments leading to the removal of Ra and Ac occurred more than once, these considerations are no longer valid.

2.2. Method for determining equilibrium activity by two measurements

This method is a general one but, in order to have a specific example and because of the ^{238}U importance, determination of its activity by gamma-spectrometric measurements is considered here.

TABLE IV. ^{288}Ac AND ^{228}Th ACTIVITY AND ACTIVITY RATIOS IN THE TIME AFTER REMOVING ^{228}Ra FROM A THORIUM SAMPLE

t(a)	K_{Ac}/K_{Th}	K_{Ac}	K_{Th}	t(a)	K_{Ac}/K_{Th}	K_{Ac}	K_{Th}
0.5	0.0697	0.0585	0.839	13	1.129	0.791	0.701
1.0	0.159	0.114	0.715	14	1.113	0.815	0.732
1.5	0.267	0.165	0.620	15	1.099	0.836	0.761
2.0	0.391	0.214	0.549	16	1.086	0.855	0.787
2.5	0.523	0.260	0.497	17	1.075	0.871	0.810
3.0	0.657	0.303	0.462	18	1.066	0.886	0.831
3.5	0.784	0.344	0.439	19	1.058	0.899	0.850
4.0	0.897	0.383	0.426	20	1.050	0.910	0.867
4.5	0.991	0.419	0.422	22	1.039	0.929	0.895
5.0	1.066	0.453	0.425	24	1.030	0.945	0.917
5.5	1.122	0.485	0.432	26	1.023	0.956	0.935
6.0	1.161	0.515	0.443	28	1.018	0.966	0.949
6.5	1.187	0.543	0.458	30	1.014	0.973	0.960
7.0	1.202	0.570	0.474	32	1.011	0.979	0.968
7.5	1.209	0.595	0.492	34	1.008	0.983	0.975
8.0	1.210	0.619	0.511	36	1.007	0.987	0.980
8.5	1.207	0.641	0.531	38	1.005	0.990	0.985
9.0	1.201	0.662	0.551	40	1.004	0.992	0.988
9.5	1.194	0.682	0.571	45	1.002	0.996	0.993
10.0	1.185	0.700	0.591	50	1.001	0.998	0.996
10.5	1.176	0.718	0.611	55	1.001	0.999	0.998
11.0	1.116	0.734	0.630	60	1.000	0.999	0.999
11.5	1.156	0.750	0.649	70	1.000	1.000	1.000
12.0	1.147	0.765	0.667	65	1.000	1.000	0.999

The instrumental gamma spectrum of U is shown in Fig.3. From the viewpoint of the suitability of emitted gamma rays (their energies and intensities), the place of different isotopes in the ^{238}U chain, and their decay periods, $^{234}Pa^{m}$ activity is the most suitable for ^{238}U measurements.

With sufficient accuracy the $^{234}Pa^{m}$ decay rate, after destroying the equilibrium (after extracting ^{234}Th and/or Pa from the measured material), can be expressed by Eq.(5):

$$\frac{dN_3(t)}{dt} = \lambda_3 N_3(t) = \lambda_1 N_1 \left(1 - e^{-\lambda_2 t}\right) \quad (5)$$

where λ_1, λ_2 and λ_3 are the corresponding decay constants of ^{238}U, ^{234}Th and $^{234}Pa^m$; and N_1 and N_3 are the numbers of ^{238}U and $^{234}Pa^m$ nuclei at the time of measurement. As the decay time ($T_{1/2}$) of ^{238}U is equal to 4.51×10^9 years its number of nuclei, N_1, can be considered as constant. The difference in the decay rates of $^{234}Pa^m$ at two different times (t_1 and $t_2 = t_1 + \Delta t$) can be determined as

$$\Delta I_3 = I_3(t_2) - I_3(t_1) = \lambda_1 N_1 e^{-\lambda_2 t_1} \left(1 - e^{-\lambda_2 \Delta t}\right) \quad (6)$$

or, in respect to the decay rate, at the first measurements

$$\frac{\Delta I_3}{I_3(t_1)} = \frac{1 - e^{-\lambda_2 \Delta t}}{e^{\lambda_2 t_1} - 1} \quad (7)$$

From Eq.(7) it is easy to determine the time interval, t_1, elapsed between the moment when the radioactive equilibrium was destroyed and the moment of the first measurements.

$$t_1 = \frac{1}{\lambda_2} \ln \left[\frac{1 - e^{-\lambda_2 \Delta t}}{\frac{\Delta I_3}{I_3}} + 1 \right] \quad (8)$$

Equation (8) is used to find Eq.(9) for the equilibrium activity of $^{234}Pa^m$:

$$I_3(t = \infty) = \lambda_1 N_1 = \frac{I_3(t_2) - I_3(t_1)\, e^{-\lambda_2 \Delta t}}{1 - e^{-\lambda_2 \Delta t}} \quad (9)$$

So, using Eq.(9) and knowing the decay rates of $^{234}Pa^m$ at two essentially different moments (when the number of $^{234}Pa^m$ nuclei differ by 2%) during the time of unrestored equilibrium, it is possible to determine its equilibrium activity and finally the ^{238}U activity.

The time dependences of

$$\frac{\Delta I_3}{I_3(t_1)}$$

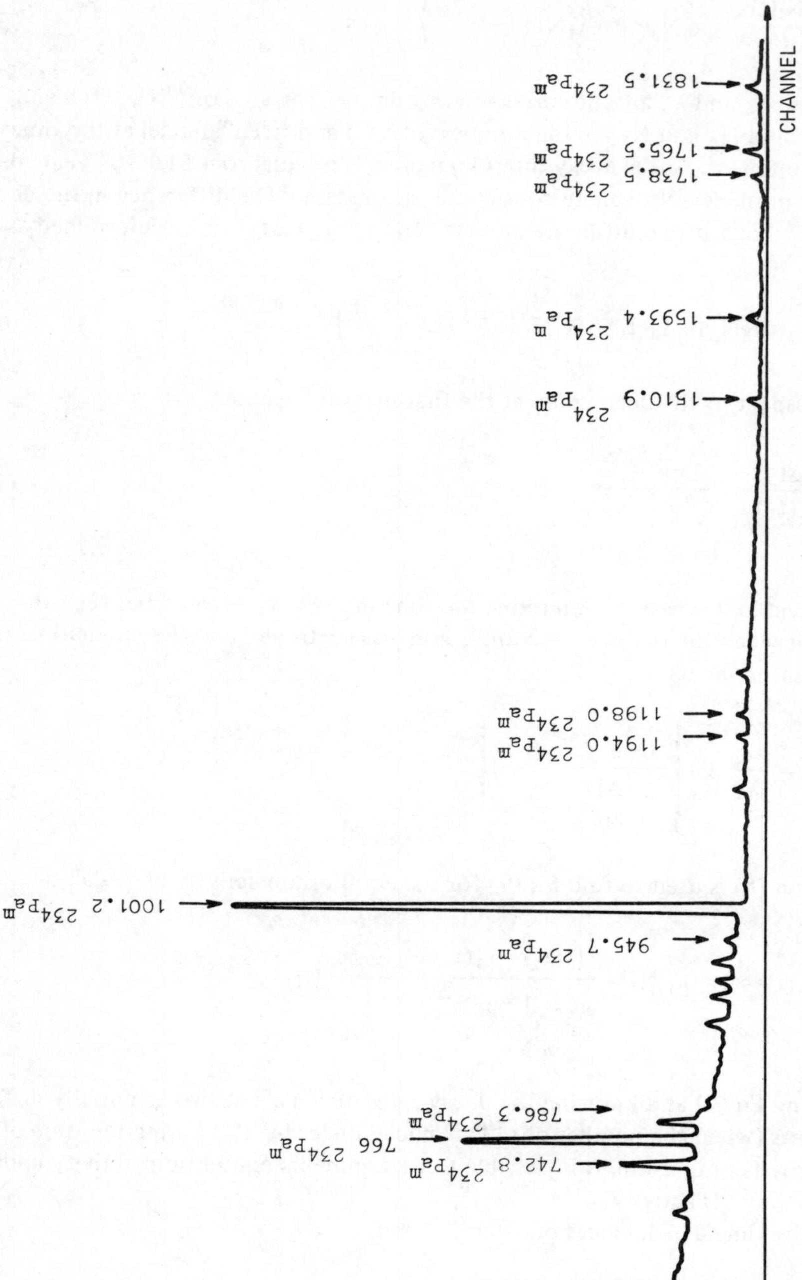

FIG.3. ^{238}U radioactive chain gamma spectrum.

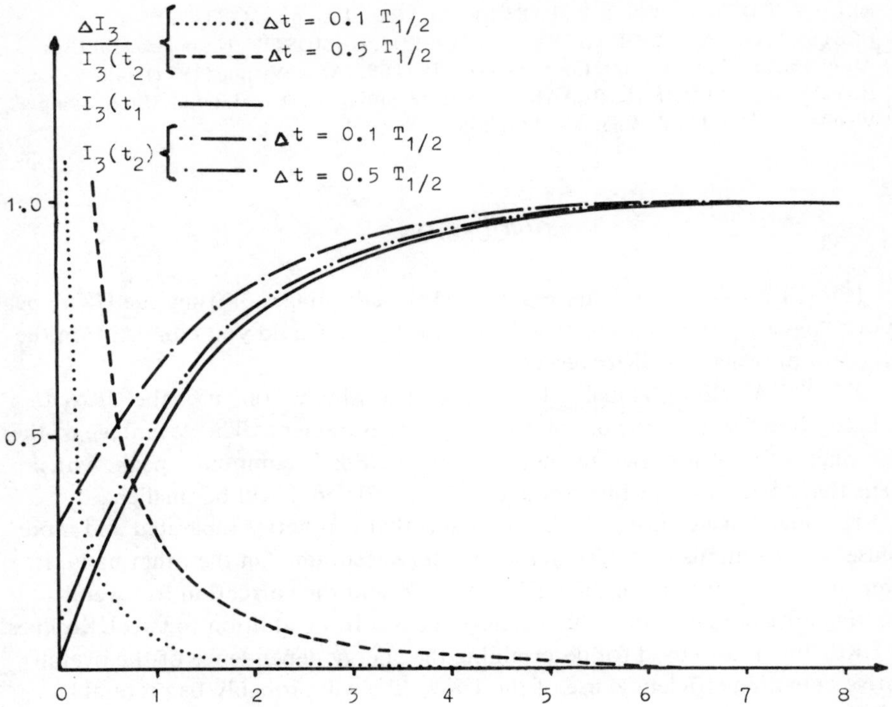

FIG.4. The time dependence of $\frac{\Delta I_3}{I_3}$ *and* $I_3(t_2)$ *for two different* Δt *intervals.*

and $I_3(t_2)$ are shown in Fig.4. The time is expressed in decay period units. From the data shown in Fig.4 it was easy to conclude that, to determine the equilibrium activity correctly during the whole period of its change, the time interval between two measurements should be at least 0.5 $T_{1/2}$. If this time interval is shorter, e.g. 0.1$T_{1/2}$, then it will be very difficult and even impossible to determine the equilibrium activity precisely during the time intervals for t_1 (3–6) $T_{1/2}$ – 40–70 days. Outside this time interval there will be no problems in determining precisely the equilibrium activity. This peculiarity should be considered for ^{238}U gamma measurement when the equilibrium of ^{234}Pa^{m} activity is not restored.

REFERENCES

[1] ZUMWALT, L.R., US AEC-Rep. No. AECU-567 (1949).
[2] BACHVAROV, N.S., et al., in Safeguarding Nuclear Materials (Proc. Symp. Vienna, 1975) 2, IAEA, Vienna (1976) 347.

[3] REILLY, T.D., WALTON, R.B., PARKER, J.L., Rep. LA-4605-MS (1970).
[4] BEETS, C., GOENS, J., DRAGNEV, T., GOOSENS, H., MOSTIN, N., in Peaceful Uses Atom. Energy (Proc. UN Int. Conf., Geneva, 1971) **9**, IAEA, Vienna (1972) 449.
[5] HARRY, R., AALDIJK, J., BRAAK, J., in Safeguarding Nuclear Materials (Proc. Symp. Vienna, 1975) **2**, IAEA, Vienna (1976) 235.

DISCUSSION

F.V. FRAZZOLI: You have indicated the advantages of using the UKα_2 peak in your "passive X-ray fluorescence" measurements. Could you comment on the possibility of using the UKα_1 peak?

T.N. DRAGNEV: It makes little difference whether one uses the UKα_1 or the UKα_2 line for this type of concentration measurement. The advantage of the UKα_2 line is that it is closer, in energy, to the 92.8-keV compound peak, which means that correction for the overall detecting efficiency will be smaller.

The advantage of using the UKα_1 line is that it is better separated and more intense, and its intensity determination is more accurate. On the other hand, it differs more in energy from the 92.8-keV peak, and the correction for overall detecting efficiency will introduce a bigger error. If, in addition to two UKα lines, the UKβ_1 line is also used for determining the energy dependence of the overall relative detecting efficiency, use of the UKα_1 line will probably be preferable.

INVENTORY VERIFICATION USING HIGH-RESOLUTION GAMMA SPECTROSCOPY (HRG)

W.C. BARTELS
Department of Energy,
Washington, DC,
United State of America

Abstract

INVENTORY VERIFICATION USING HIGH-RESOLUTION GAMMA SPECTROSCOPY (HRG).

Substantial improvement has been accomplished in high-resolution gamma spectroscopy (HRG) since 1970. This paper reviews the present status of inventory verification in the field of international safeguards. IAEA Consultant and Advisory Groups have considered the usefulness of many non-destructive assay techniques and recommended HRG for certain applications. These include the plutonium content and isotopic composition (except ^{242}Pu) of small samples; and plutonium isotopic composition of larger containers of bulk plutonium. HRG should be used for uranium enrichment measurement where there is a possibility of radiation interference from extraneous gammas. Another recommended use is in the verification of spent fuel. Improvements have been made in the capability at IAEA Headquarters for the analysis of HRG data from inspections. In addition, detection equipment and pulse-height analysers have been improved, and transportable equipment is being used for IAEA and Member State activities. The IAEA is increasing its capability for the calibration of HRG by the use of physical standards. The IAEA staff is receiving training in HRG applications to nuclear material accountability. There have been substantial progress and increasing use in HRG for nuclear material accounting and verification of inventories.

INTRODUCTION

Non-destructive assay of nuclear materials is being used for verification by the IAEA, for nuclear materials accounting by States, and for controls exercised by facility operators. The development of similar measurement techniques for these different applications has been mutually helpful. Results are evident in progress with high-resolution gamma spectroscopy (HRG) since the 1970 IAEA symposium on Safeguards Techniques [1].

High-resolution gamma spectroscopy has been used successfully for inventory verification by the IAEA because it enables the inspector to make determinations using a physical characteristic of the nuclear material itself, using equipment which the inspector demonstrates to be trustworthy, and because the inspector obtains results which are accurate and timely enough to satisfy the goals of safeguards.

This paper reviews the uses now considered appropriate for HRG in verifying inventories, as well as the difficulties encountered and progress made in implementing such uses of HRG.

RECOMMENDED USES

A 1978 IAEA Advisory Group reviewed the usefulness of many non-destructive assay techniques and recommended HRG for measuring plutonium content and isotopic composition of small samples, excluding ^{242}Pu [2]. Determination of isotopic composition by HRG [3], coupled with calorimetric measurement can provide for the timely verification of plutonium content [4]. For assay of containers with large quantities of plutonium, the Advisory Group recommended [5] a measurement procedure with two steps – one the HRG measurement to determine the plutonium relative isotopic composition, including ^{240}Pu; and the second, a neutron coincidence count to determine the "effective ^{240}Pu content" of the entire container. The "effective ^{240}Pu content" is the content of ^{240}Pu which would give the same response as that obtained by counting the spontaneous fission neutrons from the $^{238, 240, 242}$Pu in the sample. The Advisory Group acknowledged that uncertainties are expected to be in the range from 10% at present, to 5% in the future.

For uranium enrichment measurements by the so-called "enrichment meter" method, an Advisory Group recommended that HRG be used where there is a possibility of radiation interference from extraneous gammas [6], such as from radioactive decay products of uranium sometimes found to accumulate on the inner surface of shipping or storage containers for uranium hexafluoride. Non-volatile decay products accumulate in such containers if the containers are emptied by vaporizing rather than by draining the contents. However, where an inspector can be sure that interfering gamma rays do not introduce significant uncertainties, enrichment measurement by low-resolution gamma spectroscopy is adequate and more easily made.

The verification of spent fuel by HRG has been recommended [7, 8] and should be particularly helpful to the IAEA when a "starting inventory" is to be established for spent fuel coming under safeguards. Selected fission-product ratio measurements permit quantitative determination of fuel assembly burnup. Development and demonstration efforts were recommended. In my opinion, quantitative determination of fuel assembly burnup can and will reach accuracies of five per cent in the future. Information on burnup permits calculation of plutonium content by application of reactor physics principles. Information on burnup also helps to verify records of the reactor operating history. If the irradiation history of the fuel is postulated from the plant records then fission-product ratios can be used to verify cooling time.

Measurement of thorium-uranium fuel cycle materials must deal with gamma radiation which is more intense, more penetrating, and different from the radiation from uranium-plutonium fuel-cycle materials. Experience has been gained with HRG of fresh thorium-uranium fuel materials. Because of radiation, commercial-scale fuel recycling is expected to be carried out by remote control in the future [9]. I believe that bulk measurements of weight or volume will then be remote and tamper-indicating, but samples will be removed from remote work areas for assay, including HRG measurements.

DIFFICULTIES

Inspectors using HRG for the applications described above, have met difficulties. Only a limited number of key measurement points have HRG equipment. Therefore, inspectors must often bring portable or transportable detection and data-handling equipment to measurement points. Such equipment is cumbersome to transport and may have limitations where it has been designed for portability. As noted by an Advisory Group [10], by far the most significant need is for improvement of data analysis capability. A laboratory-based capability is needed, including software and access to adequate computer capacity. There is also a need to provide inspectors with at least a limited capability to determine in the field whether or not an adequate measurement has been made and an item is generally as it has been declared. Determinations in the field can be accomplished by providing additional computing power and a graphic display capability in the field. There is also a need to increase the number of channels available for spectrum analysis in the field. IAEA inspectors are using the portable pulse-height analyser provided by Silena of Italy with 1000 channels; and Silena is delivering improved units with 2000 channels to the IAEA. However, as noted by an Advisory Group [11], 4000 channels will be required for more accurate isotopic determinations involving spectral analysis of poorly resolved multiple peaks of various plutonium isotopes.

Determination by HRG of burnup, plutonium content, or cooling time of spent fuel is more difficult than identification of fuel assemblies by readily observed attributes and fuel-assembly serial numbers. Therefore, verification by using HRG to determine fission-product ratios is used only where required.

RECENT IMPROVEMENTS

There have been improvements in the technological status of HRG, starting with the measurement of plutonium content and isotopic composition of small samples. The precision of HRG has been demonstrated in an at-line system used at Los Alamos Scientific Laboratory for assay of ^{239}Pu in solutions from leached ash. With half-hour count times, a standard deviation of one per cent was obtained at a concentration of one-tenth gram per litre, improving to 0.50% at 5 g/litre. By comparison, isotopic dilution mass spectrometry had a standard deviation of 0.50%.

The measurement precision of the relative content of $^{238,239,240,241}Pu$ depends on the nature of the sample. For a sample of solution from recently separated reactor-grade plutonium, a standard deviation of less than 0.50% was obtained at Lawrence Livermore Laboratory for each of these isotopes with counting times of 10–20 min [3].

After the improvement of HRG data analysis capability was recognized as a most significant need, IAEA arranged for the assistance of four technical experts from an experienced group at Idaho National Engineering Laboratory (INEL), who have been working in Vienna and at Idaho. They have provided the IAEA Department of Safeguards with specially adapted computer programs for spectral analysis of inspector HRG data. The programs have been documented with written instructions concerning their use, and IAEA staff have been trained in the use of the programs. The IAEA had previously recognized the need for analysis of HRG data and had purchased a Nuclear Data laboratory computer (an ND-6620) for this use at IAEA headquarters in Vienna. The experts from INEL have helped to bring this computer into service. Full information on the capability of the ND-6620 and available software was obtained from the vendor. This capability was demonstrated to the IAEA staff in their analysis of a backlog of data, including spent-fuel assay data.

Specialized programs were written for use of the ND-6620 in the analysis of spectral data from uranium enrichment measurements, from measurements to determine the isotopic composition of plutonium, and from measurements for the assay of spent fuel, as well as the analysis of statistical means and variances. A new program was developed and is now available for inspectors who return to Headquarters after using the Silena multi-channel analyser to record their data on a cassette in the field. With this program they can transfer their data from the cassette of the Silena to the disk and display equipment of the ND-6620. This operation formerly required up to seven steps, but with the new program the operation is reduced to one step, which reduces opportunities for operator error. Another program in FORTRAN provides for the automatic calculation from spectral data of the enrichment of uranium unknowns, including the construction of a summary table of the results of a series of enrichment calculations. The printout permits final reports to be prepared by inspectors with minimal manipulation of numerical results. Spectra on magnetic tapes written by the ND-6620, have now been read into the main IAEA computer (an IBM 370), and additional programs now enable trained IAEA staff to use this more powerful computer when needed for more sophisticated spectral analysis.

As recommended [7, 8], the IAEA arranged for the development of improved methods for the rapid assay and verification of spent fuel assemblies. LASL is using HRG as the baseline against which the performance of other systems is compared. Experiments at commercial power reactor sites on irradiated PWR and BWR fuel have now confirmed the applicability of isotopic ratios for burnup determination.

As noted earlier, inspectors must often transport cumbersome equipment for HRG. To ease this burden, progress has been made in reducing the size and weight of detection equipment. IAEA has procured several intrinsic germanium detectors because they are more portable than lithium-drifted germanium detectors. Intrinsic germanium does not have to be cooled at all times to maintain the stability of the crystal. Therefore, intrinsic germanium can be transported and stored at room temperature. This makes it possible to avoid the presence of liquid nitrogen while transporting the detector, and to use a much smaller cryostat for liquid nitrogen to cool the detector during measurements. The IAEA now routinely uses intrinsic germanium detectors for HRG measurements. In addition, two small intrinsic germanium probes and cryostats have been obtained commercially and tested and soon will be delivered to IAEA inspectors. Each detector is mounted in a cryostat which weighs less than two kilograms, and has a cooling capacity for about six hours' use. Such portable detectors are used with portable multi-channel analysers which have become available for data retention and retrieval.

The IAEA has obtained more portable equipment for non-destructive assay of containers with larger quantities of plutonium. As noted earlier, the recommended assay technique is to determine isotopic content by HRG and "effective ^{240}Pu content", by neutron coincidence counting. As reported by Krick et al. [12], a portable, high-level neutron coincidence counter has been developed for use in IAEA inspections. The IAEA has received three such counters along with equipment manuals, and its staff members have learned to use the counters and have used them for inventory verification.

Recently IAEA introduced another application of HRG, a highly developed, transportable "segmented gamma scanner" from Los Alamos Scientific Laboratory (LASL). The system, tested by the IAEA at Casaccia, Italy, is capable of in-plant or transportable use for the assay of plutonium- or uranium-bearing scrap or waste in cans or drums which contain mostly material of low atomic number. The segmented gamma scanner is automated to make it simple to operate, and corrects for the gamma-ray attenuation within the sample. At Casaccia, the system has been applied to typical waste materials, and is able to measure the ^{239}Pu content of samples containing as little as 50 mg. The IAEA will train its staff to use the unit and is considering a possible second unit.

As reported by Sorenson [13], a mobile assay system has been mounted in a special vehicle and used since 1976 for performing inventory verification in the Department of Energy (DOE) facility at Hanford, Washington. Equipment includes intrinsic germanium detectors, multi-channel analysers, and a PDP-11 mini-computer system with a cathode-ray tube display unit.

Software provides for plutonium isotopic analysis by gamma spectroscopy and other capabilities. In addition to the DOE mobile system, other mobile systems equipped by Brookhaven National Laboratory (BNL) for HRG and other kinds of non-destructive assay are used by NRC in the northeastern United States of America [14]. IAEA has now purchased a vehicle and BNL is to equip it with mobile assay systems for IAEA's use. The vehicle is to be provided with racks for secure storage of assay and data-handling equipment. The design facilitates movement of portable equipment from the vehicle for use in facilities being inspected.

An inspector must ensure the accuracy and trustworthiness of assay equipment used in the field. When using portable or mobile equipment under IAEA control, the inspector has maximum freedom in selecting procedures to establish this assurance; when using in-plant equipment available at the inspected facility, the inspector does not have the same degree of freedom. It is expected that in-plant instrumentation will play a greater role in future inspection activities [15]. The inspector will depend most heavily for assurance of accuracy and trustworthiness on an independent calibration of the equipment using acceptable calibration standards.

In recognition of the need for such calibrations of transportable and in-plant instruments, the IAEA has arranged for technical experts from Science Applications, Incorporated, to work at IAEA Headquarters in Vienna and equip the IAEA with a standards program for assay of element and isotope. The experts are developing specifications and procedures for the use of physical standards in calibrations of HRG and neutron counting equipment.

Calibration for HRG measurement of plutonium should introduce minimal uncertainties due to differences, including matrix effects between samples and standards. Process materials, which have been assayed calorimetrically and isotopically, are now being used in DOE facilities as calibration standards with minimal differences between samples and standards, as will be discussed in the paper by Strohm [16]. By using this approach, DOE has significantly reduced inventory differences (or MUF) for plutonium with more than 90% ^{239}Pu. It is technically feasible to use the same approach to calibrate HRG measurements and reduce inventory differences for reactor-grade plutonium.

Inspectors who are to use HRG may need training in equipment and data-handling procedures. In 1977 and 1978, the IAEA arranged for such training in one-week courses at LASL. The training consisted of lectures and laboratory experience with quantitative assay of plutonium and uranium, spent-fuel assay and techniques for assay at enrichment plants. To supplement instruments available at LASL for inspector training, additional instruments normally used by IAEA inspectors have been obtained, including four of the portable pulse-height analysers made by Silena. IAEA also requested, and LASL provided, specialized training for individual members of the IAEA staff including, on occasion, specialized training in the assay of spent fuel by HRG.

CONCLUSION

HRG has long been used successfully by facility operators for nuclear material measurements for accounting purposes, and the possible usefulness of HRG for inventory verification by IAEA inspectors has long been recognized [1]. Until recently, the analysis of data has been too laborious, and available equipment and procedures have been too bulky and complex for HRG to be widely used by IAEA inspectors. These difficulties have been discussed by the IAEA and Member States in efforts to improve the usefulness of HRG. Progress has been substantial in the past few years and HRG has been used successfully by the IAEA. Further technical progress is expected and will permit broader use of this powerful and advantageous non-destructive assay technique in the future. However, it cannot replace other NDA techniques where they are adequate and easier to use.

REFERENCES

[1] HIGINBOTHAM, W.A., BARTELS, W.C., "Plant instrumentation and its relation to the design of safeguards material control systems", Safeguards Techniques (Proc. Symp. Karlsruhe, 1970), IAEA, Vienna (1970).

[2] INTERNATIONAL ATOMIC ENERGY AGENCY, Advisory Group Meeting on the Development of NDA Instrumentation and Techniques for IAEA Safeguards, AG-187, IAEA, Vienna, 22–26 May, 1978. Section II, para.1.1. Internal publication.

[3] GUNNICK, R., "Gamma spectrometric methods for measuring plutonium", Analytical Methods for Safeguards and Accountability Measurements of Special Nuclear Material (Proc. Am. Nucl. Soc. Natl. Top. Meeting Williamsburg, 1978), Am. Nucl. Soc. (1978).

[4] RODENBURG, W.W., An Evaluation of the Use of Calorimetry for Shipper-Receiver Measurements of Plutonium, US Nuclear Regulatory Commission – NUREG/CR-0014. Monsanto Research Corporation – MLM-2518 (1978).

[5] Ref.[2]. Section II, para. 2.

[6] Ref.[2]. Section II, paras 3 and 5.

[7] INTERNATIONAL ATOMIC ENERGY AGENCY, Consultants Meeting on Preparation of a Research Coordination Programme on Safeguards Instrumentation Installed at Irradiated Fuel Reprocessing Facilities, MG-89, Vienna, 10–14 May 1976, para. 3.1. Internal publication.

[8] INTERNATIONAL ATOMIC ENERGY AGENCY, Advisory Group Meeting on the Non-Destructive Analysis of Irradiated Power Reactor Fuel, AG-11, IAEA, Vienna, 4–7 April 1977. Sections 3 and 6. Internal publication.

[[9] CARPENTER, J.A. et al., IAEA-SM-231/65, these Proceedings, Vol.II,

[10] Ref.[2], Section IV.

[11] Ref.[2], Section II, para. 2.9.b.

[12] KRICK, M.S., et al., IAEA-SM-231/50, these Proceedings, Vol.II.

[13] SORENSON, R.J., et al., IAEA-SM-231/82, these Proceedings, Vol.II.

[14] ZUCKER, M., Brookhaven National Laboratory, private communication.

[15] Ref.[2], Section V, para. 3.

[16] STROHM, W.W., et al., IAEA-SM-231/83, these Proceedings, Vol.II.

DISCUSSION

A.G. HAMLIN: In the United Kingdom, we are also very interested in high-resolution gamma spectrometry, but I would like to point out that the 30-minute counting time which you mentioned corresponds to about 20 verifications a day. This capability in no way approaches the verification problem faced by the United Kingdom, let alone the Agency. What, then, is the rate-determining step? It is not count-rate; we have plenty of counts. However, as the complexity of the spectrum increases, greater resolution is required. Thus we have analysers with 2000, 4000, or more, channels. But the inspector is interested only in small areas of the spectrum, so the analyser spends most of its time sorting pulses into channels that are of no interest to the inspector. What is required is an analyser that rapidly rejects pulses outside the energy blocks of interest and then sorts the residues at high resolution. This might bring the count time down to a more realistic one minute. We have been unable to find any such system. Do you know of one, or of work directed to this end?

W.C. BARTELS: You have made a penetrating observation, and you are quite right. This inefficiency has troubled me, too. In fact, I discussed the inefficiency with scientists during a visit to Los Alamos a week or so ago. We do not know how to avoid the flow of unused information into the multichannel analyser along with the information which is needed, and we would welcome suggestions as to possible technical approaches.

T.N. DRAGNEV: I would like to make a comment in connection with Mr. Hamlin's question.

Long-time measurements are required in order to prepare secondary or working standards representing many items of the same shape, size, and so on. When such standards are available, the relative measurements are much easier and faster and this approach to the IAEA nuclear material safeguards measurements can be recommended.

However, I agree with Mr. Hamlin's point that it is desirable to have instruments which operate only with a given (important) part of the spectrum in order to reduce the loading of the instrument and speed up the measurements. In principle I think it is possible to develop such an instrument, at least for that part of the instrumentation which follows the preamplifier. For the spectrometric detector and the preamplifier it will be much more difficult, if it is possible at all.

W.C. BARTELS: It is interesting to note that my answer to Mr. Hamlin's question is based on technical considerations, while the answer provided by Mr. Dragnev is based on systems analysis considerations. The two answers are complementary.

R. MARTINC: As we saw from several papers presented in Session III,[1] there are no unique burnup/isotopic ratio correlations, not even for the fuel from a single reactor.

To improve the use of gamma spectrometry in this respect, the authors of those papers recommended "in-core" interventions, such as: more precise, space-dependent, neutron spectrum calculations; spatial neutron-density distribution measurements; and the taking into account of fuel irradiation history and in-core fuel management aspects. Such information is available to reactor operators, but not to inspectors concerned with independent "out-of-pile" verifications. In this connection two questions arise. First, to what extent do you think gamma spectrometry may be considered as a real "out-of-pile" experimental tool that can be used by inspectors for independent measurements without additional "in-pile" information? And, second, is there any "in-pile" information which could be incorporated into the reactor operation records (available to the inspectors), which could help inspectors to obtain results good enough to meet all safeguards requirements in respect of spent-fuel verification?

W.C. BARTELS: I believe your question is related to determination of burnup. At present we are working on the development of suitable techniques and, as I said, we have recently performed measurements on the large fuel assemblies for pressurized and boiling water reactors in the United States. While it is my opinion that the quantitative determination of fuel assembly burnup can and will reach accuracies of five per cent in the future, we all know that such accuracy cannot be attained without taking into account the spatial and spectral variations in neutron flux. Fuel which has been adjacent to a water gap or control-rod absorber can have big internal variations in burnup. Information on fuel-element location and movement of nearby control rods will be helpful, but not sufficient, for good burnup determination. Ideally, the inspector should be able to rely mostly or totally on various measurements he makes himself.

M.CUYPERS: I should just like to remark that the results obtained by inspectors with HRG could be improved by (a) technical improvements in all aspects of the instrumentation (including increased measuring speed) and (b) adequate inspector training programmes. It is my opinion that insufficient attention has been paid to the latter by the technical people. Training needs to be optimized by careful evaluation.

---

[1] Vol.I of these Proceedings.

CALIBRATION OF PLUTONIUM NON-DESTRUCTIVE ANALYSIS (NDA) BY CALORIMETRIC ASSAY

W.W. STROHM, W.W. RODENBURG,
J.F. LEMMING, D.R. ROGERS, C.L. FELLERS
Monsanto Research Corporation, Mound*,
Miamisburg, Ohio,
United States of America

Abstract

CALIBRATION OF PLUTONIUM NON-DESTRUCTIVE ANALYSIS (NDA) BY CALORIMETRIC ASSAY.

The calorimetric assay of plutonium is an established and documented technique used extensively in the US Department of Energy facilities for accountability measurements. Multilaboratory studies have quantified an average bias of < 0.2% for the calorimetric assay of a wide range of plutonium-bearing materials and plutonium isotopic compositions. This average bias can be reduced to < 0.1% using new half-live values provided in independent studies by the US Half-Life Evaluation Committee. An inspector's verification programme, utilizing calorimetric assay, has resulted in the increased use of calorimetric assay for the calibration of plutonium non-destructive assay systems. The use of calorimetric assay as a standard reference methodology for improved plutonium NDA measurement control is recommended.

INTRODUCTION

Calorimetric assay has been demonstrated to be a practical method for providing traceable measurement control of plutonium NDA measurements. Most NDA systems are presently calibrated using physical standards as described in ANSI N15.20 "Guide to Calibrating Nondestructive Assay Systems." With these techniques, the calibration is strictly valid only for the assay of samples which do not differ from the calibration standards with respect to any property to which the instrument is sensitive. D. Smith, in a definitive discussion of NDA calibration [1], wrote "It is the measurement biases, caused by differences between the material being assayed and the physical standards used for instrument calibration, which ultimately constitute the limit of our ability to control special nuclear material." He also observed that these biases may go completely unrecognized if the standard/sample differences are not identified and corrected.

---

* Mound Facility is operated by Monsanto Research Corporation for the US Department of Energy under Contract No. EY-C-04-0053.

In addressing the same problem from another view, Bingham, Yolken, and Reed [2] consider the chief impediment to the implementation of NDA to be a lack of demonstrable traceability of the measurements to a national or international measurement system. Traceability includes defining the limits of uncertainty of the measurement (random and systematic errors). They emphasize that having a series of traceable calibration standards is not, in itself, sufficient to meet traceability requirements; it is necessary to correct for differences between the samples and calibration standards or to include the effect of these differences in defining the limit of measurement uncertainty.

The traceability of calorimetric assay permits calorimetry to be used as a standard reference methodology for plutonium NDA calibration and measurement control [3]. When process material is calorimetrically assayed and used to calibrate the NDA measurement, sample/standard differences are minimized. This facilitates determination of NDA measurement uncertainties and demonstration of traceability.

This paper summarizes several multilaboratory experiments which have qualified calorimetric assay for plutonium bearing materials. The bias between calorimetric assay and chemical assay is quantified and the source of that bias discussed. Applications of calorimetric assay for NDA calibration and measurement control are presented.

TRACEABILITY OF THE CALORIMETRIC ASSAY OF PLUTONIUM

The calorimetric assay of plutonium consists of a measurement of sample power (watts) arising from radioactive decay and a determination of the plutonium effective specific power (watts/g of plutonium) which is used to convert the sample power measurement to plutonium content [4,5]. Mathematically, the plutonium content (in grams) of the sample is given by:

$$M = \frac{W}{P_{eff}} \qquad (1)$$

where W = sample power, and P_{eff} is the effective specific power of the sample.

The standard of the American National Standard Institute "Calibration Techniques for the Calorimetric Assay of Plutonium Bearing Solids" (ANSI N15.22-1975), describes the traceability of the component measurements of the calorimetric assay.

Calorimeter Power Measurement

The calorimeter power measurement uses standard four-terminal electrical power measuring techniques documented in ANSI N15.22-1975 [4]. Both the standard resistors and standard EMF cells are calibrated against standards certified by the National Bureau of Standards (NBS).

Calibrated ^{238}Pu heat standards are available from Mound. These heat standards have been calibrated in several calorimeters against electrical standards with direct traceability to NBS. In addition, two of these heat standards were measured independently by NBS in an entirely different type of calorimeter [6]. There were no significant differences between Mound certificate values and the NBS ice calorimeter measurements. A program is now in progress to certify and distribute heat standards under NBS auspices in a manner similar to plutonium assay and plutonium isotopic standards.

Effective Specific Power Determination

The effective specific power determination is traceable to the national measurement system by two alternate routes. The most direct is the empirical method of determining the effective specific power (P_{eff}). This method involves the calorimetric measurement of a small aliquot of plutonium and the dissolution and chemical analysis of the plutonium content of the aliquot. These measurements are combined to provide a direct measure of the P_{eff} value in watts/g of plutonium. The traceability of the empirical method is established through the use of NBS certified plutonium assay standards and the previously mentioned heat standards.

The alternate method of determining P_{eff} is called the computational method. This method involves measuring the relative weight fractions, R_i, of all plutonium isotopes and ^{241}Am relative to total plutonium. These weight fractions are then multiplied by the specific power, P_i, of each isotope to give the effective specific power of the particular plutonium isotopic composition, and summed over all plutonium isotopes and ^{241}Am.

$$P_{eff} = \sum_i R_i P_i$$

The traceability of this method relies on the use of NBS certified plutonium isotopic standards to calibrate the isotopic measurements (i.e., mass spectroscopy, gamma-ray spectroscopy, and/or alpha pulse height analysis) The techniques and nuclear constants are those recommended in ANSI N15.22 and are thus open to scientific scrutiny.

THE QUALIFICATION OF THE CALORIMETRIC ASSAY OF PLUTONIUM

For a measurement system to be considered as a standard reference methodology, it is necessary to demonstrate measurement performance following standard procedures and using available standard reference materials for the material types which it is to measure. Several experimental studies have been performed and are under way which qualify calorimetric assay for a wide variety of material types and plutonium isotopic composition. All the experiments described are multilaboratory studies. In each case, the bias between calorimetric assay and chemical assay was determined.

TABLE I. MEASURED BIAS BETWEEN CALORIMETRIC ASSAY OF PLUTONIUM AND CHEMICAL ASSAY FROM THREE MULTILABORATORY STUDIES

Attribute	Plutonium Metal A "H" Metal	Plutonium Metal B "R" Metal	Process Materials and Scrap C	PuO_2 D	PuO_2 E	PuO_2 F
Pu-240 (wt %)	5.95	5.91	∿6%	11.71	22.58	24.06
Number of Samples	11	21	16	6	6	6
Pu Content (g)	8	8	24 to 2200	4	2.5	1.2
Relative Bias[a] (%)	-0.17	-0.25	-0.17	-0.11	-0.22	-0.07
Standard Error of the Bias	±0.03	±0.02	±0.41	±0.02	±0.02	±0.02

Average Bias = -0.17%

Standard Deviation of the Biases = ±0.07%

[a] Calorimetric Assay - Chemical Assay

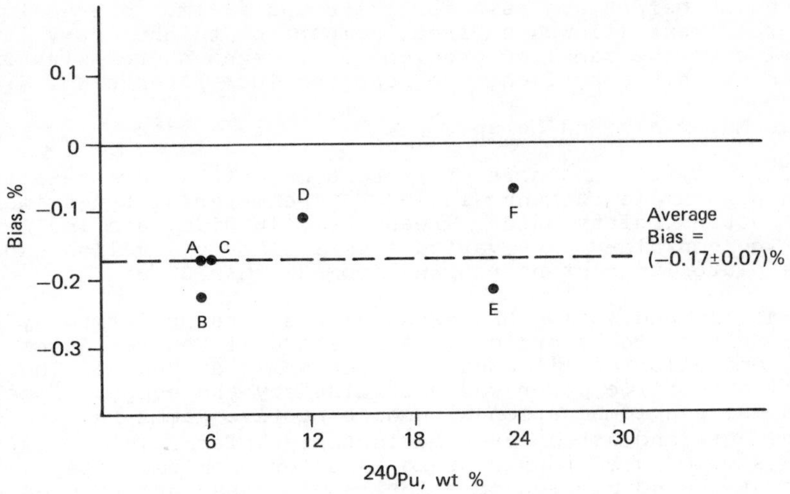

FIG.1. Bias of calorimetric assay compared with chemical assay from three multilaboratory studies.

Plutonium Metal

The U. S. Department of Energy (DOE) laboratories participate in the Plutonium Metal Exchange as part of their measurement quality control program. Eight gram plutonium metal samples, nominally 6 wt % ^{240}Pu, are distributed quarterly for chemical assay, impurity analysis, and isotopic analysis. In addition to chemical assay, Mound performed calorimetric assay on these samples for several years. The effective specific power has been determined by the computational method using isotopic compositions measured by mass spectrometry for plutonium isotopes and alpha pulse height analysis for ^{238}Pu and ^{241}Am.

The results, labeled A and B in Table I and Figure 1, are for calorimetric assay compared to by-difference assay (sample weight minus weight of impurities) for the two different metal samples distributed simultaneously in the exchange.

Reactor grade PuO_2

This experiment was part of a program conducted by Mound for the U. S. Nuclear Regulatory Commission (NRC) to evaluate the performance of different methods of determining effective specific power (P_{eff}) for the range of plutonium isotopic compositions expected in the commercial nuclear fuel cycle. Mound, Argonne National Laboratory (ANL), New Brunswick Laboratory (NBL), Savannah River Plant (SRP), and Lawrence Livermore Laboratory (LLL) participated in this program. The results, labeled D, E, and F in Table I and Figure 1, are based on the comparison of P_{eff} by the empirical method to the computational method, using mass spectrometry for plutonium isotopes and alpha pulse height analysis for ^{238}Pu and ^{241}Am. The design of this experiment allowed a direct comparison to chemistry (coulometry) without the sampling problems often encountered with oxides. Results of this experiment are reported in reference No. 7.

Process Materials and Scrap

Sixty-five containers of process materials were selected for an NDA sample exchange in 1972. The material types included: metal buttons, dirty oxide, greencake, fluoride, and incinerator ash. The container size varied from 1 pint to 1 gallon cans with a plutonium content ranging from 22 to 2200 g.

Passive and active NDA measurements were performed by Los Alamos Scientific Laboratory (LASL) and Gulf General Atomic (GGA), and calorimetric assay was performed at Mound. The effective specific power was determined by the computational method using a combination of nondestructive gamma-ray isotopic measurements and stream average isotopic information. Sixteen of the samples were then sent to LASL for chemical assay. The result of the comparison of calorimetric assay and chemical assay is labeled C in Table I and Figure 1. A more complete discussion of the results of this NDA sample exchange has been given by Reilly and Evans [8].

Summary of the Qualification Studies

These multilaboratory experiments have qualified calorimetric assay as a standard reference methodology for a wide range of plutonium bearing materials and isotopic compositions. These results reflect the insensitivity of calorimetric assay to matrix material and geometry, which affect most other NDA measurements.

Based on these studies, calorimetric assay exhibits a small negative bias which may be attributed to errors in the half-life values in ANSI N15.22. While the committee writing ANSI N15.22 considered the half-lives used to be the best available, it was recognized that large discrepancies appeared in the literature and that other compilations of nuclear data contained different values. As a result, the U. S. Half-Life Evaluation Committee (HLEC)[1] was organized under the auspices of the Department of Energy to evaluate these discrepancies.

As its first task, HLEC initiated half-life measurements of ^{239}Pu to resolve an approximate 1% discrepancy between reported measurements by alpha particle counting techniques, calorimetry, and mass spectrometry. Based on this multilaboratory half-life determination [9], the HLEC recommends a half-life of 24,119 $\pm$26 years for the half-life of ^{239}Pu (compared to 24,082 ± 46 years in ANSI N15.22). Using this new half-life for ^{239}Pu for the data shown in Table I, the average bias is reduced from -0.17% to -0.08%. This effect is shown graphically in Figure 2.

[1] Member laboratories are Los Alamos Scientific Laboratory, Argonne National Laboratory, Lawrence Livermore Laboratory, National Bureau of Standards, Rockwell Rocky Flats, and Mound.

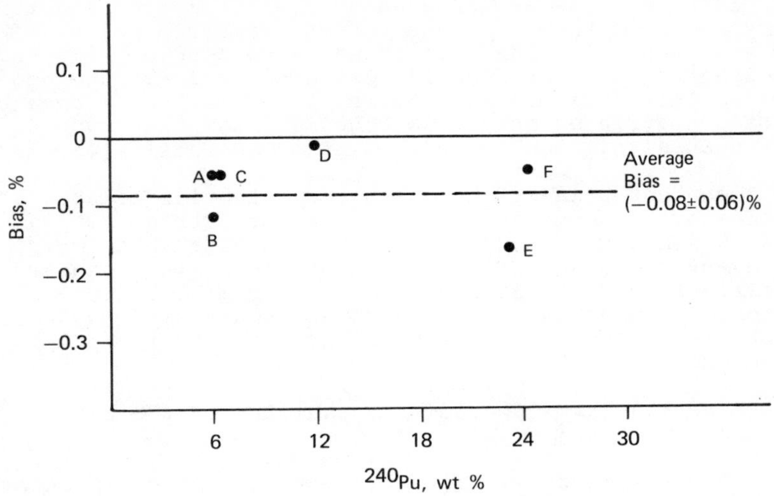

FIG.2. Bias of calorimetric assay compared with chemical assay using $T_{1/2}$ *= 24.119 years for* ^{239}*Pu.*

The half-lives of ^{238}Pu, ^{240}Pu, and ^{241}Pu are being investigated by the HLEC. New values for the half-lives of these isotopes may further reduce the bias in the calorimetric assay of plutonium.

The Half-Life Evaluation Committee is reporting the results of its efforts to the writing group responsible for a required 1980 revision of ANSI N15.22. This is an important step in the continuing effort to improve this written standard.

APPLICATIONS OF CALORIMETRIC ASSAY FOR PLUTONIUM NDA MEASUREMENT CONTROL

Inspectors' Verification Program

For six years, the DOE/Albuquerque Operations Office (ALO) has used calorimetric assay to verify contractors' accuracy statements for plutonium NDA measurements [10,11]. Because of its insensitivity to matrix effects, calorimetric assay has been used by the inspectors to measure plutonium content in a variety of feed materials and scrap categories:

- Oxides
- Metals
- Mixed Oxides
- Mixed Nitrides
- Incinerator Ash
- Ash Heel
- Fluorides
- Sand, Slag, and Crucibles
- Graphite Scarfings
- Greencake
- ^{238}Pu Scrap

ALO inspectors select assayed samples from a contractor's inventory and send them to Mound for calorimetric assay. The results are then used by ALO to identify measurement biases. One important result has been increased emphasis by the contractors on their measurement control programs and, increased use of calorimetric assay. Also, the material unaccounted for, MUF, has been significantly reduced at these facilities.

An important spin-off has been the retention of some of the samples by the contractors, after calorimetric assay at Mound, as NDA assay standards. Using standards which originated in the inventory is an effective way of minimizing standard/sample differences and the resulting measurement biases. The use of these samples as assay standards thus satisfies the two important requirements for establishing traceability of plutonium NDA measurements:

1) The standards provide traceability of the NDA calibration;

2) The standards minimize standard/sample differences.

The inspectors' verification program has proved to be a practical demonstration of the use of calorimetric assay for plutonium NDA measurement control by providing: 1) verification of prior measurements, 2) quantification of measurement biases, and 3) calibration of plutonium NDA.

Dynamic Calibration of NDA

Measurement control programs for NDA instruments are designed to ensure the quality of the measurements. These programs are extensive and include, for example, preparation of standards, calibration procedures, and operator training programs. For high-throughput plants, a dynamic calibration procedure using calorimetry as a measurement control technique is being developed at Mound for NRC.

The dynamic calibration technique selects current samples from the inventory of items being assayed to become calibration samples. The samples selected for calibration are standardized using reliable methods traceable to the national measurement system. Dynamic calibration offers several advantages: it provides continuous recalibration of the measurement system; it automatically accounts for intra- and inter-batch process process variations; it provides a linkage between measurement systems to reduce net systematic error; and it reduces the need for storage of calibration standards.

An example of dynamic calibration for the plutonium fuel cycle is shown in Figure 3. Calorimetry has been introduced as the control measurement for the NDA device.

In this example, the entire process flow is measured by NDA; a portion of the flow is measured by using calorimetric assay and returned to the process. In this measurement system, calorimetric assay is used to monitor the process and to provide

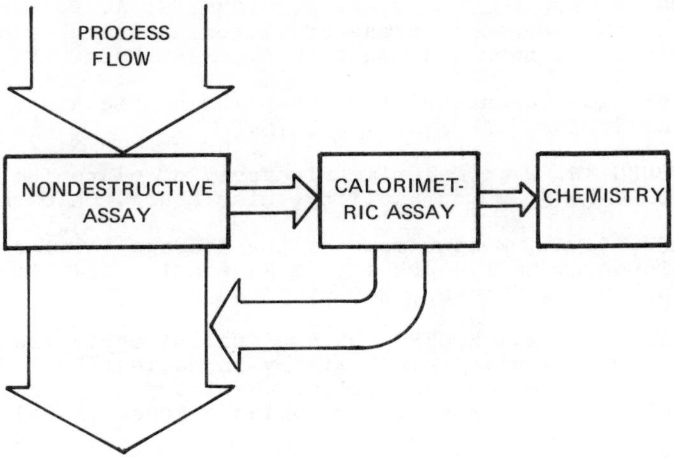

FIG.3. Dynamic calibration of plutonium NDA is being developed using calorimetric assay for measurement control.

calibration data. Chemistry is used in this system as a part of the measurement control program for the calorimetric assay. The amount of chemical analysis required in this plan is much less than that required using chemistry for controlling measurement because calorimetry assumes the burden of monitoring the NDA measurement and providing recalibration.

Calorimetric assay is an effective control measurement because it can be traced to a national measurement system, it can give absolute assay from first principles since isotopic composition is known as a part of process control information, and it is nondestructive.

CONCLUSION

Since the calorimetric assay of plutonium has been demonstrated to provide effective measurement control for plutonium NDA, it should be used more for that purpose. With an appropriate measurement control program, calorimetric assay can provide traceability of plutonium NDA.

REFERENCES

1. SMITH, D. B.,"Physical Standards and Valid Calibration" in Proceedings of the IAEA International Symposium on Safeguarding Nuclear Material, Vienna (1975) 63.

2. BINGHAM, C. D., YOLKEN, H. T., REED, W. P., Nondestructive assay measurements can be traceable, J. Inst. Nucl. Mat. Mang., V 2 (1976) 32.

3. STROHM, W. W., RODENBURG, W. W., LEMMING, J. F., Traceability of the nondestructive assay of plutonium using calorimetry for measurement control, J. Inst. Nucl. Mat. Mang., VI 3 (1977).

4. "Calibration Techniques for the Calorimetric Assay of Plutonium Bearing Solids," ANSI N15.22-1975.

5. RODENBURG, W. W., "Calorimetric assay of plutonium," NUREG-0228 MLM-NUREG-2404, Mound Facility, Miamisburg, Ohio (May 1977).

6. DITMARS, D. A., Measurement of the average total decay power of two plutonium heat sources in a Bunsen ice calorimeter, Int. J. Appl. Radiat. Isot., V (1976) 469.

7. RODENBURG, W. W., ROGERS, D. R., "Calorimetric assay of reactor grade PuO_2," Analytical Chemistry in Nuclear Fuel Reprocessing, (Proc. 21st Conf. on Analytical Chemistry in Energy Technology, Gatlinburg, Tennessee, 1977) Science Press (1978) 176.

8. REILLY, T. D., EVANS, M. L., "Measurement Reliability for Nuclear Material Assay," Los Alamos Scientific Laboratory Report LA-6574 (1977), Chapters 4 and 7.

9. STROHM, W. W., The measurement of the half-life of plutonium-239 by the U. S. Half-Life Evaluation Committee, Int. J. Appl. Radiat. Isotopes (Oct.1978) in press.

10. CROUCH, R. B., "Calorimetric Verification of Plutonium Inventories for Safeguards Surveys," Proceedings of the Symposium on Calorimetric Assay of Plutonium, MLM-2177 (1974) 96.

11. GEORGE, R. S., CROUCH, R. B., Inspector measurement verification activities, J. Inst. Nucl. Mat. Mang., IV III (1975) 327.

DISCUSSION

P. DE BIEVRE: I should like to make a number of comments which might be of interest in connection with your paper. First, a CBNM (Central Bureau for Nuclear Measurements) determination of ^{239}Pu half-life gave a figure of 24 120 years. Second, a CBNM determination of ^{241}Pu has been under way for a few years on a highly enriched ^{241}Pu sample and has yielded a provisional figure of 14.4 years which agrees with the results of the United States programme. The well-known spread in the literature is on its way to being resolved. Third, it is true that there is a great need for Pu isotope reference materials to put Pu isotope measurements on an absolute basis. In fact it is surprising that the nuclear community did not come up with such reference materials ten years ago. However, I can now announce that an agreement has been concluded between the Central Bureau for Nuclear Measurements in

Geel, Belgium and the National Bureau of Standards in Washington for a joint attack on the problem and for joint certification of such reference materials.

W.W. STROHM: The excellent agreement in half-life values obtained independently in the United States and at CBNM is very gratifying and will, it is hoped, allow an international consensus to be reached on half-life values to be used for safeguards.

Perhaps the nuclear community can take satisfaction in the fact that precise measurement technology has been developed in response to increased safeguards interest. As a consequence of this, the safeguards need for improved nuclear data and reference materials can be clearly identified.

W. BEYRICH: As you already mentioned, the ^{238}Pu makes the major contribution to the total heat. This means that an exact knowledge of the ^{238}Pu abundance is of particular improtance. In the AS-76 intercomparison experiment on ^{238}Pu determination by alpha spectrometry, which was organized by the Nuclear Research Centre at Karlsruhe and is now in the evaluation stage, we found a spread of around several per cent in the results from different laboratories. Am I correct in assuming that this will be the limiting factor on the accuracy of the method you described?

W.W. STROHM: In the PuO_2 exchange discussed in the paper, two laboratories used alpha spectrometry to measure the ^{238}Pu concentration of ~ 1.2 wt%. The two measurements agreed within 0.1% and also agreed well with mass spectrometry, which is accurate for this concentration. Also, several speakers at this conference have described gamma-ray techniques which can measure the ^{238}Pu concentration ratio to better than 1%. Accurate measurements of the ^{238}Pu concentration are thus available. Perhaps the improved isotopic reference materials mentioned by Mr. De Bievre will result in better agreement in exchange programmes.

H. KRINNINGER (*Chairman*): You mentioned that you intend to use calorimetric assay for the calibration of NDA equipment. On what types of NDA equipment will you use this calibrating procedure?

W.W. STROHM: Calorimetric assay can be used to calibrate any plutonium NDA of containers of solid materials. There is currently a lower limit of about 20 grams of plutonium in a one-gallon can. As far as very large samples are concerned, calorimeters for 55-gallon drums are not currently feasible but calorimeters for fuel bundles are.

POSSIBILITIES FOR INTERNATIONAL CO-OPERATION IN STANDARDIZING MEASUREMENT METHODS FOR NUCLEAR SAFEGUARDS

H.T. YOLKEN
National Bureau of Standards,
Washington, DC,
United States of America

Abstract

POSSIBILITIES FOR INTERNATIONAL CO-OPERATION IN STANDARDIZING MEASUREMENT METHODS FOR NUCLEAR SAFEGUARDS.

The need to determine accurately the amount of fissionable materials in nuclear fuel cycle facilities is of clear importance to both international and domestic safeguards activities. However, we have not fully developed and utilized all the measurement standardization procedures available to ensure accurate and compatible nuclear safeguards measurement results on an international scale. As the international nuclear safeguards measurement system has increased in size and sophistication, the need for effective standardization has increased. To meet this current need and projected needs from possible new fuel cycles, bilateral, multinational and international co-operation in the development of standardization capability should be encouraged. The International Atomic Energy Agency (IAEA) can and should play a key role in co-ordinating this activity. This standardization activity should encompass destructive chemical and isotopic analysis, non-destructive assay by a number of methods, bulk measurements involving mass, flow, volume, density etc., and associated sampling of materials and statistical treatment of data. A suggested international measurement and standards system is described. Examples of existing co-operative efforts in standardization of measurements for nuclear safeguards are presented. Finally, a number of recommendations for implementing and carrying forward an international co-operative effort in standardization of measurements for nuclear safeguards are presented. The author's recommendations are grouped with three time scales – those that could be implemented within a year or two; those that will most likely require several years; and long-term recommendations. The recommendations will include possible activities for both the IAEA and Member States.

INTRODUCTION

The need to rapidly and accurately determine the amount of fissionable materials in nuclear fuel cycle facilities is of clear importance in nuclear safeguards activities. The agreements between the International Atomic Energy Agency (IAEA) and States in connection with the Treaty on the Non-Proliferation of Nuclear Weapons require that the Agency shall make full use of the State's system of accounting in carrying out its verification activities. The agreements further require that the States' system of accounting shall make provision as appropriate for the evaluation of precision and accuracy of measurements and the estimation of measurement uncertainty. The words that should be stressed in the previous sentence are "evaluation of precision and accuracy of measurements and the estimation of measurement uncertainty". In order for States to fully carry out this portion of the agreement and in order for the Agency to fully utilize the States' system of accounting, there needs to be a completely developed and utilized international measurement and standardization system.

The measurement and standardization system for nuclear materials measurements is an important ingredient in assuring accurate and compatible nuclear safeguards measurement results on an international scale. However, the international measurement and standardization system for nuclear safeguards is not fully developed and implemented. This paper describes an international standards system and possible mechanisms for international cooperation in improving the standardization of measurement methods for nuclear safeguards.

INTERNATIONAL MEASUREMENT AND STANDARDS SYSTEM

Accurate and compatible measurement results on an international scale have been achieved in a number of areas utilizing several different means. For example, time measurements[1] enjoy an excellent reputation for accuracy and compatibility between countries. Atomic clock time measurements made by individual countries are constantly intercompared and coordinated with other countries' time measurements in a cooperative effort. Time measurements are disseminated via exchange of atomic clocks and by radio broadcast. Similarly, the measurement of mass[1] takes place smoothly since weights traceable to those maintained by the International Bureau of Weights and Measures, are utilized by many countries. The mass standardization methodology has been developed, to a large extent, by joint cooperative undertakings between countries.

There are a number of operational items one may select from in designing and implementing an effective international measurement and standardization system to help assure harmonization of measurement results. These items include:

- An international coordinating body
- Base and derived measurement units
- Reference measurement methods
- Reference materials and artifacts
- Evaluated reference data
- Instrument calibration services
- Field measurement methods and instruments
- Written procedural standards
- Measurement assurance programs

As the international nuclear safeguards measurement system increases in size and sophistication, the need for effective standardization based on the above building blocks will increase. In order to meet the current needs and projected needs that will be an outgrowth of new fuel cycle facilities and possible new fuel cycles, bilateral, multinational and international cooperation in the development of standardization capability should be encouraged.

---

[1] It should be stressed that these examples are intended only to illustrate international cooperation. They are not intended to illustrate the accuracy levels required for nuclear safeguards.

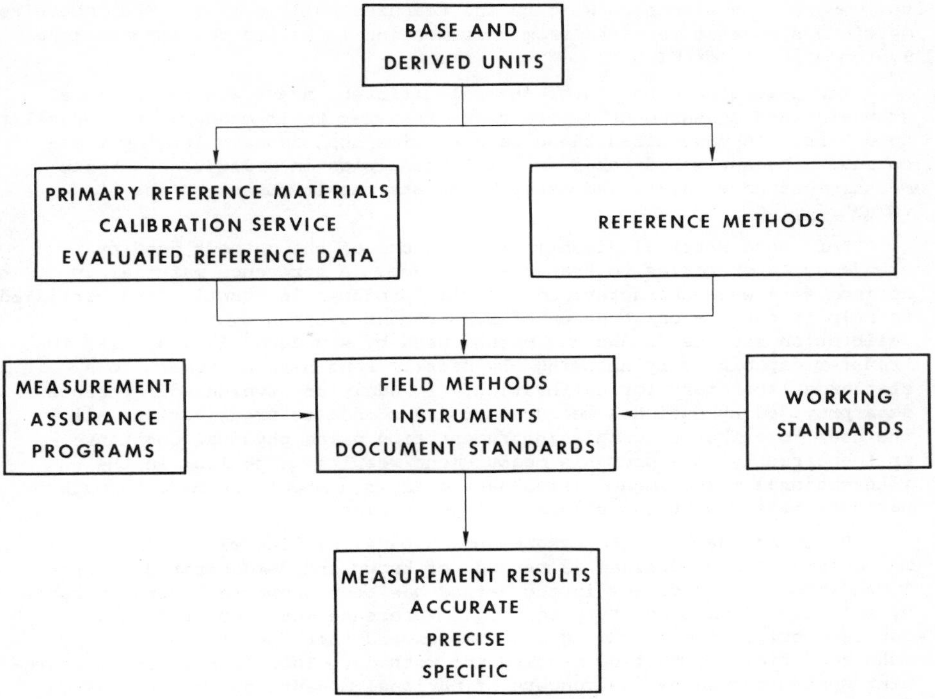

FIG.1. Proposed international measurement standards system for nuclear safeguards.

The standardization building blocks may be used to form an effective standardization system as shown in Figure 1. An international coordinating body plays an essential role in forming and maintaining the system. The coordinating body provides the administrative and management umbrella that is necessary to develop plans and priorities, initiate and carry out cooperative activities, and disseminate information and services. The IAEA can potentially provide and should assume this key role.

The standardization activity should be an integrated one encompassing all of the measurement areas needed for effective materials accountability. These measurement areas include: destructive chemical and isotopic analysis; non-destructive assay by such techniques as gamma and neutron spectrometry, calorimetry and isotopic correlation; bulk measurements involving the determination of mass, flow, volume, pressure, density, temperature, etc.; and associated plans for sampling of materials and statistical treatment of data.

Base and derived units, which form the first link in the measurement system, are a common foundation for most quantitative measurement systems. Fortunately, the base and derived units have been agreed upon internationally with the latest revisions and agreements being among forty countries at the General Conference on Weights and Measures held in 1960. The modernized metric system that resulted from this meeting is called the International System of Units or SI.

The next link consists of several different means whereby national standards and measurement laboratories transfer their measurement capability to others. In many cases these laboratories need to make lengthy time-consuming measurements based on first principles in order to determine measurement uncertainty and assure accurate results based on the "true value".

Reference material dissemination is one of the methods used by standards laboratories to transfer accuracy. A reference material is defined as a well-characterized material produced in quantity and certified to help in on-site calibration of measurement systems by the user. A calibration service is another method used by standards laboratories to transfer capability by allowing the user to send his instrument to a standards laboratory for calibration. In addition, evaluated reference data provided by national measurement and standards organizations allow the user to either check his instrumentation using physical constants or to correctly interpret his measurement results. The link in the international measurement system just described should be made through national measurement and standards laboratories.

Reference methods of measurement, a parallel link with transfer mechanisms, can be defined as methods of known and demonstrated accuracy. In addition, the accuracy in the method has been shown to be transferable by selective interlaboratory testing. Reference measurement methods can be costly, time-consuming and complex and therefore do not usually make good field or routine measurement methods. This link in the measurement system should be the concern of national measurement and standards laboratories.

Certified reference materials, instruments calibrated by national laboratories, and reference methods, in general, do not find wide-spread daily use in field measurement laboratories, but link directly to the next part of the measurement system. This part of the measurement system, which is concerned with the ultimate user, consists of field methods of measurement, document standards concerned with how to make measurements, instruments, and working or everyday calibration material and services. The responsibility for this link lies with many organizations, not only with national measurement and standards laboratories.

From a system such as this, measurements are produced that are meaningful in terms of accuracy, precision, and specificity. However, one more part is needed to insure that the system works smoothly and produces the desired results over a long period of time. A measurement assurance program, involving interlaboratory testing with unknown samples and analysis of the results as evaluated against reference method results obtained by national measurement laboratories, fulfills this need. In addition, it provides a mechanism for self-help to the participants. Measurement assurance programs that are not based on accurate results obtained by first-class reference laboratories give only a measure of between-laboratory precision, not accuracy.

SOME EXAMPLES OF CURRENT COOPERATION

There are several examples of cooperative efforts involving countries, groups of countries or international organizations that are helping to strengthen the harmonization of measurements for nuclear safeguards. In the reference materials area, the U.S.A. and ESARDA are jointly working on a series of U_3O_8 materials of varying enrichments that will be used to calibrate gamma-ray spectrometry instrumentation used to non-destructively determine enrichment. When this effort is completed, the U.S.A. and Euratom will have compatible and accurate certified reference materials available for distribution, and the IAEA will have these NDA reference materials for its use. The details of this cooperative effort will be presented in another paper at this meeting.

In another example, the U.S. National Bureau of Standards and the Euratom Bureau Central De Mesures Nucleaires have recently agreed to work cooperatively on reference materials for calibrating the mass spectrometric determination of plutonium assay and isotopic composition. In addition, many countries participate and/or cooperate in various ways in measurement assurance programs for chemical and isotopic assay of nuclear materials.

Two international organizations, the International Organization of Legal Metrology and the International Organization for Standardization, are currently encouraging international cooperation in the area of reference materials. As yet another example of cooperation, Euratom is coordinating an effort on gathering and disseminating information related to sources of currently available nuclear reference materials.

SOME POSSIBILITIES FOR COOPERATION

There are several possibilities for action to facilitate cooperation that may be considered. The items listed below are arranged in a sequence ranging from easily implemented to implemented with difficulty and are based in part on recommendations made by Cali [1] in the reference materials areas.

- Information gathering and dissemination
- Assessment of worldwide standardization needs
- Assessment of worldwide standardization priorities
- Worldwide distribution network for standard services (e.g., reference materials, calibrations)
- Harmonization of national standards programs
- Establishment of world wide network of cooperative standards laboratories
- Cooperative production of standards services at the international level

The first item, information gathering and dissemination, can build on the Euratom coordinated effort in nuclear reference materials that is currently under way. Similar efforts needs to be carried out for calibration services, document standards on how to make measurements, evaluated reference data and reference methods. The information from all of these activities then needs to be combined and disseminated.

The next two items, assessment of needs and setting of priorities, will require highly-knowledgeable and technically-trained scientists. The IAEA has made a start in this area through its technical experts committees on the need for calibration materials for both the non-destructive assay and destructive assay of nuclear fuels.

Setting up a worldwide distribution network for nuclear reference materials, calibration services, and other standards services will be extremely difficult. Several important non-technical difficulties exist such as: nuclear materials export-import regulations, currency exchange, and customs requirements. In fact, these very practical difficulties may be the most difficult to overcome.

Moving down the list, harmonization of national standards programs will involve obtaining international agreement on standards services such as reference material certification procedures and reporting, document standards and calibration services. Agreement in most of these cases will require careful and lengthy negotiations.

The final two items on the list, establishment and utilization of a worldwide network of standards laboratories, will of course be difficult. However, we do have a positive example concerning laboratory cooperation that has been set by the IAEA Safeguards Analytical Laboratory network. In addition, the willingness of a very large number of countries in the world to work together as signatories to the NPT provides some hope that cooperative production of standards services is a realistic goal.

RECOMMENDATIONS

Listed below is a series of recommendations that might facilitate the implementation of an international cooperative effort in standardization of measurements for nuclear materials safeguards. It needs to be stressed that the IAEA should play a strong leadership role in activities.

The recommendations are grouped into three time scales--those that can be implemented now or in the near future; those that require two to four years time to implement; and long-range recommendations.

Recommendations (now)

- The IAEA should establish a working group to gather information on :
  - Currently available international and national nuclear materials measurement standards for all types of nuclear fuel measurements.
  - Laboratories now performing standards work
  - National standardization procedures
- The IAEA should encourage the International Organization for Standardization to expand its activities on writing nuclear fuel document standards to include nuclear safeguards measurement standards.

Recommendations (2-4 years)

- The IAEA should establish formal technical working groups to:
  - Establish worldwide needs for additional nuclear materials measurement standards and services not currently available
  - Set priorities on these needs
  - Encourage cooperation among nations to avoid needless duplication and to encourage harmonization
  - Maintain and expand information network

The longer term recommendations may well have to be modified as the earlier work progresses. Clearly the longer the time period, the greater the uncertainty, and the greater the need to maintain flexibility.

Recommendations (4 years-onward)

- Establish under the aegis of the IAEA a formal network involving member states to provide nuclear materials measurement standard services for the Agency and member States. This includes:

- Recruitment of national measurement and standards laboratories (provision of reference materials, calibration services for NDA, etc.)
- Qualification procedures for participating laboratories
- Procedures to assure quality and integrity of work (measurement assurance programs)
- Provision for services distribution network
- Provision for financial support

If we can agree on the immediate and short-term goals and effectively attain them, then the foundation for a comprehensive program for the future will have been laid. The time is right to implement this needed program that will draw on the cooperation of many nations in the world.

REFERENCES

[1] Cali, J. P., "Standard Reference Materials and Meaningful Measurements," NBS Special Publication 408, 57-67, U.S.A. (1975).

DISCUSSION

S. DERON: Following the recommendations of the Advisory Group Meeting on Standard Reference Materials for the Nuclear Fuel Cycle[2], held in November 1977, the Agency is supporting an updating of the list of nuclear SRMs published earlier by Euratom's Central Bureau for Nuclear Measurements. An inquiry is to be addressed to Member States to establish their needs for nuclear SRMs and their capacity for preparing, characterizing and distributing them. It is hoped that the results of the inquiry will be available for discussion and evaluation by another Advisory Group Meeting to be held in 1980. The further steps which you have recommended would require the establishment of a specialized working group within the Agency.

M. CUYPERS: I should like to make a comment in this connection. The Agency has organized several Advisory Group Meetings on the subject of SNM measurements. These meetings were concerned mainly with the needs of inspectors, and the question of facility measurements was not studied. Such studies are certainly necessary to reach a consensus in the field of SNM measurements, particularly in view of the increasing use of the nuclear fuel cycle. The concept and proposals put forward in your paper are ambitious, but they ought to be given serious consideration at once, in order to avoid increasing disparity in the measurement practices and data used in safeguards and the management of fissile material.

W. FRENZEL: I'm afraid I must disagree with this point of view. The purpose of safeguards, from the operational side, is not to improve measurements and propagate exact measurement methods. Our objective is to detect diversion, and I don't see how we can undertake a task which will be tremendously expensive

[2] IAEA, Advisory Group on Chemical Standards Related to Nuclear Fuel Analysis for Fuel Characterization and Safeguards Purposes, AG-102, IAEA, Vienna (1977).

just to improve the accuracy of the digit in the third decimal place. I do not know where the Agency will get the money to do this.

H.T. YOLKEN: The cost of standardization will certainly be a great deal less than the amount we spend on developing the techniques. Secondly, many of the methods employed at present have 3–5% measurement uncertainties. Looking at the current requirements of the Agency and the recommendations contained in the SAGSI reports, we are certainly going to have to do better than that. We are not talking about a measuring accuracy of one part in 10^5; we are talking about accuracies of 0.1, 0.5, or 1%, which will enable us to assure the public that the measurements are reliable. That is the main thing.

W.C. BARTELS: The concern expressed by Mr. Frenzel is related to the potential burden on the IAEA safeguards programme of having to support a world-wide effort on nuclear standards. As Mr. Yolken knows, I am of the opinion that the benefits of such an effort would in addition be felt in the economics of nuclear power and in the conduct of international trade and commerce related to nuclear materials such as uranium hexafluoride. Therefore, the Agency should promote such a world-wide effort as a part of its technical assistance programme rather than its safeguards programme. Furthermore, the Agency would find wide support by Member States for such an approach.

A. PETIT: I agree with the remarks made by Mr. Frenzel and would like to comment on the reply given by the author. As yet no absolute values have been laid down by SAGSI which would necessitate an increase in measuring precision. Such values have been proposed, but there is at the moment no consensus regarding them. The only rule one can apply is that up-to-date measuring techniques should be used. As regards the financial aspect, we all know that, although the funds required for a certain purpose may be relatively small, they are not necessarily easy to obtain.

RESULTS OF INTERNATIONAL COLLABORATION IN THE ESARDA WORKING GROUP ON NON-DESTRUCTIVE ANALYSIS

R.J.S. HARRY*
Netherlands Energy Research Foundation ECN,
Petten, The Netherlands

Abstract

RESULTS OF INTERNATIONAL COLLABORATION IN THE ESARDA WORKING GROUP ON NON-DESTRUCTIVE ANALYSIS.

This paper concerns the main results of the collaboration in the ESARDA Working Group on Techniques and Standards for Non-Destructive Analysis. After a short introduction attention is paid to the ESARDA reports on non-destructive assay (NDA) techniques in use, and on reference materials, the reports being the first results of a collaborative effort of all participants. Then a description is given of the first project to produce an internationally certified reference material for gamma-spectrometric enrichment determination. Details of the samples to be produced are given. Further, a description of a small intercomparison experiment to determine the isotopic ratios of plutonium by means of gamma spectrometry is given, together with some preliminary results. Finally some other activities and future plans are mentioned.

INTRODUCTION

In the paper "Activities of the European Safeguards Research and Development Association (ESARDA)", presented at the IAEA symposium of 1975, a thorough description of ESARDA is given [1]. Further reports on the ESARDA activities are published in the ESARDA Newsletter, contained in the European Safeguards Bulletins [2]. In these publications the ESARDA Working Group on Techniques and Standards for Non-Destructive Analysis is introduced also. Therefore, it may suffice to note here that the principal objective of the working group is: "to establish and recommend criteria to be used as a basis for agreement between plant operators and the safeguards authorities on the acceptability of non-destructive measurements of nuclear materials".

Specialists among the ESARDA partners and an observing expert from the IAEA meet in the regular sessions twice a year, while informal meetings on specific topics are also attended by other experts with appropriate experience on these topics.

---

* At present convenor of the ESARDA Working Group on non-destructive analysis.

The working group started with the collection of information on the NDA techniques in use and on the standard reference materials available. So a clear view has been obtained on the possibilities and needs in the field of NDA. On the basis of the collected information further actions are defined and undertaken.

In this paper results of the international collaboration are described, together with continuing projects and future prospects.

NDA-TECHNIQUES IN USE

One of the first activities of the working group was the collection of information on NDA techniques in current use, or which will be put into use in the near future. An investigation has been made within all States represented in ESARDA. The compilation differs from the list of data sheets in the Safeguards Technical Manual of the IAEA [3], in so far as it also includes applications for plant management control. Further, the list concentrates on methods in actual use.

The concise and comprehensive compilation is published as an ESARDA paper entitled "Established applications of non-destructive techniques for nuclear materials measurements, control or verification, reported to ESARDA" [4]. The paper is structured in subsections relating to techniques, with a table giving cross-references to types of material. Where possible the responsible organization is identified and the range of application, the accuracy and precision achieved and the major contributions to error in the specific application have been stated. During the compilation it appeared that many techniques are in use without good knowledge of their accuracy and precision in that application. Much effort has been spent to obtain this particular information. Many of the techniques use a calibration against working standards or reference samples, whose composition is considered sufficiently well known by the operator to meet his particular needs, but which the operator does not consider sufficiently well characterized to be considered suitable for general application.

In total 57 applications are summarized, distributed over the following techniques: low-resolution gamma-ray measurements (20), high resolution gamma-ray measurements (8), passive neutron counting (9), active neutron interrogation (12) and miscellaneous techniques (10). The ^{235}U content or enrichment determination accounts for 14 applications of gamma-ray measurements and six of active neutron interrogation. For the determination of the isotopic ratios in Pu the high-resolution gamma-ray detection method is the only NDA method reported. For the determination of U and Pu in more difficult geometries such as waste, leached hulls, process liquors and solid materials, all techniques are used with, however, somewhat more emphasis on passive neutron counting (nine cases). The fuel rod scanner is mentioned four times, based on active neutron interrogation three times, and once on low-resolution gamma-ray counting.

LIST OF STANDARDS

To define what kind of reference material is lacking at present, and to decide where a particular effort should be applied, a detailed list of available NDA reference material has been established. This inventory list has been published as an ESARDA report, List of Reference Materials for Non-Destructive Assay Existing within Six Countries of the E.C., the Commission and the IAEA [5]. The list is not exhaustive. It is intended to update it regularly, to include the large number of materials under preparation. The nominal specifications of the reference materials are reproduced. Reference materials listed are used for calibration of NDA methods. It is clear that this list has to be combined with the list of NDA methods in use in order to get a clear picture of the current situation in the field of NDA applications.

In the report a general table is given in which the information is arranged according to the fuel cycles involved. For the materials test reactors – plates, rods, fuel elements and billets containing highly enriched uranium are available, in total 66 items. For a thorium fuel cycle – standards containing thorium and uranium are all in the form of graphite spheres with a diameter of 60 mm; 60 items are mentioned. For the light water reactor – fuel-cycle rods, pellets and powder of uranium oxide are available, divided over more than 400 items.

Also, mixed-oxide fuel pins are available for the light-water reactor cycle (9 items) and for the fast reactor (1 item).

Plutonium and plutonium oxide standards are available in the form of powder, platelets, pellets in pins, and simulated waste (in total 52 items).

Apart from this general information in the table, which also lists chemical composition, geometries, cladding, range of the main parameter of interest and ownership, additional information is given in a separate set of tables that will be reviewed regularly.

Prospective users of the standards should contact the ESARDA secretariat directly, or the owner of the material, for further details.

ENRICHMENT STANDARD FOR GAMMA-RAY MEASUREMENTS

The report on NDA techniques [4], and the report on standards [5], both indicate that gamma-ray measurements of the uranium enrichment are the most widely applied NDA techniques. To create a sound basis for the acceptability of these measurements by all parties concerned, one necessary step will be the supply of a certified reference material. The conclusion that there is a need for this kind of standards was also reached on the basis of a more general approach to the problem of traceability of the various NDA measurements applied in the fuel cycle [6]. A limited set of calibration material should be prepared applicable to the most accurate NDA techniques and the most commonly measured materials. One has to be aware that the NDA techniques are sensitive to various sample parameters, which means that standards and techniques have to be compatible.

There exist several standards for low-enriched uranium oxide with accurately known and documented isotopic composition, obtainable, for example, from the United States National Bureau of Standards and from British Nuclear Fuels Limited, Capenhurst. But the sample mass of these reference materials is limited to a few grams, while for the NDA measurements with gamma-ray spectrometry the samples should be a hundred times larger.

The gamma-ray measurement of the uranium enrichment is now a well-established technique, and a high precision can be obtained on uranium oxide powder materials. A precision of 0.5% can be met easily, and even better results are reported for routine measurements [7a, b]. It is also demonstrated that this NDA method can be used to reduce the need for large numbers of mass spectrometric analyses, which, however, remain the basic reference for enrichment measurements.

Other reasons to start this first approach towards the procurement of a uranium oxide standard for enrichment determination by gamma-ray spectrometry are: the simplicity of the NDA method, the large number of working group members that could contribute their experience with this measurement technique, the reasonable cost of the material, and the relative ease of fabricating the samples.

It is planned to produce standards of the following nominal enrichments: 0.4, 1.2, 1.7, 2.55 and 2.95%. Of each enrichment 100 samples of 200 g U_3O_8 will be prepared in specially designed and identified aluminium containers. In the paper "Procurement and use of reference materials for the calibration of NDA equipment" [8], details of the procedure to be followed are given.

The aim is to produce so-called Certified Reference Material, according to the definition adopted at the IAEA Advisory Group Meeting on NDA Standards in Vienna, 22–26 August 1977 [9]. Co-operation with the United States of America National Bureau of Standards, Los Alamos Scientific Laboratory and New Brunswick Laboratory has been established for the preparation, characterization and certification of the material. Also the Commission of the European Communities Joint Research Centre, Geel (Central Bureau of Nuclear Measurements), is executing part of the project. Observers from IAEA and the Euratom Safeguards Directorate are invited to witness all key steps of the fabrication, in order to enable the inspectorates to accept these multinational Certified

Reference Materials for their own use. Further, all steps of the project are fully documented in order to ensure traceability to the fundamental chemical and physical analyses data and samples of the starting materials. Finally the certification will be performed by both the Joint Research Centre of the Commission and the National Bureau of Standards. This is the first time that such standards will have been prepared in the field of non-destructive analysis of nuclear material.

SOME SPECIFICATIONS OF THE U_3O_8 STANDARD

The parameter of interest is the ^{235}U enrichment, i.e. the ratio of $^{235}U/U$. The Joint Research Centre, Geel (Central Bureau for Nuclear Measurements), is ready to certify the ^{235}U enrichment with a total uncertainty of 0.2 to 0.25% of the nominal value. The basis for this certification is mass spectrometry with NBS standards which are accurate within 0.1%. Better accuracies could only be achieved by investing an extraordinary large analytical effort. It was therefore agreed that the $^{235}U/U$ ratio should be specified within 0.25% relative.

The samples will contain 200 g U_3O_8, which is chosen for its stability, without deviating too much from the material encountered in everyday practical circumstances (UO_2). The 186-keV gamma-ray emission rate of a bulk uranium sample is proportional to the inverse of the term

$$\left(1 + \sum_i \frac{\rho_i \mu_i}{\rho_u \mu_u}\right)$$

where ρ denotes the mass density and μ the gamma-ray mass attenuation coefficient. For pure uranium oxide materials this term equals 1.0115 for UO_2 and 1.0154 for U_3O_8. Thus, the 186-keV emission rate from the two oxides differs by 0.38%. To avoid a systematic error the stoichiometry of the standard materials must therefore also be specified.

In contrast to the stoichiometry of the sample material, the impurities have a negligible influence on the measured 186-keV emission rate when they occur within the specified limits of LWR fuel. When all impurities occur at the upper limit of concentration, the enrichment analysis result changes by only 0.04%. The same applies to the "allowed" moisture content, which influence also remains below 0.03% in the worst case.

For the gamma-ray measurements a vertical counting arrangement is foreseen, where the sample is placed above the detector. A NaI(Tl) detector or Ge detector can be applied. A collimator with 5-cm-diameter max. and at least 1 cm length is proposed to avoid edge effects. Special attention has been paid to the question whether the density of the source material could influence the counting result. Theoretical studies, undertaken to assess the effect, also came to opposing results, so a simple counting experiment was performed in which UO_2 powder of a density of 2.9 Mg/m^3 was compared with a sintered pellet of the same material with a density of 10.3 Mg/m^3. A 2-cm-long collimator of 6 mm diameter was used and within the limits of the counting statistics ($1\sigma = 0.8\%$) a non-significant difference is observed of 0.3%. It is planned to repeat this counting experiment with a larger amount of uranium oxide to allow for a collimator with larger diameter, which also will result in a higher counting rate and better counting precision.

Further attention has been given to the problem of possible effects of collimation and axial density variations on the measurements. The containers will have an inner diameter of 70 mm, and a flat bottom with a thickness tolerance of less than 0.02 mm and a flatness better than 0.4 mm in order to guarantee that the the relative variation of the gamma-ray-emission rate is less than 0.1%. The sizes of the container and the collimator determine the powder density range that can be accomodated.

In a so-called "pre-experiment" a small number of samples will be prepared and tested by different laboratories. The preparation of this small set of samples is organized by the Commission of the European Communities, Joint Research Centre, Geel. This pre-experiment will facilitate a final decision about sample specifications and counting geometry to be used.

A content of 200 g U_3O_8 will guarantee an "infinite thickness" geometry: this amount of material will produce 99.9% of the 186-keV gamma-ray intensity emitted from a sample of infinite thickness (in the measurement direction).

Of course further parameters such as grain size, homogeneity, density, impurities, moisture, will also be subject to a careful control, to ensure that these parameters will not significantly influence the accuracy of the gamma-ray measurement.

INTERCOMPARISON OF Pu ISOTOPIC RATIO MEASUREMENTS BY GAMMA-RAY SPECTROMETRY

The gamma-spectrometric determination of the enrichment of uranium is in principle a simple problem of measuring similar spectra. Therefore, the accuracy and precision obtained are mainly determined by the available standards and the reproducibility of the measurement. If peak areas from several photon energies have to be compared, the measurement problem becomes more complex. In the energy region of 300 to 1500 keV the gamma-ray emission rate can be determined to an accuracy of about 1% (confidence level 68%) as was concluded, for example, in an intercomparison experiment with sources of ^{152}Eu [10]. Another conclusion of this study was that the peak area evaluation method

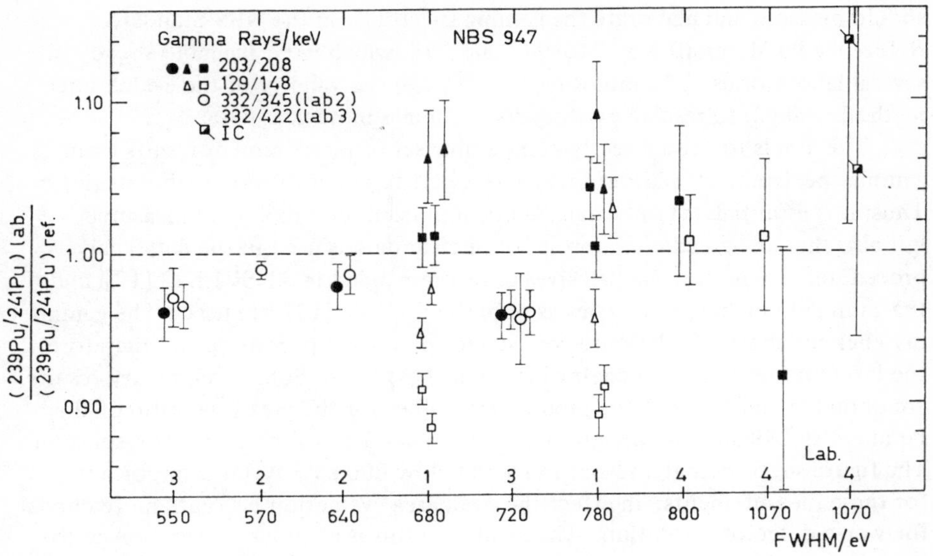

FIG.1. Preliminary results of the intercomparison of gamma-spectrometric determination of isotopic ratios in Pu sample SRM 947. The ratio of ^{239}Pu and ^{241}Pu measured is divided by the value stated by the USA National Bureau of Standards (based on mass spectrometry). The double ratios are plotted in order of detector resolution.
Legend to symbols referring to peak area determination method:
○: *channel by channel summation; background – smoothed step function*
□: *peak fitting with Gaussian; background – straight line*
Δ: *peak fitting with Gaussian; background – smoothed step function.*
Full or open symbols refer to peaks used, as indicated in the upper corner of the figure; laboratory 4 also used all main gamma peaks of ^{239}Pu and ^{241}Pu for the case indicated by IC, i.e. intrinsic calibration.

applied seems to be of minor importance as long as single peaks are considered. But the report very well recognizes that particular problems arise for lines where neighbouring lines on a non-linear background complicate the peak evaluation. A similar conclusion is reached by the Intercomparison Study of Methods for Processing Ge(Li) Gamma-Ray Spectra conducted by the IAEA [11], where the quality of the results is found to be more dependent upon the "finite tuning" by the user than on the choice of the particular evaluation method. However, it was not clear whether this conclusion also applies to real spectra elaborated in the course of the normal routine.

In a special meeting on the subject of Pu isotopic ratio determination by gamma-ray spectrometry, it appeared possible to execute a small-scale intercomparison exercise as a first step towards common actions in this field. To avoid problems with the transport of Pu, it was decided that all participants should measure and elaborate the gamma spectra from the NBS Standard Reference Pu Materials Nos 946, 947 and 948, which were available already in several laboratories. The intention is (1) to test the validity of the evaluation method, and (2) to resolve existing discrepancies in nuclear data.

The aim is to get a directly comparable set of measurement results from gamma-spectrometric plutonium analyses performed at different laboratories. Thus, it is essential not only that identical plutonium samples are measured, but also that an agreed common set of nuclear data is used in the data analysis procedure. Therefore, the half-lives as recommended in ANSI 15.22 [12], and the gamma-branching intensities as published in Ref. [13], are used. The gamma-ray energies above 120 keV are considered to be most promising and therefore the intercomparison is concentrating upon these lines. Some preliminary results are obtained and in Fig. 1 the preliminary values for the measured ratio of the Pu nuclides 239 and 241 are given, ordered according to the detector resolution. The figure suggests that a resolution well below 800 eV FWHM is important for these measurements; in effect the peak area evaluation becomes more crucial for worse detector resolution. The results, in this preliminary stage, proved to be not as good as had been hoped for, especially when one considers that most measurements were performed under good conditions and that long counting times (up to two days) were applied. Different peak area evaluation procedures applied to the same spectrum give clearly different results. Particularly with regard to the ^{238}Pu measurement results were unsatisfactory on material with a high ^{241}Pu content.

Biases in the analysis procedures presumably are the main cause of the observed scattering of the data points, and not the errors in the nuclear data. At this stage it is believed that the branching ratios of Ref. [13] are generally correct within 2%.

To identify further the systematic differences introduced by the different peak area analysis techniques (and possibly also from the intrinsic overall relative efficiency calibration, as described, for example in Ref. [14]), identical spectra recorded in one of the laboratories have been distributed to other participants. Results of these analyses are not available yet.

OTHER ACTIVITIES

In 1976 a meeting was held especially to discuss the problems of measurements of Pu in wastes. The safeguards interest in the subject is moderate as only an unimportant fraction of the total material of the plant should go into the wastes. The Commission of the European Communities, Joint Research Centre Ispra, has prepared a guide [15] in the context of the programme on radioactive waste management. The programme to prepare this guide was presented at that meeting. Further, the participants were confronted with a large variety of waste types and containers. For operational reasons only little accuracy is needed. Therefore, at the moment, only a small basis for co-operative effort in this field exists. The intercomparison of measurement results could be established after exchange of the details on the preparation of the working standards used at the different establishments.

On exploring the possibilities of an intercomparison of fuel-rod scanners, the working group noted the operation of six rod scanners of four different makes. Measurement principles applied are: active interrogation with gamma-ray measurement after irradiation (four times), active interrogation with measurement of prompt gamma and neutron radiation (once), and passive gamma-ray measurement (once). An intercomparison of the different systems would have to be based upon measurements of identical pins, or of pins which would have to be transported between the plants and laboratories. At that time also the operation of rod scanners was still hampered by some specific problems, for example the non-linear response of these instruments. Therefore, an intercomparison of rod scanners is scheduled at a later stage, when in addition more test rods and standard rods will be available.

Apart from the common actions of the group also some collaboration arose between some partners from contact made in the group. One example of such co-operation is the test of a variable dead-time counter as described in another report to this Symposium [16].

One of the members of the working group now has five batches of Pu available with isotopic compositions ranging from 5% ^{240}Pu to 21% ^{240}Pu, due to different irradiation histories and not due to blending. It has been proposed to distribute 0.5-g quantities of each to interested laboratories which would then arrange to carry out both mass-spectrometric and gamma-spectrometric measurements on the material, communicate the results and probably ultimately take part in a meeting to discuss the topic. On the basis of this exchange it would be possible to check the validity of the gamma-ray spectrometric methods. At this stage manpower limitations do not allow for a further extension of this ESARDA-inspired collaborative exercise.

In the gamma-ray spectrometric determination of the Pu isotopic composition project the results of the ^{238}Pu content were unsatisfactory. As this isotope will play an important role in the calorimetric determinations (with high ^{238}Pu content material it accounts for about 1/3 of the heat generated) it seems justified to improve the assay of this nuclide. Therefore, one of the members of the group is now undertaking to obtain better specifications of the amount of ^{238}Pu in the standards used. Alpha-spectrometric determination will yield an accuracy of 1% for this nuclide. Then, also the precision of gamma-ray spectrometry can be explored to get better results for this nuclide. At a first estimate it seems that the branching ratios in Ref. [13] for the nuclide ^{238}Pu are about 3% too high.

The regular meetings of the ESARDA Working Group on NDA form a basis for regular contact between specialists working on different NDA techniques. Individual publications of the members, e.g. Refs [17–19], indicate the wide area of applications represented in the group. The policy of the group has been, however, to limit its common actions to a few specific subjects, in order to concentrate the effort and available manpower until a result is obtained. The informal exchange of information on other related topics is useful and the available competence together with the knowledge on the applications and possibilities as reflected in the reports, Ref. [4] on techniques and Ref. [5] on standard reference material, will guide the group in designing further common actions in future. It is also expected that the ESARDA ad hoc working group on safeguards problems related to particular plants will give an input and guidance to set priorities for further efforts on specific NDA subjects.

ACKNOWLEDGEMENTS

This paper is written on behalf of the ESARDA Working Group on Techniques and Standards for Non-Destructive Analysis. The author wishes to stress that he prepared this paper in his capacity of convenor of this group. He wishes to thank all those who contributed to the work of the group and to this paper.

REFERENCES

[1] ANDERSON, A.R., "Activities of the European Safeguards Research and Development Association (ESARDA)", Safeguarding Nuclear Materials (Proc. Symp. Vienna, 1975) **2**, IAEA, Vienna (1976) 17.

[2] European Safeguards Bulletin, Office for Official Publications of the European Communities, Luxembourg, CK-PH-77-001-EN-C and CD-AH-78-001-EN-C.

[3] INTERNATIONAL ATOMIC ENERGY AGENCY Safeguards Technical Manual, **E** "Methods and Techniques", Rep. IAEA-174, IAEA, Vienna (1975) Section E.6.

[4] ADAMSON, A.S., Established Applications of Non-Destructive Techniques for Nuclear Materials Measurements, Control or Verification; Reported to ESARDA, UKAEA Rep. AERE-R-9167, ESARDA Rep. ESARDA 6 (1978).

[5] BIGLIOCCA, C., CUYPERS, M., LEY, J., List of Reference Materials for Non-Destructive Assay Existing Within Six Countries of the E.C., the Commission and the IAEA, Commission of the European Communities, Rep. EUR 6089 EN, ESARDA Rep. ESARDA 5 (1978).

[6] BINGHAM, C.D., YOLKEN, H.T., REED, W.P., Non-destructive assay can be traceable, J. Inst. Nucl. Materials Management **5** 2 (1976) 32.

[7a] MATUSSEK, P., OTTMAR, H., "Gamma-ray spectrometry for in-line measurements of ^{235}U enrichment in a nuclear fuel fabrication plant", Safeguarding Nuclear Materials (Proc. Symp. Vienna, 1975) **2** IAEA, Vienna (1976) 223.

[7b] BEETS, C., et al., "Acquisition and verification of reference light-water reactor rods", IAEA-SM-231/24, these Proceedings, Vol.II.

[8] BUSCA, G., et al., "Procurement and use of reference materials for calibrating non-destructive assay equipment", IAEA-SM-231/120, these Proceedings, Vol.II.

[9] INTERNATIONAL ATOMIC ENERGY AGENCY, Advisory Group Meeting on the Use of Physical Standards in Inspection and Measurements of Nuclear Materials by Non-Destructive Techniques, IAEA Rep. AG-112 (1977).

[10] DEBERTIN, K., Intercomparison of Gamma-Ray Emission Rate Measurements by Means of Germanium Spectrometers and ^{152}Eu Sources, PTB Report PTB-Ra-7 (ICRM-S-3) (1978) 32.

[11] PARR, R.M., HOUTERMANS, H., SCHAERF, K., Private communication.

[12] AMERICAN NATIONAL STANDARDS INSTITUTE, American National Standard Calibration Techniques for the Calorimetric Assay of Plutonium-Bearing Solids Applied to Nuclear Materials Control, American National Standards Institute Inc., New York, N.Y. (1975).

[13] GUNNINK, R., EVANS, J.E., PRINDLE, A.L., A Re-evaluation of the Gamma-Ray Energies and Absolute Branching Intensities of ^{237}U, ^{238,239,240,241}Pu and ^{241}Am, Lawrence Livermore Laboratory, Rep. UCRL-52139 (1976).

[14] HARRY, R.J.S., AALDIJK, J.K., BRAAK, J.P., "Gamma-spectrometric determination of isotopic composition without use of standards", Safeguarding Nuclear Materials (Proc. Symp. Vienna, 1975) 2, IAEA, Vienna (1976) 235.

[15] BIRKHOFF, G., NOTEA, A., Monitoring of Plutonium Contaminated Solid Waste Streams, Commission of the European Communities, Reps EUR 5635 e, EUR 5636 e, EUR 5637 e.

[16] LEES, E.W., RODGERS, J.G., IAEA-SM-231/51, these Proceedings, Vol.II.

[17] EBERLE, H., et al., Elemental and Isotopic Concentration Analyses on Nuclear Fuels Using Non-Destructive Assay Techniques, J. Nucl. Materials, to be published.

[18] STEIN, G., et al., Zerstoerungsfreie Messsysteme zur Kernmaterialueberwachung in der KFA Jülich, KFA Rep. Jül-1473 (1977).

[19] ADAMSON, A.S., Summary of safeguards in the United Kingdom, J. Inst. Nucl. Materials Management **6** 2 (1977) 38.

DISCUSSION

R. NILSON: Will participation in the sample measurement exchange programme be limited to laboratories in the Community countries or will United States laboratories, including those in fuel supplier facilities, be permitted to participate as well?

R.J.S. HARRY: The ESARDA Working Groups are in principle open to any observers or participants who can contribute to the progress of the work.

R. NILSON: Are any U_3O_8 standards planned?

R.J.S. HARRY: It is intended to make the uranium oxide standards for gamma-spectrometric determination of enrichment commercially available, in particular from the National Bureau of Standards in the United States of America and the Central Bureau of Nuclear Measurements of the Commission of the European Communities at Geel in Belgium.

Requests to use the characterized reference material for other purposes can be considered only in so far as such use will not delay the progress of the project. An amount of spare material will be reserved for future needs.

ACQUISITION AND VERIFICATION OF REFERENCE LIGHT-WATER REACTOR RODS

C. BEETS*, G. BUSCA**, J. COLARD*, M. CORBELLINI†, M. CUYPERS†, B. FONTAINE*, F. FRANSSEN*, D. REILLY†, F. VAN CRAENENDONCK§, P. VAN LOO**

* CEN/SCK, Mol, Belgium

** CEC Safeguards Directorate, Luxembourg

† CEC Joint Research Centre, Ispra, Italy

§ FBFC, Dessel, Belgium

Abstract

ACQUISITION AND VERIFICATION OF REFERENCE LIGHT-WATER REACTOR RODS.

In manufacturing LWR fuel assemblies, the UO_2 pellet homogeneity is being tested using a rod scanner device, which is fundamentally intended to detect in very short time (about 1 min) off-specification pellets, i.e. pellets with an enrichment departing x% from the nominal enrichment, the value of x being currently between 10 and 15%. Furthermore, this device is able to verify the ^{235}U content and the weight of the individual rods; therefore, it could be used in nuclear material accountancy, either directly at the rod KMP, or indirectly at the pellet KMP, giving further assurance of the homogeneity of the pellets, or better confidence in the representative sampling. To test the rod scanner device, the operator produces a set of 13 reference rods. It is evident that a high degree of assurance for these reference rods is required, and this is best achieved by collaboration between the operator and the safeguards authorities. Such collaboration was established between the operator, the safeguards authorities (Euratom and IAEA) and nuclear centres (CCR Euratom, Ispra and CEN/SCK Mol), to agree on the way in which characterization should be performed so as to be adequate for both management and verification purposes. Procedures followed by the operator and the safeguards authorities are described. These procedures include: (a) chemical analysis; (b) mass-spectrometric measurements performed by three laboratories; (c) non-destructive measurements of enrichment using NaI and Ge detectors effected by three teams working in plant environment; and (d) surveillance and containment measures. An estimate of the cost for safeguards verification (manpower, samples to be measured, surveillance) is given.

INTRODUCTION

When manufacturing LWR fuel assemblies, the UO_2 pellet homogeneity is tested by a rod scanner, the basic purpose of which is to detect in a short time (about 1 min) off-specification pellets, i.e. pellets whose enrichment deviates from the nominal enrichment by x%, the value of x lying usually between 10 and 15%, with a 5% false rejection rate.

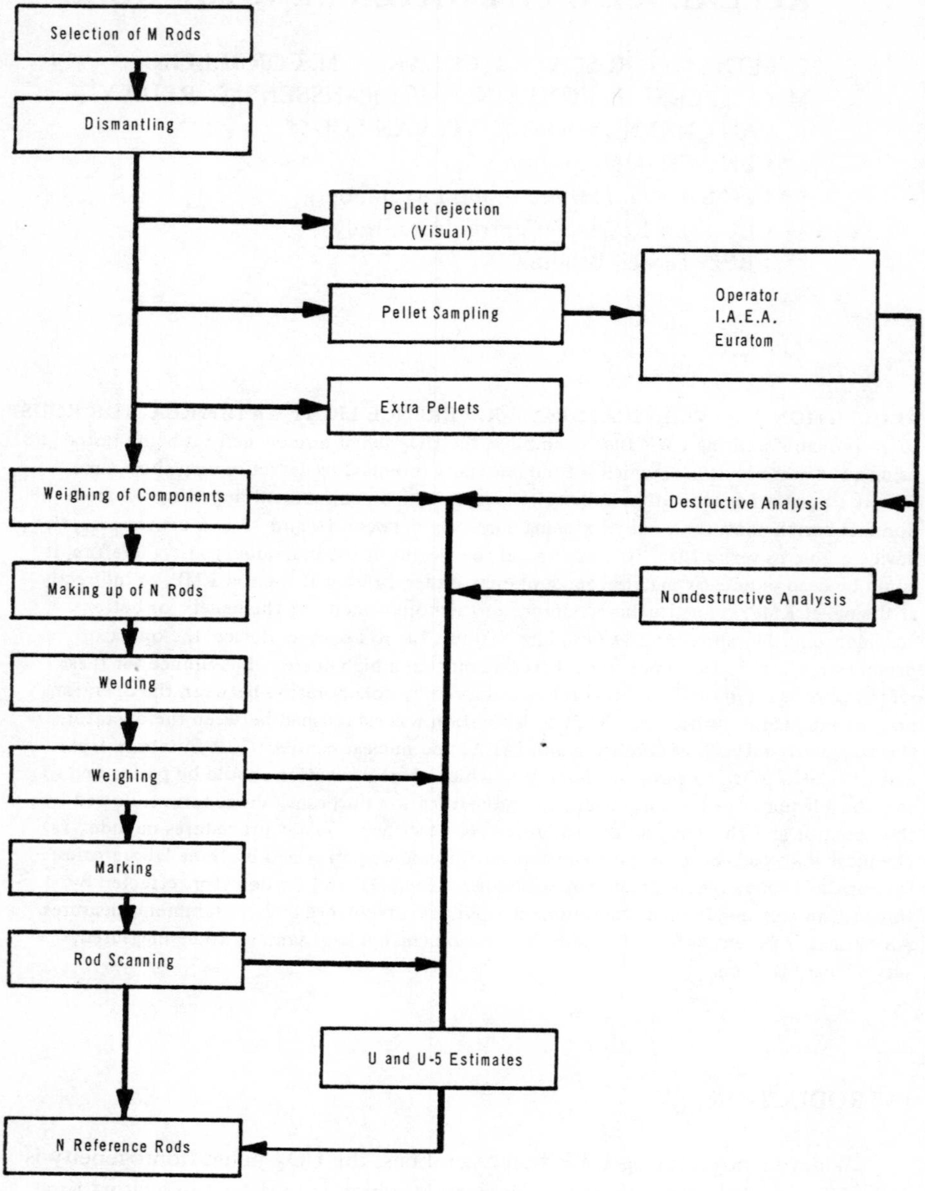

FIG.1. Procedure for the preparation of reference LWR rods.

TABLE I. SAMPLING PLAN

		m	n	p Operator	p Euratom	p I.A.E.A.
A	2.1 %	2	1	30	28	4
B	2.6 %	2	1	14	28	4
C	3.1 %	3	2	37	14	2

m : number of selected fabrication rods
n : number of reference rods
p : number of selected pellets

Furthermore, as this device is able to verify the ^{235}U content and the weight of each rod, it could be used in nuclear material accountancy, either directly at the rod KMP, or indirectly at the pellet KMP, giving better confidence in representative sampling.

The rod scanner is calibrated by means of a set of reference rods. It is obvious that a high degree of assurance for these reference rods is required, and this is best achieved in collaboration between the operator and the safeguards authorities. Such a collaboration has been established between the operator, the safeguards authorities (Euratom, Luxembourg and IAEA), and Nuclear Research Centres (Euratom, Ispra and CEN/SCK, Mol).

The procedure for preparing reference UO_2 rods followed by the operator and the safeguards authorities is outlined in Fig. 1.

For each enrichment, a total of m production rods was dismantled in order to obtain n reference rods. The numbers m and n are given in Table I for each enrichment, together with the number p of sampled pellets taken by the operator, Euratom and the IAEA.

To test the rod scanner sensitivity, the operator produces a set of 13 UO_2 powder samples of different enrichments, from which off-specification pellets were made. For each enrichment, 100 g UO_2 powder samples and about 20 pellets were taken by the operator and Euratom for destructive and non-destructive analysis. Off-specification pellets were introduced into dismantled nominal enrichment rods in order to get "special non-homogeneous rods".

This paper describes the procedures outlined in Fig. 1 in order to obtain a set of homogeneous rods; however, measurements related to "special non-homogeneous rods" are also presented because they enhance confidence in the experimental results. For clarity, the procedures and the measurements are referred to as type A (homogeneous rods) and type B (non-homogeneous rods).

The division of the work among the participants was decided according to the type of tasks (rod dismantling, pellet sampling, rod reassembly and sealing, etc.; the physical form of the items (powder, pellet, rod); and the chosen methods (destructive analysis) and equipment.

1. SEQUENCE OF THE TASKS

The task sequence was as follows:

(1) Enrichment measurements of 12 powder batches with an enrichment varying from 1.9 to 4% (type B)
(2) Enrichment measurement of selected pellets from these batches (type B)
(3) Dismantling of a set of homogeneous rods (type A)
(4) Post-dismantling tasks under continuous containment and surveillance measures (type A):
  (i) Random sampling of pellets
  (ii) Weighing of all components (pellet, spring, can, end-plug)
  (iii) Calibration verification of the scales
  (iv) Closing of the rods
  (v) Marking and punching of the rods
  (vi) Identification by macrophotography
(5) Pellet sampling for destructive analysis (type A)
(6) Sealing of containers for powders or pellets (type B) and pellets (type A)
(7) Destructive and non-destructive analyses of the pellets (type A).

2. ENRICHMENT MEASUREMENTS OF TWELVE UO_2 POWDER BATCHES WITH AN ENRICHMENT VARYING FROM 1.9 TO 4%

2.1. Destructive analysis: Mass-spectrometric measurements

The 1-g samples were analysed by three different laboratories – A, B, and C. Independent estimates of the precision were deduced by applying Grubb's analytical method to the results (see Table II) [2]. The results of Lab. A are more precise. However, the A-B and A-C differences are all negative, while the B-C differences are randomly distributed about zero. An estimate of the systematic difference between Lab. A and Lab. B and C could be expressed by

$$= \frac{1}{2}(|m_{AB}| + |m_{BC}|) = 0.003 \pm 0.001 \text{ or } 0.10 \pm 0.04 \text{ rel.\%}.$$

TABLE II. COMPARISON OF MASS-SPECTROMETRIC DATA BY USING GRUBB'S ANALYSIS

	GRUBB'S ANALYSIS	3 METHODS 10 SAMPLES	
	A	B	C
	1.899	1.9035	1.900
	2.320	2.3226	2.320
	2.439	2.4426	2.440
	2.670	2.6747	2.671
	2.800	2.8048	2.803
	2.915	2.9177	2.922
	3.020	3.0227	3.024
	3.428	3.4310	3.434
	3.601	3.6030	3.602
	3.779	3.7820	3.781
	A-B	A-C	B-C
	-0.0045	-0.001	0.0035
	-0.0026	0.000	0.0026
	-0.0036	-0.001	0.0026
	-0.0047	-0.001	0.0037
	-0.0048	-0.003	0.0018
	-0.0027	-0.007	-0.0043
	-0.0027	-0.004	-0.0013
	-0.0030	-0.006	-0.0030
	-0.0020	-0.001	0.0010
	-0.0030	-0.002	0.0010
	-0.00336	-0.00260	0.00076
	9.760E-07	5.600E-06	7.567E-06
	A	B	C
σ^2	4.956E-07	1.472E-06	6.096E-06

2.2. Non-destructive analysis: Gamma-spectrometry

2.2.1. Data acquisition

The measurements were done under field conditions by using

Two NaI + MCA equipment (No. 1 and No. 2)
One intrinsic Ge detector + MCA (No. 3)
One Ge(Li) detector (No. 6)
Two SAM-2 instruments (No. 4 and No. 5)

The 100-g UO_2 powder samples were measured through a collimator.

Concerning experiment No. 3, two small ^{154}Eu sources were added at the bottom of the collimator; the 244-keV peak of the ^{154}Eu was used as a reference for the counting time given by the MCA.

The technical characteristics of these instruments are given in BLG522 [1]. The net response attributed to the 185.7-keV gamma rays is the accumulated counts in the peak region minus a multiple of the counts accumulated in a neighbouring background region. In all the measurements, only an upper background region was monitored.

2.2.2. Data reduction and analysis

The enrichment z is determined by the relation

$$z = ax + by$$

where x = gross counts in the 185.7-keV peak region and y = counts in the background region in a given time interval.

For each sample i, the data set is the following:

(1) The shipper's mass-spectrometer data, $z_{\mu,i}$, considered as the true value of the sample enrichment; and
(2) A number q_i of $x_{i,j}$, $y_{i,j}$ pairs.

The parameters a and b defined by linear regression analysis, are based on counts, and the calibration error σ_c can be estimated from these counts [1].

To test the coherence of gamma measurements,

$$z_{\gamma,i,j} = ax_{i,j} + by_{i,j}$$

the $t^*_{i,j} = \dfrac{s_i - r_i}{\sigma(r_i)}$ statistics is used [3] in which

TABLE III. ENRICHMENT MEASUREMENTS BY GAMMA SPECTROMETRY
Experiment No. 3. Instrumentation: Intrinsic Ge + M.C.A.

DATA SET					SR DIFFERENCES		
Ident.	s_i	X_i	Y_i	r_i	$s_i - r_i$	$\sigma(r_i)$	t*
1	1.899	249500	55590	1.887	0.01226	0.00571	2.146
2	2.320	292500	56070	2.315	0.00463	0.00609	0.761
3	2.439	303700	55720	2.432	0.00659	0.00618	1.066
4	2.670	328000	54560	2.691	-0.02091	0.00636	-3.286
5	2.800	342100	56450	2.812	-0.01176	0.00649	-1.811
6	2.915	352200	56290	2.916	-0.00054	0.00657	-0.082
7	3.020	364400	57270	3.028	-0.00756	0.00667	-1.133
8	3.196	379100	56120	3.189	0.00697	0.00677	1.029
9	3.428	404500	56840	3.437	-0.00926	0.00697	-1.328
10	3.602	421700	57430	3.604	-0.00218	0.00710	-0.307
11	3.779	436500	56770	3.761	0.01791	0.00720	2.488
12	3.972	458000	57620	3.968	0.00353	0.00736	0.480

[a] SR = Shipper-receiver.

$$s_i = z_{\mu,i}$$

$$r_i = z_{\gamma,i,j}$$

$$\sigma_{i,j} = \frac{\sqrt{a^2 x_{i,j} + b^2 y_{i,j}}}{a x_{i,j} + b y_{i,j}} \cdot z_{\gamma,i,j}$$

$$\sigma(r_i) = \sqrt{\sigma_c^2 + \sigma_{i,j}^2}$$

Detailed analyses of experiments Nos 1–5 are given in Ref. [1], a typical table is presented (Table III) and the t* statistics are summarized in Table IV. The t* statistics sign distribution is acceptable for all the samples; however, the ranges $|t^*_{max}| + |t^*_{min}|$ differ appreciably, which means that the actual standard deviations are not only due to counting statistics, but also to other causes, i.e. electronic instability, container wall thickness variations, etc. With good estimates of the standard deviation, the t* distribution ranges would be the same for all the experiments, since the same samples are used throughout. This is our working hypothesis to get estimates of the actual standard deviation. Range 4, corresponding to Pr(0.95) so that a result lies in that range, was arbitrarily chosen as a reference; the estimates of the actual standard deviation were calculated by

TABLE IV. 100 g POWDER SAMPLES *(t* statistics)*

Ident.	Experiments 1	2	3	4	5
1	0.17	-0.47	2.1	-2.2	0.84 -0.72
2	-0.45	-0.02	0.76	-0.57	-2.6
3	-0.01	0.99	1.1	0.86	1.4
4	0.92	-0.32	-3.3	-0.15	0.73
5	1.7	0.48	-1.8	1.0	0.66
6	-0.99	0.67	-0.1	4.1	2.1
7	1.4	0.42	-1.1	3.1	-0.92
8	-2.6	-0.56	1.0	0.27	-2.3
9	-1.1	-1.0 -1.4	-1.3	-0.29	1.6
10	-3.1	2.0	-0.3	-2.3 -1.1 -1.8	-0.95
11	1.4	1.6	2.5	-0.91	-0.38 -0.03 -1.7
12	2.6	0.05	0.48	0.28 0.16	0.97 2.2
Range	5.7	3.4	5.8	6.4	4.8
$\sigma(r_i)$	0.0078	0.0220	0.0066	0.0092	0.0096

multiplying $\overline{\sigma(r_i)}$ by 1/4 of the observed range. The relative standard deviations are 0.4, 0.7, 0.3, 0.5 and 0.4% for experiments 1–5, with measurement times of 600, 600, 10 000 s, 720, and 720 s, respectively.

The enrichment measurements of the experiment No. 6 (10% efficiency Ge(Li) detector) were calculated directly from the ratio between the 185.7-keV net peak area of the sample and the same area of a reference 100-g UO_2 powder with a known enrichment in an identical container. The Compton subtraction was done routinely using a microprocessor. The range is 3.63 and $\sigma(r_i) = 0.015$; therefore, the relative standard deviation is 0.5% for a 600-s measurement time interval.

TABLE V. 100 g UO_2 POWDER SAMPLES *(Enrichment: NDA results)*

Ident.	Experiments						
	1	2	3	4	5	6	Weighted Means
1	1.898	1.909	1.887	1.907	1.899	1.896	1.896 ± 0.003
2	2.323	2.320	2.315	2.325	2.343	-	2.324 ± 0:010
3	2.439	2.418	2.432	2.431	2.427	2.446	2.433 ± 0.003
4	2.663	2.677	2.691	2.671	2.663	-	2.675 ± 0.006
5	2.787	2.789	2.812	2.790	2.794	2.827	2.802 ± 0.006
6	2.963	2.900	2.916	2.877	2.895	2.923	2.909 ± 0.012
7	3.009	3.011	3.028	2.991	3.029	3.026	3.019 ± 0.006
8	3.217	3.208	3.189	3.193	3.219	3.178	3.200 ± 0.007
9	3.437	3.450	3.437	3.431	3.412	3.414	3.430 ± 0.005
10	3.627	3.557	3.604	3.607	3.612	3.570	3.603 ± 0.007
11	3.767	3.745	3.761	3.788	3.786	3.778	3.771 ± 0.006
12	3.950	3.971	3.968	3.970	3.956	3.958	3.962 ± 0.04
r.s.d.	0.4 %	0.7 %	0.3 %	0.5 %	0.4 %	0.5 %	0.23 %
w	0.19	0.06	0.32	0.12	0.19	0.12	

The weighted means of the enrichments with their relative standard deviations are given in Table V.

Grubb's analysis gives the comparison between all the NDA devices. (See Table VI.)

The random variances as deduced from Grubb's analysis are in agreement with the estimated relative standard deviations.

3. NON-DESTRUCTIVE MEASUREMENTS OF PELLETS WITH ENRICHMENT FROM 1.9 TO 4%

Data analysis

The procedure described in Section 2.2.2 was applied, the t* statistics is given in Table VII. The weighted means are calculated and the r.s.d. are 0.3, 1.4, 1.0 and 1.3% for 1800, 600, 720, and 720 s counting times, respectively. (Table VIII).

TABLE VI. COMPARISON OF NDA METHODS BY GRUBB'S ANALYSIS

Identif. \ Experiment	No. 1	No. 2	No. 3	No. 4	No. 5	No. 6
1	1.898	1.909	1.887	1.908	1.899	1.896
2	2.323	2.320	2.315	2.325	2.343	-
3	2.439	2.418	2.432	2.431	2.427	2.446
4	2.663	2.677	2.691	2.671	2.663	-
5	2.787	2.789	2.812	2.790	2.794	2.827
6	2.923	2.900	2.916	2.877	2.895	2.923
7	3.009	3.011	3.028	2.991	3.029	3.026
8	3.217	3.208	3.189	3.193	3.219	3.178
9	3.437	3.455	3.437	3.431	3.412	3.414
10	3.627	3.557	3.604	3.607	3.612	3.570
11	3.767	3.745	3.761	3.788	3.786	3.778
12	3.950	3.971	3.968	3.970	3.956	3.958

$m_{12} = 0.0104$ $s_{12} = 0.0294$ $v_{12} = 8.652\ 10^{-4}$	$m_{13} = 0.0033$ $s_{13} = 0.0229$ $v_{13} = 5.230\ 10^{-4}$	$m_{14} = 0.0083$ $s_{14} = 0.0286$ $v_{14} = 8.162\ 10^{-4}$	$m_{15} = 0.0038$ $s_{15} = 0.0247$ $v_{15} = 6.091\ 10^{-4}$	$m_{16} = 0.0078$ $s_{16} = 0.0306$ $v_{16} = 9.353\ 10^{-4}$
	$m_{23} = -0.0071$ $s_{23} = 0.0199$ $v_{23} = 3.979\ 10^{-4}$	$m_{24} = -0.0022$ $s_{24} = 0.0234$ $v_{24} = 5.494\ 10^{-4}$	$m_{25} = -0.0067$ $s_{25} = 0.0257$ $v_{25} = 6.584\ 10^{-4}$	$m_{26} = -0.0058$ $s_{26} = 0.0267$ $v_{26} = 7.153\ 10^{-4}$
		$m_{34} = 0.0024$ $s_{34} = 0.0207$ $v_{34} = 4.272\ 10^{-4}$	$m_{35} = 0.0004$ $s_{35} = 0.0210$ $v_{35} = 4.417\ 10^{-4}$	$m_{36} = 0.0018$ $s_{36} = 0.0174$ $v_{36} = 3.020\ 10^{-4}$
			$m_{45} = -0.0045$ $s_{45} = 0.0172$ $v_{45} = 2.974\ 10^{-4}$	$m_{46} = -0.0031$ $s_{46} = 0.0281$ $v_{46} = 7.874\ 10^{-4}$
				$m_{56} = 0.0013$ $s_{56} = 0.0253$ $v_{56} = 6.391\ 10^{-4}$

$\Sigma v_{i,j} = 8.9646\ 10^{-3}$

$S_1 = 3.7488\ 10^{-3}$	$S'_1 = 5.2158\ 10^{-3}$
$S_2 = 3.1862\ 10^{-3}$	$S'_2 = 5.7784\ 10^{-3}$
$S_3 = 2.0918\ 10^{-3}$	$S'_3 = 6.8728\ 10^{-3}$
$S_4 = 2.8776\ 10^{-3}$	$S'_4 = 6.0870\ 10^{-3}$
$S_5 = 2.6457\ 10^{-3}$	$S'_5 = 6.3189\ 10^{-3}$
$S_6 = 3.3791\ 10^{-3}$	$S'_6 = 5.5855\ 10^{-3}$

$\sigma_1^2 = 4.89\ 10^{-4}$	$\sigma_4^2 = 2.71\ 10^{-4}$
$\sigma_2^2 = 3.48\ 10^{-4}$	$\sigma_5^2 = 2.13\ 10^{-4}$
$\sigma_3^2 = 0.747\ 10^{-4}$	$\sigma_6^2 = 3.97\ 10^{-4}$

TABLE VII. PELLET SAMPLES *(t^* statistics)*

Ident.	Experiments 1	2	4	5
1	-0.11	-0.37	-0.748	-1.1
2	-0.20	2.2	2.5	2.9
3	0.28	-1.0	0.70	4.2
4	0.05	0.86	-0.77	-1.3
5	2.2	1.7	-0.72	-0.52
6	-1.1	0.05	0.48	-2.0
		-0.45		
7	-1.1	-0.84	-1.6	-3.4
8	0.67	-2.9	-2.3	-2.6
9	-1.5	0.62	-2.6	-1.5
10	0.78	-2.34	-3.7	1.2
11	-1.2	-0.46	2.1	3.7
			-0.4	3.2
			0.14	2.3
			-0.80	-0.20
			-0.63	-0.56
			0.34	-1.1
			0.40	-1.4
			-0.07	0.58
			-2.3	-1.5
			1.3	-3.2
			1.2	-1.4
12	0.24	-3.20	-	-
13	1.0	1.2	-4.1	0.56
		0.28		
		0.70		
Range	3.8	5.5	6.6	7.6
$\sigma(r_i)$	0.009	0.031	0.019	0.021

TABLE VIII. PELLET SAMPLES *(Enrichment: NDA results)*

Ident.	Experiments				
	1	2	4	5	Weighted Means
1	1.900	1.910	1.912	1.921	1.902 ± 0.003
2	2.322	2.252	2.277	2.262	2.314 ± 0.012
3	2.437	2.470	2.427	2.354	2.434 ± 0.011
4	2.670	2.643	2.684	2.697	2.671 ± 0.005
5	2.781	2.746	2.813	2.811	2.783 ± 0.008
6	2.925	2.941	2.906	2.957	2.925 ± 0.006
7	3.030	3.046	3.049	3.094	3.035 ± 0.009
8	3.190	3.288	3.240	3.253	3.200 ± 0.015
9	3.442	3.409	3.477	3.460	3.444 ± 0.007
10	3.595	3.677	3.678	3.575	3.604 ± 0.016
11	3.790	3.793	3.778	3.777	3.789 ± 0.005
12	3.970	4.075	-	-	3.975 ± 0.005
13	4.182	4.169	4.277	4.179	4.189 ± 0.005
r.s.d.	0.3 %	1.4 %	1.0 %	1.3 %	0.3 %
w	0.83	0.04	0.08	0.05	

TABLE IX. COMPARISON BETWEEN MASS-SPECTROMETRIC AND GAMMA-SPECTROMETRIC MEASUREMENTS

Identification	Weighted values of mass spec.	Weighted values of gamma spec.
1	1.900	1.898
2	2.321	2.321
3	2.440	2.433
4	2.671	2.674
5	2.801	2.797
6	2.916	2.913
7	3.021	3.023
8	3.197	3.200
9	3.429	3.434
10	3.602	3.603
11	3.780	3.776
12	3.973	3.966

TABLE X. COMPARISON OF MASS-SPECTROMETRY AND GAMMA-SPECTROMETRY RESULTS

Grubb's Analysis
3 Methods
10 Samples

	B	C	Y
	1.90350	1.90000	1.89800
	2.32260	2.32000	2.32100
	2.44260	2.44000	2.43300
	2.67470	2.67100	2.67400
	2.80480	2.80300	2.79700
	2.91770	2.92200	2.91300
	3.02270	3.02400	3.02300
	3.43100	3.43400	3.43400
	3.60300	3.60200	3.60300
	3.78200	3.78100	3.77600
	B-C	B-Y	C-Y
	0.00350	0.00550	0.00200
	0.00260	0.00160	-0.00100
	0.00260	0.00960	0.00700
	0.00370	0.00070	-0.00300
	0.00180	0.00780	0.00600
	-0.00430	0.00470	0.00900
	-0.00130	-0.00030	0.00100
	-0.00300	-0.00300	0.00000
	0.00100	0.00000	-0.00100
	0.00100	0.00600	0.00500
	0.00076	0.00326	0.00250
	7.567E-06	1.636E-05	1.606E-05
	B	C	Y
σ^2	3.934E-06	3.633E-06	1.242E-05

γ = weighted values of all the gamma-spectrometry measurements.

4. BATCH ENRICHMENTS – DESTRUCTIVE AND NON-DESTRUCTIVE MEASUREMENTS COMPARISON

The non-destructive measurements on powder and pellets were weighed for each batch. These batch enrichment estimates are compared with the mass-spectrometric measurements in Table IX.

Grubb's analysis applied to these results (see Table X) shows that the overall precision of the non-destructive measurements carried out in plant conditions is quite satisfactory. As can be seen, the Lab. A mass-spectrometry results are not included in that analysis because these measurements were used to calibrate the NDA techniques.

5. DISMANTLING OF A SET OF HOMOGENEOUS RODS

The operator dismantled three 3.1% nominal enrichment rods under continuous surveillance by IAEA and Euratom inspectors until the two reference rods were completed.

The inspectors selected a number of pellets for verification purposes, and the operator rejected pellets with small visual defects.

The reference rod components were weighed by the operator using a balance of 0.01 g accuracy for springs and plugs, and of 0.1 g accuracy for tubes and rods.

The calibration, loading, welding, helium test, radiographical inspection of the welds, and visual and dimensional inspection, were all done by the operator under continuous surveillance, following the same specifications and conditions as in a normal rod production; furthermore, the marking with an electrical marker, punching with an Euratom tool, and macrophotography of the ends were done by the inspectors.

The required time for the entire procedure was about half a day.

The same procedure was followed for two 2.1% and two 2.6% nominal enrichment rods to obtain one reference rod of these enrichments, with an equivalent time of half a day for the two reference rods.

Data relating to chemical and isotopic analysis are carefully collected and conserved, so as to guarantee the traceability of the used UO_2.

6. REFERENCE RODS: ENRICHMENT MEASUREMENTS OF SELECTED PELLETS

6.1. Destructive analysis

The pellets were analysed by three different laboratories, A, D and E. Independent estimates of the random variances were deduced by applying Grubb's analytical method and the weighted means were calculated, giving to each observation a weight proportional to the reciprocal of the independent estimate of its variance.

The results are 2.089, 2.563 and 3.104 for the A,B and C enrichments, respectively.

TABLE XI. COMPARISON OF ESTIMATED ENRICHMENTS AND RESULTS

Methods	Enrich. A	Enrich. B	Enrich. C
Mass spec. A	2.087 ± 0.005	2.562 ± 0.006	3.101 ± 0.007
Mass spec. D	2.089 ± 0.001	2.563 ± 0.001	3.104 ± 0.0065
Mass spec. E	2.091 ± 0.003	2.566 ± 0.003	3.113 ± 0.005
Exp. No. 7	2.085 ± 0.002	2.560 ± 0.002	3.096 ± 0.002
Exp. No. 1, 2, 6	2.106 ± 0.007	2.569 ± 0.006	3.102 ± 0.003
Estimated enrich.	2.088 ± 0.002	2.564 ± 0.003	3.103 ± 0.003

6.2. Non-destructive analysis

The procedure described in Section 2.2.2 was followed.

Detailed results of experiments 1, 2 and 6, done under field conditions, are given in Ref. [1]. The enrichment determinations are based on the Lab. A mass-spectrometric determinations of the thirteen UO_2 batch enrichments varying from 1.9 to 4.2%, and of the three nominal enrichments, and therefore constitute a first set of NDA results.

In experiment No. 7, a known and fixed amount of uranium (about 0.5g) was dissolved in nitric acid-sulphuric acid mixture and transferred into a Baird atomic 1-inch polyethylene tube as a 5-ml solution. The amount of ^{235}U isotope in the tubes was determined by comparing their gamma activity at 186 keV to the activities of reference sources prepared similarly from a set of NBS isotopic standards. The total uranium in the source is known from chemical titration and serves to calculate the ^{235}U enrichment [4]. The results of this experiment constitute a second set of NDA results.

6.3. Reference rod enrichments: Destructive and non-destructive measurements

The means of the two sets of NDA measurements and of the three mass-spectrometric measurements were chosen as the best estimates of the enrichments. As the precision of the NBS standards is about 0.1%, and as the random errors of the different sets are of the same order of magnitude, the overall precision has been estimated at 0.12%.

Therefore, the estimates of enrichment are

Enrichment A: 2.088 ± 0.002
Enrichment B: 2.564 ± 0.003
Enrichment C: 3.103 ± 0.003

TABLE XII. PELLET URANIUM CONCENTRATIONS *(in wt. %)*

Enrichment	Lab. 1	Lab. 2	Lab. 3	Theoretical value
A				
2.1 %	88.116 ± 0.002	88.11	87.82	88.145
2.088 %	88.105 ± 0.002	88.11		
		88.12		
		88.14		
		88.17		
		88.13		
		88.12		
		88.16		
B				
2.6 %	88.109 ± 0.0016	88.14	87.80	88.145
2.564 %	88.098 ± 0.0013	88.12		
		88.13		
		88.13		
		88.12		
		88.16		
		88.14		
		88.11		
C				
3.1 %	88.118 ± 0.005	88.16	87.80	88.144
3.103 %	88.116 ± 0.013	88.12		
	88.125 ± 0.010	88.15		
		88.18		
		88.15		
		88.08		
		88.13		
		88.14		
		88.13		
		88.14		
		88.15		
		88.14		

These estimates are compared with the different sets in Table XI. The last NDA set was not taken into account for the mean of enrichment A, owing to the random error, being rather high compared with the other set.

All the measurements are in agreement with the stated errors.

7. REFERENCE RODS: URANIUM AND ^{235}U CONTENT ESTIMATES

The uranium concentrations were measured by three laboratories. The results are given in Table XII. The stated errors are one order of magnitude lower than the weighing errors and can be neglected in the estimates of the U content by rod. The weight, the U content and the ^{235}U content are tabulated as follows:

	Weight (g)	U (g)	^{235}U (g)
Enr. A	1972.96	1739.07	36.312
Enr. B	1979.95	1745.23	44.748
Enr. C	1972.27	1738.44	53.944
	1999.93	1762.82	54.700
	1996.93	1760.17	54.618

8. CONCLUSIONS

The accuracy and precision of the U content of a reference rod are essentially those of the weight measurement. A precision of 0.05% has been achieved.

The accuracy and precision of the ^{235}U content result from the combined errors on the U content and on the enrichment. Because NBS enrichment standards are certified to ± 0.1%, the combined errors on the U content and on the enrichment standard limit the accuracy of the ^{235}U content to ± 0.15%.

The objective of the experiment is to obtain the best ^{235}U estimates. Different techniques were applied with a view to eliminating systematic errors to the greatest extent possible. The random errors of the different techniques, as they derive from the present experiment, are given in Table XII.

To assess the relative merits of the techniques, the number of duplicates required to attain the level of, say, 0.1% (comparable to the other components), can be deduced from the quoted r.s.d. as the square of r.s.d./0.1. Because

TABLE XIII. TOTAL MEASUREMENT EFFORT TO REACH PRECISION OF OF 0.1%

		RSD (%)	Measurement time, (incl. prepar.) (s)	No. of duplicates to reach 0.1%	Total measurement effort to reach 0.1% (s)
Lab.	MS	0.15	10 000	2.25	22 500 (20 000)
	NaI, titration	0.25	5 000	6.25	31 250 (30 000)
Plant	Ge	0.3	5 000	9	45 000 (45 000)
	NaI with MCA	0.4	600	16	9 600 (10 000)
	SAM-2	0.5	720	25	18 000 (20 000)

man-power is the main factor in the cost of any measurement, the total time of measurements gives a rough indication of the effort required by each technique.

The cost of applying continuous containment and surveillance measures can be roughly estimated by the length of inspector presence necessary in order to verify the weight measurements and to apply sealing and identification techniques. In this experiment, the total time amounted to about two days for five reference rods.

ACKNOWLEDGEMENT

In the characterization of the standards described in this paper, the assistance of D.R. Terrey, S. Deron, H. McKown, E. Kuhn, and M. Miguel of the IAEA is gratefully acknowledged.

REFERENCES

[1] BEETS, C., et al., BLG. 522 (to be published).
[2] INTERNATIONAL ATOMIC ENERGY AGENCY, IAEA Safeguards Technical Manual: Part F. Statistical Concepts and Techniques, 1 IAEA-174, IAEA, Vienna (1977) 48.
[3] JAECH, J.L., Statistical Methods in Nuclear Material Control, TID-26298 (1973) Chap. 8.
[4] MIGUEL, M., DERON, S., "Performance of an Am-241 stabilized NaI scintillation gamma spectrometer for the determination of the U-235 isotope abundance in small safeguards samples", Rep. IAEA-RL/31, IAEA, Vienna (1975).

A NEW APPROACH FOR SAFEGUARDING ENRICHED URANIUM HEXAFLUORIDE BULK TRANSFERS

L.W. DOHER
Rockwell International,
Rocky Flats Plant,
Golden, Colorado

P.E. PONTIUS, J.R. WHETSTONE
Center for Mechanical Engineering and
Process Technology,
National Bureau of Standards, Washington DC,
United States of America

Abstract

A NEW APPROACH FOR SAFEGUARDING ENRICHED URANIUM HEXAFLUORIDE BULK TRANSFERS.

The unique concepts of American National Standard ANSI N15.18–1975 "Mass Calibration Techniques for Nuclear Material Control" are discussed in regard to the establishment and maintenance of control of the mass measurement of uranium hexafluoride (UF_6) both within and between facilities. Emphasis is placed on the role of the control of the measurements between facilities, thus establishing decision points for detecting measurement problems and making safeguards judgements. The unique concepts include the use of artifacts of UF_6 packaging cylinders, calibrated by a central authority, to introduce the mass unit into all the industries' weighing processes. These are called Replicate Mass Standards (RMS). This is accomplished by comparing the RMS to each facility's in-house standards (IHS), also artifacts, and thence the usage of these IHS to quantify the systematic and random errors of each UF_6 mass measurement process. A recent demonstration, where UF_6 cylinders were exchanged between two facilities, who used ANSI N15.18-1975 concepts and procedures, is discussed. The discussion includes methodology and treatment of data for use in detecting measurement and safeguards problems. The discussion incorporates the methodology for data treatment and judgements concerning (1) the common base, (2) measurement process off-sets, (3) measurement process precision, and (4) shipper-receiver bulk measurement differences. From the evidence gained in the demonstration, conclusions are reached regarding the usefulness of the realistic criteria for detecting mass measurement problems upon acceptance of the concepts of ANSI N15.18-1975.

Introduction

In 1970 American National Standards Institute (ANSI) Committee N15 (Nuclear Materials Control), with The Institute of Nuclear Materials Management (INMM) as secretariat, formed subcommittee INMM-8 (Calibration Techniques). One of the tasks was the creation of a standard for calibration and control of weighing systems within the unique confines and rigorous requirements of the nuclear industry for measurement of nuclear materials for control and safeguards. The responsible writing group was designated INMM 8.1.

In 1972, the membership of INMM 8.1 conducted a study which included visits to some nuclear facilities which expressed concern with mass measurements. The most productive visits were those to the portions of the industry that weigh massive cylinders of uranium hexafluoride (UF_6). It was discovered that one area of mass measurement, that which controls the interchange of material between various facilities, is of greatest concern to the nuclear industry. INMM 8.1 dealt with this problem and published ANSI N15.18-1975 "Mass Calibration Techniques for Nuclear Materials Control."[1] This document provides guidelines for establishing and maintaining adequate mass measurement processes for nuclear material control both within a facility and for transfer of material between facilities. Specific guidelines include: (1) Selecting weighing instruments and mass standards; (2) Evaluating performance of mass measurement processes, which includes determining the magnitude of random errors, systematic errors, and error limits; (3) Assigning mass values to test objects; (4) Initiating and maintaining a program using replica mass standards of large cylinders of uranium hexafluoride, (UF_6) and (5) Establishing and maintaining control of mass measurement both within and between facilities.

This paper discusses the implementation of these guidelines particularly those relating to UF_6 Replica Mass Standards (RMS) and the demonstration of realistic uncertainties to be used to maintain control of shipper/receiver differences in mass measurement of UF_6 cylinders.

ANSI N15.18-1975 describes in detail the concepts of the mass measurement process and makes recommendations relative to the mass measurements involved in the transfer of uranium hexafluoride (UF_6). These recommendations are based on the assumption that a regulatory function will (1) accept the uncertainty limits as demonstrated and monitored by the various facilities as adequate evidence of facility process control; (2) monitor the exchange of material from one facility to another by a review of shipper/receiver measurement data relative to the uncertainty statement of the facilities involved and (3) test the system on occasion by requiring each facility to furnish measurement data on selected objects circulated to all of the facilities within the system. The Standard details characterization of the mass measurement process to include error limit evaluation, test data, process parameter determination and updating together with demonstration of process control. Assignment of mass values to test objects and the uncertainty determination associated with those values is discussed in detail. These discussions rely heavily on the work of Pontius[2,3].

In 1975 two decisions were mutually made by INMM 8.1 and the U. S. Nuclear Regulatory Commission (NRC). They were: (1) the inauguration of a pilot Measurement Assurance Program (MAP) involving a select group of facilities who routinely measure UF_6, and (2) an overview of how the concepts of ANSI N15.18-1975 could be incorporated in a nuclear materials control system. It was emphasized that a large portion of ANSI N15.18-1975 had been directed toward establishing realistic, as to opposed to historically arbitrary, uncertainties of the mass measurement processes concerned with the transfer of UF_6.

A status report by Doher and Pontius[4] of the first portion of the UF_6 pilot MAP was presented at the 18th Annual Meeting of the INMM in

TABLE I. PROVISIONAL NBS VALUES FOR UF_6 RMS

Cylinder Designation	Value	Uncertainty	Displacement Volume
RMS-1	1296.238 (lb)	± .127 (lb)	29.63 (ft^3)
RMS-2	6355.999	± .127	29.89
RMS-3	4463.810	± .127	120.75
RMS-4	25332.076	± .127	121.08
RMS-5	5284.925	± .127	154.67
RMS-6	32507.424	± .127	154.67

Note: 1 ft^3 = 0.0283 m^3 (S.I. units)

June of 1977. Key portions of that report have been included in this paper for clarity and continuity.

The UF_6 Measurement Assurance Demonstration

There are two aspects of the UF_6 measurement assurance demonstration. One aspect is concerned with the performance of the weighing processes within a given facility, as discussed in detail in ANSI 15.18-1975. The other is concerned with the agreement between facilities, which is also the concern of those interested in safeguards. In the discussion that follows, the latter aspect is treated first.

There are three essential elements for realistic evaluation of the agreement of the results between measurements made on the same object by two different measurement processes. First, the measurements must be referred to a common base. This is not to say that all facilities must use this common base routinely but rather that they must be able to establish the offset between the basis for the results in each facility with reference to a common base. The second and third elements are facility measurement offset (relative to the common base) and process precision. These are closely related. There must be evidence that each measurement process involved is indeed operating in a state of control, and there must be an index of precision for that portion of the measurement process which produced the results which are to be compared.

The Common Base

A common base for stating the mass values is established through the use of artifacts which are replicate UF_6 cylinders. The two reasons for choosing replicate cylinders are: (1) the ease of handling by the facilities as compared to the conventional summations of mass standards classically used to calibrate and (2) the minimization of bias associated with apparent mass values assigned to product cylinders relative to the summations of mass standards maintained above. These artifacts are the Replica Mass Standards (RMS), owned by the NRC. They are stainless steel facsimilies (two each) of the Models 30B, 48X, and 48Y as specified in ANSI N14.1-1971[5]. One cylinder of each size has been filled to provide an RMS at two mass levels, full and empty.

The RMS artifacts have been calibrated by the National Bureau of Standards (NBS). The results of this work are shown in Table I. Detailed accounting of the buoyant effect was done in the initial calibration so

TABLE II. IN-HOUSE STANDARDS CALIBRATION

Facility		30B Mass lb	Provisional Unc lb		48X Mass lb	Unc lb		48Y Mass lb	Unc lb
A	F_1	----		F_1	25327.62	±.92	F_1	32497.46	±1.25
	F_2	----		F_2	25370.79	±.92	F_2	32488.67	±1.25
	E_1	----		E_1	4428.81	±.92	E_1	5287.27	±1.25
	E_2	----		E_2	4419.06	±.92	E_2	5289.20	±1.25
B	F_1	6341.52	±.48	F_1	25322.73	±.42	F_1	32497.27	±.42
	F_2	6348.50	±.48	F_2	25338.42	±.42	F_2	32493.80	±.42
	E_1	1523.15	±.48	E_1	4474.29	±.48	E_1	------	
	E_2	1555.18	±.48	E_2	4460.33	±.48	E_2	------	
C*	F_1	6356.74	±.26	F_1	-----		F_1	------	
	F_2	6357.35	±.28	F_2	-----		F_2	------	
	E_1	1396.23	±.26	E_1	-----		E_1	------	
	E_2	1396.23	±.26	E_2	-----		E_2	------	

Where F = Full Cylinder
E = Empty Cylinder

*NOTE: Facility C was unable to participate in any further aspects of the demonstration.

that the values assigned are true mass values. Since the displacement values of the artifacts are essentially the same as that of the product cylinders, the bias introduced by the buoyant effect of the atmosphere has been minimized.

The common base is introduced into each facility by In-House-Standards (IHS) which are also fascimilies of UF_6 cylinders.

For each model cylinder, each participating facility has four "In-House-Standards," (IHS), i.e. two "Full" and two "Empty." There are two reasons for using pairs of IHS's: (1) to permit the facility to monitor the constancy of one with respect to the other, and (2) to permit greater flexibility in the use of these cylinders to control the product weighing process.

Each facility is responsible for establishing the mass values for its set of IHS. These values are established relative to those assigned to the RMS, following the carefully prescribed procedures of ANSI N15.18-1975 which incorporates a weighing design and substitution measurement methods. In this work, the various weighing devices are used as comparators. Therefore, the uncertainty associated with the mass values assigned to the IHS artifacts is only a function of random variability. Since all IHS artifacts in the system are calibrated relative to the RMS artifacts, the uncertainty associated with the values assigned to the RMS by NBS are common to the entire system and need not be accounted for in comparing results. Table II summarizes the results of this effort.

The use of two "full" and two "empty" IHS artifacts at each facility provides a means of monitoring the constancy of the artifacts, one relative to the other. The results of this comparison at the beginning and at the end of the demonstration for one facility proved to show

TABLE III. TYPICAL IN-HOUSE STANDARD DATA FOR PROCESS OFFSET $(F_1 + b_i)$ and $(E_1 + b_i)$

b_i (lb)	$F_1 + b_i$ (lb)	May 31, 1978	
		Observation (W_i)	Difference (OS_i)
0	25,371	25,364[a]	-7
30	25,401	25,395	-6
65	25,436	25,430	-6
95	25,466	25,460	-6
	$E_1 + b_i$		
0	4,429	4,423[b]	-6
30	4,459	4,454	-5
65	4,494	4,489	-5
95	4,524	4,519	-5

[a] Scale zero was checked and adjusted, as necessary, prior to obtaining this weight.

constancy (less than one lb[1] difference). Such comparisons can be made at any time and if the results are not in order, the RMS artifacts can be requested for the purpose of establishing new values for the IHS artifacts.

Facility Offset Relative to the Common Base

Two IHS artifacts in each facility, designated F_1 and E_1, (full and empty) were used in "calibration" or to determine the offsets of the processes involved. Two summations were used $(F_1 + b_i)$ and $(E_1 + b_i)$, where b_i are known small weights used for incremental changes over a range of approximately 200 lbs. With the mass values of the summation known, the process offset (OS_i) of full and empty is defined as:

$$OS_i = W_i - (F_i + b_i) \text{ and } OS_i = W_i - (E_i + b_i), \text{ respectively} \qquad (1)$$

Where: W_i is the observed weight for the various summations. Several sequences of measurements were made during the course of the demonstration, with the average of the total collection, OS_i, defined as the process offset. Typical data relating to process offset are shown in Table III and the offsets for the various processes are superimposed on precision data in Figure 1.

Factors which contribute to the magnitude of the offsets shown are related to the methods and procedures used in each facility, the largest being associated with the air buoyancy and essentially eliminated as discussed before. These are on the order of 8 lb. for the empty and 5 lb. for the full 48X cylinders. Other elements include the methods in which the scale calibration results were used and procedural differences.

[1] 1 lb = 0.4535 kg (S.I. units). The unit lb is used throughout this paper.

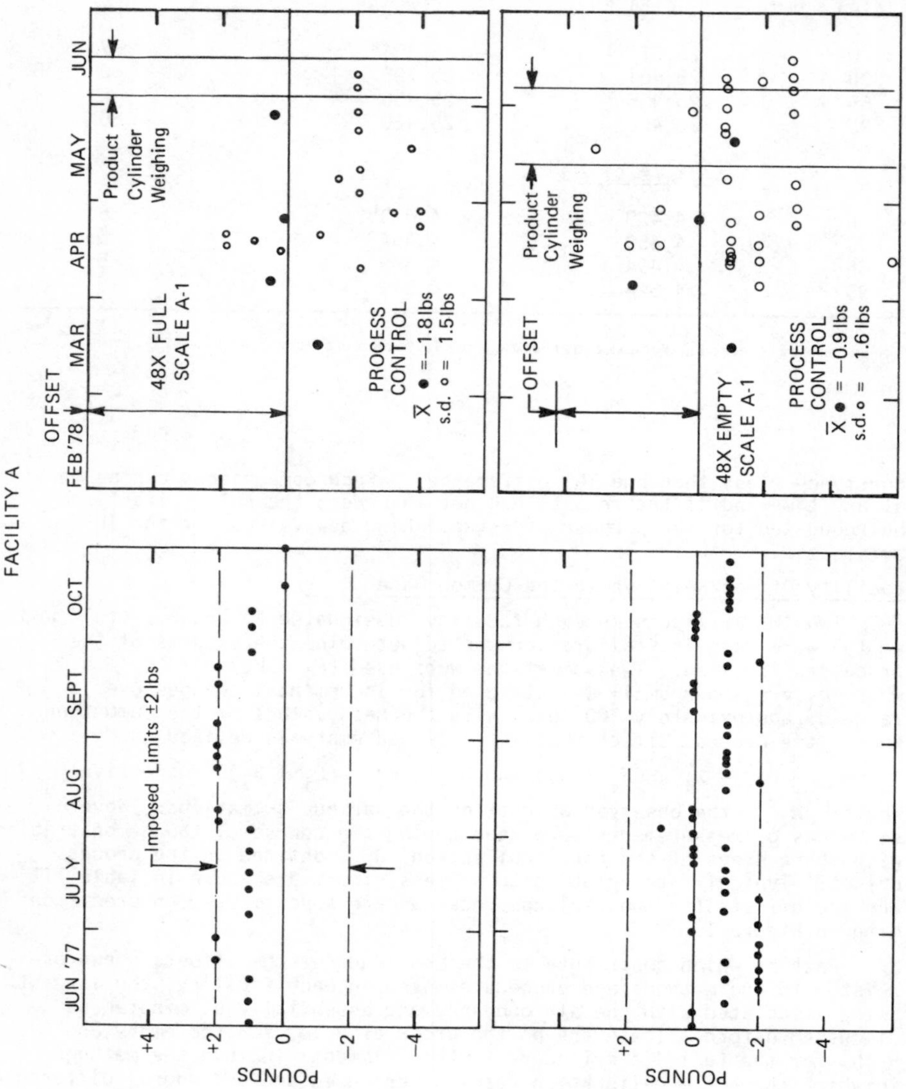
FACILITY A
JUN '77
JULY
AUG
SEPT
OCT
Imposed Limits ±2 lbs
POUNDS
+4
+2
0
-2
-4
FEB '78
MAR
APR
MAY
JUN
OFFSET
48X FULL SCALE A-1
Product Cylinder Weighing
PROCESS CONTROL
X̄ ● = -1.8 lbs
s.d. ○ = 1.5 lbs
OFFSET
48X EMPTY SCALE A-1
Product Cylinder Weighing
PROCESS CONTROL
X̄ ● = -0.9 lbs
s.d. ○ = 1.6 lbs

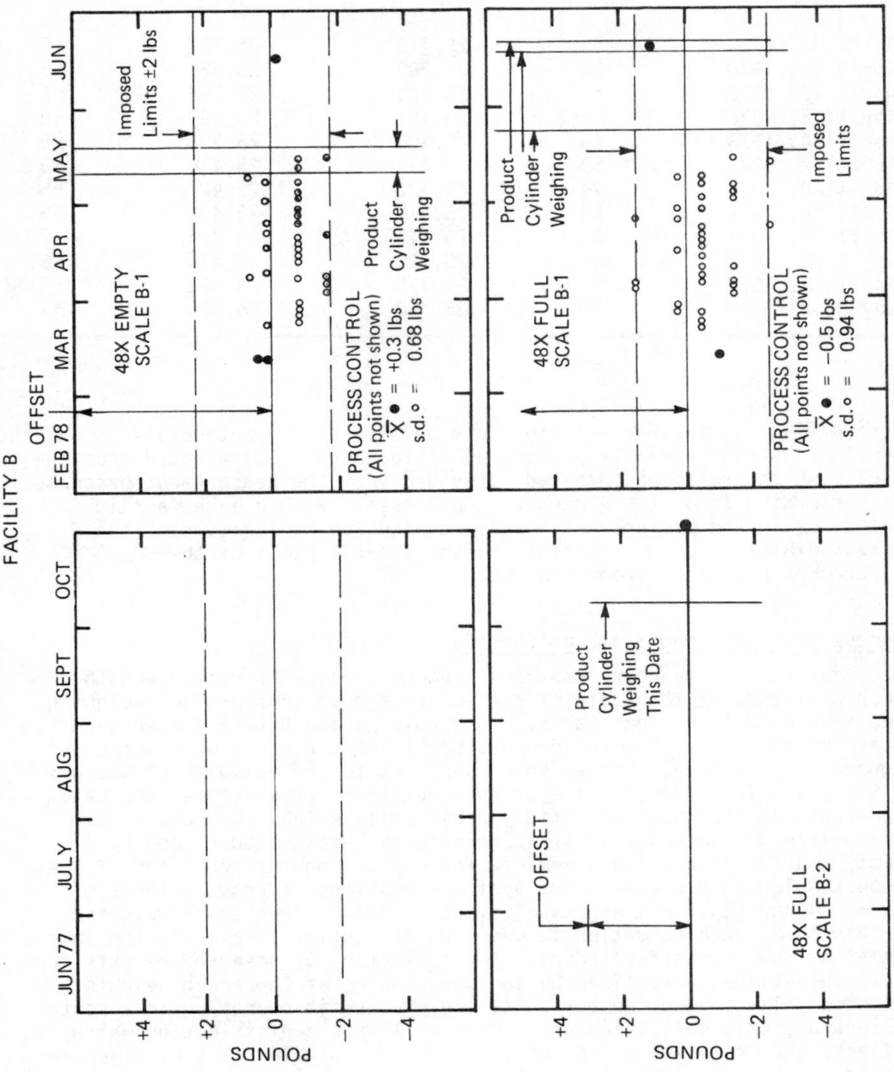

FIG.1. Process offset and precision (Facility A and Facility B).

TABLE IV. TYPICAL IN-HOUSE STANDARD DATA FOR PROCESS PRECISION ($F_2 + c_i$)

Date	Added Weight of c_i in lb	$F_2 + c_i$ lb	Observation lb	Difference lb
April 7, 1978	55	25,413	25,407	-6
April 11, 1978	28	25,386	25,381	-5
April 12, 1978	45	25,403	25,398	-5
April 13, 1978	45	25,403	25,393	-10
April 14, 1978	45	25,403	25,398	-5
April 17, 1978	59	25,417	25,414	-3
April 20, 1978	74	25,432	25,427	-5
April 24, 1978	53	25,411	25,406	-5
April 25, 1978	53	25,411	25,404	-7
April 26, 1978	64	25,422	25,416	-6
April 28, 1978	17	25,375	25,372	-7
May 8, 1978	30	25,388	25,318	-7

By referring all measurements to a common base by the method discussed, the effect of between facility procedural differences is eliminated from the results which are to be compared, provided that the measurement processes are performing in a stable manner. Conversely, within each facility confidence in the measurement results depends on explanation of the offsets exhibited. Such studies are the other aspects of the UF_6 measurement assurance demonstration.

Process Precision and State of Control

Most regulatory agencies require each facility to generate data which will exhibit the state of control of the various product weighing processes on a continuous basis. As a rule, these data are also used to determine an estimate of the process precision. A study of a number of sequences of such data raised some doubts as to the validity of the use of the precision index for the product weighing process, i.e. the data appears to be truncated to satisfy arbitrarily established bounds for variability. A second pair of IHS artifacts, designated F_2 and E_2 (empty and full), provide a vehicle for evaluating the precision of the product weighing process. Once again, summations of known weights, ($F_2 + c_i$) and ($E_2 + c_i$) are made, c_i again being known small weights. In this case, each summation is weighed frequently, in exactly the same manner as the product cylinders. The collection of observed weights for these summations, normalized by the known value of the added weights, define a collection of numbers. If the process is operating in a state of control, this collection of numbers will have a distribution which reflects the precision of the process. Each new value in the collection provides information about the state of control, either "in control" with respect to past performance, or "out of control." "In control" data reinforces the precision estimate i.e., it lies within the bounds of the previous distribution. "Out of control" data signals possible process change. Table IV shows typical data for ($F_2 + c_i$) when the summation has been weighed by the same procedures used to weigh product

TABLE V. BETWEEN FACILITY COMPARISON OF PRODUCT CYLINDER WEIGHTS (A – B)

48X Empty		48X Full	
Cyl.	(A - B) lb.	Cyl.	(A - B) lb.
1 - E	.5	1 - F	.4
2 - E	-3.5	2 - F	.4
3 - E	-.5	3 - F	2.4
4 - E	-1.5	4 - F	5.4
5 - E	-1.5	5 - F[a]	.3
6 - E	-4.5	6 - F[a]	-.7
7 - E	-1.5	7 - F[a]	+2.3
8 - E	-3.5	8 - F[a]	-1.7
9 - E	-.5	9 - F[a]	1.3
10 - E	-1.5	10 - F[a]	.3
11 - E	+.5	11 - F[a]	1.3
12 - E	-.5	12 - F	-.6
13 - E	-3.5	13 - F	-4.6
14 - E	-2.5	14 - F	-3.6
15 - E	-1.5	15 - F	-2.6
16 - E	+.5	16 - F	-3.6
17 - E	-.5	17 - F	-.6
18 - E	-.5	18 - F	-6.6
19 - E	-3.5	19 - F	-1.6

[a]48X Full weighings for these cylinders were made on Scale B - 2. All other weighings were made on Scales A - 1 and B - 1 respectively.

cylinders. It should be noted that this collection of values is also a measure of the process offset previously discussed. Figure 1 summarizes the precision estimates of the processes numbers in the demonstration.

The top left hand control chart of Figure 1 shows the data generated to comply with the requirements of a regulatory agency in facility A. The top right hand chart, superimposed over the process offset data, reflects the performance of the measurement process when the IHS ($E_2 + c_i$) is treated as a product cylinder. Likewise, the treatment of ($F_2 + c_i$) is shown in Figure 1, for both the full and empty cylinders. These support the suspicion that the data shown on the left of Figure 1 may be truncated. The two outliners in the 48X full data are marginally in control relative to the computed precision.

Facility B elected to treat the appropriate IHS's in the same manner used to satisfy the regulatory agency's requirements. Therefore, it can be postulated that the limits associated with these measurements do not properly describe the product weighing process. The "blank" chart of Figure 1 indicates the imposed limits.

The last part of the demonstration was the actual exchange of sets of full and empty product cylinders between facilities A and B. Figure 1 also shows the time intervals where product cylinder weighings were made.

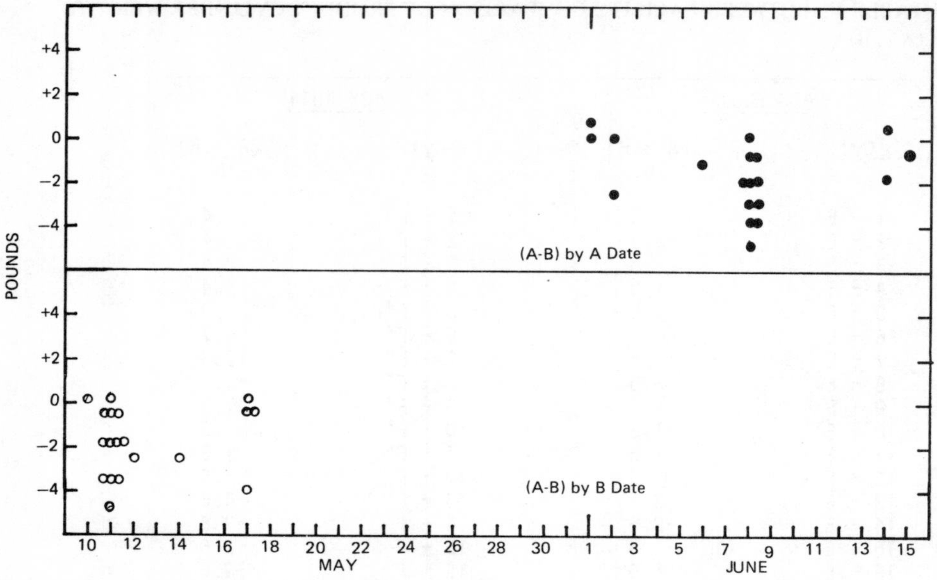

FIG.2. 48X empty product cylinder comparisons (S/R Diff) by date.

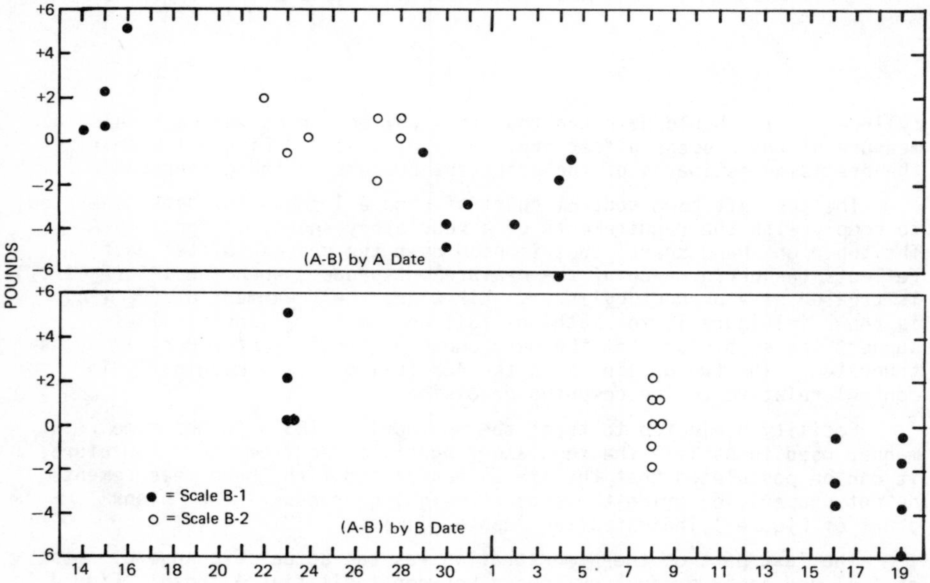

FIG.3. 48X full product cylinder comparisons (S/R Diff) by date.

Between Facility Comparisons

The results of the comparisons are summarized in Table V where the differences shown, Facility A-Facility B, are computed as follows:

$$(A - B) = (W_A + OS_A) - (W_B + OS_B) \qquad (2)$$

Where: W_A and W_B are the observed weights for each cylinder, at facilities A and B and OS_A, OS_B are the respective process offsets.

Diagnostic analysis of the above set of differences is the basis for the second aspect of the UF_6 measurement assurance demonstration; i.e. the evaluation of the within facility process performance. From the parameters developed in the demonstration for each facility, the expected error bounds for the differences from both the "full" and "empty" product cylinders, in pounds, are:

$$EB = \pm 3\sqrt{(s.d._A)^2 + (s.d._B)^2} \pm (SE_A + SE_B) \qquad (3)$$

$$EEB = \pm 3\sqrt{(1.5)^2 + (.68)^2} \pm (.92 + .42)$$

$$= \pm 3(1.64) + (1.34)\ \text{lb} = \pm 5.04 \pm (1.34)\ \text{lb}$$

And Similarly

$$FEB = \pm 3\sqrt{(1.6)^2 + (.94)^2} \pm (.92 + .42)$$

$$= \pm 3(1.84) + (1.34) = \pm 1.52 \pm (1.34)\ \text{lb}$$

Where:

EB = Total error bounds
EEB= Total error bounds of empty weights
FEB= Total error bounds of full weights

$s.d._A$ = Standard deviation of facility A process

$s.d._B$ = Standard deviation of facility B process

SE_A = Systematic error of A process

SE_B = Systematic error of B process

In Figure 2 the differences, (A - B), between the exchanged 48X Empty Cylinders are plotted by the date the measurements were made in each facility. Both processes appear to be stable over the time intervals shown, with a standard deviation (s.d.) = 1.54 lb which compares favorably with the expected s.d. of 1.64 lb. The offset, $\bar{X}$ = -1.5 lb, indicates that the values obtained from the empty cylinders in facility A are lower than obtained in facility B somewhat in excess of the expected limits for offset of facility A with respect to B, 1.34 lb. It is important to note that the expected offset may be either + or - and of any magnitude equal to or less than the propagated error bounds of A and B.

In Figure 3 the differences, (A - B) for the 48X Full Cylinders are plotted in the same manner. Here there is clear evidence of grouping, particularly in facility B where mutliple weighings were made on each of four days. The range of the values exceed the expected ±3 s.d. limit (±5.5 lb), thus it must be concluded that the two extremes are probably out of control. The nature of the plot suggests that the source of this difficulty is with facility B.

In order to support the tentative conclusions stated above, refer again to Figure 1. In the case of the 48X empty process at facility A, it is clear that the ($E_2 + c_i$) simulated product cylinder weights are offset by -1.8 lb from the ($E_1 + b_i$) weighings used to determine the process offset. For the 48X full cylinders at facility B, the two sets of data used to determine the offset show that the process on scale B - 1 changed over the time interval shown in the direction and by an amount which would remove the out of control differences in the full cylinder data. In both cases, the data used to determine the process offsets was generated by the procedures now in use to "calibrate" the various scales. The offset for scale B - 2 at facility B, used initially to establish the values of the IHS relative to the RMS and recently overhauled, was established by 10 independent weighings of ($F_2 + c_i$). The differences for the product cylinders weighed on this scale reflect the expected agreement between the two facilities.

Conclusions and Future Plans

This demonstration proved that the minimum detection level was a function of the combined error bounds for the two facilities involved, and thus is an example of the technique's practicability for safeguarding the transfer of UF_6.

Therefore, a data base, formulated from ANSI N15.18 guidance, for UF_6 transfer participants, provides a realistic safeguards tool for detection of loss as opposed to arbitrary criteria used in the past. Given real time data processing for maintenance of data bases, a mass value difference greater than the accepted uncertainties for the two measurement processes alerts the system to the spurious nature of a transfer. Thus, the system is able to react in a timely fashion and investigate the "out of control condition" be it safeguards problems, equipment and/or operator error, or other assignable causes.

Such a data base also provides an inspection agency with the ability to verify the mass measurement of UF_6 cylinders within a facility. For example, the inspector can require demonstration of the realism and accuracy of uncertainty statements for mass measurements by a set of witnessed weighings of the IHS and comparisons of the value obtained relative to the currently accepted uncertainties associated with that facility. This method provides the tool to verify the realism of the facility's mass measurement statements and to estimate the reliability of the facility's inventory.

Recent funding from NRC has provided the UF_6 MAP with an interim administrator. The data base from the recent demonstration will be delegated to this administrator for expansion. The task thus far accomplished has shown that the methodology and philosophy of the MAP detects problems in the mass measurement of UF_6 cylinders. That is, a UF_6 shipper-receiver mass difference greater than the propagated measurement uncertainties of the shipper and the receiver raises safeguards questions, while a weighing of the IHS resulting in a value outside the measurement process limits detects internal measurement problems.

Therefore, with the use of real time communication and computing equipment, the MAP administrator can immediately (1) update the data base, and thus (2) detect possible safeguards problems and (3) detect internal mass measurement problems. All of these actions and judgments can then be communicated to the shipper-receiver facilities and/or regulatory bodies for immediate action.

REFERENCES

[1] AMERICAN NATIONAL STANDARDS INSTITUTE, Mass Calibration Techniques for Nuclear Material Control, American National Standard N15.18-1975, New York (1975).

[2] PONTIUS, P.E., Mass and Mass Values, NBS Monograph 133, Washington, DC NBS, US Dept. of Commerce (Jan. 1974).

[3] PONTIUS, P.E., Notes on the Fundamentals of Measurement as a Measurement Process, NBSIR 74-545, Washington, DC NBS, US Dept. of Commerce (1974).

[4] PONTIUS, P.E., DOHER, L.W., "The joint ANSI INMM 8.1 Nuclear Regulatory Commission study of uranium hexafluoride cylinder material accountability bulk measurements", Proc. 18th Ann. Meeting, Inst. Nucl. Materials Management, **6** 3, (1975) 480–87.

[5] AMERICAN NATIONAL STANDARDS INSTITUTE, Packaging of Uranium Hexafluoride for Transport, American National Standard N14.1-1971, New York (1971).

AN APPROACH TO ENSURE THE RELIABILITY OF NON-DESTRUCTIVE ASSAY TECHNIQUES FOR THE CONTROL AND ACCOUNTING OF SPECIAL NUCLEAR MATERIAL

P. TING, W.B. BROWN
United States Nuclear Regulatory Commission,
Washington, D.C.,
United States of America

Abstract

AN APPROACH TO ENSURE THE RELIABILITY OF NON-DESTRUCTIVE ASSAY TECHNIQUES FOR THE CONTROL AND ACCOUNTING OF SPECIAL NUCLEAR MATERIAL.

This paper summarizes those development programmes that the United States Nuclear Regulatory Commission (NRC) has under way to ensure the reliability of non-destructive assay (NDA) results for material control and accounting. The NRC development programmes described in this paper are intended to provide a better understanding of the principles of NDA techniques, to define those parameters that affect measurement quality, and to establish a basis for measurement quality control for NDA.

INTRODUCTION

In recent years non-destructive assay (NDA) has been used with increasing frequency for the quantitative measurement of special nuclear material (SNM). Having recognized the potential utility of NDA for material control and accounting, the nuclear community has made a substantial effort to improve the technical level of the equipment and to develop various applications for the techniques. As the technology matures, it is desirable to establish a body of standard practices for the use of NDA within the nuclear industry.

Currently, the US Nuclear Regulatory Commission (NRC) supports a programme designed to provide a sound technical foundation for the broad-based implementation of NDA technology. The programme encompasses several major efforts directed towards (1) elucidating the principles of NDA measurement techniques; (2) determining those parameters that affect measurement quality; (3) developing measurement quality control systems; (4) generating calibration materials for various applications; and (5) establishing national standards (ANSI) for the proper practice of NDA methodology.

To provide a better understanding of the principles and parameters of NDA measurement techniques, efforts are under way to publish a Handbook of Safeguards

Measurement Methods, which is to provide an up-to-date, reliable reference document for use by the nuclear community in designing and/or evaluating SNM measurement systems.

As part of a plan for ensuring NDA measurement accuracy, the NRC is sponsoring a programme at the National Bureau of Standards (NBS) for characterizing and calibrating assay techniques for nuclear materials. The objectives are (1) to develop calibration standards, reference measurement methods, sampling schemes, statistical methods, and reference data; (2) to disseminate the standards, reference methods, and data to users, and (3) to establish means to assist the nuclear industry to ensure that their measurement results are of sufficient accuracy to provide adequate control over nuclear material. In addition, a study is under way to investigate dynamic calibration techniques for the control of NDA systems. Such techniques are designed to minimize errors caused by any differences that exist between reference materials and unknowns.

With respect to the development of bases for ensuring measurement uniformity, studies are under way to standardize SNM containers, to develop the use of calorimetry for shipper-receiver measurements of plutonium, and to provide guidance to licensees for the selection of proper gamma-ray spectrometers for the quantitative determination of SNM content of materials and the qualitative determination of radionuclide abundances.

The following paragraphs briefly describe the current individual studies run by NRC for ensuring the reliability of NDA techniques for the control and accounting of special nuclear material.

HANDBOOK OF SAFEGUARDS MEASUREMENT METHODS

At present the literature describing methods suitable for measuring special nuclear material consists of a large collection of technical journal articles, reports, and books. Some excellent topical compilations exist that summarize the present status of a few measurement methods, or of all measurement methods appropriate for a specific material type. None of these compilations cover the full range of measurement methods for all material types. There is a need for a digest that describes the capability and performance characteristics of all available SNM measurement methods for all types of SNM bearing materials. Such a digest will be provided as a part of the NRC Safeguards Measurement Handbook project. It will be an up-to-date, reliable reference document for use in designing and evaluating SNM safeguards measurement systems.

The Safeguards Handbook will be based on a user's survey and a literature survey. Information gathered from the first will include measurement methods in routine use by NRC licensees and Department of Energy contractors. Results from the latter will include both current level of technology and routine measurement methods and their application, and a current bibliography.

The Handbook will describe methods for measuring ^{235}U, ^{233}U, and plutonium for material control and accounting, and will be organized into five parts to include (1) summary of methods applications and performance, (2) bulk measurement methods, (3) chemical assay methods, (4) active NDA methods and, (5) passive NDA methods. Applications will include material types and compositions routinely encountered in the nuclear industry, i.e. feed and source material, product material, waste and scrap material, intermediate product and recycle material, and hold-up material.

Part 1, "The summary of methods applications and performance" will be tabular in nature for ready reference. Information will include applicable sample size or concentration, level of performance (i.e. estimates of random error and systematic error), availability of standards, number of routine users included in the survey, and cross-references to Parts 2, 3, 4, or 5 for details. Parts 2 through 5 will contain detailed information describing each method, its scope of application, major sources of error and interferences, description of instrumentation, measurement, measurement control and data analysis requirements, specific applications, and references.

MEASUREMENT QUALITY ASSURANCE PROGRAMME

Currently, the NRC is sponsoring a major effort at the National Bureau of Standards to develop written standards and reference materials for calibrating and controlling measurement systems used for SNM accounting across the nuclear fuel cycle. The NBS programme will ensure the availability of reference materials, reference measurement methods, and quality assurance methodology for adequate control of safeguards measurements. The programme will be structured to assist licensees in identifying and correcting their measurement problems and in reducing the magnitude of their measurement errors. A programme of intercomparison measurements will be used to improve measurement consistency across the nuclear industry. Also, some special measurement techniques, such as the use of infra-red thermometry for measuring material hold-up quantities in pipes and equipment, will be evaluated for reliability.

The NBS programme is divided into five tasks:

(1) The Development and Implementation of Measurement Assurance Programmes and Statistical Support for Nuclear Materials Safeguards

(2) Standardization of Destructive Analytical Chemistry Methodology for Nuclear Materials.

(3) Standardization of Non-Destructive Assay Methodology For Nuclear Materials

(4) Feasibility Study on the Use of Accurate Infra-red Thermography Techniques For Estimating Nuclear Material Hold-up, and

(5) Standardization of Bulk Measurement Methodology in Nuclear Fuel Cycle Plants.

Task (1) will result in a comprehensive approach to measurement assurance for materials accounting in the nuclear industry. Activities will include: development of techniques for calibrating and monitoring the performance of individual measurement methods to ensure long-term stability and control; identification and characterization of sources of error for specific measurement systems and material types; investigation of the relative magnitudes of measurement errors and sampling errors for a wide variety of instruments and material types under field conditions; and development of efficient statistical sampling plans for testing field measurements and for controlling measurement performance.

Task (2) will be directed towards the precise and accurate characterization of concentration and isotopic composition of the various secondary reference materials now used in the US nuclear industry. Means to establish necessary traceability to the National Measurement System will be developed. This task, in addition, will provide for the production and certification of Standard Reference Materials (SRMs) in accordance with the needs of the nuclear industry.

The objective of Task (3) is to provide standard procedures and calibration materials to ensure that the desired accuracy can be attained by NDA systems under field conditions. Areas where immediate efforts are needed to tie measurement results to the National Measurement System include gamma- and X-ray spectrometry, spontaneous fission rate measurements, calorimetry, and active assay techniques.

Task (4) is to determine if infra-red thermography can be applied as a diagnostic tool in locating and evaluating radioactive "hot-spots" in nuclear fuel cycle piping and equipment after clean-out. The problem of estimating material hold-up could be reduced if the latest infra-red thermometry equipment could be used to locate pockets of residual which either could be removed by special efforts or which could be measured in situ.

Task (5) consists of six sub-tasks which will cover all significant aspects of bulk measurement methodology. Complete evaluations will be made of every bulk measurement technique being used in existing fuel-cycle facilities.

DYNAMIC CALIBRATION FOR URANIUM NDA

Currently, most NDA systems are calibrated using the "fixed standards technique" described in ANSI N15.20, Guide to Calibrating Non-destructive Assay Systems. This technique is valid only for the assay of items which do not differ from the calibration standards with respect to any property to which the instrument is sensitive. With the use of the "fixed standards technique", any measurement inaccuracy may go unrecognized if potential sources of bias are not eliminated, controlled, or evaluated to permit corrections to be made. Unidentified sources of error make traceability of NDA to the National Measurement System difficult to verify.

Dynamic calibration is a measurement control technique designed to minimize error effects caused by differences between calibration standards and unknowns. The calibration procedure calls for the selection of unknowns from the process line to be standardized by a method known to be traceable to the National System.

This technique has been applied successfully to the NDA of plutonium-bearing materials. The NRC sponsored study will extend the application of the concept to uranium-bearing materials. First, control measurements that could be used for dynamic calibration of the primary NDA system will be identified. The control techniques should be rapid in comparison with chemistry, responsive to parameters different from those that stimulate the primary NDA system, non-destructive in mode of operation, accurate, precise, traceable to the National Measurement System, amendable to measurement of various types of unknown, and readily monitored for variability. Based on these criteria, potential options for control measurement systems could be gamma-ray assay coupled with transmission measurements, or an active assay system or, potentially, calorimetry coupled with an enrichment meter.

STANDARD CONTAINERS FOR SNM STORAGE, TRANSFER AND MEASUREMENT

A successful non-destructive measurement programme greatly depends on the selection of instruments, the optimization of measurement conditions and the matrix of the SNM to be measured. Many parameters that have been identified as potential sources of error in the observed response of an NDA measurement system can be minimized through the proper selection of containers to be employed when assaying special nuclear material.

It is generally believed that procedural, not technical, criteria inhibit the broad-based implementation of non-destructive assay techniques within the nuclear industry. Difficulty in comparing test results has been noted to be a major shortcoming in the NDA approach. Elimination of sources of variability in unknowns, other than SNM content, is an important prerequisite for solving the comparability problem. The role and the relationship of such sources of variability have been identified. The standardization of SNM containers for key process applications within the nuclear industry will eliminate a significant source of variability and contribute towards attaining measurement uniformity. The NRC has sponsored a study to establish specifications for the selection of containers for storage and transfer of SNM throughout the nuclear industry in an attempt to minimize this source of variability on NDA measurements.

The study consists of three basic tasks: (a) to survey present container usage, (b) to determine the effect of container variability on NDA results, and (c) to develop specifications for standardized containers.

The first task includes determining

The number and description of containers currently in use
The range of the quantity of SNM stored in the containers
The plant and process requirements that affect the selection of containers
The safety requirements that affect the selection of containers
The source and magnitude of any economic impact that would result from changes in containers
The data generated by users on the effect of container variability on NDA accuracy

The second task will involve defining a set of empirical tests to evaluate the effect of container characteristics on NDA analysis and the effectiveness of container sealing mechanisms with regard to SNM leakage. This task will provide a basis for establishing the relationships between container characteristics and NDA results that must be taken into account in generating specifications for container standardization.

The third task is intended to relate the results of tasks (a) and (b). Specifications will be prescribed for standardizing SNM containers and an assessment will be made of any technical and economic impact. In addition to the cost of standardized containers, the assessment will estimate any economic impact on plant operations including process and safety considerations that could be pertinent to the development of meaningful specifications.

SPECIFICATIONS FOR Ge(Li) DETECTORS

To assay SNM effectively in its various forms and concentrations, the nuclear fuel cycle needs the most up-to-date instrumentation. A major tool in current use for such assay is gamma-ray spectroscopy. To ensure the acquisition and implementation of efficient and effective gamma-ray spectroscopy systems, an understanding of the capabilities and trade-offs available in the instrumentation is necessary. Based on such an understanding, it is possible to define specific performance requirements and specifications for equipment, especially currently available germanium detector systems.

An NRC sponsored study is under way to establish the operational requirements and limitations of commercially available gamma-ray assay techniques and to investigate the effects of different germanium detector parameters on the assay procedure. Recently developed combinations of detector geometries and coincidence circuitry will be considered.

The results of this study will be published as recommendations for germanium detector specifications for various types of gamma-ray assay. The report will include parameter sensitivity data, information on system availability, and guidelines for specifying and purchasing such a system.

CALORIMETRY FOR SHIPPER-RECEIVER MEASUREMENT

It has been recognized that errors in shipment and receipt measurements can contribute significantly to the limit of error of material unaccounted for (LEMUF) in throughput dominated plants. In addition, shipper-receiver differences can be a valuable source of information about undetected measurement bias that may exist in either a shipper's or in a receiver's plant, or both. Therefore, accurate measurement of receipts and products is essential to a good material control and accounting programme.

A study has been sponsored by NRC to evaluate the use of calorimetric techniques for shipper-receiver verification of plutonium-bearing materials. Three different potential techniques have been examined: (1) Calorimetry combined with gamma-ray spectroscopy to obtain isotopic ratios for plutonium; (2) a direct comparison of the shipper's and receiver's calorimetry measurements; and (3) calorimetry combined with the results of chemical analysis. A final report on this topic, An Evaluation of the Use of Calorimetry for Shipper-Receiver Measurements of Plutonium, NUREG/CR-0014, has just been published by NRC. The results of this study will contribute to later studies concerning methods of performing shipper-receiver measurements on uranium-bearing materials.

CONCLUSION

To ensure the reliability of non-destructive assay techniques for the control and accounting of special nuclear material, the NRC is supporting major studies, carried out by national and private laboratories, for the assessment of current NDA measurement technology. In addition, NRC is participating actively in the National Standards writing committees in the area of NDA. Through efforts of this kind, i.e. sharing expertise within the nuclear community, the unique capabilities of NDA can be exploited and fully utilized to improve safeguards for special nuclear material.

THE ROLE OF CERTIFIED REFERENCE MATERIALS IN MATERIAL CONTROL AND ACCOUNTING

S.P. TUREL
United States Nuclear Regulatory Commission,
Washington, DC,
United States of America

Abstract

THE ROLE OF CERTIFIED REFERENCE MATERIALS IN MATERIAL CONTROL AND ACCOUNTING.

One way of providing an adequate material control and accounting system for the nuclear fuel cycle is to calculate material unaccounted for (MUF) after a physical inventory and to compare the limit of error of the MUF value (LEMUF) against prescribed criteria. To achieve a meaningful LEMUF, a programme for the continuing determination of systematic and random errors is necessary. Within this programme it is necessary to achieve *traceability* of all Special Nuclear Material (SNM) control and accounting measurements to an International/National Measurement System by means of Certified Reference Materials. SNM measurements for control and accounting are made internationally on a great variety of materials using many diverse measurement procedures by a large number of facilities. To achieve valid overall accountability over this great variety of measurements there must be some means of relating all these measurements and their uncertainties to each other. This is best achieved by an International/National Measurement System (IMS/NMS). To this end, all individual measurement systems must be compatible to the IMS/NMS and all measurement results must be traceable to appropriate international/national Primary Certified Reference Materials. To obtain this necessary compatibility for any given SNM measurement system, secondary certified reference materials or working reference materials are needed for every class of SNM and each type of measurement system. Ways to achieve "traceability" and the various types of certified reference material are defined and discussed in this paper.

1. Background

Special nuclear material (SNM) measurements for control and accounting are made on a large variety of material types and concentrations, employing a diverse number of measurement procedures by many industrial, R&D, and academic facilities. Some means of linking all these measurements and their uncertainties to an International or National Measurement System (IMS/NMS) are necessary to achieve valid overall SNM accountability. To do this, all measurement systems must be compatible with the IMS/NMS, and all measurement results must be traceable to the appropriate international/national (primary) reference standards or Primary Certified Reference Materials. "Traceability" is defined as the ability to relate individual measurement results to international/national standards or internationally/nationally accepted measurement systems

TABLE I. CATEGORIZATION OF REFERENCE MATERIALS

RM Type and Abbreviation	Definition	Examples
Reference Material (RM)	A general term that is recommended as a substitute for that which previously has been referred to as a standard or standard material. [A material or substance one or more properties of which are sufficiently well established to be used for the calibration of an apparatus or for the verification of a measurement method ISO - Guide 6 - 1977(E)]	Any or all of the materials listed below.
Certified Reference Material (CRM)	A general term for any PCRM or SCRM or these materials as a group. [A RM accompanied by, or traceable to, a certificate stating the property value(s) concerned, issued by an organization, public or private, which is generally accepted as technically competent ISO - Guide 6 - 1977(E)].	Any PCRM or SCRM or these materials as a group. See examples below.
Primary Certified Reference Material (PCRM)	A stable material characterized, certified, and distributed by a national or international standards body.	Standard Reference Materials of the United States National Bureau of Standards (NBS SRMs) and Standard Materials of the International Atomic Energy Agency (IAEA) bearing the IAEA classification, S.
Secondary Certified Material (SCRM)	A RM characterized against PCRMs, usually by several laboratories. Unlike PCRMs, SCRMS can be typical impure materials.	Reference Materials available from U.S. Department of Energy New Brunswick Laboratory (NBL) or from IAEA. Those from the latter bear the IAEA classification, R.

Table I (continued)

RM Type and Abbreviation	Definition	Examples
Working Reference Material (WRM)	A RM derived from CRMs or characterized against CRMs, used to monitor measurement methods, to calibrate and test methods and equipment, and to train and test personnel.	Process stream materials and any RM prepared according to this and related reports; Materials prepared and distributed in the Safeguard Analytical Laboratory Evaulation (SALE) Program; IAEA's Inter-comparison exchange samples.

through an unbroken chain of comparisons. "Reference standard" is defined as "a material, device, or instrument whose assigned value[1] is known relative to international/national standards or internationally/nationally accepted measurement systems." To obtain this necessary compatibility for any given SNM measurement task, Secondary Certified Reference Materials appropriate for each SNM type and measurement system are nearly always required. Table I defines the various types of reference materials.

Traceability is a property of the overall measurement process, including all Certified Reference Materials, instruments, procedures, measurement conditions, techniques, and calculations employed. Each component of a measurement contributes to the uncertainty of the measurement result relative to an Internationalor National Measurement System (IMS/NMS). The IMS/NMS itself comprises a number of components, including Primary Certified Materials, national or international laboratories, calibration facilities, and standards-writing groups. If the IMS/NMS is viewed as if it were an entity capable of making measurements without error, then traceability can be defined as the ability to relate any measurement made by a local entity (e.g., facility) to the "correct" value as measured by the IMS/NMS. If it were possible for the IMS/NMS to make measurements on the same item or material as the local entity, then this relationship--and hence traceabilty--could be directly obtained. Since the IMS/NMS is largely an intangible reference system, not a functioning body, such direct comparisons are not ordinarily possible, and alternative means for achieving traceability must be employed. This necessary linkage of measurement results and their uncertainties to the international/national measurement system can be achieved by:

A. Periodic measurements of Primary Certified Reference Materials whose assigned values and uncertainties have been certified by a national standards organization or the International Atomic Energy Agency (IAEA). This option applies only if the materials to be measured have either a substantially identical effect upon the measurement process as do the reference materials, or where the difference is relatively small and

[1] The term "value" includes instrumental response and other pertinent factors.

easily correctable by means of the known effects of all interfering parameters. Also, of course, the measurement of the Reference Materials must be performed in a manner identical to that employed for the SNM measurements.

B. Periodic measurements of well-characterized process materials or synthesized artifacts that have been shown to be substantially stable and either homogeneous or having small variability within known limits. The uncertainties (relative to the IMS/NMS) associated with the values assigned to such process materials or artifacts are obtained by direct or indirect comparisons with Certified Reference Materials.

C. Periodic submission of samples for comparative measurement by a recognized facility having established traceability in the measurement involved by employing one or both of the above procedures, and involving only samples not subject to change in their measured values during storage or transit. ("Round-robin" sample exchanges between facilities can be useful in confirming or denying compatability of results, but such exchanges do not of themselves constitute the establishment or maintenance of traceability.)

Valid assignment of an uncertainty value to any measurement result demands a thorough knowledge of all of the observed or assigned uncertainties in the measurement system, including an understanding of the nature of the sources of these uncertainties--not just a statistical measure of their existence. <u>The valid determination of the uncertainty of a measurement relative to the IMS/NMS, and thus of the degree of traceability, is not a rigorous procedure but is the result of sound judgment based on thorough knowledge and understanding of all factors involved.</u>

<u>Every measurement must be considered, in all aspects, as an individual determination subject to error from a variety of sources, none of which may be safely ignored</u>. The all-too-natural tendency to treat successive measurements as routine must be rigorously avoided. Physical Reference Materials such as weights and volumes in particular, tend to be mistakenly accepted as true and unvarying; but they may be subject to changes in effective value (measured response), as well as unrepresentative of the calibrated items unless wisely selected and carefully handled.

The characteristics required of Reference Materials include:

- Small and known uncertainties in the assigned values. The uncertainties of the Reference Materials should contribute only a small fraction of the total uncertainty of the measurement.

- Predictability in the response produced in the measurement process. (Ideally, the measurement process will respond to the reference material in the same way as to the unknown to be measured. If there is a difference in measurement response to the measured parameter arising from other measure-affecting factors, these effects must be known and quantifiable).

- Adequate stability with respect to all measurement-affecting characteristics of the Reference Material. (This is necessary to avoid systematic errors due to changes in such properties as density, concentration, shape, and distribution).

- Availability in quantities adequate for the intended application.

It cannot be assumed that Reference Materials will always remain stable as seen by the measurement system employed, that Working Reference Materials will forever remain representative of the measured material for which they were prepared or selected, or that the measured material itself will remain unchanged in its measurement characteristics. Therefore, it is essential that these Reference Materials as well as the measurement instrumentation and procedures be subject to a program of continuing confirmation of traceability. Many of the factors involved in such a program will be discussed in U.S. Nuclear Regulatory Commission (USNRC) Guides.[1,2,3,4]

It is doubtful that WRMs can ever be exact representations of the material under measurement in any given instance, even for highly controlled process materials shown to be substantially uniform in both composition and measurement-affecting physical characteristics (e.g., density or shape, for NDA measurements), such as formed fuel pieces or uniform powdered oxide, etc. However, in most cases Reference Materials can be either prepared or obtained which yield measurement uncertainties within the selected limits for the material in question. The errors resulting from mismatch of the RM with the measured material will be largest in heterogeneous matter such as waste materials, but in these cases the SNM concentrations normally will be low, and the limits of uncertainty correspondingly less critical.

2. Physical Measurements

National systems of mass and volume measurement are usually so well established that Certified Reference Materials meeting the above criteria are readily available. Where necessary, the facility can use the Certified Reference Materials to calibrate Working Reference Materials which more closely match the characteristics of the measured material in terms of mass, shape, and density, in the case of mass measurements, or are more easily adapted to the calibration of volume-measurement equipment.

Specific procedures for the use of mass and volume Certified Reference Materials for the calibration of measurement processes and equipment are given in the corresponding American National Standards Institute (ANSI) Standards.[7,8,9]

3. Elemental and Isotopic Measurements

Methods for chemical analysis and isotopic measurement are often subject to systematic errors caused by the presence of interfering impurities, gross differences in the concentrations of either the measured component(s) or of measurement-affecting matrix materials, and other compositional factors. Traceability in these measurements can be obtained only if such effects are recognized and either are eliminated by adjustment of the Reference Materials (or sample) composition or, in some cases, are compensated for by secondary measurements of the measurement-affecting variable component(s) and corresponding correction of the measured SNM value. The latter procedure involves additional sources of uncertainty and, therefore, should be employed only if it has either a substantial economic or time advantage, if the interferences or biasing effects are small and limited in range, if the method is reliable, and if the correction itself is verifiable and is regularly verified.

3.1 International/National Standards

US NBS Standard Reference Materials or IAEA Class S Primary Certified Reference Materials are not recommended for use directly as Working Reference

Materials, not only because of cost and required quantities but also because of differences in composition (or isotopic ratios) compared to the process materials to be measured. Primary Certified Reference Materials are more often used to prepare synthesized intermediate Reference Materials of composition and form matching the process material, or to evaluate (and give traceability to) non-Primary Certified Reference Materials of substantially identical material from which matching Working Reference Materials are then prepared. This is necessary because of the wide diversity of process materials encountered and the very small quantities and variety of Primary Certified Reference Materials available. These intermediate Reference Materials can be used directly as Working Reference Materials, if appropriate, but should be reserved for less frequent use in the calibration of suitable synthetic or process-material Working Reference Materials of like characteristics, as well as to verify instrumental response factors and other aspects of the measurement system. However, each level of subsidiary Reference Materials adds another level of uncertainty to the overall uncertainty of the SNM measurement.

Primary Certified Reference Materials also can be used to "spike" process samples or Working Reference Materials to determine or verify the measurability of incremental changes at the working SNM level. However, because of possible "threshold" or "zero error" effects and/or nonlinearity or irregularity of measurement response with concentration, this does not of itself establish traceability.

3.2 Synthesis and Use of Working Reference Materials

Working Reference Materials which closely match the effective composition of process material, or a series of such Reference Materials which encompass the full range of variation therein, serve as the traceability link in most chemical analyses and isotopic measurements. The Working Reference Materials derive traceability through calibration relative to either Primary Certified Reference Materials or, more often, synthesized intermediate Reference Materials containing either Primary Certified Reference Materials or other material evaluated relative to the Primary Certified Reference Material.

The characteristics required of a Working Reference Material are that it is chemically similar to the material to be measured, including interfering substances, is sufficiently stable to have a useful lifetime, and has sufficiently low uncertainty in its assigned value to meet the overall desired accuracy.

Working Reference Materials can be prepared from process materials having the characteristics of the material to be measured or by synthesis using known quantities of pure Primary Certified Reference Material. The former offers the advantage that the Working Reference Material will include that properties that can affect the measurement, such as impurities, SNM concentration level, and chemical and physical form. It suffers from the disadvantage that the assigned value is determined by analyses which are subject to uncertainties that must be ascertained. The latter method involves preparations using Primary Certified Reference Materials (not usually economical unless small amounts are used) or Secondary Certified Reference Materials with the appropriate combination of other materials to simulate the material to be measured. The advantage of the latter method include a more accurate knowledge of the SNM content and better control of other variables such as the level of impurities and matrix composition. The chief disadvantage is that the preparation of Working Reference Materials from Certified Reference Materials will be substantially more costly than Working Reference Materials prepared from process material. Also the use of the Primary Certified Reference Material on a large scale for synthesizing Working Reference Materials is generally an inefficient use of a scarce commodity and therefore is not recommended. Detailed procedures for preparing plutonium and uranium Working Reference Materials are described in U.S.N.R.C. reports. [10,11]

The primary concern in using Working Reference Material to establish traceability in SNM measurements is the validity of the assigned value and its uncertainty. Considerable care is necessary to ensure that the Working Reference Materials are prepared with a minimum increase in the uncertainty of the assigned value above that of the Primary Certified Reference Material upon which the Working Reference Material value is based. If the assigned value of a Working Reference Material is to be determined by analysis, the use of more than one method of analysis is necessary to strengthen confidence in the validity of the assigned value. The methods should respond differently to impurities and to other compositional variations. If the Working Reference Material has been synthesized from a Primary Certified Reference Material or from an intermediate reference material, the composition and SNM content can be verified by subsequent analyses.

The composition of a Working Reference Material can change with time, as by changes in oxidation state, crystalline form, hydration, adsorption. These changes and their effects on measurement are minimized by appropriate packaging and proper storage conditions. Additional assurance is attained by distributing premeasured amounts of the material into individual packets at the time of preparation. These can be appropriately sized so that the entire packet is used for a single calibration or test. Even between such subsamples there may be variability in SNM content, and this factor must be taken into account in determining the uncertainty of the assigned value.

3.3 The Use of Other Laboratories and Sample Exchange Programs

Traceability of chemical assay and isotopic analysis values also may be obtainable through comparative analyses of identical samples under parallel conditions. A comparative-measurement program may take either or both of two forms.

- Periodic submission of process samples for analysis by a recognized facility having demonstrated traceability in the desired measurement.
- Interfacility exchange and measurement of well-characterized and representative materials with values assigned by a facility having demonstrated traceability in the measurement.

Sample exchange programs in which samples are only analyzed by a number of laboratories do not establish traceability but can only indicate interlaboratory agreement or differences, unless traceability of one or more of the samples in a set has been established as above.

3.4 Nondestructive Assay Methods

In nearly all NDA methods, the integrity and traceability of the measurements depend upon the validity of the Certified Reference Materials by which the NDA system is calibrated. Calibrations generally are based on Working Reference Materials which are, or are intended to be, well-characterized and representative of the process material or items to be measured. While the matching of Reference Materials to process items is not difficult to achieve for homogeneous materials of substantially constant composition (e.g., alloys) having fixed size and shape (e.g., machined pieces), such ideal conditions are not obtained for most NDA measurements. Many of the materials and items encountered are nonhomogeneous, noncomforming in distribution, size or shape, and highly variable in type of material and composition. In order to assure traceability of the measurement results to the International/National Measurement System, variations in the physical characteristics and composition of process items and their effects upon the response of the NDA measurement system must be evaluated and carefully considered in the selection or design for Working Reference Materials and measurement procedures.[12,13]

Working Reference Materials usually are prepared either from process materials which have been characterized by measurement methods whose uncertainties have been ascertained relative to the IMS/NMS (i.e., are traceable) or are artifacts synthesized from well-characterized materials to replicate the process material.[2] However, calibration of the NDA method by means of such Reference Materials does not automatically establish continuing traceability of all process item measurement results obtained by that method. The effects of small variations in the materials being assayed may lead to biased results even when the Working Reference Material and the material under assay both were obtained from nominally the same process material. It therefore may be necessary to either establish traceability of process item measurement results by comparing the NDA measurement results with those obtained by means of a reliable alternative measurement system of known traceability, e.g., by total dissolution and chemical analysis or to establish adequate sample characterization to permit the selection of a similarly characterized WRM for method calibration.

4. <u>Characterization by a Second Method</u>

If the process items or materials being measured are subject to non-SNM variations which affect the SNM measurement, it may be possible to employ one or more additional methods of analysis to measure these variations, and thus to characterize process materials in terms of such analysis results. If the secondary analyses are also by an NDA method, they often can be performed routinely with the SNM measurements. In many cases, the results of secondary analyses can be used to derive simple corrections to the SNM measurement results. Corrections also can be obtained, and traceability preserved, by the judicious modification of Reference Materials so as to incorporate the same variable factors, i.e., so that they can produce the same relative effects in the SNM and non-SNM measurements as does the process variable(s).

Alternatively, it may be advantageous to prepare Working Reference Material which span the normal range of variability of the measurement-affecting non-SNM parameter(s) (and also the SNM-content range, if appropriate). These Working Reference Materials can then be characterized on the basis of their non-SNM measurement results, or of some function(s) of SNM and non-SNM measurement results, and can be assigned a corresponding "characteristic figure." If this procedure can be carried out with adequate sensitivity and specificity relative to the interfering parameters factors, and within acceptable limits of uncertainty, the process material can be routinely characterized in like manner, and the appropriate Working Reference Material selected on the basis of such characterization.

---

[2] The advantages of similarily derived WRMs also apply here (3,2).

5. Traceability Is a Continuous Effort

Initial or occasional demonstration that a laboratory has made measurements which are compatible with the International/National Measurement System is not sufficient to support a claim of traceability. Measurement processes are by their nature dynamic. They are vulnerable to small changes in the skill and care with which they are performed. Deterioration in the reliability of their measurement results can be caused by changes in personnel performance, deterioration of or the development of defects in Reference Materials, instrumentation or other devices, or variation in the environmental conditions under which the measurements are performed. The techniques discussed in preceding sections ensure traceability only if they are used within a continuing program of measurement assurance. This should include planned periodic verifications of the assigned values of all Reference Materials used for calibrations.

5.1 Verification of Calibrations

A formal program should be established which fixes the frequency at which calibrations and calibration checks are performed. The required frequencies are strongly dependent on system stability and should be determined and modified for each case using historical performance experience. Current performance of the measurement system based on measurement control program data may signal the need for more frequent verifications. Also, the effects of changes in process parameters, such as composition of material or material flows, should be evaluated when they occur to determine the need for new calibrations.

Working Reference Materials that are subject to deterioration should be recertified or replaced on a predetermined schedule. The frequency of recertification or replacement should be based on performance history. If the integrity of a Reference Material is in doubt, it must be discarded or recalibrated.

5.2 Recertification or Replacement of Certified Reference Materials

Objects, instruments or materials calibrated by the USNBS, IAEA, or other authoritative laboratories and used as Certified Reference Materials by the facility can be monitored by intercomparisons with other Certified Reference Materials to assure their continued validity. In any case, the values can be redetermined periodically according to the following schedule.

Test Objects & Devices	Maximum Periods
Mass	5 a
Length	5 a
Volumetric Provers	5 a
Thermometers & Thermocouples	3 a
Calorimetric Standards	2 a
Certified Reference Materials	
Plutonium Metal (after unpacking)	3 months
U_3O_8 (after unpacking)	1 a

This schedule is based on long term experiences of the U.S. National Laboratories in the calibration of a large variety objects, devices, and materials.

5.3 Interlaboratory Exchange Programs

The facility operator can participate in interlaboratory exchange programs when such programs are relevant to the types of measurements performed in his laboratory. The data obtained through this participation and other comparative measurement data (such as shipper-receiver difference and inventory verfication analyses) can be used to substantiate the uncertainty statements of his measurements.

When significant deviations in the results of the comparative measurements occur, indicating lack of consistency in measurements, the operator should carry out an investigation. The investigation should identify the cause of the inconsistency and, if the cause is within his organization, the operator should initiate corrective actions to remove the inconsistency. The investigation may involve a reevaluation of the measurement process and the Reference Materials to locate sources of bias or systematic error, or a reevaluation of the measurement errors to determine if the stated uncertainties are correct.

ACKNOWLEDGEMENTS

This paper is based in part on a report prepared for the U.S. Nuclear Regulatory Commission on traceability by R. J. Brouns and F. P. Roberts, Battelle, Pacific Northwest Laboratories, Richland, Washington, U.S.A., and on contributions by Willard B. Brown. The author gratefully acknowledges their efforts and the portions of their report contained herein.

REFERENCES

1. Measurement Control Program for Special Nuclear Material Accounting U.S.N.R.C. Report (in preparation).

2. Considerations for Determining the Systematic Error of Special Nuclear Material Accounting Measurements U.S.N.R.C. Report (in preparation).

3. Considerations for Determining the Random Error of Special Nuclear Material Accounting Measurements U.S.N.R.C. Report (in preparation).

4. Nondestructive Assay of Special Nuclear Material Contained in Scrap and Waste, U.S.N.R.C. Guide 5.11, (1973).

5. Mass Calibration Techniques for Nuclear Material Control, ANSI Standard N15.18, American National Standards Institute, 1430 Broadway, New York, New York (1975).

6. Volume Calibration Techniques for Nuclear Material Control, ANSI Standard N15.19, American National Standards Institute, 1430 Broadway, New York, New York (1975).

7. Design Considerations for Minimizing Residual Holdup, U.S.N.R.C. Guide 5.25, Special Nuclear Material in Equipment for Wet Process Operations (1974).

8. Design Considerations for Minimizing Residual Holdup, for Special Nuclear Material in Equipment for Dry Process Operations, U.S.N.R.C. Guide 5.42 (1975).

9. Design Considerations - Systems for Measuring the Mass of Liquids U.S.N.R.C. Guide 5.48, (1975).

10. SWANSON G. C., MARSH S.F., REIN J.E., TIETJEN G. L., ZEIGLER R. K., and WATERBURY G. R., "Preparation of Working Calibration and Test Materials-- Plutonium Nitrate Solution," U.S.N.R.C. Report, LA-NUREG 63.48-1976.

11. YAMAMURA S. S., SPRAKTES F. W., BALDWIN J. M., HAND R. L., LASH R. P., and CLARK J. P., Preparation of Working Calibration and Test Materials: Uranyl Nitrate Solution, U.S.N.R.C. Report, NUREG-0253.

12. Guide to Calibrating Nondestructive Assay Systems, ANSI Standard N15.20, American National Standards Institute, 1430 Broadway, New York, New York (1975).

13. Qualification, Calibration, and Error Estimation Methods for Nondestructive Assay U.S.N.R.C. Guide 5.35, (1975).

14. NDA Physical Standards - Standard Verification Technique ANSI/INMM 9.3 Standard, (in preparation).

PROCUREMENT AND USE OF REFERENCE MATERIALS FOR CALIBRATING NON-DESTRUCTIVE ASSAY EQUIPMENT

G. BUSCA*, M. CUYPERS**, A. PROSDOCIMI**,
T.D. REILLY§, L. STANCHI**
* CEC Safeguards Directorate, Luxembourg
** CEC Joint Research Centre,
Ispra, Italy
§ Los Alamos Scientific Laboratory,
United States of America

Abstract

PROCUREMENT AND USE OF REFERENCE MATERIALS FOR CALIBRATING NON-DESTRUCTIVE ASSAY EQUIPMENT.

The effort made jointly by the Euratom Safeguards Directorate, the Joint Research Centres of the CEC and the operators in the preparation, characterization and use of reference materials for non-destructive assay (NDA) is presented. General guidelines and some practical solutions for the procurement and use of reference materials (RM) or normalization samples (mostly in fuel fabrication and conversion plants) are reported. A consensus on the acceptance of some RMs widely used in the fuel cycle by national and international organizations is recognized as a necessary condition for improving the quality of the fissile materials assay on routine basis. The initial stages in this direction are mentioned. Furthermore, an optimization of the effort in the procurement of RM on the basis of a broad collaboration between Safeguards Authorities and operators is proposed as a solution to solve the problem of high-quality measurements in a continuously increasing fuel cycle where a large variety of materials is encountered.

1. INTRODUCTION

Since 1968 the Euratom Safeguards Directorate of the Commission of the European Communities (CEC) has used non-destructive assay (NDA) techniques extensively for the measurement of fissile materials. Particularly in fuel fabrication plants passive gamma and neutron methods and active neutron interrogation techniques are applied routinely on input and product materials for flow measurements, and on intermediate products during physical inventory takings. Experience gained over several years in calibrating NDA equipment under field conditions, allows us to propose a number of guidelines for the choice and procurement of adequate reference material (RM).

The Commission is now dedicating substantial effort to the standardization of DNA methods and to the definition, preparation, and characterization of reference materials.

In many cases NDA methods have now reached a state of development that in future it will be primarily the availability of adequate reference materials and the standardization of the measurement methods which may be expected to improve the quality of the fissile material assay under routine conditions. This is now largely recognized by several national and international organizations active in the field of NDA. For example, ESARDA has promoted specific actions for the procurement of reference materials for safeguards and management purposes [1]. In addition, a list of RMs existing within most of the EEC countries for NDA has also been established [2]. During the last two years the IAEA has organized several advisory group meetings on the calibration of NDA methods and the use of RMs [3–5]. In the USA action is being pursued by INMM, national laboratories and NBS.

Almost no certified reference materials (CRM) for NDA are available in the world. The preparation and certification of RM is generally a long procedure, but practical solutions have to be found in a short time for safeguards purposes. Some more pragmatic solutions are presented here.

2. GENERAL GUIDELINES

Particular attention is paid in this paper to the RMs used for NDA performed in fuel fabrication and conversion plants. An examination of these plants in this respect allows the fissile materials to be roughly divided into input materials, intermediate products and finished products. Waste and scrap materials are not considered in this paper because the measurement complexity requires studies that cannot easily be generalized. In these cases the procurement of reference materials for calibrating NDA instruments is often very specific.

The measurement and calibration of these materials have been considered by the Joint Research Centre at Ispra within the framework of other studies [6]. Instrument calibration standards, which serve to verify the correct operation of the NDA equipment, are also not considered in this paper.

2.1. Cost, traceability

Owing to the wide variety of materials to be assayed (process material or finished products) by the operators and/or inspectors, the cost of procuring RMs to cover the NDA field may be very high. This cost and inspection time could be reduced substantially, taking into account the following considerations:

(a) The selection and/or preparation of RMs should be made in such a manner that they have the largest possible application for different NDA methods and different measurement conditions. The number of samples to be prepared may thus be reduced. However, it should be remembered that the specifications for the selection of RMs for NDA are closely linked to the technique used (for example, container dimensions and material).

(b) Plant operators and inspectors often perform measurements on the same product in the production line with the same or a different NDA technique. In this case Safeguards Authorities and operators should combine their effort as much as possible in the preparation and characterization of the fissile material, which will result in the preparation of a common standard.

(c) When the RMs are of interest to many users a special effort should be made by national or international normalization organizations to prepare and certify these materials; an attempt should be made to reach as broad a consensus as possible on the acceptance of these RMs, a consensus which can be obtained through a joint certification process.

The Safeguards Authorities are unable to assist in the detailed preparation of all RMs used by plant operators for their measurements on which Safeguards declarations are based. The harmonization of the procedures for procuring RMs by individual laboratories should largely facilitate their acceptance by the Safeguards Authorities. These procedures should contain general guidelines concerning the characterization of the materials, including sampling procedures, statistical treatment of the measurement data and the traceability criteria.

RMs can be traced by referring back to international or multinational CRMs (practically unavailable for NDA), or to a combination of destructive and/or non-destructive analyses performed in specialized laboratories with experience in this field. A network of internationally accepted NDA laboratories does not yet exist.

Recommendations should be elaborated and made available by the Safeguards Authorities to obtain this harmonization in the procedures. Safeguards criteria to be considered is discussed in Section 2.2.

2.2. Safeguards requirements and measures

A few safeguards considerations relevant to the use and procurement of RMs are presented. Reference materials are used by Safeguards Authorities for the verification of fissile materials based generally on attribute or variable sampling. The acceptable measurement error is very different in the two cases of the characterization of the material. The effort may be reduced substantially in the case of attribute verification. During the procurement and use of RMs, Safeguards Authorities should be enabled to apply a certain number of independent measures such as:

(i) Surveillance by inspectors during the preparation of the RMs in the laboratories or plants. For this reason, an important factor is that the preparation time (or time when the basic material used for preparing RMs is found in an unsealed form) should be kept to a minimum;
(ii) Sampling of basic material, if necessary at the different steps of the production, for the performance of independent measurements in agreed laboratories;
(iii) Means to ensure that the RMs have not been altered, and to prove the integrity for its content (including the container in certain cases) using, for example, ultrasonic or X-ray techniques.

These last measures have particular importance when the RMs are kept at the plant or are also used by the plant operators.

2.3. Materials classification

Some materials in a fuel fabrication plant are homogeneous, of well-defined composition, and generally encountered in different plants (UF_6, UO_2, etc.). Here reference materials of general application may be prepared and certified by specialized laboratories and a broad consensus can be obtained on these CRMs. An example of this is the U_3O_8 gamma-ray enrichment standard at present being prepared and characterized jointly by the Commission, European research laboratories (ESARDA), NBS and other US laboratories. More details are presented in Section 3.1.

The preparation of reference materials for product material is generally plant-specific and very costly, and accounts for a significant amount of the plant inventories. The procurement of a high-quality RM can be achieved in most cases only by close collaboration between operator, Safeguards Authorities and specialized measurement laboratories. For this purpose, procedures have to be developed in order to ensure that the specifications, the traceability and integrity requirements are satisfied for all interested parties. Examples of such a scheme applied for the preparation of mixed-oxide fuel pins is given in Section 3.2.

In cases of process materials which are also specific to a plant, the preparation of RMs is not suited for industry-wide standardization. In the case of bulk material measurements after an initial calibration of the NDA instrument with the appropriate material, obtained from the operator and verified by the Safeguards Authorities, a normalization sample has to be selected and kept with the instrument. It is too expensive for the Safeguards Authority to keep all the material used for the instrument calibration. As a practical example of this procedure, the calibration of an active neutron interrogation system (Sb-Be) for measuring various products in a fabrication plant is described in Section 3.3.

2.4. Practical use of RMs

The safeguards measurements carried out by the CEC inspectors are mostly performed with instruments installed for long periods in a plant. This means that the RMs are always kept at the same plant. Transport is only needed in a few cases, where the confirmation of the basic data of the reference material has to be obtained by measurement, for example at the JRC-Ispra laboratory.

With the increase of the number of RMs, a more systematic effort will have to be made in the future to verify also the long-term integrity and validity of the RMs. In a few instances, fuel pins now produced will be used more than five years from now. The validity of RMs depends on the life of the product within the fuel cycle.

3. PRACTICAL EXAMPLES

Practical examples for the procurement of RMs for the three above-mentioned categories of materials are now presented.

3.1. Reference material for uranium enrichment determinations

The ESARDA NDA working group has studied the procurement of reference materials for the measurement of the ^{235}U enrichment (range depleted to 4%) by gamma-ray spectrometry [1]. The measurement of the enrichment is mostly performed on UO_2 powder or pellets and is of practical interest to numerous plants and control authorities. As a result, it was proposed that 30-kg batches of U_3O_8 material of five different enrichments be prepared.

The plan is to prepare 200 g U_3O_8 samples in metallic boxes (diameter 7 cm, height 2 cm). The samples are intended for use in a collimated geometry so as to ensure an infinitely thick geometry in respect of the gamma-ray emission of ^{235}U at 185.7 keV.

An extensive characterization programme of the powder by DA and NDA is foreseen, in collaboration with several European and US laboratories. The detailed specifications for the preparation and characterization of the material and the container of this RM are now being examined. Special attention will also be paid to the characterization and identification of the metal container by ultrasonics, in order to be able to verify the integrity of the samples in a function of time. The final preparation of the RM samples will then be performed at the JRC-Geel (CBNM).

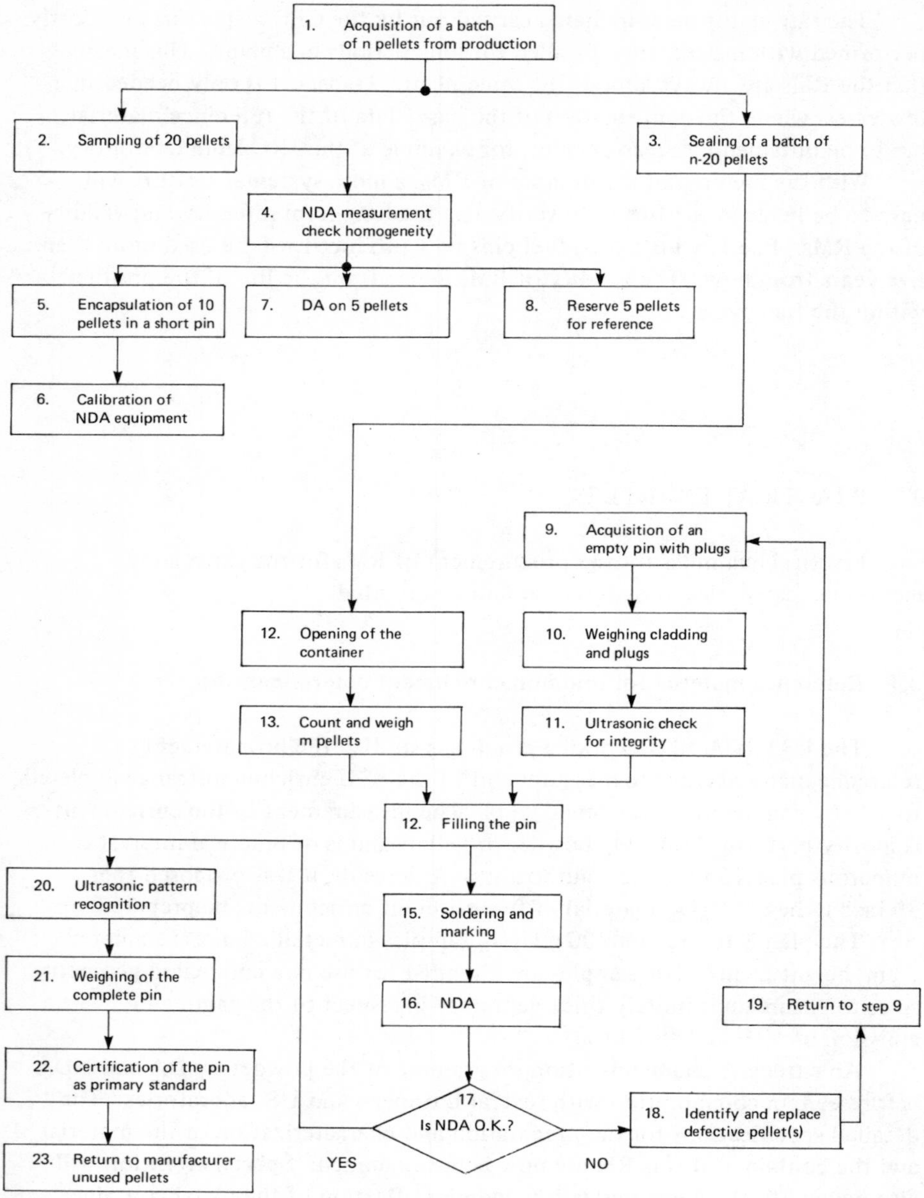

FIG.1. Procurement of CRM for mixed-oxide fuel pins.

3.2. Mixed-oxide fuel pins

Some characteristics of finished fuel pins of mixed oxides (MOX) are measured by the plant operator with a rod scanner, based on active neutron interrogation and prompt or delayed gamma counting. At the same time the safeguards inspectors apply NDA methods to verify the plutonium content of these pins. For the calibration of both techniques reference fuel pins are required, which have the nominal value for the normal production, in addition to a few "off-spec." pins. The physical characteristics of these pins are very plant specific and thus no CRMs are available outside the plant. So as to limit the effort in the preparation and characterization of reference fuel pins, the JRC-Ispra, in collaboration with the Safeguards Directorate of the CEC and concerned plant operators, has developed procurement schemes for RMs, which take into account:

The requirements of the NDA methods of operator and inspector (all parameters which may influence the measurement response of both methods have to be fully characterized);
The safeguards requirements as mentioned earlier (surveillance, independent measurements and guarantee of integrity); and
Requirements of the operator and inspector to use freely the reference samples (a primary standard only used in the presence of both parties and several secondary standards used individually, have to be created to facilitate their actual use).

Figure 1 shows a typical flow chart for the procurement of a certified reference material (primary standard). The example is made for a mixed-oxide fuel pin containing "m" pellets. With minor modifications concerning the fertile material the flow chart holds for fast breeder reactor pins. The detailed explanation of all the steps is reported elsewhere [7]. Here the main points are summarized. A batch of "n" pellets ($n > m$) is acquired from the normal production. The first check is only a visual control of the integrity. Then a reduced number of pellets, for instance 20, is sampled from the batch under control of the Safeguards Authorities. Suitable statistical studies are needed to determine the number of samples and the accuracy of the method. The sampled pellets are controlled by NDA and a small number is checked with DA. NDA gives information about homogeneity of the batch while DA provides the full information about the chemical and isotopic composition and is the reference for the certification. The entire operation, starting from the opening of the batch (step 12) to final recognition (step 20), must be made under direct surveillance of the inspectors. The pin is filled with the normal number of pellets as for the production (m pellets) and is welded at both ends.

An ultrasonic check of the cladding and the determination by ultrasonic techniques [8] of the pattern of the welded areas permit the later check of the integrity and uniqueness of the pin. A final NDA check ensures that no abnormal pellets are inside the pin. If lack of uniformity is found, the pin must be opened and the cladding replaced. A new pin must be prepared by substituting the defective pellet or pellets. The returning loop from steps 17 to 9 escapes only if full uniformity is proved by NDA.

Several RM procurement schemes of this kind have already been applied in the past by Commission inspectors and the JRC on other types of finished product, such as MTR cores and MTR elements. Recently, an experiment was also performed for the preparation of 16 LWR fuel pins [9]. This experiment was carried out with the collaboration of FBFC plant, CEN (Mol) and IAEA.

The implementation of this approach is based on the fact that a competent NDA measurement team is available to the inspectors for the characterization of the RMs at the plant itself. The measurements required have to be performed with more sophisticated instrumentation than that normally available during inspections, and then a combination of NDA techniques in addition to DA techniques have to be applied.

3.3. Normalization samples

When measurements have to be performed on products for which no CRMs are available and on which the plant operator does not perform NDA measurements, a different procedure should be adopted by the inspectors. Two cases may occur:

(a) If the material to be measured is well defined and does not contain large quantities of fissile material (typical examples are pellets, platelets, pebbles), representative samples of the normal production are selected by the inspectors. A first calibration curve is established using the nominal values of those samples, based on operators' data. These normalization samples are kept at the plant until the end of the production campaign of that particular material. One or several samples are then sent to an NDA laboratory for full characterization of the sample with respect to NDA and finally element and isotopic contents are determined by destructive techniques. At this point the sample composition is known precisely and all the earlier data taken must be modified accordingly.
The procedure adopted also ensures the traceability back to destructive analysis and CRMs for these methods. This procedure has often been applied by CEC inspectors for LEU pellets, powders, MOX platelets and pellets and U-Th pebbles.

TABLE I. TYPICAL APPLICATION LIMITS FOR THE Sb-Be ACTIVE NEUTRON INTERROGATION SYSTEM

Uranium compound	^{235}U enrichment[a]	^{235}U mass upper limit
U metal fragments	Medium	3 kg
	High	5 kg
UF_4 (powder)	High	1 kg
U oxide	Low	0.5 kg
(powder or pellets)	Medium	4 kg
	High	4 kg
U carbon-coated particles	Low	20 g
U + Al cores	High	1 kg

[a] Low: less than 20%; medium: between 20 and 70%; high: more than 70%.

(b) If bulk quantities (100 g to several kilograms of fissile material) have to be assayed by the inspector, an initial calibration is performed with a number of samples which cover the range to be measured. These samples should be obtained from the operator. Each of these samples is carefully weighed by the inspectors and small samples are drawn for the independent verification of the isotopic and element composition of the basic material. Several kilograms of fissile material are needed to establish the basic calibration curve and these quantities are generally not available at a later time for a recalibration of the NDA instrument. To overcome this difficulty, a normalization sample is kept (if possible a sample corresponding to one of the points of the calibration curve) by the inspector at the plant. This procedure was applied for calibrating the active neutron interrogation system (Sb-Be), used in a HEU fabrication plant, during physical inventory takings [10].

Calibrations were established for the following families of nuclear materials and within the upper limit of the available fissile material quantities given in Table I.

The response reproducibility of the whole system between the conditions of the calibrations and those of the assay carried out for inventory purposes, is ensured by a "normalization sample". This one has been established by a set of five U + Al cores of high ^{235}U enrichment, and the reference to it enables the correction for the effect of the source strength decay, undesired drifts and variation of experimental conditions. Indeed it may be necessary to modify the detector geometry or container type and size, in order to adapt it to the sample assay, or to readjust some electronic conditions, i.e. HV supply, amplifier gain or discrimination thresholds.

4. CONCLUSIONS

As mentioned earlier, the harmonization of the measurement methods and the availability of suitable RMs can definitely improve the quality of the NDA techniques. Intercomparison of results obtained by NDA techniques in different laboratories will constitute a major guide to decide on the validity of methods, to establish application details and to propose standard procedures. In addition, suitable intercomparison campaigns between laboratories allow the acquisition of sufficient results to establish beforehand the specifications of RMs. Efforts in this direction have been promoted in Europe by the ESARDA NDA working group.

The acceptance of NDA techniques as a valid safeguards measure, when applied by the operator or by the Safeguards Authorities, is finally conditioned by the availability of CRMs, ensuring the traceability of the measurements to a national or multinational system. Up to now no international standards in NDA exist.

The cost of the procurement of RMs for NDA measurement in the fuel cycle is very high. This means that whenever RMs are prepared, the broadest consensus on their validity should be sought, in other words a close collaboration between the operator and the Safeguards Authorities, or between specialized measurement laboratories, is highly recommendable in the future. A special effort is still to be dedicated to define clearly the safeguards requirements during procurement, and in the practical use of CRMs.

This paper is an attempt to contribute some guidelines towards an efficient procurement of RMs, and reports a few examples encountered by the CEC (Euratom) inspectors.

REFERENCES

[1] HARRY, R.J.S., IAEA-SM-231/23, these Proceedings, Vol.II.

[2] BIGLIOCCA, C., CUYPERS, M., LEY, J., List of Reference Materials for Non-Destructive Assay of U, Th and Pu Isotopes, Rep. ESARDA 5 and EUR 6089e (1978).

[3] INTERNATIONAL ATOMIC ENERGY AGENCY, Qualification of Non-Destructive Analysis for Application in IAEA Safeguards Verification Activities, Rep. Advisory Group Meeting, IAEA, Vienna, Rep. AG-80 (1976).

[4] INTERNATIONAL ATOMIC ENERGY AGENCY, The Use of Physical Standards in Inspection and Measurements of Nuclear Materials by Non-Destructive Techniques, Rep. Advisory Group Meeting, IAEA, Vienna, Rep. AG-112 (1977).

[5] INTERNATIONAL ATOMIC ENERGY AGENCY, The Development of NDA Instrumentation and Techniques for IAEA Safeguards, Rep. Advisory Group Meeting, IAEA, Vienna, 22–26 May 1978.

[6] BIRKHOFF, G., et al., "Monitoring of plutonium contaminated solid waste streams", Rep. EUR 5635e, 5636e, 5637e (1976) Chaps 1–3.

[7] STANCHI, L., Typical Examples for the Procurement of Certified Reference Materials for Mixed-Oxide Fuel Pins, Int. Rep. JRC, Ispra, FMM 42 (1978).

[8] CRUTZEN, S.J., HAAS, R., JEHENSON, P.S., LAMOUROUX, A., "Application of tamper-resistant identification and sealing techniques for safeguards", Safeguarding Nuclear Materials (Proc. Symp. Vienna, 1975) 2, IAEA, Vienna (1975) 305.

[9] BEETS, C., et al., IAEA-SM-231/24, these Proceedings, Vol.II.

[10] BIRKHOFF, G., et al., ^{235}U Measurements by Means of an Sb-Be Photoneutron Interrogation Device, Rep. EUR 3327e (1977); and PROSDOCIMI, A., A Neutron Active Interrogation Device for the Fissile Material Assay, Int. Rep. JRC-Ispra, FMM 23 (1977).

Session VII (Part 2)

SAFEGUARDS DATA EVALUATION

Chairman: H. KRINNINGER
(Federal Republic of Germany)

Rapporteur summary: *Problems of safeguards data evaluation*
Papers IAEA-SM-231/8, 60, 71, 103, 104 were presented by
P.T. GOOD as Rapporteur

STRATIFICATION OF NUCLEAR MATERIAL AS A PART OF THE SWEDISH STATE SYSTEMS FOR ACCOUNTANCY AND CONTROL (SSAC)

A. NILSSON, L. EKECRANTZ
Swedish Nuclear Power Inspectorate,
Stockholm, Sweden

Abstract

STRATIFICATION OF NUCLEAR MATERIAL AS A PART OF THE SWEDISH STATE SYSTEMS FOR ACCOUNTANCY AND CONTROL (SSAC).

Nuclear material stratification plays an important role in the Swedish SSAC. How stratification of low-enriched uranium is performed in a fuel fabrication facility is described. The stratification principle is based upon knowledge of random and systematic errors for each step in the measurement of quantities of uranium-total and ^{235}U. Stratification of nuclear material is implemented in the NPT safeguards agreement between Sweden and the International Atomic Energy Agency. The roles of an SSAC and an NPT safeguards agreement are discussed.

1. INTRODUCTION

There is a continuous discussion going on about how state systems for accountancy and control (SSAC) should be designed in order to obtain its various objectives. The objectives for an SSAC are variable with the prerequisites of each state designing its SSAC.

One of the purposes of this paper is to provide a general impact of what objectives Sweden has considered to be relevant in the design of its SSAC. Another purpose is to describe the solution Sweden has found to the specific parts of its SSAC that concerns measurement requirements and stratification of material.

2. LEGAL BACKGROUND TO THE SSAC

All nuclear activities are subject to the requirements in the Atomic Energy Act of 1956. As a general piece of information it may be mentioned, that for the loading of a new reactor with fuel, an additional permit according to a law of 1977 regarding the safe deposits of waste is needed.

Since the major part of the nuclear material used in Sweden is supplied by other countries there are a number of bilateral agreements between Sweden and the supplying countries e.g. USA, Canada, Great Britain and USSR. In these agreements the peaceful use of the nuclear material is asserted.

Sweden signed the Treaty on the Non-Proliferation of Nuclear Weapons (NPT) in 1970. An NPT-safeguards agreement with the International Atomic Energy Agency (the Agency) was signed and taken into force 1975. During 1972-1975 Sweden had a Safeguards Transfer Agreement with the Agency and USA covering all US supplied material.

The Swedish Nuclear Power Inspectorate has been given the responsibility to regulate the safe use of source and special nuclear material according to the Atomic Energy Act, including safety aspects, nuclear material accountancy and control and physical protection. The Inspectorate has therefore got the responsibility to assure that all undertakings regarding the peaceful use of nuclear material Sweden has made in international agreements are fulfilled.

Licenses to use source nuclear material and small quantities of special nuclear material are given by the inspectorate, as well as licenses to export small quantities of nuclear material. Other licenses are given by the Government.

For the fulfillment of all responsibilities undertaken, a "National System for the Accountancy and Control" was issued 1970 by the Inspectorate. The system has been revised twice, 1975 and 1978 due to stricter national policy, new agreements and development of safeguards techniques. The requirements in the Swedish SSAC are operational conditions for licenses according to the Atomic Energy Act.

3. OBJECTIVES AND GENERAL SCOPE OF THE SWEDISH SSAC

The Swedish prerequisites in nuclear material accountancy and control have determined the design of the Swedish SSAC. The following general objectives for the SSAC can be distinguished

- to provide means to assert the licensed use of nuclear materials
- to provide means for the fulfillment of undertakings in bilateral safeguards and multilateral agreements
- to provide means for the fulfillment of undertakings in the safeguards control agreement with the Agency.

From a facility´s point of view the SSAC implies a set of conditions concerning the nuclear materials control, such as organizational responsibilites, the facility´s internal nuclear materials control, frequency and scope of physical inventories, internal and external reporting systems, program for the measurement of nuclear materials, statistical treatment of data and program for the training of operational personnel assigned to nuclear material control functions.

Each facility shall determine how to meet the requirements in the SSAC. The solution each facility has adopted to the requirements shall be described in a "Facility Safeguards Description" and be approved by the Inspectorate. Included in the Facility Safeguards Description is the Design Information that is to be provided to the Agency.

According to NPT-safeguards agreement [1], a national system of accounting for and control of nuclear material must be established in a state entering into an agreement with the Agency. The Agency´s verification activities shall take due account of e.g. the technical effectiveness of the state´s system. The SSAC will therefore contribute to the content of technical parts, i.e. the Subsidiary Arrangements, of the Agreement.

In the SSAC, independent verification inspections by the Inspectorate play an important role. Inspections are often performed jointly by the Inspectorate and the Agency. Even if the purpose of inspection may vary between the Inspectorate and the Agency, this arrangement can reduce the disturbance for the operator. Since all Agency inspections are performed together with national inspectors according to the SSAC, there are provisions to follow the Agency´s implementation of the safeguards agreement.

4. FACILITIES SUBJECT TO THE SSAC

The main nuclear facilities in Sweden are

- A facility for the conversion of low enriched UF_6 into UO_2 and the subsequent fabrication of fuel assemblies.
- Five lightwater reactors of BWR-type and another three under construction.
- One lightwater reactor of PWR-type and another two under construction.
- A research facility with a number of laboratories, including a low-active waste treatment area and one research reactor fueled with highly enriched uranium.
- A mining facility.
- A number of locations where small quantities of nuclear material are handled.

5. PRINCIPLES FOR THE MEASUREMENT SYSTEM AT A BULK MATERIAL FACILITY

5.1 <u>Some remarks about the measurement of nuclear material</u>

When the SSAC is implemented, the fulfillment of the requirements in the SSAC will give knowledge about the quantities of nuclear material held on inventory and involved in inventory changes in facilities where nuclear material is handled. In this respect, accurate values of material quantities must be derived from the use of a measurement system.

A system for the measurement of nuclear material has been in operation since 1970. However, before the NPT-safeguards agreement was taken into force, it was realized that the existing measurement system had to be reworked in order to avoid unnecessary overlapping and duplication between the SSAC and the Agency´s safeguards system. A measurement system designed to fit both the national requirements on accountancy and control at a facility and the Agency´s reguirements on sufficient information, will enable both parties to perform adequate safeguards according to their respective objectives.

5.2 Objectives for the measurement system

At the fuel fabrication facility the measurement system in force until 1975 was designed to measure uranium in the production of fuel assemblies with UO_2-powder as feed material. When conversion of UF_6 to UO_2 was added to the process, the facility had to revise the measurement system in order to have it applicable also for uranium in the conversion process. The objectives for the new system were to

- provide information necessary for effective implementation of safeguards
- be applicable to all nuclear material in any form present at the facility
- provide information about random and systematic errors for each step in the measurement system
- be easily understood by the facility personnel that will perform the measurements
- to the extent possible, combine measurements needed for production control and safeguards purposes
- fit into both the national and the Agency´s systems for accountancy and control.

Some measurement requirements in the SSAC go beyond what is needed for production control. Additional measurements must therefore be performed. As examples can be mentioned uranium and fissile content in some intermediate products, scrap and waste material. But to the extent possible, measurements for safeguards purposes and production control should be combined.

The determination of uranium and fissile content of material items will include the following steps:

- determination of net weight or volume
- determination of uranium concentration (% U)
- determination of enrichment (% U-235).

5.3 Determination of net weight

5.3.1 Procedure

In order to obtain the net weight of an item of nuclear material, two weights must be determined - gross weight and tare weight of the container used. For the determination of gross weight, several different scales can normally be used. However, only the gross weight of the item has influence on the choice of scale. The scale used must have a design capacity suitable to the actual weight of the item. The container tare weight can be determined at the weighing operation but is normally predetermined and recorded for each container.

A limited number of different types of containers are used. Safety rules regulate the maximum quantity for each type of container while the practice regulates the minimum quantity.

Scales of different design and with different capacity give different random and systematic errors in the weighing. From safeguards point of view it is important to know the error contributions from each step in the measurement system to the total limit of error in stated content of uranium and fissile isotope.

5.3.2 Random and systematic errors

Each scale used in materials accountancy and control is thoroughly tested with standard weights according to a requirement in the SSAC. Standard weights used at the facility must be crowned (certified to its weight) by the appropriate Swedish authority (statens anläggningsprovning). From the various calibration series, random and systematic errors have been determined according to procedures in [2]. Scales with detected biases shall normally be adjusted for the bias. If not, the bias shall otherwise be accounted for.

Scales used are periodically checked by the Inspectorate according to the performance of independent calibration. Hereby also quantities for random and systematic errors can be controlled.

All net weights shall be determined within prespecified limits of errors. In practice this means that material items with a gross weight in a prespecified interval shall be weighed only on scales assigned to the weighing of such items.

Scales used and associated limits of errors are described in the Facility's Safeguard Description.

5.4 Determination of volume

5.4.1 Procedures

Liquids containing uranium can be stored in a variety of tanks. The tanks must be equipped or constructed so that the volume of a contained liquid is easily determined, even if the tank is only partly filled. Normally this is done by a transparent scale, graded for direct reading of the volume.

5.4.2 Random and systematic errors

The volume of tanks must be "calibrated" with predetermined volumes and/or control measurements of dimensions according to SSAC requirements. The results were used to estimate quantities for random and systematic errors.

5.5. Determination of uranium concentration

5.5.1 Principles

In the determination of weights only the gross weight of the item has influence on the method (scale) used. In the determination of uranium content a number of factors can influence on the method used and the result obtained.

In this respect, items of nuclear material at a fuel fabrication plant can be divided into two "problem-groups":

- homogeneous material with a very low rate of impurities
- heterogeneous impure material.

Determination of uranium concentration is normally made by the analysis of a representative material sample. All sampling must be made carefully. To sample homogeneous material is relatively straight forward because of the very small variation of uranium concentration within the material. The sampling of heterogeneous material imposes quite different problems. Since the uranium concentration can vary considerably within one item of heterogeneous material, a true representative sample may not be possible to get. When the uranium concentration is determined the sampling error must be accounted for.

The analytical method used must be suitable for the chemical form of the material. Different analytical methods may have to be used for e.g. UF_6 and UO_2. Impure material may need special chemical treatment before analysis can be performed.

5.5.2 Random and systematic errors

Each analytical method is associated with a set of random and systematic errors which must be determined according to the SSAC requirements. To determine the quantities of errors calibration series and designed tests were performed. In calibration, well-known standard material such as NBS-standards were used. In the various tests analysis of material with different chemical and physical form were made. From the results obtained, random and systematic errors were determined according to [2].

To the systematic error variance several factors contribute, such as sampling, analyst, treatment of sample, sample container used etc. So far only total systematic error variances have been estimated. Splitting up the systematic error variances remains, at present, to be done.

In some calculations standardfactors are used for uranium concentration, e.g. for sintered pellets. These factors have been determined experimentally. Since this analysis is not made in the safeguards control thereafter, the derived error is treated as being a systematic error variance.

The SSAC requires the facility to check analytical methods used periodically by the analysis of standard material. In this respect, secondary standard material may be used. Since the Inspectorate samples material in its verification activities, independent analysis of these samples, made by another laboratory, are used in the Inspectorate´s control of measurement errors specified.

Uranium concentration shall be determined within prespecified limits of errors. The best analytical method shall be chosen if more than one method is available. Methods that contribute equally to the total limit of error in the stated uranium concentration can be used alternatively.

Analytical methods used and associated limits of errors are described in the Facility´s Safeguards Description.

5.6 Determination of enrichment

5.6.1 Principles

The determination of enrichment has been considered in the same way as the determination of uranium concentration. Evaluation revealed that one method for enrichment analysis could be assigned to each type of material present at the facility. The enrichment of major quantities of uranium shall be determined by use of the best available technique.

Enrichment shall be determined within specified limits of errors. Methods that contribute equally to the total limit of error in the stated enrichment can be used alternatively.

Uranium in process is accounted for under a "nominal enrichment". The nominal enrichment can be regarded as a label enrichment for true enrichments in an interval 0.05 %.

5.6.2 Random and systematic errors

Well-known standard material was used to calibrate the analytical methods. Quantities for random and systematic errors were calculated according to [2].

6. NUCLEAR MATERIAL STRATA AT A FUEL FABRICATION FACILITY

6.1 Stratification principles

The study of the measurement methods revealed that the nuclear material at a fabrication facility could be stratified according to the following criteria:

TABLE I STRATIFICATION IN A FUEL FABRICATION FACILITY

STRATUM CODE	CHEMICAL AND PHYSICAL FORM OF MATERIAL	ANALYTICAL METHOD AND EQUIPMENT USED FOR THE DETERMINATION OF WEIGHT OF COMPOUND, U-CONCENTRATION AND ENRICHMENT	MEASUREMENT ERROR STANDARD DEVIATIONS SYSTEMATIC	RANDOM	REMARKS
A	URANIUM AS POWDER OR PELLETS NOT SINTERED IN CONTAINERS CONTAINING <100 kg	WEIGHING: SCALE A1 U-CONCENTRATION: BI1 ENRICHMENT: BII1	10 g 0.01 % 0.006 %	14 g 0.03 % 0.008 %	
B	URANIUM AS POWDER IN CONTAINERS CONTAINING >100 kg	WEIGHING SCALE A2 U-CONCENTRATION: BI1 ENRICHMENT: BII1	0.1 kg 0.01 % 0.006 %	0.3 kg 0.03 % 0.008 %	
C	URANIUM AS SINTERED PELLETS KEPT ON TRAYS IN CONTAINERS CONTAINING <50 kg OR IN CUPBOARDS FOR EACH TRAY:	WEIGHING: SCALE PE 11 U-CONCENTRATION: ENRICHMENT:	0.1 g 0.03 % 0.03 %	1.4 g	FOR URANIUM CONCENTRATION STANDARD ELEMENT FACTOR 88.14% IS USED. NOMINAL VALUES ARE USED FOR ENRICHMENT
D	SMALL QUANTITIES OF NUCLEAR MATERIAL IN VARIOUS FORMS KEPT IN CUPBOARDS	WEIGHING: SCALE 12, 12A 13, 4 U-CONCENTRATION: BI2 ENRICHMENT:	0.1 g 0.01 % 0.03 %	1.4 g 0.1 %	NOMINAL VALUES ARE USED FOR ENRICHMENT
E	URANIUM IN VARIOUS CHEMICAL FORMS AS DRY SCRAP FROM THE PROCESS AREA	WEIGHING: SCALE A1 U-CONCENTRATION: ENRICHMENT: BII1	10 g 1.0 % 0.006 %	14 g 0.008 %	FOR URANIUM CONCENTRATION STANDARD FACTORS AS SPECIFIED IN NOTE 3 ARE USED. THEY ARE ESTIMATED TO HAVE A SYSTEMATIC ERROR STANDARD DEVIATION OF 1.0%

TABLE I (CONT.)

STRATUM	CHEMICAL AND PHYSICAL FORM OF MATERIAL	ANALYTICAL METHOD AND EQUIPMENT USED FOR THE DETERMINATION OF WEIGHT OF COMPOUND, U-CONCENTRATION AND ENRICHMENT		MEASUREMENT ERROR STANDARD DEVIATIONS SYSTEMATIC	RANDOM	REMARKS
F	URANIUM AS SCRAP FROM THE WET RECOVERY. THE MATERIAL IS HETEROGENIOUS	WEIGHING: U-CONCENTRATION: ENRICHMENT:	SCALE A1	10 g 10 % 1.0 %	30 g	HETEROGENOUS MATERIAL CAUSES A SAMPLING ERROR. THIS IS ACCOUNTED FOR IN THE SYSTEMATIC ERROR FOR ANALYSIS OF U-CONCENTRATION AND ENRICHMENT
G	URANIUM IN SOLID WASTE	VOLUME DETERMINATION: URANIUM CONTENT: CONTENT OF FISSILE ISOTOPE:		$0.5\ m^3$ $190\ g/m^3$ $3.8\ g/m^3$		FOR CONTENT OF TOTAL URANIUM AND FISSILE ISOTOPE STANDARD FACTORS ARE USED WITH ESTIMATED ERROR STANDARD DEVIATIONS AS SPECIFIED
H	FUEL RODS	WEIGHING: U-CONCENTRATION: ENRICHMENT:	SCALE PE 11 BII1	0.1 g 0.03 % 0.006 %	1.4 g 0.008 %	SEE STRATUM C FOR U-CONCENTRATION
I	FUEL ASSEMBLIES	WEIGHING: U-CONCENTRATION: ENRICHMENT:	 BII1	6.3 g 0.03 % 0.006 %	11.1 g 0.008 %	ONE FUEL ASSEMBLY CONTAINS 63 FUEL RODS
K	URANIUM IN CYLINDERS CONTAINING UF_6	WEIGHING: U-CONCENTRATION: ENRICHMENT:	SCALE K1 BI5 BII2	NOT EVALUATED 0.02 % 0.001·E %	0.6 kg 0.02 % 0.001·E %	E= ENRICHMENT
L	URANIUM SOLUTIONS CONTAINING ≥10 g U/lt	VOLUME DETERMINATION: U-CONCENTRATION: ENRICHMENT:	PAGE 31 BI2 BII1	$0.003 \cdot V\ m^3$ 0.01 % 0.006 %	$0.003 \cdot V\ m^3$ 0.1 % 0.008 %	V= VOLUME
M	SOLUTIONS CONTAINING <10 g U/lt	VOLUME DETERMINATION: U-CONCENTRATION: ENRICHMENT:	PAGE 31 BI3 BII1	$0.01 \cdot V\ m^3$ NOT EVALUATED 0.006 %	$0.01 \cdot V\ m^3$ 0.02·U % 0.008 %	U= U-CONCENTRATION V= VOLUME

a) systematic and random error variance in the determination of weight = type of scale used

b) systematic and random error variance in the determination of uranium concentration

c) systematic and random variance in the determination of enrichment.

Material with the same specified systematic and random error variances in a - c above will be defined as one material stratum.

The uranium at the fuel fabrication facility is stratified according to Table I. In the table, scales and analytical methods used are coded. Table I is included in the Facility's Safeguards Description and in the Design Information provided to the Agency. Elsewhere in these documents, scales, analytical methods, calibration procedures and statistical methods used are described in its entirety.

6.2 Some remarks about the material stata

The only difference between stratum A and B, according to Table I is the scale used for weighing. Items belonging to stratum A are feed-uranium powder normally stored in containers, each containing approximately 50 kg of uranium. To the production lines, however, uranium is fed in containers containing about 1 000 kg stratum B. Obviously the same scale cannot be used in weighings. The chemical form of the uranium powder is not specified, the criteria being that the analysis can be performed as specified.

At a fuel fabrication facility, an almost infinite number of items exist in quantities less than or equal to 10 effective grams. These small quantities belong to stratum D irrespective of chemical or physical form. Most parts of the items are well-known production reference samples.

Strata E and F consist both of scrap material. Scrap material collected directly from the production process belongs to stratum E. This material has a high content of uranium but is impure and heterogeneous. Products from chemical recovery processes belong to stratum F. This material has a low content of uranium and is very heterogeneous. For this stratum a considerable sampling-error has to be considered. Uranium in strata E and F will be recovered for later use in the production process. This is not the case for solid waste material, stratum G, which will be disposed as "measured discard".

A great variety of liquids exist in a conversion facility. For the accountancy and control, however, the only distinction made is between liquids with a high uranium content, stratum L, and liquids with a low uranium content, stratum M, depending upon the analytical method used. Analysis is made with a high reliability for material in stratum L.

7. STRATIFICATION USED IN ACCOUNTANCY AND CONTROL

In the national system there is a requirement that nuclear material in inventory changes shall be measured and the result reported with a 95 %

confidence interval specified. Measurement methods specified in Table I shall be used when nuclear material quantities in inventory changes are determined. Specified limits of errors for each stratum shall be used when the confidence interval is determined.

Physical inventory of all nuclear material held at the facility shall be taken periodically. All nuclear material quantities at the facility shall be determined as to net weight, uranium concentration and enrichment. The items shall be listed on an inventory list with material stratum given. Evaluation of the result of inventory is based on error propagation of the total inventory for each material stratum. The error propagation includes calculation of LEMUF for the facility's material balance period.

In the Inspectorate's verification of the physical inventory a statistical sampling plan is used, based on the material strata. For the items chosen according to the sampling plan, the Inspectorate will independently determine net weight, uranium concentration and enrichment. Conclusions about stated physical inventory can hereby be made by the use of statistical evaluation techniques.

8. STRATIFICATION OF NUCLEAR MATERIAL IN OTHER FACILITIES

In Sweden there are two consecutive steps in the nuclear fuel cycle, fuel fabrication and irradiation in a nuclear reactor. For a later step in the nuclear fuel cycle it is important to have control of the quantities of nuclear material present at a nuclear power plant. Adopting the same philosophy as in the fuel fabrication facility will give the following nuclear material strata for a nuclear power plant

- fuel assemblies
- fuel rods
- small quantities of nuclear material kept at the nuclear power plant.

Initial pre-irradiation data are used in the accountancy. Of course, all relevant fuel assembly data, such as position in core, integrated burn up, plutonium decay etc. are kept simultaneously.

In the reactor uranium is consumed and plutonium is produced. Quantities of nuclear loss and production are calculated by means of a computer program. For the calculation initial uranium and fissile content of the fuel assemblies and reactor thermal power etc. are used as input values. For the input values, associated limits of errors must be known. For initial uranium and fissile content this is obtained from the fuel fabrication facility. Evaluation and error propagation of the computer program is a requirement in the SSAC, in order to get values of nuclear loss and production specified with limits of error. If the fuel assemblies are to be reprocessed it is important to have good knowledge about what quantities of uranium and plutonium are shipped for reprocessing.

TABLE II KMP:s FOR PHYSICAL INVENTORY IN A FUEL FABRICATION FACILITY

KMP	DESCRIPTION OF TYPICAL BATCH	DESCRIPTION OF TYPICAL ITEM	SOURCE DATA	MATERIAL DESCRIPTION	MEASUREMENT BASIS
A	URANIUM AS POWDER OR PELLETS NOT SINTERED IN CONTAINERS CONTAINING LESS THAN 100 kg: ANY NUMBER OF SUCH CONTAINERS	ONE CONTAINER	(1) INVENTORY NUMBER (2) WEIGHT OF COMPOUND (3) CHEMICAL FORM, INCLUDING URANIUM CONCENTRATION AND ISOTOPIC COMPOSITION (4) WEIGHTS OF TOTAL AND FISSILE URANIUM, INCLUDING THE MEASUREMENT ACCURACY	PHØF	M
B	URANIUM AS POWDER IN CONTAINERS CONTAINING MORE THAN 100 kg: ANY NUMBER OF SUCH CONTAINERS	ONE CONTAINER	AS ABOVE	PHØF	M
C	URANIUM AS SINTERED PELLETS IN CONTAINERS CONTAINING LESS THAN 50 kg: ANY NUMBER OF SUCH CONTAINERS	ONE CONTAINER	AS ABOVE	CPØF	M
	URANIUM AS SINTERED PELLETS INTENDED FOR LOADING INTO RODS, STORED TEMPORARILY IN CUPBOARDS: ANY NUMBER OF TRAYS WITH SUCH PELLETS	ONE TRAY	FOR THE CUPBOARD: INVENTORY NUMBER FOR EACH TRAY: (1) WEIGHT OF COMPOUND (2) CHEMICAL FORM, INCLUDING URANIUM CONCENTRATION AND ISOTOPIC COMPOSITION (3) WEIGHTS OF TOTAL AND FISSILE URANIUM, INCLUDING THE MEASUREMENT ACCURACY	CPØF	M

KMP	DESCRIPTION OF TYPICAL BATCH	DESCRIPTION OF TYPICAL ITEM	SOURCE DATA	MATERIAL DESCRIPTION	MEASUREMENT BASIS
D	FOR SMALL QUANTITIES OF NUCLEAR MATERIAL (EACH LESS THAN 0.01 EFF **kg**): ANY NUMBER OF SUCH QUANTITIES	NOT APPLICABLE	FOR THE CUPBOARD OR ROOM: INVENTORY NUMBER FOR EACH TRAY: (1) WEIGHT OF COMPOUND (2) CHEMICAL FORM, INCLUDING URANIUM CONCENTRATION AND ISOTOPIC COMPOSITION (3) WEIGHTS OF TOTAL AND FISSILE URANIUM, INCLUDING THE MEASUREMENT ACCURACY	SSØF	M
E	URANIUM AS DRY SCRAP FROM PROCESS AREA: ANY NUMBER OF SUCH CONTAINERS	ONE CONTAINER	(1) INVENTORY NUMBER (2) WEIGHT OF COMPOUND (3) CHEMICAL FORM INCLUDING URANIUM CONCENTRATION AND ISOTOPIC COMPOSITION (4) WEIGHTS OF TOTAL AND FISSILE URANIUM, INCLUDING THE MEASUREMENT ACCURACY	SNØF	M
F	URANIUM AS SCRAP FROM WET RECOVERY: ANY NUMBER OF SUCH CONTAINERS	ONE CONTAINER	AS ABOVE	SNØF	M

TABLE II (CONT.)

KMP	DESCRIPTION OF TYPICAL BATCH	DESCRIPTION OF TYPICAL ITEM	SOURCE DATA	MATERIAL DESCRIP-TION	MEASURE-MENT BASIS
G	URANIUM AS SOLID WASTE: ANY NUMBER OF SUCH CONTAINERS	ONE CONTAINER	(1) INVENTORY NUMBER (2) AVERAGE ISOTOPIC COMPOSITION (3) WEIGHTS OF TOTAL AND FISSILE URANIUM	AMØF	M
H	FOR FUEL RODS: A NUMBER OF FUEL RODS	ONE ROD	FOR EACH RACK OR COFFIN: INVENTORY NUMBER FOR EACH TYPE OF FUEL RODS ON THE RACK OR IN THE COFFIN: (1) NUMBER OF ELEMENTS (2) ISOTOPIC COMPOSITION (3) AVERAGE WEIGHTS OF TOTAL AND FISSILE URANIUM, INCLUDING THE MEASUREMENT ACCURACY	ERØF	M
I	FOR FUEL ASSEMBLIES: ONE FUEL ASSEMBLY	ONE ASSEMBLY	FOR EACH FUEL ASSEMBLY: (1) INVENTORY NUMBER (2) IDENTIFICATION NUMBER OF THE FUEL ASSEMBLY (3) AVERAGE ISOTOPIC COMPOSITION (4) WEIGHTS OF TOTAL AND FISSILE URANIUM, INCLUDING THE MEASUREMENT ACCURACY	EAØF	M

KMP	DESCRIPTION OF TYPICAL BATCH	DESCRIPTION OF TYPICAL ITEM	SOURCE DATA	MATERIAL DESCRIPTION	MEASUREMENT BASIS
K	URANIUM IN CYLINDERS CONTAINING UF_6 ANY NUMBER OF SUCH CYLINDERS	ONE CONTAINER	(1) INVENTORY NUMBER (2) WEIGHT OF COMPOUND (3) CHEMICAL FORM, INCLUDING URANIUM CONCENTRATION AND ISOTOPIC COMPOSITION (4) WEIGHTS OF TOTAL AND FISSILE URANIUM, INCLUDING THE MEASUREMENT ACCURACY	LØØF	M
L	URANIUM IN SOLUTIONS CONTAINING MORE THAN OR EQUAL TO 10 **g U/lt**: ANY NUMBER OF SUCH CONTAINERS	ONE CONTAINER	(1) INVENTORY NUMBER (2) VOLUME OF SOLUTION (3) CHEMICAL FORM, INCLUDING URANIUM CONCENTRATION AND ISOTOPIC COMPOSITION (4) WEIGHTS OF TOTAL AND FISSILE URANIUM, INCLUDING THE MEASUREMENT ACCURACY	LØØF	M
M	URANIUM SOLUTIONS CONTAINING LESS THAN 10 **g U/lt**: ANY NUMBER OF SUCH CONTAINERS	ONE CONTAINER	AS FOR KMP L ABOVE	LØØF	M

9. IMPLEMENTATION OF MATERIAL STRATIFICATION IN THE NPT-SAFEGUARDS AGREEMENT

In NPT-safeguards agreement with the Agency the reporting requirements normally include concepts such as Key Measurement Point (KMP) for flow and inventory of nuclear material, Material Description and Measurement Basis. It is tacitly understood by us that what is reported under these "headlines" shall enable the Agency to get such information about the nuclear material on inventory, inventory changes, etc. that can be used in statistical treatment of data, in order to draw relevant conclusions. Normally inventory-KMPs are equivalent to the location of material within a facility. The Material Description consists of four keywords associated with physical form (fuel elements, ceramics, solids, waste etc.), chemical form (hex, oxides, al-alloys etc.), containment (fuel units, flask etc.) and irradiation status. The key-words exist in many combinations.

In the agreement between Sweden and the Agency, nuclear material strata are substituted for inventory KMPs. To each KMP/material stratum, one set of key-words for Material Description is defined. Only those defined for the nuclear material strata are used in reporting inventory and changes. This means that no information is given in the reports about location at the facility or container used. On the other hand, with this arrangement, the Agency gets information about the material, its chemical and physical form, measurement methods used and associated limits of errors. KMP codes for the flow of material are defined according to the "model"-agreement. For Measurement Basis the choice is only "measured at the facility" (M) or "measured elsewhere" (N). Table II shows how this is solved in the Facility Attachment for the fuel fabrication facility.

The same philosophy is introduced in Facility Attachments for nuclear power plants. Inventory KMPs correspond to the nuclear material strata mentioned above. The Material Description Code will show if the fuel assemblies or fuel rods are irradiated or fresh. At a nuclear power plant, the nuclear material will become "measured" only upon time of shipment, the event which will initiate reporting of nuclear loss and production.

10. FINAL REMARKS

It must be emphasized that what is presented here is only one part of the Swedish SSAC and how this part has been implemented into the safeguards-agreement with the Agency. Furthermore, only the basic concepts of the measurement system is presented. Statistical treatment of data and verification activities are not included but are essential in the national safeguard.

Continuous evaluation of measurement methods must be made as well as periodic updating of the SSAC, entirely or partly, in order to account for new development in the field of safeguards.

REFERENCES

[1] INTERNATIONAL ATOMIC ENERGY AGENCY, The Structure and Content of Agreements Between the Agency and States Required in Connection with the Treaty on the Non-Proliferation of Nuclear Weapons, INFCIRC 153 (1972).

[2] JEACH, J.L., Statistical Methods in Nuclear Material Control, TID-26298, USAEC (1973).

DISCUSSION

A.G. HAMLIN: You mentioned that you have difficulty in controlling human error. At the United Kingdom Atomic Energy Authority we regard this as probably the dominant error in accountancy statements. It is totally unpredictable and can affect not only quantity, but also quality, chemical nature and even the account eventually posted. Our current approach in the UKAEA is to minimize human intervention in the data gathering chain.

L. EKECRANTZ: We have the same approach in Sweden, but we think it is impossible to completely eliminate the effect of the "human factor".

G. HOUGH: I would like to confirm that the method of stratification used in this paper can be very useful to the Agency because it groups together items which have a common measurement basis and makes it much easier to determine sample plans and make inferences about the whole population of items sampled. Stratification schemes based on locations where several different material types and sizes of items are grouped together make it very difficult to implement sample plans, estimate errors and make valid inferences.

A.M. BIEBER: Is the stratification used in the facility reflected in the inventory KMP structure used in the IAEA facility attachment for the facility?

L. EKECRANTZ: Yes. The facility uses the same inventory KMP structure as that given in Table II.

PRACTICAL APPROACH TO A PROCEDURE FOR JUDGING THE RESULTS OF ANALYTICAL VERIFICATION MEASUREMENTS

W. BEYRICH*, G. SPANNAGEL
Kernforschungszentrum Karlsruhe GmbH,
Karlsruhe, Federal Republic of Germany

Abstract

PRACTICAL APPROACH TO A PROCEDURE FOR JUDGING THE RESULTS OF ANALYTICAL VERIFICATION MEASUREMENTS.

For practical safeguards a particularly transparent procedure is described to judge analytical differences between declared and verified values based on experimental data relevant to the actual status of the measurement technique concerned. Essentially it consists of two parts: Derivation of distribution curves for the occurrence of interlaboratory differences from the results of analytical intercomparison programmes; and judging of observed differences using criteria established on the basis of these probability curves. By courtesy of the Euratom Safeguards Directorate, Luxembourg, the applicability of this judging procedure has been checked in practical data verification for safeguarding; the experience gained was encouraging and implementation of the method is intended. Its reliability might be improved further by evaluation of additional experimental data.

INTRODUCTION

Judging the deviations of analytical results provided by a verification laboratory from values declared by the plant operator is an important task of practical safeguards. Since this has to be performed quickly and reliably for numerous data on a routine basis, simplicity and transparency of the applied method are fundamental requirements.

As shown by analytical intercomparison programmes such deviations are mainly induced through laboratory related systematic error components (biases) whereas the laboratory internal reproducibility of the measurements (precision) is of minor importance.

As is known, it is possible in principle to determine the bias of an individual laboratory by the analysis of calibrated sample material. However, not only might

---

* Delegated from Euratom, CEC.

considerable difficulties be encountered[1] in practice, but it must also be taken into consideration that the bias of a laboratory is subject to time-dependent fluctuations. By contrast, if a group of laboratories analysing the same sample material is considered, the spread of their measurement results around the mean (interlab-deviation) might be considered as essentially independent of time.

Furthermore, when conceiving a judging procedure applicable in practice it has to be taken into account that complete information on the error limits of the declaration value is not always available; this concerns estimates of its systematic error component and the measurement precision as well as uncertainty on the number of repetitions performed in the individual analytical steps.

Considering these particularities, it seemed reasonable to base a judging procedure for the differences between declared and verified values directly on the interlaboratory differences observed in practice. Extensive data on these differences are already available from international analytical interlaboratory comparisons, verification measurements, and shipper-receiver differences for most of the analytical methods used in safeguards.

1. DESCRIPTION OF THE JUDGING PROCEDURE PROPOSED

Already in 1975 an empirical method was described [1] for evaluating the results of analytical intercomparison programmes with respect to the interlaboratory differences to be expected in practice.

The method consists of calculating the modulus of the relative difference between the results of two laboratories; this is done for all possible combinations of two laboratories among the participants in the intercomparison programmes. Then, after appropriate grouping and counting of these differences, a distribution curve is plotted as shown schematically in Fig. 1. The ordinate value is a measure of the probability P to observe an interlaboratory difference greater than or equal to the corresponding abscissa value $|d|$ under the analytical method investigated[2].

As shown in Fig. 1 two abscissa values, D_0 and $k \cdot D_0$ ($k > 1$), are chosen as so-called "threshold values", generating in this way three ranges of difference,

---

[1] For example, if the material for the calibration sample is taken from a production process, its "true" composition has to be determined by extended characterization measurements that are restricted to special cases. On the other hand, blending synthetic materials results in precisely calibrated reference samples; however, owing to their high purity analysis, they are often no longer representative for practical conditions.

[2] A detailed description of the features of these distribution curves and their general applicability to the evaluation of analytical intercomparison studies is in preparation [2].

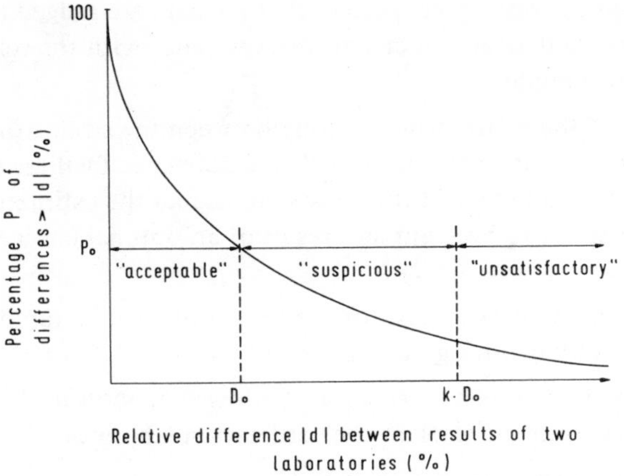

FIG.1. Distribution curve of interlaboratory differences and classification ranges for judging (schematic).

which may be used for classifying and judging deviations observed between declared and verified values as follows:

(a) If the first verification measurement yields $|d| \leqslant D_0$, the agreement with the declared value is considered as "acceptable" and a repetition measurement by the verification laboratory can be avoided. If $|d| > D_0$, repetition measurements are performed at the verification laboratory in order to endorse the reliability of its first determination.

(b) If it is confirmed that $|d| > D_0$ while $|d| \leqslant k \cdot D_0$ (thus $D_0 < |d| \leqslant k \cdot D_0$), the deviation is considered as "suspicious". In this case it is not mandatory that the control authority intervenes in a direct manner, but it is recommended that the frequency of occurrence of this situation in the respective facility be surveyed and that this finding be taken into account when planning the sampling in the course of subsequent inspections. Together with tests on the sign of such "suspicious" differences, systematic deviations between declared and verified values in the control of a plant become visible.

(c) If $|d| > k \cdot D_0$ is obtained, the deviation is considered as "unsatisfactory" and a direct action of the control authority is recommended to elucidate the situation.

The main features of this procedure can be summarized as follows:

It is based exclusively on the experimental results so far obtained in practice;

It is flexible – for example, it can be easily adapted at any time to current developments in measuring technology;

For the plant operator it guarantees that his data are judged according to uniform criteria derived directly from experience with the relevant measuring technique;

Knowledge of the relationship existing between the size of the differences to be expected and the probability of their occurrence facilitates, for the control authority, the planning of future inspections and the estimation of the verification effort to be spent as a result of an optimal choice on the value of D_0;

It provides an auxiliary means for the inspector to get a rapid survey of the situation prevailing during plant control;

By partly avoiding replicate analyses it allows expenditure to be reduced in terms of measurement work at the verification laboratory;

It might be expected that, with time, a consequent application of this judging procedure will result in reaching the least dispersion of interlaboratory differences achievable in the field application of the measurement techniques involved.

2. TESTING OF THE PROPOSED JUDGING PROCEDURE

Empirical distribution plots needed for applying this method were determined two years ago for some of the analytical problems relevant to safeguards. In particular, results were used which had been obtained with the analytical inter-laboratory comparison programmes JEX-70 [3] and IDA-72 [4].

The analytical requests were subdivided into uranium and plutonium concentration determinations using chemical methods and the mass-spectrometric isotope dilution technique, respectively, as well as the determination of isotopic compositions of these elements with thermionic mass spectrometers[3]. For the distribution curves related to concentration determinations it was assumed that there is no dependence on the absolute value of the concentration; for mass-spectrometric isotopic composition determinations the dependence of measuring uncertainty on the isotopic abundance was taken into account.

Figure 2 presents distribution curves for the chemical determination of the uranium and plutonium concentrations derived from data obtained in the JEX-70 experiment. According to these curves the occurrence of interlaboratory differences greater than or equal to 0.2% (uranium) and 0.5% (plutonium) had to be expected with about 50% probability[4].

---

[3] Regarding the determination of the ^{235}U isotope in UF by different analytical techniques, see Refs [5, 6].

[4] Assuming the distribution curve of the interlaboratory differences is derived from a group of laboratory results with a Gaussian distribution and a standard deviation σ, the difference d exceeded by about 50% of the values just corresponds to 1σ.

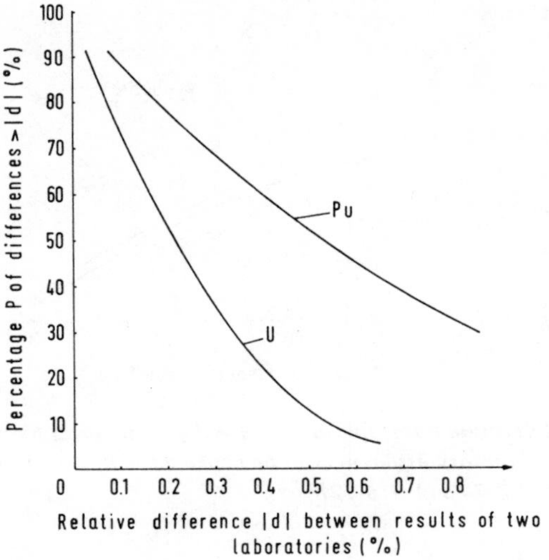

FIG.2. Distribution curves of interlaboratory differences for uranium or plutonium concentration determinations by chemical methods (derived from data of JEX-70).

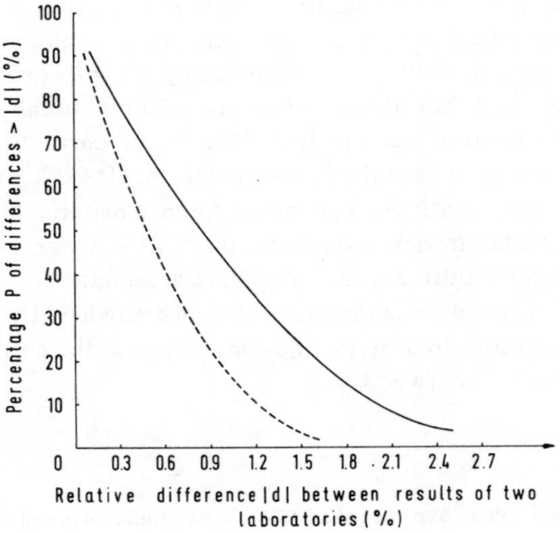

FIG.3. Distribution curves of interlaboratory differences for uranium concentration determinations by isotope dilution mass spectrometry (derived from data of IDA-72); solid line: ——: individual spiking by laboratories; . . . . : measurement of commonly prespiked samples.

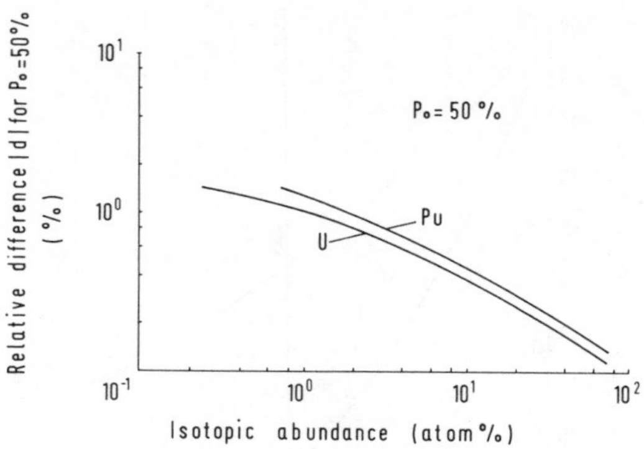

FIG.4. Uranium and plutonium isotopic abundance determinations by mass spectrometry: Minimal relative interlaboratory differences to be expected with a probability of P_0 = 50% (derived from data of JEX-70 and IDA-72).

Figure 3 displays such curves based on experimental data from IDA-72 concerning uranium concentration determinations by the mass-spectrometric isotope dilution techniques as used with the analysis of highly active input solutions of reprocessing plants.

With this analytical method the special means of sample treatment preceding mass-spectrometric measurement has to be taken into account in order to obtain results relevant to practical conditions: the solid curve is valid in case the spiking procedure is performed at the individual laboratory using its own calibration procedure[5]; therefore, at best it approaches the situation usually encountered in plant control by verification measurement. The dotted curve displays results obtained with the so-called "standard experiment" of IDA-72, where the sample material had been spiked with one common reference material before shipping it to the participating laboratories. A comparison of these two curves exhibits the considerable influence resulting from the spiking procedure.

Owing to the relatively small number of values provided by IDA-72 for plutonium determinations by isotope dilution technique the corresponding distribution curve will be derived later[6].

---

[5] So-called "self-spike" experiment of IDA-72. It should be noted, however, that, contrary to practical conditions, fission-product-free synthetic sample solution had to be used in this part of IDA-72 in order to avoid the influence of possible aging effects on the results.

[6] For this purpose extended data material can be expected from PAFEX-II [6] and the IDA-78 evaluation programme initiated at the beginning of 1978.

As already mentioned a dependence of the measurement accuracy on the isotopic abundances has to be taken into account in the mass-spectrometric determination of isotopic compositions. Therefore, the distribution curves were derived for different isotopic abundances; then the related d-values were obtained from these curves for arbitrarily chosen values of the probability P and plotted as a function of the isotopic abundance. Figure 4 presents such curves for uranium and plutonium with the probability P fixed at 50%. The areas below the curves now represent the range of "acceptable" differences if the d values related to the probability chosen are considered as threshold values D_0. The main data were derived from the IDA-72 experiment, part of the data were taken from the JEX-70 experiment.

The experimental situation allowed the curve for plutonium to be based on data material covering the isotopic abundance range between about 1% and about 70%. With respect to the curve for uranium, experimental data were available from the interlaboratory programmes mentioned for the isotopic abundance range between 0.3 and 3%. For the isotopic abundance range above 3% an extrapolation was performed by essentially following the shape of the curve for plutonium.

Based on these plots, which had been derived from the results gained with international analytical intercomparison programmes, the practical application of the judging procedure was subsequently field-tested by courtesy of the Euratom Safeguards Directorate.

In this test, several hundred verification values were judged, which resulted from the control of 10 different plants. These judgements concerned Pu and U concentration determinations as well as isotopic abundance determinations of ^{235}U, ^{239}Pu, ^{240}Pu, ^{241}Pu and ^{242}Pu.

For all analytical methods considered, the threshold values D_0 resulted from $P_0 = 30\%$ probability[7], the latter value being chosen arbitrarily. Figure 5 displays similar curves as already described in Fig.4, however, for $P_0 = 30\%$.

For the constant k, a rather high value of three was applied in order to classify in this test only such differences as "unsatisfactory" which were caused by error sources of high significance.

With a value of $P_0 = 30\%$ chosen for establishing the threshold values D_0, it could be expected that about 70% of the differences observed between declared and verified values would be classified as "acceptable". This prediction was fulfilled approximately with an average of 63% "acceptable" differences found for the whole test. They varied only between 61% and 64% for the different analytical problems considered. This finding shows that the actual field situation

---

[7] Should the distribution curve for the interlaboratory difference be derived from a group of laboratory results with a Gaussian distribution and a standard deviation σ, the threshold value D_0 related to $P_0 = 30\%$ is given by about 1.5σ.

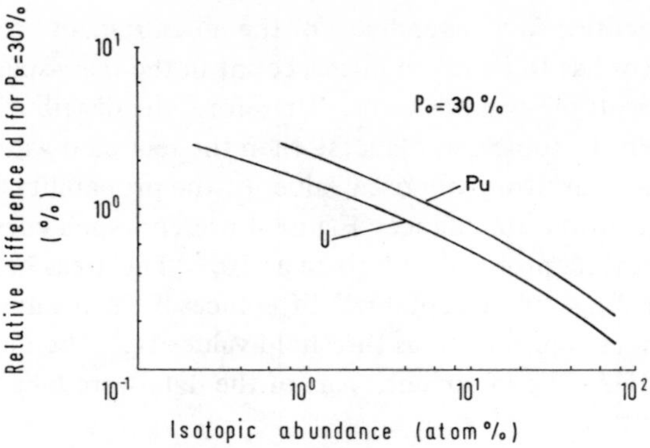

FIG.5. Uranium and plutonium isotopic abundance determinations by mass spectrometry: Minimal relative interlaboratory differences (D_0 values used for testing the judging procedure) to be expected with a probability of P_0 = 30% (derived from data of JEX-70 and IDA-72).

encountered in the safeguarded area does not differ from that characterized by the results of the international intercomparison experiments evaluated for this test.

In agreement with this finding, regarding the verification effort there was about 30% reduction in the analytical work compared with the customary performance of duplicate determinations: according to the judging procedure repetition measurements are performed only in the case when a non-acceptable deviation from the declared value resulted with the first determination.

More detailed investigation of the test results provided the following information. Although verification measurements on the isotopic composition of plutonium were corrected for the reference data of the declaration value, relatively high differences were more often observed for ^{241}Pu than for the other isotopes. This might indicate that, in some cases with the data reporting, not enough attention is paid to the time dependence of the isotopic composition of plutonium or to the completeness of americium separation; relatively often these effects were the reason that a deviation had to be classified as "unsatisfactory" in addition to inhomogeneity of sample materials or errors in data transmission.

An over-representation of differences not belonging to the acceptable range might indicate that the deviations between declaration values of a specific plant and verification values carry contributions from systematic deviations.

To check the capability of the procedure in this respect, the distributions of differences obtained in three plants for uranium isotope abundance determinations were compared; the results are presented in columns 2 to 3 of Table I. Obviously, for plants A and B the distribution between acceptable and non-acceptable deviations is even better than expected for P_0 = 30%, the probability chosen for the

TABLE I. DISTRIBUTION OF DIFFERENCES BETWEEN DECLARED AND VERIFIED VALUES OBSERVED FOR THREE PLANTS WITH ^{235}U ABUNDANCE DETERMINATIONS

1	2	3	4
	LEU		LEU + HEU
Classification	Plant A	Plant B	Plant C
"Acceptable"	86%	83%	48%
"Suspicious" and "Unsatisfactory"	14%	17%	52%
Data evaluated	21	30	60

TABLE II. DISTRIBUTION OF DIFFERENCES BETWEEN DECLARED AND VERIFIED VALUES OBSERVED FOR PLANT C WITH ^{235}U ABUNDANCE DETERMINATIONS FOR DIFFERENT SUBGROUPS

1	2	3	4	5
Classification	LEU + HEU	LEU	HEU	
"Acceptable"	48%	67%	38%	69%[a]
"Suspicious" and "Unsatisfactory"	52%	33%	62%	31%[a]
Data evaluated	60	21	39	39

[a] Results of judgements according to threshold values derived for isotopic determinations of plutonium.

selection of a D_0. However, for plant C, non-acceptable differences are distinctly over-represented.

The measurements from plants A and B were performed on low-enriched uranium (LEU); the measurements from plant C comprise those on both low- and on high-enriched uranium (HEU). Therefore, for further investigations the latter measurement results were split into two groups, one group concerning LEU and the other concerning HEU.

The distributions so obtained are presented in columns 3 and 4 of Table II. The distribution of the results received with the low-enriched material now fulfils expectations related to $P_0 = 30\%$ now for plant C also; the over-representation of large differences obviously concerns only the results obtained from the highly enriched material. When explaining this finding related to HEU two reasons should be considered:

For this type of measurement a laboratory bias might be indicated.

The curve derived for uranium with respect to the threshold value D_0 as a function of the abundance had been extrapoalted (Fig. 5). In doing so some uncertainty might have been introduced.

To investigate the first reason a test on the signs of the deviations observed was made; a ratio of 1:2 was found pointing towards a one-sided deviation. Detailed examination on the actual values of the differences judged as non-acceptable revealed that this laboratory bias, if existing at all, would probably amount to less than 0.1%.

Considering the second reason, the curve given in Fig. 5 for plutonium should be an upper limit for the uncertainty induced while extrapolating the curve for uranium. Therefore, the values in column 5 of Table II were also derived according to the threshold values valid for plutonium. Obviously, the distribution obtained now corresponds to that found for LEU. The difference in the threshold values D_0 used for uranium and plutonium also amounts to about 0.1%.

For the time being it is not possible to decide which of these two possible reasons dominates because of the limited data base. Nevertheless, it should be pointed out that for this analytical problem the procedure is sensitive to deviations as small as 0.1%.

3. CONCLUSIONS

The experience gathered so far in testing the judging procedure proposed confirms expectations regarding its applicability and its capability. The procedure is based on data relevant to the actual international status of measurement techniques and it is felt that it meets the demands of practical safeguards. To endorse its reliability and to support the implementation of this method as envisaged by Euratom Safeguards Directorate, the performance of further analytical intercomparison programmes and the evaluation of existing data material are recommended.

ACKNOWLEDGEMENTS

The authors thank the Euratom Safeguards Directorate, Luxembourg, for making it possible to test the applicability of the procedure described with actual field data. Especial thanks are due to Mr. E. Van der Stricht for his continuous engagement and encouragement. They also acknowledge the helpful discussions with Mr. R. Avenhaus on special subjects. One of us (W.B.) is also indebted to his colleagues of the Euratom Joint Research Centres who contributed to the testing of the procedure.

REFERENCES

[1]] BEYRICH, W., "The problem of analytical interlaboratory differences in practical safeguards" Safeguarding Nuclear Materials (Proc. Symp. Vienna, 1975) **2**, IAEA, Vienna (1976) 175.

[2] BEYRICH, W., SPANNAGEL, G., Distributions of analytical interlaboratory differences, in preparation.

[3] Joint Integral Safeguards Experiment (JEX-70) at the EUROCHEMIC Reprocessing Plant, Mol, Belgium, (KRAEMER, R., BEYRICH, W., Eds) KFK 1100, EUR 457e (1971).

[4] BEYRICH, W., DROSSELMEYER, E., The Interlaboratory Experiment IDA-72 on Mass Spectrometric Isotope Dilution Analysis, KFK 1905, EUR 5203e (1975).

[5] BEYRICH, W., SPANNAGEL, G., The mass spectrometric determination of U-235 in uranium hexafluoride – discussion of two interlaboratory evaluation programs, Nucl. Tech. (to be published).

[6] SZABÓ, E., BUSH, W., HOOGH, C., Results of the IAEA Process Analysis Field Experiment II for Safeguards (PAFEX-II), IAEA-STR-71, IAEA, Vienna (1978).

DISCUSSION

P. DE BIEVRE: I think you have developed an excellent way of "describing" and "handling" actual differences in measurement results, i.e. differences between declared values and verification measurement results. It is probably the best practical system in existence. I wonder whether you could comment on the next step, which is how to judge, handle and evaluate differences between declared values and verification measurement results on the one hand and "true" fissionable material contents on the other. This is what we are really after in safeguards. We should realize that in your system, as well as in all other existing systems for "handling" data, both the declared values and the verification measurement results could be wrong.

W. BEYRICH: I think it is the dream of every analyst to know the "true" value for comparison of his results. There are various specialized laboratories in existence which establish very accurate data, for example those operated by the Central Bureau for Nuclear Measurements at Geel, Belgium, and the National Bureau of Standards in the United States of America. However, this characterization of materials is extremely time-consuming, complicated and costly, as you know, and for these reasons it is feasible only in special cases such as for the characterization of materials used in analytical intercomparison experiments. In daily practical safeguards operations, for which the reported procedure is proposed, the situation is very different, however, and often one does not even know whether a declared value is based on a single determination or whether it is the average of many results.

You mentioned the possibility of both declared and verified values being biased in the same direction, causing significant deviations from the "true" value. To reduce this possibility, I think two measures can be taken. First, in the case of intercomparison experiments in which sample material is characterized as exactly as possible, it should be established whether the specific analytical method under investigation contains error sources which tend to cause a bias of the same sign for all laboratories. An example of this situation is the memory effect in the isotopic analysis of UF_6 by gas mass spectrometry; once this effect is known, it can be overcome, e.g. by using the double-standard technique. Secondly, every laboratory should run – and probably already does run – an internal measurement assurance programme involving the frequent analysis of samples whose composition is accurately known and as similar as possible to the material subjected to routine analysis. In this respect, it might be of advantage if more suitable materials could be made available in sufficient quantities and at a reasonable price by such specialized laboratories as those I mentioned.

W.L. ZIJP: The procedure described appears very interesting. It assumes that the distribution curve of interlaboratory differences can be established satisfactorily, and this calls for a high incidence of these differences. A large number of differences may result from a large number of participating laboratories analysing a few samples or, conversely, from a small number of laboratories analysing a large number of samples. Can you suggest "good" values for the number of laboratories and for the number of samples from the point of view of satisfactory application of your method?

W. BEYRICH: To get a representative distribution, a large number of laboratories analysing a few samples is certainly better than a few laboratories analysing a large number of samples. It is our experience that ten participating laboratories is the lower limit when only one or two samples are measured. We compared the distribution of differences for the uranium concentration determination by isotope dilution mass spectrometry, obtained from the measurements of 23 laboratories in the IDA-72 experiment, with the results from 12 laboratories in PAFEX-II. The agreement was very satisfactory, showing that both curves are representative. There is, however, another feature of these curves which is of interest in connection with their representativeness or relative independence of strong regularities, namely the way they are influenced by "outliers" of the laboratory results on which they are based. As long as the relative number of such outliers is small, their influence on the distribution curve is restricted mainly to the tail part. It is thus possible to derive from the middle part of the curve a meaningful estimate of the variance in the laboratory results without the prior rejection of outliers, which is not possible when using statistical methods in the conventional manner. This is of interest for the evaluation of analytical intercomparison experiments.

W.L. ZIJP: In an interlaboratory test, the participating research and development laboratories generally have a good "capability". Do you think the differences found in such a test are representative of operator-inspector differences where operator results are obtained by routine procedures? Perhaps the solution lies in the choice of your parameters D_0 and $k \cdot D_0$.

W. BEYRICH: In field testing of the judging procedure with operator-inspector data, the Euratom Safeguards Directorate at Luxembourg obtained nearly the same ratio between "acceptable" and "non-acceptable" differences as expected from the threshold values applied, which were derived from interlaboratory intercomparisons only. This means that experience has not so far indicated any significant effect of the type you mention.

A STATISTICAL APPROACH TO THE VERIFICATION OF LARGE STOCKS OF FISSILE MATERIAL IN DIVERSE FORMS

P.T. GOOD
NMACT, Harwell

J. GRIFFITH
SNPDL, Springfields

A.G. HAMLIN
NMACT, Harwell,
United Kingdom

Abstract

A STATISTICAL APPROACH TO THE VERIFICATION OF LARGE STOCKS OF FISSILE MATERIAL IN DIVERSE FORMS.

An examination of the total safeguards problem shows that by far the greatest proportion of material to be controlled at any one time is in store or a similar static condition. This static material may be in very diverse forms and containers, and on any one site may be too large in quantity for 100% verification to be possible. It is therefore necessary to use sampling systems to obtain a quantifiable assurance regarding its verification. Conventional stratification approaches to defining sampling systems may fail for a variety of reasons, and may frequently fail most for the most sensitive material. The UKAEA has developed a sampling scheme based upon the monetary unit accounting system of financial audit, which avoids the need for stratification and which can allow weighting of sensitive material so that its probability of selection for verification can be enhanced in a controlled manner. The advantages and disadvantages of the scheme are discussed and examples of its practical application described.

1. INTRODUCTION

1.1 Verification of large stocks of nuclear materials that may be held at a particular location in a diversity of forms and containers poses particular problems. If the stock runs to hundreds or thousands of items, the auditor or inspector generally has not time to verify every item, and must rely upon some form of statistically based sampling plan that will enable him to minimise the number of items he must examine before he can make a statement, qualified by appropriate uncertainty, regarding the accuracy of the accountancy.

1.2 The traditional approach to such a problem of sampling has been to divide the stock into "strata" of similar items and then to apply simple random sampling plans within the strata, finally recombining the observations on the strata to give a conclusion on the whole [1,2]. If the number of strata is large, as is frequently the case, the preparatory work is tedious and, since each stratum requires to be verified to an appropriate degree, the total number of samples may be rather high.

1.3 A similar problem has existed in the field of financial auditing, where an auditor may be faced with judging the validity of accounts by sampling a population of thousands of transactions, represented by prime documents ranging in value from a few units to perhaps hundreds of thousands of monetary units. In financial accounting, the material represented by the documents, monetary units, is not in doubt and the auditor is concerned to check for errors of recording and transcription. Clearly any conceivable error in a document for a few monetary units will have less effect on the overall validity of the accounts than conceivable errors in documents for hundreds or thousands of units, and should attract less of the auditor's attention. This has led to the development of monetary value sampling [3], based on the theory of sampling with unequal probabilities [4], in which the sampling plan is arranged so that the probability of choosing a document for examination is proportional to the monetary value that it represents. This theory produces a very substantial reduction in the number of samples that must be drawn in order to make a quantified statistical statement on the validity of the account of the whole population and does not require a preliminary stratification.

1.4 In nuclear materials accounting, the material represented by the document is in doubt, but this does not reduce the validity of the above argument. An error in correspondence between a small batch and its document is less serious as regards the validity of accounts than a similar error in a large batch, and this error in samples selected for verification would be no different from a clerical or transcription error in financial terms.

1.5 This paper demonstrates the practicability of applying such a technique to the verification of stocks of nuclear material in diverse forms. For convenience, and by analogy with monetary value sampling it might be termed "mass value sampling".

1.6 The lack of an invariant relationship between a document and the nuclear material it represents could be a further advantage in auditing stocks of material containing items of mixed strategic value[1]. By weighting the mass of these items according to their strategic value, it would be possible to obtain an assurance of the stock in terms of strategic value accountancy and to avoid any strategic stratification prior to sampling. This aspect is not however considered further in this paper.

2. THE OPERATION OF MASS VALUE SAMPLING

2.1 The basic requirement for mass value sampling is a simple list of the items comprising the stock, assembled to give a cumulative total in terms of mass units. The items do not have to be in any particular order, and can conveniently be taken from a Physical Inventory Listing or from a computer print-out of stock. The mass units may be arbitrary, but are conveniently either the units normally used, or the appropriate multiples or sub-multiples of these by factors of ten.

2.2 This list of mass units then becomes the population to be checked for attributes i.e. for correctness or incorrectness according to normal attribute theory. The fact that items occupy larger or smaller parts of

---

[1] Strategic value may be interpreted in any way appropriate to the situation value to a divertor, ease of diversion, sensitivity in the safeguards or commercial terms, etc.

TABLE I. DISTRIBUTION OF MATERIAL IN THE STORE

Weight per item	Items	Weight per item	Items	Weight per item	Items
0 - 10	62	71 - 80	9	141 - 150	5
11 - 20	21	81 - 90	10	151 - 160	9
21 - 30	21	91 - 100	4	161 - 170	2
31 - 40	1	101 - 110	1	171 - 180	1
41 - 50	8	111 - 120	2	181 - 190	2
51 - 60	9	121 - 130	5	191 - 200	5
61 - 70	14	131 - 140	3	961 - 970	3

the list according to the number of mass units they contain gives them automatically the required higher or lower probability of being sampled. The samples may be drawn by choosing a series of random numbers within the total of mass units, or more simply, since the original listing is random, by drawing at regularly spaced intervals from a random starting point.

2.3 Where two or more sample points fall within the span of mass units allocated to an item they take the attribute determined for that item and re-allocation is not required. For an example of this see Table I.

2.4 It is important to note that the change from variable to attribute sampling inherent in the mass value sampling approach has an important consequence in that the objective of the audit, and through this the attribute to be examined, needs to be determined and specified before the sampling scheme can be established. Thus one scheme may show with equal probability that there is a surplus or deficit in a stock but give only a low degree of confidence in the actual amount of surplus or deficit. Another scheme may give much more confidence that a deficit is or is not present but may blind the auditor to the presence of a surplus.

3. OBJECTIVE OF THE VERIFICATION

3.1 There are a number of possible objectives which involve verification of nuclear material stocks by the auditor. These include:-

(a) The determination of whether material is missing from a declared total stock.

(b) Verification that either the total book stock or the book values of individual items agree with the auditor's measured values.

(c) A check that some characteristic, e.g. stated gross weight, enrichment, presence of plutonium, is in accordance with book values. (This approach could be used if the necessary equipment for quantitative measurement was not available or if it was considered that the material was not sufficiently important to justify more extensive characterisation, e.g. for stocks of low enriched or natural uranium).

(d) Comparison of independent measurements with those carried out by local staff in order to come to a conclusion regarding the effectiveness (accuracy, precision) of the measuring system in use.

This paper is primarily concerned with the first and second objectives given above.

3.2 The simplicity of the procedure, once the objective has been determined is illustrated by means of step by step instructions for application.

4. METHODS OF ACHIEVING VERIFICATION

4.1 Determination of Losses

4.1.1 Suppose the auditor's instructions are 'check to 95% confidence that not more than Y kg are missing from a store with stated total content X kg'.

An important implication in this case is that the auditor is solely concerned with checking that no item or material is missing from those given on the list presented to him by the store holder. The auditor will not, except possibly by accident, discover if some items are present but have not been included on the list.

4.1.2 A meaningful mass unit is perhaps the minimum discrepancy which can be detected by the inspector's quantitative technique, for example if this can distinguish a difference of 5 g from the stated value with certainty but not a difference of 4 g then the mass unit may be taken to be 5g.

4.1.3 The subsequent steps in the mass value sampling scheme are to:

(1) List individually the amounts of nuclear material stated to be in each item of the group to be audited and form a cumulative sum for the total amount of material present ,Y.

(2) Set the amount of loss out of the total Y which is to be detected at X, and decide the degree of confidence required.

(3) Decide the mass unit, Z, and express the total amount Y in terms of mass units. This gives the population size P = Y/Z.

(4) Express the item contents, individually and cumulatively, in terms of mass units.

(5) Calculate the sample size, n, required to detect a loss of X/Y% of the total inventory by using binomial or hypergeometric tables, depending on the population size as expressed in mass units. For P less than three thousand use Hypergeometric tables or the equation quoted by Jaech [1,4]. For Y/Z greater than three thousand, Binomial tables give a sample size which is only slightly larger than that given by Hypergeometric.

(6) Select n different random numbers up to Y/Z, and order them. The production and ordering of the random numbers is extremely tedious if a computer is not available, and normally very little element of randomness in the selection of the mass units to be examined will be lost if they are selected on a regular interval basis, i.e. units are selected for examination at intervals of Y/Zn from a random starting number which is less than Y/Zn.

(7) Use the cumulative list of mass units to select, as the sample to be measured, the items which include these numbers. If in selecting the units of account to be verified a particular item

TABLE II. PARTIAL STOCK LIST SHOWING HOW ITEMS ARE SELECTED FOR EXAMINATION

Item No.	Item Weight	Cumulative Mass Unit Sum	Selected Mass Units contained - 100 g detection level [a]	Selected Mass Units contained - 200 g detection level [a]
1	158.9	1589	1 - 4	1, 2
2	135.2	2941	5 - 8	3, 4
3	157.8	4519	9 - 13	5, 6
4	112.3	5642	14 - 16	7, 8
5	61.4	6256	17, 18	9
6	71.2	6968	19, 20	10
7	156.0	8528	21 - 25	11, 12
8	20.6	8734	26	13
9	3.6	8770		
10	165.2	10422	27 - 31	14, 15
11	121.3	11635	32 - 35	16, 17
12	4.1	11676		
13	0.8	11684		
14	2.1	11705		
15	5.6	11761		
16	197.2	13733	36 - 41	18 - 20

[a] For clarity, serial numbers are given instead of the actual mass unit numbers selected by the sampling scheme

includes more than one of the selected numbers there is no need to select an alternative mass unit or an alternative item.

(8) Measure the chosen items and check that any apparent discrepancies are not due to clerical or other reasonable human errors by local staff or by the auditors. Note that such clerical errors by local staff may, or may not, affect the total amount stated to be in the store.

(9) Calculate the best estimate of the proportion defective in terms of mass units i.e. number of defective mass units in the total number of mass units examined. For this purpose the term in the denominator is the total number of mass units said to be present in the items measured and not the sample size n. This arises from the fact that in examining a particular mass unit contained

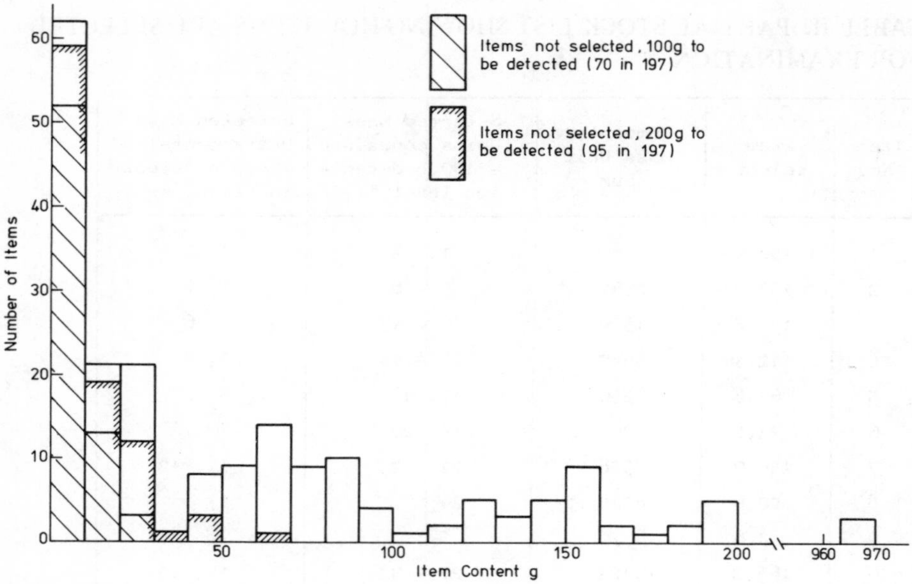

FIG.1. Distribution of the Pu content of the items in the store and of the items not selected for examination.

in an item one automatically examines all the mass units contained in that item.

(10) Calculate the upper 95% confidence limit of the proportion defective, using binomial or hypergeometric confidence limits.

4.1.4 If as a result of the measurements only losses are discovered, the interpretation of the results in terms of the weight of material is straightforward. If, however, some items are found which contain more than the stated number of mass units then provided that the number of excess or deficient defects is small in comparison with the number examined it seems intuitively reasonable to compute the proportion of 'loss' defects and also the proportion of 'gain' defects. The net gain or loss so determined can be used as a guide as to whether the scheme should be redesigned as below. The auditor may wish to reconsider his strategy since he has started by defining as "defective" only those items which have less than the stated amount of material. Items having an excess are then "normal".

4.2 Determination of Total Stock

4.2.1 A difference between this case and the one where the auditor is looking for a loss is that, in determining the total stock, the auditor will need first to take positive steps to ensure that the declared stock is not understated. He should thus count the number of items present and ensure that they are all recorded in the stock lists presented to him by the local staff. If there are any not included on the list his first task would be to determine their contents and add them to the cumulative mass unit list.

The auditor can then proceed to estimate the content of the listed items using a procedure similar as that used in detecting a loss, but with an appropriately different criteria under Step 5, since the test is now double sided.

4.2.2 While an estimate of the total stock can be obtained quickly from the proportion defective, a more accurate method is to estimate the discrepancy between each item examined and its stated value, and then to calculate a mean value of the discrepancy which can be applied to the stated total stock.

5. EXAMPLE OF THE METHOD

5.1 As an example of the application of the method used to select the items to be measured, data on items in an actual fissile material store is listed. The total holding amounted to 13,130 g contained in 197 items of various sizes. The distribution of material throughout the items is given in Table II and is shown on Figure 1.

5.2 We wished to be 95% confident of detecting a loss of 100 g and a mass unit of 0.1 g was chosen because this was the content of the smallest item in the store.

$$\text{Loss to be detected} = \frac{100}{13130} \times 100\% = 0.76\% \text{ of the total holding.}$$

$$\text{Population size} = 13130/0.1 = 131300.$$

Therefore we used the Binomial table to select the sample size and found that the sample size to detect a 0.76% loss is 395. Using the simple procedure of selection at regular intervals, the sampling interval needed was 131300/395 = 332, and we worked down the cumulative list selecting every 332nd mass unit, starting with a random number, less than 332.

5.3 The application of this procedure to the items in the store is illustrated in Table I which shows the data for first 16 items. The first 4 mass units to be measured occur in the 1st item, the second 4 occur in the 2nd item etc, and items number 9, 12, 13, 14 and 15 are not examined at all. Proceeding in this way through the complete list produces the result that, of the 197 items in the store, 127 are selected for examination. The 70 items not selected (35.5% of the total), however, contain only 457.7 g of material (3.48% of the total), and all contain less than 30 grams of material per item, the majority containing less than 10 g per item. Their distribution is shown in Fig. 1.

5.4 A similar result is obtained if the amount to be detected is raised to 200 gr. In this case 198 mass units need to be examined. This results in 95 items not being selected for examination, 48.2% of the number present. These 95 items contain 1057.6 g of material, 8.1% of the total. Again the items not examined are grouped at the lower end of the weight distribution; nearly all of the items containing less than 10 grams of material, see Fig. 1.

5.5 Thus an appreciable saving in time will be achieved by using a planned random sampling technique when verifying items in stores or similar areas.

5.6 It could be argued that in the examples given a straightforward decision not to examine any item containing less than 20 g of material would have led to a similar time saving. This is of course correct, in these examples, but such a decision would have removed any possibility of detecting diversions from items in this class. The structured sampling procedure does permit the possibility of the inspector detecting such diversions albeit with a small probability proportional to their importance - three items containing less than ten grams were examined - and this is embodied in the final statement made regarding the confidence in the audit. The straightforward removal of the smaller items would not permit the statement to be made.

5.7 It is worth remarking that in this example,apart from the group of very small items,the size distribution of items is roughly continuous and it would have been very difficult to make any rational stratification.

6. CONCLUSIONS

6.1 Monetary value sampling has been proposed for financial audit purposes as a means of reducing verification workloads on heterogeneous populations of financial documents.

6.2 The technique is based on the theory of sampling with unequal probabilities, and directs the attention of the auditor preferentially to the documents likely to have the most significant affect on the accounts.

6.3 The method can be applied equally well to verification of nuclear material accounts by converting the material under investigation to a population of mass units.

6.4 Application of the methods to this population enables quantitative statements to be made regarding the uncertainty of the accountancy with considerable saving in effort compared to 100% examination of the material and without the need to stratify heterogeneous populations.

6.5 The technique could conveniently be referred to as mass value sampling. It could be extended to recognise strategic value as an alternative.

REFERENCES

(1) BROWN, F., GOOD, P.T., PARKER, J.B. "Sampling of Stores for Safeguards Purposes" COS 4A, A.W.R.E. Aldermaston, June 1970.

(2) BROWN, F. et al, "Progress in the Development of a System for the Safeguards Inspection of the Fuel Store of a Zero-Energy Fast Reactor" COST 20, A.W.R.E., Aldermaston, October 1970.

(3) ANDERSON, R.and TEITLEBAUM, A.D. "Dollar-unit sampling" Canadian Chartered Accountant,April 1973, p.30.

(4) JAECH, J. "Statistical Techniques in Nuclear Material Control" TID 26298, p.321.

DISCUSSION

T.M. BEETLE: The unequal probability sampling technique described in your paper might also be useful in the sampling of records in records audit work. For example, mistaken addition of a zero in recording a number will result in an increase in the probability of selection of that record in a sample, so that the mistake is more likely to be detected. A difficulty in applying the technique, however, is that manual location of the sample of records would require a large amount of arithmetic. If the records are computerized, this difficulty is not present.

STATISTICAL ANALYSES IN NUCLEAR ACCOUNTABILITY
A simulation approach

M.E. JOHNSON, G.L. TIETJEN, M.M. JOHNSON
Los Alamos Scientific Laboratory,
Los Alamos, New Mexico,
United States of America

Abstract

STATISTICAL ANALYSIS IN NUCLEAR ACCOUNTABILITY: A SIMULATION APPROACH.

This paper describes a computer simulation approach to modelling material balances and to deriving the limits of error attributable to measurement procedures. A new probability distribution is presented which is useful in the computer simulations. This distribution permits the investigator to assess the sensitivity of initial distributional assumptions on the computed limits of error. The simulation approach is illustrated with a case study example.

1. INTRODUCTION

In this paper we discuss the statistical treatment of the numbers arising from the process of nuclear accountability. Our goal is to decide whether a given amount of material unaccounted for (MUF) is actually missing from the facility or is apparently missing because of combined measurement errors.

If the MUF falls within certain computed limits, we conclude that it is within measurement error. If outside these limits, we conclude that some material is missing. Our approach in calculating these limits on measurement error is first to model the given process. This involves analyzing the flow of material in the process and the associated measurement instruments and practices (including calibration techniques). We model each measurement in the process with a random variable whose expected value is the true value to be measured and whose probability distribution reflects the likely variability in the observed value. We then employ a computer program to simulate the process and to generate many realizations of the MUF. Given the simulated MUFs from a model which assumes no missing material, we readily can see the variability which can be expected in the normal course of events. Intervals containing the middle 95% and 99% of the generated MUF values yield reasonable estimates of the "warning" and "out of control" limits, respectively.

A much simpler approach to estimating these limits is to assign a standard deviation (or precision) to each measurement in the process and then to assimilate this information in an overall standard deviation by propagation of error. While this gives an estimate of the variance, it is not known how to use such an estimate to form a confidence interval for the mean. The usual practice of taking 2 or 3 estimated standard deviations on either side of the mean as "warning" or "out of control" limits depends heavily on the assumption of normality. Although this approach is easy to carry out, the resulting limits may be poor estimates of the overall measurement error. Frequently, measurements are the product of two values (for example, weight and concentration) which can lead to non-normal probability distributions. Another major difficulty with this approach is its inability to handle calibration errors. Since calibration curves are estimated from the measurement of standards (material with a "known" value), the mere assignment of standard deviations to individual measurements does not accurately incorporate calibration errors.

The simulation approach requires considerable expertise in modeling a given process, but leads to reasonable estimates of the overall measurement error. A desirable feature of this approach is that we can test the effect of our distributional assumptions on our estimates of measurement error. In particular, we can investigate the effects of departures from the normal distribution assumption. This test is performed by exercising our computer model for a variety of assumed probability distributions. For each computer run, the estimated measurement error is obtained. The complete set of these estimates indicates the effect of the distributional assumptions. In the desirable situation, the set of estimates do not vary dramatically so that we can conclude that the results are not sensitive to the initial assumptions. Sensitivity analysis is an essential tool in evaluating the simulation model and assessing the appropriateness of the estimates of measurement error.

In section 2 we describe a new family of symmetric univariate probability distributions which can enhance sensitivity analysis studies, as described above. This family is particularly useful in analyzing quantitatively the effect of departures from normality on the estimates of measurement error. The proposed family includes as special cases the uniform and normal probability distributions, which are commonly used in nuclear accountability. The kurtosis of the family (i.e., the fourth standardized moment) which is an indicator of tail weight, varies from 1.8 (the uniform) to 3.0 (the normal) to 5.4 (a heavy-tailed distribution). Hence, the family includes a broad spectrum of probability distributions. Random variates from the proposed family are easy to generate, and thus, they can be used in the computer simulation model.

The simulation approach together with the new family of distributions leads to robust estimates of the overall measurement error. In section 3 we describe in detail a case study in which measurement errors for a particular process were estimated by simulation. We conclude that our approach leads to reasonable estimates of overall measurement error.

2. PROPERTIES OF THE NEW DISTRIBUTION

The proposed distribution has probability density function

$$f(x) = \frac{\sqrt{\alpha}\Gamma(\alpha-1/2)}{2\sigma\Gamma(\alpha)\sqrt{3}} \quad [1 - H(\frac{2\alpha(x-\mu)^2}{3\sigma^2})],$$

for $\alpha > 1/2$, $-\infty < x < \infty$. H is the distribution function of a gamma random variable with shape parameter $\alpha - 1/2$ and scale parameter 2. Numerous properties of this distribution are derived in [1, 2]. Properties of importance to nuclear materials simulation applications are enumerated below.

1. A random variable X with the density f is symmetric and all moments exist. In particular, the mean of X is μ, the variance is σ^2, the skewness is 0, and the kurtosis is $1.8(\alpha + 1)/\alpha$.

2. The kurtosis can assume any value in the interval $[1.8, 5.4)$. For a specified kurtosis, say β_2, set $\alpha = 1.8/(\beta_2 - 1.8)$.

3. A range of distributional properties is obtained by appropriate choice of parameters. For $\alpha = 1.5$, a normal distribution is obtained. As α tends to infinity, f approaches a uniform distribution. More generally, the probability in the tails can be regulated by the choice of α: large α gives light tails, α near 1.5 gives medium tails, and heavy tails are obtained for α near 0.5.

4. The proposed distribution can be easily generated on a digital computer. One algorithm is as follows: Generate a gamma variate x_1 with shape parameter α and scale parameter 2. Then, generate conditionally a uniform variate x_2 on the interval $(-\sqrt{x_1}, \sqrt{x_1})$. A random variate with density f is $\sqrt{(1.5\alpha)}\sigma x_2 + \mu$. Recommendations for the appropriate gamma generation algorithm are given in [3].

5. By using a computer simulation program for a range of α values, one can assess the effects of almost any type of symmetric non-normality on the simulated results.

3. CASE STUDY

In this section we describe the methodology for computing limits of error (LE) in a process for recovering uranium from metal scrap. We first describe the physical material. We then discuss the material balance areas and the measurement devices and practices. Finally, we present results from a computer simulation model which is used to estimate LE.

3.1. Physical Material

A part of the uranium reprocessing operations at the Los Alamos Scientific Laboratory consists of recovering uranium from turnings created in machining uranium metal. After burning to eliminate oily residues, this material is stored in cans in a vault for possibly several months. Periodically, several cans are removed and their contents dissolved. Ultimately, fairly pure uranium oxide is precipitated.

3.2. Material Balances and Measurements

We concentrate on two material balances in the recovery process. Before a can is placed in the vault, its contents are burned to an oxide and a non-destructive assay is performed using a random driver device. This device is calibrated with standards of 250g, 500g, 1000g, 1500g and 2000g uranium per can. There are five replications per can during the calibration run. Each can in the vault usually contains 1500g to 2000g uranium. One material balance area is defined by considering a processing batch of 4 or 5 cans. The corresponding MUF is the difference between the total uranium assays at the times of putting the cans in the vault and taking them out of the vault.

The batch can usually be completely dissolved in a nitric acid solution. The volume of the solution is typically 30 to 40 liters and is obtained from reading graduated cylinders (especially designed for radioactive solutions). The concentration is determined from a non-destructive uranium solution assay device (USAD) using a 20 ml sample. The calibration standards used for this device are 150, 250, 300 and 350 grams uranium per liter solutions with five replications each. The product of the volume and the concentration yield an estimate of the uranium in solution. The second material balance area is defined by the material as it leaves the vault and the uranium in the solution.

3.3. Computer Simulation Model

The problem is to derive LE for each of the material balance areas defined in section 3.2. Under the assumption of no hold up or diversion, the corresponding MUFs can be modeled, as follows:

$$\text{MUF1} = \sum_{i=1}^{n} \left(\frac{x_i - a_1}{b_1}\right) - \sum_{i=1}^{n} \left(\frac{y_i - a_2}{b_2}\right)$$

$$\text{MUF2} = \sum_{i=1}^{n} \left(\frac{y_i - a_2}{b_2}\right) - V \cdot \left(\frac{C - a_3}{b_3}\right),$$

where

x_i = random driver measurement for i^{th} can as it enters the vault

y_i = random driver measurement for i^{th} can as it leaves the vault

n = number of cans

a_i, b_i = estimated calibration constants assuming a linear relationship $y_i = b_i x_i + a_i$

TABLE I. MUF1 RESULTS

β_2	Est. 95% Limits	Est. 99% Limits
1.8	±319g	±419g
2.5	±331g	±435g
3.0	±328g	±436g
4.0	±341g	±452g
5.3	±313g	±443g

TABLE II. MUF2 RESULTS

β_2	Est. 95% Limits	Est. 99% Limits
1.8	±273g	±355g
2.5	±272g	±360g
3.0	±274g	±356g
4.0	±279g	±370g
5.3	±280g	±378g

V = volume measurement

C = concentration measurement.

We can treat each of the measurements and estimates as random variables, with a variance derived from historical or designed experimentation. For investigating particular MUFs, we use the observed measurements as the means of the random variables. For the estimated calibration "constants," we simulate readings for the standards and fit a line to them. The slope of the fitted line is b_i; the intercept is a_i. Since the sets of random driver measurements are taken months apart, different calibration constants are simulated for the repeated measurements on a batch of cans.

The next step in the methodology is to simulate in a computer program the models for MUF1 and MUF2. Naturally, we use the proposed distribution of section 2 to model the individual random variables. From the previous paragraph, the values of μ and σ^2 are determined, and the parameter α gives us a degree of freedom in a sensitivity analysis. In particular, we can select, say, five kurtosis values 1.8, 2.5, 3., 4. and 5.3 with corresponding α values ∞, 2.57, 1.5, 0.818 and 0.514. Eventually, we compare five sets of estimates of LE. The details are apparent from the subsequent example.

Consider, for illustration, four cans with initial random driver measurements 1688g, 1676g, 1723g and 1705g and with later random driver measurements 1735g, 1719g, 1719g, and 1682g, respectively. The MUF for this material balance area is 63g gain. The solution assay is 6584g with a volume measurement of 26.2ℓ. Thus, the second material balance area has a MUF of 271g loss. Are these MUFs within their limits of error?

Our approach to the question is to simulate five replications of 1000 samples of MUF1 and MUF2. Each set of 1000 values is sorted, and the 5th, 25th, 975th and 995th observations provide estimates of the 0.5, 2.5, 97.5 and 99.5 percentiles. Denote the four estimates as p_i, q_i, r_i and s_i, respectively, where i is the replication. Since the limits are symmetric, we can justify estimates of the 95% warnings limits as $\pm$ [median $|q_i|$ + median r_i]/2 and the 99% out of control limits as $\pm$ [median $|p_i|$ + median s_i]/2. Certainly, other estimates could be proposed, but our experience indicates these to be robust.

The resulting estimates from the simulation run are given in Tables I and II.

From these simulation results, we can observe that the 63g MUF gain and the 271g MUF loss are within their respective LE for all distributions sampled. We conclude the MUFs represent material apparently missing because of combined measurement errors. We also notice that the LE estimates are reasonably stable over the range of distributions sampled.

4. CONCLUSIONS

In this paper we have presented a computer simulation methodology for determining limits of error for material unaccounted for. This approach is straightforward, leads to reasonable LE estimates, and can incorporate measurement errors induced by calibration. The new probability distribution can be used effectively to assess the impact of non-normal distributional assumptions. This facilitates the analysis of computed warning and out of control limits. An example has been given which illustrates the methodology.

REFERENCES

[1] JOHNSON, M. E., JOHNSON, M. M., "A New Probability Distribution with Application in Monte Carlo Studies," Los Alamos Scientific Laboratory Report No. LA-7095-MS, (1978).

[2] JOHNSON, M. E., TIETJEN, G. L., BECKMAN, R. J., "A New Probability Distribution with Applications to Monte Carlo Robustness Studies," submitted to the Journal of the American Statistical Association (1978).

[3] TADIKAMALLA, P. R., JOHNSON, M. E., "A Survey of Methods for Sampling from the Gamma Distribution," to appear in the Proceedings 1978 Winter Simulation Conference, Miami Beach (1978).

TWO-STAGE SAMPLING IN SAFEGUARDS WORK

T.M. BEETLE
Department of Safeguards,
International Atomic Energy Agency, Vienna

Abstract

TWO-STAGE SAMPLING IN SAFEGUARDS WORK.

There are several situations in safeguards work where two-stage sampling is more natural than one-stage sampling. That is, instead of drawing one set of random numbers to identify the items to be measured in each stratum, the items within strata are clustered into groups and groups are drawn at random followed by a selection of items within groups drawn at random. Some mathematical results for two-stage sampling are presented in this paper. An expression for the probability of non-detection of diversion of special nuclear material is derived from a description of the sampling plan under mild constraints. This expression is shown to be useful for determining the best strategy available to a diverter and determining a strategy for an inspector to combat this best strategy. A sample-size formula for fixing an upper bound on the maximum probability of non-detection of a chosen amount of diverted special nuclear material is presented. The constraints on the problem are primarily in the stratification of the containers of items. These constraints can be made less restrictive by a procedure which is conservative in the sense that increased sample sizes decrease the probability of non-detection of diversion.

INTRODUCTION

Random sampling of units of nuclear material for measurement is frequently employed in safeguards inspections. The simplest procedure (<u>simple random sampling</u>) is to number the units from 1 to N and to select a random sample of $n < N$ numbers from a random number table to identify the units which will be measured. This procedure is easy to execute when N is "small" and the storage arrangement of the units allows easy access (e.g. UF_6 cylinders, drums of UO_2 powder). However, there are safeguards inspection situations where the number and storage arrangements of units suggests the use of <u>two stage sampling</u>. That is, when units are clustered in subgroups it may be easy to use random numbers to select some subgroups and then to use random numbers again to select units within subgroups. Examples are fuel rods in boxes, pellet trays in stacks, spent fuel bundles in baskets, and plutonium coupons in drawers. The mathematics for describing the two stage sampling procedure is, of course, different from the mathematics for describing simple random sampling. This paper presents some results of a study of two stage sampling.

RESULTS

It is assumed that containers of items can be grouped into strata where containers in a stratum contain equal numbers of items. When an inspector selects random numbers with replacement for identifying some containers to be sampled in each stratum and then selects equal numbers of random numbers with replacement for identifying items within the selected containers within strata to be measured, then the best strategy for a diverter is to allocate the desired amount of special nuclear material (SNM) to be diverted to strata in inverse relationship to the stratum proportions of the total number of random numbers to be selected for containers and to divert the SNM from all items in some of the containers in each stratum.

In order to combat the best strategy for a diverter, the inspector should choose the amount of SNM which is to be considered significant if diverted, M, and choose the maximum probability of non-detection of the diversion, β, and calculate sufficient sample sizes. The number of random numbers for identifying containers in stratum i can be calculated as follows:

$$n_i = \ln\beta \left\{ \ln \left\{ \left[\sum_{i=1}^{N} (\overline{X}_i N_i R_i) - M \right] \left[\sum_{i=1}^{N} (\overline{X}_i N_i R_i) \right]^{-1} \right\} \right\}^{-1} \left\{ \sum_{i=1}^{N} (\overline{X}_i N_i R_i) \right\}^{-1} (\overline{X}_i N_i R_i)$$

$\overline{X}_i$ = average quantity of SNM per item in stratum i (as declared by the operator)

N_i = number of containers in stratum i

R_i = number of items in each container in stratum i

N = number of strata

For each random number identifying a container, the inspector should select one random number to identify an item which is to be measured. Selection of more than one item per random number identifying a container would be a waste of effort.

When the number of items per container are very variable, the inspector may construct strata consisting of containers that contain approximately the same number of items and calculate sample sizes as if each container had the number of items contained in the container with the largest number of items. This would be conservative in the sense that the probability of non-detection of diversion would be at least as small as the probability of non-detection of diversion when the number of items per container are all equal to the number of items per container for the container with the largest number of items.

DISCUSSION

Situation

The inspection situation considered here is that we wish to inspect items (subunits) of SNM distributed in containers (primary units) with a two stage sampling plan. Primary units are stratified so that each primary unit within a stratum contains the same number of subunits as every other primary unit in that stratum. The inspector wishes to verify the data on amounts of SNM in subunits as presented by the operator by measuring several subunits and comparing his values to the presented values. Within each stratum he plans to select several primary units and then to select the same number of subunits from each of those primary units. For each of the two stages of selection he plans to sample at random with replacement. That is, each random number selected in the first stage will be used to designate a primary unit from the whole group of primary units in the stratum, and each random number selected in a second stage will be used to designate a subunit from the whole group of subunits in a primary unit. With this sampling technique each primary unit and each subunit may be selected more than once. The mathematical results associated with this sampling technique are easier to derive and understand than the mathematical results associated with a random sampling without replacement technique which would not allow a selection of a primary unit or a subunit more than once.

Notation

N = number of strata

N_i = number of primary units in stratum i

R_i = number of subunits in each primary unit in stratum i

$\overline{X}_i$ = average quantity of SNM per subunit in stratum i (as declared by the operator)

M_{ij} = total quantity of SNM that might be diverted from j subunits in each of $M_{ij}/(j\overline{X}_i)$ primary units in stratum i, $j = 1, 2, \ldots, R_i$ (as decided by a diverter)

$M = \sum_{i=1}^{N} \sum_{j=1}^{R_i} M_{ij}$ = total quantity of SNM that might be diverted

n_i = number of random numbers to be selected for designating primary units in the sample for stratum i

r_i = number of random numbers to be selected for designating subunits in the sample for each designated primary unit in stratum i.

The N, N_i, R_i, $\overline{X}_i$, n_i and r_i symbols are defined in accordance with the inspection situation described above. The definitions of the M_{ij} and M symbols indicate a diversion strategy. It is assumed that a diverter will remove all of the SNM from some subunits in some primary units. It is also assumed that if the inspector selects in his sample a subunit which has had the SNM diverted, he will detect this with his measurement procedure. With these situations in mind, it is useful to consider the probability that the subunits designated for inspection will not include subunits from which SNM has been diverted.

Probability of Non-Detection (P)

In stratum i we have $\sum_{j=1}^{R_i} M_{ij}(j)^{-1}\overline{X}_i^{-1}$ primary units from which SNM has been diverted. Since there are N_i primary units in the stratum, the probability of not selecting one of the primary units from which SNM has been diverted in a single random selection is

$$1 - \left[(\overline{X}_i N_i)^{-1} \sum_{j=1}^{R_i} M_{ij}(j)^{-1}\right]$$

The probability of selecting one of the primary units containing j subunits from which SNM has been diverted in a single random selection is

$$(\bar{X}_i N_i)^{-1} M_{ij}(j)^{-1}.$$

Given that such a primary unit is selected, the probability of not selecting one of the j subunits from which SNM has been diverted out of a total of R_i subunits in a single random selection at the second stage is

$$1 - jR_i^{-1}.$$

Since selections with replacement are independent of each other, the probability of not selecting one of the j subunits in r_i random selections with replacement is

$$(1 - jR_i^{-1})^{r_i}.$$

Hence, the probability of selecting a primary unit containing j subunits from which SNM has been diverted in a single first stage random selection and not selecting one of the j subunits in r_i second stage random selections with replacement is

$$[(\bar{X}_i N_i)^{-1} M_{ij}(j)^{-1}][(1-jR_i^{-1})^{r_i}].$$

The ways in which a single first stage random selection and r_i second stage random selections can be made without selecting a subunit from which SNM has been diverted are the first stage selection of a primary unit from which no SNM has been diverted, or the first stage selection of a primary unit from which j subunits have had SNM diverted and a second stage selection of r_i random numbers which do not designate one of those j subunits for $j = 1, 2, \ldots, R_i$. The probability of this event is

$$[1-(\bar{X}_i N_i)^{-1} \sum_{j=1}^{R_i} M_{ij}(j)^{-1}]+[(\bar{X}_i N_i)^{-1} \sum_{j=1}^{R_i} M_{ij}(j)^{-1}(1-jR_i^{-1})^{r_i}].$$

Since there are n_i first stage random selections with replacement the selections are independent of each other, and the probability of making n_i first stage random selections and r_i second stage random selections for each first stage selection without selecting a subunit from which SNM has been diverted is

$$\left\{[1-(\bar{X}_i N_i)^{-1} \sum_{j=1}^{R_i} M_{ij}(j)^{-1}]+[(\bar{X}_i N_i)^{-1} \sum_{j=1}^{R_i} M_{ij}(j)^{-1}(1-jR_i^{-1})^{r_i}]\right\}^{n_i}.$$

Finally, sampling is carried out independently in each stratum, so the probability of non-detection of a subunit from which SNM has been diverted is

$$P = \prod_{i=1}^{N} \left\{[1-(\bar{X}_i N_i)^{-1} \sum_{j=1}^{R_i} M_{ij}(j)^{-1}]+[(\bar{X}_i N_i)^{-1} \sum_{j=1}^{R_i} M_{ij}(j)^{-1}(1-jR_i^{-1})^{r_i}]\right\}^{n_i}.$$

Maximum P

A diverter would want to use a strategy which would maximize P. Consider the case of diversion of SNM from all subunits contained in several primary units in each stratum. That is, the diverter sets $M_{ij} = 0$ for $j = 1, 2, \ldots, R_{i-1}$ and $M_{iR_i} = M_i$ for all i. Then,

$$P = \prod_{i=1}^{N} [1-(\bar{X}_i N_i R_i)^{-1} M_i]^{n_i}$$

$$= \prod_{i=1}^{N} \{[1-(\bar{X}_i N_i)^{-1} \sum_{j=1}^{R_i} M_{ij}(j)^{-1}]+[(\bar{X}_i N_i)^{-1} \sum_{j=1}^{R_i} M_{ij}(j)^{-1}(1-jR_i^{-1})^{1}]\}^{n_i}$$

$$\geq \prod_{i=1}^{N} \{[1-(\bar{X}N_i)^{-1} \sum_{j=1}^{R_i} M_{ij}(j)^{-1}]+[(\bar{X}_i N_i)^{-1} \sum_{j=1}^{R_i} M_{ij}(j)^{-1}(1-jR_i^{-1})^{r_i}]\}^{n_i}.$$

The strategy of diverting all of the SNM in several primary units in each stratum has the same probability of nondetection as any strategy with the same amounts of SNM diverted from the strata when the inspector selects single second stage samples ($r_i = 1$), and has a probability of non-detection greater than or equal to that of any strategy when the inspector selects larger second stage samples.

We find Max P under the constraint $\sum_{i=1}^{N} \sum_{j=1}^{R_i} M_{ij} = \sum_{i=1}^{N} M_i = M$ as follows:

$$\frac{\partial}{\partial M_i} \{\prod_{i=1}^{N} [1-(\bar{X}_i N_i R_i)^{-1} M_i]^{n_i} + \lambda[\sum_{i=1}^{N} M_i - M]\} =$$

$$= n_i[1-(\bar{X}_i N_i R_i)^{-1} M_i]^{n_i-1}[-(\bar{X}_i N_i R_i)^{-1}] \prod_{k\neq i} [1-(\bar{X}_k N_k R_k)^{-1} M_k]^{n_k} + \lambda =$$

$$= -n_i[(\bar{X}_i N_i R_i)-M_i]^{-1} P + \lambda = 0$$

$$M_i = (\bar{X}_i N_i R_i) - n_i P \lambda^{-1}$$

$$M = \sum_{i=1}^{N} M_i = \sum_{i=1}^{N} (\bar{X}_i N_i R_i) - P\lambda^{-1} \sum_{i=1}^{N} n_i$$

$$\lambda = P[\sum_{i=1}^{N} (\bar{X}_i N_i R_i) - M]^{-1} \sum_{i=1}^{N} n_i$$

$$M_i = (\bar{X}_i N_i R_i) - n_i[\sum_{i=1}^{N} (\bar{X}_i N_i R_i) - M][\sum_{i=1}^{N} n_i]^{-1}$$

$$\text{Max } P = \prod_{i=1}^{N} \{n_i[\sum_{i=1}^{N} (\bar{X}_i N_i R_i) - M][\sum_{i=1}^{N} n_i]^{-1}[\bar{X}_i N_i R_i]^{-1}\}^{n_i}$$

The next to last equation indicates that a diverter should set the amount of SNM to divert from stratum i in an inverse relationship to the proportion of primary units sampled which are selected by the inspector in the first stage sampling in stratum i. Hence, the best strategy for a diverter is to determine the amount of SNM to divert from each stratum according to the inspectors choice of the n_i values, and to divert the SNM from all subunits in some primary units.

Sample Sizes

We have shown that a diverter can eliminate the effectiveness of additional second stage sampling ($r_i > 1$) by diverting SNM from all subunits in some of the primary units. So, the inspector should set $r_i = 1$ for $i = 1, 2, \ldots, N$. If he sets the number of first stage sample selections, n_i, proportional to the amount of SNM in stratum i, $(\bar{X}_iN_iR_i)$, and chooses Max $P = \beta$ and M, then we derive a sample size formula as follows:

$$\text{Max } P = \beta = \prod_{i=1}^{N} \left\{ n_i \left[\sum_{i=1}^{N} (\bar{X}_iN_iR_i) - M \right] \left[\sum_{i=1}^{N} n_i \right]^{-1} \left[\bar{X}_iN_iR_i \right]^{-1} \right\}^{n_i}$$

$$n_i = p(\bar{X}_iN_iR_i)$$

$$\beta = \prod_{i=1}^{N} \left\{ \left[\sum_{i=1}^{N} (\bar{X}_iN_iR_i) - M \right] \left[\sum_{i=1}^{N} (\bar{X}_iN_iR_i) \right]^{-1} \right\}^{p(\bar{X}_iN_iR_i)}$$

$$= \left\{ \left[\sum_{i=1}^{N} (\bar{X}_iN_iR_i) - M \right] \left[\sum_{i=1}^{N} (\bar{X}_iN_iR_i) \right]^{-1} \right\}^{p \sum_{i=1}^{N} (\bar{X}_iN_iR_i)}$$

$$p = \ln\beta \left\{ \ln \left\{ \left[\sum_{i=1}^{N} (\bar{X}_iN_iR_i) - M \right] \left[\sum_{i=1}^{N} (\bar{X}_iN_iR_i) \right]^{-1} \right\} \right\}^{-1} \left\{ \sum_{i=1}^{N} (\bar{X}_iN_iR_i) \right\}^{-1}$$

$$n_i = \ln\beta \left\{ \ln \left\{ \left[\sum_{i=1}^{N} (\bar{X}_iN_iR_i) - M \right] \left[\sum_{i=1}^{N} (\bar{X}_iN_iR_i) \right]^{-1} \right\} \right\}^{-1} \left\{ \sum_{i=1}^{N} (\bar{X}_iN_iR_i) \right\}^{-1} (\bar{X}_iN_iR_i$$

Thus, the inspector chooses the amount of SNM which would be considered significant if it were diverted, M, and chooses the maximum probability of non-detection, β, and calculates the first stage sample sizes as above. These sample sizes may not be optimum in the sense that the total sample size, $\sum_{i=1}^{N} n_i$, is minimized.

Example

Suppose that we have two types of primary units each containing two types of subunits. Let the 100 primary units of type 1 constitute strata 1 and 2, and let the 200 primary units of type 2 constitute strata 3 and 4. Assume the following information:

i	$\bar{X}_i$	N_i	R_i	$(\bar{X}_iN_iR_i)$
1	3	100	5	1500
2	1	100	10	1000
3	2	200	5	2000
4	1	200	15	3000

$M = 500$ $\beta = 0.05$

1. Calculate

$$\ln(0.05)\left\{\ln\left\{\left[\sum_{i=1}^{4}(\overline{X}_iN_iR_i)-M\right]\left[\sum_{i=1}^{4}(\overline{X}_iN_iR_i)\right]^{-1}\right\}\right\}^{-1}\left\{\sum_{i=1}^{4}(\overline{X}_iN_iR_i)\right\}^{-1} =$$

$$= (-3)\left\{\ln\left\{(7000)(7500)^{-1}\right\}\right\}^{-1}\left\{7500\right\}^{-1}$$

$$= 0.00579$$

2. Calculate

$$n_1 = (0.00579)(\overline{X}_1N_1R_1) = 8.7 \sim 9$$

$$n_2 = 5.79 \sim 6$$

$$n_3 = 11.58 \sim 12$$

$$n_4 = 17.37 \sim 18$$

3. Check the sample size calculations by calculating

$$\text{Max } P = \left\{9\,[7000]\,[1500]^{-1}\,[45]^{-1}\right\}^{9} + \left\{6\,[7000]\,[1000]^{-1}\,[45]^{-1}\right\}^{6} + \left\{12\,[7000]\,[2000]^{-1}\,[45]^{-1}\right\}^{12} + \left\{18\,[7000]\,[3000]^{-1}\,[45]^{-1}\right\}^{18} =$$

$$= 0.045 \sim 0.05 = \beta$$

4. Draw random numbers with replacement

Stratum 1				Stratum 2			
Random Number	Primary Unit	Random Number	Subunit	Random Number	Primary Unit	Random Number	Subunit
71	71	6	1	17	17	0	10
94	94[b]	1	1	86	86	9	9
50	50	1	1	94	94[b]	4	4
54	54	4	4	53	53	4	4
42	42[a]	7	2	49	49	3	3
21	21	3	3	61	61	8	8
34	34	0	5				
42	42[a]	6	1				
78	78	4	4				

Stratum 3

Random Number	Primary Unit	Random Number	Subunit
638	38	2	2
357	157	2	2
105	105	6	1
084	84	5	5
724	124[c]	2	2
524	124[c]	3	3
922	122	1	1
342	142[d]	0	5
680	80	7	2
611	11	6	1
745	145	0	5
696	96	2	2

Stratum 4

Random Number	Primary Unit	Random Number	Subunit
482	82	51	6
051	51	85	10
740	140	14	14
790	190	61	1
154	154	04	4
339	139	88	13
588	188	61	1
617	17	09	9
481	81	25	10
846	46	62	2
942	142[d]	55	10
512	112	81	6
028	28	04	4
812	12	26	11
441	41	41	11
870	70	61	1
526	126	18	3
895	95	29	14

In primary unit type 1 (strata 1 and 2) we have primary units 42 and 94 each drawn twice. We measure type 1 subunits 1 and 2 in primary unit 42, and type 1 subunit number 1 and type 2 subunit number 4 in primary unit number 94. In primary unit type 2 (strata 3 and 4) we have primary units units numbered 124 and 142 each drawn twice. We measure type 3 subunits numbered 2 and 3 in primary unit 124, and type 3 subunit number 5 and type 4 subunit number 10 in primary unit 142.

Numbers of Subunits Unequal

The above sampling problem involves stratification of prime units into strata where primary units contain equal numbers of subunits of the same type. When this leads to an excess number of strata because of a wide variability in the numbers of subunits, we might reduce the number of strata by combining those strata with primary units containing approximately the same number of subunits. The sample sizes can then be calculated by using subunit numbers, R_i, equal to the largest R_i for each stratum. We can then consider some mythical subunits containing the declared amount of SNM for primary units which do not have the maximum R_i, and proceed with the selection of random numbers. This may preclude the best strategy for a diverter since he cannot obtain SNM from mythical subunits. That is, he might have to divert SNM in a way which is equivalent to diverting less than the complete number of real plus mythical subunits in some primary units. This would result in a smaller probability of non-detection than Max P.

REFERENCES

[1] JAECH, J.L., Statistical Methods in Nuclear Material Control, USAEC (1973).

[2] FELLER, W., An Introduction to Probability Theory and Its Applications, Vol. 1, 2nd Ed., John Wiley & Sons, New York (1958).

DISCUSSION

R. AVENHAUS: Your paper and IAEA-SM-231/60[1] give simple, yet practical formulae for inspection sample sizes. Now, ideally, the sample sizes should be such that the probability of detection is minimized with respect to all diversion strategies for a total diversion, and is maximized with respect to all inspection strategies for a total inspection effort. Usually, this will lead to very complicated solutions, if closed solutions can be obtained at all. Do you have any idea how far your practical solutions are from the ideal solution?

T.M. BEETLE: I have minimized the probability of detection over all diversion strategies, and thus found the best strategy for a diverter. The inspector sample size is then derived to provide adequate protection against that strategy. I have not derived the optimum sample size, or allocation to strata, by considering the maximum probability of detection over all inspection strategies. However, that is an interesting question for future investigation.

[1] GOOD, P.T. et al., these Proceedings, Vol. II.

SAFEGUARDS MEASUREMENT ERRORS

W. BUSH, C. HAGERTY, G. HOUGH
Department of Safeguards,
International Atomic Energy Agency, Vienna

Abstract

SAFEGUARDS MEASUREMENT ERRORS.

Safeguards quantitative measurement data collected by the International Atomic Energy Agency (IAEA) from facility inspections, facility design information, analytical interlaboratory measurement programmes, and standard reference material characterizations, form the bases by which the IAEA confirms an operator's stated material balance. Statistical analyses of these data result in estimates of material unaccounted for, shipper-receiver differences, operator-inspector differences, and their associated uncertainties reflected in random and systematic error variance components. Knowledge of random and systematic error variances is used, in addition to assessing the quality of safeguards measurements, in planning inspections and in analysing inspection data. This paper includes some definitions of random and systematic error variances as used by the IAEA and a collection of recent measurement error experience on various parts of the nuclear fuel cycle under IAEA safeguards.

INTRODUCTION

For several years, the Department of Safeguards has been developing procedures for the estimation of measurement errors and error propagation, statistical evaluation of safeguards data and definition of statistical terms such as random and systematic errors. The objective is to develop as uniform a set of definitions and procedures as possible which can be used by facilities under safeguards in many parts of the world. Considerable progress has been made in this direction in spite of the complexity of the subject, mainly owing to the publication of Volume 1, Part F of the Safeguards Technical Manual [1]. Volume 1 contains a brief introduction to statistics, procedures for estimating random and systematic errors and error propagation procedures for MUF and operator-inspector comparisons (D statistic). It is designed primarily as a guide to facilities for setting up their internal quality control programmes for safeguards material balance measurements, Volume 2 (soon to be published [2]) contains statistical procedures for planning inspections, evaluating inspection data, preparing statements of results and conclusions and a section on interlaboratory comparisons. It also includes some important definitions which are presented below. Volume 2 is primarily designed for inspectors and those who are actively involved in evaluating inspection data. A third volume is planned for completion some time in 1979.

Many difficulties have been met in evaluating inspection data. Principal among these are (1) meaningful stratification of material balance data into groups of items that have a common measurement basis [3]; (2) definition of facility records that make it easy for inspectors to calculate sample plans, carry out counting and identification of items, randomly select items for measurement and make valid inferences about the population of items in each stratum; and (3) collection of the data in the form of working papers that can be input to the computer for statistical evaluation. The actual evaluation of the data has been relatively easy compared with these three activities. At present very little non-destructive (NDA) and weight (scale) measurements data can be summarized owing to data-collection problems. However, the following sections do include summaries of analytical measurements of reference materials, standards and inspection samples and volume calibrations of some process vessels. One NDA application is included.

1. DEFINITIONS OF SYSTEMATIC AND RANDOM ERRORS

The terminology of measurement errors is best understood in the context of an example. Suppose that six sintered UO_2 pellets of nominally the same composition are distributed to two laboratories (three to each laboratory) for analysis of per cent uranium. Both laboratories use the same analytical method. Laboratory 1 analyses all three pellets on the same day; laboratory 2 performs the analyses over three days. Let x_i be the measured per cent uranium for pellet i; (i = 1, 2, . . . , 6). Then a model *may* be written as follows for the six observations:

$$x_1 = \mu + \Delta + \ell_1 + t_1(1) + p_1 + \epsilon_1$$

$$x_2 = \mu + \Delta + \ell_1 + t_1(1) + p_2 + \epsilon_2$$

$$x_3 = \mu + \Delta + \ell_1 + t_1(1) + p_3 + \epsilon_3$$

$$x_4 = \mu + \Delta + \ell_2 + t_1(2) + p_4 + \epsilon_4$$

$$x_5 = \mu + \Delta + \ell_2 + t_2(2) + p_5 + \epsilon_5$$

$$x_6 = \mu + \Delta + \ell_2 + t_3(2) + p_6 + \epsilon_6$$

In the model, the symbols have the following meanings:

μ is the nominal value

Δ is the value common to all measurements made using the analytical technique in question. It is called a *systematic error* or a *bias* and these terms are generally used interchangeably. (Some users make a distinction between these two terms as follows: If the quantity Δ is estimated by making measurements on standards, and if values are then corrected on the basis of these standard measurements, then Δ is referred to as a *bias*. However, the effect of the bias cannot be corrected for completely because one never knows Δ; one only estimates Δ. There is a *residual bias* that remains even after the bias correction is applied, and this is called a *systematic error* by some users.) Since Δ, which may be either positive or negative, affects *all* observations in the data set exactly the same, it is also called a *long-term systematic error*. This term, which is meaningful only with respect to a given data set, is used to make a distinction between it and a *short-term systematic error* illustrated in the following paragraph. Finally, if the quantity Δ is regarded as a random variable with zero mean and variance σ_Δ^2, then σ_Δ^2 is called a *systematic error variance*, or, again to make the distinction noted above, a *long-term systematic error variance*.

ℓ_j is some value common to all measurements performed by laboratory j, with j = 1, 2 in the example. The quantity ℓ_j may be either positive or negative. It is referred to in various ways. It may be called a *laboratory effect*, or it may be called a *short-term systematic error* with reference to the data set in question since it is a common effect for *more than one* member of the data set, *but not for all*. If the data set had consisted of only the first three observations, then ℓ_1 would be called a *systematic error*, a *bias*, or a *long-term systematic error*, because it then affects *all* members of the data set, and its effect would be indistinguishable from that of Δ. If ℓ_j is regarded as a random variable with zero mean and variance σ_ℓ^2, then σ_ℓ^2 is called a *short-term systematic error variance*, or a *between-laboratory variance*. σ_ℓ^2 and σ_Δ^2 are called *variance components*.

$t_{k(j)}$ is some value common to all measurements performed at time k by laboratory j. For laboratory 1, since all measurements are performed in the same time frame, one cannot distinguish between this *time effect* and the laboratory effect. This is an important point because one professed aim of interlaboratory experiments is to remove the effects of differences between laboratories by correcting all results to some base value, that is, by obtaining estimates of the ℓ_j's and correcting for the laboratory effects. However, this approach does not recognize the importance of the time effect, $t_{k(j)}$, which is usually "confounded with" or indistinguishable from the laboratory effect ℓ_j. Thus, when one attempts to remove laboratory biases in this way, the results are only applicable to the given time frame that existed at the time of the interlaboratory experiment. The *between-time* variance, σ_t^2, may well be a dominant effect when compared with σ_ℓ^2, in which case it would be misleading to conclude that one can correct for differences between laboratories. Rather, in most instances, one would use interlaboratory data to obtain the combined estimate, $\sigma_\ell^2 + \sigma_t^2$, which becomes a *systematic error variance, long-term,* when applied only to a given laboratory, and *short-term* otherwise. In this instance, one usually calls this simply a between-laboratory variance, it being understood that the time effect is implicitly included in that variance component. Note that for laboratory 2, the measurements are made at three different times, but this is not the usual mode in interlaboratory experiments.

p_i is a deviation from the nominal value, μ for pellet i. Since this differs for each observation, p_i is called a *random error.* Further, since it is a pellet or sample-oriented effect, it is more specificially called a *sampling random error.* With p_i regarded as a random variable having zero mean and variance σ^2, this variance is called a *random error variance* or, in this instance, a *sampling random error variance.*

ϵ_i is a deviation due to analytical measurement i. Since this differs for each observation, ϵ_i is called a *random error* or, more specifically in this instance, an *analytical random error.* The quantity σ_ϵ^2 is called a *random error variance* or an *analytical random error variance.* In this example, one cannot distinguish between the *sampling* and *analytical* effects and the model might be rewritten to use one symbol, say m_i, to denote the combined effect. σ_m^2 might then be called the *measurement random error.* With respect to this last point, it is recognized by the modellers that there are many potential sources of error that affect a given result, some identifiable and some not. It is common practice to group potential effects in a model, especially when the effects cannot be distinguished. Thus, for example, the systematic error, Δ, represents the combined effects of many factors that would, in total, be represented by Δ.

Other terminology is used in characterizing measurement errors. *Precision* is related to *random errors.* When the *random error variance* is small (in a relative sense), the measurement is said to have *high precision,* or *good precision.* A precise measurement method is one that exhibits little scatter in the data, or good *repeatability.*

The term accuracy is a general one that varies widely in use depending on the purpose of the experiment or test. Some authors define accuracy, bias and systematic error as synonymous. Others define accuracy as a combination of bias and random-error standard deviation, and the definitions may vary depending on the specific application. The definition agreed upon by past IAEA advisory groups is:

Accuracy is the error in the measurements of a quantity of nuclear material in a batch as for the sum of many batches and include *random* and *systematic errors* due to bulk material (weight, volume) sampling and analytical sources or calibration of NDA systems. As the number of batches increases, the random error reduces to the *limit accuracy*, i.e. the systematic errors tend to become controlling.

The effect of a systematic error cannot be reduced by making additional measurements; the average result is offset from the true value by the same amount as each individual result. Of course, the effects of short-term systematic errors can be reduced by enhancing the data set in such a way that the average reflects more levels of the effects involved, i.e. more time periods, more laboratories, more analysts, etc.

The primary technique for estimating these variance components is the hierarchical analysis of variance. This is the method used in the following sections on the calibration of physical standards and interlaboratory comparisons.

2. CALIBRATION OF PHYSICAL STANDARDS

Non-destructive assay (NDA) techniques are used extensively for inspection measurements of nuclear material under safeguards. Associated with each NDA determination are the components of systematic and random error. Included in the random error uncertainty are operator effects, geometry effects from re-positioning of the sample in the instrument, and counting statistics. The systematic error component includes the errors due to varied container dimensions, varied container material compositions, calibration, and the error in the assigned value of the calibrating standard.

The IAEA maintains an inventory of approximately two hundred NDA calibrating standards of various materials (highly enriched uranium, low enriched uranium, natural uranium), and in various physical forms and containers (powder, pellets, fuel rods, fuel plates, billets, glass vials, plastic bags). These standards are used throughout the world to calibrate NDA equipment utilized by IAEA inspectors in the verification of nuclear material subject to international safeguards. Table I itemizes diverse NDA standards recently characterized by the IAEA and its Network of Analytical Laboratories (NWAL). For each standard contained within the table are separate estimates of nuclear material content and major isotopic abundances. Together with each assigned value are estimates of between-laboratory, between-sample, and within-laboratory standard deviations plus the standard deviation of the mean or assigned value. These error variance estimates do not demonstrate the present status of analytical chemistry measurements but rather reflect the intended accuracy of the assigned values for use in the calibration of NDA instruments in the field by the IAEA inspectors.

TABLE I. IAEA RECENTLY CHARACTERIZED PHYSICAL STANDARDS

Material Type	Container or Form	U-235 Certified Enrichment	Between Laboratory Absolute SD	Between Laboratory Percent RSD	Sampling Error Absolute SD	Sampling Error Percent RSD	Random Error Absolute SD	Random Error Percent RSD	Standard Deviation of Certified Values Absolute SD	Standard Deviation of Certified Values Percent RSD
U_3O_8	Plastic Bags/Powder	0.6804	0.0040	0.588			0.00054	0.079	0.0024	0.353
U_3O_8	Plastic Bags/Powder	0.7115	0.00073	0.103			0.0010	0.141	0.00057	0.080
U_3O_8	Plastic Bags/Powder	1.1763	0.0016	0.136			0.0022	0.187	0.0012	0.102
U_3O_8	Plastic Bags/Powder	1.3702	0.0	0.0			0.0019	0.139	0.00065	0.047
U_3O_8	Plastic Bags/Powder	1.5629	0.0	0.0			0.0021	0.134	0.00085	0.054
U_3O_8	Plastic Bags/Powder	1.9627	0.0032	0.163			0.0020	0.102	0.0020	0.102
U_3O_8	Plastic Bags/Powder	2.4476	0.0	0.0			0.0041	0.168	0.0015	0.061
U_3O_8	Plastic Bags/Powder	3.4314	0.0022	0.064			0.0064	0.187	0.0026	0.076
U_3O_8	Plastic Vials/Powder	3.8239	0.0066	0.173			0.0029	0.076	0.0041	0.107
U_3O_8	Plastic Vials/Powder	3.9217	0.0090	0.229			0.0025	0.064	0.0053	0.135
U_3O_8	Plastic Vials/Powder	4.9010	0.0054	0.110			0.0072	0.147	0.0042	0.086
U_3O_8	Plastic Vials/Powder	6.8777	0.011	0.160			0.011	0.160	0.0076	0.111
U_3O_8	Plastic Vials/Powder	7.8598	0.022	0.280			0.0048	0.061	0.013	0.165
U_3O_8	Plastic Vials/Powder	9.8163	0.023	0.234			0.0035	0.036	0.014	0.001
U_3O_8	Plastic Vials/Powder	93.1488	0.014	0.015			0.012	0.013	0.0094	0.010
UO_2	Fuel Rod/Pellets	1.415	0.0021	0.148	0.0025	0.177	0.0035	0.247	0.0019	0.134
UO_2	Fuel Rod/Pellets	1.7498	0.00067	0.038	0.0021	0.120	0.0013	0.074	0.00065	0.037
UO_2	Fuel Rod/Pellets	1.7522	0.0	0.0	0.0010	0.057	0.0037	0.211	0.00079	0.045
UO_2	Fuel Rod/Pellets	2.049	0.0022	0.107	0.0037	0.180	0.0010	0.049	0.0020	0.098
UO_2	Fuel Rod/Pellets	2.500	0.0047	0.188	0.0039	0.156	0.0027	0.108	0.0032	0.128
U/Zr	Billet	92.901	0.020	0.022			0.026	0.028	0.012	0.013

TABLE II. PAFEX II

SAU-SAMPLE: CONCENTRATION AND ISOTOPIC COMPOSITION MEANS AND STANDARD DEVIATIONS OF UNSPIKED INPUT SOLUTION

	Including all reported results						Excluding outliers					
	LM	RL	ATOT	RBL	RWL	RTOT	LM	RL	ATOT	RBL	RWL	RTOT
U_1	2.0664[a]	2.061	0.016	0.37	0.66	0.75	2.0656	2.061	0.010	0.48	0.18	0.51
Pu_1	12.223[b]	12.07	0.25	2.0	0.36	2.1	12.124	12.07	0.14	1.1	0.32	1.2
U_2	2.0683[a]	2.053	0.016	0.47	0.64	0.80	2.0677	2.053	0.014	0.62	0.25	0.67
Pu_2	12.234[b]	2.06	0.31	2.5	0.73	2.6	12.101	12.06	0.058	0.41	0.25	0.48
^{234}U	0.01266	0.01140	0.0012	8.7	3.3	9.3	0.01237	0.01140	0.00078	5.3	3.3	6.3
^{235}U	1.1901	1.1899	0.0070	0.45	0.38	0.59	No Outliers					
^{236}U	0.24799	0.24830	0.0035	1.0	0.96	1.4	No Outliers					
^{238}U	98.550	98.550	0.0091	0.0070	0.0060	0.0092	98.551	98.550	0.0073	0.0042	0.0061	0.0074
^{238}Pu	0.64661	0.64676	0.027	3.4	2.4	4.1	0.63834	0.64676	0.010	1.5	0.72	1.7
^{239}Pu	67.768	67.826	0.17	0.22	0.12	0.25	67.808	67.826	0.092	0.11	0.075	0.14
^{240}Pu	21.234	21.256	0.059	0.0	0.31	0.28	21.230	21.256	0.035	0.12	0.11	0.16
^{241}Pu	8.1858	8.0990	0.11	1.2	0.56	1.3	8.1624	8.0990	0.070	0.79	0.34	0.86
^{242}Pu	2.1668	2.1726	0.030	1.1	0.89	1.4	No Outliers					

[a] mg/ml.
[b] μg/ml.

LM = Mean of labs.
RL = Reference laboratory
ATOT = Absolute total standard deviation
U_1 and Pu_1 = Element concentration calculated from SAS-1

RBL = Relative per cent between-laboratory standard deviation
RWL = Relative per cent within-laboratory standard deviation
RTOT = Relative per cent standard deviation
U_2 and Pu_2 = Element concentration calculated from SAS-2

The error in the assigned value of the standard, itself composed of both random and systematic errors, then becomes a systematic error component in an NDA measurement result based on a calibration utilizing that standard. Depending on the NDA instrument and its accessories, type and form of nuclear material, and operating environment in the field, the effect on the final measured value of the error due to the standard can range from negligible to very significant. In the case of Table I calibrating standards, the contribution to the total NDA measurement error is negligible.

3. ANALYTICAL MEASUREMENTS OF AN INPUT SOLUTION TO A REPROCESSING PLANT

Because of the importance in safeguarding reprocessing facilities, and because the input solution is the first analytical measurement point of nuclear material content in spent fuel, the IAEA conducted the PAFEX II experiment as an investigation of sample-handling requirements, and as a determination of the quality of plutonium and uranium content measurements in an input solution.

Table II is abstracted from the PAFEX II Report [4] and contains a summary of plutonium and uranium input solution sample concentration and isotopic abundance means and standard deviations as calculated from the results of thirteen participating laboratories. The table is divided into two primary sections. The left half reports the means and standard deviations based on all reported data. The right half contains the same parameter estimates based on exclusion of those laboratories whose mean or within-laboratory standard deviation was determined to be an outlier. These values, based on the reduced data set, probably represent the best state of measurement quality among this group of laboratories at the time of the experiment in early 1975.

Adjacent to each mean is the reference laboratory's estimate of concentration and isotopic abundance. In addition, the absolute total, relative total, relative between, and relative within-laboratory standard deviations are tabulated.

4. ANALYTICAL MEASUREMENTS OF REPROCESSING PLANT PRODUCT MATERIALS

For reasons allied with the PAFEX II experiment, the PAFEX I experiment was conducted for assessment of PuO_2, Pu nitrate and PuO_2-UO_2 measurements. Table III is abstracted from the PAFEX I report [5] and presents estimates of between and within-laboratory standard deviations.

TABLE III. PAFEX I CONCENTRATION RESULTS

Material	Mean wt%	Between laboratory (% RSD)	Within laboratory (% RSD)
Pu nitrate	1.766	0.23	0.31
PuO_2	86.58	0.12	0.22
Pu-mixed oxide	3.462	0.26	0.41
U-mixed oxide	84.562	0.46	0.29

5. OPERATOR-INSPECTOR PAIRED ANALYTICAL SAMPLE MEASUREMENTS

For each physical inventory of a bulk-handling facility under IAEA Safeguards, a random sample of inventory items is selected by an inspector to be measured. Among this set of items, total net weight, element concentration, and fissile isotopic determinations are needed to verify total element and isotopic weights. Inspectors collect samples to be distributed to one or more of the IAEA's Network of Analytical Laboratories (NWAL) where element concentration and fissile isotopic determinations are made. At the time of collection, the inspector simultaneously records in his inspection working papers the operator's values for the sampled item. Subsequently, the laborator's analytical results are forwarded to the inspector, the Safeguards Analytical Services Officer, and the Data Evaluation Services (DES) Section. Then, within DES, the sample results are matched to the operator's values, are prepared for input to a computer data base, and are statistically analysed by computer programs.

Inherent in both the operator and inspector's measurement values are errors due to material heterogeneity, possible contamination, hygroscopic absorption, and the effects of laboratory errors, reflected in significant between-laboratory components of variance found in interlaboratory quality control programmes. In addition, the operator's value may not be the result of a measurement, but an average factor assigned to a large batch of process input material, or a historical value based on a priori knowledge of the process, or a theoretical value based on stoichiometry. Each of these contribute to operator-inspector differences and need to be quantitatively determined or eliminated, such as inter-laboratory quality control programmes for between-laboratory errors and sealing material in inert gas for absorption prevention.

TABLE IV. OPERATOR-INSPECTOR DIFFERENCES

Material description	Element concentration (wt%)					% ^{235}U enrichment				
	Number of pairs	Nominal value	Mean operator-inspector difference	Systematic error (% RSD)	Random error (% RSD)	Number of pairs	Nominal value	Mean operator-inspector difference	Systematic error (% RSD)	Random error (% RSD)
UO_2 powder	153	87.4	0.130	0.26	0.23	153	2.19	−0.012	0.32	2.91
UO_2 sintered pellets	228	88.1	0.042	0.13	0.20	227	2.54	−0.005	0.30	0.61
U_3O_8 powder	38	85.1	0.237	0.48	0.59	40	2.16	−0.003	0.038	3.24
ADU powder						17	2.18	0.006	0.0	0.98
UO_2 scrap (pure)	7	87.1	0.532	0	1.20	7	1.49	0.008	0.0	3.90
UO_2 green powder	21	87.6	0.026	0.011	0.08	21	2.67	0.005	0.12	0.24
UO_2 green pellets	18	87.6	0.020	0.086	0.64	16	1.92	−0.001	0.0	0.38
UO_2 green scrap	3	82.5	1.802	1.227	2.81	3	2.15	0.019	0.449	2.00
UO_2 contaminated scrap	4	87.5	9.693	5.9	10.33	4	1.85	0.015	0.0	2.99
AUC	4	46.4	0.435	0.0	2.40	4	0.22	0.028	8.8	3.78
UO_2 scrap powder	3	22.3	−0.373	0.0	15.86	3	2.28	0.170	0.0	23.90
HNO_3 liquid waste	3	11.6	4.762	0.0	53.42	3	2.11	0.060	0.0	15.63
UO_2 scrap – dry waste	4	38.3	−1.837	0.0	8.82	4	2.34	0.515	0.0	22.92
U_3O_8 waste	8	52.8	1.646	0.0	7.03	6	1.57	−0.083	0.0	9.00
U_3O_8 recycle	3	84.7	0.067	0.049	0.05	3	3.21	−0.015	0.29	0.29
U_3O_8 scrap	7	71.8	0.142	0.048	0.39	7	2.14	0.050	1.08	5.00
UO_2 scrap pellets/recycle	6	88.0	0.299	0.196	0.56	6	2.99	−0.017	0.442	0.09

For a variety of material categories Table IV presents the IAEA's experience during the last two years with operator-inspector differences for low-enriched (<5%) uranium concentration and ^{235}U isotopic abundance. The random and systematic error components are estimated using a randomized block analysis-of-variance (ANOVA) technique. The ANOVA table for randomized blocks is as follows:

Effect	Mean square	Degrees of freedom	Expected mean square
Mean	$(\Sigma\Sigma X_{ij})^2/2n = G = MSM$	1	$\sigma_R^2 + 2\,\sigma_B^2 + n\sigma_T^2 + 2n\sigma_M^2$
Treatment	$\sum_{i=1}^{2} (\Sigma X_{ij})^2/n - G = MST$	1	$\sigma_R^2 + n\sigma_T^2$
Blocks	$\sum_{j=1}^{2} (\Sigma X_{ij})^2/2 - G = MSB$	(n−1)	$\sigma_R^2 + 2\sigma_B^2$
Residual	$\Sigma\Sigma X_{ij}^2 - MST - MSB - MSM$	(n−1)	σ_R^2

If the following identifications are made

Treatment → shipper-receiver → operator-inspector
Blocks → items
Residual → random measurement variance

then it is apparent that differences based on paired measurements are equivalent to randomized blocks as far as variance components estimation is concerned. Thus

$$\hat{\sigma}_R^2 = \text{RMS (residual mean square)}$$

$$\hat{\sigma}^2_{\text{BLOCKS}} = \hat{\sigma}^2_{\text{PROCESS VARIANCE}} = \text{MBS} - \text{RMS}/2$$

$$\hat{\sigma}^2_{\text{TREATMENTS}} = \hat{\sigma}^2_{\text{BIASES}} = (\text{MST} - \text{RMS})/n$$

If d_i is the difference between operator/inspector measurements then

$$\text{MST} = n\bar{d}^2/2$$

$$\text{RMS} = s_d^2/2 \qquad (1)$$

$$\hat{\sigma}^2_{\text{TREATMENTS}} = \frac{\bar{d}^2}{2} - \frac{s_d^2}{2n} \qquad (2)$$

TABLE V. NDA CALIBRATION ACCURACY

% fissile content	Element	Container content (kg)	% RSD
75	Pu	1.5	0.52
81	Pu	1.5	0.22
92	Pu	1.5	0.36

Thus, in Table IV the random error %RSD is based on (1) and the systematic error %RSD is based on (2). The latter is not a precise estimate when based on one comparison, but such estimates accumulated over different comparisons (inspections) give valuable information on the variance of biases in practice. Note also that the random error %RSD estimates are valid for the case when the operator and inspector use comparable measurement methods that have approximately the same random error. Although the selection is limited, it is possible to compare the results in Table IV with those in Tables I–III, recognizing that the materials in Table IV are process materials while the materials in Tables I–III are relatively stable reference materials.

6. NON-DESTRUCTIVE ASSAY OF PLUTONIUM METAL PLATES

Plutonium metal plates have been verified under inspection conditions using non-destructive equipment. The plates are stored in containers and the containers were measured directly with the SAM-2 Mark 3 electronic unit, four 2.54-cm-dia. by 13-cm active length ^{3}He proportional counters, and a flexibly designed moderator assembly. The counting time is one minute.

Calibration curves were constructed by measuring several full containers and then consecutively fewer plates per container. Individual plates were compared with a standard plate and fabricator's data were available for the content of each plate. The relative standard deviation (RSD) of the calibration varied depending on the fissile concentration and on the number of plates per container. Typical values for full containers are shown in Table V. However, the systematic error standard deviations were actually larger than the calibration errors because of systematic effects due to different plate sizes, material origin, drift in the instrument from day-to-day and variation in background from day-to-day.

Using analysis of variance methods it was determined that the best estimate of the systematic error standard deviation is 21 g and the random error standard

TABLE VI. VOLUME MEASUREMENTS

	Tank 1	Tank 2	Tank 3
Function	Input accountability	Pu product accountability	Pu product storage
Shape	Right circular cylinder	Rectangular slab	Right ciruclar cylinder
Capacity (litre)	4000	45	120
Slope (litre · mm^{-1})	Variable from 0.6 to 4.0	0.05	0.08
Heel (litre)	15	3	7
Systematic error	0.5 at 20 litre	0.1 litre	0.02 litre
Standard deviation of calibration	1 litre/2000 litre 1.5 litre/4000 litre		
Random error standard deviation (one measurement)	0.9 litre/20 litre 4 litre/2000 litre 5 litre/4000 litre	0.04 litre	0.1 litre

deviation is 7.3 g for one full container. Based on a population of 236 containers and a sample size of 145 containers, the projected total bias between operator and inspector was 0.42% and the limit of error (2 sigma) was 1.34%, i.e. 0.42 ± 1.34%. This was a very satisfactory result. With certain improvements in the calibration procedure, even better results could be obtained.

7. CALIBRATION OF ACCOUNTABILITY TANKS

Tank calibrations are verified by observation of the calibration procedure and statistical evaluation of the calibration data using the methods of Chapter 5 in Ref. [1]. Reference [6] gives very useful procedures for collection and evaluation of tank calibration data.

Indirect volume measurements are made, based upon differential pressure measurements between a pair of pneumatic bubbler dip tubes, one extending to the near bottom of the tank and the other situated above the solution in the tank atmosphere. Then the differential pressure is proportional to the height of the solution above the extended dip tube with the constant of proportionality depending on the density of the solution.

If the cross-sectional area of the tank spanning the height of the solution is constant, volume is then a linear function of pressure. If the cross-sectional area varies greatly as a result of tank geometry or numerous internal pipes such as cooling coils, then volume estimates should be made as segmented linear functions or a curvilinear function.

The initial determination of the calibration function or functions is made from paired measurements of incrementally added volumes and the corresponding pressure readings. Typically, volume in litres is based on accurate weight and density measurements, and pressure, in millimetres of water, is measured by a high precision water manometer.

Table VI contains summary information concerning three functional tank types found in a reprocessing plant. The descriptors, function, shape and capacity, are self-explanatory. Slope is unit change in volume per unit change in manometer height. For the first tank, the cross-sectional area is not constant and thus the slope varies as a function of the solution height in the tank. The range of the variation in slope is given. The heel is the residual volume of solution left in the tank below the jet-out nozzle.

The estimates of the coefficients of the calibration curve contain those errors attributable to both volume and pressure system measurement errors during the initial calibration. The error contribution of the calibration curve is given by its standard deviation. In the case of the variable slope tank, the standard deviation varies directly with the slope of the tank. The random error standard deviation is that attributable to the error in the pressure measurement system at the time a volume estimate is made.

ACKNOWLEDGEMENTS

We gratefully acknowledge K.B. Stewart and J.L. Jaech for their assistance with the statistical models used in the preparation of this paper.

REFERENCES

[1] INTERNATIONAL ATOMIC ENERGY AGENCY, Statistical Concepts and Techniques, IAEA Safeguards Technical Manual 1, Part F (1977).

[2] INTERNATIONAL ATOMIC ENERGY AGENCY, Statistical Concepts and Techniques, IAEA Safeguards Technical Manual 2, Part F (to be published).

[3] BIEBER, A.M. Jr., et al., IAEA-SM-231/98, those Proceedings, Vol.

[4] SZABO et al., Results of the IAEA Process Analysis Field Experiment II (PAFEX II) for Safeguards, IAEA-STR-71, IAEA, Vienna (1978).
[5] SZABO, E., BEETLE, T.M., Results of the First IAEA Process Analysis Field Experiment for Safeguards (PAFEX I), IAEA-STR-70, IAEA, Vienna (1978).
[6] AMERICAN NATIONAL STANDARDS INSTITUTE, Volume Calibration Techniques for Nuclear Materials, N.15.19 (1975).

Session VIII (Part 1)

ADVANCED MATERIALS CONTROL CONCEPTS AND SYSTEMS

Chairman: G. Robert KEEPIN
(United States of America)

Rapporteur summary: *Dynamic control of nuclear materials*
Papers IAEA-SM-231/62, 66, 67, 76 were presented by
A.W. DE MERSCHMANN as Rapporteur

OPTIMAL ESTIMATES OF INVENTORY AND LOSSES USING NUCLEAR MATERIAL ACCOUNTANCY DATA

D.H. PIKE, G.W. MORRISON
Computer Sciences Division,
Union Carbide Corporation,
Nuclear Division,
Oak Ridge, Tennessee,
United States of America

Abstract

OPTIMAL ESTIMATES OF INVENTORY AND LOSSES USING NUCLEAR MATERIAL ACCOUNTANCY DATA.

State-estimation theory has been shown to be applicable in obtaining minimum variance unbiased estimates of inventory and loss of special nuclear materials. The loss estimates are far superior to those obtained by the ID/LEID approach and the CUSUM approach. The technique may be extended to other situations where more complex models are appropriate. For example, one can add additional state variables to the system model to represent holdup at various points in the MBA. One could also model several interactive unit processes. So long as the system can be described by a linear system model, it is possible to construct a Kalman Filter and Linear Smoother to obtain minimum variance unbiased estimates of the state variables.

CURRENT APPROACH

The system of material accounting used by the U. S. nuclear industry is comprised of three basic subsystems [1]:

1) Measurement

2) Accounting

3) Management.

As an integral part of an overall safeguards program, material accounting acts as: (1) a deterrence to theft or diversion of special nuclear materials by providing the capability of detecting a loss, isolating the area of the loss, and identifying suspects and (2) a prevention of thefts by increasing the likelihood that an alarm will be sounded before the diverted material leaves the area.

The measurement system is concerned with measuring the inputs and outputs from a Material Balance Area (MBA) and providing an estimate of the uncertainty (via a measurement error variance) of the measured quantities. Shipments, receipts, and wastes will be treated as a single entity which will be referred to as measured net transfers:

$$U(t) = R(t) - S(t) - W(t) \quad (1)$$

where

$U(t)$ = Measured net transfers during period t,
$R(t)$ = Cumulative measured receipts during period t,
$S(t)$ = Cumulative measured shipments during period t,
$W(t)$ = Cumulative measured wastes and discards during period t.

The measured net transfers, $U(t)$, the measured inventory at the start of the period, $Y(t)$, and measurement error variances, $\sigma^2_{U(t)}$ and $\sigma^2_{Y(t)}$, respectively, are the data generated by the measurement system. The accounting subsystem is concerned first with maintaining the records of data generated by the measurement subsystem. The data is maintained by means of a double-entry accounting system similar to that used in the business community. Periodically the data maintained by the accounting subsystem is used to reconcile the "book inventory" with the "physical inventory". This reconciliation which is also referred to as constructing a "material balance" is the second function of the accounting subsystem. The "book inventory" is that inventory predicted by the data maintained in the accounting records. Specifically:

$$B(t) = Y(t) + U(t) \quad (2)$$

where

$B(t)$ = book inventory at the end of period t,
$Y(t)$ = measured inventory at the beginning of period t,
$U(t)$ = measured net transfers during period t.

Ideally, the book inventory should be equal to the end-of-period physical inventory $Y(t+1)$. Because of a variety of errors, the two will rarely be equal. The variable $M(t)$ is constructed according to:

$$M(t) = B(t) - Y(t+1) = Y(t) + U(t) - Y(t+1) \quad (3)$$

where $M(t)$ is the inventory difference (ID). The variable $M(t)$ is often referred to as the Book-Physical Inventory Difference (BPID) or historically as Material-Unaccounted For (MUF). The final function of the material accounting subsystem is to analyze the magnitude of $M(t)$ to ascertain whether or not the magnitude of $M(t)$ is significantly different from zero. This is done by means of a statistical test of hypothesis. From the measurement control program, one obtains an estimate of the variance of $M(t)$, $\sigma^2_{M(t)}$. If all measurement errors are independent then

$$\sigma^2_{M(t)} = \sigma^2_{Y(t)} + \sigma^2_{U(t)} + \sigma^2_{Y(t+1)} \, . \quad (4)$$

The limit-of-error for the inventory difference is defined as:

$$\text{LEID} = 2\sigma_{M(t)} \; . \tag{5}$$

If $|M(t)| \leq$ LEID then the calculated inventory difference is assumed to be the result of measurement errors alone. If $|M(t)| >$ LEID then it is assumed that some assignable cause is responsible for the value of M(t) (i.e., misplaced inventory, loss of material, etc.). The reconciliation and analysis are performed each time a physical inventory is measured. This usually occurs every two months.

The management subsystem is concerned with administrative procedures and processes to implement other aspects of the material accounting system and will not be considered herein.

LIMITATIONS

There are several limitations associated with the current approach. First, the limit of error for inventory difference (LEID) may be large. If a plant has a high throughput (resulting in large $\sigma_{U(t)}$ or if large inventories are experienced (resulting in large $\sigma_{Y(t)}$ and $\sigma_{Y(t+1)}$) then the LEID will be large. It can be shown that if one assumes normally distributed measurement errors there is a 0.5 probability of detecting a loss whose magnitude is equal to the LEID. Hence if the LEID is large, relatively large material losses can occur without detection. If this situation occurs, both the detection and prevention functions of material accounting will be degraded. The Rosenbaum Report [2] was highly critical of the ID/LEID approach primarily because of this limitation. Several alternatives have been proposed to reduce the sources of error and improve the timeliness of data (see [3] and [4] for example).

The second limitation of the ID/LEID approach is a conceptual limitation. The approach analyzies the data from each individual balance period without utilizing the data from previous periods. Thus, trends and patterns which might be indicative of material loss are not utilized. Jaech [3] has suggested that more information could be extracted from accounting data if the history of ID's were analyzed as a time series rather than as isolated data points.

MINIMUM VARIANCE UNBIASED ESTIMATES OF LOSS

This paper is concerned with utilizing the history (or time series) of inventory measurements, net transfer measurements, and the measurement variances to determine the magnitude of loss of material from an MBA. The criterion used will be to obtain a Minimum-Variance-Unbiased Estimate (MVUE) of loss and inventory. An estimate will be a statistic since, by necessity, it will be constructed as a function of several random variables (i.e., net transfer measurements and inventory measurements). An unbiased estimate is an estimate whose mean is equal to the true value of the variable. A minimum-variance unbiased estimate is therefore an unbiased estimate with as small or smaller variance than any other unbiased estimate. Such an estimate is generally <u>not</u> unique.

The MVUE estimate of loss will be contrasted to the Cumulative-Summation (CUSUM) approach which although is not an MVUE loss estimate approaches in limit the MVUE under certain conditions.

KALMAN FILTER APPROACH

The mechanism which will be presented for obtaining a MVUE estimate of loss and inventory is by use of the Kalman Filter. This procedure was developed by R. E. Kalman [6] and has gained widespread acceptance in navigational systems, guidance and control systems, and orbital estimation. The technique is designed to use measurement data, a system model, and *a priori* information to obtain estimates of the state of a linear system. The technique produces MVUEs of the system state by sequentially processing the measurement data.

The state of the system is a two-component vector consisting of inventory and loss. The system model shows how the states evolve in a dynamic environment. One state equation is necessary for each state variable. The inventory equation is:

$$I(t+1) = I(t) + T(t) - L(t) \tag{6}$$

where

$I(t)$ = True inventory of start of period t,
$T(t)$ = True net transfers during period t,
$L(t)$ = Material loss during period t.

The equation for loss describes the loss mechanism. Experience has shown [7] that a desirable loss equation is:

$$L(t+1) = L(t) + \varepsilon(t) \tag{7}$$

where

$L(t)$ = loss during period t,
$\varepsilon(t)$ = zero-mean random variable denoting modeling error,
σ_ε^2 = variance of modeling error.

The loss mechanism modeled by Equation (7) assumes that true loss is a constant ($L(t+1) = L(t)$). The modeling error, $\varepsilon(t)$, accounts for deviations of the actual loss from a constant value. It should be noted that the variables $I(t)$ and $T(t)$ represent true values and not the measured variables used in calculating inventory difference. The relation between the true and measured variables is:

$$Y(t) = I(t) + V(t) \tag{8}$$

$$U(t) = T(t) + Z(t) \tag{9}$$

where

V(t) = zero-mean measurement error for inventory,
Z(t) = zero-mean measurement error for transfers.

The variance of V(t) is $\sigma^2_{Y(t)}$ and the variance of Z(t) is $\sigma^2_{U(t)}$. Equations (6)–(9) constitute the linear system model.

The Kalman Filter algorithm for this system is:

1) Initialize the state variables:

a) $\hat{I}(0) = Y(0)$ (10a)

b) $\hat{L}(0) = 0.$ (10b)

2) Supply error covariance matrix for initial state estimate:

$$\underline{G}(0) = \begin{bmatrix} \sigma^2_{\hat{I}(0)} & \mathrm{Cov}[\hat{I}(0), \hat{L}(0)] \\ \mathrm{Cov}[\hat{I}(0), \hat{L}(0)] & \sigma^2_{\hat{L}(0)} \end{bmatrix}$$

$$= \begin{bmatrix} \sigma^2_{Y(0)} & 0 \\ 0 & \infty \end{bmatrix}$$

3) Predict State at time t:

$$\tilde{I}(t) = \hat{I}(t-1) + U(t-1) - \hat{L}(t-1) \tag{12a}$$

$$\tilde{L}(t) = \hat{L}(t-1) \tag{12b}$$

4) Calculate error covariance matrix for predicted state:

$$\underline{P}(t) = \begin{bmatrix} \sigma^2_{\tilde{I}(t)} & \mathrm{Cov}[\tilde{I}(t), \tilde{L}(t)] \\ \mathrm{Cov}[\tilde{I}(t), \tilde{L}(t)] & \sigma^2_{\tilde{L}(t)} \end{bmatrix} \tag{13a}$$

where

$$\sigma^2_{\tilde{I}(t)} = G_{11}(t-1) - 2G_{12}(t-1) + G_{22}(t-1) + \sigma^2_{U(t-1)} \tag{13b}$$

$$\mathrm{Cov}[\tilde{I}(t), \tilde{L}(t)] = G_{12}(t-1) - G_{22}(t-1) \tag{13c}$$

$$\sigma^2_{\tilde{L}(t)} = G_{22}(t-1) + \sigma^2_{\varepsilon} \tag{13d}$$

5) Calculate gains:

$$K_1(t) = P_{11}(t)/(P_{11}(t) + \sigma^2_{Y(t)}) \quad (14a)$$

$$K_2(t) = P_{12}(t)/(P_{11}(t) + \sigma^2_{Y(t)}) \quad (14b)$$

6) Update state estimate

$$\hat{I}(t) = \tilde{I}(t) + K_1(t)\,\{Y(t) - \tilde{I}(t)\} \quad (15a)$$

$$\hat{L}(t) = \tilde{L}(t) + K_2(t)\,\{Y(t) - \tilde{I}(t)\} \quad (15b)$$

7) Calculate error covariance matrix for state estimate:

$$\underline{G}(t) = \begin{bmatrix} [(1-K_1(t)]\,P_{11}(t) & [(1-K_1(t)]\,P_{12}(t) \\ [(1-K_1(t)]\,P_{12}(t) & P_{22}(t)-K_2(t)P_{22}(t) \end{bmatrix} \quad (16)$$

8) Set t = t+1 and return to step (3).

The particular algorithm given above assumes that there is no *a priori* knowledge of the loss (since $\sigma^2_{\hat{L}(0)} = \infty$). If some knowledge is available, $\hat{L}(0)$ should be set equal to the user's best estimate, $\sigma^2_{\hat{L}(0)}$ should be set equal to the variance of the user's estimate of $\hat{L}(0)$.

The Kalman Filter inventory estimate, $\hat{I}(t)$, is a better estimate of the true inventory than either the measured value or the book value. The loss estimate, $\hat{L}(t)$, can be used with its variance $G_{22}(t)$ to test for significance in a manner analogous to the ID/LEID approach.

LINEAR SMOOTHER

One of the desirable features of the state variable approach is the ability to "smooth" the data. Smoothing refers to improvement of previous state estimates. Immediately after receipt of the observation Y(t) and U(t-1) at time t, the state variables are updated to obtain $\hat{I}(t)$ and $\hat{L}(t)$. All available information has been utilized in calculating $\hat{I}(t)$ and $\hat{L}(t)$. However, new information is now available which can be used to update state variable estimates which were made at time periods, t-1, t-2, etc. This process of updating previous state estimates given new data is referred to as smoothing. There are various forms of smoothing. The particular smoother used in this study is the Rauch-Tung-Striebel smoother [8]. The algorithm is:

1) Initialization (starting with period $\hat{L}(T)$:

a) $I^*(T) = \hat{I}(T)$, $L^*(T) = \hat{L}(T)$, (17a)

b) $\underline{G}^*(T) = \underline{G}(T)$, (17b)

c) Set $t = T - 1$. (17c)

2) Obtain the predicted state $[\tilde{I}(t+1), \tilde{L}(t+1)]$. These can be saved during the filter phase. Alternatively, one can save $[\hat{I}(t), \hat{L}(t)]$ and use Equations (12a) and (12b) to calculate the predicted state.

3) Obtain error covariance matrix $\underline{P}(t+1)$. As with the predicted state, $\underline{P}(t+1)$ may be saved during the filter phase or calculated from Equation (13a)–(13d).

4) Calculate gain: (Using $\underline{G}(t)$ and $\underline{P}(t+1)$)

$$\underline{D}(t) = \frac{1}{\Delta}\begin{bmatrix} (G_{11}-G_{12})P_{22}-G_{12}P_{12} & P_{11}G_{12}-P_{12}(G_{11}-G_{12}) \\ (G_{12}-G_{22})P_{22}-G_{22}P_{12} & P_{11}G_{22}-P_{12}(G_{12}-G_{22}) \end{bmatrix} \tag{18}$$

where $\Delta = P_{11}P_{22} - P_{12}^2$.

5) Update State Estimate:

$$I^*(t) = \hat{I}(t) + D_{11}[I^*(t+1) - \tilde{I}(t+1)] + D_{12}[L^*(t+1) - \tilde{L}(t+1)] \tag{19a}$$

$$L^*(t) = \hat{L}(t) + D_{21}[I^*(t+1) - \tilde{I}(t+1)] + D_{22}[L^*(t+1) - \tilde{L}(t+1)]. \tag{19b}$$

6) Calculate error covariance matrix:

$$\underline{G}^*(t) = \underline{G}(t) + \underline{D}(t)[\underline{G}^*(t+1) - \underline{P}(t+1)]\underline{D}(t)^T \tag{20}$$

7) Set $t = t - 1$ and return to step 2.

The smoother should be used when σ_ε^2, the variance of the modeling error, is positive. If $\sigma_\varepsilon^2 = 0$ then the smoother results in losses which set $L^*(t) = \hat{L}(T)$ for $t = T-1, T-2, \ldots, 0$. One must save the state estimates $\hat{I}(t)$ and $\hat{L}(t)$ and the error covariance matrix $\underline{G}(t)$ during the filter phase if smoothing is desired.

CUSUM APPROACH

The average cumulative sum of IDs divided by time is defined as:

$$C(t) = \frac{1}{t}\sum_{j=0}^{t} M(j) . \tag{21}$$

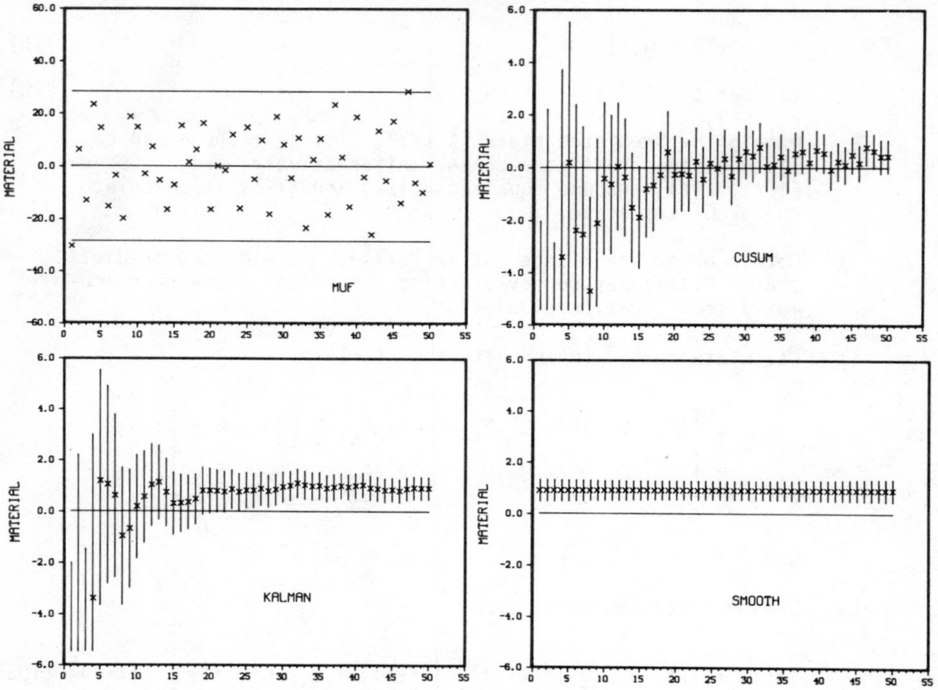

FIG.1. Results of simulation 1.

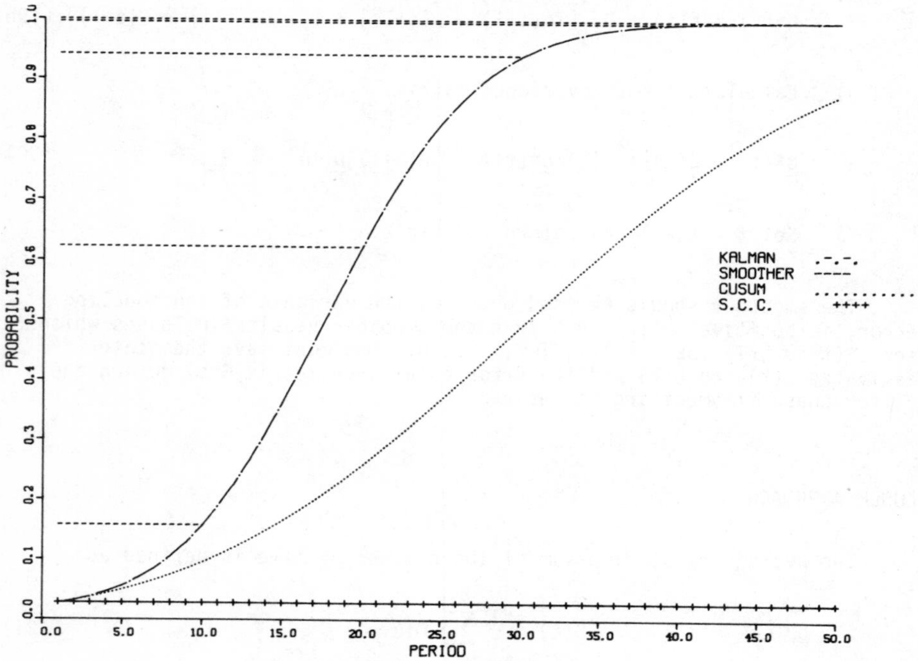

FIG.2. Detection probabilities – Case 1.

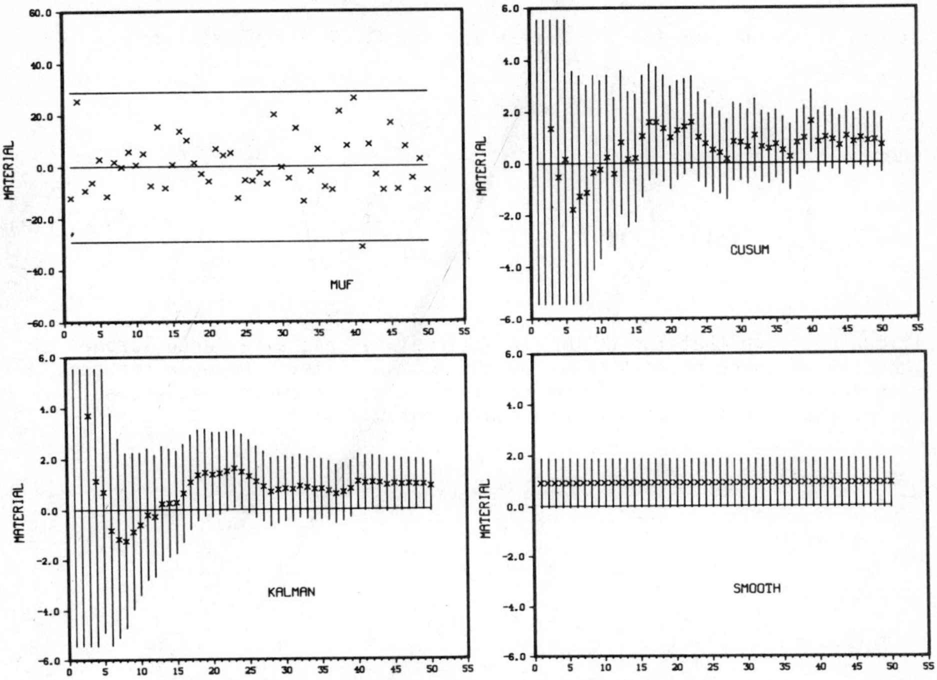

FIG.3. Results of simulation 2.

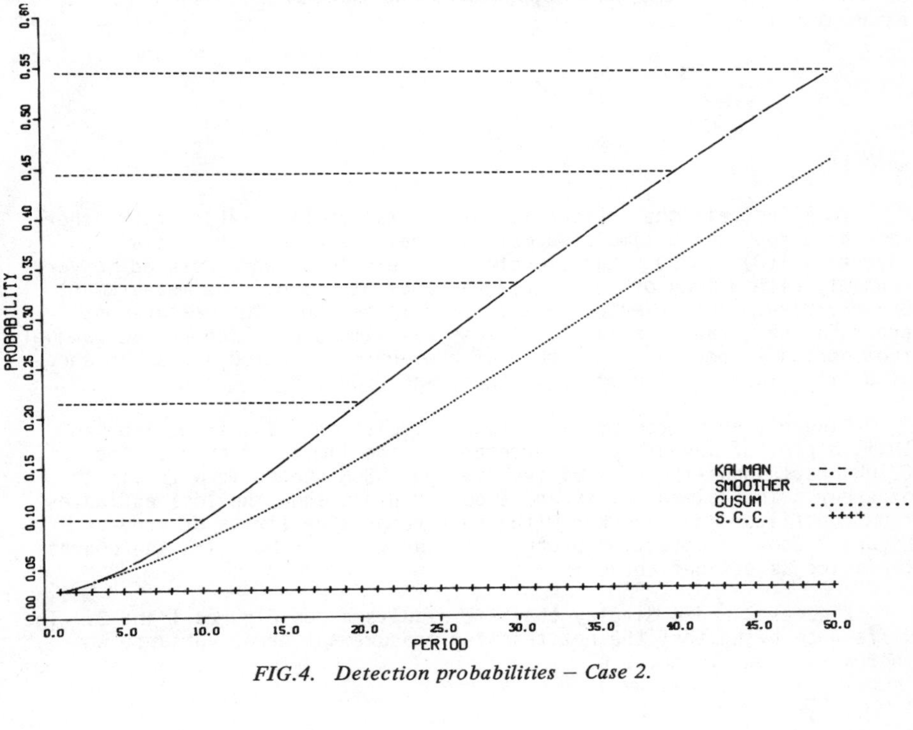

FIG.4. Detection probabilities – Case 2.

It can be shown that the limit-of-error for CUSUM (LECUSUM) is:

$$\text{LECUSUM} = 2\,\sigma_{C(t)}\ .$$

where

$$\sigma^2_{C(t)} = \left(\sigma^2_{\hat{Y}(0)} + \sigma^2_{\hat{Y}(t+1)} + \sum_{j=0}^{t} \sigma^2_{\hat{U}(t)}\right) \Big/ t\ . \tag{22}$$

It can be shown that the CUSUM statistic always has a variance larger than the variance of loss given by the Kalman Filter. However, as the transfer variance becomes large relative to the inventory measurement error, the loss estimates become equivalent. That is:

$$\sigma^2_{C(t)} > G_{22}(t) \tag{23}$$

but

$$\sigma^2_{C(t)} \rightarrow G_{22}(t) \text{ as } \sigma_{U(t)}/\sigma_{Y(t)} \rightarrow \infty\ . \tag{24}$$

Thus, in plants where measurement error is dominated by net transfer measurement error, the CUSUM approaches the two state Kalman filter loss estimator.

EXAMPLES

To illustrate the application of the Kalman Filter/Linear Smoother approach, consider a simulated material balance area with initial inventory $\hat{I}(0) = 2200$. Net transfers (true values) were assumed to vary randomly with a mean of zero and a variance of 10000. The measurement error variance for inventory was assumed to be 100. The measurement error for net transfers was 1. A loss was simulated which varied randonly from period to period with a mean of one unit per period and a variance of 0.111. The modeling error variance was 0.001.

Figure 1 shows the results of the simulation. The graph labeled ID is a plot of inventory differences and the limits of error. The CUSUM chart shows the plot of the average CUSUM statistic with limits of error. The Kalman Filter and Smoother plots show the loss estimates from the filter and smoother with their respective limits of error. Figure 2 shows a detection probability curve for an MBA with the characteristics as defined above.

Figures 3 and 4 display the same results as do Figures 1 and 2. The difference being that the net transfer measurement error variance was increased from 1.0 to 10.0.

REFERENCES

[1] U. S. Nuclear Regulatory Commission, "Report of the Material Control and Material Accounting Task Force", NUREG-0450, 2 (April 1978) IV:1–IV:17.

[2] Rosenbaum, D. M., Googin, J. N., Jefferson, R. M., Klectman, D. J. and Sullivan, W. C., Special Safeguards Study, USAEC, Report T-201.

[3] Bain, E. E. Jr., et. al., An Evaluation of Real-Time Material Control And Accountability in a Model Mixed-Oxide Fuel Plan, Lawrence Livermore Laboratory Report, Sept. 15, 1975.

[4] Senbaugh, P. W., Rogers, D. R., Woltermann, H. A., Fushimi, F. C. and Ciramella, A. F. Application of Controllable Unit Methodology to a Realistic Model of a High-Throughput Mixed-Oxide Fabrication Process, Mound Laboratory, Miamisburg, Ohio (1977).

[5] Jaech, J. L., "Control Charts for MUF's", J. of Inst. of Nucl. Mater. Mang. II 4 (1974) 16–28.

[6] Kalman, R. E., "A New Approach to Linear Filtering and Prediction Problems", J. of Basic Eng. (ASME) 82D (1960) 33–45.

[7] Pike, D. H., Morrison, G. W. and Downing, D. J., Time Series Analysis Techniques Applicable to Nuclear Material Accountability Data, ORNL/NUREG/CSD-10 (1978) Draft Report.

[8] Rauch, H. E., Tung, F. and Striebel, C. T., "Maximum Likelihood Estimates of Linear Dynamic Systems", AIAA Journal, 3 8 (August 1965) 1445–50.

DISCUSSION

R. AVENHAUS: Several years ago K.B. Stewart from Battelle Pacific Northwest Laboratory in the United States of America developed his theory of a minimum-variance unbiased estimate (MVUE) for the starting inventory of a new inventory period based on the ending book and physical inventories of the foregoing inventory period. Is there any difference between Stewart's procedure and yours?

D.H. PIKE: The method of K.B. Stewart produces a minimum-variance unbiased estimate of the inventory only. Stewart's technique is an MVUE only when the true loss is zero. When the true loss is zero, the Kalman filter method and the Stewart method should give identical estimates of inventory. When the true loss is not zero, Stewart's method is not an MVUE and the Kalman filter method must be used.

R. AVENHAUS: I should also like to make a comment. You showed a diagram where the probability of detection was plotted versus the number of inventories. This diagram indicated that the use of your technique led to a continuously increasing probability of detection. I think that this diagram is somewhat misleading in relation to international nuclear material safeguards, as one normally specifies a reference time (e.g. one year) after which a final statement has to be made as to whether or not there was a diversion. This means that the estimate of losses should not be extended beyond this reference time.

G. HOUGH: The Agency is required to confirm that there has been no diversion of nuclear material at least once a year. Under present safeguards agreements the inventory frequency varies from one to four times per year. Do you think that the technique you describe, or any other technique, will be very useful based on one to four data points?

D.H. PIKE: Probably no technique will provide good estimates of loss with only four data points. This is because the limit of error will be excessively large and it is doubtful whether one could reject the null-hypothesis (i.e. true loss is zero) with only four data points. Depending on the magnitude of measurement errors, it might be possible to detect loss with 16–20 data points.

R.H. AUGUSTSON: If the Agency moves towards detection times of the order of days or weeks, there will be far more material balance closures than four a year. This being so, the Kalman filter techniques may be applicable to Agency inspection data.

D.H. PIKE: This is true. It should also be noted that even though one is limited to four data points per year, the information contained in the previous year's data will also be of use in analysing the current year's data.

THE POTENTIAL VALUE OF DYNAMIC MATERIALS CONTROL IN INTERNATIONAL SAFEGUARDS

G. Robert KEEPIN
Los Alamos Scientific Laboratory,
Los Alamos, New Mexico,
United States of America

J.E. LOVETT
Department of Safeguards,
International Atomic Energy Agency, Vienna

Abstract

THE POTENTIAL VALUE OF DYNAMIC MATERIALS CONTROL IN INTERNATIONAL SAFEGUARDS.

The difficulties inherent in conventional materials accountancy based on semi-annual or annual shutdown cleanout physical inventories have been recognized for many years. The increasing importance of international nuclear materials safeguards, coupled with the availability of advanced non-destructive measurement technology which could be installed on or near process lines, has led to the development of the concept of advanced or dynamic materials control. The potential benefits of dynamic materials control in terms of significantly improved detection capabilities (ranging from a few kilograms of plutonium down to perhaps a few hundred grams, even for large-scale bulk processing facilities), and even more dramatically improved detection timeliness (typically a few days, and potentially only a few hours, in advanced facilities), are reviewed. At least twelve major dynamic material control systems already in existence or in the process of being installed are noted, and some of the essential characteristics are discussed. Some currently unresolved questions are explored, and future prospects for the concept of dynamic material control in international safeguards are reviewed.

I. INTRODUCTION

Strengthening and fostering the worldwide growth of nuclear power while at the same time minimizing or eliminating the risks of nuclear proliferation or mis-use of nuclear materials; that is the dual challenge facing the international nuclear community. It was the subject of the International Conference on Nuclear Power and Its Fuel Cycle [1]. It is the subject of the International Fuel Cycle Evaluation Program (INFCE). It is the subject of national programs in at least several Member States of the IAEA. It is the subject of this symposium.

Certain conclusions seem inescapable. The world needs nuclear power, at least for the next several decades. This nuclear power requires not only the nuclear power reactors themselves, but also the supporting bulk processing facilities. Indeed if nuclear power is to be generated on a large scale, then the nuclear process facilities must also operate on a large scale, and must process large quantities of nuclear material. Nevertheless, this nuclear power must not and cannot be purchased at the cost of nuclear proliferation.

Safeguards remain the central element of any combination of measures taken against nuclear proliferation. We find no alternative. Nuclear safeguards considerations must be a major factor both in the selection of processes and in the subsequent design of nuclear facilities. Whatever fuel cycle or combination of fuel cycles may be implemented in various parts of the world, we cannot afford to treat safeguards as an afterthought.

It is in this context that we review here the concept of dynamic materials control, and the role that we believe such systems can and must play in safeguarding nuclear materials and in preventing nuclear proliferation.

II. CONVENTIONAL MATERIALS ACCOUNTANCY

In conventional safeguards practice the accountability of nuclear materials within a facility, and the detection of unauthorized removals from that facility, have relied almost exclusively on the counting of discrete physical items or containers, coupled with periodic shutdown, cleanout physical inventories for purposes of material balance accounting. These conventional material balances have been drawn for the most part around entire nuclear facilities, or at least around some major portion of a nuclear facility such as a complete process line. During the periods between physical inventories material control, if it exists at all, has been vested in administrative and process controls.

The problems with actual inplementation of this traditional concept of materials accountancy are well known. In the context of today's demands for stringent international safeguards monthly material balance closings for facilities handling large quantities of strategic nuclear material (e.g., separated Pu) are only marginally acceptable, and the more common annual or semi-annual material balances are totally unacceptable. In those cases where more frequent (i.e., monthly) material balances have been attempted, moreover, the inventory quality usually suffered, and in most cases it is not clear that there was any true net gain in diversion detection capabilities.

Material balance uncertainty is also a major problem in conventional materials accounting. Even when actual quantity measurements are carefully made and statistical uncertainty data are carefully generated, the overall material balance uncertainties propagated over a six month or twelve month material balance period are rarely acceptable. The combined measurement uncertainty associated with MUF is seldom less than 0.5% of total throughput and may reach or exceed 1.0%. In the spent fuel reprocessing facility even 1.0% control is difficult to achieve using conventional methods of materials accountancy.

These periodic "shutdown and cleanout" operations will always have an important role in international safeguards. Nevertheless, it seems clear that this procedure alone, as employed in the past, has serious limitations in sensitivity and timeliness. Moreover, these limitations are inherent in the basic concept of conventional accountancy. Sensitivity is limited by measurement uncertainties, which are not likely to be improved greatly, and by the inherent characteristics of the Gaussian error propagation statistical model, for which there does not appear to be a viable alternative, while timeliness is limited by the practical difficulties and economic penalties associated with shutdown physical inventories, which difficulties likewise are not likely to be reduced to any significant extent. Efforts to improve the capabilities of conventional materials accountancy should be continued, but conventional materials accountancy cannot solve today's safeguards problems.

What are the alternatives? The one most commonly discussed is an increased reliance on containment measures, optical, electronic, and human surveillance, and continuous inspector presence in critical nuclear facilities. These measures, like conventional materials accountancy, have an important role in international safeguards. Yet they too have their limitations, and at least some of these limitations would appear to be inherent in the concept of containment as a basic safeguards measure. Existing containment/surveillance equipment has a demonstrated record of being failure-prone, and cannot be used as an independent safeguards measure. Improvements in reliability seem probable, but are likely to prove costly, and in any case will stop significantly short of 100% reliability. When equipment failure does occur, it is often difficult to re-establish the physical inventory so that meaningful containment/surveillance can be resumed.

Containment and surveillance safeguards measures are also difficult to quantify. They can be 100% effective, but it is sometimes difficult to know whether they have been effective in a given application. Containment safeguards, like conventional materials accountancy, when patched onto a nuclear facility in which international safeguards requirements were never a serious design criterion, are not likely to safisfy currently discussed criteria.

The other alternative, the one discussed in this paper, is dynamic materials accountancy.

III. DYNAMIC MATERIALS ACCOUNTANCY

We have made several attempts to define dynamic materials control or accountancy, and have not been fully satisfied with any of them. The reason, we conclude, is that dynamic materials accountancy is in reality not greatly different from conventional materials accountancy except that:

a) the uncertainty in a conventional material balance is too large, primarily because the material balance unit is too large, both in time and space. Dynamic materials control divides the process into numerous small units and prepares material balances at frequent intervals. Not only is the uncertainty in any single material balance greatly reduced, but the large quantity of unit material balance data available permits the use of decision analysis methods and statistical techniques which previously either were invalid or could not be used for lack of sufficient input data.

b) conventional material balances require costly and time-consuming shutdown physical inventories. This is partly because the process equipment was not designed to facilitate the taking of in-process inventories and partly because the uncertainty in those inventories only adds to an already unacceptably large total uncertainty. By proper design of process equipment, coupled with recently developed non-destructive assay (NDA) techniques, dynamic materials accountancy utilizes physical inventory data obtained either without process interruption or in some cases with only momentary interruptions which do not significantly affect process operations. Since material balances are prepared for small process units and at frequent time intervals, and since in general different statistical techniques are used for data evaluation, comparatively large relative uncertainties (in individual measurements) can be tolerated.

c) conventional measurement and data handling techniques often introduced delays ranging from days to weeks between the closing of the material balance and the calculation of the resulting material unaccounted for (MUF). By utilizing computerized data handling systems, coupled where possible with in- or on-line measurement systems, dynamic materials accountancy can reduce data processing delays to minutes or hours instead of days.

Stated in broad terms, the overall objective of dynamic materials control is to demonstrate, in a more or less continuously operating bulk processing facility, that an undetected diversion or theft of W weight units within an interval of T time units, by any combination of scenarios, has not occurred. Dynamic materials control thus may be viewed as an advanced form of conventional materials accountancy, one in which recently developed NDA technology, state-of-the-art conventional measurement methods, improved facility design, and computerized data processing are combined to generate useful material balance data about small portions of the total facility on a near real-time time scale.

It is difficult to assign values to W and T except in terms of general capabilities. In conventional materials accountancy T is fixed by the frequency of shutdown physical inventories, and W is fixed by the propagated uncertainty in measurements for a single material balance period. In dynamic materials control, however, there are trade-offs between W and T which remain to be studied on a specific facility. There appear to be no inherent obstacles to the achievement of T=8 hours, or even less if required. On the other hand, T=1 to 10 days should satisfy most requirements for timeliness, and this may permit the use of better measurement techniques which in turn reduce W. Paper studies on large scale facilities suggest that W=1 to 3 kgs Pu should be attainable, and here again cost-effectiveness considerations remain to be studied.

A. Material Balance Areas vs. Unit Process Areas

In conventional materials accountancy the basic material balance unit is termed the material balance area (MBA). In many cases the definition of the MBA is synonymous with "facility", or at least with "process line". There is normally little advantage to any finer sub-division, and the disadvantage most frequently cited is the need for increased inter-MBA measurements if multiple MBAs are created.

In dynamic materials control it is important to divide the total nuclear facility into discrete accounting envelopes, each of which involves, insofar as possible, either only material storage, or only batch processing, or only continuous processing. Although often loosely referred to as MBAs, a better terminology is to refer to these process units as unit process areas (UPAs). A unit process area can still be one or more chemical or physical processes, but the UPA boundaries now are drawn on the basis of process logic, residence time of material, ability to perform quantitative measurements using in- or on-line NDA techniques, need to identify and localise possible unrecognized loss mechanisms, etc.

B. Measurement Techniques and Measurement Quality

In conventional materials accountancy inter-MBA and physical inventory measurements are performed using the most accurate and precise methods available. There is no other alternative. The combined uncertainty in a six or twelve month material balance usually is uncomfortably large under the best of circumstances, and making it larger through compromises in measurement quality only makes a bad situation worse.

In dynamic materials control the accuracy of measurement techniques is still an important factor, but the significant improvement in material balance timeliness, the ability to use other statistical techniques (discussed in a later section), and other factors usually more than compensate for the loss of accuracy which may result from the use of non-destructive measurement techniques. While NDA measurements often involve measuring the entire bulk quantity in a given batch or process operation, this is certainly not essential. Many excellent NDA measurements can be performed on samples taken from large bulk quantities. As with all measurements - destructive or non-destructive - which are based on sampling, a potential sampling error is introduced, but the resulting combination of a moderate sampling uncertainty with virtually no analytical delay usually is preferable to other feasible alternatives.

Implementation of dynamic or "near real-time" materials measurement requires the rapid, quantitative measurement of nuclear materials locally (i.e., in-line or at-line) at each process unit. Modern non-destructive assay techniques are well suited to this rapid, direct in-line measurement, and NDA instruments are being developed, adapted and applied to process measurement requirements in several different ways:

1) as the primary measurement technique at unit process boundaries;

2) as part of a complementary set consisting of frequent or even "continuous" NDA measurements that may be updated by periodic analytical chemistry data (e.g., in connection with periodic cleanout physical inventories); and

3) to assay or verify the contents of discrete items such as sealed containers, fabricated pieces, finished components, etc., where the non-destructive feature is virtually essential.

C. Physical Inventory

It is in the area of physical inventory, perhaps more than any other single area, that dynamic materials control differs from conventional materials accountancy. In the latter it is extremely important that all nuclear material be included in the physical inventory listing. Any material not included, for whatever reason, becomes part of the MUF for that material balance period. Since the statistical model is, of necessity, a single-period model in which the expected value of MUF is zero, failure to achieve completeness in the physical inventory only tends to suggest the existence of unknown losses when in fact none have occurred.

In dynamic materials accountancy, on the other hand, the ability to perform multiple-period statistical evaluations permits the deliberate omission of minor inventory quantities. Such perturbations normally can be handled by using operational experience or "historical" data to interpret trends in apparent holdup. In many processes the added control obtained by measuring small, relatively constant inventory quantities may be negligible, and may not justify the difficulty and expense.

Dynamic materials control, however, places considerable emphasis on the ability to measure all significant inventory quantities without disturbing process operations. Sometimes this means installed measurement equipment which truly measures material quantities "as is, where is". In other cases it may require that the physical inventories be scheduled to coincide with specified process operations, such as the discharging of a product solution evaporator, which occur at sufficiently frequent intervals. In still other cases the system may be designed to maintain more or less continuous book inventory records on a container-by-container or batch-by-batch basis, with the dynamic physical inventory being taken by verifying the existence of all batches and the contents of randomly selected batches. Such judgements must be made on an individual process basis, taking into account the nature of the processes to be inventoried, together with the strategic value and safeguards vulnerability of the materials.

D. Evaluation

In the statistical evaluation of MUF as calculated by the conventional material balance, the assumption is that the MUF should be zero, and the statistical hypothesis is that the observed non-zero MUF occurred solely through a chance combination of measurement uncertainties. In practice, it is recognized that an observed MUF may be non-zero through any combination of three causes; bias, random measurement error, and unknown losses (including possible diversions). It is further recognized that the ability to measure bias is limited, and that there is a range, commonly termed systematic uncertainty, within which efforts to measure or eliminate bias may not have been successful. Since this residual bias, if it exists, is a cumulative effect, the magnitude of the systematic uncertainty usually defines the overall uncertainty in the material balance.

In dynamic materials control the material balance period is short, i.e., day or hours rather than weeks or months. This permits the use of statistical tests which ask not whether the observed MUF is different from zero, but whether it is different from previous experience. Indeed, since there is usually a small unmeasured in-process inventory, the expected value of MUF is not zero. It is, or should be, constant.

The result of this difference in evaluation philisophy is that while measurement bias in input or output data is still very important, the uncertainty in this bias, the previously all-important systematic uncertainty, is no longer of overriding concern. Further, since changes in the magnitude of the physical inventory from one period to another usually are small, even bias is of secondary importance in the inventory measurements.

The availability of relatively large quantities of material balance data underscores the need for an organized framework of techniques to ensure efficient and complete extraction of information. The discipline of decision analysis [2], combining techniques from estimation theory, decision theory, and systems analysis, provides such a framework. Stated realistically, in terms of actual practice, decision analysis is used to provide factual and quantitative assurance that diversion has in fact not occurred. Augmented by computer-display and pattern-recognition techniques, such as the CUSUM plot and the alarm sequence chart, decision analysis can be used to reduce errors caused by subjective data evaluation, and to condense large collections of data to smaller sets of more descriptive statistics. The use of these techniques makes the decision process more timely and efficient, as well as more consistent and more objective [3].

IV. EXISTING SYSTEMS AND PROCESS SIMULATIONS

The development of dynamic materials accountancy systems has been underway for a number of years, and is now worldwide. Table I lists a number of systems presently in existence or in various stages of development. Some encompass only some of the aspects of true dynamic materials control, as for example where emphasis is placed on computerized data processing without a corresponding emphasis on in-process inventory determination, but the number and diversity of the systems is indicative of the current high level of interest in advanced material control systems.

In addition, modeling and simulation techniques [4] have proved extremely valuable in the design, evaluation and comparison of the relative effectiveness of alternative processes, measurement systems and materials accounting strategies. These techniques permit the prediction of the dynamics of nuclear material flow under a wide range of operating parameters, and the rapid accumulation of data for relatively long (simulated) operating periods. For each facility, this approach requires:

1) a detailed dynamic model of the process;

2) simulation of the model process on a digital computer;

3) a dynamic model of each measurement system;

4) simulation of accountability measurements on nuclear material flows and in-process inventories generated by the model process; and

5) evaluation of the simulated measurement data from each accounting strategy.

As a specific example of the performance capabilities of dynamic materials control, we cite simulation studies [5] on a reference conversion process based on plutonium oxalate precipitation and calcination. In this conversion process, the key measurement points were located at the receipt tank, the output of the precipitator, and the product loadout stage. The estimated plutonium detection sensitivity levels are presented in Table II. Diversion sensitivity is given for periods of one material balance (one batch), one day (approximately 20 batches), one week (approximately 125 batches), and one month (approximately 530). For comparison Table II also lists conventional accounting sensitivity, based on current U.S. regulations which require that material balancing in conversion plants be performed every two months with a material balance uncertainty (2σ) of less than 0.5% of the facility throughput. This limit of error corresponds to 33 kg of plutonium for the reference conversion process, which has a design throughput of 6600 kg of plutonium over a two month period. A recent estimate [25] of the two sigma limit of error that should be achievable using conventional

TABLE I
SOME DYNAMIC MATERIALS ACCOUNTING SYSTEMS

Facility, Location, Function	System Name	System Functions & Comments	Ref.
General Electric Wilmington, North Carolina; UF_6 conversion to fuel-bundle assembly	MICS	Material Inventory Control System. Diversion detection, information quality, loss localization, system management & control. NDA used.	[6, 7]
General Electric Vallecitos, California; Pu fuel-development laboratory for LMFBR	GERTA	Material distribution, diversion detection. Primarily automated record-keeping; no NDA at this time.	[8]
Mound Laboratory Miamisburg, Ohio; MOX fuel fabrication	CUA	Controllable Unit Accounting. Conceptual system for accounting, diversion detection. No NDA used.	[9, 10]
Combustion Engineering Windsor, Connecticut; Nuclear fuel manufacturing	FACS	Fuel Accounting & Control System. Timely & accurate reporting on SNM status & flow; no NDA as yet.	[11]
AECL Chalk River, Canada; Fuel materials development & fabrication	INMACS	Integrated Nuclear Material Accounting & Control System. On-line material accounting, data base mgmt.; no NDA as yet.	[12]
Y-12 Plant Oak Ridge, Tennessee; Enriched uranium processing facility	DYMCAS	Accountability, diversion detection, physical inventory, NDA verification. Incorporates on-line or keyboard verification of weights.	[13, 14]
Rocky Flats Plant Golden, Colorado; Pu processing facility	NMC COMSAC	Accountability, criticality control, NDA calibration, NDA measurements.	[15, 16]
Karlsruhe Research Center Karlsruhe, Federal Republic of Germany; Research, processing, handling, storage facilities	---	Generalized SNM accounting & data handling system for variety of SNM processing, handling functions.	[17]
ARHCO Richland, Washington; Storage & processing facility	---	Accountability, process monitoring, laboratory bookkeeping, monitoring & control of storage locations.	[18]
PNC Tokai-mura, Japan; MOX fuel fabrication	PINC	Plutonium Inventory Control system using on-line NDA sensors & computerized inventory, process control.	[19]

TABLE I (continued)

Facility, Location, Function	System Name	System Functions & Comments	Ref.
AGNS Barnwell, South Carolina; Fuel reprocessing plant	AGMAC	Laboratory data system & materials accounting & control system. Some process monitoring.	[20]
LASL Los Alamos, New Mexico Pu processing	DYMAC	Accountability, in-plant NDA instrumentation; computerized near-real-time inventory control; data-base management, unit process SNM localization.	[21–24]

TABLE II
COMPARISON OF CONVENTIONAL AND DYNAMIC MATERIALS ACCOUNTING DIVERSION SENSITIVITIES FOR THE CONVERSION PROCESS

	Average Diversion per balance (kg Pu)	Total Diversion Sensitivity (kg Pu)	Detection Time
Dynamic Materials[a] Accounting	0.13	0.13	1 batch (1.35h)
	0.03	0.63	1 day
	0.01	1.24	1 week
	0.005	2.65	1 month
Conventional Materials Accounting	33	33	2 months

a For a single-unit-process accounting strategy as described in Ref. 14.

materials accountancy and two month material balances for the same model facility is ± 0.38%, or approximately 25 kgs Pu.

Dynamic materials control is not solely for the large facility. A study is currently underway concerning the feasibility of back-fitting dynamic control systems into existing reprocessing facilities in the 250 t/a or under range. The study is still in its early stages, and it is too early to quote quantitative data, but the basic feasibility has been shown to exist. Two models are being studied, one based on weekly inventories and one based on daily inventories. In terms of effectiveness both models are expected to be capable of detecting an abrupt diversion of 2 kgs Pu, and of detecting a long-term small-quantity diversion before the total diverted has reached 5 kgs Pu.

V. CURRENT QUESTIONS AND FUTURE PROSPECTS

Dynamic materials control, like nuclear power itself, has its apostles and its critics. A growing number of safeguards experts, we are pleased to note, are studying its potentialities, but there are still many sceptics. In part perhaps this is our own fault. Elaborate paper studies have been prepared on truly dynamic systems for facilities many of which likewise exist only on paper, at a time when true demonstrations of practical systems were hard to find. We have, metaphorically speaking, claimed that we could win an Olympic race at a time when our ability to walk was still in question.

We believe that ultimately we can win that race. In fact, we believe that dynamic materials control is the only candidate in sight that has a chance of winning. Nevertheless, we freely acknowledge that the race is still far from won, and in this section we discuss some of the current questions and future prospects of dynamic materials control.

A. Cost Factors

Some critics express the opinion that dynamic materials control systems will be expensive, and ask who will pay the cost. They also point out that dynamic materials control systems must be installed and implemented by the facility operator, and ask where is the mechanism requiring the use of dynamic materials control.

There is, clearly, no current mechanism whereby the IAEA can require that such systems be installed, and we would make no proposals for changes which would give them that right. There are, at the moment, two potential mechanisms. One is that the State adopts regulations which require dynamic materials controls, or better, which impose performance standards which can best be met via dynamic materials controls. The IAEA does not have such authority, but individual Member States, for their own reasons, might invoke such requirements.

A better and ultimately more compelling reason, we think, is economics. There is a growing body of evidence that overall effective costs can be very significantly reduced through the improvements in operational efficiency, process and quality control, criticality safety, greater knowledge of process status, elimination of unrecognized loss mechanisms, and the reduced need for costly shutdown/cleanout physical inventories which result from the increased and more timely availability of process data. (Reference, for example, the General Electric facility near Wilmington, N.C.) It should also be remembered that the total cost of fuel fabrication and reprocessing is small compared to the total cost of nuclear energy, and the cost of dynamic materials control should, in turn, be only a small part of the cost of operating bulk processing facilities. If, for example, fuel cycle costs are increased by 5%, which we believe is probably an extreme upper limit, the cost of nuclear energy will be increased by only about 0.3%, which should be well within profitability margins.

More importantly, we again ask, "What are the alternatives?". Most use-denial concepts, spiking, pre-irradiation, etc.) can greatly complicate operational procedures, personnel shielding, safety, fission product retention etc., all of which can in turn lead to increased capital and operational costs. Many use-denial concepts can also introduce severe safeguards problems. If the material is less accessible for processing, then by the same token it is less accessible for verification. We suggest that the alternative which the critic usually has in mind is a minimum quality conventional system capable, at best, of achieving W and T values which are an order of magnitude larger than is needed to meet even minimum safeguards goals. Compared to other alternatives which achieve similar W and T values, we believe the cost of dynamic materials control will not be excessive.

B. Measurement Technology

It is sometimes alleged that the on-line measurement technology assumed in dynamic materials control studies does not truly exist. In part this is true. Although impressive progress has been made in the field of measurement technology - both destructive and non-destructive - there are many important materials measuring problems still to be solved, and in-line or at-line measurement instruments still to be developed. Certainly also, much of the newly-developed technology remains to be demonstrated in actual in-plant usage. There is much work still to be done.

In many cases, however, what is needed is not more laboratory experiments, but actual installation and in-plant evaluation. The demonstration of in-plant capability can only come from practical experience, in an operating plant environment, and over an extended period of time.

C. Verification

Then finally, we acknowledge an important question mark, "How is the international safeguards authority to verify a dynamic materials control system?" This verification clearly is an essential feature of any dynamic materials control system if it is to function as an effective international safeguards guarantee.

In answering this question, it is important first to separate out two important aspects of dynamic materials control, but which are not unique to it. The first is the question of detection time. If safeguards systems are to be designed to detect the possible abrupt diversion of significant quantities within a few days after the event (and current controversies seem to relate more to the definition of "a few days" rather than to the concept) then the facilities to be so safeguarded will require continuous or near-continuous inspector presence. The verification of a dynamic materials control system will require a continuous inspector presence, but so will any other system which achieves T = 10 days or less. There may be a potential need for a significant increase in inspection staff capabilities and professional requirements, but that need is related to the definition of safeguards goals, not to dynamic materials control.

Second, the verification of a single NDA measurement is not a simple task. With destructive measurements one can pretend that he has an independent calibration of the bulk measurement system, send a portion of the facility operator's sample to an independent laboratory, and claim that he has performed an independent verification. With NDA measurements there is no sample to be sent off for independent analysis. Nearly all NDA instruments now have some sort of associated data processing, whether a built-in mini-processor or a connection to a convenient central processor. The only verification commonly discussed requires the inspector to have independently prepared calibration standards, and since these must be facility, material, and geometry specific, their preparation is not an insignificant task.

As with continuous inspection, however, the verification of NDA measurements is a function of current developments in safeguards measurement technology, not of dynamic materials control. The same independent calibration standards, if that is the approach which must be used, are required for one NDA measurement as for one hundred.

If the safeguards authority has arranged for the necessary continuous or near-continuous inspection effort and has solved the problems inherent in the verification of NDA measurements, we believe that they have solved the major portion of the dynamic materials control verification problem. We do not pretend that this problem is solved, and indeed we acknowledge the existence of a cooperative effort to explore potential solutions in greater detail. We do believe, however, that the major portion of the problem is independent of dynamic materials control itself, and must be solved in any case.

VI. CONCLUSION

There is today an increasing awareness and appreciation of the global nature of the international safeguards problem and the vital importance of effective international safeguards. Clearly the over-all goal for safeguarding the world nuclear industry is an ensemble of effective national systems meeting certain broad concensus standards, with an overlay of truly effective international safeguards inspection and verification. Hopefully those "concensus standards" can be stated in terms of performance criteria - in terms of W and T as defined in this paper. We are, perhaps, a long way from this goal, and its achievement will require much dedication and hard work. Nevertheless, as was stated early in the paper, nuclear energy is essential, and international safeguards are essential to nuclear energy. As a conclusion to this paper, it seems appropriate to review briefly the coming dramatic changes in international safeguards implementation over the years ahead [26].

1) As more facilities come under IAEA safeguards, there will be further large increases in the volume of information that must be gathered, assimilated and analyzed.

2) Pursuant to "NPT safeguards", new types of large, high-throughput facilities located in large industrial nations will come under IAEA safeguards for the first time. Such key fuel cycle facilities include isotope separation plants, spent fuel reprocessing plants, conversion and fuel fabrication plants producing mixed-oxide fuels, and large critical assembly facilities.

3) With the rapid increase in the number and size of facilities under international safeguards, the IAEA will be required to deal with complete nuclear fuel cycles within individual nations, or with closely coupled international/regional groups of nations.

4) For a variety of technical and economic reasons, including - in addition to international safeguards - operational efficiency, quality and process control, radiological and criticality safety, and economics, these large future facilities will turn more and more to timely, on-line materials measurement and accounting systems.

Given the growing trend toward automation and increased sophistication in nuclear materials measurement, processing and handling systems in today's competitive worldwide nuclear industry, effectively safeguarding that industry is clearly a challenge. Along with the challenge comes an important new opportunity, inasmuch as these same advanced materials accountancy systems can provide far more incisive knowledge - both in time and in space - of plant inventory. This knowledge must, of course, be fully available to the inspector as well as to the plant operator. To further strengthen independent verification capabilities, new techniques and procedures must be and are being investigated.

Whatever nuclear fuel cycles are to be pursued in various countries, groups of countries, or regions of the world, it seems certain that the already large demands on IAEA inspection and safeguards verification will continue to increase, both in volume and in complexity. Under these conditions, full exploitation of the inherent capabilities and advantages of modern materials control systems, properly deployed and independently verified, may ultimately prove to be the only practical means of implementing effective international safeguards.

REFERENCES

[1] INTERNATIONAL ATOMIC ENERGY AGENCY, Nuclear Power and its Fuel Cycle (Proc. Conf. Salzburg, 1977) **1–8**, IAEA, Vienna (1978).

[2] SHIPLEY, J.P., Trans. Am. Nucl. Soc. **27** (1977) 178.

[3] SHIPLEY, J.P., in Proc. Nineteenth Annu. Meeting of Inst. Nucl. Materials Management, Cincinnati, J. Inst. Nucl. Mater. Manage. **8** 3 (1978).

[4] COBB, D.D., SMITH, D.B., in Proc. Eighteenth Annu. Inst. Nucl. Mater. Manage. Meeting, Washington D.C., J. Inst. Nucl. Mater. Manage. **7** (1977).

[5] DAYEM, H.A., COBB, D.D., DIETZ, R.J., HAKKILA, E.A., KERN, E.A., SHIPLEY, J.P., SMITH, D.B., BOWERSOX, D.R., Los Alamos Rep. LA-7011 (1978).

[6] STEWART, J.P., in Safeguarding Nuclear Materials (Proc. Symp. Vienna, 1975) **1**, IAEA, Vienna (1976) 341–45.

[7] VAUGHN, C.M., WALKER, H.F., "A safeguards and cost effective alternate to biannual uranium physical inventories", Proc. Nineteenth Annu. Inst. Nucl. Materials Management, Cincinnati, J. Inst. Nucl. Mater. Manage. **8** 3 (1978).

[8] ELDRED, D., General Electric, Sunnyvale, Personal communication to G.R. Keepin, Aug. 1975.

[9] SEABAUGH, P.W., ROGHERS, D.R., WALTERMANN, H.A., FUSHINI, S.C., CINAMELLA, A.F., Mound Laboratory Rep. MLM-MU-77-68-0001 (1977).

[10] SEABAUGH, P.W., WHITE, L.S., "Application of CUA to analyzing safeguards measurement systems", Proc. Nineteenth Annu. Inst. Nucl. Mater. Manage. Meeting, Cincinnati, J. Inst. Nucl. Mater. Manage. **8** 3 (1978).

[11] KERSTEEN, C., "Summary of combustion engineering fuel accountability and control system (FACS)", *ibid.*

[12] PAUL, R.N., YAN, G., "INMACS: A computer assisted on-line fissile material inventory and criticality control system," *ibid.*

[13] WILSON, W., "Y-12 program overview", in Transcript, Safeguards Topical Workshop, LASL, 18–19 Jan. 1977.

[14] BARNES, G.L., DARBY, D.M., DUNIGAN, T.H., "Dynamic SNM control and accountability system prototype station", Proc. Nineteenth Annu. Inst. Nucl. Mater. Manage. Meeting, Cincinnati, J. Inst. Nucl. Mater. Manage. **8** 3 (1978).

[15] BARTLETT, J., FEDERICK, J., "Rocky flats program overview" in Transcript, Safeguards Topical Workshop, LASL, 18–19 Jan. 1977.

[16] CHANDA, R., "Rocky flats NDA system", *ibid.*

[17] JARSCH, V., ONNEN, S., DOLSTER, F.J., WOIT, J., "An approach to a generalized real-time nuclear materials control system", Proc. Nineteenth Annu. Inst. Nucl. Mater. Manage. Meeting, Cincinnati, J. Inst. Nucl. Mater. Manage **8** 3 (1978).

[18] SARICH, J.P., Nucl. Mater. Manage. **4** 3 (1975) 464–67.

[19] MISHIMA, T., AOKI, M., SHIGA, K., MUTO, T., AMSNUMA, T., in Proc. Am. Nucl. Soc. Topical Meeting on Back End of the LWR Cycle, Savannah, 19–23 March 1978.

[20] WORKMAN, G.D., Nucl. Mater. Manage. **6** 3 (1977) 302–11.

[21] AUGUSTSON, R.H., in Proc. Seventeenth Annu. Inst. Nucl. Mater. Manage. Meeting, Seattle, J. Inst. Nucl. Mater. Manage **5** 3 (1976) 302–316.

[22] KEEPIN, G.R., MARAMAN, W.J., in Safeguarding Nuclear Materials (Proc. Symp. Vienna, 1975) **1**, IAEA, Vienna (1976) 305–20.

[23] AUGUSTSON, R.H., LASL Rep. LA 7126-MS (1978).

[24] AUGUSTSON, R.H., BARON, N., FORD, R.F., FORD, W., HAGEN, J., LI, T.K., MARSHALL, R.S., REAMS, V.S., SEVERE, W.R., SHIRK, D.G., IAEA-SM-231/101, these Proceedings, Vol. II.

[25] McSWEENEY, T.I., JOHNSTON, J.W., SCHNEIDER, R.A., GRANDQUIST, D.P., BNWL-2098 (1975).

[26] KEEPIN, G.R., in Pacific Basin Fuel Cycle Conference (Proc. 2nd. Int. Conf. Tokyo, 1978), Am. Nucl. Soc., to be published; see also J. Nucl. Mater. Manage. **7** 3 (1978).

DISCUSSION

D. GUPTA: Your paper contains valuable information on this interesting system. The principle behind the system is very important for the attainment of IAEA safeguards objectives, but I should like to mention a number of aspects which seem to require further clarification:

(1) Containment/surveillance (C/S) measures: your paper gives the impression that C/S measures may be useless for international safeguards, whereas a number of other papers presented by the Agency give an exactly opposite impression;

(2) Normally, the influence of the systematic error on material balance uncertainty persists regardless of the frequency of inventory taking. How could a diversion strategy based on this error component be detected by the system you describe?

(3) The verification of this system may present the Agency inspectors with an enormous problem. They will have to be perfectly familiar with the facility operation, the functioning of the NDA systems and the data processing system; and they will have to know the implications of these systems in regard to the IAEA safeguards system.

J.E. LOVETT: These are very important questions, especially the one concerning verification. I will try to give the best answers I can in the time available.

First, C/S measures are by no means useless for international safeguards. They are extremely valuable as a means of preserving the integrity of previously verified data. However, when used in this manner they do not *improve* safeguards, they only make safeguards *easier*. The uncertainty in any quantitative statement prepared by the Agency is exactly the uncertainty in the "previously verified data" whose integrity has been preserved. In the last year or so it has been suggested that C/S measures could be used in other ways that would in fact improve safeguards. This matter is still under study, but I personally am sceptical, and it is this scepticism which is reflected in our paper.

With regard to your second question, one must distinguish very carefully between bias itself and systematic uncertainty, which is the extent to which one does not know whether a bias exists. In dynamic materials accountancy each MUF is compared against past history rather than against zero. Since MUFs are obtained daily or weekly and recalibrations occur much less frequently, systematic uncertainty is not an element in the evaluation equations. Bias very definitely is an element in the equations, and one of the ultimate tests of the concept will be our ability to keep bias under control.

The verification of *any* safeguards system which achieves detection times of the order of days or weeks will present the Agency with a significant problem, and I acknowledge that we do not at the moment have any ready answers. The feasibility of dynamic materials accountancy is still being studied and demonstrated, and although the verification problem has been recognized, it has not as yet been given the study it unquestionably requires. Mr. Keepin and I do believe, however, that the problem is a general one, not one which is unique to dynamic materials accountancy. We also believe that ways must be found to substitute technology for sheer numbers of inspectors, and that our concept offers considerable promise of being able to do that.

A. PETIT: What we have just heard illustrates the risk of error due to confusion between the two meanings of the word "safeguards", i.e. whether it is taken to signify controls exercised by an operator or Government, or international safeguards related to non-proliferation. The Agency is competent in regard to the latter only. In the case of operator or Government controls, it is easy to appreciate the value of dynamic systems for detecting thefts by individuals or sub-national groups. In the case of international safeguards, however, dynamic systems may be completely useless, since any State which had decided to carry out diversion would only need to make an intentional programming error, or to feed false data into the system, in order to make it ineffective.

J.E. LOVETT: This question again relates to the problem of verification, and I do not agree that dynamic materials accountancy is likely to be of value only as an operator or State control system. Conventional materials accountancy is installed and maintained by the facility operator, and verified by the international safeguards inspector. Dynamic materials accountancy likewise is installed and maintained by the facility operator, and must be verified by the international safeguards inspector. Thus the two systems, while significantly different in detail, are in fact not different in terms of fundamental concepts. The quality of safeguards today is largely determined by the quality of the accountancy system maintained by the facility operator, and the same will be true of any dynamic materials accountancy system. Moreover, I believe that the much larger quantity of accountancy data generated, and the almost infinite number of cross-checks which are possible, will in fact make the falsification of computerized dynamic record systems more difficult rather than easier.

G.Robert KEEPIN *(Chairman)*: In consideration of the points just made by Mr. Petit regarding confusion of terminology, it should be noted that the term *accountancy* has been used in our paper interchangeably with *control* (See Section III entitled "Dynamic materials accountancy"). Thus, an advanced materials control system is not in any way "fed back" into the control of plant processes, but rather is designed to provide information on material inventory status in a facility on as accurate and timely a basis as possible. This information is then used by the responsible nuclear safeguards and materials management people in the plant for continuous assessment of the safeguards status of the plant and to advise plant management of appropriate response options and recommended actions. Needless to say, a major consideration in the design of an acceptable dynamic materials control system is the scrupulous avoidance of any significant interference with process operations and plant production. The workability and effectiveness of any system can be convincingly demonstrated only through extensive in-plant operation and evaluation, and a number of the dynamic systems shown in Table I of our paper are currently undergoing this process.

It is certainly time, as Mr. Petit has indicated, that advanced material control systems, such as DYMAC, are responsive to the needs of a national system of safeguards and control. However, as has been stated many times, an effective international safeguards system must ultimately be based on an ensemble of effective national systems, which in turn are placed under an umbrella of credible and independent verification by the international safeguards authority.

S.V. KUMAR: You say that it is not enough to perform about two physical inventories annually. In that case, don't you think that verification will have to be carried out more frequently, as in the dynamic inventory concept? This would require considerable effort on the part of the operator and the inspector in order to reconcile the values obtained, and it could possibly mean that the operator would be almost continuously engaged on this exercise.

J.E. LOVETT: I have said that annual or semi-annual physical inventories are not acceptable to anyone, and I believe that. They clearly are not acceptable to those who argue for detection times of the order of one or two weeks. Even those who argue against such short detection times would agree, I believe, to a compromise of the order of one or two months, if they thought that such an inventory frequency could be achieved without virtually crippling process operations. Annual or semi-annual physical inventories are necessary as base points, but they cannot be the first line of defence in international safeguards.

As I stated in answer to the question by Mr. Gupta, *any* safeguards system which achieves short detection times implies an increased inspection effort. The means must be found, however, to make this an increased technological effort, not a simple increase in the number of inspectors. The problem is fully recognized. The answers, unfortunately, are not so obvious.

AN APPLICATION OF THE CONTROLLABLE UNIT APPROACH (CUA) TO THE ANALYSIS OF SAFEGUARDS MEASUREMENT SYSTEMS

P.W. SEABAUGH, D.R. ROGERS, H.A. WOLTERMANN,
F.C. FUSHIMI, A.F. CIRAMELLA
Mound Facility*,
Miamisburg, Ohio,
United States of America

Abstract

AN APPLICATION OF THE CONTROLLABLE UNIT APPROACH (CUA) TO THE ANALYSIS OF SAFEGUARDS MEASUREMENT SYSTEMS.

The controllable unit approach (CUA) is a material control and accountability methodology that takes into account the system logic and statistical characteristics of a plant process through the formulation of closure equations. The methodology is adaptable to plant processes of varying degrees of design and operational complexity. No alteration or modification of a process is required to apply the methodology. Cost/benefits of refinements in, or changes to, the proposed measurement system are obtained as incremental cost. To encourage improved safeguards accountability, the United States Nuclear Regulatory Commission (NRC) has been considering the use of performance-oriented regulations to supplement those currently used. The study, sponsored by NRC/Office of Standards Development, evaluated CUA methodology to meet performance-oriented regulations. For this study, the criterion is defined as the detection of a material loss of two kilograms of SNM with 97.5% confidence. Specifically investigated were the timeliness of detection, the ability to localize material loss, process coverage, cost/benefits, and compatibility with other safeguards techniques such as diversion path analysis and data filtering. The feasibility of performance-oriented regulations is demonstrated. To use the system of closure equations fully, a procedure was developed to integrate formally the effect of both short-term and long-term closure equations into an overall systems criterion of performance. Both single and multiple diversion strategies are examined in order to show how the CUA method can protect against either strategy. Quantitative results show that combined closure equations improve the detection sensitivity to material loss, and that multiple diversions provide only diminishing returns.

INTRODUCTION

The controllable unit approach [1] (CUA) is a material control and accountability methodology that takes into account the system logic and statistical characteristics of a plant process

---

* Mound Facility is operated by Monsanto Research Corporation for the US Department of Energy under Contract No. EY-76-C-04-0053.

through the formulation of closure equations. These material balance equations model inputs, outputs, inventories, holdups and possible losses or diversions. They depend upon fixed measurement points in the process. The methodology is adaptable to plant processes of varying degrees of design and operational complexity, exemplary of present and future facilities. Application of the method does not require alteration or modification of an applicant's process. Because base-case calculations are a natural first step in the evaluation scheme, the cost/benefits of refinements in, or changes to, the proposed measurement system for purely safeguards purposes are easily obtained as incremental cost.

To encourage improved safeguards accountability, the Nuclear Regulatory Commission (NRC) has been considering the use of performance oriented regulations to supplement those currently used. In the area of material control and accountability, for instance, one such performance oriented criterion could be the assurance that a given loss of material be detected within a specific time frame. The present study, sponsored by NRC/Office of Standards Development, evaluated CUA methodology to meet performance oriented regulations. For purposes of this study, the criterion is defined as the detection of a material loss of two kilograms of SNM with 97.5% confidence. Specifically investigated were: the timeliness of detection, the ability to localize material loss, process coverage, cost/benefits, and compatibility with other safeguards techniques such as DPA (diversion path analysis) and data filtering. In addition, this study was undertaken as a first step in providing the NRC with the methodology and information to:

1. Support development of safeguards regulations that emphasize performance requirements,

2. Assess license applications, and

3. Inspect processes.

Like many successful management systems, the CUA methodology iteratively compares the actual situation to the need. In this study, the performance of the proposed or existing measurement system is compared to the material control criterion. Then, additions or refinements to the measurement system or process are iteratively compared to the criterion until it has been met. This systematic comparison can efficiently ensure that a complicated process measurement system will perform to the level as specified by the need. Furthermore, because the existing or proposed system is mathematically modeled with the CUA method, modifications to the process for any reason can be tested quickly for their effect on material control before implementation. A summary flow diagram of the CUA methodology is shown in Figure 1.

PROCESS MODEL

A process model [1] was developed to provide a severe test of controllable unit methodology. The process model was based primarily on a commercial high-throughput (200 MT) mixed-oxide

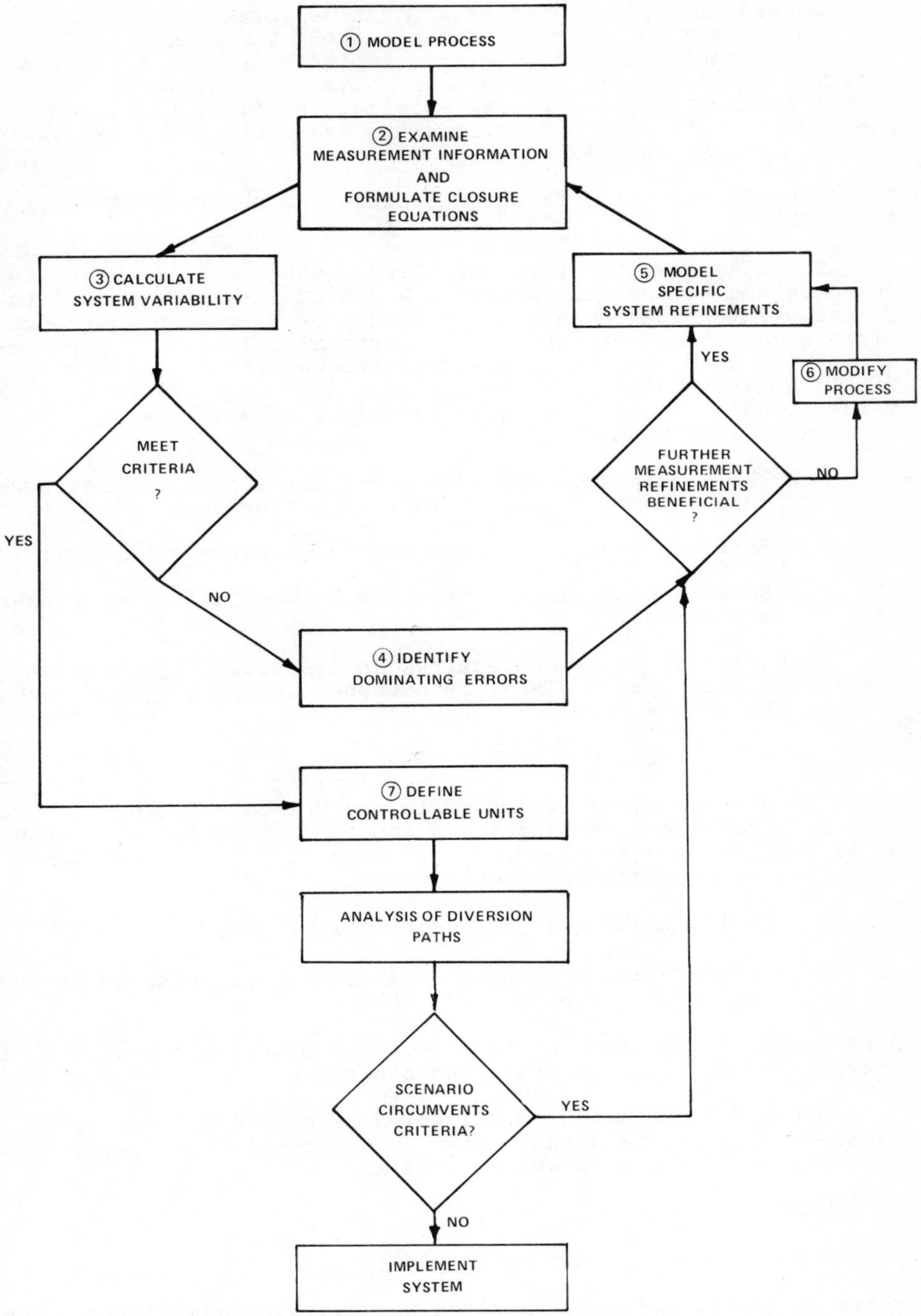

FIG.1. CUA methodology is a systematic approach to material control.

fuel fabrication plant similar to that proposed by Westinghouse [2] and further described by Science Applications Inc. [3]. Modeling techniques were developed to include as much realism into the model process as possible. Simultaneous operation modes, randomly varied process streams, and flow, time, and event-dependent hold-up functions are examples of the realistic features of the process model.

EFFECTIVENESS OF MULTIPLE CLOSURE EQUATIONS AND VULNERABILITY ASSESSMENT

To fully use the system of closure equations generated to describe and control the process, a procedure was developed to formally integrate the effect of both short-term and long-term closure equations into an overall systems criterion of performance. The objective is to maximize the detection sensitivity within a given detection time period. In this assessment of the value of using multiple closure equations, the following situations were accounted for:

1. The combination of independent non-overlapping closure equations to obtain an overall performance criterion;

2. Possible overlap between several closure equations;

3. Possible correlated variables between different closure equations.

Each closure equation requires an associated hypothesis test procedure. For explanatory purposes assume a closure equation of the simple form

$$M_1 = X_1 - X_2,$$

with the distributions of errors in measurements X_1 and X_2 normal with zero means and variances σ_1^2 and σ_2^2 respectively. Then M_1 is $N(0, \sigma_1^2 + \sigma_2^2)$ assuming no losses in this control area. Next form a one-sided hypothesis test:

H_0: Null Hypothesis (No Diversion) $M_1 \leq L_1$

H_1: Alternative Hypothesis (Diversion has occurred at the Q_1 level) $M_1 = Q_1$

A threshold C_1 is obtained to effect the hypothesis test. H_0 is <u>accepted</u> if $M_1 < C_1$, and rejected otherwise.

Let M_1 be the measured CEI (closure equation imbalance), whereas CEI_1 is the <u>actual material diverted</u>; then Type-1 (α_1) and Type-2 (β_1) errors are defined as:

$$P(M_1 > C_1/CEI_1 = L_1) = \alpha_1$$

$$P(M_1 > C_1/CEI_1 = Q_1) = (1 - \beta_1);$$

where α_1 is the probability given no diversion has occurred that the measured CEI (M_1) exceeds the threshold C_1, and is sometimes

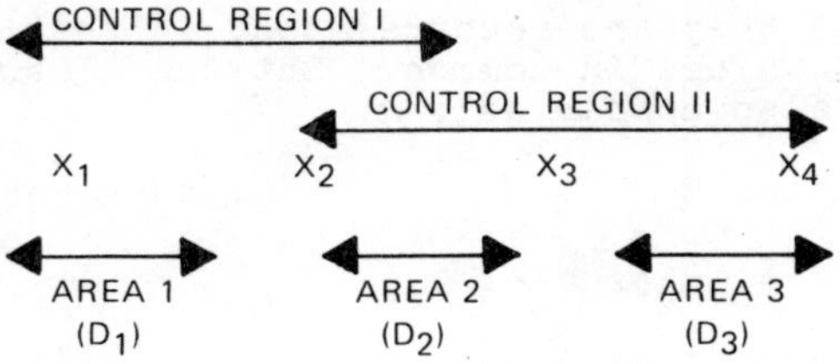

FIG.2. A schematic for the overlapping but uncorrelated closure equation model shows the control regions and diversion areas.

called the "false-alarm" error probability; and $(1 - \beta_1)$ is the probability given a diversion at the Q_1 level that the measured CEI (M_1) exceeds the threshold C_1, and is referred to as the "power" of the test. The probability that a diversion has occurred but the hypothesis test has failed to detect it is given by β_1.

Assume for each control region that α, L, Q, and measurement error variances are given, so that the error β and the required threshold C can be obtained from the two equations presented above. It will be assumed for the analyses that each control region hypothesis test is designed independently, and thus all of these parameters are determined before computing overall error probabilities.

For the case of n non-overlapping independent closure equations (situation one), the overall probability that a diversion at a given level has occurred, but has not been detected by any of the closure equations, can be obtained as

$$\text{Overall Probability of nondetection} = \prod_{i=1}^{n} \text{Prob} \begin{bmatrix} \text{nondetection in the } i^{th} \\ \text{closure equation} \end{bmatrix}$$

$$= \prod_{i=1}^{n} \beta_i$$

The product of the β_i follows from the observation that all n potential independent diversions must go undetected for overall nondetection.

A schematic for overlapping closure equations is shown in Figure 2. If a diversion D_2 should occur in area 2, it could be detected by either closure equation. Yet closure equations I and II may or may not be correlated depending upon their specific form. For example, if they are of the form:

Model A	Closure Equation
$M_1 = X_1 - X_3$	I
$M_2 = X_2 - X_4$	II

it is clear that they are uncorrelated, in that they do not have any measurement values in common. But the closure equations in Figure 2 might also appear as:

Model B	Closure Equation
$M_1 = X_1 - X_2 - X_3$	I
$M_2 = X_2 + X_3 - X_4$	II

in which case M_1 and M_2 are clearly correlated when they share measurements X_2 and X_3 in common. In such cases, if the underlying measurement errors are Gaussian the resulting closure equation imbalances M_j can be described by multivariate Gaussian distributions. Probability calculations are considerably more complicated, but can be evaluated numerically.

On the other hand, whether the closure equations are correlated or not, the effect of overlap can be handled simply. In Figure 2, assume diversions D_1, D_2 and D_3 have occurred in areas 1, 2, and 3 respectively. Then since it is assumed that all measurement errors have zero mean Gaussian distributions, the distributions for M_1 and M_2 have means of $(D_1 + D_2)$ and $(D_2 + D_3)$ respectively. Thus the controlling effect of the overlap shows up in a relatively simple way in the imbalances, and a diversion in area 2 shows up in the means of <u>both</u> M_1 and M_2. Because this diversion D_2 occurs in both means, it is much more liable to be detected.

As an example consider the determination of the overall probability for Model A given diversions D_1, D_2, and D_3 in areas 1, 2, and 3 respectively shown in Figure 2. A worst case analysis requires examination of all possible partitions of Q into D_1, D_2, and D_3 where

$$Q = D_1 + D_2 + D_3$$

and $$D_1 + D_2 \geq Q_1$$

and $$D_2 + D_3 \geq Q_1.$$

If either of these inequalities is <u>not</u> valid, then we will consider that <u>no diversion has occurred in the corresponding region at all</u>, and a much simpler problem results.

With these assumptions the maximization of the probability of nondetection is given by

$$Z = \max_{\substack{D_1, D_2, D_3 \\ D_1 + D_2 + D_3 = Q}} \left[\int_{-\infty}^{C_1} f_1(M_1)\, dM_1 \int_{-\infty}^{C_2} f_2(M_2)\, dM_2 \right]$$

which rewritten in terms of the usual normal distribution notation using normalized variables becomes

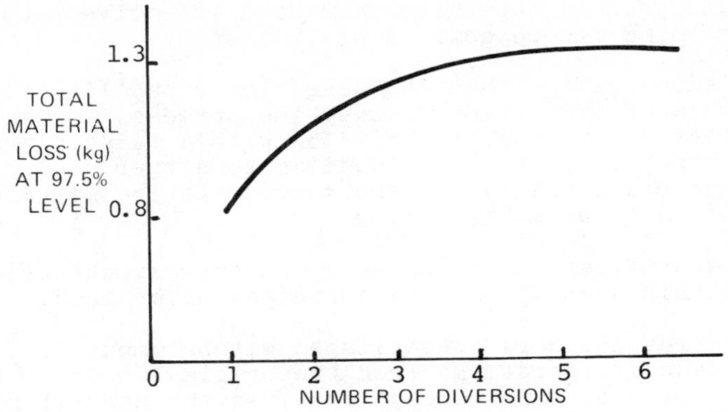

FIG.3. Computations show that multiple diversions reach a point of diminishing returns.

$$Z = \Phi\left(\frac{C_1-(Q-Q_2)}{\sigma_I}\right) \cdot \Phi\left(\frac{C_2-Q_2}{\sigma_{II}}\right)$$

where for brevity $\sigma_I^2 = \sigma_1^2 + \sigma_3^2$

$$\sigma_{II}^2 = \sigma_2^2 + \sigma_4^2.$$

A numeric example is selected using data from the first stage of the mixed oxide process, controlling this process from the initial weighing of incoming nuclear material until material enters the MO_2 subblend silo. Only short term closure equations [1]

S-1, S-3, S-4, S-5, S-9, and S-10

are considered. For the specific data, see Table 4.8 and Table I.1 in Reference 1.

These results are shown in Figure 3 where it is clear that multiple diversions provide only diminishing returns for the potential divertor even without the increased risk and logistic difficulty taken into account.

RESULTS

Comparative results for CUA [1] and MUF/LEMUF [4] as applied to the mixed-oxide process are given which show that CUA provides an improvement factor of 3 for detection sensitivity and a greater improvement for timeliness of detection.

Results to date indicate that the methodology will be highly effective in timely detection of SNM material loss and in material control. Specifically through the CUA methodology accountability and process data have been used effectively to meet the following principle objectives of this study.

- Demonstration that the detection capability for material loss of SNM in the mixed-oxide process is 2 kg at a detection probability of 97.5% with a false alarm rate of three per year. This applies to either a single-event material loss or to random accumulative material losses up to a two-month period.

- Identification of the area and approximate time (generally within a shift) of the suspected diversions.

These results were accomplished without modification of the plant process or operations from the original model. Furthermore, the application of this concept provides the user with the added benefit of estimating the cost and effectiveness of additional measurements or measurement points anywhere in the process. Currently, CUA is being applied to an existing high-throughput operating plant.

REFERENCES

[1] SEABAUGH, P. W., ROGERS, D. R., WOLTERMANN, H. A., FUSHIMI, F. C., CIRAMELLA, A. F., The Controllable Unit Approach to Material Control: Application to a High Through-put Mixed-Oxide Process, Mound Facility Rep. MLM-NUREG-2532 (1978).

[2] "Westinghouse License Application for the Recycle Fuels Plant at Anderson, S. C.," USAEC Docket No. 70-1432 (July 1973).

[3] "An Evaluation of Real-Time Material Control and Accountability in a Model MO_2 Fuel Plant," Science Applications, Inc., SAI-75-648-1J (September 15, 1975).

[4] GLANCY, J. E., "Safeguards implementation practices for a mixed oxide recycle fuel fabrication facility," Science Applications, Inc., La Jolla, California, BNL-21409, Brookhaven National Laboratory, Upton, N.Y. (May 1976), 109 pp.

DISCUSSION

E.R. MORGAN: I was interested to hear that the controllable unit approach (CUA) is not limited to any particular measurement method. Would you please further explain the differences between the controllable unit approach and other methods of material control which have been discussed.

P.W. SEABAUGH: First of all, in contrast to a series of MBAs, the use of CUA multiple closure equations allows material balances to be closed in both time and space senses. An example is the use of two equations which are overlapping but which close in different time periods to monitor the same area. Secondly, as indicated in our paper, the multiple closure equations allow more flexibility in the use of the data. You can formulate equations that are non-overlapping and use uncorrelated data; non-overlapping and use correlated data; overlapping and use uncorrelated data; and overlapping and use correlated data.

Also in contrast to the method just presented by Mr. Lovett[1], CUA – by using both long-term and short-term closure equations – provides a straightforward way of protecting against both abrupt and non-random trickle diversions.

Before proceeding, I should like to correct an error in Table I ("Some dynamic materials accounting systems") of paper IAEA-SM-131/133[1] because it illustrates another difference. Contrary to what is stated in that Table, CUA can and does include NDA methods. Their use, or for that matter the use of any other technique, is evaluated in a systematic manner. CUA starts with what is already proposed or attained and systematically evaluates improvements in a practical, cost-effective way. By identifying *controlling errors,* CUA focuses on the areas of major concern so that efforts and techniques and/or methods such as NDA can be directed to areas of major concern and not misdirected to those areas already well covered or of minor impact. It thus enables you to selectively upgrade and clearly establish a set of priorities.

Another difference is that the emphasis of MBA methods is on accountancy whereas the emphasis of CUA is on diversion detection, which is the mission of Agency inspectors. With the CUA methodology an inspector can use his own set of closure equations which can be different from the facility's closure equations and which can be further varied as the Agency chooses.

[1] KEEPIN, G. Robert, LOVETT, J.E., IAEA-SM-231/133, these Proceedings, Vol.II.

A DEVELOPMENT, TEST AND EVALUATION PROGRAMME FOR DYNAMIC NUCLEAR MATERIALS CONTROL*

R.H. AUGUSTSON, N. BARON, R.F. FORD,
W. FORD, J. HAGEN, T.K. LI, R.S. MARSHALL,
V.S. REAMS, W.R. SEVERE, D.G. SHIRK
University of California,
Los Alamos Scientific Laboratory,
Los Alamos, New Mexico,
United States of America

Abstract

A DEVELOPMENT, TEST, AND EVALUATION PROGRAMME FOR DYNAMIC NUCLEAR MATERIALS CONTROL.*

A significant part of the Los Alamos Scientific Laboratory Safeguards Program is directed towards the development, test, and evaluation of dynamic nuclear materials control. The building chosen for the prototype system is the new Plutonium Processing Facility in Los Alamos, which houses operations such as metal-to-oxide conversion, fuel pellet fabrication, and scrap recovery. A DYnamic MAterials Control (DYMAC) system is currently being installed in the facility as an integral part of the processing operation. DYMAC is structured around interlocking unit-process accounting areas. It relies heavily on automatic non-destructive assay measurements made in the process line to draw dynamic material balances in near real time. In conjunction with the non-destructive assay instrumentation, process operators use interactive terminals to transmit additional accounting and process information to a dedicated computer. The computer verifies and organizes the incoming data, immediately updates the inventory records, monitors material in transit using elapsed time, and alerts the nuclear materials officer in the event that material balances exceed the predetermined action limits. The DYMAC system comes within the jurisdiction of the United States safeguards programme, and is under control of the facility operator. The system's advanced features will oblige the IAEA to upgrade its inspection capability. The central issue is how the IAEA can make use of the system's features yet maintain independent verification. This is the subject of a current study sponsored by the US-IAEA Technical Assistance Programme.

INTRODUCTION

A number of programs world-wide are examining techniques and technology for the purpose of up-grading nuclear materials accountability systems for safeguards purposes.(1) The Los Alamos Scientific Laboratory (LASL) in the United States has one such program.(2) The program's goals are three-fold: to develop features that improve the effectiveness of a nuclear

---

* Sponsored by the United States Department of Energy, Office of Safeguards and Security.

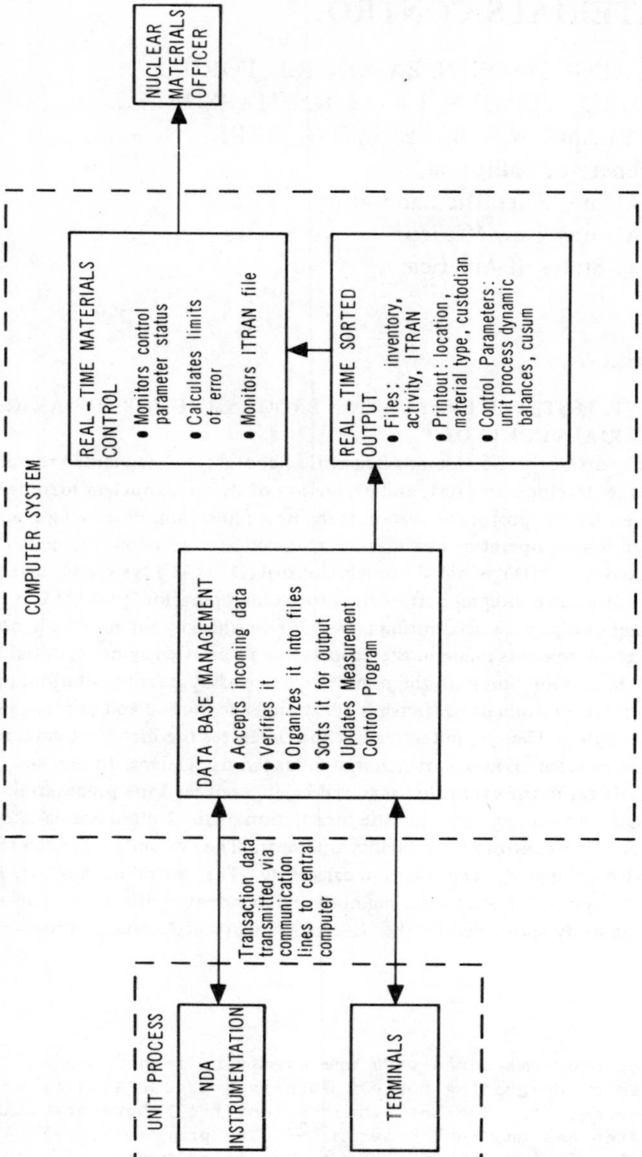

FIG.1. DYMAC system configuration.

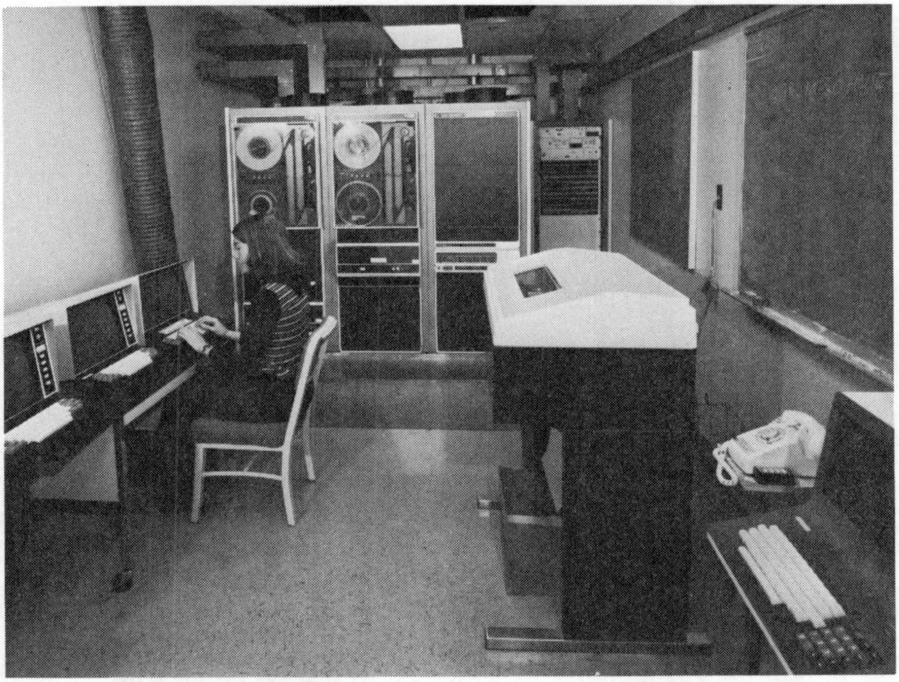

FIG.2. Data General Eclipse C330 computer with system consoles and line printer.

materials accountability system, to identify the technology necessary to implement such improvements, and to test and evaluate how a system built on these principles functions in a real processing environment.

Certain features can increase the effectiveness of a nuclear materials accountability system: (a) real-time updating of the inventory records, (b) incorporation of clock-time as part of the information contained in the records, (c) data verification upon entry into the records system, (d) unit-process structuring[3], (e) dynamic materials balancing around the unit-process structure, (f) monitoring of control limits based on these dynamic balances, and (g) incorporation of decision analysis[4] techniques in examining the balances with respect to the control limits.

To incorporate these features into a viable accountability system, we must turn to technology for in-line nondestructive assay instrumentation, interactive terminal data communication, modern computer hardware, and data base management software.

Real time implies that the information generated in the processing area describing the movement or change in form of nuclear material should be incorporated into the computer data base in as timely a fashion as possible. Thus, the "book" inventory is an up-to-date, accurate accounting of

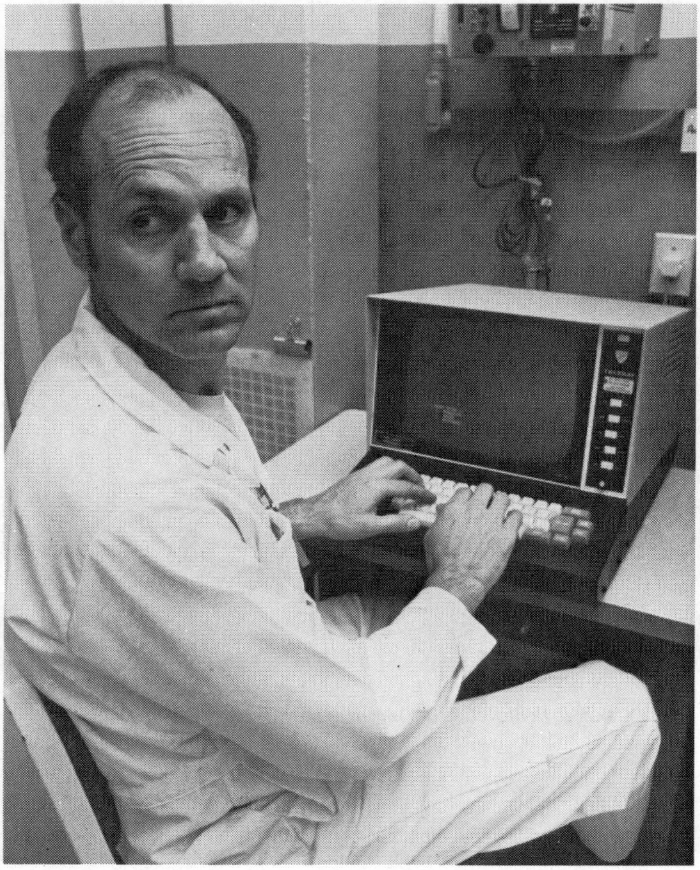

FIG.3. Teleray video terminal: the system displays messages and prompts on the screen to which the operator responds via the keyboard.

the material within the facility, and can be used for material control. Incorporation of clock time as part of each record in the data base permits such features as the monitoring of nuclear material in transit within the facility. For example, when material is sent from the vault to the processing area, the system can monitor the time it is in transit and notify the safeguards office if it fails to arrive in a predetermined time interval. By incorporating time as part of each record, an audit trail can be followed in chronological order. Drawing on the data base, it is possible to follow the movement of material through the facility.

Unit-process structuring consists of dividing the facility into material balance areas, which serve to localize the material within the facility both in space and time. As with any balance area, material is measured as it

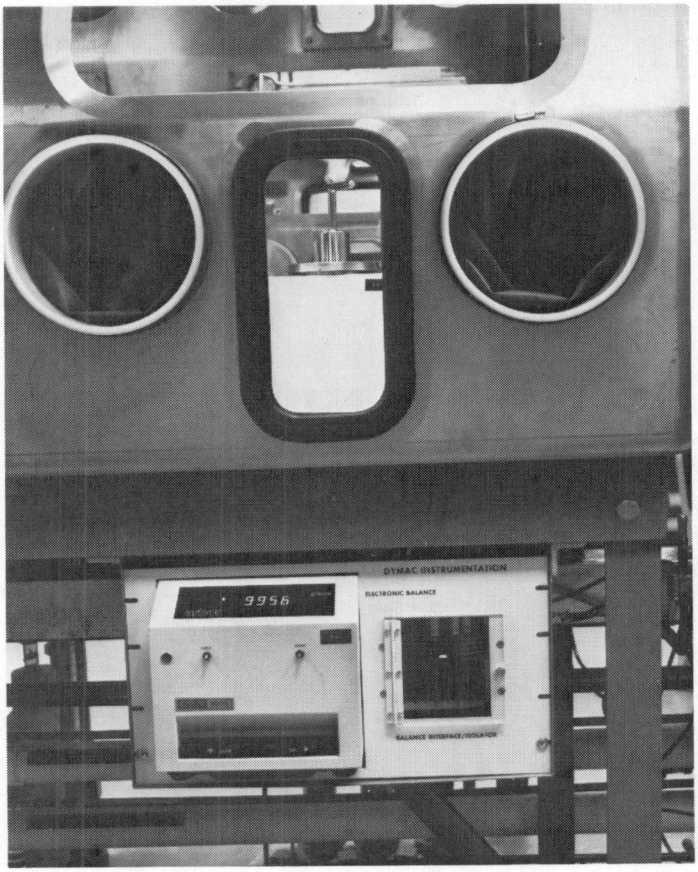

FIG.4. Digital electronic balance: the weighing unit is located inside the glovebox and the readout unit is installed beneath.

crosses the boundaries. At certain points in time, material balances are drawn and quantitatively examined for indications of material loss. In a facility based on batch balancing, these points in time are linked to the complete processing of a batch in that particular unit process.

The term "dynamic balancing" implies that a unit-process balance is drawn without stopping the process.[5] Thus the dynamic balance may not necessarily be zero, but contain contributions from material in process, unmeasured scrap and residues, and holdup, as well as the measurement uncertainties. Material control is maintained by comparing this unit-process balance with predetermined control limits. Control limits are set for the individual balances and for the cumulative sum (cusum) of these balances. Processing is allowed to continue as long as the balances and cusum values stay below the limits. When either limit is exceeded, the

FIG.5. Thermal-neutron coincidence counter installed on top of the glovebox line. An elevator extends down into the glovebox to convey material up into the counting chamber.

operator must stop and identify, on a measured basis, the cause of the problem. He can, for example, measure his scrap and residues, clean out the process area, or measure the residual holdup. If, after he takes these steps, the balance is not completely accounted for, then a MUF is declared.

Decision analysis techniques are being developed to analyze dynamic material balance data to provide the maximum sensitivity in detecting material loss, and to give a formal framework in which computer monitoring of the data might take place, eventually leading to a decision as to whether or not material is missing. These techniques, such as Kalman filtering, are described in Ref. 4.

TABLE I.

DYMAC EQUIPMENT

Equipment	Explanation
Data General Eclipse C330 computer	196,000 16-bit-word memory; two 10-megabyte disks; two 9-track tape units; one line printer; communications interface for 128 lines
27 Teleray video terminals	Cathode-ray tube devices with keyboards
6 Texas Instruments hardcopy terminals	
Texas Instruments label printer	
30 Arbor 5.5-kg capacity electronic balances	0.1 g precision; 0.3 g accuracy
5 Arbor 5.5-kg capacity electronic balances	0.01 g precision; 0.03 g accuracy
3 Mettler 15.5-kg capacity electronic balances:	0.1 g precision; 0.2 g accuracy
20 thermal-neutron coincidence counters	
5 solution assay instruments	gamma-ray spectroscopy
2 segmented gamma scanners	gamma-ray spectroscopy
1 random driver	fast coincidence; active interrogation

OPERATIONAL TEST OF DYNAMIC NUCLEAR MATERIALS ACCOUNTING

A dynamic materials control system, called DYMAC, is currently being installed at the Los Alamos Scientific Laboratory's new Plutonium Processing Facility. An operational test of the system is under way. DYMAC serves not only as the working materials accountability system for the facility, but also incorporates the features mentioned earlier. The new facility houses a variety of processing operations, including metal fabrication, electro-refining, fuel pellet fabrication, nitrate-to-oxide conversion, plutonium oxide powder preparation, and a number of cold scrap recovery operations. As of December 31, 1978, it will contain about 10,000 inventory items. As with any system, DYMAC is being installed in a specific facility and must

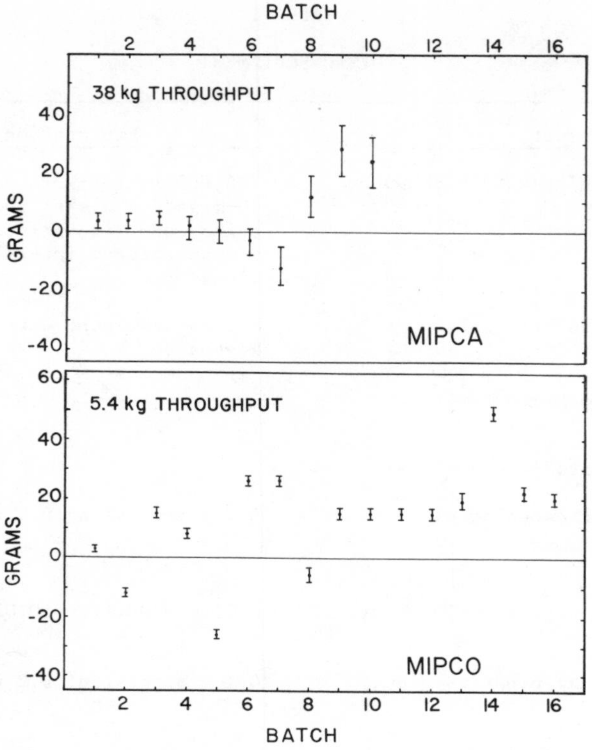

FIG.6. Cumulative sum (cusum) charts for dynamic material balances for two unit processes during the period 17 April through 11 June 1978.

accommodate itself to the design features of the processing areas. Although some of the system features are specific to the LASL facility, the system contains the generic principles necessary to improve materials accountability as a safeguards tool. Although the facility contains many different operations, with no one operation having a throughput comparable to fuel cycle facilities of the future, modeling and simulation techniques can extrapolate the operating experience obtained there to future facilities.(6)

DYMAC incorporates a number of features for an effective nuclear materials accountability system. As shown in Fig. 1, it has a unit-process structure, with nondestructive assay measurements made at the boundaries of those unit processes; information is communicated from the process area to a central computer system via interactive video terminals located in the process area; a mini-computer assembles the data, maintaining certain real-time files; and an accountability system structure focusses on unit-process balances so as to provide the material status indicators necessary for control. Figures 2-5 show various features of the DYMAC system: computer,

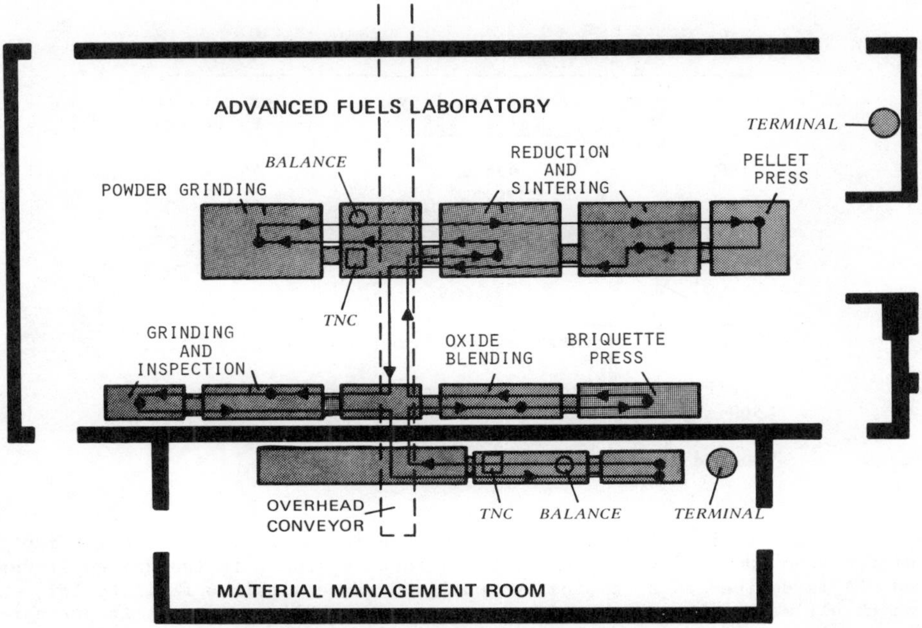

FIG.7. Pellet fabrication process in the advanced carbide fuels laboratory.

video terminal, electronic balance, and in-line thermal-neutron coincidence counter. Table I summarizes the DYMAC data processing and communications equipment and enumerates the installed NDA instrumentation.

The accountability system focusses its attention on three types of material balance. To begin with, the station balance gives the total quantity of nuclear material contained within the physical walls of the facility, broken down by each type of nuclear material. There are two independent ways to determine the station balance. The first is to take the difference between shipments and receipts of material into the facility. The Nuclear Materials Officer keeps these records in the form of shipping and receiving documents that accompany the items as they enter or leave the building. The second method of calculating the station balance is to add up the individual inventory items within the facility as recorded by the DYMAC accountability system. These two methods produce independently calculated values that should agree exactly. The Nuclear Materials Officer reconciles the two values at the end of each business day.

Coincidentally, a station balance is also available on the international level, in that the IAEA inspector has access to the shipping and receiving documents for each facility that he inspects. Thus, for facility inspection, one check is to add up all its inventory items and compare that value to what the shipping documents predict as the total inventory.

TABLE II.

VERIFICATION MEASUREMENTS OF PUO_2 FEED MATERIAL

Lot	Accountability Value (g)[a]	Verification Value (g)[b]
100	496.2	497 ± 10[c]
200	495.8	488 ± 10[c]
300	496.0	490 ± 10[c]

[a]Determined by total weight times a chemical factor.

[b]Verified using thermal neutron coincidence counting (500-second count time).

[c]Quoted uncertainties are 2σ values.

The second kind of material balance area is the formal MBA, which may differ from the national to the international system. In the United States, an MBA is defined as a physically contiguous area within a facility for which all material is measured as it crosses the MBA boundary. At periodic intervals, a closed material balance is drawn around that particular physical area. At the new LASL facility, there are 15 designated MBAs.

Under the DYMAC scheme, material balance areas are further subdivided into unit processes, for the third type of material balance. A unit process coincides with the actual physical process going on within its boundary; often several processes may be combined into a single unit-process area. Material flowing across the physical boundaries of the unit process is measured on a batch basis, and dynamic material balances are drawn for the unit process.

These material balances are not closed, in the sense that the balance can be drawn while there is still material left within the unit process. For example, generated scrap may exist in the unit process and, hence, will be included in the balance. Obviously, the holdup of material within the unit process will also constitute part of this balance. Hakkila et al.(7) have examined how dynamic balancing can work for a continuous process.

The DYMAC approach to materials control works well in a facility, such as the new one at LASL, that processes material on a batch basis. The unit-process balance is drawn only when a batch has completely passed through the unit process. Thus, the balance becomes a primary indicator of material status. The system monitors individual batch balances for each unit process as well as their cumulative sum. Such a cusum plot is shown in Fig. 6 for two unit processes at the new facility. The chart plots the individual points contributing to the cusum as a function of time. It shows a process upset that occurred in the MIPCA unit process for batch 8. Thus we see that the cusum plot not only gives information about the material status, but also about the process operation. This information, which is readily available to the process operator, enables him to run the process efficiently because he can quickly detect problems.

```
<<< SELECT OPTION >>>
     1...RECEIVE FROM MANAGEMENT ROOM
     2...TAKE SAMPLE
     3...RETURN FEED STOCK TO MANAGEMENT ROOM
     4...SEND SAMPLE OR SCRAP
     5...PREPARE NEW BATCH
     6...ADD CARBON AND UO2 TO PUO2 BATCH
     7...TRANSFER IN GLOVEBOX LINE (WITH MEASUREMENT)
     8...REDESCRIBE BATCH
     9...WEIGH BATCH
    10...SEND BRIQUETTES TO REDUCTION FURNACE
    11...RECEIVE AND VERIFY BRIQUETTES
    12...REDUCE BRIQUETTES
    13...BLEND LOT
    14...SEND SINTERED PELLETS TO GRINDING
    15...SEND SINTERED PELLETS
    16...WEIGH AND SEND FINISHED PELLETS
?____
```

FIG.8. Option list of transactions available to the operator in the advanced carbide fuels laboratory.

Such detailed information is what process and safeguards people find extremely useful for their own purposes. From the information in the data base, a running time history of material within the facility can be constructed. It is also possible to make use of the unit-process balances by combining them to help close material balances around the MBAs. Problems are quickly identified using dynamic balancing, and are easily traced to their point of origin. Thus, at physical inventory time when the process is closed down, the time spent in closing material balances should be significantly less than under current accountability procedures.[5]

A good example of how the DYMAC system actually works is the pellet fabrication process in the advanced carbide fuels laboratory. Figure 7 shows the flow of material among the six unit processes: oxide blending, briquette press, reduction and sintering, powder grinding, pellet press, and grinding and inspection. Using this unit-process structure, the material is localized to within a single glovebox. Feed material from a storage vault is introduced into the glovebox line in the material management room. The first processing step blends plutonium oxide, uranium oxide, and carbon powder. The powder is pressed into low-density briquettes, which are then sent to a furnace for reduction from oxide to the carbide chemical form. These low-density carbide briquettes are ground into powder, to which binder is added, and this mixture is sent to the pellet press. The low-density pellets are sintered into a high-density form and sent to grinding and final inspection.

The DYMAC system follows the movement of material through each unit process by means of transactions. A transaction is the mechanism for changing inventory information stored in the data base. The process operator enters information at a video terminal keyboard, like the one shown in Fig. 3, which is located in his process area. In answer to successive prompts that appear on the screen, he identifies the process operation he

INVENTORY IN ACCOUNT 711 6/12/78, 9:59

LOC	MT	ITEM ID	UP	SNM VALUE	BULK VALUE	SHLF	DESC	SEAL
G137	1C	3353A01	RS	22. G	23.83 G		CH3	
G137	1C	3353A02	RS	57. G	60.00 G		CH3	
G137	1C	3353A03	RS	29. G	30.75 G		CH3	
G137	1C	3353A04	RS	16. G	17.26 G		CH3	
G137	1C	3353B01	RS	44. G	47.00 G		CH3	
G133	1C	FS 2770-1	OB	1263. G	1445.58 G		701	
G133	1C	FS 2770-2	OB	4092. G	4681.70 G		701	
G133	1C	FS 2770-3	OB	2027. G	2319.70 G		701	
G133	1C	MIPCO	CO	118. G	126.35 G		CH9	
G133	1C	MIPOB	OB	19. G	22.11 G		CJ9	
G132	1C	MIPOP	OP	7. G	8.13 G		CJ9	
G126	1C	MIPPP	PP	1. G	.59 G		CH9	
G136	1C	MIPRS	RS	35. G	38.24 G		CH9	
G137	1C	MIPSF	SF	19. G	470.02 G		CH9	
G126	1C	SCP 126	PP	25. G	25.80 G		CH9	
G132	1C	SCP 132	OP	2. G	2.00 G		CJ9	
G133	1C	SCP 133	OB	2. G	1.80 G		CJ9	
G136	1C	SCP 136	RS	33. G	34.30 G		CH9	
G137	1C	SCP 137	SF	215. G	224.15 G		CJ9	
G138	1C	SCP 138	CO	182. G	191.10 G		CH9	

FIG.9. Terminal display of inventory by account.

has just completed and the particular batch of material in question. As part of the transaction, he will include a measurement at the unit-process boundary.

DYMAC transactions for the pellet fabrication process begin following the feed material as it leaves the vault. The vault custodian makes a transaction to notify the system that material is on its way to the material management room. The system notes that the material is "in transit" and establishes an expected arrival time. If the material does not arrive within the specified time frame, the system alerts the Nuclear Materials Officer who begins an investigation. When the material arrives at its destination, the receiving operator makes a transaction to notify the system. Part of the transaction verifies that (a) the seal on the container is intact, and (b) the seal number agrees with the number in the data base, which was entered at the time the container was stored in the vault. In addition, the transaction allows for verifying the material by nondestructive assay. In the advanced fuels case, this verification measurement consists of counting material in a thermal-neutron coincidence counter such as the one shown in Fig. 5. The system compares the value obtained from the 15-minute count with the value in the data base. If the two values agree within statistical limits, the original value is kept as the accountability value. The most recent verification value for that material is also entered elsewhere in the data base. If they fail to agree, then the operator must make further measurements. Table II shows the verification results for the first plutonium feed material introduced into the pellet fabrication line. For this case, the verification value agrees with the data base value.

INTERNAL ACTIVITY OF ITEM -- 711/1C/MIPOB

										TRANSACTION NO.
TRANSFER INTO ITEM										
0.	G	FROM	711/1C/	3213CC	UP:	OB	LOC:G133	1/25		001D2
0.	G	FROM	711/1C/	3213DC	UP:	OB	LOC:G133	1/25		001D3
29.	G	FROM	711/1C/	3217	UP:	OB	LOC:G133	3/16		001H3
-7.	G	FROM	711/1C/	3224	UP:	OB	LOC:G133	4/19		001L7
-1.	G	FROM	711/1C/	3353	UP:	OB	LOC:G133	5/24		039G3
TRANSFER FROM ITEM										
2.	G	FROM	711/1C/SCP	133	UP:	OB	LOC:G133	5/24		039G2

FIG.10. Terminal display of the transaction activity for the oxide blending unit process material balance.

From the material management room, the feed material is sent to the oxide blending unit process via the conveyor system, and the corresponding transactions are written to update the data base. When the plutonium oxide, uranium oxide, and the carbon powder are blended on a weighed basis, this information is entered into the DYMAC system. From this blended lot, sub-lots of powder are sent to the briquette unit process, each on a weighed basis. The DYMAC system can relate the weight to the amount of nuclear material by means of factors previously determined by chemical analysis. Briquettes are moved out of the pressing unit process on a weighed basis, and at this point it is possible to draw a dynamic balance for that batch of material. The briquettes can be verified using the thermal-neutron coincidence counter located in the glovebox line.

As each sublot of material moves through the pressing unit process, balances are drawn on a batch basis, and a cumulative sum for the successive batches is kept in the data base. This cusum now becomes the material status indicator, which is used for material control. Accountability personnel set limits for the value of the cumulative sum, as well as for each individual batch that goes into that sum. The same procedure holds for the other unit processes in the advanced fuels laboratory. In Fig. 6, the MIPCO chart shows the behavior of material in the powder grinding unit process for the first two months of operation.

To expedite transaction-making, the system displays an option list on the terminal screen from which the operator selects the transaction corresponding to the process step he has just completed. Figure 8 shows the 16 transaction options from which the operator in the advanced fuels laboratory selects. A transaction only requests information necessary for that particular process step. The system checks the operator's entries for validity as he enters them. For example, when he identifies the item he has just processed, the DYMAC system searches the data base to see whether it currently exists on the books; if it is a new item, the system notifies the operator that a new entry has been made for it in the data base.

In addition to making transactions, the operator can recall information from the data base on his video terminal. For example, in Fig. 9, the operator has asked for an MBA level inventory of all the items in the advanced fuels process line, account 711. Note that the display shows which unit process (UP) each item is in. Besides the various items currently being processed, the display shows the items associated with the material balance for each of the unit processes: MIPCO, MIPOB, MIPPP, MIPRS, and MIPSF. To determine how each MIP represents the cumulative sum of the batch balances that have been processed through that unit process (i.e., to determine the individual components of a given MIP) the operator asks for the item's activity internal to the facility, as shown in Fig. 10. The activity report lists all the transactions that have been written for that item in the last 45 days. It is extremely useful in that it enables both process operators and safeguards officers to examine the flow of materials in detail through a given process line.

ROLE OF DYNAMIC NUCLEAR MATERIALS CONTROL IN INTERNATIONAL SAFEGUARDS

LASL and the IAEA are jointly studying the role of a DYMAC-like system in the implementation of IAEA safeguards. The following are a few preliminary thoughts on how an international inspection agency, such as the IAEA, might independently verify an inventory maintained by a DYMAC system.

Basically, the IAEA divides a facility into three MBAs: one at the receiving area where feed material arrives and is stored until it is ready to be processed. The second is a processing MBA, which, in a given facility, might be more than one MBA. The third MBA is the shipping area, where products from the processing area are stored before shipment to other facilities.

Shipping and receiving MBAs are primarily item control areas in which inspection takes the form of ensuring that every item on the inventory listing is indeed present. Some of the items may be verified by portable NDA equipment, for example, and by conventional sample-taking. In these areas, a real-time accountability system can be an asset in that it provides accurate, up-to-date listings of every item. The listings enable the inspector to perform his item check quickly because he has information concerning the location of each item, as well as up-to-date information about its content. Hence, in the shipping and receiving areas, a DYMAC-like system could provide timely records to aid in the inspection process.

The main focus of a DYMAC system, however, is to improve accountability in the processing MBA. The system can localize material by subdividing the MBA into unit processes. It can keep up-to-date, accurate information on the status of the material in each unit process, drawing balances at the appropriate time as material crosses the boundaries.

At the present time, the IAEA inspector has the greatest difficulty making measurements in the processing MBA, and must limit his inspections primarily to times when the facility is shut down for a physical inventory and cleanout of the material. With material under control of a DYMAC-like system, facilities will tend to have fewer shutdowns and cleanouts. Such a system will enable them to keep track of the material in the facility between physical inventories. At inspection time, the IAEA representative will be able to reconcile the physical inventory quickly without having to resolve errors in the accounting records.

However, fewer facility shutdowns and cleanouts could degrade the IAEA inspection capability unless it can find a method to make use of the computer repository of information. The IAEA may wish to use the unit-process structure, perhaps treating each unit process as an item. When inspecting the processing MBA, the IAEA representative could verify the dynamic balance in a similar fashion to the way he verifies item control areas. He could choose to sample the unit processes and verify the items in one particular process. Even though the number of unit processes is small, it might be possible to consider this a statistical sampling. The crux is to find techniques that enable the inspector to independently verify a particular unit process.

To perform an independent verification, an inspector could use in-line NDA instrumentation in such a fashion that the IAEA could guarantee instrument performance, perhaps by the use of independent standards. The interlocking structure of unit processes and the associated flow of material from one unit process to another can guarantee that a particular unit process is being correctly accounted for, thus enabling the IAEA to verify the process. Inspection would become less burdensome because there is less need to shut down the processing in the facility. In principle, it may be possible to inspect a facility with dynamic nuclear materials control essentially at any time, because the inspections need not coincide with actual plant shutdowns. This gives both the IAEA inspectors and facility operators more flexibility in scheduling inspections.

Should international fuel cycle centers become a reality, the dynamic materials control approach can readily transform a facility safeguards system into an international system. The IAEA would assume control of the facility's material control system, drawing on the accountability information to safeguard the material in the facility.

Throughout the nuclear community, countries are developing accountability systems that exhibit some, or all, of the features I have described in this paper. Once such systems are implemented, we can expect tighter material control both at the facility and national level. These tighter controls will be a distinct asset to international safeguards. As the IAEA continues to gain experience with dynamic material control and assimilates the full implication of improved levels of control, it can develop commensurate inspection techniques.

REFERENCES

(1) KEEPIN, G. R., "Safeguards Implementation in the Nuclear Fuel Cycle," Pacific Basin Fuel Cycle Conference (Proc. 2nd Int. Conf. Tokyo, 1978), Am. Nucl. Soc.

(2) AUGUSTSON, R. H., "Dynamic Materials Control Development and Demonstration Program," (Proc. 19th Annual Meeting, Cincinnati, 1978), J. INMM VII III (1978).

(3) AUGUSTSON, R. H., "DYMAC Demonstration Program: Phase I Experience," LASL report LA-7126-MS (1978), Appendix A.

(4) SHIPLEY, J. P., "Efficient Analysis of Dynamic Materials Accounting Data," (Proc. 19th Annual Meeting, Cincinnati, 1978), J. INMM VII III (1978).

(5) LOVETT, J. E., "In-Plant Material Controls: an International Perspective," (Proc. 17th Annual Meeting, Seattle, 1976), J. INMM V III (1976).

(6) SHIPLEY, J. P. et al., "Coordinated Safeguards for Materials Management in a Mixed-Oxide Fuel Facility," LASL report LA-6536 (1977).

(7) HAKKILA, E. A. et al., "Coordinated Safeguards for Materials Management in a Fuel Reprocessing Plant," LASL report LA-6881 (1977).

DISCUSSION

A.G. HAMLIN: An important aspect of a nuclear materials control system is the speed with which it can react to an external statement that in fact control has been lost. Given an external claim that material had been diverted, how fast could the DYMAC system described establish a physical inventory?

R.H. AUGUSTSON: A complete inventory listing is available immediately, the only delay being the time necessary to print it out. The installed NDA instrumentation may be used to verify any items containing large quantities of strategic materials. Ordinarily, assay times are around 20 minutes. Under the proposed conditions, this time would most likely be considerably reduced (~ 1–2 minutes). Most of the time will be spent transporting the material to the instruments.

J.E. LOVETT: In connection with Mr. Hamlin's question I should like to add that in dynamic systems one does not "gather everything together" for inventory purposes. Take a pellet press as an example. A known batch of material has been fed to that press, and processing of the batch will be completed within a period of around 24 hours, at which time batch MUF data become available. Not only is clean-out for inventory purposes unnecessary but the required data become available routinely on a time scale that may often be shorter than clean-out time.

S. SANATANI: I have never had a chance to see your DYMAC system at Los Alamos and I look forward to seeing it in operation during my next visit to LASL. However, I should like to ask whether it isn't all too cumbersome, time-consuming and complicated? It seems that many people would be so busy tracking and keeping record of small amounts of material and feeding the computer with data that they would have little time for other work.

R.H. AUGUSTSON: An accountability system such as DYMAC does indeed generate a large amount of information. However, the system has the inherent computer capability for organizing and screening the data in an efficient way, freeing accountability people from routine report generation. These people can now concentrate on material control. When properly implemented, DYMAC focuses the control activities on real problems.

A. PETIT: Let us imagine that tomorrow Los Alamos becomes a purely civil installation under IAEA safeguards. Would the entire DYMAC system then be put at the inspectors' disposal or would only a part of it be made available owing to considerations of commercial and industrial secrecy?

R.H. AUGUSTSON: The DYMAC data base does contain detailed process information from which information such as process efficiency could be derived. If this is sensitive information, the facility operator would want assurances that the Agency will treat the data as confidential. Since the Agency normally does this anyway, I see no reason why the entire system should not be open for inspection.

R. NILSON: We have an accounting system in which inventory movement transactions are recorded by plant personnel on paper forms. One of our biggest problems, encountered during reconciliation, is the apparent non-recording of a transaction, not just transaction errors. Does your system have interlocks which require the operator to record a transaction before or during the physical movement? If not, could such a feature be added?

R.H. AUGUSTSON: There are no physical interlocks. However, the system is so designed that the transactions must follow in logical order. If an operator fails to make a transaction, or makes it out of order, the system will at some point detect it.

H.G. STURMAN: While it appears that the system described is of assistance to a facility operator in confirming his material account, it is by no means clear that the system is verifiable by an independent inspectorate wishing to check that the operator has not diverted.

J.E. LOVETT: Perhaps I can offer a reply here. Mr. Sturman's concern is valid. The verification of dynamic material control systems is still being studied, and I cannot claim to have a complete set of answers. I will make the following observations, however:

(1) The IAEA can neither require a facility operator to adopt a dynamic system, nor require him *not* to adopt one. Since the advantages of such a system to the facility operator are beginning to become apparent, it seems likely that the IAEA will have to verify dynamically generated data whether it wants to or not.

(2) The effort required to verify a dynamic system is probably comparable to that required to implement any other system which meets currently discussed short detection time and stringent detection sensitivity goals. That is to say, the effort is more a function of the goals than of the method used to achieve those goals.

(3) Whether one thinks in terms of dynamic materials accountancy, increased containment and surveillance, or any other safeguards approach, it is clear that we must learn how to substitute a higher level of technology for sheer numbers of inspectors.

G.Robert KEEPIN (*Chairman*): I should like to comment here on the "transparency" and "obscurity" of materials accountancy systems. In the conventional "paper" system of nuclear materials transactions, the plant operator records all pertinent aspects of the transaction (material identification, location "from", "to", etc.) including hand tabulation of any associated measurements (NDA or destructive) with, of course, the attendant possibility of human error, unintentional or otherwise. By contrast, in advanced material control systems, such as DYMAC, all relevant transaction data (including measurement results) that can be handled automatically are so handled, thus minimizing the likelihood of human error (or even the possibility of attempted circumvention with intent to divert material).

Thus, the conventional paper system often requires considerable input by plant personnel under conditions of relative "privacy". The automated system, on the other hand, reduces the degree of human intervention required and, through the interactive terminal ("paperless") system, makes individual transaction information immediately available to other plant personnel, including the plant safeguards/materials accountancy authority and, as needed, to the national and/or international safeguards inspection authority at the same time. Further, the advanced system enables a number of important additional consistency and continuity checks, detailed inventory and safeguards-status analyses (cusum, alarm-sequence charts, etc.) to be performed, with resultant greatly increased incisiveness of control in both space and time. In general, such incisive analyses are not feasible using conventional materials accountancy.

It is in this sense that we contrast the "transparency" of advanced materials control systems with the comparative "obscurity" of conventional accountancy systems.

This does, of course, imply the need for international (and national) inspectors with an understanding of general plant operations, processes and functions as well as of the nuclear material accountancy and control system in the plant. It may also imply resident inspectors, as appropriate, in large complex facilities.

Given the trend towards on-line measurement and accountancy, together with automated processing, remote-handling equipment etc., in the design of large-scale facilities of the future, it is essential that appropriate methods and techniques be available for effective inspection and verification by both IAEA and national inspectors, and these problems are now being tackled by the safeguards technical community. Full exploitation of the inherent capabilities and advantages of modern materials control system, properly deployed with minimum intrusion into plant operations and independently verified by the safeguards inspectorate, can be expected to play a key (and probably indispensable) role in implementing effective national and international safeguards in the large, complex fuel-cycle facilities of the future.

DEVELOPMENTAL ASPECTS OF A DYNAMIC SPECIAL NUCLEAR MATERIALS CONTROL AND ACCOUNTABILITY SYSTEM

G.L. BOWERS, D.M. DARBY, T.H. DUNIGAN,
T.E. SAMPSON, J.L. COCHRAN
Union Carbide Corporation,
Nuclear Division,
Oak Ridge, Tennessee,
United States of America

Abstract

DEVELOPMENTAL ASPECTS OF A DYNAMIC SPECIAL NUCLEAR MATERIALS CONTROL AND ACCOUNTABILITY SYSTEM.

To improve the accountability of special nuclear materials (SNM) and provide a timely integrated item inventory of SNM in each material balance area (MBA), a Dynamic Special Nuclear Materials Control and Accountability System (DYMCAS) is proposed for the Oak Ridge Y-12 Plant. This paper describes three aspects of the DYMCAS effort in the Y-12 Plant. A broad overview of the system's features is followed by a discussion of the DYMCAS prototype work done in support of the final DYMCAS design and, finally, by a discussion of the non-destructive assay (NDA) effort in determining specific NDA requirements, operating procedures, standards development, and instrumentation precision.

PRESENTATION OF EXPERIMENTAL WORK

A prototype of the DYMCAS was designed and evaluated. The prototype station provides for pre-evaluation of computer hardware, software, non-destructive assay techniques and equipment, and operating procedures required for the complete DYMCAS. An SNM machining area was chosen for the installation of the prototype station. To simulate the complete DYMCAS more fully, a remote station host computer concept was implemented. Control and accountability data generated in the SNM machining area are transmitted to the central base where an integrated item inventory is maintained on all SNM in the MBA.

Interface hardware for the video terminal, line printer, and the communications facilities was purchased with the computer and peripheral equipment. Three non-standard data entry devices were interfaced to the central processing unit (CPU). A static card reader, capable of reading a single 80-column tab card, was interfaced to provide for input of operator and part identification. Individual parts are weighed on a digital scale with a 20-kg load cell with a resolution of ±1 g. A stabilized assay meter was also interfaced to the CPU.

A multi-tasking operating system (RSX-11S) was chosen for the remote station. Software written in-house consists of the device drivers and terminal control routines. The device drivers are system software routines to make the badge reader, assay device, and digital scale look like normal RSX data entry devices.

To gain access to the system, the operator inputs a two-component identification number. If these "match", he is a legitimate operator, and the system authorizes one of the varying degrees of entry into the control program. Once the operator is logged on, he is presented a menu of options. Some options are for material transfer, others are various report-form requests. At the top of every CRT page is the MBA number, the date, the operator's name, and the status of the host computer. As the SNM transfer is being made, extensive error checks are made both at the remote station and at the host computer.

The software developed for the host processor of the DYMCAS prototype system was coded entirely in FORTRAN, and provides three distinct functions: maintenance and diagnostic services, report generation and information retrieval, and data base and interprocessor communication. The primary objective of the host processor software is to provide an interface with, and support of, an on-line inventory for each MBA. The major program development effort was directed towards providing transaction management for one or more MBAs. The transaction management program controls six disk-based files: (1) MBA index files, (2) inventory-accountability file, (3) daily log file, (4) items in transit file, (5) error log, and (6) console log. Programs were also developed to provide inventory and accountability reports. These reports are requested by the MBA terminal and listed on the MBA line printer.

One of the most critical phases of the DYMCAS prototype is reliable communication between the remote station and the host computer. DECNET communications were used for the prototype facility. This software package provides error-free intercomputer communication over long distances in industrial environments.

RESULTS

The purpose of DYMCAS is to increase the sensitivity and timeliness of the SNM control and accountability system in detecting diversion of SNM from the controlled process stream. Two very basic and important capabilities must be provided by the plant-wide DYMCAS if this goal is to be achieved. These are: (1) the ability to produce an accurate integrated item inventory at any point in time for a particular MBA within a matter of minutes, and (2) the ability to maintain a more accurate and timely accountability for the total facility and local MBAs.

The prototype facility has been in operation for six months. Comparisons were made each month between the interim accountability system and the DYMCAS prototype. This included a comparison of both the item part count (physical inventory) and the total accountable SNM transfer weight for the MBA. From the results obtained during the operating period, the DYMCAS prototype has made possible a timely integrated item inventory and improved the accountability data figures for the MBA.

DYMCAS NDA TECHNIQUES

To realize the optimum possible benefits from DYMCAS, one step that must be taken is to determine NDA requirements and establish operating procedures for NDA instrumentation. NDA techniques are needed that will ensure accurate and timely assay data input to DYMCAS. At present, several NDA methods are available, and studies are now under way to determine which of these is best suited for each of the processing areas at Y-12.

Four specific NDA techniques to be evaluated in support of DYMCAS are: (1) segmented gamma scanner, (2) solution assay meter, (3) portable enrichment monitor (stabilized assay meters), and (4) neutron interrogator (random driver). Currently, the enrichment monitor and neutron interrogator are being evaluated.

The SAM-2 enrichment monitor is in widespread use at Y-12 for verification of ^{235}U enrichment. In all but a relatively few instances, the SAM-2 has performed well. Problems with the SAM-2 measurements have arisen when the material contains higher than normal ^{232}U-daughter-product content or fission products. In these cases, a Ge(Li) detector can be used to verify the ^{235}U enrichment correctly. Even though the SAM-2 is stabilized, long-term gain shifts have been observed. These are corrected by periodic recalibration.

A random driver neutron interrogation system is currently under evaluation at Y-12. This CAMAC-controlled, computer-based unit was designed and constructed at the Los Alamos Scientific Laboratory.

Results to date indicate that the calibration is linear up to 10 kg U for impure U_3O_8. Assay accuracy and precision is in the 1 to 3% range for 300-s counts with 8-kg U samples. Blending and sampling problems that currently limit the accuracy of standards are under investigation.

The unit has also been tested on unleached incinerator ash. A lower practical assay limit of about 100 g U has been found with a 2000-s count time for this material. The statistical accuracy at this limit is about 10%. These numbers apply for the fast neutron interrogation operational mode. Thermal neutron interrogation, which would give higher sensitivity, has not been utilized to date.

ACKNOWLEDGEMENTS

Development of the DYMCAS in the Oak Ridge Y-12 Plant has been a team effort. C.W. Holland, J.W. Strohecker, J.N. Treadwell and J.M. Younkin formulated the system criteria and features. G.L. Bowers, D.M. Darby and T.H. Dunigan designed, fabricated and made the prototype station operational. G.D. Ellis, T.E. Sampson and W.H. Tipton have been responsible for evaluating the non-destructive assay techniques.

DYNAMIC MATERIALS ACCOUNTING IN CHEMICAL SEPARATIONS, CONVERSION, AND FUEL FABRICATION FACILITIES*

H.A. DAYEM, D.D. COBB, R.J. DIETZ,
E.A. HAKKILA, J.P. SHIPLEY, D.B. SMITH
Los Alamos Scientific Laboratory,
Los Alamos, New Mexico,
United States of America

Abstract

DYNAMIC MATERIALS ACCOUNTING IN CHEMICAL SEPARATIONS, CONVERSION, AND FUEL FABRICATION FACILITIES.

Concepts for advanced materials accounting and control systems for future industrial-scale, uranium-plutonium fuel cycle facilities are reviewed. The concepts are developed in terms of specific mixed-oxide fuel fabrication, chemical separations, and nitrate-to-oxide conversion facilities and, when combined with advanced physical protection elements, provide effective safeguards for the back end of the uranium-plutonium fuel cycle. Modelling and simulation techniques are used to evaluate the sensitivity of proposed materials accounting systems to single and multiple thefts and to compare safeguards options. Design criteria to improve the safeguardability of future plants are identified.

I. INTRODUCTION

Effective safeguards control of special nuclear materials (SNM) in a nuclear fuel cycle facility requires the ability to draw materials balances about the facility or portions of the facility. In the past, the accountability of nuclear materials and the detection of unauthorized removals have relied, almost exclusively, on discrete-item counting and materials-balance accounting following periodic shutdown, cleanout, and physical inventory. The classical materials balance associated with this system usually is drawn around the entire facility or a major portion of the process, and is formed by adding all measured receipts to the initial measured inventory and subtracting all measured removals and the final measured inventory. Although conventional materials-balance accounting is essential to safeguards control of nuclear materials, it has inherent limitations in sensitivity and timeliness. Sensitivity is limited by measurement uncertainties that can obscure the diversion of a trigger quantity of SNM in a large throughput plant. Timeliness is limited by the infrequency of process shutdown, cleanout, and physical

* Work performed under the auspices of the US Department of Energy, Office of Safeguards and Security.

inventory; i.e., a loss of material could remain undiscovered until the next inventory is taken.

Two ways of improving materials-accounting sensitivity and timeliness are to increase the physical inventory frequency or to augment conventional accounting procedures with dynamic materials accounting. Two to four weeks typically are required to make a physical inventory in a large-throughput (1500 t/a) chemical separations plant. Therefore, increasing the frequency of physical inventories drastically reduces the plant throughput. Using dynamic materials accounting, materials balances can be drawn perhaps as often as every hour, and parts of the process as well as the entire process can be examined. Thus, timely materials balances are obtained about relatively small amounts of nuclear material.

Conceptual designs of advanced materials management systems for safeguarding SNM in the three major components of the back end of the uranium-plutonium fuel cycle (chemical separations [1], nitrate-to-oxide conversion [2], and mixed-oxide fuel fabrication [3]) have been developed and evaluated. When combined with advanced physical protection elements, they provide concepts for effective safeguards in future industrial-scale, uranium-plutonium fuel cycle facilities.

In this paper, the philosophy, the conceptual designs, and the evaluation techniques of dynamic materials-accounting and control systems are discussed briefly. The application of these techniques is illustrated by highlighting the results of the three studies.

II. DYNAMIC MATERIALS ACCOUNTING AND CONTROL

Dynamic materials management systems are developed with the philosophy that SNM accounting and control can be improved significantly in large-throughput facilities only if the facility is divided into accounting areas that permit closing materials balances about relatively small amounts of in-process material. These systems are designed for near-real-time control of SNM through the concepts of unit process accounting, dynamic materials balances, and graded safeguards.

Unit process accounting is accomplished by dividing the entire process area into unit process accounting areas (UPAAs). A UPAA is one or more chemical or physical processes and is chosen on the basis of process logic and the ability to draw a materials balance. Dividing the process into UPAAs and measuring the flow of in-process material permits control of quantities of SNM much smaller than the entire process inventory. Furthermore, discrepancies are localized to that portion of the process contained in the UPAA.

Materials balances drawn around such unit processes during plant operation are called dynamic materials balances to distinguish them from balances drawn after a cleanout and physical inventory. Ideally, the dynamic materials balances would all be zero unless nuclear material had been diverted. In practice they never are for two reasons. First, measured values are never exact because of the errors inherent in any measuring procedure. Second, constraints on cost or effects on materials processing operations may dictate that not all components of a materials balance be measured equally often; therefore, even if the measurements were exact, the material-balance values would not be zero until closed by additional measurements. In the interim, it is sometimes possible to use historical data to estimate unmeasured material, and then to update the estimates when more measurements become available.

The various forms and concentrations of SNM in each facility dictate that these ideas be applied flexibly. Graded safeguards relies on the fact that the strategic value and safeguards vulnerability of the material depend on its location and form within the process, and within the fuel cycle. It is a major factor in the choice of measurement points and UPAAs.

III. MODELING AND SIMULATION APPROACH

The design and evaluation of the materials measurement and accounting system are based on computer simulations of a reference facility because neither the facility nor its safeguards system presently exists in readily modifiable form. Usually we are working with the design of nuclear facilities that are expected to be built some years in the future, and rarely can we expect to change or to experiment with facilities already literally cast in concrete. Furthermore, the use of simulation techniques permits prediction of the dynamic behavior of materials flows under a wide range of operating parameters and the rapid accumulation of data for relatively long operating periods. Alternative measurement strategies can be readily compared and safeguards data-analysis algorithms can be tested. In principle, the necessary data could be obtained from experiments on test loops and on mockups of plant operation, but these are time consuming and expensive. Carefully selected test loops can be used more effectively to validate the computer models and to test portions of the final design of the system.

Modeling and simulation are essential parts of designing safeguards systems. Design concepts are developed by identifying key measurement points and appropriate measurement techniques, comparing potential materials-control strategies, developing and testing appropriate data-analysis methods, and evaluating the capability to detect diversion quantitatively.

The modeling and simulation approach has been used extensively in the three studies. This approach [4] requires a detailed dynamic model of the process based on actual design data and simulation of the model process on a digital-computer; a dynamic model for each measurement system and simulation of accountability measurements applied to SNM flow and in-process inventory data generated by the model process; and evaluation of simulated data from various materials accounting strategies.

IV. DECISION ANALYSIS

The most promising measurement and accounting strategies are combined with powerful statistical data-analysis and sequential decision techniques in comparative studies of diversion sensitivity. A general framework of data-analysis and decision methods has been developed, and is described in Ref. 1, Vol. II, App. E and in Refs. 5-9. Within this framework, the cusum, Kalman filter, and other data-analysis techniques are combined naturally with techniques from decision theory such as the sequential probability ratio test. This framework is called decision analysis. Interpretation of results is aided by computer-display and pattern-recognition tools such as the alarm-sequence chart [5]. This package is an important first step in software development for dynamic accountability systems.

The effectiveness of the materials measurement and accounting system is evaluated by applying the decision-analysis algorithms to the simulated accounting data. Sensitivity to diversion and effective false-alarm and detection probabilities are estimated by examining test results from many sets of materials-accounting data derived from the process and measurement models.

V. ALARM-SEQUENCE CHARTS

The decision tests must examine all possible sequences of the available materials balance data because, in practice, the times at which diversions occur are never known beforehand. Furthermore, to ensure uniform application and interpretation, each test should be performed at several levels of significance. Thus, a graphical display that indicates those sequences that cause alarms, specifying each by its length, time of occurrence, and significance, is essential. One such tool is the alarm-sequence chart [5], a type of pattern-recognition device that has proven very useful for summarizing the results of the various tests and for identifying trends.

VI. RESULTS

The performance of the conceptual dynamic accountability systems for each of the processes is discussed in the following sections. Each system was developed for a specific reference facility so that realistic, quantitative conclusions could be reached. Each system is based on measurement and control technology that has been demonstrated or that can be reasonably projected for the early 1980's. The results are discussed in the order in which the studies were completed, mixed-oxide fuel fabrication, chemical separations, and nitrate-to-oxide conversion. For complete descriptions of the conceptual safeguards systems, see Refs. 1-3.

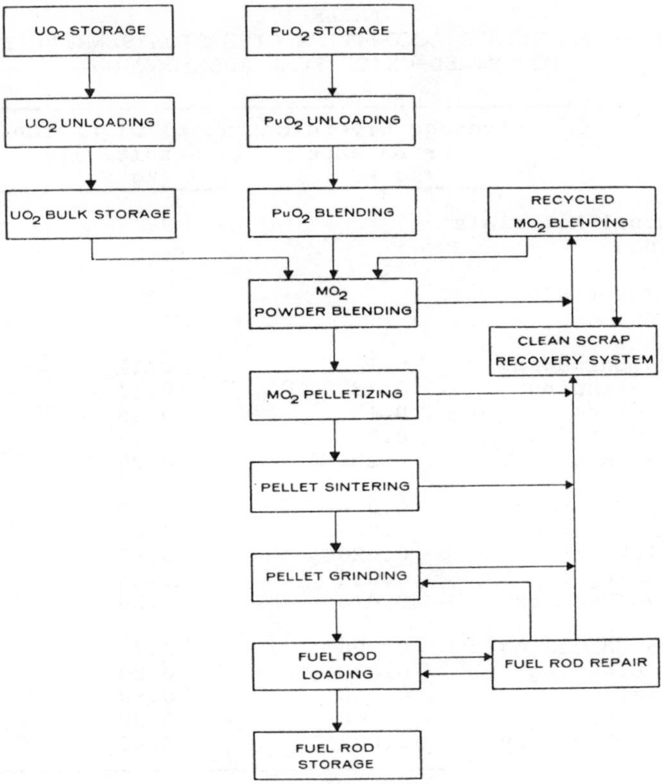

FIG.1. Mixed-oxide fuel-fabrication process.

A. Mixed-Oxide Fuel Fabrication

The model mixed-oxide fuel-fabrication process [3] is based on the Westinghouse design of the proposed Anderson, South Carolina, plant [10]. This is a batch process with an annual throughput of 200 metric tons of mixed-oxide fuel containing 4% plutonium oxide by weight (approximately 100,000 fuel rods annually). Figure 1 is a block diagram of the model process. For dynamic accountability the mixed-oxide fuel-fabrication process line is divided into five UPAAs: plutonium oxide unloading, mixed-oxide blending, pelleting, grinding, and scrap recovery. Measurements are made in each UPAA using digital balances, neutron counters, and gross gamma counters.

The results of closing dynamic-materials balances on a once-per-shift basis are given in Table I. The single-shift diversion quantities detected after one shift, one week, and

TABLE I
DYNAMIC MATERIALS ACCOUNTING DIVERSION SENSITIVITIES FOR MIXED-OXIDE FUEL FABRICATION

	Average Diversion Per Balance (kg Pu)	Total Diversion Sensitivity (kg Pu)	Detection Time
Conventional Materials Accounting	7	7	2 months
Dynamic Materials Accounting			
PuO_2 Unloading	0.11	0.11	2 hours
MO_2 Blending	0.13	0.13	8 hours
Pelleting	0.15	0.15	8 hours
Grinding	0.08	0.08	8 hours
Scrap Recovery	0.20	0.20	8 hours
PuO_2 Unloading	0.016	0.35	1 week
MO_2 Blending	0.020	0.40	1 week
Pelleting	0.020	0.40	1 week
Grinding	0.012	0.25	1 week
Scrap Recovery	0.010	0.20	1 week
PuO_2 Unloading	0.008	0.70	1 month
MO_2 Blending	0.010	0.80	1 month
Pelleting	0.010	0.80	1 month
Grinding	0.006	0.50	1 month
Scrap Recovery	0.005	0.40	1 month

one month of monitoring are shown along with the total material missing at the time of detection. These quantities can be compared with the 7 kg limit of error for conventional materials-balances accounting. Single-theft sensitivity for one shift in these unit processes ranges from 80 to 200 grams. A complete description of the accounting strategies is given in Ref. 3, Chap. IV.

B. Chemical Separations

The Barnwell Nuclear Fuel Reprocessing Plant (BNFP) at Barnwell, South Carolina,[11] was selected as the reference chemical separations facility [1]. Schematics of the BNFP are shown in Figs. 2 and 3. The BNFP design is based on Purex technology and has an annual throughput of 1500 MTHM.

Dynamic materials accounting is restricted to the plutonium purification portion of the process (Fig. 3). Prior to this, the plutonium is in a very dilute and highly radioactive solution and the process is confined to heavily-shielded, high-level radiation cells.

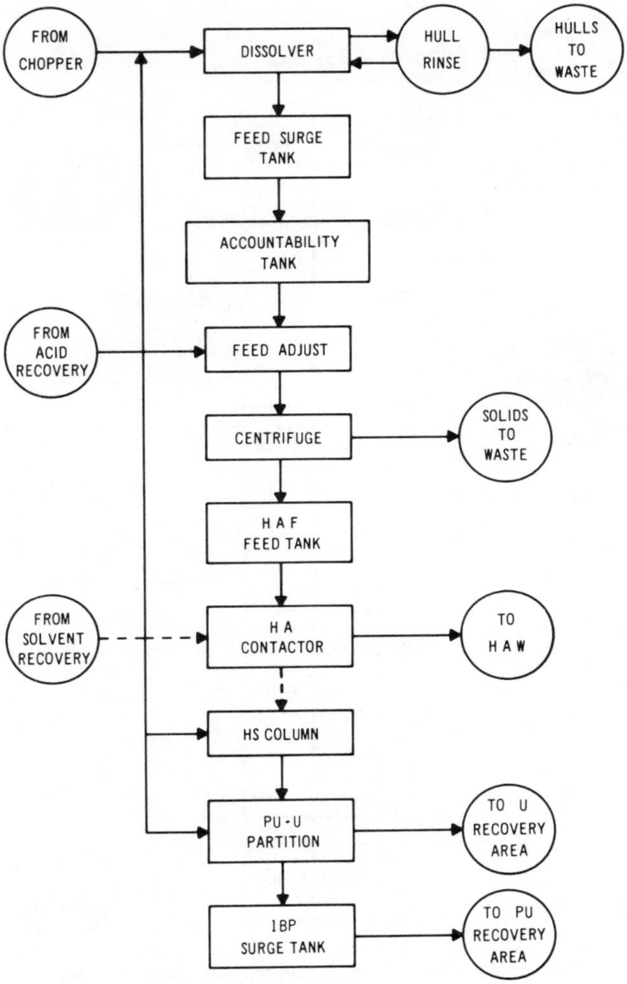

FIG.2. Dissolver-separations process block diagram.

The plutonium purification process is treated as a single unit process accounting area. A dynamic materials balance could be closed perhaps as often as once every hour by combining measurements of flow and plutonium concentration from the 1BP tank, the 3P concentrator, and the waste and recycle streams (2AW, 3AW, 3PD, 2BW, and 3BW). Concentrations are measured by absorption edge densitometers in the feed and product streams and by alpha monitors in the waste streams. The feed-stream flow could be measured by an ultrasonic or electromagnetic flow meter. A bubble-transit flow meter is used in the product stream. In the waste streams air lifts or orifice-plate flow meters are used.

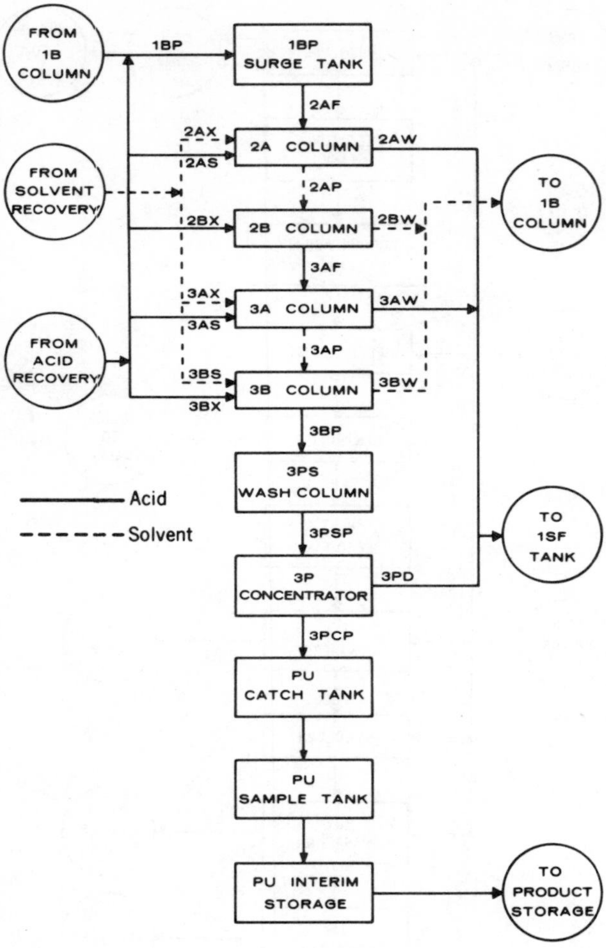

FIG.3. Plutonium purification process block diagram.

Diversion sensitivities for one hour, one day, and one week of one-hour balances are given in Table II. Note that a single-theft diversion sensitivity of 2.6 kg is obtained. If material is diverted in many small thefts, then approximately 4.2 kg (accumulated at a rate of about 25 g/balance) of missing material can be detected after one week. This can be compared to a sensitivity of 75 kg per six months for conventional materials accounting.

C. CONVERSION PROCESS

A conversion-facility design provided by the Savannah River Laboratory and Savannah River Plant [12,13] was used as

TABLE II
DYNAMIC MATERIALS ACCOUNTING DIVERSION SENSITIVITIES FOR THE PLUTONIUM PURIFICATION PROCESS

	Average Diversion Per Balance (kg Pu)	Total Diversion Sensitivity (kg Pu)	Detection Time
Conventional Materials Accounting	75	75	6 months
Dynamic Materials Accounting	2.6	2.6	1 hour
	0.075	1.8	1 day
	0.025	4.2	1 week

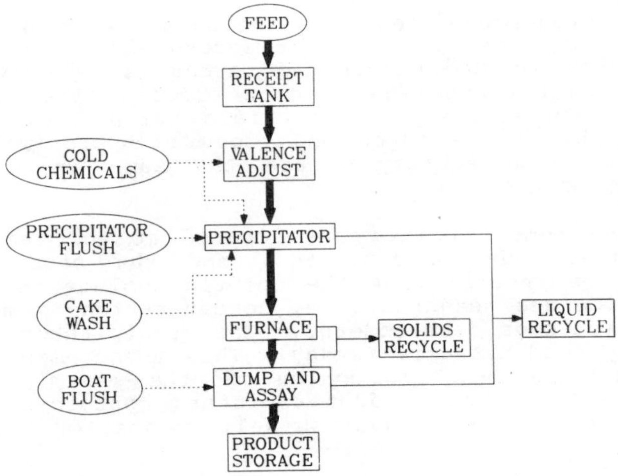

FIG.4. Simplified block diagram of the plutonium nitrate-to-oxide conversion process.

a reference in the third study. The process converts plutonium-nitrate solution to plutonium-oxide powder using plutonium (III)-oxalate precipitation and calcination. The design capacity is 116 kg of plutonium per day.

The conversion process comprises four parallel process lines, and dynamic accountability is applied to each line. The block diagram (Fig. 4) represents a single process line. Two accounting strategies were investigated for each process line. In the first strategy, each process line is divided into two UPAAs. The precipitation-accounting area includes the receipt, valence adjust, and precipitation operations.

TABLE III

DYNAMIC MATERIALS ACCOUNTING DIVERSION SENSITIVITIES FOR THE CONVERSION PROCESS

	Average Diversion Per Balance (kg Pu)	Total Diversion Sensitivity (kg Pu)	Detection Time
Conventional Materials Accounting	33	33	2 months
Dynamic Materials Accounting	0.13	0.13	1.35 hours
	0.03	0.63	1 day
	0.01	1.24	1 week
	0.005	2.65	1 month

The calcination-accounting area includes the furnace and the dump and assay station. In the second accounting strategy, the precipitation and calcination areas are combined into a single accounting area; i.e., each process line is a single UPAA. In each strategy, a materials balance is closed for every 2-kg batch. A batch is processed approximately every 1.3 hours. Both strategies use the same instruments and measurement points.

Key measurement points are located at the receipt tank, the output of the precipitator, and the product loadout area. At the receipt tank the solution volume and plutonium concentration are measured. The concentration is measured by an absorption-edge densitometer. The precipitator product is measured by nondestructive assay. This measurement is difficult because the moisture content of the wet plutonium oxalate is unknown. The product canisters containing plutonium oxide are measured by a neutron well counter or by a calorimeter.

The analysis clearly showed that the second strategy (i.e., a single accounting area) to be more sensitive to low-level, long-term diversion, because of the inaccuracy of the wet oxalate cake measurements. The estimated diversion sensitivities for the second accounting strategy (single process line) are given in Table III. Sensitivities are given for periods of one materials balance (one batch), one day (about 20 batches), one week (about 125 batches), and one month (about 530 batches). The number of batches is not constant because the schedule for daily flushing of precipitators varies. The sensitivity of conventional accounting applied to such a facility is included for comparison.

VII. CONCLUSIONS

The three dynamic accountability studies provide concepts for advanced materials accounting and control systems for future industrial-scale, uranium-plutonium fuel cycle facilities. When combined with advanced physical protection elements, they provide concepts for effective safeguarding of the back end of the uranium-plutonium fuel cycle.

Examination of the uranium-plutonium fuel cycle shows that the conversion process provides a unique safeguards challenge. Unlike the situation that exists in other elements of the fuel cycle, the feed, in-process inventory, and product materials are all attractive, high-purity, concentrated targets for diversion, unhampered by high-level radiation, heavy shielding barriers, or impractically low concentration levels. Also, the conversion process naturally tends to become the process buffer between the loosely coupled functions of plutonium separation (chemical reprocessing) and plutonium use (fuel fabrication) that characterize the commercial fuel cycle. Its buffer function causes the conversion process to be bracketed physically by significant inventories of extremely attractive feed and product materials representing the greatest diversion potential in any domestic fuel cycle and a serious nuclear proliferation risk in the nuclear fuel cycle of any nonweapons state.

The pivotal role of the conversion facility in a safeguarded fuel cycle suggests that any enhanced safeguards or nonproliferation strategy first be applied directly to the conversion plant and then be expanded to include the adjacent functions of separation (or coseparation) and fuel fabrication, starting with the critical areas of product and feed storage and inventory control. This could be done best in future facilities by expanding the conversion facility to include product storage for the separations plant; solution blending or early dilution, if plutonium partitioning is used; custom blending of mixed-oxide powders or coconversion; and feed storage for the fuel-fabrication plant. Colocation of these crucial functions under a single controlling authority in a facility inside or contiguous to the separations plant has been suggested in the "Bonded Crucial Facility" (BCF) concept, proposed as a nonproliferation strategy for fuel-cycle facilities [14]. In this way, the safeguards controlling authority could monitor and verify production and consumption rates and could maintain cognizance of the disposition of all fissile products produced by the complex, thus ensuring that no significant quantities of undiluted plutonium could leave the complex undetected.

Finally, the following points are apparent from the three studies:

- Materials accounting and control in batch processes is generally easier than in continuous processes.

- Redundant accounting information and overlapping unit process boundaries provide added assurance and decreased vulnerability both for diversion and for equipment failure.

- Processes that minimize in-process inventories usually improve safeguards sensitivity.
- Parallel process lines are preferred to a single high-throughput line.
- Safeguards system design should be incorporated at an early stage in facility design, when integration of the functions of dynamic materials accounting and control, process control, and containment and surveillance can provide the most effective safeguards system.
- The conversion plant is a unique safeguards challenge as well as a unique opportunity to address the proliferation problem in a positive way, consistent with production and fuel-cycle requirements and projected institutional arrangements.

REFERENCES

1. HAKKILA, E. A., COBB, D. D., DAYEM, H. A., DIETZ, R. J., KERN, E. A., SCHELONKA, E. P., SHIPLEY, J. P., SMITH, D. B., AUGUSTSON, R. H. and BARNES, J. W., "Coordinated Safeguards for Materials Management in a Fuel Reprocessing Plant," Los Alamos Scientific Laboratory report LA-6881 (September 1977).

2. DAYEM, H. A., COBB, D. D., DIETZ, R. J., HAKKILA, E. A., KERN, E. A., SHIPLEY, J. P., SMITH, D. B., and BOWERSOX, D. F., "Coordinated Safeguards for Materials Management in a Nitrate-to-Oxide Conversion Facility," Los Alamos Scientific Laboratory report LA-7011 (April 1978).

3. SHIPLEY, J. P., COBB, D. D., DIETZ, R. J., EVANS, M. L., SCHELONKA, E. P., SMITH, D. B. and WALTON, R. B., "Coordinated Safeguards for Materials Management in a Mixed-Oxide Fuel Facility," Los Alamos Scientific Laboratory report LA-6536 (February 1977).

4. COBB, D. D. and SMITH, D. B., "Modeling and Simulation in the Design and Evaluation of Conceptual Safeguards Systems," Proc. 18th Annual Meeting of the Institute of Nuclear Materials Management, Washington, DC, June 29-July 1, 1977.

5. COBB, D. D., SMITH, D. B. and SHIPLEY, J. P., "Cumulative Sum Charts in Safeguarding Special Nuclear Materials," unpublished Los Alamos report LA-UR-76-2749 (December 1976).

6. SHIPLEY, J. P., "Decision Analysis in Safeguarding Special Nuclear Material," Trans. Am. Nucl. Soc. 27, 178 (1977).

7. SHIPLEY, J. P., "Decision Analysis and Nuclear Safeguards," in "Nuclear Safeguards Analysis - Nondestructive and Analytical Chemical Techniques," E. A. Hakkila, Editor, Am. Chem. Soc., Washington, DC (1978).

8. SHIPLEY, J. P., "Decision Analysis for Dynamic Accounting of Nuclear Material," paper presented at the American Nuclear Society Topical Meeting, Williamsburg, Virginia, May 15-17, 1978.

9. SHIPLEY, J. P., "Efficient Analysis of Materials Accounting Data," paper presented at the 19th Annual Meeting of the Inst. Nucl. Mater. Manag., Cincinnati, Ohio, June 27-29, 1978.

10. "Westinghouse License Application for the Recycle Fuels Plant at Anderson, SC," USAEC Docket 70-1432 (July 1973).

11. "Barnwell Nuclear Fuels Plant Separations Facility--Final Safety Analysis Report," Docket 50-332, Allied-General Nuclear Services, October 10, 1973.

12. Savannah River Laboratory document DPSTP-LWR-76-5 (November 1976).

13. Savannah River Laboratory document DPSP-LWR-77-29 (March 1977).

14. DIETZ, R. J., in "Non-Proliferation Strategies--Foreign Spent-Fuel Reprocessing Plants," Draft Discussion Paper, prepared for US ERDA-DSS (June 20, 1977). (Abstracted in LA-7011, Ref. 2.)

DISCUSSION

R.J. SORENSON: Is plutonium nitrate shipped to the Bonded Crucial Facility (BCF)? In the United States of America, plutonium nitrate would be converted to plutonium oxide at the reprocessing plant. Only plutonium oxide could be shipped.

R.J. DIETZ: We attempted to make this analysis as general as possible so that it would be most useful for international applications. The proscription of plutonium nitrate solution shipment is unique to the United States regulatory requirements, and in our paper we covered this by suggesting that the nitrate conversion modules be located adjacent to, or within, the reprocessing plant complex, as they currently are.

ADVANCED ACCOUNTABILITY TECHNIQUES FOR BREEDER FUEL FABRICATION FACILITIES

S.I. BENNION, R.L. CARLSON, A.W. DeMERSCHMAN,
W.F. SHEELY
Hanford Engineering Development Laboratory,
Richland, Washington,
United States of America

Abstract

ADVANCED ACCOUNTABILITY TECHNIQUES FOR BREEDER FUEL FABRICATION FACILITIES.

The United States Department of Energy (DOE) has assigned the Hanford Engineering Development Laboratory (HEDL), operated by the Westinghouse Hanford Company, the project lead in developing a uniform nuclear materials reporting system for all contractors on the Hanford Reservation. The Hanford Nuclear Inventory System (HANISY) is based upon HEDL's real-time accountability system, originally developed in 1968. The HANISY system will receive accountability data either from entry by process operators at remote terminals or from non-destructive assay instruments connected to the computer network. Nuclear materials will be traced from entry, through processing to final shipment through the use of mini-computer technology. Reports to DOE will be formed directly from the real-time files. In addition, HEDL has established a measurement program that will complement the HANISY system, providing direct interface to the computer files with a minimum of operator intervention. This technology is being developed to support the High Performance Fuels Laboratory (HPFL), which is being designed to assess fuel fabrication techniques for proliferation-resistant fuels.

INTRODUCTION

Since 1968, HEDL has been developing and applying computerized nuclear materials accountability systems with the ultimate goal of achieving real-time accountability. Development has been a continuing evolution in the application of new technology to meet changing requirements.

The initial system, which was installed in 1968, maintained an on-line file of special nuclear material (SNM) quantities and locations using a dedicated minicomputer. Data were entered manually by the process operators on remote terminals. Status of SNM was available in real-time from either the computer center or at the remote terminals.

HEDL is presently embarking upon a major revision to the initial system in cooperation with the other major contractors at Hanford, i.e., HEDL (Project Leader), Rockwell Hanford Operations (RHO), United Nuclear Industries (UNI), and Battelle Pacific Northwest Laboratories (PNL). The purpose of this revision is to provide uniform safeguards accountability for all nuclear material on the Hanford Reservation, which cuts across administrative boundaries and achieves economy of operation by eliminating diverse systems aimed at the same objectives. The new HANISY system will provide accurate and protected records on SNM movement that not only supply the process operator with necessary data, but provide information for reporting to the Department of Energy (DOE), Nuclear Regulatory Commission (NRC) and the International Atomic Energy Agency (IAEA) directly from real-time files.

The HANISY system must interface with the special and diverse tasks being performed by the various Hanford contractors. One specific example is the HPFL designed by HEDL to demonstrate that remote fabrication and processing of high gamma activity fuels is a viable and economic alternative that will achieve the necessary nonproliferation goals of the nuclear industry. In support of this mission, HEDL is developing advanced accountability techniques that reply upon remote data collection and will test these techniques through application to the fuels processing now in operation. Nondestructive assay (NDA) instruments are being adapted to provide the remote sensing that will supply data directly for the real-time files. This will reduce operator intervention to an absolute minimum with consequent enhancement of safeguards and data protection.

This paper describes the HANISY system basis and features, its application in HEDL laboratories, its support of the HPFL mission, and its application for a Hanford-wide standardized accountability reporting system.

SYSTEM SCOPE AND DESCRIPTION

The HANISY safeguards system is designed to meet or exceed specifications outlined in DOE Manual Chapter 6104, Appendix A. The capability to comply with NRC regulations will be incorporated where no comparable DOE regulation exists. To these ends, the system for nuclear materials accountability and inventory is a minicomputer-based integrated safeguards system. The objective is to maintain records on the location and amounts of all SNM in each of the laboratories.

Two computer centers will operate PRIME 400 computers with similar software to provide backup capability for report production. One will be operated by HEDL, and the other will be operated by RHO. Both systems will have the capability for protection of classified data and production of classified reports. Both PNL and UNI will utilize the HEDL facility for production of their reports. A diagrammatic representation of the hardware configuration for the HEDL system is shown in Figure 1.

HANISY is being designed to provide near-real-time material balancing as a minimum service to all contractors. Daily updates to the data base will be performed by all contractors even if one computer is inoperable. Real-time material balancing will only be available for the on-line portions of the system. The ability of a contractor to provide rapid data on location of SNM in response to a safeguards alert will be in conformance with the graded safeguards approach defined in DOE Appendix 6104.

REAL-TIME ACCOUNTABILITY SYSTEM DESCRIPTION

HANISY will maintain records on SNM at the material balance area (MBA) level in real-time on a 24-hour, 7-days-per-week basis. Records based upon item control will be maintained on the smallest identifiable containment

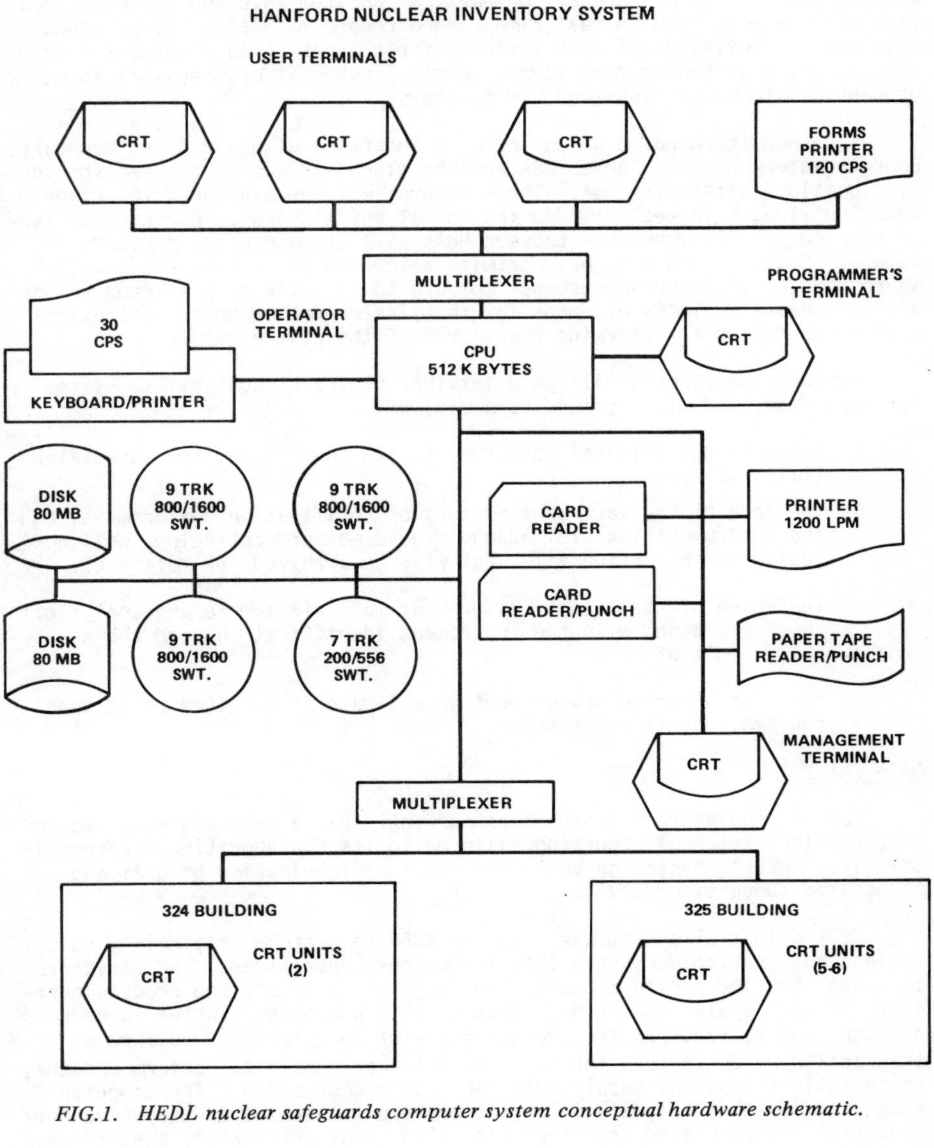

FIG.1. HEDL nuclear safeguards computer system conceptual hardware schematic.

that houses SNM. This includes such items as cans, pins, processing equipment, glovebox enclosures, and waste barrels. SNM quantities and accountable isotopes will be recorded to the nearest gram. Each item will be assigned to a location, and all movement of an item into and out of a location will be reflected in real-time, and a record maintained of all such movements. Provisions will be included for changes to an item caused by consolidation of two or more items, chemical composition, physical form, processing, sampling, loss, or use category.

Each MBA is composed of one or more locations within a fixed boundary. Book inventory records for an MBA are the algebraic sum of the records for all locations within the MBA. These records will be available to authorized individuals in real-time during normal shift hours or during a declared emergency. All movements between MBAs will be made using measured values. Measurements will be made with automated and remote equipment where possible. Measurements must compare to the file or predicted values within specified limits of error for that instrument or an investigation will be conducted to determine the source of the discrepancy.

Measurement records will be maintained by the accountability system for each item, and will include as a minimum:

- Value of the original measurement, its source, and the associated limits of error.
- The date of the last confirmatory or verification measurement, and the instrument identification. These measurement values will not reside in the active file, but will be archived for reference.
- The value of the latest measurement made via sample and analytical chemistry along with the instrument identification, and the associated limit of error.
- The most accurate measurement or estimate of the item based upon the smallest limit of error.

DATA ENTRY

Each of the Hanford contractors currently has a data entry system and a specialized internal reporting tailored to its own operational responsibilities. HEDL's system serves a breeder fuel fabrication program and associated supporting services.

HEDL's initial system was built in 1968 to perform criticality calculations and provide accountability for process operations. The basic premise was that the person actually handling the material, the process operator, should supply the raw data input. This philosophy, which is carried over to HANISY, requires the process operator to enter data to form a transaction which in turn interacts with the data base to perform updates, write audit files, and supply data for report generation. The computer then verifies that all material transfers are authorized by proper authority, that the requested transfers are valid, that criticality specifications are not exceeded by the proposed transfer and that the inventory is updated for each transfer. The master data base is maintained by the central processor and is updated upon completion of each transfer. Access to the system is through either the computer operator's terminal or remote terminals located throughout the plant operations area. Physical access to the terminals and equipment is restricted.

HEDL's safeguards program is developing instrumentation interfaces to allow direct collection of data from NDA instrumentation by the computer. Such instruments include electronic balances, label scanners, neutron counters and gamma scanners. Automatically extracted data from these on-line instruments will replace manual entries in the transaction that would have to be entered by the operator in the earlier system. Not only will this reduce input errors, but it will also provide a greater measure of safeguards protection during SNM movement.

Processing equipment for HPFL will provide automated and remote handling of SNM. A major portion of the accountability data for material in process will be received automatically and directly from the instrumentation used to control the process. NDA will be used as overchecks on reported data in places and at times that will impose a minimum impact upon the process. In HPFL a minimum of operator intervention is expected for data collection.

TRANSACTION PROCESSOR

The heart of the HANISY system is a transaction processor (TP) which routes transactions to and from the appropriate software sections within the system. The TP writes a recovery file upon receiving a transaction and sends it to the proper application program. On completion, the TP writes an output file, an audit file, a transaction input file and routes the result back to the appropriate output driver. These various files, in conjunction with the data base management software package, provide the information necessary for proper restart by the applications program in the event of a software or hardware failure.

Input and output software drivers handle all interaction with the system user. A transaction request is constructed in a front end driver and the data are verified to the extent possible without actually accessing the data base. If the transaction fails due to bad input data, the operator is returned his original input and allowed the opportunity to correct the bad data and retransmit without the problem of reentering all of the data. He is, of course, allowed only a limited number of attempts before the software rejects his input on the basis of a potential security violation.

Upon completion of a transaction, the data are sent from the remote terminal through dedicated wireways to the central processing unit (CPU). Modular software is being developed as part of the HANISY project to provide additional data protection, traceability and recovery.

Application routines, which perform the logic processing for the transaction and modify the data base, will write the transaction number to the record just prior to ending the modification. This allows the recovery logic to determine the status of interrupted modifications. After processing, the results are written to an output file, audit information is written to the audit file, and finally the recovery file is changed to a permanent transaction input file.

The data base management software incorporates before- and after-imaging of the data base to assist in recovery processing. Modifications to the data are written to an after-image file rather than directly to the data base. In the event that the system fails during a modification, the after-image file is not closed. When the system is restarted, the modifications which were only partially completed are not incorporated into the

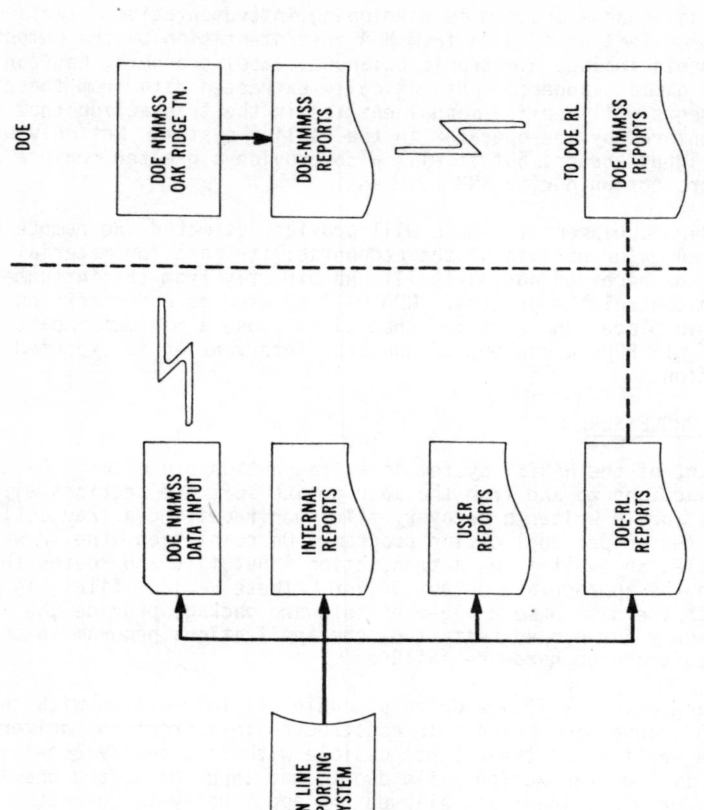

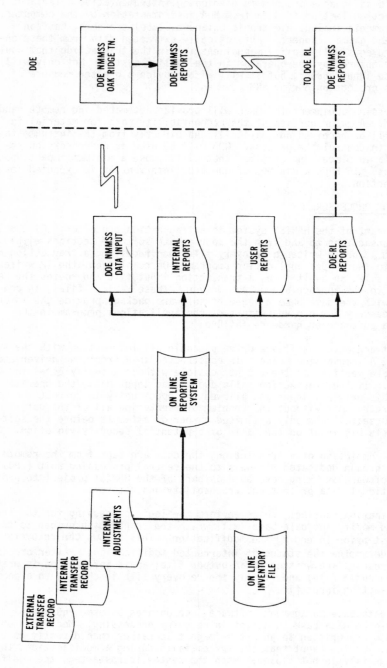

FIG.2. Typical nuclear material reporting.

data base, thus allowing the operations to be repeated without danger of a double update and destroying the integrity of the data base.

AUDIT CAPABILITIES

The audit file which is written by the TP contains summary data on all operations performed on the data base. Authorized personnel can search the audit file for specific transactions and display them on a cathode ray tube (CRT) terminal as they appeared to the original user. The displays can also be sent to a line printer to obtain a permanent copy of the transaction.

All transactions for a specified date and period of time can be recalled to provide a complete history of the transactions over a period of interest. All transactions performed by a specified user can be retrieved to provide a history of that user's actions. It is also possible to extract transactions involving specified material, specific locations, or changes in a material composition. These search keys can be used in combinations to provide very specific searches. This capability provides an invaluable tool for resolving discrepancies as well as for responding to unusual incidents.

RECOVERY PROCESSING

One of the important features of the system is a recovery that is as rapid as possible after a failure, while maintaining the integrity of the data base. Recovery logic in the TP makes use of both the audit files and the before- and after-imaging of data base software to answer the questions --what data were being processed, and what stage of processing had been reached when the failure occurred? Once these questions have been answered the recovery logic in the TP restarts the incomplete transactions. The user is notified that his transaction is complete, but he must identify himself before output data are sent to the terminal. This is to prevent inadvertent display of sensitive information to a possibly unauthorized individual.

If data are physically destroyed by the failure, the computer operator must rebuild the data files before transactions can be processed. Due to system design, this can normally be accomplished in a matter of minutes, except when two physical data storage devices are simultaneously destroyed. Even in this unlikely case, the system can be restored in a few hours, thus providing a high degree of reliability for the user.

REPORTS

DOE requires each Hanford contractor to submit periodic reports to Richland Operations Office and to the Oak Ridge Nuclear Materials Safeguards Systems (NMMSS). These reports include data on material status, project activity, material composition and SNM location. HANISY will provide a package for production of uniform reports.

Each contractor will maintain a data base with a standard format. Each of the required reports will be generated from the contractor's data base by a software package common to all the users. This is the first attempt to standardize reporting from several contractors' on-line computer systems. Detail of the report flow is shown in Figure 2.

In addition, real-time reports will be available to authorized individuals at the remote terminals. This system will provide process operators with the following needed data:

- An inventory of the material holdings of an operator or a work station, ICA or MBA
- The location of a specified item
- The composition and physical status of a specified item
- Internal inventory reports based upon both physical areas and owners.

MEASUREMENT PROGRAM

In addition to HANISY, HEDL has established a measurement program to complement the accountability system. Its purpose is to supply basic input data for feed material; substantiate accountability data on scrap, waste, and material in processes; and provide final assay of product.

A combination of active and passive neutron counting, calorimetry, and gamma scanning NDA techniques are being used, in addition to sampling and wet chemical analysis.

Accuracy is being maintained as high as practical, and conforms to the concept of graded safeguards which places greatest emphasis on strategic quantities of SNM. NDA is used where techniques have been developed to provide reliable accuracy. Measurements by sampling and destructive analysis are made where applicable to assure adequate accountability, consistent with operational needs and the graded safeguards approach.

A continuing quality assurance program is being established to (1) ensure measurement accuracy meets established standards for each instrument, and (2) where possible, provide more accurate measurements using physical, procedural and statistical techniques.

This program currently has two major objectives--experimental determination of limits of error, stratified by instrument, SNM composition, and form; interfacing the measuring instruments directly to the computer. Thus the goal of HEDL's system is a combination of HANISY (which is a software product), and a measurement system.

DEMONSTRATION

Upon completion of HEDL's integrated system, all transfer of material will be accomplished under full safeguards through remote terminal containing microprocessor intelligence and NDA measuring devices. The operators will be identified to the system and their level of authorization will be assessed by the microprocessor. When proper authorization has been received by the terminal, a transaction may be requested. If that transaction is permitted, the computer will gather all the data needed to complete the transaction, using NDA instruments as appropriate, provide the edit, safety checks and validate the transfer.

As the material is moved throughout the process, the computer will automatically balance the inventory records. A file will be kept of all transactions to provide not only traceability but information for audit trail construction. Inventory reports will be produced directly from the computer data base. The location of all SNM will be available to authorized individuals from on-line files.

Systems capability will be established throughout the demonstration phase. The security system will be linked to the accountability system. Communications will originate in either system to provide material protection. If a breach of physical security is indicated, the criticality and inventory system will be placed in a wait state to hamper the efforts of an intruder until the situation is resolved. If unauthorized individuals attempt to tamper with the safeguards equipment, or attempt illegal operations, the same lockup and alarm sequence will be generated.

One of the major objectives will be to demonstrate safeguards systems which prevent, deter, or respond to diversion threats. Many of the measuring devices to be tested will be geared to routine process operation requirements so that a dual function of process control and inventory verification is served. However, in the event of a diversion threat, a very rapid physical inventory must be carried out to establish whether or not diversion has taken place and how much material may have been diverted. Measurement techniques used for such a physical inventory may differ somewhat from those associated with routine process control. HEDL's safeguards program will establish the combination of measurement techniques, real-time records, and item verification necessary to counter diversion threats in minimum time periods.

SCHEDULE

Development of the coordinated HANISY system is currently underway. One PRIME 400 computer has been received and has been installed in a staging area for software development. Without the operating constraints required of an on-line system, software development can proceed at a faster rate. The initial implementation is scheduled for completion expected in October 1979.

Interfaces between the process equipment for HPFL, remote handling devices, NDA instruments and the safeguards computer are keeping pace with the hardware and software design. Accountability system component testing will coincide with process equipment evaluation. On the present schedule, required equipment and the accountability system technology will be available for transfer to the HPFL in mid-1983.

DYNAMIC MATERIAL ACCOUNTANCY IN AN INTEGRATED SAFEGUARDS SYSTEM

J.S. MURRELL
Goodyear Atomic Corporation,*
Piketon, Ohio,
United States of America

Abstract

DYNAMIC MATERIAL ACCOUNTANCY IN AN INTEGRATED SAFEGUARDS SYSTEM.

The nuclear material safeguards system at the Portsmouth Gaseous Diffusion Plant is currently being improved. A new material control system will provide computerized monitoring and accountability, and a new physical protection system will provide upgraded perimeter and portal entry monitoring. The control system incorporates remote computer terminals at all processing, transfer and storage areas throughout the plant. Terminal equipment is interfaced to a computer through teletype equipment. A typical terminal transaction would require verification that the particular activity (material movement or process operation) is authorized, identifying the container involved, weighing the container, and then verifying the enrichment with non-destructive assay instrumentation. The system, when fully operational, will provide near real-time accountability for each eight-hour work shift for all items in process.

INTRODUCTION

The uranium enrichment plant at Piketon, Ohio, U.S.A., more commonly known as the Portsmouth Gaseous Diffusion Plant, is operated by the Goodyear Atomic Corporation for the United States Department of Energy (hereafter referred to as DOE or the state). Under the toll enrichment program the plant provides enrichment services to greater than 97 weight % ^{235}U. The plant facilities consist of three process buildings containing the gaseous diffusion equipment and several buildings for auxiliary operations. A site map (Figure 1) shows the process buildings as well as the auxiliary operation buildings in a fenced area of approximately 4.0 square kilometers.

Processing and/or storage operations are performed in various areas throughout the plant. Auxiliary processing activities conducted at the plant include the conversion of uranium oxides to UF_6 by fluorination, decontamination of process equipment, and recovery of uranium from solutions and solid wastes. The conversion facility is used to process all scrap uranium materials returned to the DOE. The facility is designed for direct fluorination of enriched uranium oxides and is nuclearly safe for all enrichments. The decontamination facility is used for cleaning all contaminated equipment removed from the process. The recovery facility is operated in conjunction with the decontamination operation and produces oxide from the solutions

---

* Acting under contract with the US Department of Energy.

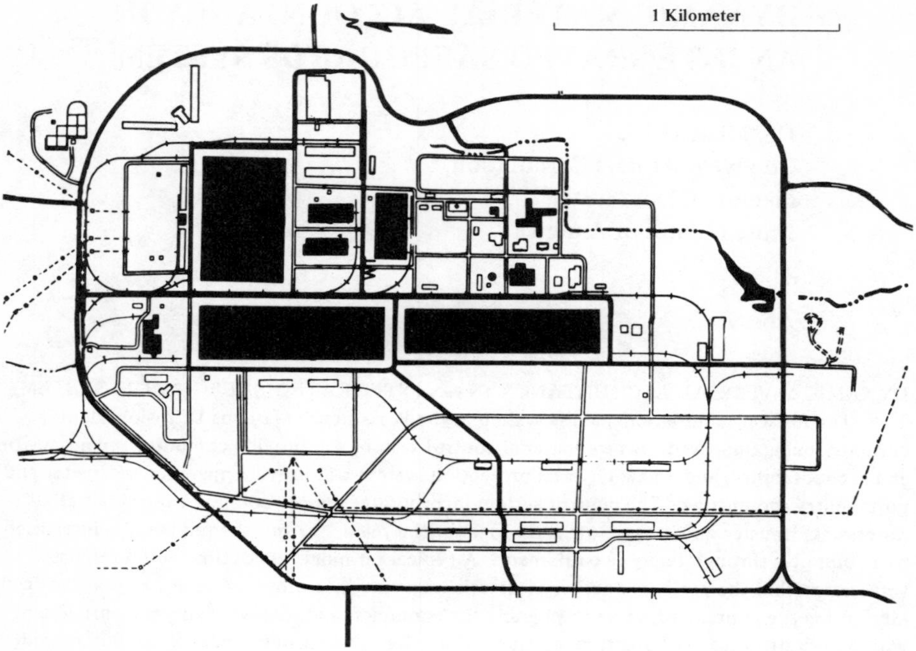

FIG.1. Portsmouth gaseous diffusion plant site map – shaded areas contain DYMCAS equipment.

generated by the decontamination of process equipment and the dissolution of waste solids. Typically, the waste solid materials include trapping materials from the process such as magnesium fluoride, alumina, sodium fluoride and incinerator ash. These various waste solutions and solids which contain uranium-bearing compounds span the entire isotopic range.

In the plant record system the process and auxiliary operations are documented by three accountability reporting stations comprised of nine material balance areas and 57 separate accounts. The accountancy work load centers around the record system required for these activities and focuses on an average inventory of 18,000 items, of which approximately 20% are items of material greater than 20 weight % ^{235}U or otherwise referred to as strategic special nuclear material (SSNM). Plant activities associated with the toll enrichment program include annual receipts of approximately 8.5 million kilograms of uranium as feed material and shipments of approximately 2 million kilograms of uranium of low enriched product. There are also significant quantities of highly enriched uranium received and shipped annually for foreign as well as domestic nuclear activities. The monthly internal transfers average 1250 documents, representing 5000 transactions. An additional 1500 documents are required for internal sample transactions. SSNM is involved in 10 to 30% of these transactions, depending upon operating schedules of plant facilities. Requirements for the state system include daily input of all shipments and receipts to the central state computer.

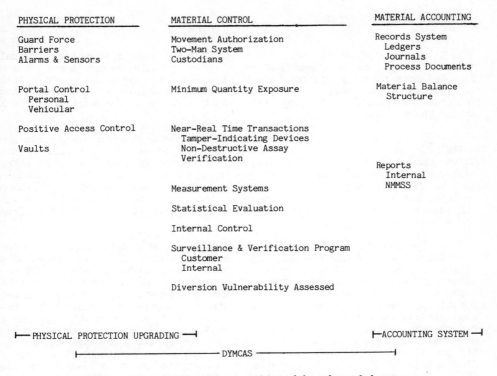

PHYSICAL PROTECTION	MATERIAL CONTROL	MATERIAL ACCOUNTING
Guard Force Barriers Alarms & Sensors	Movement Authorization Two-Man System Custodians	Records System Ledgers Journals Process Documents
Portal Control Personal Vehicular	Minimum Quantity Exposure	Material Balance Structure
Positive Access Control Vaults	Near-Real Time Transactions Tamper-Indicating Devices Non-Destructive Assay Verification	Reports Internal NMMSS
	Measurement Systems	
	Statistical Evaluation	
	Internal Control	
	Surveillance & Verification Program Customer Internal	
	Diversion Vulnerability Assessed	

├── PHYSICAL PROTECTION UPGRADING ──┤ ├─ACCOUNTING SYSTEM ─┤

├──────── DYMCAS ────────┤

FIG.2. Characteristics and composition of the safeguarded system.

Additionally, the monthly transactions are completed and transmitted to the computer by the tenth day of the following month. All plant inventories are transmitted by the fifteenth day of the following month. The state requirement, the volume of transactions and documents, as well as a large inventory spanning the entire ^{235}U isotopic range in a variety of process systems, make the plant's safeguard problems complex if not somewhat unique. This necessitates a strongly integrated safeguards system which includes the many aspects of physical protection, material control, and material accountancy. Activities incorporated in each of these functional areas are outlined by subject and shown in Figure 2.

Recent changes in state regulations have necessitated improvement to the existing physical protection system for facilities processing or storing SSNM. Similarly, reporting requirements with improved material control and measurement techniques, including measurement redundancy, prompted implementation of a near real-time control system, which at Portsmouth is referred to as the Dynamic Material Control Accountancy System (DYMCAS). The two projects which implement the requirements for the integrated safeguard system are the Physical Protection Project and the DYMCAS Project (see Figure 2). Since there is some overlap of the physical protection and the material control aspect of safeguards, an overview of the physical protection project would be appropriate.

PHYSICAL PROTECTION PROJECT

Implementation of new and revised state regulations for physical protection resulted in identification of four material access areas. These include three existing process facilities and one storage building which will be constructed to meet the safeguards vault concept. Process buildings required modifications ranging from vault enclosures to isolation zones around the building.

One of the more significant segments of the project is the storage building. The single-story reinforced concrete structure is divided into three sections, one each for UF_6, miscellaneous uranium compounds (UNH, oxide, UF_4, etc.), and staging and sampling operations. All storage locations in the facility are vertical cylindrical holes below ground level. Each position has an adjacent smaller diameter hole which can be used for obtaining either a gamma or neutron signature without removing the container from its sealed location.

The process facilities will require extensive modifications including a perimeter fence around the building. These facilities will be equipped with the following protection systems: perimeter and intrusion alarm systems, closed circuit TV monitoring, and most importantly, portal control for both personnel and vehicles. All of the material access areas will contain manned portals and secured and alarmed vehicle gates. The manned portals will contain surveillance equipment, radiation and metal detection capability, and a card access system to monitor ingress and egress. All material access areas will have alarms and an intrusion detection system which are communicated to the portal and the communication center of the guard headquarters building. Closed circuit TV will be installed to cover the portal hallways, vehicle traps, feed and withdrawal locations in the process material access areas, and buffer zones which are surrounded by fencing. Additional intrusion detection systems are being evaluated for buffer zones.

All manned portal card-key access systems will be interfaced with a dedicated computer system and will verify card data as well as individual code numbers. The computer card access system will assure that only authorized personnel are admitted and in addition maintain computer control for the two-man rule in material access areas.

An individual attempting access at a portal must present a plastic card that will be electronically scanned. He must also manually enter a personal code number into the system. Access will be granted only if the card and code number match on system base data. Additional control such as anti-card-pass-back, door-open sensing, anti-tampering and remote override will also be included in the system. The computer will control ingress and egress so that it administers the two-man rule and sounds an alarm if only one individual remains in the material access areas beyond a predetermined time. The computer will also assure that individuals are authorized into the material access area in the appropriate time frame and in accordance with assigned status levels.

The communication center in the guard headquarters is being expanded to include two control consoles; one devoted to the SSNM monitoring, and one to other plant systems. A new computer-based integrated security system will monitor security alarms and the personnel entering and exiting the material access areas. The system will be able to annunciate, display, record both video and audio, and produce hard copy records of all alarm activity and personnel ingress and egress on a daily basis.

DYMCAS EQUIPMENT

The physical protection project and the material control project overlap in such areas as the two-man rule and custodial activities. These and most other facets of the material control project are incorporated into the DYMCAS program, which provides two significant improvements when compared with the present material control system. The first is a near real-time item accountability and the second a prompt verification procedure for all transactions. Initially the system was directed toward verification of SSNM. As the concepts developed it was evident that independent material control systems for above and below 20% ^{235}U were not practical and the scope of the system was expanded to include all uranium transactions into a common data gathering system inputting to the main nuclear material accounting system.

To implement the data gathering system, high capacity multi-use computers in parallel are utilized and data are transmitted or accessed by 35 remote terminal stations distributed throughout the site for on-line time-sharing capability. These remote stations are of two types; communication terminals with computer access, and terminals with measurement instrumentation in addition to the teletype equipment for material transactions. The instrumentation at these remote stations is controlled by micro-processor and transmits data to the control computer for processing. Terminal equipment includes a standard teletype terminal (TTY) and/or a video display unit (CRT) with a keyboard. These communication terminals will be used by the operations personnel, laboratory personnel, and nuclear material control personnel. Field station terminals to be used for transactions will include a CRT or teletype and the equipment described below.

a) Bar code label readers will provide reliable container identification information to insure the integrity of container identity in the computer data structure.

b) Electronic digital scales will verify the container's gross weight against data base values. Containers in process will be weighed to determined process weights. This procedure will provide interim accountability information.

c) Ultrasonic thickness gauges will measure the wall thickness of UF_6 cylinders so that attenuation corrections may be made to the gamma-ray enrichment measurements.

d) Stabilized sodium iodide gamma-ray spectrometers will make enrichment measurements of pure non-UF_6 materials and UF_6 in cylinders where applicable.

e) Shielded neutron assay probes will verify the enrichment of UF_6 where the use of gamma-ray techniques is inappropriate.

Application of additional instrumentation for both special and abnormal transactions will be processed through a special terminal equipped like the other terminals but including the equipment described below.

a) Random driver active neutron interrogation will measure or verify the fissile content of nonhomogeneous materials where other techniques are not reliable.

b) A segmented gamma scanner with a high resolution gamma-ray detector will measure ^{235}U content in low density scrap and waste materials, and provide technical spectral information in cases where abnormalities prohibit the normal use of other gamma-ray techniques.

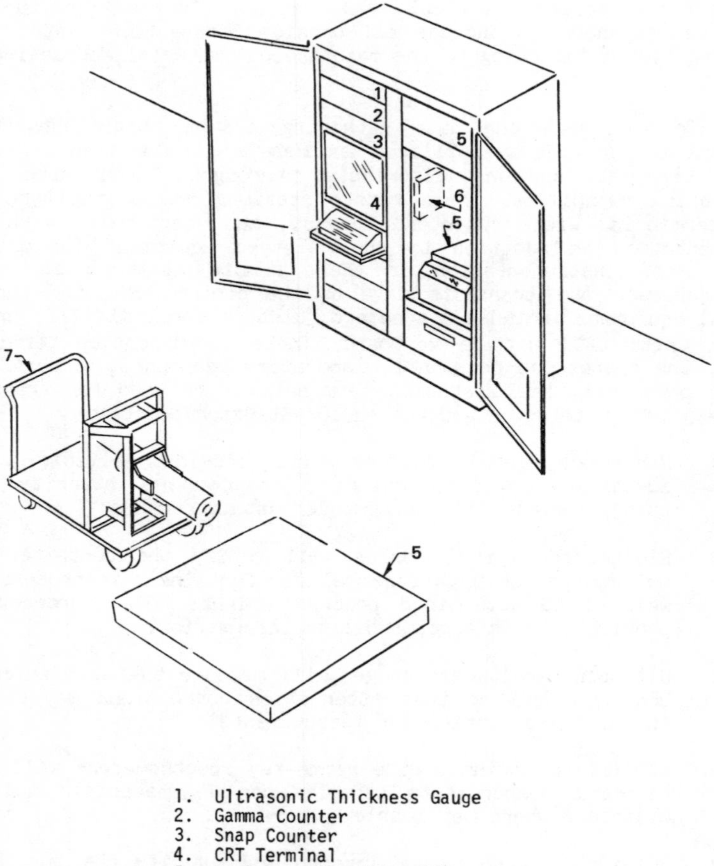

1. Ultrasonic Thickness Gauge
2. Gamma Counter
3. Snap Counter
4. CRT Terminal
5. Scales: 5kg, 15kg, 50kg, etc.
6. Bar Code Reader
7. Detector Cart

FIG.3. DYMCAS remote data entry station.

DYMCAS STATION OPERATION

A typical field terminal with the instrumentation described is shown in Figure 3. Operationally, each field terminal will be activated using a computer identification code similar to codes used on other time-sharing systems. Once the terminal is activated for the day's operations, the operator will enter a station standardization transaction into the system and perform calibration checks of the station's instruments. Using the container identification device (a bar code reader) the check weights for the station will be used to determine the performance of the weighing device which will be audited by the computer verification program. Using previous calibration data, control limits will be maintained on current performance. Capacities of the weighing devices vary according to activity for the field stations. An indication of the range and expected accuracy of the electronic digital weighing devices are as follows:

Maximum Load, kg	Accuracy, ± g	± Percent
5	0.01	0.0002
15	0.1	0.007
50	5	0.01
100	10	0.01
500	100	0.02
4,536 (10,000 pounds)	2.268 (5 pounds)	0.05

Calibration checks will also be performed for the ultrasonic thickness gauge. Design specifications for accuracy of the gauge are ± 0.001 inch.

Enrichment measurements will be made using primarily a gamma-ray assay meter. Specially prepared standards will be used for continued calibration checks with data comparison by the computer for actual versus standard values. A stabilized NaI gamma-ray assay meter will be used to measure the following compounds with the use of the calibration procedure. The estimated accuracies are:

Compound	Enrichment, wt. % ^{235}U	Accuracy, %
UF_6	>20	2 to 5
	5 to 20	5 to 10
	<5	about 10
Oxide	>20	about 5
	<20	about 2
UNH	>20	about 5
	<20	about 5

Additional verification of instrumentation for the low enriched UF_6 (<5 wt. % ^{235}U) will be available using a shielded neutron assay probe. The accuracy of this instrument is expected to approach ±5% for material less than 5 wt. % ^{235}U.

In the storage areas mobile stations will be provided and equipped with a bar code reader and a weighing device. In addition, gamma and neutron instrumentation will be provided to obtain signature data for containers in storage using the adjacent hole concept. The method provides access to material for NDA measurement while it is stored in a sealed position. The method is expected to provide signature data relative to the ^{235}U mass with a reproducibility estimated at ± 10%. The equipment will be used for routine inventory as well as verification procedures. Special instrumentation for measuring and verifying various plant waste materials as well as any scrap containers requiring measurement reconciliation will be permanently located in the storage building staging area. These commercially available instruments include a segmented gamma scanner and a random driver. A summary of the equipment to be installed in the various field stations is shown in Table I by type of application and in Figure 1, by shaded area, with respect to plant location.

DYMCAS COMPUTER EQUIPMENT

The basic computer system for the DYMCAS project requires a dedicated computer, communications equipment, and storage capabilities for processing data transmitted from the field stations. All data obtained at the stations are transmitted using multiplexer equipment. The DYMCAS computer requirements are being implemented along with uprating of other plant computer requirements so that the DYMCAS will have access to one of two large frame multi-user computers, either of which will access multiple disk storage

TABLE I. DYMCAS TERMINAL EQUIPMENT

Terminal Application	Terminal CRT	Terminal TTY	Container Identifier	Digital Scale	Thickness Gauge	NDA	Special NDA
Field Stations							
Operations	23	2	34	40	19	23	
Support	4	2	7	8	1	2	
Communication							
Administrative	7	6	-	-	-	-	-
Material Control	6	-	-	-	-	-	2
	40	10	41	48	20	25	2

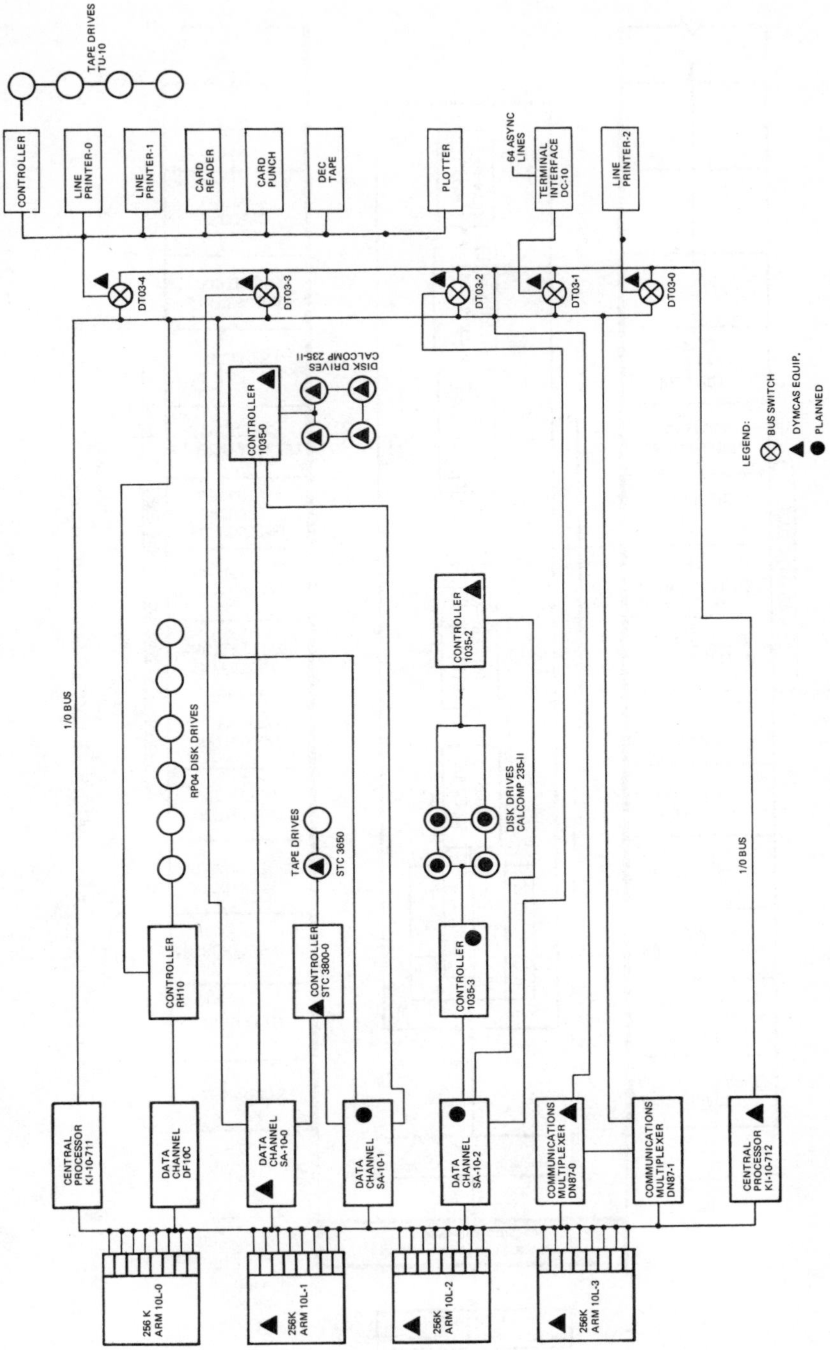

FIG.4. Computer configuration.

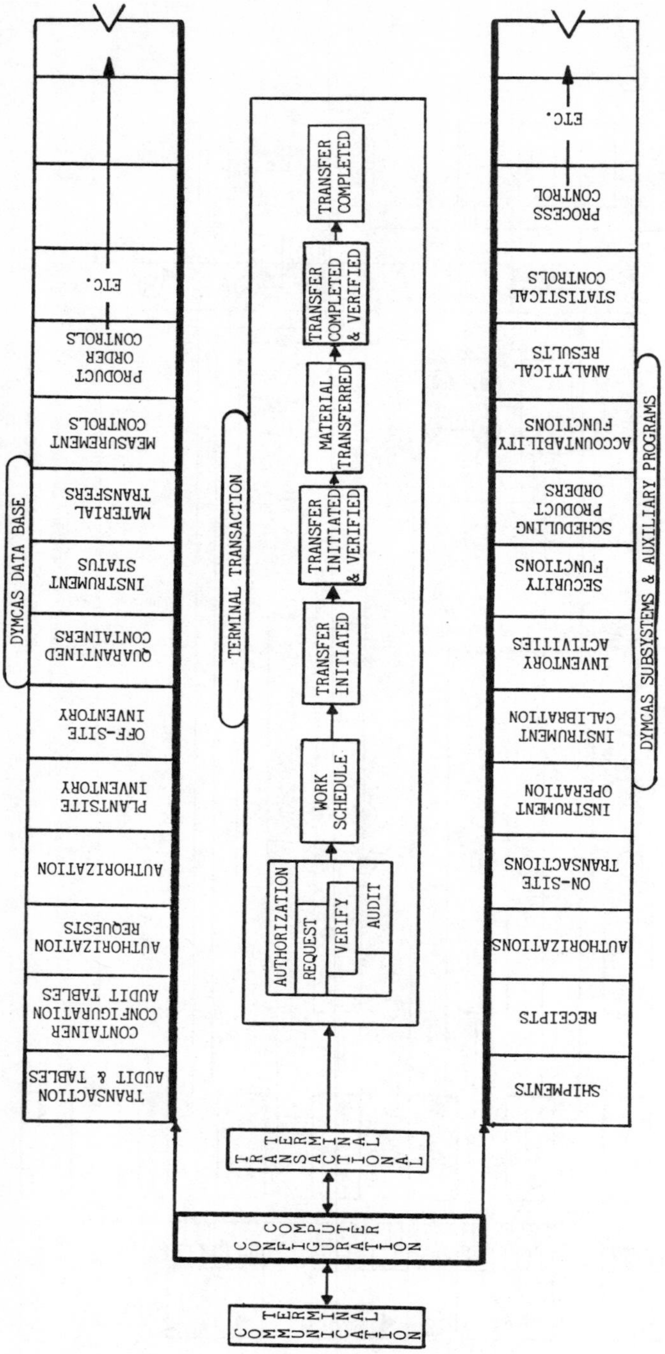

FIG.5. DYMCAS functional diagram.

units. The system uses a controller, with disk drive units from the communication multiplexer, which interfaces with the field stations. Because the existing plant computer system was being uprated, the DYMCAS design incorporated sharing of common equipment resulting in a higher degree of reliability, at least from a project cost-effectiveness standpoint. Thus, essentially all equipment systems are redundant with corresponding improvement in flexibility for all access modes from the terminal interface, through the central processing unit, to the disk storage. A significant part of the computer system, not commercially available, is a specially designed multiplexer system used at the field terminals. Each station multiplexer is capable of interfacing twelve input signals. For reference, the computer equipment configuration for the DYMCAS is shown in Figure 4 with the existing and planned plant equipment.

DYMCAS SYSTEM CONCEPTS

The programming aspects of the DYMCAS consists of three sections. The first is the major system functions and their interactions. The second consists of the authorization program defining what activities are available to approved users. The third is the station transaction which triggers the system with updated (new) data. When data concerning a transaction is input at a field station, the DYMCAS will perform authorization verification and accounting activities.

Figure 5 is a functional diagram of the DYMCAS. The system can accommodate several activities simultaneously; including station transactional data, scheduling activity and/or operational inquiries. The versatility and capacity of this system provides for easy program modification as well as the addition or deletion of functional programs directly or indirectly affecting material safeguards.

Transactions are programmed into DYMCAS computers before being performed. Authorizations for all material movement must be obtained from the DYMCAS which confirms that individual transactions correspond to the approved program. This procedure performs both authorization and auditing functions and is expected to accomplish a two-fold improvement of the present system. First, it increases the assurance that only materials for approved project activity will be accessed, and second, it reduces the number of transactional errors and permits corrections when errors occur.

Figure 6 is a generalized diagram of typical DYMCAS transactions. Figure 7 illustrates the steps for a specific container movement. The procedural system illustrated in Figure 7 is applied to all transactional activities. The most frequent transactional activities and the procedures for authorization and verification (audit) are illustrated by these two figures. Receipts of materials from off plantsite are authorized prior to arrival. When the material is received, the material verification procedure is completed and the system is updated. All shipments off site are first authorized, then verified; the shipment is then initiated, and the system is updated.

The example of Figure 7 applies to a SSNM movement, which requires both field station measurements and security, that is plant guard, activity.

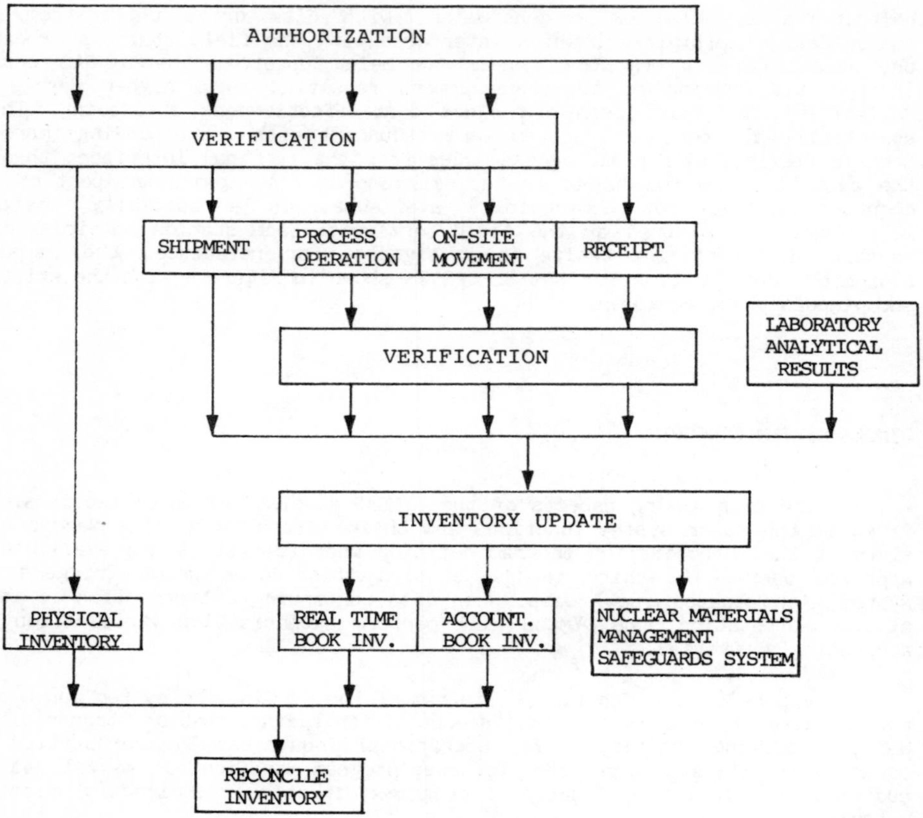

FIG.6. Typical DYMCAS transactions.

The example includes the authorization check, which initiates the field work and locates the container for movement to the measurement station for verification. Measurement data is taken before and after the material movement with the security notification made prior to movement. Not shown on the diagram is an additional safeguard requirement; the guard headquarters must notify the guard force at the transaction site that the movement is authorized and must alert the receiving area of the transaction.

DYMCAS EXTENDED UTILIZATION

The large data base necessary for safeguards and material control functions prompted a transition of many related activities to DYMCAS. Some of these functions are directly related to plant material control. One example

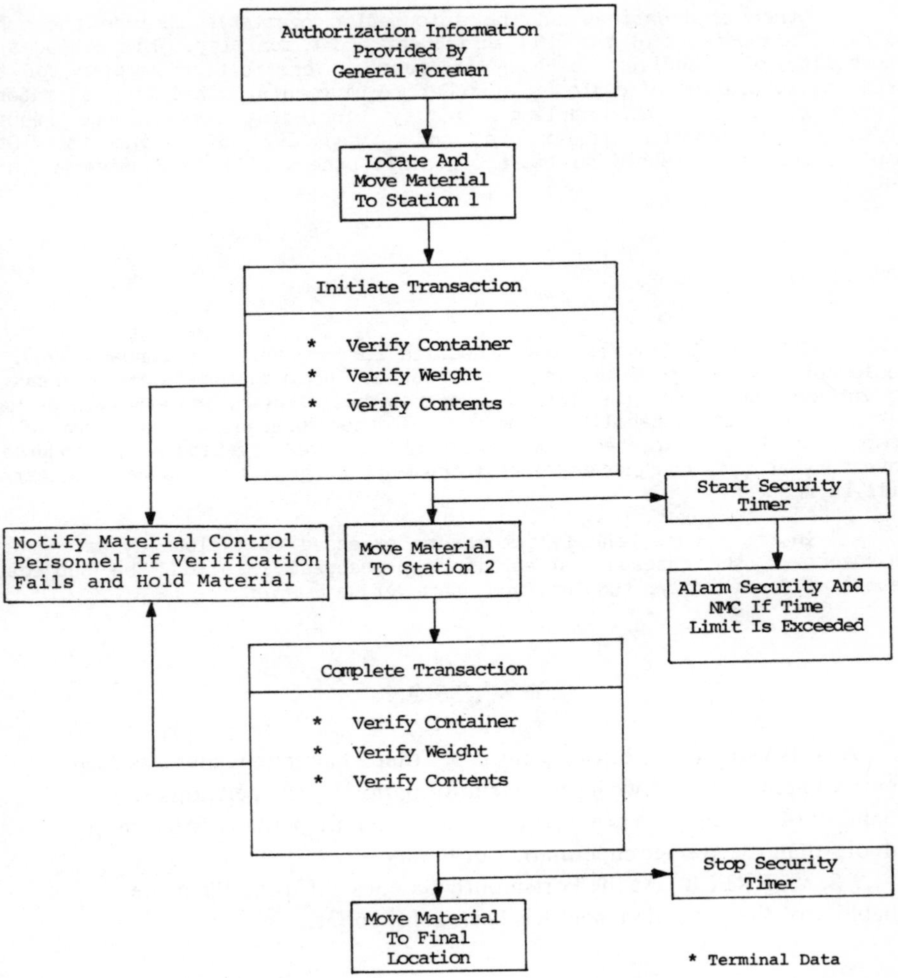

FIG. 7. Field terminal transaction.

is the cascade daily material balance activity. This is now performed using daily log sheets (for feed and withdrawal data), adjustments for power level, and operating variables using a programmable calculator. The extension could be to compute the differences on an hourly basis with all necessary data inputted from the various areas. Further, but possibly with greater uncertainty, measured isotopic values could be added to the data and thus provide both a uranium and uranium-235 balance (feed minus withdrawals) on a daily or shift basis.

Other applications of the information available include the scheduling of all aspects of the toll enrichment order activity. This includes the capability of schedling cascade withdrawals, verification sampling of toll product, scheduling of customer cylinder requirements, scheduling of material transfers, control of sampling activity (including surveillance requirements), verification of orders and scheduling of shipments. Other applications are planned and most certainly others will be considered in the future.

CONCLUSION

The DYMCAS is presently scheduled for operation in January 1980. All equipment has been received or is on order and programming is in progress. It is intended that when the system is operational SSNM in process will be under 8-hour shift accountability. Because of the substantial capacity of the computer, it is expected that material control, particularly scheduling, inventory status, and process evaluation will be greatly enhanced, and errors will be reduced.

In the future, the system may be expanded to include dynamic process inventory of the cascade and auxiliary processes. This will be subject to technical feasibility, funding, and state authorization.

DISCUSSION

A.G. HAMLIN: I see from your paper that a bar code reader was used to identify batches. Could we have some information on the performance reliability of this equipment under conditions that are perhaps rather more arduous than the average supermarket or library?

J.S. MURRELL: At the Portsmouth Gaseous Diffusion Plant the reliability of the bar code reader has been 100% so far.

Session VIII (Part 2)

THORIUM/URANIUM-233 FUEL CYCLES

Chairman: G. Robert KEEPIN
(United States of America)

IAEA-SM-231/41

THE NUCLEAR MATERIALS CONTROL SYSTEM IN THE THTR-300 HIGH-TEMPERATURE PEBBLE-BED REACTOR

H. BÜKER, H. ENGELHARDT
Kernforschungsanlage Jülich GmbH,
Jülich, Federal Republic of Germany

Abstract

THE NUCLEAR MATERIALS CONTROL SYSTEM IN THE THTR-300 HIGH-TEMPERATURE PEBBLE-BED REACTOR.

The paper describes a safeguards system for the 300-MW(e) THTR pebble-bed reactor being now built in the Federal Republic of Germany. A continuous loading and unloading of the core during operation is typical for this reactor. As much as possible the system will be based on accounting measures, especially on item-counting. As "flanking measures", containment and surveillance measures will also be applied. It is also typical for this reactor that physical inventory verification in the reactor core is impossible. Balancing of the nuclear material will be carried out by automatic counting of the fuel-element pebbles to be introduced into the core and the loading facility and those leaving the core. Counting of the fuel elements will be performed with the aid of the pebble counters, which are also used for operational reasons. The installation of the counter indications of the individual counters can be described by more than 25 balance-equations, so that it seems almost impossible for an element to be removed unnoticed from the fuel element flow through the loading facility and the core. Building construction measures will ensure that all fuel elements must pass the pebble counter at the entrance and exit of the reactor. The continuous operation of the facility and the counters will provide a continuous supervision of the physical inventory. The proposed safeguards system demonstrates that a secure and complete supervision of the flow of the fissile material of the THTR-reactor should be possible.

I. THE PRINCIPLE OF THE PEBBLE BED REACTOR

The characteristic property of a pebble bed reactor is that its fuel-elements are composed of graphite balls having a diameter of 6 cm and containing the fuel in the form of coated particles. The coated particles are distributed homogeneously in a graphite matrix in the centre of the ball (Fig. 1). Each ball contains 0.96 g U-235 and 10.2 g Th. The prototype of this reactor is being built in Fed. Rep. Germany. It is the THTR-reactor. The general lay-out data of the reactor are shown in table I.

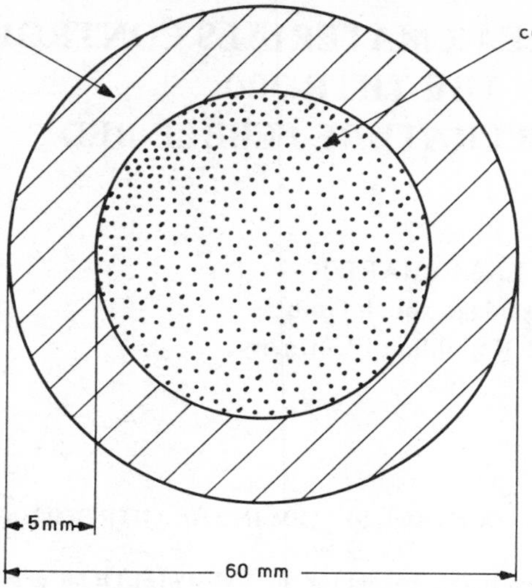

FIG.1. Fuel element of the THTR prototype reactor.

Table I: General Lay-out Data of THTR Prototype

Electrical Power	300 MW
Thermal Power	750 MW
Average Power Density	6 MW/m^3
Core Diameter	5.6 m
Average Core Height	5.1 m
Number of Fuel Elements	675,000
Number of Absorber Rods	42
Number of Control Rods	36
Cooling Gas Pressure	40 atm (a)
Primary Circuit Pressure Drop	1 atm (a)
Blower Power	6 × 2.7 MW
Core Inlet Temperature	270°C
Core Outlet Temperature	750°C
Number of Heat Exchangers	6

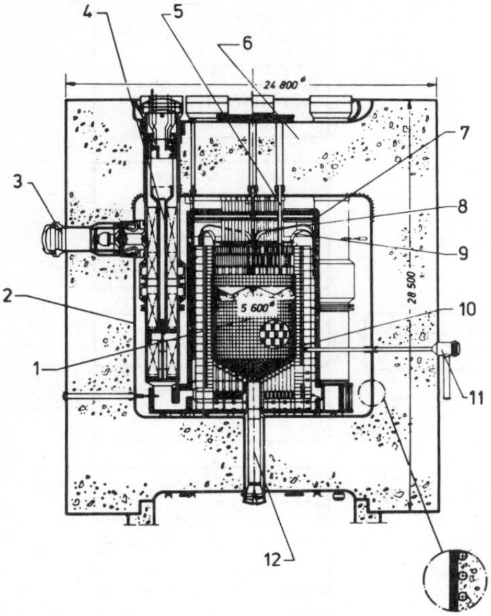

Key:

1 Pebble Bed
2 Liner
3 Blower
4 Steam Generator
5 Absorber Rod
6 PCPV
7 Thermal Barrier
8 Fuel Element Loading
9 Pneumatic Pebble Buffer
10 Reflector
11 Start-up Instrumentation
12 Fuel Element Discharge Pipe

FIG.2. Prestressed concrete pressure vessel with components of THTR prototype facility.

The core of the reactor consists of a statistical pebble fill (Fig. 2) containing 675,000 pebbles. The absorber rods move freely into the pebble bed. Control of reactor will be effected by means of control rods moving in shafts in the reflector of the reactor. The steam-generators are arranged in vertical positions concentric around the core while the cooling gas blowers are installed horizontally around the core in the wall of the prestressed concrete pressure vessel. The reactor possesses 6 cooling gas blowers.

The fuel element cycle is operated by means of a loading facility in which the pebbles are thrown pneumatically in vertical transport-tubes and, after having been decelerated pneumatically, fall onto the pebble bed from above. Below the core, in the discharge pipe, the pebbles will be withdrawn continuously and will first of all be examined for their mechanical condition. Damaged fuel elements will be passed out of the cycle. In the next step the burn-up of each fuel element will be determined. These data will be fed into a process-computer which will decide whether the element concerned should be removed from the cycle or be once more fed into the core. The loading facility is constructed in such a way that it permits the fuel elements to circulate continuously during reactor operation.

II. THE FUEL ELEMENT FLOW

The flow of the fuel elements through the THTR-300 power plant is shown in fig. 3.

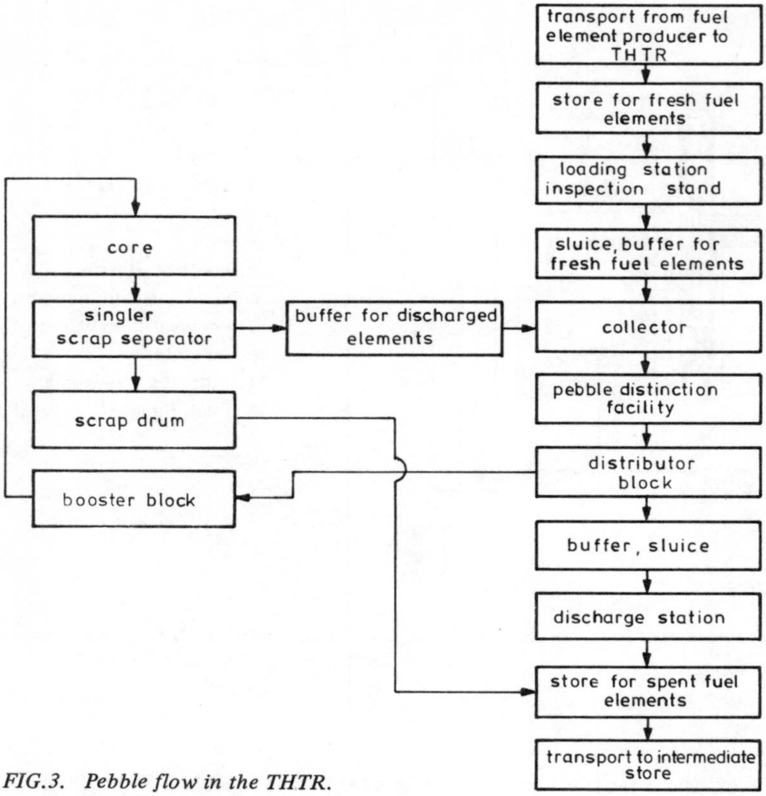

FIG.3. Pebble flow in the THTR.

The fuel elements are supplied in welted barrels containing approximately 1,000 elements each and are stored in the storage facility for fresh fuel elements. The fuel element barrels will have been sealed at the fabrication plant by the controlling organization (EURATOM). Following flange-mounting of the barrels for fresh fuel elements on the loading station of the loading facility, the fuel elements are transferred from the storage stock to the inventory of the loading facility. After the fuel elements have left the calibration facility in the inspection stand of the loading station they remain in a closed cycle during normal reactor operation, until they are discharged at the discharge station.

The elements flowing through the core are led through a scrap separator. Destroyed or damaged elements are sorted out and collected in the scrap drum. This barrel is constructed in such a way that it will not need to be replaced during the life-time of the reactor. The barrel will contain a maximum of 2,100 fuel elements. In case of need a second scrap barrel will, however, be available. It is not possible to state the fissile material content of the scrap barrel. This constitutes an unavoidable inaccuracy for the fissile-material balancing of the THTR which will amount at the maximum to approximately 4 ‰ of the total inventory of the high-pressure part of the reactor. As the scrap barrel will not be removed from the reactor-material-balance area a "tight containment" can, however, be guaranteed at this location from the safeguards point of view. Undamaged elements are stored in the buffer for undamaged pebbles. The buffer pipes of the fresh elements and of those withdrawn from the core are connected in the collector block. The collector is controlled via the central computer and control unit. The collector releases individual elements towards the measurement reactor which distinguishes between fuel elements and non-fuel elements (absorber and graphite elements).

The information obtained is passed on to the process computer. The computer will decide with the aid of the burn-up figure whether the fuel element can be introduced into the reactor core once more or whether it should be removed. Spent elements are discharchged; elements designated for the core are transported to the core via the booster pipe determined by the computer.

At the discharge station 2,100 fuel elements are filled in a discharge drum which is then sealed and transported to the storage facility for spent fuel elements. After a cooling time of approx. one year the fuel element drums are loaded in a transport container and taken to an external storage facility (leaving the reactor MBA).

The described pebble flow applies the run-in phase and the equilibrium core. In contrast to this, during first core loading, the individual types of elements are fed into the core by means of a temporary loading facility. This phase which has not yet been fixed in detail for the THTR-300 cannot be incorporated in an automated and instrumental nuclear material safeguards system. According to the present state of knowledge, it seems to be appropriate to envisage a full-time permanent inspection for the duration of this phase (approx. three months, once at the beginning of the reactor operation).

III. THE SAFEGUARDS-SYSTEM[1]

From the explanation so far made it can be seen that at any location of the reactor facility the nuclear material does not occur in smaller units than in the spherical fuel element. This fact suggests that the fuel element be defined and accounted as an item. In other words, if accounting on the basis of pebble counting can prove that each pebble was not used for any purpose other than the one stated, proof has also been furnished that no nuclear material was diverted. This definition does not imply that at certain locations of the reactor facility accounting may not take place in a higher aggregated form (e.g. fuel element drum, fuel element barrel), if an identity and integrity control can ensure that the containers are exactly definable and their integrity is clearly controllable. The nuclear material inventory of the containers then amounts to an integral multiple of the item inventory. Nuclear material accounting in a higher aggregated form has advantages not only for the operator but also for the control authority: for the operator during inventory taking and for the control authority during inventory verification. Therefore, it is suggested, that accounting on the basis of an integral multiple of the item iventories be carried out in cases where the reactor concept renders it possible.

[1] A detailed description of this safeguards system is given in the report: ENGELHARDT, H., "Development of a Nuclear Material Safeguards System for the THTR-300 Pebble Bed Reactor", JÜL-1522 (1978). This is the final report of IAEA Research Contract No.1877/RB.

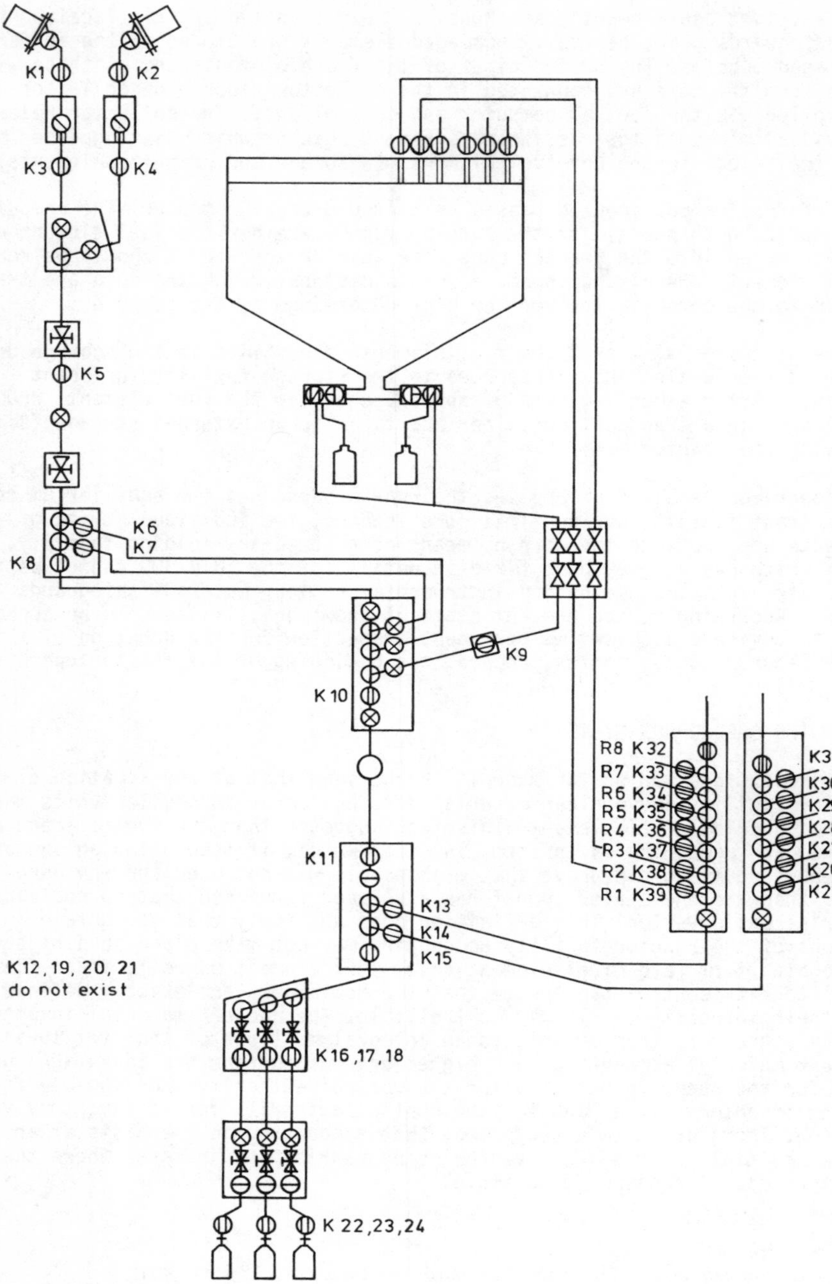

FIG.4. Indication of the pebble counter locations.

This is the case not only in the storage facility for fresh elements but also in the facility for spent elements. Accounting in the storage facility for fresh elements can be carried out via a fuel element barrel (≙ 1,000 elements) and in the storage facility for spent elements via a fuel element drum (≙ 2,100 elements). The nuclear material content of the barrel and drum is then accounted as charge. The nuclear material content of the charge amounts to approx. 1,000 g U-235 and approx. 10 kg thorium per barrel and to approx. 670 g U-235-equivalent per drum.

Metrological determination of the nuclear material within the framework of physical inventory and verification may be superfluous in the reactor facility, if a precise identification method and a possibility for determining the integrity of the barrels in connection with the nuclear material safeguards system can be found. In this case accounting and physical inventory in the two storage facilities is carried out on the basis of "charge counting".

The maximum nuclear material inventory for the storage facility for fresh elements, thus, amounts to 171 barrels, what is equivalenet to 171 kg U-235 and 1,710 kg thorium. The storage facility for spent fuel elements shows a maximum inventory of 234 drums, what is equivalent to 163 kg U-235-equivalent and 4,848 kg thorium. Accounting of the core of the reactor facility, including the loading facility, is based on element counting. In order to obtain rough estimates on the nuclear material of this part of the nuclear power plant, the number of elements, type of element and nuclear material share per element as a function of the burn-up must be known. All this information is necessary for operational reasons as input-data for the process-computer, so that they are also available for safeguards purposes. The nuclear material inventory of the core amounts to approx. 350 kg $U_{equivalent}$.

The fuel element charge (1 barrel of fresh fuel elements with 1,000 elements) is broken up in the loading facility and converted into "individual items". This procedure is repeated at full power of the facility at approx. 1 1/2 day intervals. After the loading station pebble counting is carried out by automatic pebble counters and recording devices accounting takes place via a process computer. The reasons why these components of pebble counting and accounting which the operator requires for his loading strategy are also used for nuclear material safeguards purposes, must be seen in the comprehensive redundancy of the counting devices, the high degree of tamper resistance of the total system and in the great reliability of all components as shown in the following.

The schematic diagram of the loading station (fig. 4) contains all pebble counter positions (number and location). This diagram shows that the elements are not only counted in the loading station and the filling facility but that there are further counters within the loading facility that permit tracking of the pebble flow through the loading facility. A total of 36 bridge pebble sensors have been installed outside the prestressed concrete reactor vessel for control of the loading facility. If all 36 counters are also employed for element accounting within the framework of the nuclear material control system, an extremely high redundancy of the instrumented control systems is provided which operates almost tamper-proof due to the numerous control calculations and accounting of deliberately chosen component systems.

It is possible to show that with the aid of different counters the same element inventories for deliberately chosen balance areas can be determined.

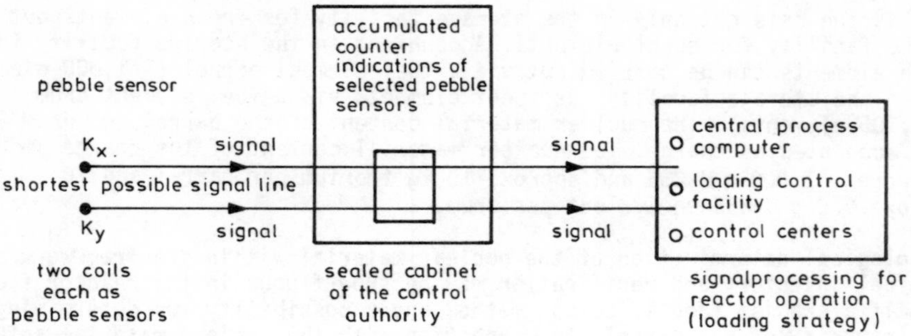

FIG.5. Signal processing of the pebble sensor signals.

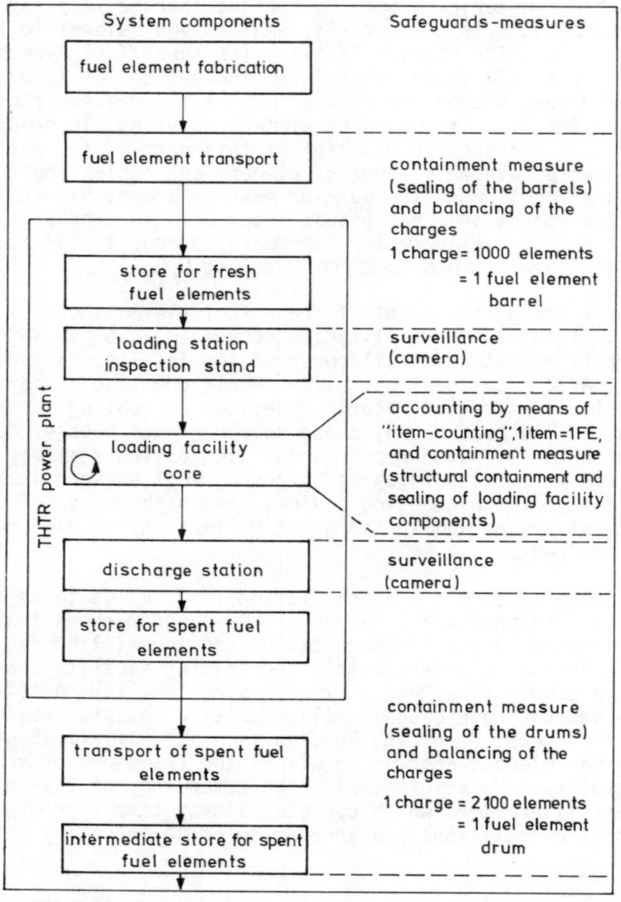

FIG.6. Safeguards measures for the different system components of the HTR fuel-element cycle.

To demonstrate the interrelation of the counter indication of the individual counters, one can define more than 25 balance-equations, which have to be satisfied all together at the same time so that it almost seems to be impossible to remove unnoticed an element from the element flow through the loading facility. Generally speaking, it is possible to carry out with the installed counters a great number of accounting and control calculations on a deliberately chosen balance area so that apart from the favourable control efficiency also an extremely high degree of redundancy of the system is guaranteed. Nevertheless, it must be said that not all operationally required counters have to be employed for nuclear material safeguards. Also the selection of a smaller number of counting devices will permit satisfactory information on non-diversion.

Beyond this, there are some possibilities to improve the control efficiency of nuclear material accounting in the THTR. One possibility for instance is a modification of impulse processing of the pebble counting facilities. The complete counting facility consists of a casing and the evaluator unit, whereby the latter can be subdivided into the analog part and the logic part with its output stage. The tamper-resistance of the counter indication of a counter facility can be increased, if the evaluator unit of the pebble sensing signals for each counter location which is also to be employed for nuclear material control purposes is designed in twin units. The part of the counter facility exclusively serving control purposes may be accommodated in a (sealed) cabinet accessible to inspectors only. The basic circuit diagram of this proposal is shown in fig. 5. The increased tamper-resistance of this signal processing is a result of the fact that during uninterrupted loading operation there are two parallel counter indication registrations for one pebble counting location, whereby it is possible to recognize manipulations in the (accessible) electronic equipment of the evaluator unit. It is not necessary to mention the increased security of this circuit in case of defects of the evaluator units of the operator.

Apart from accounting, containment and surveillance measures may also be incorporated into the THTR-safeguards-system. Fig. 6 shows the safeguards-measures for the different system components of the HTR fuel element cycle. For the storage facility for fresh fuel elements, for instance, containment measures can be envisaged. It is suggested that the barrels for fresh fuel elements be sealed at the fuel element fabrication plant to store the barrels sealed in the fresh fuel element storage facility of the power plant and not to remove the seal until shortly before they are flanged to the loading station. This measure brings advantages not only for the power plant operator, i.e. that the periodical physical inventory in the store for fresh elements represents a simple proof of non-diversion, but also for the overall system. The fresh elements (at least in the bigger unit "fuel element barrels") are clearly identifiable on their way from the fuel element producer to the nuclear power plant.

For the drums with spent fuel elements a containment measure flanking accounting in the store for spent fuel elements should also be applied for the above reasons. Containment measures for the store itself would not seem reasonable due to the great number of operationally required openings in the store and to the many required nuclear material movements through these openings. Sealing of the filled barrels in the discharge station also has the advantage that this measure is not only exclusively effective in the store for spent fuel elements, but also remains effective via the transport sector for the intermediate store and for the storage in the reprocessing plant. Sealing of the drums can be carried out in such a way that the locking covers of the drums are provided with a tamper-proof mark. Detailed investigations for developing a practical sealing procedure for the drums are at present on the way.

Furthermore, at the loading and discharge station surveillance measures can be applied. At these points the seals have to be removed or attached. If the above described measures of accounting, containment and surveillance become an integral part of the safeguards-system for the THTR, this reactor type may be regarded as reliably safeguardable within the meaning of the NPT.

DISCUSSION

M. HONAMI: What is the justification for regarding spent fuel elements (pebbles) as individual items which can be counted, when you have burnup information from the pebble distinction facility and therefore could maintain quantity records?

H. BÜKER: The operator uses only an average burnup for discharged fuel elements (pebbles).

M. HONAMI: But a spent fuel drum contains absorber elements, graphite elements, and fuel elements.

H. BÜKER: Once the equilibrium operating phase is reached, the reactor will contain only fuel elements, and therefore only fuel elements will be discharged into spent fuel drums.

NON-DESTRUCTIVE MEASUREMENTS ON NUCLEAR MATERIAL FROM THE URANIUM-THORIUM CYCLE

G. STEIN, P. CLOTH, P. FILSS, M. HEINZELMANN
Kernforschungsanlage Jülich GmbH,
Jülich, Federal Republic of Germany

Abstract

NON-DESTRUCTIVE MEASUREMENTS ON NUCLEAR MATERIAL FROM THE URANIUM-THORIUM CYCLE.

At the KFA Jülich fresh and irradiated nuclear material of the U/Th cycle was subjected to non-destructive measurements. Passive and active measuring methods were used. The passive method used was a measuring system on the basis of Ge(Li) technology, whereas the active method comprised a system on the basis of fission coincidence, a system on the basis of delayed neutrons and, finally, one on the basis of selective neutron transport. After positive test results have been obtained, these measuring methods will be used in safeguards systems for facilities of the HTR fuel cycle.

1. Introduction

In the Federal Republic of Germany the U-Th fuel cycle is being developed within the framework of the application of high-temperature reactors. So far, a fabrication facility for fresh fuel and a 15 MW_e AVR experimental reactor facility are in operation. A 300 MW_e THTR reactor facility is presently being built which is to start operation in approx. 1980. An experiment for reprocessing HTR-elements is being carried out at KFA Juelich. In addition, the KFA Juelich is developing HTR-reactor facilities for the production of nuclear process heat (PNP) and direct gas turbines (HHT). Apart from the development of these facilities, special attention is attached to the closing of the fuel cycle and in this connection to the development of suitable nuclear fuels, mainly consisting of uranium and thorium mixtures in graphite matrixes.

Within the framework of the development of safeguards systems for facilities of the HTR-fuel cycle, that is being carried out at the KFA Juelich, different systems for the non-destructive assay of nuclear material of the U/Th cycle are being developed and/or are in operation. It is envisaged, following successful testing, to employ these systems in suitable stations of the fuel cycle for nuclear material safeguards and quality control.

2. Description of the Nuclear Material from the U-Th Cycle

The nuclear material presently used for the AVR-operation and for the THTR has an enriched uranium content of 93 %. Thorium admixtures as breeder material are added to the fuel element at a ratio of 5:1 and/or

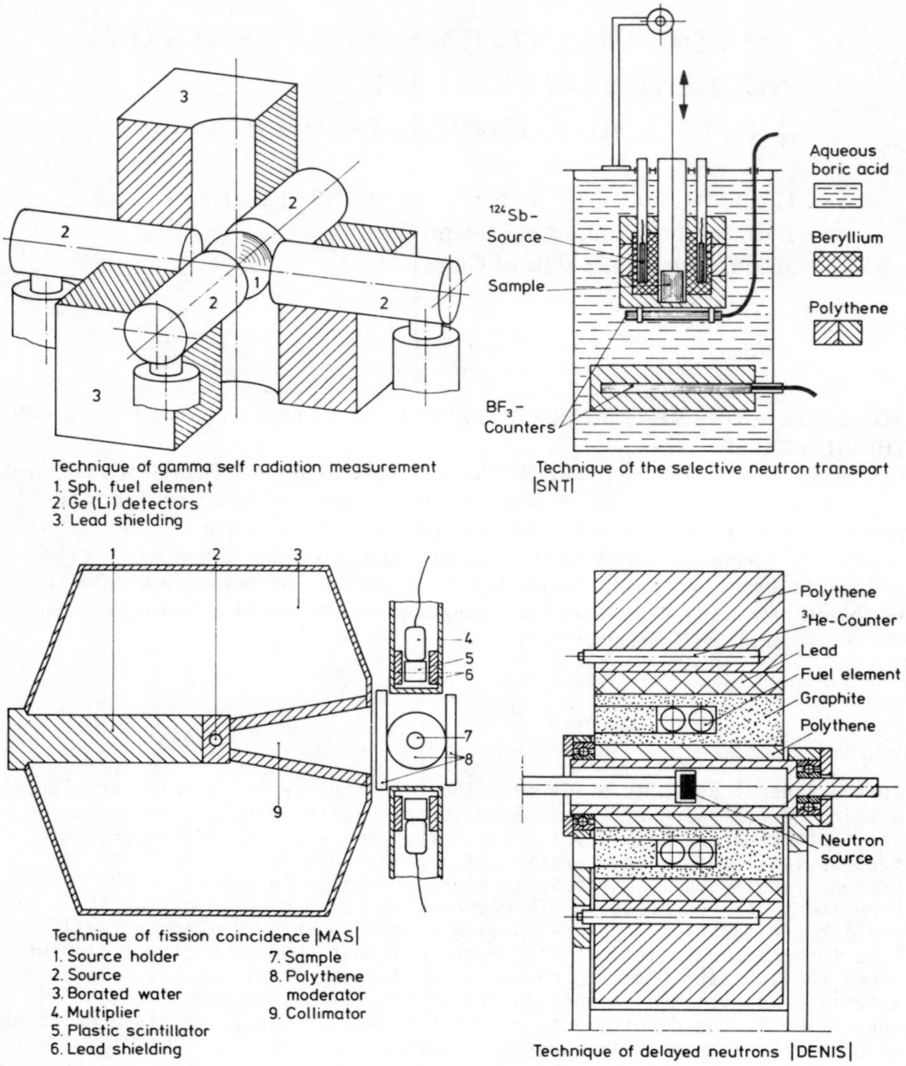

FIG.1. Measurement systems for non-destructive assay.

10:1. During the fuel element production UO_2 and THO_2 kernels as mixed oxides with pyrolytic carbon and/or silicon-PyC layers are shaped to coated particles. The UO_2 and/or THO_2 kernels have a diameter of $\sim$ 400 μm and the coated particles of $\sim$ 800 μm.

The final fuel element has a spherical shape with an outside diameter of 6 cm. The graphite matrix of this pebble contains 15000 coated particles whose uranium weight amounts to 1.1 g of 93 % enriched U-235. The fuel-free graphite shell of the pebble has a diameter of 0.5 cm.

When assuming a burn-up of 100 000 MW d/t for the fuel element, one obtains a U-233 content of 400 mg and a U-235 content of 70 mg at a thorium/uranium mixture ratio of 5:1 in the fresh fuel element.

The configurations occuring for the nuclear material in the HTR-cycle for non-destructive analysis are: kernels, particles, pebbles and solutions which are generated during reprocessing.

3. Description of the Measuring Systems for the Non-Destructive Assay of Nuclear Material.

3.1 Technique of the selective neutron transport /SNT/

A low energy neutron flux is induced by an antimony-beryllium neutron source in the samples containing fissile materials. The neutrons from the fission events are detected by a BF_3 counter after traversing a water layer containing neutron poisons. The primary source neutron flux is selectively absorbed in this region. The contribution of the neutron source corresponds to 30 mg of U-235. For a spherical fuel element with 1 g U-235 and a measuring time of 1 000 sec the statistical error is less than 1 % (1σ).The measuring system operates is a γ-field of up to 10^4 R/h. This method was developed with special regard to recycling of HTR-fuel. Fig. 1 shows a schematical diagram of this measuring system [1].

3.2 Technique of delayed neutrons /DENIS/

Non-destructive measurements of fissile material contents can be performed by detection of delayed neutrons after irradiation of the samples in a low ernergy neutron field. Somewhat different time behavior of these delayed neutrons in the case of U-235 and U-233 makes it possible to distinguish between the samples. The measuring system is shown in fig. 1 [2]. The target of the 14-MeV neutron generator is surrounded by a 20 mm-thick iron cylinder followed by a layer of 30 mm polyethylene as neutron moderator. The iron cylinder is the hollow axis of a wheel and made of graphite. It bears a circle of 16 holes parallel to the axis, each capable of holding two pebbles of 6 cm in diameter. The wheel is rotated at 250 rpm in order to work against inhomogeneous loading of fissile material and inhomogeneous efficiency of the detectors, which are at positions in the static part of the apparatus. This static part consists of a lead cylinder of 50 mm thickness surrounding the rotating wheel and acting as a γ-shield in the case of irradiated fuel. The lead cylinder is embedded in a polyethylene block containing bore holes close to the lead in which up to 16 detectors can be inserted. The number of detectors can be increased considerably if necessary to increase counting efficiency. During the test measurements usually He-3 counters were used for efficiency reasons; however, BF_3-counters were also tested.

3.3 Technique of fission coincidence /MAS/

The MAS measuring system is shown in fig. 1 [3]. Fission processes are generated with neutrons of a Cf-252 source in the fissile material and/ or with fast neutrons in the fertile materials of the samples. The fast neutrons and gammas generated during this fission process are determined by means of a four-detector arrangement. The 40-μg Cf-252 neutron source is accommodated in a cylindrical container which, for radiation reasons, is filled with boric water. The pertinent neutron collimator is conically shaped and made of lead and artificial resin. The four-detector arrange-

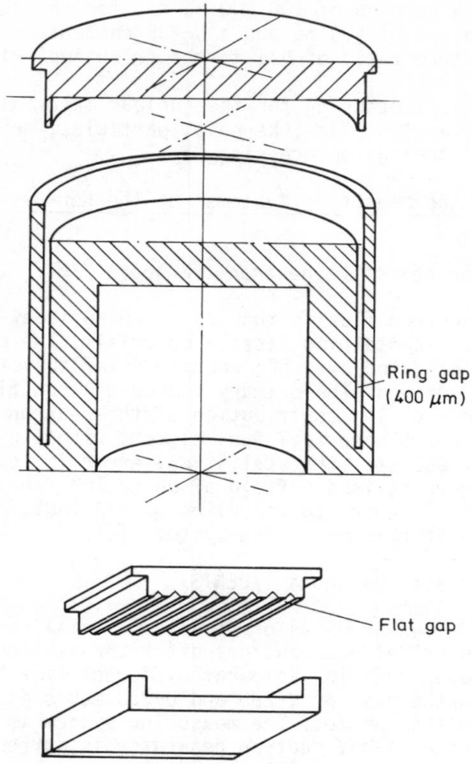

FIG.2. Sample container.

ments consists of two boxes, each of which contains two plastic scintillators on top of each other that are equipped with fast photo-multipliers and embedded in lead and/or a steel housing. The electronic counting unit for the four-detector arrangement consists of 4 discriminators, a coincidence unit by means of which 3/4 and 2/4 coincidences can be pre-selected, as well as of four counters. The counting time is pre-selectable by means of a timer unit and the measured number of impulses is automatically printed via a printer.

3.4 Technique of gamma self-radiation measurement

By means of Ge (Li) detectors with high resolution the 185.7 keV line of the U-235 and the 238.6 keV line of the Pb-212 are used for the determination of U-235 and/or Th-232. Fig. 1 shows the cross-shaped arrangement of four Ge (Li) detectors for analyzation of fresh HTR-fuel element pebbles [4]. The samples are moved into the measuring position by a stepping motor. By turning the fuel during the measuring process and with the aid of lead shields a location-independent determination of self-radiation is achieved. Gamma spectra were analyzed with a view to their applicability for burn-up or fissile material determination using Ge (Li) detectors with high resolution on spent fuel elements.

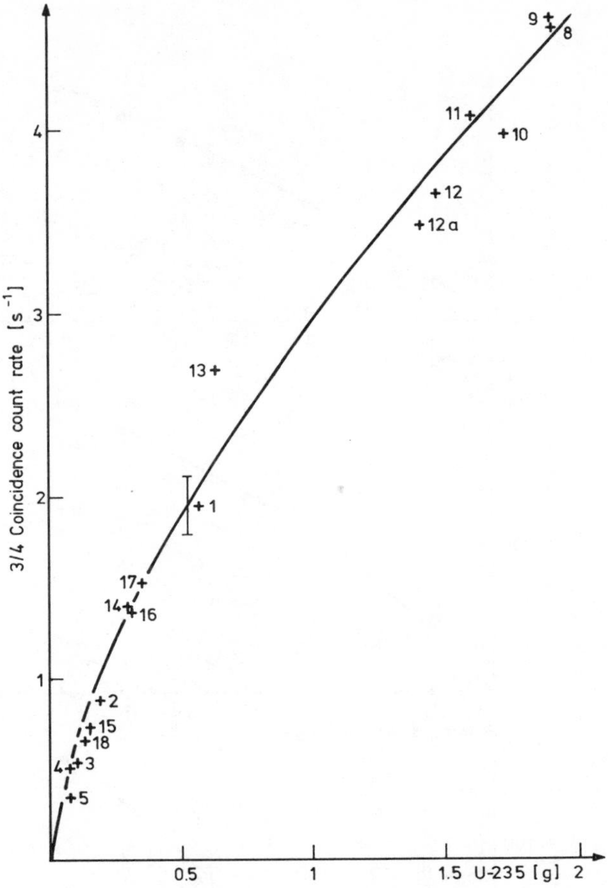

FIG.3. 3/4 coincidence count-rate for different ^{235}U contents in the flat container.

4. Results of the Measurements on Nuclear Material from the U-Th Fuel Cycle

4.1 Fresh nuclear material

The kernels and coated particles of different enrichment, thorium admixtures and different diameter were analyzed by means of the MAS-system. The aim of the experiments was to obtain a universal calibration curve of the U-235 content for the different samples.

In order to keep the effect of self-shielding of the samples on the measuring results as low as possible, sample containers were constructed by means of which it is possible to determine the nuclear material in an as thin as possible layer thickness. Fig. 2 shows the sample containers used here. The nuclear material to be examined is distributed on a cylinder plane with the ring gap container. During the analyses it became

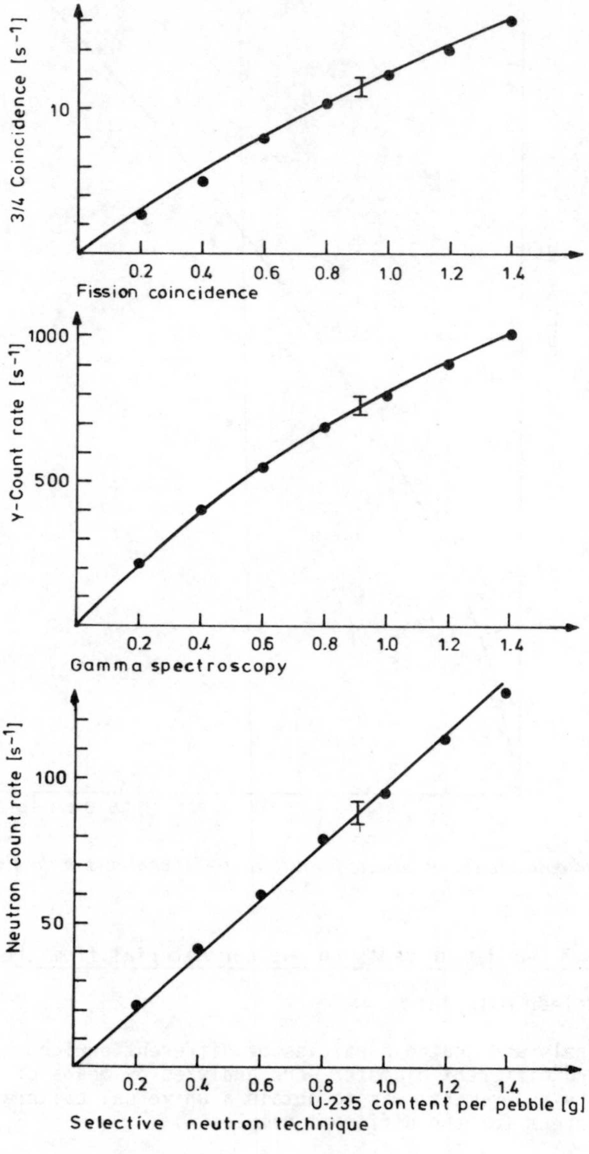

FIG. 4. Calibration curves for THTR-type pebbles.

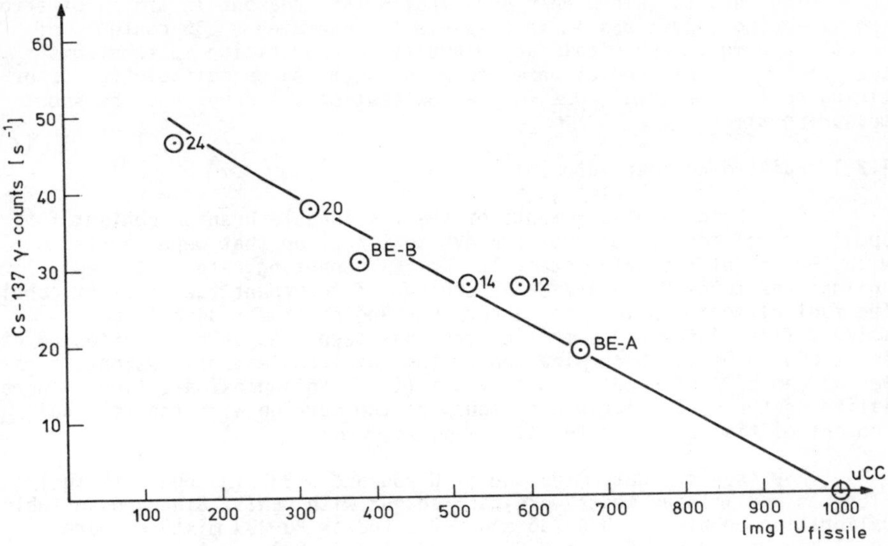

FIG.5. Measurements on pebbles with different burnup.

apparent that it was not possible to fill the ring gap of all sample configurations in the same manner. In the case of the flat gap container the coated particles and/or kernels are filled in the grooves of a plexiglass pane. In this container the material is distributed even more widely than in the ring gap container. Fig. 3 shows the measuring results gained with the flat gap container. It becomes evident that the flat gap container produces the best results with regard to the linearity and deviation of the measuring points from the curve. The numbers 1-18 in figure 3 represent kernels and coated particles of different enrichment, thorium composition coating etc. Summarizing, it may be said that the flat gap container is well suited for the assay of nuclear material samples gained in processes and that it permits determination of the U-235 content also in the case of unknown sample compositions with an accuracy of up to ± 1o % for a 2o-minute measuring time, except for result of sample 13 which consists of Al/Si-mantled particles.

Fig. 4 shows the results of measurements of the THTR-type fuel elements. The U-235 content varies between 0.2 g and 1.4 g. The thorium admixture amounts to a 10:1 ratio vis-à-vis the respective uranium content. The figure shows the U-235 content per pebble in g and the response signal of the respective measuring system. During the self-radiation measurement the max. statistical error is ± 0.5 % (1 σ) for a measuring time of 200 sec. per pebble. For the SNT-system the statistical error amounts to ± 1 % (1 σ) at a measuring time of 1 000 sec. When applying the MAS-system, the statistical error varies between 2.5 % and 14 % (1 σ) for a 20 minute measuring time.

The results gained here show that within reasonable limits of error and measuring conditions it is possible to determine U-235 contents of up to 1 400 mg in spherical fuel elements. For precision measurements the SNT and/or self-radiation measurement seems to be suitable for determining nuclear material, due to the low statistical error and the short measuring times.

4.2 Irradiated nuclear material

Fig. 5 shows measurements of the fissionable uranium content for spent fuel element pebbles of the AVR-test reactor that were carried out with the aid of the SNT-system. The Cs- 137 counting rate is traced against the total U-235 and U-233 content of the spent fuel element pebbles. The fuel element pebbles show a burn-up ranging from ∿ 34 % fifa to ∿ 120 % fifa. A fresh UCC fuel element has been used as calibration point for 1 000 mg U_{fiss}. Determination of the total content of fissionable material can be performed of up to ± 3 % (1 σ). An approximate linear correlation of the Cs-137 curve and, thus, of the burn-up with the residual content of fissionable material can be assumed.

For separate determination of U-235 and U-233 in spent HTR-fuel elements preliminary tests were carried out with DENIS. Since no suitable calibration pebbles with U-235 and U-233 and/or Pu-239 mixtures were available, analyses of pure U-235, U-233 and Pu-239 samples were performed which are compared with computed values. For determination of the short-lived group of the delayed neutrons irradiation and measuring times of 1 sec were applied, whereas the long-lived group was subject to 30 sec. These irradiation and measuring procedures are repeated in a suitable manner. Practical problems came from the pulsing and running of the counting system immediately after the output of the very intense fast neutron source of 10^{11} neutrons/sec. The background of the system was found to be equivalent to less than 0.05 g U-235. The intensity ratios between the short-lived and long-lived group show the following results contained in table I.

TABLE 1. MEASURED AND COMPUTED INTENSITY RATIOS BETWEEN THE SHORT-LIVED AND LONG-LIVED GROUP OF THE DELAYED NEUTRONS [5]

Isotope	$V_{meas.}$	V_{cal}
U-233	1.97 ± 0.16	2.23
U-235	1.35 ± 0.12	1.79
Pu-239	1.24 ± 0.10	1.48

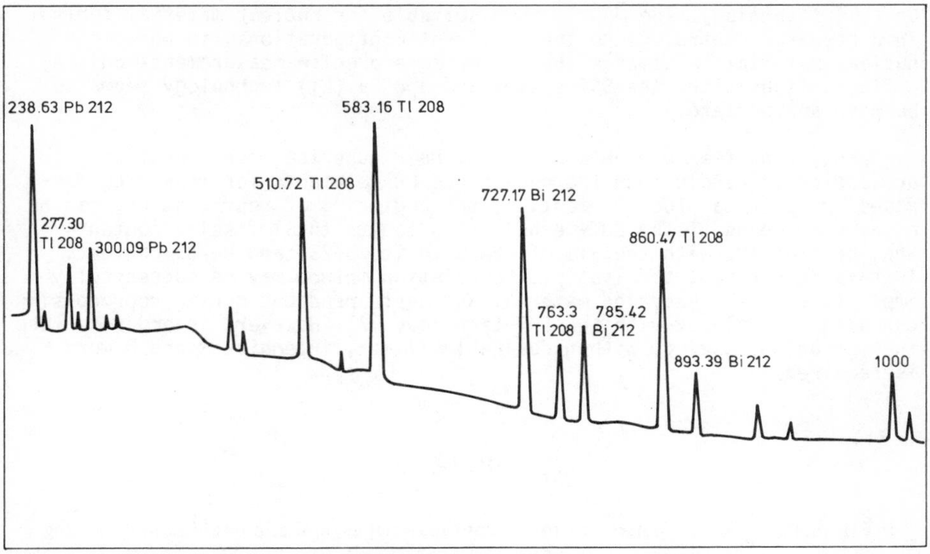

FIG.6. Gamma-spectrum of reprocessed ^{233}U.

For the U-235 measurements fresh HTR-fuel element pebbles were used, for the Pu-239 coated particles and for the U-233 measurements oxide powder with a comparatively very high densisty. The results show that mixtures of U-233 and U-235 as they may occur in spent HTR-elements can basically be determined. For further detailed studies suitable calibration pebbles will be produced containing U-233 and U-235 as coated particles.

The isotopic correlation method is also a suitable means of analyzing nuclear material in irradiated fuel elements. Therefore, the irradiation history of a fuel element in a HTR pebble bed reactor with U/Th fuel was simulated and investigated using a modified ORIGEN program for determination of isotopic correlation of fission product and heavy metal content [6]. Only few of the fission products are suitable for analyzation among them Cs-137, Cs-134 and Eu-154. Interesting because of their linear correlation, were the following correlations which were evaluated, especially: [U-234/U-236] : [Cs-137], [U-234/U-236] : [Eu-154].
Fig. 6 shows a γ-spectrum of a U-233 sample which has been reprocessed from spent U/Th nucelar material. It becomes evident that the daughter products of the U-232 and particularly the 2.6 MeV line of the Tl-208 seems to be suitable for qualitative nuclear material control. The γ-activity of the U-233 itself is fully covered by the very intensive γ-lines of the U-232 daughter products.

5. Conclusion

Fresh nuclear material from the U-Th fuel cycle can be assayed with systems like SNT or MAS as well as with passive systems on the basis of

Ge (Li) technology. The MAS is more suitable for nuclear material control in a research centre due to the different configurations, in which the nuclear material can emerge there. For more precise measurements on specific configurations the SNT system and the Ge (Li) technology seems to be more appropriate.

The spent fuel elements - due to their spherical configuration - are easier to handle than the rod-shaped LWR-elements for measuring purposes. A spherical fuel element, at the beginning of reprocessing, can be assayed by means of the SNT-technique as to its total fissile content and, separately, with the aid of DENIS to its U-233 and U-235 content. In this connection, the isotopic correlation method may be successfully employed as well. Recycled material and waste produced during reprocessing can also be analyzed with the SNT-technique [7]. However, in order to keep the measuring accuracy within desirable limits, intensive R and D work is required.

REFERENCES

[1] FILSS, P., "Non-destructive control of fissile material in solid and liquid samples arising from a reactor and fuel reprocessing plant," Safeguarding Nuclear Materials (Proc. Symp. Vienna, 1975) **2**, IAEA, Vienna (1976) 471.

[2] CLOTH, P., KIRCH, N., KRING, F.J., "Non-destructive measurements of ^{235}U and ^{233}U content in HTR fuel elements by delayed neutron analysis, ibid. p.533.

[3] STEIN, F., et al., Zerstörungsfreie Meßsysteme zur Kernmaterialüberwachung in der KFA Jülich, Jül-1473 (1977).

[4] LANG, H., MEIXNER, Ch., Kerntechnik **8** (1975) 351.

[5] ROSSENBACH, M., KFA Jülich, Diplomarbeit (1978).

[6] SUSANTI, K., KFA Jülich, Diplomarbeit (1978).

[7] FILSS, P., "Non-destructive assay techniques for irradiated fissile material in extended configurations", IAEA-SM-231/17, these Proceedings, Vol.II.

DISCUSSION

F. SCHINZER: You said that for the measurement of thorium with the Ge(Li)-detector you used the gamma-emitting daughter ^{212}Pb. How do you overcome the difficulty arising from the fact that the daughters of thorium are usually not in equilibrium?

G. STEIN: With the method described it is only possible to determine thorium of the same age.

MATERIAL CONTROL AND ACCOUNTABILITY ASPECTS OF SAFEGUARDS FOR THE UNITED STATES URANIUM-233/THORIUM FUEL RECYCLE PLANT*

J.A. CARPENTER, Jr., S.R. McNEANY, P. ANGELINI
Oak Ridge National Laboratory,
Oak Ridge, Tennessee

N.D. HOLDER, L. ABRAHAM
General Atomic Company,
San Diego, California,
United States of America

Abstract

MATERIAL CONTROL AND ACCOUNTABILITY ASPECTS OF SAFEGUARDS FOR THE UNITED STATES URANIUM-233/THORIUM FUEL RECYCLE PLANT.

The material control and accountability aspects of reprocessing and refabrication in a large-scale high-temperature gas-cooled reactor (HTGR) fuel recycle plant are discussed. Two fuel cycles are considered. The highly enriched uranium (HEU) cycle uses uranium enriched 93% in ^{235}U as the initial fuel. The medium-enriched uranium (MEU) cycle uses uranium with a ^{235}U enrichment less than 20% as its initial fuel. In both, ^{233}U is bred from thorium. The HEU ^{235}U and the ^{233}U of both cycles are recycled. The MEU ^{235}U is retired to waste after one reactor cycle. Typical heavy metal contents of spent fuel elements from both cycles are presented. The main functional areas of the recycle plant are shipping, receiving, and storage; reprocessing; refabrication; and waste treatment. A real-time materials accountability system will manage the data provided by measurements from all four areas. Simulations of material flow used in the HTGR development programme are forerunners of such a system. The material control and accountability aspects of reprocessing and refabrication only are discussed. The proposed accountability areas are identified and the measurement techniques appropriate to various streams crossing the boundaries of the areas are identified. Special emphasis is placed on novel non-destructive methods developed for assaying solid materials containing ^{233}U-Th. The material form, total uranium and plutonium, and activity of selected reprocessing streams are listed. The isotopics and activity of the uranium input into refabrication are also presented.

INTRODUCTION

The U.S. HTGR Recycle National Program

The objective of the U.S. High-Temperature Gas-Cooled Reactor (HTGR) ^{233}U-Th fuel recycle program is the design and licensing of a large-scale demonstration recycle plant to be built and operated in the

* Research sponsored by the Nuclear Power Development Division, US Department of Energy under Contract W-7405-eng-26 with the Union Carbide Corporation.

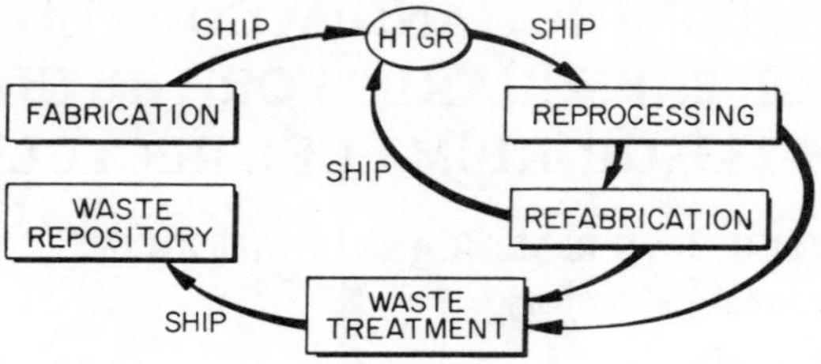

FIG.1. Fuel cycle.

time frame of 1995–2000. Heretofore, emphasis of the program [1] was on the development of the recycle technology, much of it done in cooperation with the United Kingdom and the Federal Republic of Germany. The development effort has now progressed to the stage in which almost all the process steps of reprocessing and refabrication have been demonstrated in prototypic equipment with natural or depleted uranium. While this development work progressed, conceptual design studies of such a recycle plant were conducted and included materials control and accountability.

The fuel cycle

The general flow of materials for the HTGR ^{233}U-Th fuel cycle is indicated in Fig. 1. Enriched uranium and thorium are fabricated into elements in a fresh fuel plant and sent to the reactor. The spent fuel is sent to the recycle plant, where it is reprocessed to recover the fissile ^{233}U produced from the thorium and, in some cases, the residual ^{235}U. These fissile materials are combined with fresh thorium and refabricated into recycle elements, which are shipped back to the reactor. The unrecovered fissile material and other wastes are processed in waste treatment and sent to a repository. The spent thorium is stored for later use.

This paper examines the two nuclear fuel cycles of primary interest for implementation with the HTGR. The first is called the highly enriched uranium (HEU) cycle. It has the best economic performance and resource utilization and traditionally has been the prime candidate for HTGR use. However, in response to proliferation concerns, a number of alternate cycles have been examined. A fuel cycle that may provide more proliferation resistance is the medium enriched uranium (MEU) cycle; however, its economic performance is not as good as that of the HEU cycle. A low enriched uranium (LEU) cycle has also been considered. As there would be no recycle of the fuel at all, this cycle is not considered in this paper.

Scope of paper

Shipping, receiving, and storage; reprocessing; refabrication; and waste treatment are the main functions of the recycle plant. The majority of the materials control and accountability problems are in reprocessing and refabrication and it is there that most of our efforts have been placed; therefore, only reprocessing and refabrication are discussed. Item control will be used in shipping, receiving and storage. The materials control and accountability aspects of waste treatment are being defined. Physical security aspects are not addressed, although they have been and are being considered in our studies.

GENERAL SAFEGUARDS CONSIDERATIONS FOR THE ENTIRE FUEL RECYCLE PLANT

The fuel and the fuel cycle

The HTGRs and HTGR fuel recycle have been reviewed in detail elsewhere [2–5]. The U.S. design of the General Atomic Company uses a hexagonal graphite block 0.79 m high and 0.36 m across the flats as its fuel element. The fuel and the fuel element are depicted in Fig. 2. The fuel is contained in microspheres less than 1000 μm in diameter. The fissile particle containing the initial fuel, ^{235}U or ^{233}U, is coated with three layers of pyrolytic carbon and a single layer of silicon carbide. The fertile particle, containing thorium, from which additional ^{233}U is produced, is coated with two carbon layers only. The two types of particles are bonded by a graphitic matrix to form a fuel rod about 51 to 65 mm long and 13 to 16 mm in diameter. These fuel rods are stacked end-to-end into holes drilled longitudinally through the block parallel to the coolant holes.

Selected heavy metal compositions and characteristics of spent fuel elements are presented in Table I for both the HEU and MEU fuel cycles. The spent fuel compositions are for burnups of about 70,000 MWd per tonne heavy metal (U + Th) for the HEU fuel cycle and 85,000 MWd per tonne heavy metal for the MEU fuel cycle, both cooled 180 days from reactor discharge.

The HEU fuel cycle uses three types of elements. One is the initial or makeup element, produced by the fresh fuel plant. This contains uranium highly enriched in ^{235}U (∿93%) as its initial fuel. The other two types of elements are products of the recycle plant. One type contains uranium highly enriched in ^{233}U (∿70%) produced from the thorium in previous reactor cycles. The third type uses uranium containing about 30% ^{235}U, which is the residual of the uranium from previous irradiation of the initial or makeup elements. These three types of elements are designated IM, 23R, or 25R elements to denote elements charged to the reactor or IMS, 23RS, or 25RS to denote spent elements, respectively. The uranium in the fissile particles of the IMS and 23RS elements and in the fertile particles from all three types of elements is recovered in the recycle plant. The fissile particles from the 25RS elements, containing uranium that is about 8% ^{235}U but about 70% ^{236}U, are retired to waste.

The MEU cycle uses only two types of elements, the IM element and the 23R element containing ^{233}U (70%).[1] The IM element contains uranium with an enrichment less than 20% ^{235}U. This burns down to 4% ^{235}U in the spent fuel element, so the fissile particles of the IMS

[1] Another MEU cycle, in which the recycle 23R uranium is denatured to less than 12% ^{233}U has also been considered but is not the current reference.

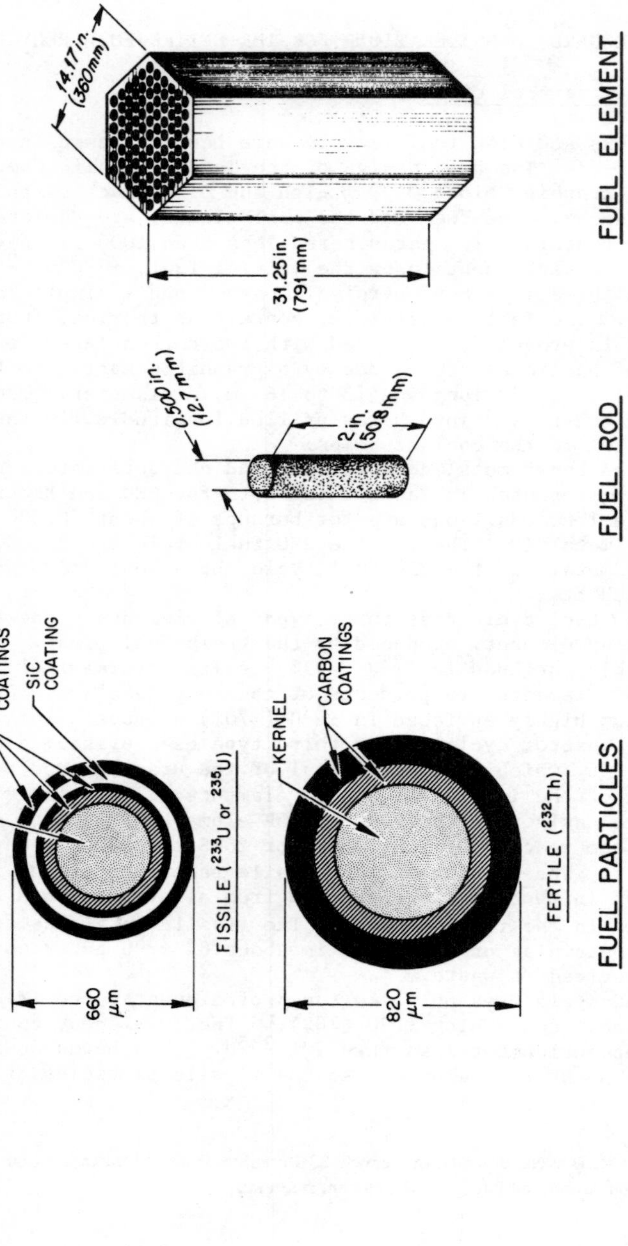

FIG.2. HTGR fuel components.

TABLE I. QUANTITIES OF SELECTED HEAVY METALS IN TYPICAL HIGH-TEMPERATURE GAS-COOLED REACTOR SPENT FUEL ELEMENTS COOLED 180 DAYS FROM REACTOR DISCHARGE

	Highly Enriched Uranium (HEU) Fuel Cycle						Medium Enriched Uranium (MEU) Fuel Cycle			
	Initial or Makeup Fuel Element		23R Recycle Fuel Element		25R Recycle Fuel Element		Initial or Makeup Fuel Element		23R Recycle Fuel Element	
	Fissile Particle	Fertile Particle	Fissile Particle	Fertile Particle	Fissile Particle	Fertile Particle	Fissile Particle	Fertile Particle	Fissile Particle	Fertile Particle
Thorium, Total wt/Fuel Element (g)	0.01017	10,550	9.76×10^{-3}	10,550	0.01984	10,550	0.0333	3216.3	2.02×10^{-3}	1,657
-228, wt% of Total Th	0.75275	1.48×10^{-5}	1.266	1.48×10^{-5}	0.2807	1.48×10^{-5}	2.81×10^{-5}	1.36×10^{-5}	46.06	1.36×10^{-5}
-229, wt% of Total Th	0.2322	1.69×10^{-5}	20.03	1.69×10^{-5}	0.0862	1.69×10^{-5}	1.72×10^{-5}	1.72×10^{-5}	15.56	1.72×10^{-5}
-230, wt% of Total Th	91.73	7.89×10^{-5}	78.23	7.89×10^{-5}	34.717	7.89×10^{-5}	1.07×10^{-5}	1.07×10^{-5}	38.38	1.07×10^{-5}
-231, wt% of Total Th	2.04×10^{-4}		1.65×10^{-5}		2.31×10^{-4}					
-232, wt% of Total Th	7.2858	99.99989	0.4731	99.99989	64.913	99.99989	99.994	99.996		99.996
-234, wt% of Total Th	3.55×10^{-4}				2.22×10^{-3}					
Protactinium, Total wt/Fuel Element (g)	6.61×10^{-6}	0.1118	4.23×10^{-4}	0.1118	0.600	0.1118	1.3×10^{-6}	0.1104	1.67×10^{-4}	0.0569
-231, wt% of Total Pa	99.224	67.57	99.951	67.57	85.2	67.57	30.72	20.98	99.9999	20.98
-233, wt% of Total Pa	0.776	32.43	0.049	32.43	14.8	32.43	69.27	79.02		79.02
Uranium, Total wt/Fuel Element (g)	198.5	270.5	174.4	270.5	730.5	270.5	1851.9	109.45	85.135	56.384
-232, wt% of Total U	4.45×10^{-5}	0.0531	6.39×10^{-4}	0.0531	3.91×10^{-6}	0.0531	4.55×10^{-8}	0.0382	5.65×10^{-2}	3.82×10^{-2}
-233, wt% of Total U	1.31×10^{-4}	77.57	8.67	77.57	1.77×10^{-5}	77.57	5.07×10^{-5}	82.956	20.17	82.95
-234, wt% of Total U	8.40	17.34	42.64	17.34	0.7894	17.34	8.66×10^{-6}	14.171	44.44	14.17
-235, wt% of Total U	30.92	4.36	23.81	4.36	8.160	4.36	3.904	2.54	19.49	2.54
-236, wt% of Total U	45.57	0.6844	24.76	0.6844	69.025	0.6844	3.469	0.2956	15.83	0.2956
-238, wt% of Total U	15.1	9.7×10^{-4}	0.1144	9.7×10^{-4}	22.026	9.7×10^{-4}	92.624	1.08×10^{-5}	1.96×10^{-3}	1.08×10^{-5}
Neptunium, Total wt/Fuel Element (g)	13.07	0.1083	5.979	0.1083	0.1589	0.1083	8.5603	0.01836	2.114	9.46×10^{-3}
-237, wt% of Total Np	99.9999	99.9999	99.9999	99.9999	99.9999	99.9999	99.9997	99.9999	99.9999	99.9999
Plutonium, Total wt/Fuel Element (g)	8.537	0.02912	3.551	0.02912	61.21	0.02912	48.647	4.07×10^{-3}	1.2093	2.10×10^{-3}
-238, wt% of Total Pu	61.773	82.546	74.46	82.546	64.55	82.546	7.07	89.80	85.52	89.80
-239, wt% of Total Pu	16.31	10.686	12.05	10.686	15.00	10.686	37.17	6.27	7.96	6.27
-240, wt% of Total Pu	9.30	4.104	6.5	4.104	8.75	4.104	16.39	1.65	2.86	1.65
-241, wt% of Total Pu	6.48	2.000	4.25	2.000	6.17	2.000	20.20	1.03	2.57	1.03
-242, wt% of Total Pu	6.14	0.6621	2.73	0.6621	5.55	0.6621	19.16	1.25	1.09	1.25
Total Weight of Fuel Element (g)	120,600		119,900		122,700		115,180		110,050	
Total Activity of Fuel Element (Ci)	56,450		53,100		61,720		67,310		22,300	

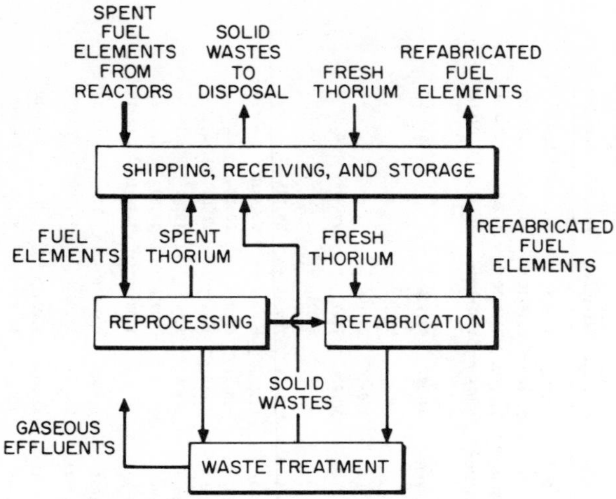

FIG.3. HTGR fuel recycle flowsheet.

elements are discarded and not recycled. The uranium in the 23RS elements and that in the fertile particles of the IMS elements are recycled.

The Fuel Recycle Plant

The overall flow of material within the recycle plant is indicated in Fig. 3. The spent fuel elements enter the plant in Shipping, Receiving, and Storage and are stored before delivery to Reprocessing.

From Reprocessing recovered fissile material is delivered to Refabrication. Retired fissile material and all wastes are sent to Waste Treatment. The spent thorium is solidified and placed in storage for 20 to 30 years to allow its radioactivity to decay before recycling.

In Refabrication the recovered fissile material is joined by fresh thorium in the form of coated particles, and these are fabricated into recycle elements, which are stored and eventually sent to the reactor. Refabrication wastes are also sent to Waste Treatment.

In Waste Treatment all solid and liquid and some gaseous wastes are processed to a repository-ready solid waste form. The remaining gaseous effluents are treated and vented to the atmosphere. Liquid effluents from the plant contain no nuclides from the fuel elements. The solid wastes are stored and ultimately delivered to a waste repository.

Because of the inherent radioactivity associated with most of these streams, virtually all process operations in all four main areas of the plant will be done remotely.

General Plant Material Control and Accountability

The HTGR Fuel Recycle Plant will incorporate the latest state-of-the-art techniques to implement a highly effective safeguards program.

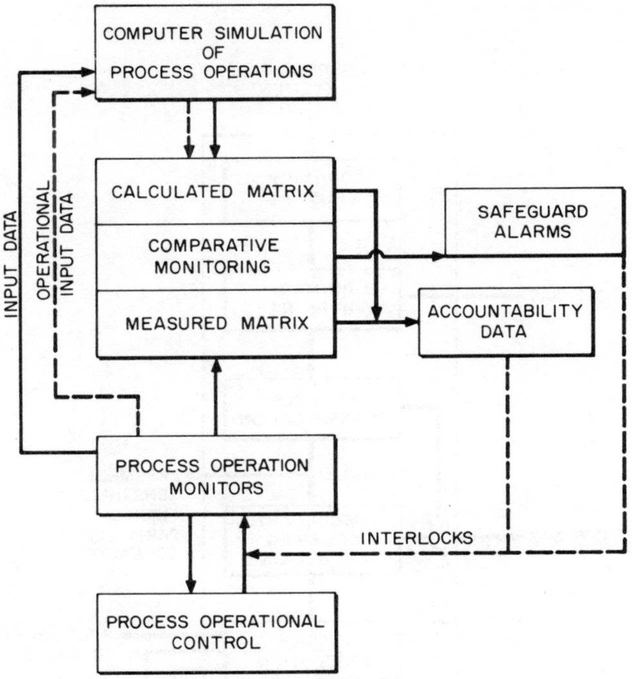

FIG.4. Real-time material accountability concept by integration of process simulation and monitoring.

Safeguards are maintained by the inherent radioactivity of the fuel, the physical barrier of the required heavy shielding, an integrated system of measurement including destructive analysis of samples and nondestructive assay, and physical security. The data collected will be managed by a real-time accounting system similar to the DYMAC Program [6,7] developed at Los Alamos Scientific Laboratory (LASL). A schematic of the proposed system is shown in Fig. 4. A central computer collects and analyzes instrument data and operator-supplied information to continuously update the recorded status of material locations throughout the plant. Simultaneous computer simulations of plant operations continuously calculate expected amounts of material in various parts of the plant. The two values are continuously compared.

In support of the development of real-time accounting capability, several material flow models [8,9] have been developed to simulate expected mass flow patterns throughout the recycle plant. One of these, which calculates average fissile mass movements, has been used to determine accuracy requirements of measurement devices [10] to meet U.S. government material accountability standards. Other studies have yielded the time-dependent variations of these flows. Simulations of reprocessing operations are under way at General Atomic Company using

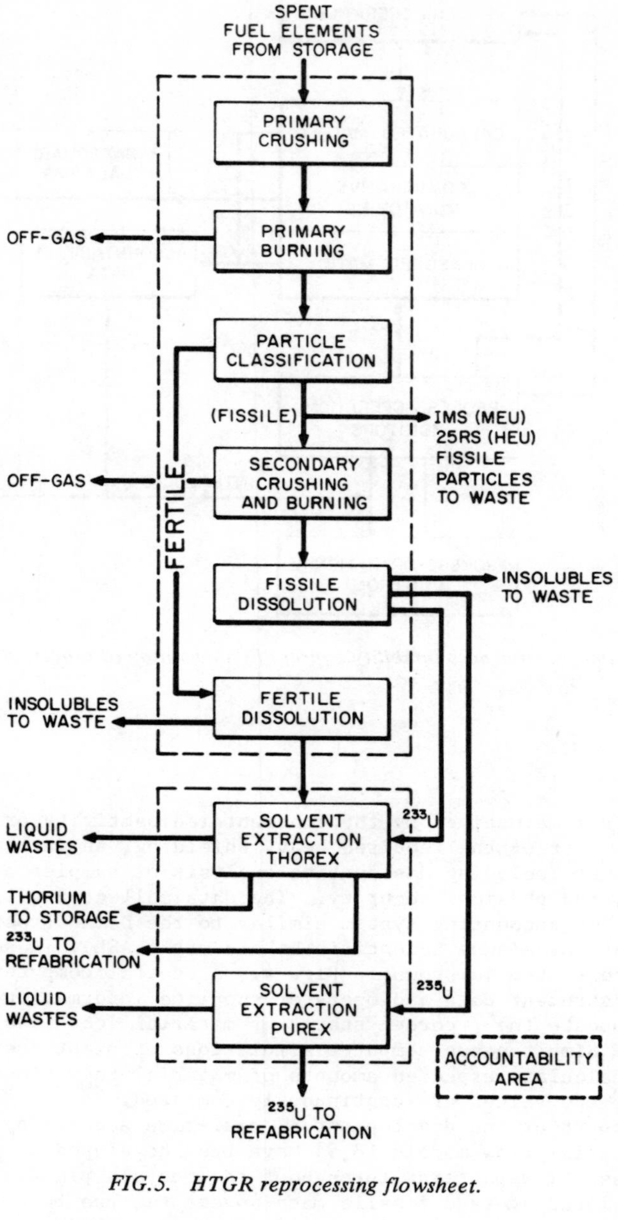

FIG.5. HTGR reprocessing flowsheet.

Table II. Characteristics of Main Nuclide Streams HTGR Reprocessing HEU and MEU Fuel Cycles Fuel Cooled 180 Days from Reactor Discharge

Stream	Description[a]	Type of Element	Uranium Content of Stream, wt. %, Based on: Fuel-Element-Derived Material	Uranium Content of Stream, wt. %, Based on: Uranium Input to Plant	Plutonium Content of Stream, wt. %, Based on: Fuel-Element-Derived Material	Plutonium Content of Stream, wt. %, Based on: Plutonium Input to Plant	Specific Activity of Nuclides in Stream (Ci/g)	Specific Neutron Emission Rate (n/s·g)
Spent fuel elements	Whole graphite blocks	HEU-IMS	0.39	100	0.007	100	4.7 E-1	1.69 E+1
		-25RS	0.82	100	0.050	100	5.0 E-1	9.73 E+1
		-23RS	0.37	100	0.003	100	4.4 E-1	2.5
		MEU-IMS	1.70	100	0.042	100	7.5 E-1	2.05 E+2
		-23RS	0.13	100	0.001	100	2.7 E-1	5.80 E+1
Retired fissile particles	Burned-back fissile particles	HEU-25RS	9.51	72.4	0.79	98.1	3.68	1.53 E+3
		MEU-IMS	20.8	92.9	0.55	98.2	6.33	2.58 E+3
Fissile dissolution Product solution	Liquid, 24 kg/m^3	HEU-IMS	15.63	43.1	0.647	97.67	1.51 E+1	1.52 E+3
		-23RS	14.66	40.1	0.286	97.2	1.39 E+1	2.15 E+2
		MEU-23RS	14.39	60.1	0.201	97.8	2.02 E+1	1.00 E+4
Insolubles	SiC hulls	HEU-IMS	0.0082	0.045	0.00339	0.102	5.4 E-1	8.0 E-1
		-23RS	0.0095	0.042	0.00018	0.101	5.4 E-1	1.4 E-1
		MEU-23RS	0.0045	0.063	0.00006	0.102	3.3 E-1	3.13
Fertile dissolution Product solution	Liquid, 240 kg/m^3	HEU-IMS	2.13	55.5	0.00024	0.38	2.59	2.28
		-25RS	2.13	26.3	0.00048	0.10	2.59	2.70
		-23RS	2.13	59.1	0.00024	0.84	2.59	2.28
		MEU-IMS	2.66	5.47	0.00072	0.059	7.36	1.64 E+2
		-23RS	2.27	38.8	0.00110	2.19	6.32	1.40 E+2
Insolubles	Noble metals, carbon, few fissile particles	HEU-IMS	5.60	0.76	0.114	1.79	6.08	8.56 E+1
		-25RS	10.70	1.31	0.896	1.79	4.04	1.74 E+3
		-23RS	5.60	0.75	0.114	1.78	6.08	8.56 E+1
		MEU-IMS	22.25	1.70	0.585	1.80	6.73	2.76 E+3
		-23RS	3.24	1.08	0.046	1.79	5.03	2.30 E+3
Solvent extraction-Thorex 1A column aqueous wastes	Liquid, 4.3 kg/m^3	HEU-IMS	0.014	0.056	0.00176	0.378	1.73 E+1	1.01 E+1
		-25RS	0.014	0.026	0.00322	0.097	1.73 E+1	1.28 E+1
		-23RS	0.014	0.099	0.135	98.0	1.89 E+1	1.07 E+2
		MEU-IMS	0.013	0.0055	0.00362	0.059	3.72 E+1	8.25 E+2
		-23RS	0.012	0.099	0.0980	98.0	2.24 E+1	5.17 E+3
^{233}U Product	Liquid, 233 kg/m^3	HEU-IMS	100	54.7	*b*	*b*	4.6 E-3	3.05 E+1
		-25RS	100	25.6	*b*	*b*	4.6 E-3	3.05 E+1
		-23RS	100	97.4	*b*	*b*	4.6 E-3	3.05 E+1
		MEU-IMS	100	5.42	*b*	*b*	1.7 E-2	2.51 E+1
		-23RS	100	97.8	*b*	*b*	1.7 E-2	2.63 E+1
Thorium product	Liquid, 464 kg/m^3	HEU-IMS	0.00694	0.15	*b*	*b*	1.3 E-4	2.3 E-1
		-25RS	0.00694	0.068	*b*	*b*	1.3 E-4	2.3 E-1
		-23RS	0.0113	0.267	*b*	*b*	1.3 E-4	2.3 E-1
		MEU-IMS	0.0093	0.147	*b*	*b*	8.2 E-4	2.1 E-1
		-23RS	0.023	0.265	*b*	*b*	6.2 E-4	1.07
Solvent extraction-Purex 5A column aqueous wastes	Liquid, 12.5 kg/m^3	HEU-IMS	0.019	0.043	0.00078	0.098	1.80 E+1	1.70 E+3
5D column aqueous wastes	Liquid, 0.57 kg/m^3	HEU-IMS	6.28	0.214	48.96	91.5	1.13	8.44 E+3
5F column aqueous wastes	Liquid, 0.007 kg/m^3	HEU-IMS	7.57	0.013	16.3	1.51	3.61	2.93 E+3
^{235}U Product	Liquid, 470 kg/m^3	HEU-IMS	100	42.24	1.00	0.149	2.3 E-3	2.42

[a] Liquids are nitrate solutions. Concentrations are of material originating from fuel element.
[b] Less than 1×10^{-6} wt. %.

the GASP IV simulation language, and similar simulations of the refabrication operations are scheduled for the near future at the Oak Ridge National Laboratory (ORNL). Total nuclide flows and the associated radioactivity through the recycle plant have been calculated at ORNL with isotope-depletion codes ORIGEN [11] and ORIGEN2 [12].

DETAILED MATERIAL CONTROL AND ACCOUNTABILITY ASPECTS OF REPROCESSING

The general flowsheet for the operations involved in reprocessing of spent HTGR fuel is shown in Fig. 5. The proposed accountability areas are indicated by the dashed lines. Material form, total uranium and plutonium, and activity for selected streams are presented in Table II. The relative attractiveness (or unattractiveness) depends more on dose-rate than simply on activity, but as the former depends upon geometry, the specific energies of the emitted radiation, and matrix, only activity is reported.

Reprocessing is divided into wet and dry head-end and solvent extraction. The entire head-end constitutes one large accountability area. Dry head-end consists of Primary Crushing, where the fuel elements are reduced to 5-mm-diam granules; Primary Burning, where the graphite fuel block and exposed fuel particle carbon coatings are burned away in a CO_2-O_2 atmosphere; Particle Classification, where the silicon-carbide-coated fissile particles are separated from the burned-back fertile kernels; and, for the ^{235}U particles of the HEU IMS elements and ^{233}U recycle fissile particles of the 23RS elements, Secondary Crushing and Burning to crack the silicon carbide coatings and burn away the remaining carbon. The fissile particles of the HEU 25RS and the MEU IMS elements are retired to waste after Particle Classification. A major safeguard advantage in the MEU flowsheet is that the residual ^{235}U and the plutonium bred from the ^{238}U remain in containment with fission products in intact fissile particles.

Input into the dry head-end consists of whole fuel elements, and output consists of fuel particles and CO_2-bearing off-gas from the burners. Item count identity is lost at the fuel element crushing stage. By well-planned administrative controls, material can be batched and total mass accountability maintained by weighing before and after each process step. The burning step will also require CO_2 measurements and calculations of the quantities of carbon removed in the burning process to combine with weight data for total mass accountability. Gross gamma activity measurements may be used to monitor various areas; however, background activity in the process cells may make such measurements of little value. From time to time, material will be removed from the process in failed equipment. There is also a steady flow of process control samples to sample analysis areas. For proper administrative control, both the decontamination and maintenance areas and hot laboratories must be included in the head-end material balance area. Differing requirements for reprocessing and refabrication imply separate maintenance and laboratory facilities, so that separate material balance areas should pose no real problem.

The wet head-end consists of fertile and fissile dissolution, where the burned-back fuel kernels are dissolved and insoluble materials such as the SiC hulls are separated from the product nitrate solutions. At this point, the material is in a form where uranium isotopic content can be measured. Several chemical methods exist; however, most are not well suited for a high throughput operation. In particular, plutonium and other fission products may interfere with uranium isotopic resolution, necessitating chemical separation before the uranium can be measured. Development work on ^{233}U measurements without chemical separation is necessary for high throughput.

The uranium content measured at the dissolving stage for an entire customer lot must be balanced against the fresh fuel uranium and thorium loadings and reactor burnup calculations for a special nuclear material balance. This balance must include the uranium content of any fissile particles and insoluble materials removed for waste disposal. Again, since the purpose of particle disposal is containment of special nuclear material and fission products in the repository, the development of non-destructive assay methods on irradiated particles is preferable to chemical dissolution and separation for assay.

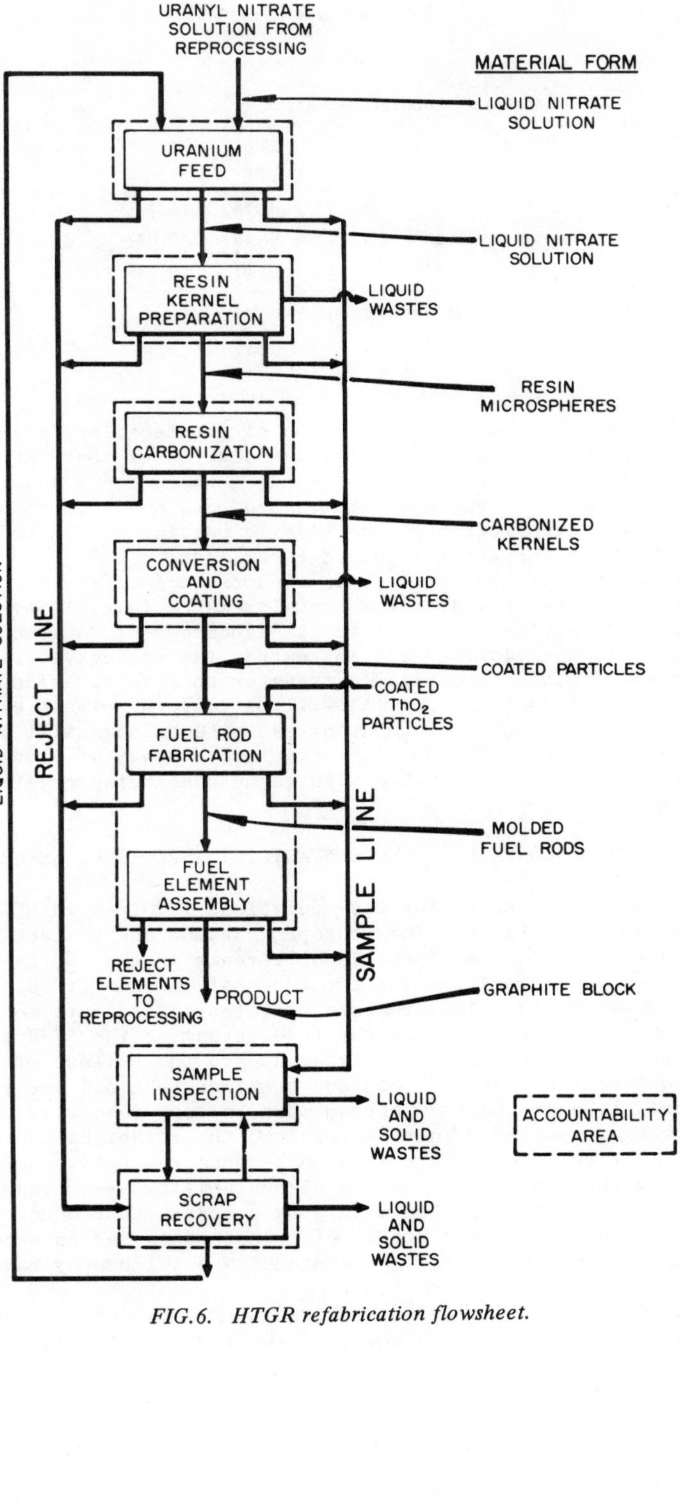

FIG. 6. HTGR refabrication flowsheet.

Table III. Characteristics of Input Streams HTGR Refabrication

Fuel Element Type	Uranium Isotope Content, %						Specific Activity (mCi/g)	Specific Neutron Emission Rate (n/s•g)
	232	233	234	235	236	238		
HEU-235	0.002	2.93	1.62	29.96	48.78	16.72	2.3	2.621
HEU-233	0.045	66.73	21.30	7.42	4.48	0.025	19.2[a]	30.97[a]
MEU-233	0.043	67.55	21.33	6.58	4.01	0.48	18.5	29.81

[a] Assumes 30 d since solvent extraction cleanup.

The solutions from the dissolution of the fertile particles are combined with the solutions from the dissolution of the 23RS fissile particles and sent through the Thorex line, where the uranium and thorium are separated from the fission products and each other. The products of the dissolution of the fissile particles from the HEU IMS elements are sent through a standard Purex line.

When the measurement of ^{233}U in the presence of plutonium and fission products is solved, the material balance initial inventory for the solvent extraction area will be straightforward. A good isotopic inventory can be provided at solvent extraction product storage. This inventory then becomes the balance transfer to refabrication. Some minimal amount of nuclear material will be lost to solvent extraction waste streams, and the development of assay techniques will include uranium measurements at very low concentrations. Assay of plutonium in the waste streams should be possible with techniques being developed for LWR reprocessing.

DETAILED MATERIAL CONTROL AND ACCOUNTABILITY ASPECTS OF REFABRICATION

The general flowsheet for the operations involved in HTGR refabrication is shown in Fig. 6. The isotopic content and the activity associate with the uranium for each of the input streams in the two fuel cycles are presented in Table III. The two types of streams (^{235}U and ^{233}U) are never mixed. The products of refabrication are separate fuel elements containing either the ^{235}U stream or the ^{233}U stream. The high activities in the ^{233}U stream due to the buildup of decay products of the inherent ^{232}U content requires that all the steps in the refabrication of the ^{233}U stream must be done remotely. Because of crossover of some ^{233}U in reprocessing, the refabrication of the ^{235}U stream of the HEU cycle must also be done remotely. Refabrication has many more measurements and accountability areas than has Reprocessing. Though dictated mainly by process and product quality control concerns, the great number of measurements serves accountability purposes as well. The sampling techniques and philosophy have been discussed in detail elsewhere [13].

Uranium, as liquid nitrate, enters the refabrication line from Reprocessing and from Scrap Recovery. The first system is Uranium Feed, where the liquid is stored, isotopically blended, and chemically adjusted. The liquid goes to Resin Kernel Preparation, where the uranium is loaded onto resin microspheres. These then go to Resin Carbonization, where the resin is decomposed (carbonized) to produce a kernel consisting of uranium dioxide in a carbon matrix. The kernel goes to Conversion and Coating, where, in conversion, the UO_2 is caused to react with the carbon matrix to produce a mixture of the UO_2 and UC_2 and where, in coating, three layers of pyrocarbons and one SiC coating are applied to produce the coated fissile particle. These coated fissile particles then go to Fuel Rod Fabrication, where they are blended to homogenize any slight differences in coating batches and mixed with carbon-coated ThO_2 particles from a fresh-fuel plant. The mixture is molded to form the "green" fuel rod held together by a pitch-base binder. The green fuel rods go to Fuel Element Assembly, where they are inserted into the graphite fuel blocks. The assembled fuel elements are heated to carbonize the pitch binder of the rods. Finally, the fuel elements are cleaned, inspected, and sent to Shipping, Receiving, and Storage. Scrap fuel elements are temporarily stored, then campaigned to Reprocessing for recovery.

Sample Inspection and Scrap Recovery are major systems in Refabrication. They receive streams from all the systems mentioned previously and each other. The material exiting Sample Inspection is routed to Scrap Recovery or to Waste Treatment. The material entering Scrap Recovery exits principally as recovered uranyl nitrate solution returned to Uranium Feed or as various forms sent to Waste Treatment.

Uranium Feed is the first accountability area in Refabrication. Uranyl nitrate solutions are received from Reprocessing or Scrap Recovery. Once in Refabrication there is no transfer of material back to Reprocessing with the exception of the whole reject blocks. Liquid samples are transferred to Sample Inspection to assess the impurity levels of the feed and to verify the results of the isotopic blending and chemical adjustments. Unacceptable liquid feed is transferred to Scrap Recovery and the product is delivered to Resin Kernel Preparation. Accountability for these input and output streams is by volume measurement and uranium assays of the samples.

Resin Kernel Preparation, Resin Carbonization, and Conversion and Coating are also separate accountability areas. Accountability for the liquid stream received by the Resin Kernel Preparation is by volume measurement and uranium determination of a liquid sample. Accountability for the product of Resin Kernel Preparation and the inputs and outputs of Resin Carbonization and Conversion and Coating is by means of automatic remote weighing devices and destructive and nondestructive analyses of samples. Special passive samplers suitable for remote handling have been developed [14]. Solid particles are conveyed pneumatically between and within systems. Besides the main product streams, solids or liquids are transferred to Scrap Recovery or Waste Treatment.

Fuel Rod Fabrication and Fuel Element Assembly together constitute another accountability area. The basic accountability approach is to nondestructively assay 100% of the acceptable fuel rods produced in Fuel Rod Fabrication. Accountability in Fuel Element Assembly depends on knowledge of the location and weight of assayed fuel rods and upon the weight of a fuel block before and after loading.

The as-coated fissile particles are pneumatically transferred from Conversion and Coating to a precision weigher and then to a batch blender, which blends up to 24 coating batches, each containing about 3 kg U. The blended particles are passed through a sampler, and the sample is nondestructively and destructively chemically analyzed in Sample Inspection to determine the uranium assay, isotopic contents, and the mass of the particles. The fertile particles are blended, sampled, and analyzed similarly to the fissile particles before being transferred into the hot cells. The mass of the incoming fertile material is determined. The particles are then molded into fuel rods. At this point in the process, the scrap is in the form of particles, fuel rods, and pieces of rods. The scrap is assayed by nondestructive methods or by total mass measurements in those cases where accurate uranium weight factors are known, before being transferred to Scrap Recovery.

The accepted rods then undergo fuel homogeneity inspections. All the rods are analyzed in a gamma scan system. The overall mass distribution is determined and the fissile and fertile content verified semi-quantitatively. A second system samples approximately 10% of the fuel rods and determines the total heavy metal, thorium, and uranium distributions in the rods. Such a system, which operates on the principle of multi-energy radiation attenuation with selective K-edge absorption, has been developed at ORNL in the HTGR program [15]. It uses the radioisotopes ^{169}Yb and ^{177m}Lu, which emit gamma rays of energies between the respective thorium and uranium K-absorption edges. Such gamma rays permit separation of the thorium and uranium contribution because the attenuation coefficients of the thorium and uranium are very different in that energy range.

The next inspection is fuel rod assay. Two nondestructive assay devices are used for this purpose. One is an on-line device capable of assaying 100% of the fuel rods produced from two machines. The identity, location, and disposition of rods are monitored by the computer in subsequent storage and fuel element loading. A second device accepts a sample from the main product line and nondestructively assays the fuel rod in a laboratory. In addition, a limited number of rods is also chemically assayed. The two nondestructive assay devices are used for product verification and for determining the ^{233}U, ^{235}U fissile contents of the fuel rods. Both devices use active neutron interrogation with a ^{252}Cf neutron source in each irradiator assembly. The on-line assay device detects the prompt fission neutrons from the irradiated sample. A device of this type has also been developed at ORNL in the HTGR program [16].

Sample Inspection is a separate accountability area. This system comprises the analytical laboratories and equipment necessary to perform analyses on samples transferred from the other systems of Refabrication. Samples are analyzed to characterize the main batches of material present in the other systems. The mass of each sample entering and leaving the system is determined. Both chemical and nondestructive methods are used for the analyses. Uranium, thorium, and the isotopic contents of samples are determined by potentiometric, volumetric, and other techniques. The nondestructive analyses are performed by gamma- and alpha-ray counting and by neutron interrogation. A nondestructive device that assays particles and fuel rod samples has been developed and is being tested at ORNL [17]. The device uses a ^{252}Cf neutron source in the irradiator assembly and detects the delayed fission neutrons emitted from the irradiated sample. The device yields accurate assay information and complements other assay devices and methods.

Scrap Recovery receives material in a variety of forms from all the other systems. This system is in effect a mini-reprocessing operation largely provided to avoid the inherent accountability problems associated with transfers of material back to Reprocessing. Incoming accountability is via techniques appropriate to the material form. The major exiting stream is the recovered uranium nitrate product, which is directed back to the front end of the refabrication line. Accountability for this stream is by volume measurement and sample analyses including isotopic analyses. The other exiting streams are various aqueous wastes for which the accountability is by volume measurement and sample analyses. A final stream consists of insolubles, mainly coated particles. Accountability for this stream is by means of weighing and sample analyses.

SUMMARY

The materials control and accountability aspects of the Reprocessing and Refabrication of a conceptual large-scale HTGR fuel recycle plant have been discussed. Two fuel cycles were considered. The traditional highly enriched uranium cycle uses an initial or makeup fuel element with a fissile enrichment of 93% ^{235}U. The more recent medium enriched uranium cycle uses initial or makeup fuel elements with a fissile enrichment less than 20% ^{235}U. In both cases, ^{233}U bred from the fertile thorium is recycled.

Materials control and accountability in the plant will be by means of a real-time accountability method. Accountability data will be derived from monitoring of total material mass through the processes and a system of numerous assays, both destructive and nondestructive.

REFERENCES

[1] LOTTS, A. L., et al., "HTGR Fuel and Fuel Cycle Technology," Nuclear Power and its Fuel Cycle (Proc. Int. Conf. Salzburg, 1977) 3, IAEA, Vienna (1977) 433–52.

[2] Project Staff, *Conceptual Design Summary and Design Qualifications for HTGR Target Recycle Plant,* General Atomic Company Rep. GA-A13365 (April 1975).

[3] Project Staff, *Preconceptual Design and Estimate Summary for HTGR Recycle Demonstration Facility (HRDF),* General Atomic Company Rep. GA-A13502 (July 1975).

[4] LOTTS, A. L., COOBS, J. H., *HTGR Fuel and Fuel Cycle Technology,* ORNL/TM-5501 (August 1976).

[5] NOTZ, K. J., *An Overview of HTGR Fuel Recycle,* Oak Ridge National Laboratory Rep. ORNL/TM-4747 (January 1976).

[6] AUGUSTSON, R. H., "Development of In-Plant Real-Time Materials Control: The DYMAC Program," *Nucl. Mater. Manage.* 5 3 (1976) 302–16.

[7] KEEPIN, G.R., MARAMAN, W. J., "Nondestructive Assay Technology and In-Plant Dynamic Materials Control — DYMAC," Symposium on Safeguarding Nuclear Materials (Proc. Int. Symp. Vienna, 1975 IAEA, Vienna (1976).

[8] McNEANY, S. R., Oak Ridge National Laboratory, unpublished computer program, September 1974.

[9] McNEANY, S. R., Oak Ridge National Laboratory, unpublished computer program, September 1976.
[10] JENKINS, J. D., McNEANY, S. R., RUSHTON, J. E., *Conceptual Design of the Special Nuclear Material Nondestructive Assay and Accountability System for the HTGR Fuel Refabrication Pilot Plant,* Oak Ridge National Laboratory Rep. ORNL/TM-4917 (July 1975).
[11] BELL, M. J., *ORIGEN — The ORNL Isotope Generation and Depletion Code,* Oak Ridge National Laboratory Rep. ORNL-4628 (May 1973).
[12] CROFF, A. G., Oak Ridge National Laboratory, personal communication, October 1977.
[13] PECHIN, W. H., et al., *Inspection of High-Temperature Gas-Cooled Reactor Recycle Fuel,* Oak Ridge National Laboratory Rep. ORNL-5165 (June 1977); also "Quality Control Tests for High-Temperature Gas-Cooled Reactor Recycle Fuel," *Nuclear Fuel Quality Assurance,* (Proc. Sym. Vienna), IAEA, Vienna (1976), 425–27.
[14] SUCHOMEL, R. R., LACKEY, W. J., *Device for Sampling HTGR Recycle Fuel Particles,* Oak Ridge National Laboratory Rep. ORNL/TM-5739 (March 1977).
[15] ANGELINI, P., et al., Oak Ridge National Laboratory Rep. ORNL-5266 (1977) 181–84.
[16] RUSHTON, J. E., Oak Ridge National Laboratory Rep. ORNL-5266 (1977) 186–89.
[17] KNOLL, R. W., Oak Ridge National Laboratory Rep. ORNL-5266 (1977) 189–90.

DEVELOPMENT AND DEMONSTRATION OF ADVANCED INSTRUMENTAL METHODS TO DETERMINE THE VOLUME OF AN INPUT ACCOUNTABILITY VESSEL AND THE RESIDUAL NUCLEAR MATERIAL RETAINED IN THE LEACHED HULLS

G. BARDONE
Comitato Nazionale Energia Nucleare,
Fuel Cycle Department,
Casaccia, Rome

T. CANDELIERI
Comitato Nazionale Energia Nucleare,
ITREC Plant, Trisaia Center,
Policoro,
Italy

Abstract

DEVELOPMENT AND DEMONSTRATION OF ADVANCED INSTRUMENTAL METHODS TO DETERMINE THE VOLUME OF AN INPUT ACCOUNTABILITY VESSEL AND THE RESIDUAL NUCLEAR MATERIAL RETAINED IN THE LEACHED HULLS.

The report summarizes the results obtained during the work carried out at the CNEN-ITREC Plant within a programme of co-operation with the IAEA concerning: (1) TDR (Time Domain Reflectometry) system installation in the input accountability vessel and comparison of TDR with the conventional dip-tube pneumatic system; and (2) Installation of a gamma spectrometric system using germanium detectros for measuring fissile material retained in the leached hulls. A short description of the systems and technique is given together with the layout of the overall installation; data obtained during installation, calibration and hot operating phases are also reported.

INTRODUCTION

The report summarizes the results obtained during the work carried out at the CNEN-ITREC plant within a programme of co-operation with the IAEA concerning the testing of systems and techniques useful for safeguarding reprocessing plants.

The ITREC plant is a multipurpose pilot plant, located in the southern part of Italy, for reprocessing thorium-uranium oxide fuels and advanced reactor fuels, with the aim of performing research and development operations under plant conditions close to industrial use [1].

It became operational in July 1975 by reprocessing thorium-uranium spent fuels from the Elk-River Reactor, using a modified Thorex process [2].
The objectives of the programme are:

(1) Installation of a complete time domain reflectometry (TDR) system in the D-30 input accountability vessel; calibration of the vessel using TDR and conventional dip-tube pneumatic systems; and comparison of accuracy and precision of data at a selected operating level including the heel.

(2) Installation of a gamma spectrometric system using Ge detectors for measuring fissile material remaining on the leached hulls; and data collection of sufficient batches of leached hulls to assess its usefulness as an accountability technique.

A complete TDR system has been installed in the D-30 ITREC input accountability tank. Calibration both with TDR and with the conventional dip-tube pneumatic system with high precision manometric reading has been carried out and the results are reported and compared. The system has been in operation since March 1975 and a considerable amount of data in various operating conditions is available. Figures showing the TDR system's accuracy, precision and reliability can also be given.

As far as item (2) is concerned, data on the evaluation of the amount of nuclear material losses in the leached hulls, of the region chosen for analysis, and the results of a series of measurements on dissolution baskets containing hulls, are also available.

The data has been collected over two years of hot operation.

1. TDR (TIME DOMAIN REFLECTOMETRY)

1.1. Generalities

The TDR (Time Domain Reflectometry) is a technique through which mismatches can be determined in a circuit or in a signal transmission line. TDR analysis starts with the propagation of an incident signal (voltage pulse) in a circuit or in a cable under examination, followed by the observation of the reflected signal coming back from the controlled system.

All conditions causing discontinuities can be physically located by recording, with respect to time, the position of the reflected wave-form. When a coaxial probe, matched to the characteristic impedance of the system (50 ohm), with air as dielectric medium, is put into a liquid phase, the change of the dielectric constant at the interface between the two fluids will give a signal reflection which will be displayed on the oscilloscope CRT, associated with the TDR system.

It is then possible to measure dielectric constant of liquids; and liquid level in stationary or moving conditions even in the presence of foam or emulsion.

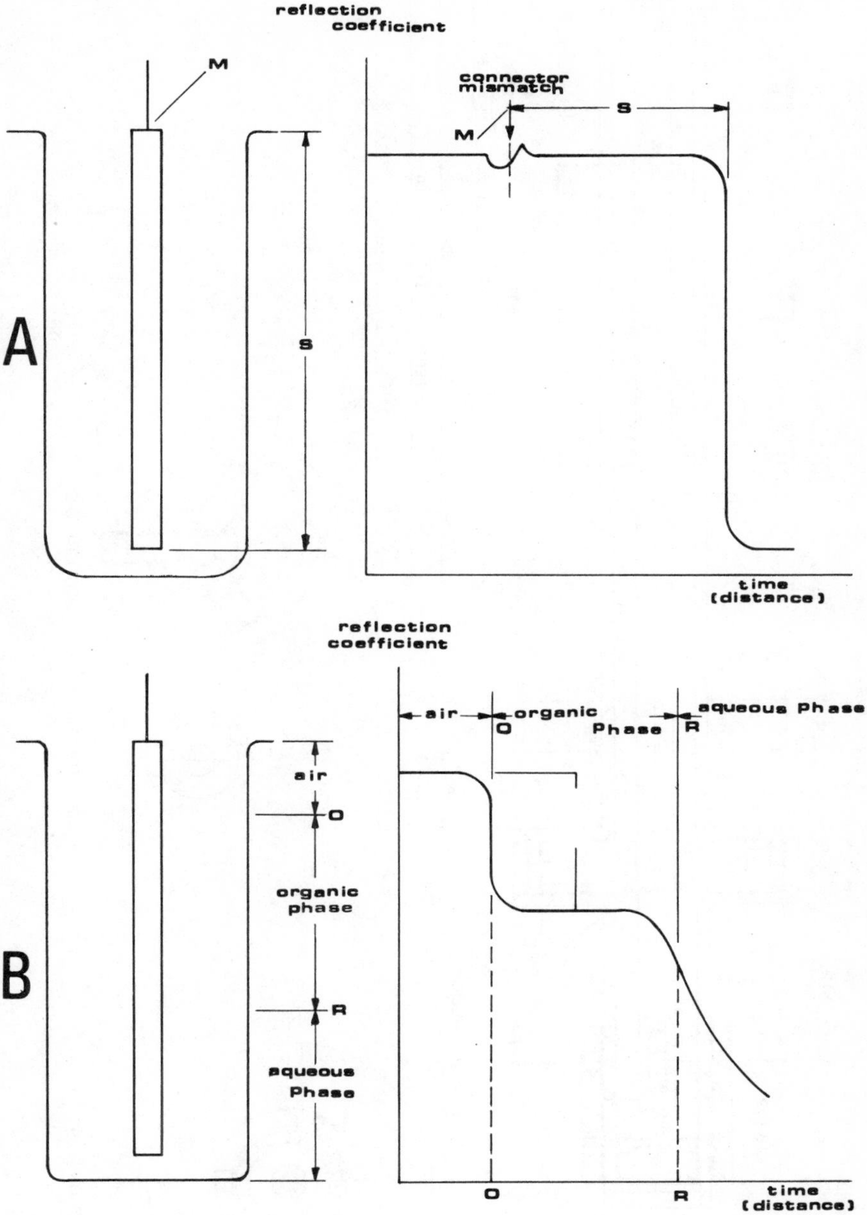

FIG.1. Schematic oscillogram display in the TDR system for liquid level and dielectric constant measurements.

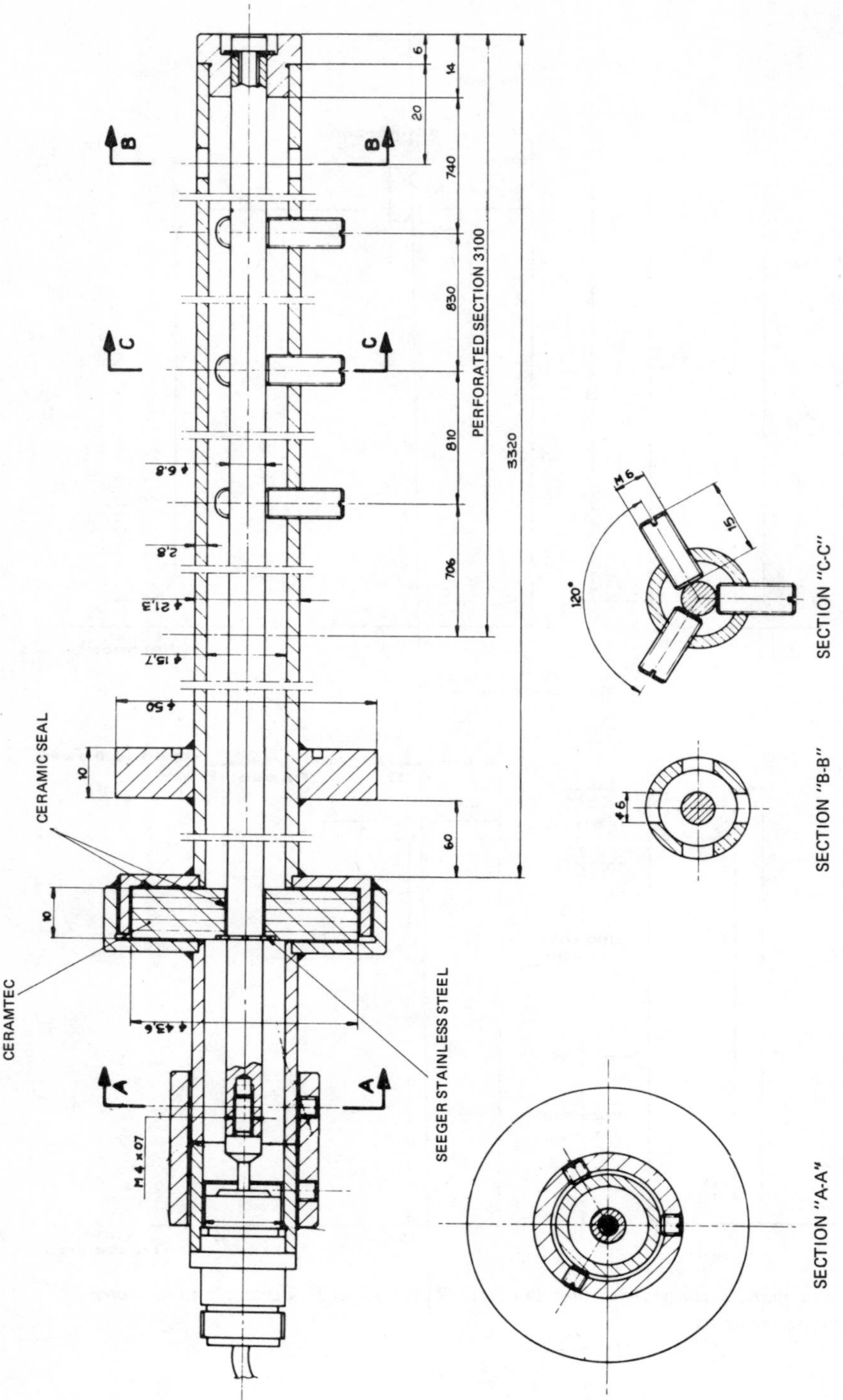

FIG.2. ITREC plant: TDR probe for the D-30 installation.

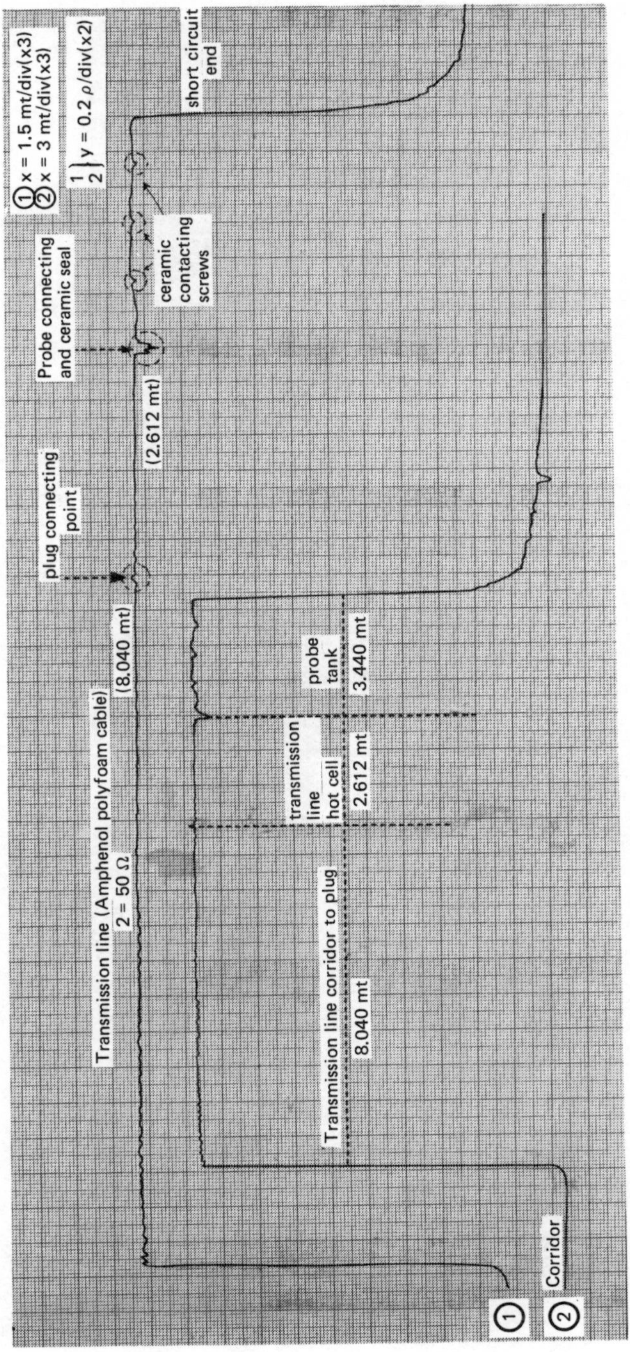

FIG.3. TDR signature of the probe and cabling installation.

FIG.4. TDR probe.

FIG. 5. TDR probe installation on the D-30, particularly of the welded flange (1) and connector.

1.2. Schematic measurement example

The practical application of this method is better illustrated by Fig. 1. A probe installed in an empty vessel is schematically represented at the left hand of part A of Fig. 1, while the resulting oscillogram, which shows that the signal does not meet any electrical mismatches, is represented at the right-hand side; the S part represents the probe length from the line connection to the probe end.

The same probe, inserted in a vessel containing two unmixable liquids, is schematically represented on the right-hand side of part B of the same figure. The corresponding oscillogram, represented on the right, shows two steps at points O and R. The first point corresponds to the upper air-liquid interface, the second to the interface between the upper and lower liquid. The horizontal O-R part corresponds to the upper liquid thickness, while the first step height corresponds to the dielectric constant of the same liquid.

1.3. Dimensioning of the TDR probes

For dimensioning the TDR probes some general rules must be kept:

(1) The diameter of the coaxial probe conductors must be calculated in order to yield an impedance close to 50 ohm ± 10% and minimize the mismatch related to the ceramic seal;
(2) The material thickness must be sufficient for safety considerations and concern for life;
(3) The welding requirement must be minimized. Taking the above considerations into account a probe was designed and constructed (see Fig.2).

The basic features of the probe are the same as those described in Ref.[3]. Owing to the height of the probe (≅ 3600 m) the inner conductor had to be centred by three series of ceramic screws. A complete TDR recorded signature of the probe and cabling installation is given in Fig.3. A full view of the probe and the probe installation on the D-30 ITREC input accountability tank are shown in Figs 4 and 5.

1.4. Transmission line

The signal from the TDR probe is transferred to the electronic TDR unit by a coaxial polyethylene polyfoam cable with low capacity and low dielectric losses, and protected by a stainless steel tubing. A schematic layout of the transmission line is shown in Fig.6.

1.5. TDR electronic and data acquisition system

The electronic equipment used is supplied by Hewlett Packard and includes:

TDR Mod.1818A with oscilloscope mod.182C
TDR Mod.1815B with: Sampler 1815A, tunnel diode mount 1106A, oscilloscope mainframe 180 A.

The signal (behaviour) was transferred from the oscilloscope display to an x-y recorder and the x-y co-ordinates were increased by a factor of 3 and 2 respectively in order to improve the measurement sensitivity and precision. The recorded TDR signatures were manually elaborated.

1.6. Experimental

1.6.1. Calibration

The calibration of the D-30 input accountability vessel was carried out, adding step by step an amount of accurately weighed demineralized water and

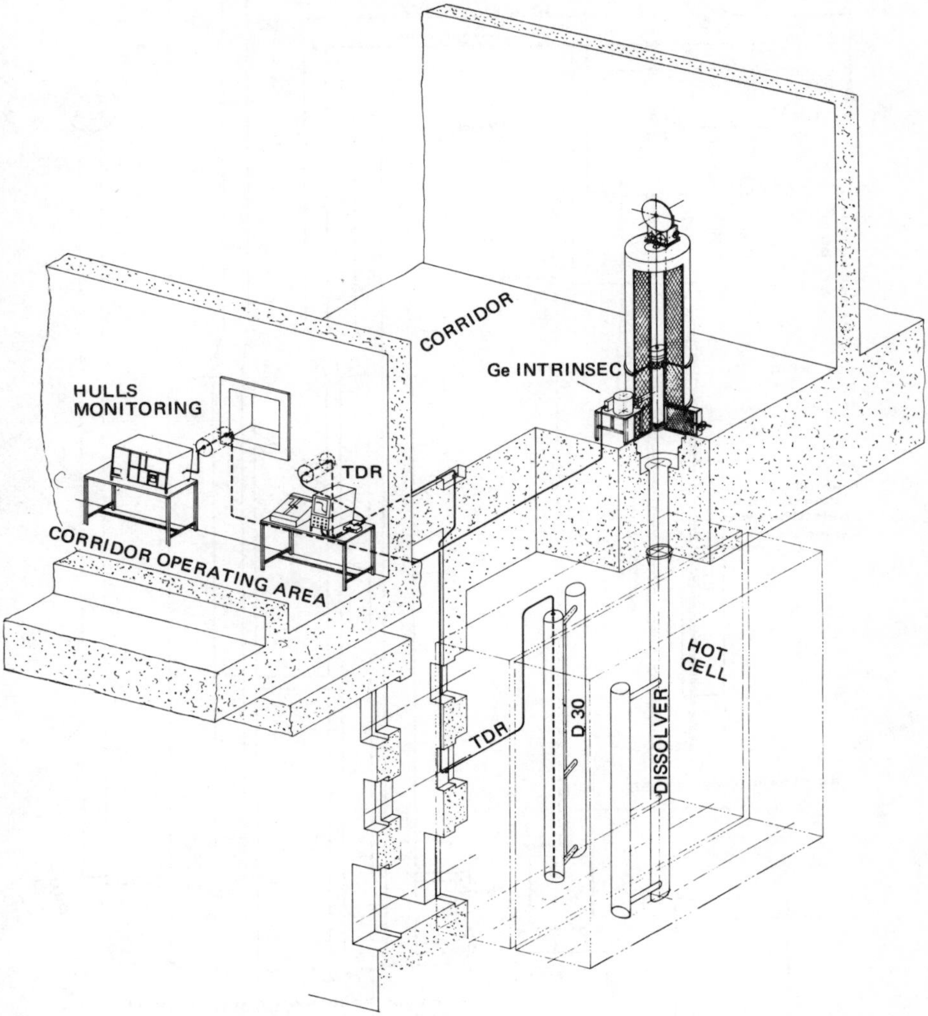

FIG. 6. Layout of TDR and hull monitoring system installation.

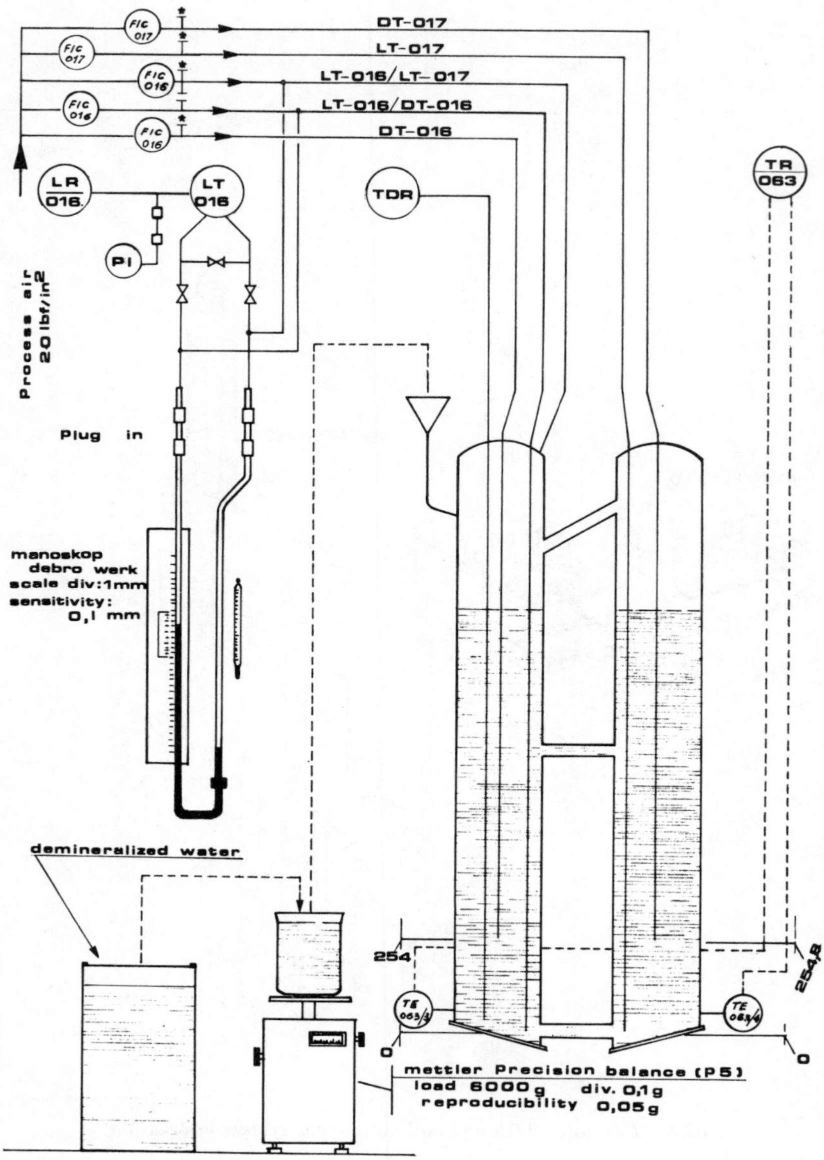

FIG. 7. Calibration set-up. D-30.

recording both the dip-tube manometric reading and the TDR signature. The calibration set-up is shown in Fig.7.

The scale used had a reading accuracy of 0.1 g and a precision of 0.05 g. The vessel containing the amount to be added was weighed before and after the additions. Accurately known check weights were used routinely during calibration to check scale accuracy. Weighed water and manometric fluid temperature were controlled with a high precision thermometer: 0–50° centigrade range; 0.1°C per division.

Before starting the calibration the tank was completely drained and dried. The "dead volume" and the minimum detectable level were accurately estimated. The calibration data are reproduced in Table I. The cumulative regression model [4] was applied for the data evaluation. The factors limiting the calibration precision for the TDR system are in our opinion essentially the ceramic-centring screws which cause discontinuity in the calibration curve, the low reading sensitivity of the electronic used for calibration, and the limited number of calibration runs. Measurements carried out recently have enabled us, however, to reduce the average error to below 2%.

1.6.2. Measurements under hot operation conditions

The measurements under plant operation conditions were taken both with TDR and the dip-tube manometric system. The two following types of measurements are made following level variations in the accountability tank:

(1) Process measurements: Every time a transfer operation of a solution from the dissolver to the D-30 input accountability vessel is involved.

(2) Input accountability measurements: Every time a transfer operation from the D-30 input accountability vessel to the process sections is carried out. Some TDR recordings in plant operating conditions are shown in Fig.8.

It should be noted that the γ-dose rate during hot plant operation in correspondence to the sensor probe and the transmission line is 2×10^3 rad/h; the integrated γ-dose absorbed after three years of installation is 2×10^7 rad. System integrity checks at scheduled time interval do not show, at the moment, any variation in the electric characteristics of the sensor probe and of the transmission line.

From the large amount of data collected (about 300) at different selected and random operating levels the following conclusions can be drawn:

(1) The TDR system data, when compared with the manometric data, show a negative, very reproducible systematic error, particularly in the high-level zone (between 90–115 litres). The average values of the systematic error

TABLE I. D-30 ACCOUNTABILITY VESSEL CALIBRATION DATA

	Manometer	TDR		TDR
		1st linear section		
	1st linear section	Uncorrected data	Corrected data	2nd linear section
$\sigma^2_{\epsilon_i}$	0.000013	0.002288	0.00065	0.000507
σ^2_a	0.000335	0.1099	0.0312	1.8469
σ^2_b	0.15×10^{-7}	0.000003	0.000001	0.000005
σ_ν	0.15	1.80	0.96	1.00
i.f.	0.4	6.16	3.3	2.55
a	2.161	1.450	1.450	0.929
b	0.03976	0.03899	0.03899	0.03879

					TDR
		3rd linear section			
	2nd linear section	Uncorrected data	Corrected data	4th linear section	5th linear section
$\sigma^2_{\epsilon_i}$	0.000082	0.001204	0.00067	0.000474	0.000658
σ^2_a	0.14564	5.0839	2.82909	2.43528	9.699
σ^2_b	0.42×10^{-7}	0.000003	0.000002	0.6×10^{-6}	0.000001
σ_ν	0.70	1.94	1.44	1.51	1.97
i.f.	0.58	3.06	2.3	1.6	1.71
a	3.454	0.506	0.506	1.582	1.468
b	0.03952	0.03992	0.03992	0.03876	0.03858

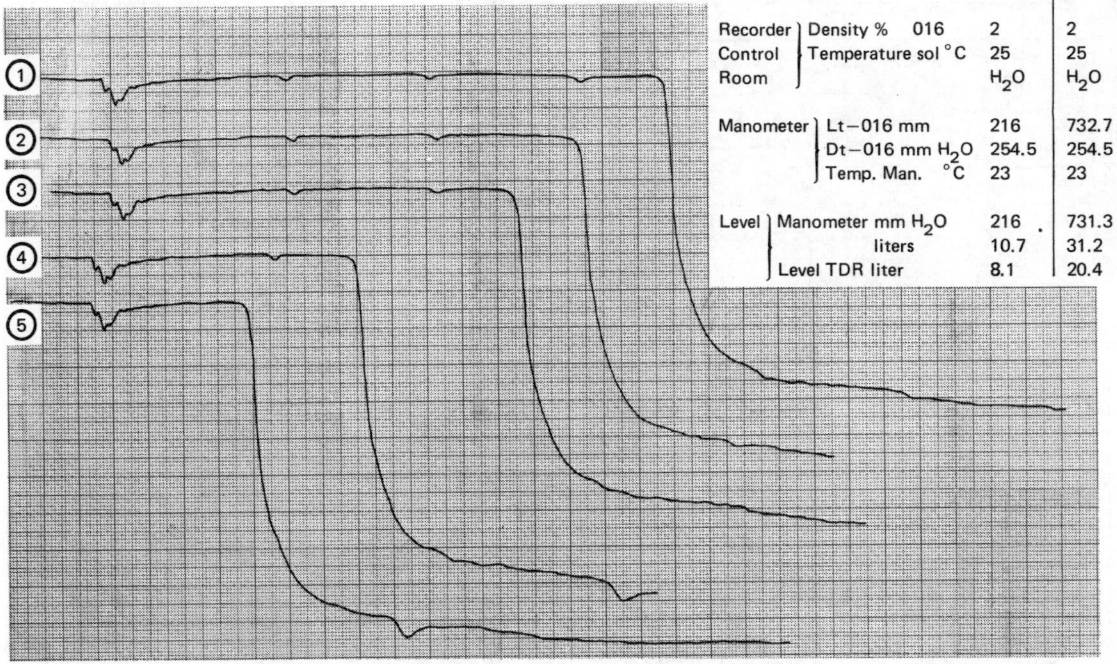

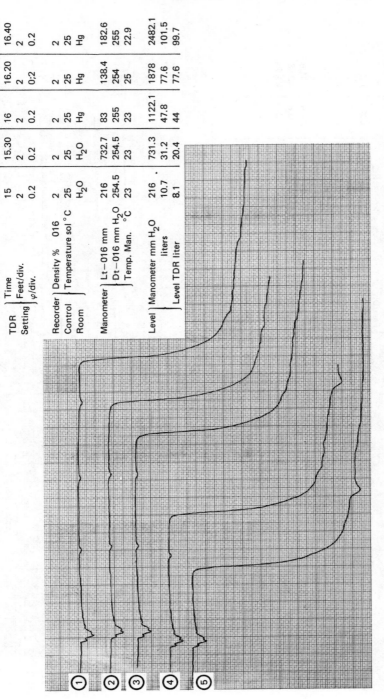

		Date 20/10/76				
		1	2	3	4	5
TDR Setting	Time	15	15.30	16	16.20	16.40
	Feet/div.	2	2	2	2	2
	φ/div.	0.2	0.2	0.2	0;2	0.2
Recorder Control Room	Density % 016	2	2	2	2	2
	Temperature sol °C	25	25	25	25	25
		H_2O	H_2O	Hg	Hg	Hg
Manometer	Lt−016 mm	216	732.7	83	138.4	182.6
	Dt−016 mm H_2O	254.5	254.5	255	254	255
	Temp. Man. °C	23	23	23	25	22.9
Level	Manometer mm H_2O	216	731.3	1122.1	1878	2482.1
	liters	10.7	31.2	47.8	77.6	101.5
	Level TDR liter	8.1	20.4	44	77.6	99.7

FIG. 8. TDR signatures in the plant under hot operating conditions. Transfer measurements D-30 (input accountability tank) to D-40 (feed adjustment tank).

			1	2	3	4
TDR Setting	Time		16.30	16.30	16.30	16.30
	Feet/Div		2	2	2	2
	φ/Div		0.2	0.2	0.2	0.2
Recorder Control Room	Level %	016	53.5	53.5	53.5	53.5
	Density %	016	77.5	77.5	77.5	77.5
Manometer	LT 016 mm Hg		194.9	194.9	194.9	194.9
	DT 016 mm H_2O		403.8	403.8	403.8	403.8
	Temperature Man.		23.5	23.5	23.5	23.5
Level	Man. Heff. mm		1163.9	–	–	–
	Man. volume ltr		69.2	id	id	id
	TDR mm		2424			
	TDR volume ltr		94.9			

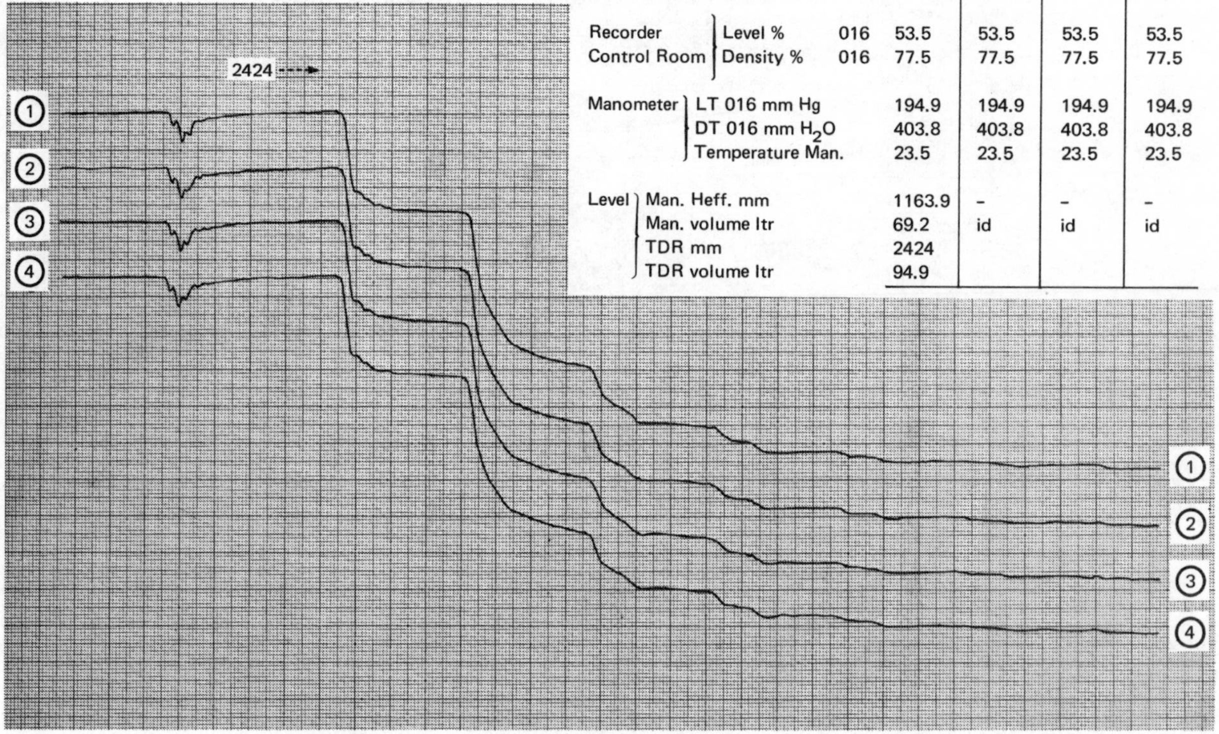

FIG.9. TDR signatures illustrating anomalies after liquid homogenization.

is about 1.5 litres. The bias, in our opinion, is probably due to the different electronic units used for calibration and for plant operating measurements.

Further measurements and tests are in progress in order to find out the bias origin, and, if possible, to eliminate it.

(2) Series of measurements ($\cong$ 100) taken at selected operating levels, fixed and variable, give a satisfactory precision (0.30% average at 95% confidence level).

(3) System anomalies have been noted when the TDR measurements were taken after the D-30 dissolver solution homogenization, required by the accountability procedure, and when emptying the same vessel. After these operations some liquid is retained by the ceramic screws, consequently giving a wrong TDR level indication (see Fig.9).

A variable time period (from few minutes to some hours), depending on the different operating conditions, is required for system normalization.

Action will be taken to study a new material for centring the screws or a new design of a self-centring probe.

2. DETERMINATION OF RESIDUAL NUCLEAR MATERIAL RETAINED IN THE LEACHED HULLS BY GAMMA SPECTROMETRY

2.1. Chopping and dissolution process [1]

The amount of nuclear material losses in the leached hulls mainly depends on:

(1) Physical and chemical properties of the spent fuel to be dissolved;
(2) Chopping system;
(3) Dissolution process.

In the ITREC plant the chopping operation is carried out pin by pin, obtaining geometrically regular hulls with a good attack surface, or high dissolution rate. The dissolution process is discontinuous, performed in four consecutive steps. Each dissolution batch is transferred to the D-30 input accountability tank. The fuel element is charged in the two first batches, while the third one should complete (98–99%) the dissolution process. The fourth batch represents the dissolution test; the dissolution is stopped when, from consecutive analyses, no nuclear material concentration increase is observed. The basket containing 25 pins (constituting a fuel element) is then washed with concentrated nitric acid solution sprayed by jets installed in the dissolver. Subsequently the basket is dried and finally extracted from the dissolver for monitoring and storing the hulls.

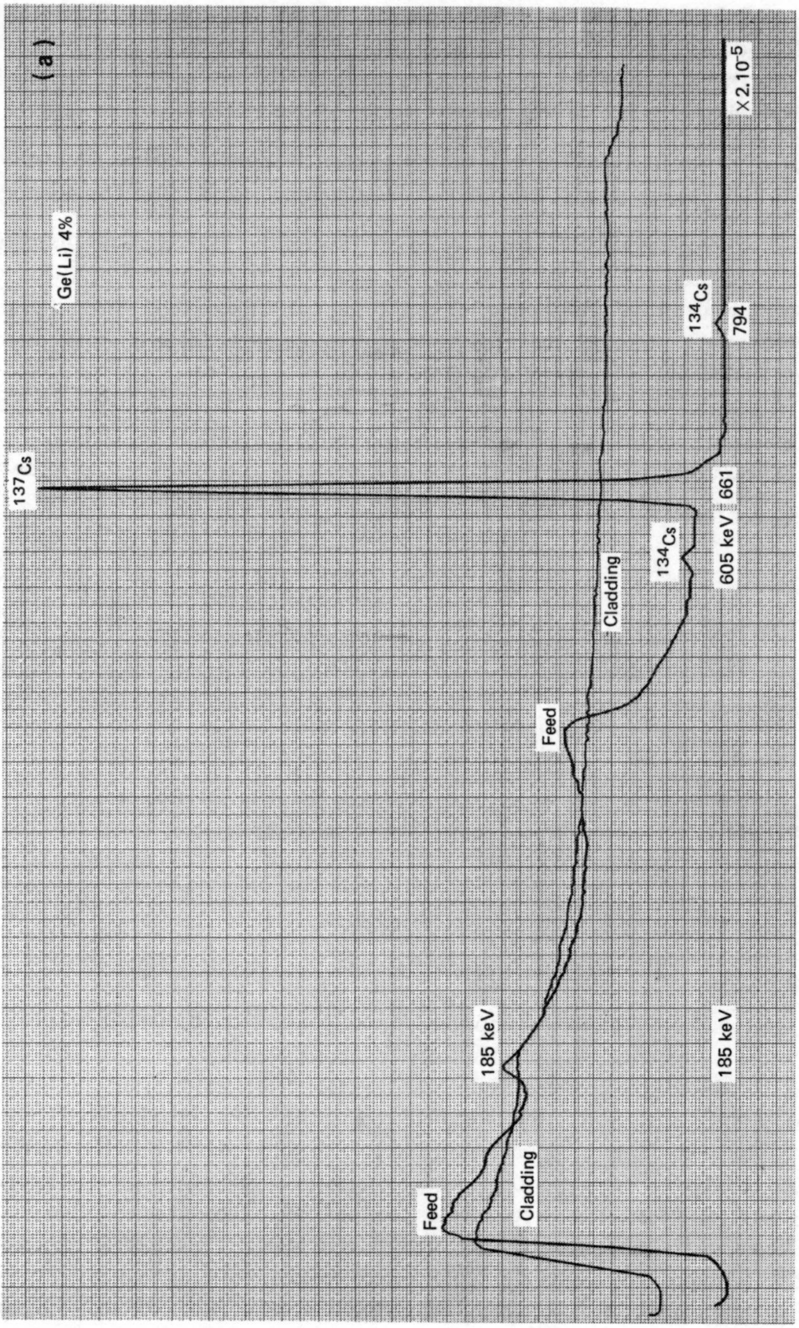
(a)
Ge(Li) 4%
×2.10^{-5}
^{137}Cs
^{134}Cs
794
^{134}Cs
605 keV
661
Cladding
Feed
185 keV
185 keV
Feed
Cladding

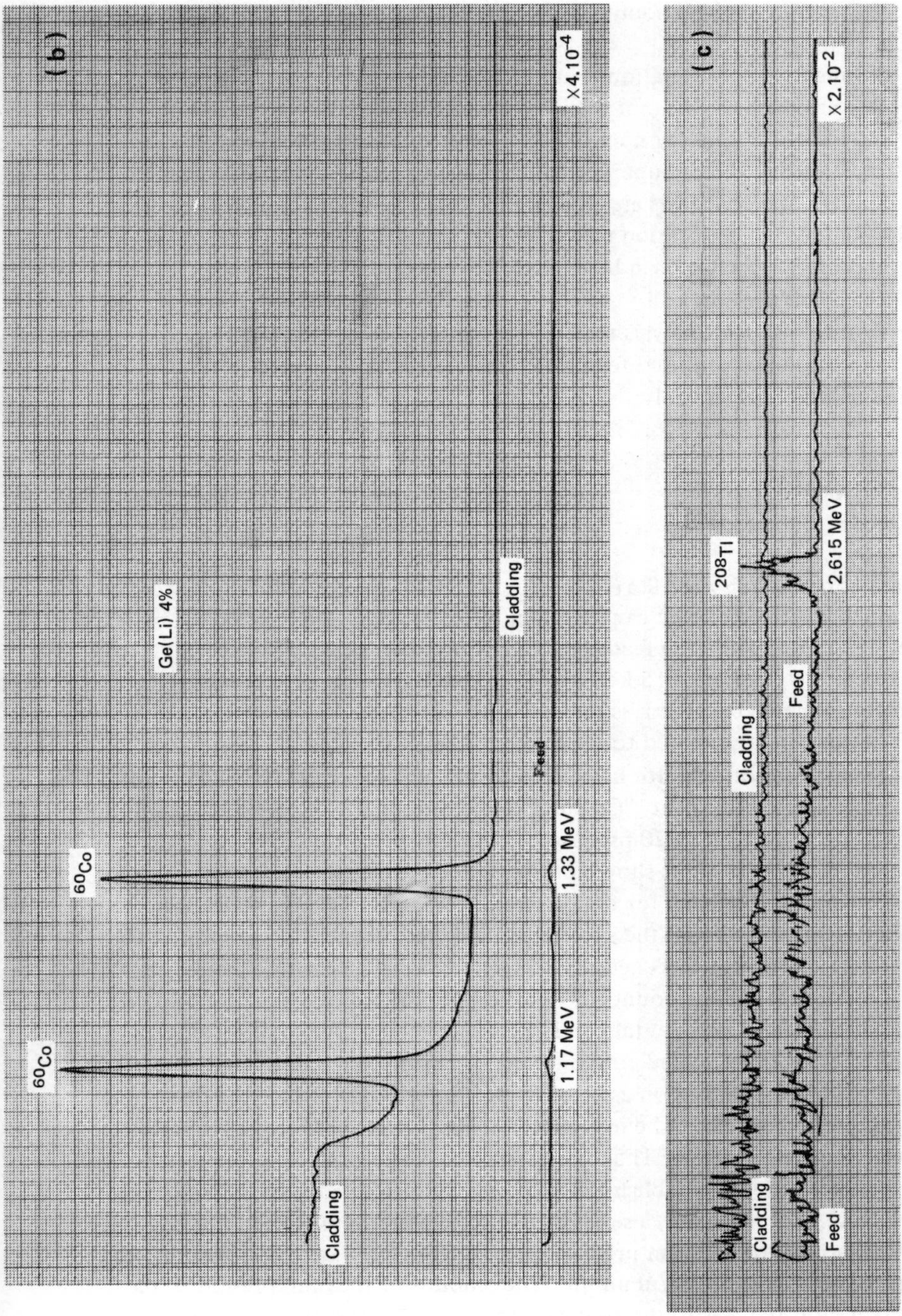

FIG.10. Gamma spectra of a hull sample and a diluted feed sample. (a) Energy range: 0–1023 keV; (b) Energy range: 1024–2047 keV; (c) Energy range: 2048–3071 keV.

2.2. Estimation of the amount of residual nuclear material (NM) in the hulls

Taking into consideration that ITREC hulls are geometrically regular and that they are washed with hot nitric acid and dried before leaving the dissolver, they should not retain any nuclear material in solid or liquid form. Tests performed during the pre-operational campaign gave no detectable residual NM in the hulls. The cold tests are valuable for hot operation, considering that the irradiated fuel dissolution rate is higher, and that the possibility of insoluble Th compound formation can be excluded because of the low burnup (6000–13 000 MW·d/t).

To establish the analysis region, a series of gamma spectra were taken on a hull sample and on a diluted feed sample. The spectra, divided in three ranges of energy were, respectively:

I 0–1023 keV
II 1024–2047 keV
III 2048–3071 keV

and are compared in Fig.10(a)–(c). The cladding spectrum does not show any radionuclide whose energy can be taken into consideration for nuclear material losses detection, while the feed spectrum shows two peaks of interest, at 185 keV (^{235}U) and at 2615 keV (^{208}Tl). This latter energy was selected for the measurements because it is located at an interference-free region and because ^{208}Tl can be directly related to thorium via the ^{228}Th scheme.

Three types of detector have been tested and finally a Ge(Li) detector, with an efficiency of 8% at ^{60}Co energy, was chosen. A series of 20 dissolution baskets corresponding to 20 processed fuel elements were monitored for ^{208}Tl emission in the conditions shown in Fig.6 (i.e. with the basket in the shielded transfer container) giving, for the different positions of the basket, a large number of spectra. From the spectra analysis all the baskets were found to contain no detectable ^{208}Tl activity.

To check if small amounts of nuclear material were present, but not detectable in the experimental conditions described above, the basket was extracted from the shielded container and monitored in a hot cell of the plant, using a standard penetration as a collimator, in order to reduce the dead time of the counting system. The measurements performed under these conditions show an increase of the ^{208}Tl activity with respect to the background of the cell.

Furthermore, to establish the amount of thorium corresponding to the above increment observed, a series of measurements were carried out adding known amounts of thorium product coming from the extraction section before separation of the thorium/uranium. The amount of thorium present in the baskets monitored was found to correspond to 3 g, equivalent to a nuclear material loss of about 0.012% (referred to 25 kg of thorium per element). This very small amount of nuclear material retained in the ITREC leached hulls is a result of the chopping system used, which produce geometrically regular hulls, and of the dissolution procedure adopted.

ACKNOWLEDGEMENTS

The work has been carried out by the Operating and the Analytical Groups of the ITREC plant. The authors thank Mr. G. Arcuri for the statistical evaluation of data, Mr. T. Conte for the gamma spectrometric measurements and Mr.G. Ossola for his helpful suggestions and interest in this work.

REFERENCES

[1] CNEN, Reprocessing and Fabricating Thoria Urania Fuel Elements, Doc. PCUT (67)10, Rome (Feb. 1967).

[2] CNEN, Programma generale prove nucleari impianto ITREC. 1st Campaign, Doc. TR (75)2/Rev.1, CRN TRISAIA (July 1975).

[3] DE CAROLIS, M., BARDONE, G., TDR Methods and Apparatus for Measurement of Levels as Physical Characteristics of Moving or Static Fluids in Pipelines or Tanks, CNEN, Rep. RT/CHI (74)/7 (1974).

[4] HOUGH, C.G., Statistical Analysis. Accuracy of Volume Measurements in a Large Process Vessel, General Electric Co., HW-62177 (Oct. 1959).

DISCUSSION

M.R. IYER: Have you tried to estimate the residual nuclear material in the hulls in hot operation using fission-product gamma lines?

G. BARDONE: As mentioned in the paper, a great many spectra were taken on feed samples and on a series of baskets containing hulls after dissolution, in order to find gamma-energy lines useful for the detection of nuclear material losses. The spectra reported in Fig.10 (a)–(b) of the paper, for example, show that there are no useful gamma reference lines except the 2.615 MeV of ^{208}Tl. This was expected because of the long cooling time of the Elk River Reactor fuel elements (more than ten years).

D. JUNG: Is there sufficient interest in transmitter-type TDR level probes to warrant commercial production in the near future?

G. BARDONE: At CNEN we have discussed the possibility of developing a commercial TDR transmitter similar to the types of transmitter used in a conventional plant (e.g. differential pressure transmitter). In our opinion, the transmitter-type TDR would be useful for process control purposes.

As far as safeguards applications are concerned, we would rather keep the present oscilloscope-type instrument because of the continuous CRT display of the transmission-line and probe signatures. This feature is very useful for checking whether any tampering has taken place. In addition, a new small high-performance portable TDR unit (weight 8 kg) is now available for operation under extreme environmental conditions; it has been designed to meet military specifications and also incorporates an x-y recorder. This instrument is very useful for inspection purposes.

Sessions IX and X

SPENT FUEL REPROCESSING

Chairman (Session IX): A. PETIT (France)

Chairman (Session X): A. von BAECKMANN (IAEA)

Rapporteur summary: *Reprocessing plant input measurements*

Papers IAEA-SM-231/4, 35, 48, 95 were presented by

L. KOCH as Rapporteur

Rapporteur summary: *Isotope correlation techniques in spent fuel reprocessing*

Papers IAEA-SM-231/20, 25, 26, 45, 142 were presented by

J. BOUCHARD as Rapporteur

PRELIMINARY CONCEPTS FOR DETECTING DIVERSION OF LIGHT-WATER REACTOR SPENT FUEL

T.A. SELLERS
Sandia Laboratories
Albuquerque, New Mexico,
United States of America

Abstract

PRELIMINARY CONCEPTS FOR DETECTING DIVERSION OF LIGHT-WATER REACTOR SPENT FUEL.

Sandia Laboratories, under the sponsorship of the Department of Energy, Office of Safeguards and Security, has been developing conceptual designs of advanced systems to detect rapidly the diversion of LWR spent fuel. Three detection options have been identified and compared on the basis of timeliness of detection and cost. Option 1 is based upon inspectors visiting each facility on a periodic basis to obtain and review data acquired by surveillance instruments and to verify the inventory. Option 2 is based upon continuous inspector presence, aided by surveillance instruments. Option 3 is based upon the collection of data from surveillance instruments with periodic readout, either at the facility or at a remote site and occasional inspection. Surveillance instruments are included in each option to ensure a sufficiently high probability of detection. An analysis technique with an example logic tree that was used to identify performance requirements is described. A conceptual design has been developed for Option 3 and the essential hardware elements are now being developed. These elements include radiation, crane and pool acoustic sensors, a Data Collection Module (DCM), a Local Display Module (LDM), and a Central Monitoring and Display Module (CMDM). A demonstration, in operating facilities, of the overall system concept is planned for the March-June 1979 time frame.

INTRODUCTION

The safeguarding of reactor spent fuel against acts of national diversion aimed at establishing a nuclear explosive capability is a principal international concern.

Sandia Laboratories, under the sponsorship of the Department of Energy, Office of Safeguards and Security, has been developing conceptual designs of advanced systems to rapidly detect diversion of LWR spent fuel.

A specific objective of international safeguards is the timely detection of diversion of significant quantities of nuclear material from peaceful nuclear activities to the manufacture of nuclear weapons or of other nuclear explosive devices or for purposes unknown, and deterrence of such diversion by the risk of early detection. Timeliness of detection is related to the time required to convert diverted material to nuclear explosive devices.

International response to a diversion must be preceded by detection and verification and for purposes of this paper are defined as follows:

- Detection is the receipt of an indication by an international authority that an undeclared transfer of spent fuel may have occurred.

- Verification is the determination by an international authority that a diversion has occurred. It requires independent assessment of the data upon which the detection was based and may be supplemented by:

  1) additional records and reports from the state or from the facility operator,

  2) additional data from the site, e.g. physical inventory, and
  3) on-site observation.

- Response is the set of institutional and political actions that may be set in motion by appropriate organization following the verification of a diversion.

TIMELINESS OF DETECTION

A reprocessing facility is essential to recover the plutonium in reactor spent fuel for use in weapons production. Reprocessing plants located throughout the world that can produce between 0.2 and 18 kg of fissile plutonium per day have been or are now operating. Since this basic reprocessing technology is well known and although design and construction is a complex and highly technical process, the potential for a clandestine reprocessing plant cannot be ignored.

It is possible to express the time from diversion to fabrication of the first weapon in the following manner. Let

T_1 = Time to transport diverted spent fuel to a reprocessing plant

T_2 = Startup time, after cold testing, for the reprocessing plant

T_3 = Time to produce enough plutonium for one weapon given an operating plant

T_4 = Time to fabricate a nuclear explosive device assuming all non-nuclear parts are available and assembled then the time to produce the first weapon will be

$$T_{first\ weapon} = T_1 + T_2 + T_3 + T_4$$

Given the broad range of estimates which may be made for each of the various times (T_1, T_2, T_3, T_4), the time to first weapon if the diverted material is spent fuel ranges from days to months.

DETECTION CONCEPTS

Three concept options have been identified for the detection of diversion of spent fuel. These concepts are defined in detail in "Preliminary Concepts for Detecting National Diversion of Spent Fuel," SAND77-1954, April 1978. They are based on the facility information contained in "Baseline Description for Reactor Spent Fuel Storage, Handling and Transportation," SAND77-1953, May 1978.

Option 1 is based upon inspectors visiting each facility on a periodic basis to obtain and review data acquired by surveillance instruments and to verify the inventory. Option 2 is based upon continuous inspector presence, aided by surveillance instruments. Option 3 is based upon the collection of data from surveillance instruments with periodic readout either at the facility or at a remote CMDM and occasional inspection. Surveillance instruments are included in each option to assure a sufficiently high probability of detection.

In Option 1, the duties of the inspector include the following:

- Reviewing accounting records for comparison with previously submitted reports
- Performing sampling tests to assure the presence and integrity of the fuel inventory
- Reviewing the recorded surveillance data and investigate anomalies detected by the safeguards instrumentation
- Observing the installation and removal of fuel assembly integrity devices

In Option 2, the duties of the on-site inspector will be similar to those of the periodic inspector, but will be performed at a much greater frequency.

In Option 3, authenticated data from on-site safeguards instrumentation would be transmitted over a tamper-indicating communication link to an off-site monitoring facility where assessment of the safeguards status of the spent fuel is made. The data could be transmitted over conventional landlines, high-frequency radio channels, or by satellite. The data transmittals could occur continuously, on a prearranged schedule,

TABLE I. MANPOWER AND COST ESTIMATES

Concept	Communication Mode	Reporting Interval	Number of Personnel	Annual Cost ($ Millions)
1-Periodic Inspections	Commercial Telephone	Every 2 Months	15	1.7
		Monthly	22	2.3
2-Resident Inspectors	Commercial Telephone	Daily	96	5.4
3-Remote Surveillance	HF Radio	Daily	22	2.7
	Leased Line	Daily	22	3.0
	Satellite	Daily	22	3.55

or on demand, and may include real-time observations and/or selective replays of the data recorded during periods of nontransmission. The frequency of data transmission is not necessarily limited by the technical system, but rather will be selected by interpretation of timeliness requirements and cost considerations. Occasional inspections would be conducted to confirm safeguards instrumentation integrity, and an inspector may be present at each reactor for the refueling period. Other duties of the inspector would be the same as those discussed for Concept 1.

CONCEPT COMPARISON

The three concepts have been compared in terms of communication mode, reporting interval, number of personnel, and rough estimates of cost. All comparisons are based on a network of 60 power reactors and 3 supporting storage facilities.

Table I compares the three facility concepts. The manpower estimates are based on 240 work days per inspector per year, at $50,000 per year per inspector. For Concept 1, each inspection including travel time requires five days. For Facility Concept 2, Resident Inspectors, each facility is monitored 365 days of the year during normal working hours. For Concept 3, Remote Surveillance, the off-site monitoring facility is manned by two inspectors, 24 hours a day, 365 days a year; a total of 10 personnel would be required to allow for weekends, holidays, and sick leave. In addition, quarterly inspections are assumed. An additional 30 days of inspector presence is assumed to be required at each reactor during the yearly refueling operation for Concepts 1 and 3. Travel expenses are estimated for each concept based on a cost of $750 per round trip.

Cost for the basic safeguards instrumentation is estimated at $100,000 per facility; this amount is used for all concepts. It is recognized that the safeguards instrumentation for Concept 2, Resident Inspectors, may not need to be as reliable or tamper resistant and, therefore, may not be as costly as the instrumentation for the other concepts. This factor is not considered due to the "rough estimate" nature of the cost estimates. Detection, assessment, fuel assembly identification, and data processing and storage equipment are included in this amount. These costs, and other capital costs, are amortized over a 10-year period to provide annual cost estimates.

For Concept 3, three communication modes are compared: high-frequency (hf) radio, leased landlines, and satellite. With this concept, an off-site monitoring facility is required; these are estimated at $500,000 for the hf radio and landline modes and $700,000 for the satellite mode. The communications cost for the three modes are based on:

TABLE II. CONCEPT COMPARISON

Concept	Communication mode	Reporting interval	Annual cost ($ × 10^6)	Time for verification and response (d)			
				6-day plant startup 6-day fabrication		6-month plant startup 21-day fabrication	
				(5 kg/d)[a]	(0.5 kg/d)[a]	(5 kg/d)[a]	(0.5 kg/d)[a]
1. Periodic inspections	Commercial telephone	Every 2 months	1.7	None (23 weapons produced before detection)	None (2 weapons produced before detection)	145	163
		Monthly	2.3	None (7 weapons produced before detection)	4	175	193
2. Resident inspectors	Commercial telephone	Daily	5.4	15	33	204	222
3. Remote surveillance	HF radio	Daily	2.7	15	33	204	222
	Land lines	Daily	3.0	15	33	204	222
	Satellite	Daily	3.6	15	33	204	222

a Assumed reprocessing plant Pu production.

- High-Frequency Radio Mode - 63 transceivers at $10,000 each and 63 digital interfaces between the instrumentation and the transceivers at $100,000 each. Five relay sites at $125,000 each are also required to provide communications over a land area similar to that of the United States.

- Leased Landline Mode - 63 lines at $6,000 per year each (based on typical foreign costs of $6 per mile and an average line length of 1000 miles) and 63 digital interfaces between the instrumentation and the landlines at $100,000 each.

- Satellite Mode - 63 transceivers at $100,000 each, 63 digital interfaces between the instrumentation and the transceivers at $100,000 each, and $300,000 per year for a dedicated nonpreemptable channel. A nonpreemptable channel may not be required and this cost could be significantly lower.

In comparing these concepts, a key factor to be considered is the time for verification and response that is provided before an initial weapon can be fabricated.

Three key elements in determining this time are the reprocessing plant startup time, the daily plutonium production rate, and the device fabrication time. A range of times has been considered that encompasses available estimates for these elements and 10 kg was used as the amount of material needed for a weapon. Table II shows the impact of these estimates on the available verification and response time.

The periodic inspection concepts, with inspection intervals of one and two months, offer the lowest costs; however, given the initiation of a diversion sequence soon after the inspector(s) departs the facility, the detection times is one to two months. Given the assumption of rapid plant startup, rapid fabrication, and a moderate production rate (5 kg/day), such a detection time may allow many weapons to be produced before detection.

Concept 2, Resident Inspectors, provides very rapid detection time at an expense of approximately $5.4 million per year--the highest of all concepts presented.

Concept 3, Remote Surveillance, also would provide very rapid detection time but with reduced inspector presence. The costs for this concept appear to be competitive with the bimonthly or monthly inspection concepts, with an improved detection time. This concept would require a comprehensive evaluation of its feasibility, the development of subsystem elements, and a capability demonstration. The communication options are straightforward adaptations of existing communication systems. Implementation of the hf radio link on an international basis may present a significant problem in the area of frequency allocations. Similarly, implementation of the leased landline link may be severely affected by the wide range of existing landline quality.

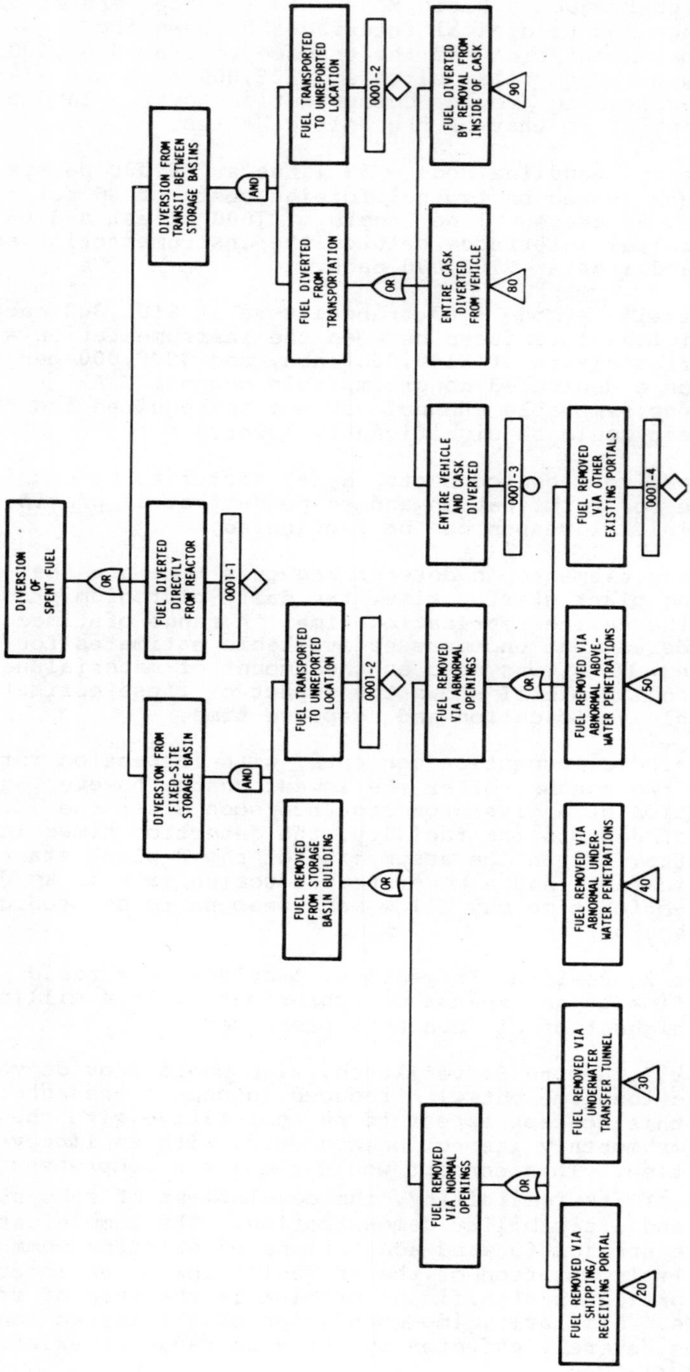

FIG.1. Example logic trees.

CONCEPTUAL DESIGN

In order to make an effective choice of the possible options available, a tradeoff analysis of hardware configurations and inspection options must be completed.

This tradeoff analysis will provide data on the effectiveness of various hardware configurations in detecting the unreported movement of spent fuel. This data will allow comparisons of relative effectiveness and cost of inspectors versus hardware in these tested configurations.

All three options require the use of surveillance hardware to assure an acceptable probability of detection. This instrumentation contributes significantly to the detection probability but does not influence timeliness of detection. Option 3 incorporates remote monitoring of surveillance equipment to increase timeliness of detection while requiring only occasional on-site inspection operations.

A conceptual design has been developed for Option 3 to meet the following performance requirements:

- Confirm in a rapid manner reported spent fuel movements and detect unreported movements including
  - Unreported transfers of fuel between a reactor and the storage pool.
  - Unreported transfers of fuel between the storage pool and a reprocessing area.
  - Unreported transfers on or off-site.
  - Nonarrival of reported transfers
- Provide capability for assessing alarms
- Provide capability for rapid inventory verification

The general principles that guided the conceptual design include:

- Assure high probability of detecting the unreported movement of spent fuel, including tamper indication
- Allow rapid reporting of detections
- Minimize need for on-site inspectors and other personnel requirements of the inspection agency
- Minimize cost and operational impact on the facility

The initial step of the conceptual system design is a detailed analysis to identify the basic activity which must be detected. The analysis that has been used is adapted from fault tree

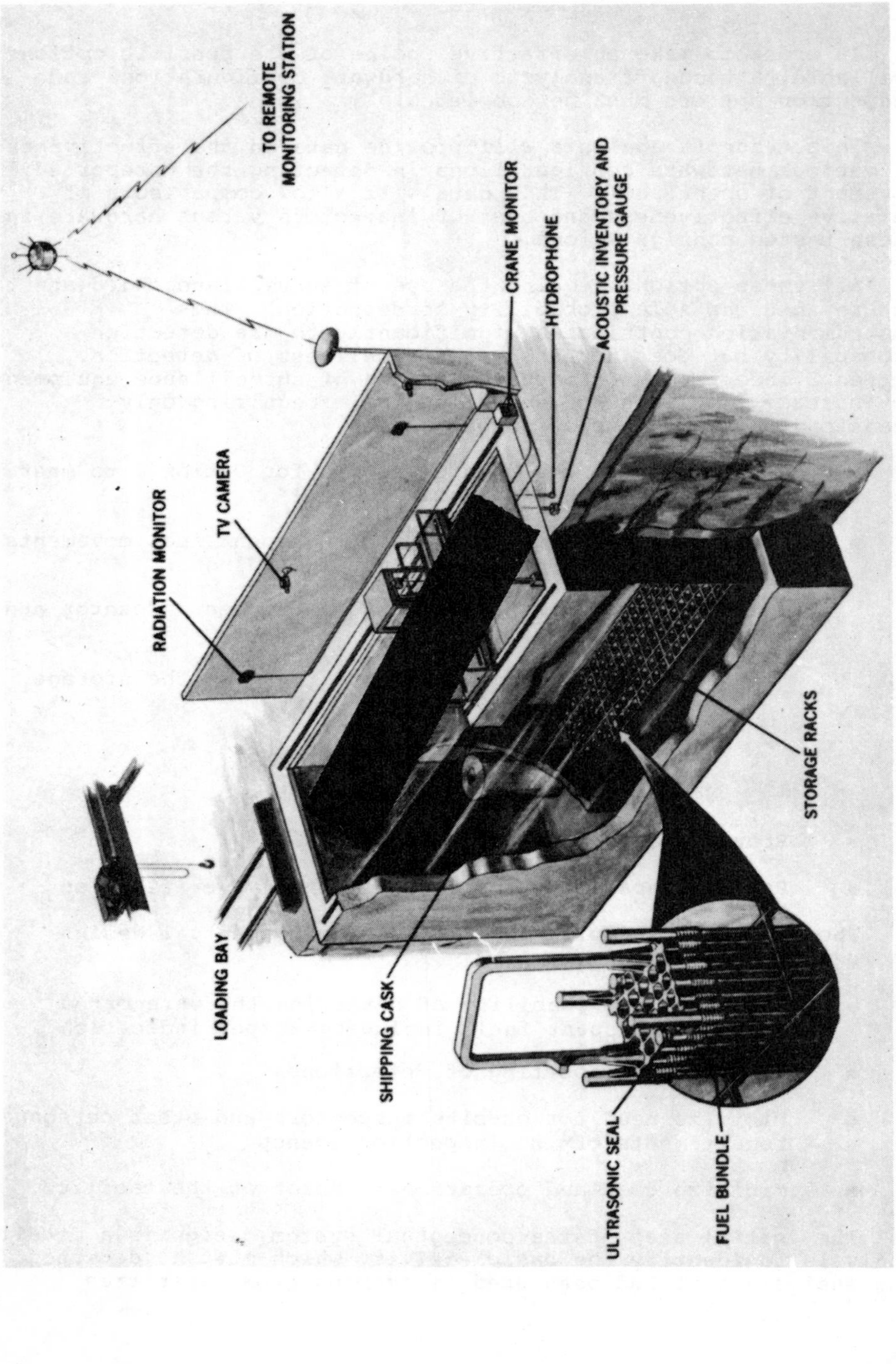

FIG.2. Spent fuel storage monitoring.

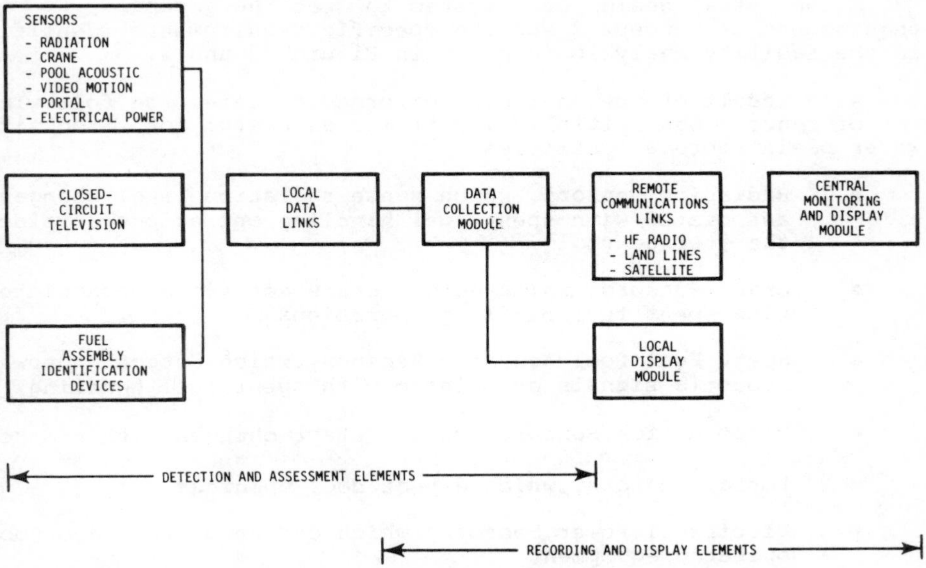

FIG.3. Conceptual aspect of spent fuel containment and surveillance system.

methodology where events related to specific concerns are connected using logic gates to show their interdependence. This approach has the following features:

- systematic
- provides visibility of logic
- provides traceability

Using this analysis a choice of surveillance elements required to detect specific activity can be identified. An example logic tree is shown in Figure 1.

When the trees are fully developed by following them along any desired set of branches, a set of basic activities is identified which must be accomplished before fuel diversion can take place along that particular path. Where these activities are inputs to an OR gate, the detection of all the activities is required to block that diversion path. For activities grouped as inputs to an AND gate, the detection of any one will block that path. In order to achieve reasonable reliability, however, a surveillance system should be capable of detecting more than one activity along any diversion path. For example, the detection of three activities may be a practical goal because it allows for single failures and still provides for redundancy.

A conceptual design of a system to meet the general requirement of Concept 3 and the specific requirements identified in the facility analysis is shown in Figures 2 and 3.

As a result of the analysis performed to date, the following set of sensors was initially identified as useful for fixed site, water basin storage facilities:

- Radiation Sensors, which sense radiation level changes associated with spent fuel handling operations outside the storage pool
- Crane Sensors, which detect crane activities associated with spent fuel handling operations
- Spent Fuel Pool Acoustic Sensors, which detect underwater acoustic signals associated with spent fuel handling
- Video Motion Sensors, which detect changes within a scene
- Portal Sensors, which detect door openings
- Electrical Power Sensors, which detect operation of motor operated equipment

In addition to the preceding detection sensors, Fuel Assembly Identification Devices (FAIDs) that could be installed on fuel assemblies to provide unique identification and integrity information, and Closed Circuit Television (CCTV) to assist in assessment activities were included in the initial conceptual design.

Communication between the sensor modules and the DCM takes place over fiber optics local data links. The fiber optics system and tamper indicating enclosures around the various on-site modules will allow detection of any tampering attempt on the system.

The DCM collects sensor data and tamper indicator status from the various sensor modules. It correlates the sensor data and checks for unusual or inconsistent operations.

As an example of such a correlation process, the relative radiation levels as seen by several spatially separated radiation monitors should correlate in a predictable manner with a radiation source, such as a cask containing spent fuel, being moved through the area. The monitor outputs can be used to generate source strength and position location solutions by a process called deconvolution, or simply, unfolding. An unusual condition could then be a solution which shows the radiation source (presumed to be spent fuel) moving in a direction indicating a movement that was not reported. An inconsistent condition could be one in which the unfolding process does not converge to a unique solution, possibly due to an attempt to mask or fool the system by the introduction of shielding around the sensors and/or the introduction of extraneous radiation sources within the area. Such unfolding techniques further enhance the tamper-indicating capability of the system, reduce false alarms and increase the probability of detection of an unreported spent fuel movement over what a system of uncorrelated sensors could provide.

The data base of normal operations will have to be generated on the basis of observations and data obtained from each facility during the initial installation and checkout phase.

In operation, the DCM will poll the sensors and tamper status indicators every few seconds and perform the unfolding calculations. Whenever an abnormal solution is obtained, sensor data will be stored, and when required, a CCTV picture will be stored in the DCM. This sequence could also be initiated at random times or an interrogation from the LDM or the CMDM as shown in Figure 1. Approximately 24 hours of data can be stored by the DCM for later interrogation by the LDM and/or the CMDM.

The system elements which allow a timely reporting of a detection of a diversion are the DCM, LDM, the communication links and the CMDM. It should be noted that the DCM is included in both the detection elements and the recording and display elements. The data links to the CMDM can be satellite, ground station RF, or hardware links. Data authentication will be applied to these links to detect tamper attempts.

The CMDM may be located many miles from the spent fuel site. It contains the necessary equipment to communicate with the DCM's, to record, to process, and to display data. It also provides the capability to accomodate inputs from system operators and aid them in assessing spent fuel status via any of the system DCM's.

HARDWARE DEVELOPMENT AND TEST

At the present time, the installation and checkout of the initial system elements are under way at the General Electric, Morris, Illinois, away-from-reactor storage facility. Preliminary design of a configuration for a pressurized water reactor is under way. The design and checkout of the initial CMDM is in the final stages and will be located at Sandia Laboratories, Albuquerque, New Mexico to support a demonstration of the system capability that will take place in the March to June 1979 time frame.

DISCUSSION

A.J. STIRLING: Has the IAEA stated that option 2 (i.e. the stationing of resident inspectors at reactors) would be acceptable to them?

T.A. SELLERS: No IAEA approval has been given for any of the three options. Our study considers all possible inspection options and tries to identify the advantages and disadvantages of each.

A.G. HAMLIN: This analysis is very interesting, but it raises the question of where do you stop. You have taken steps to close a number of routes. A diverter who attempts to use one of these is then either a fool or, politically, he has decided that he does not need to worry about detection. Neglecting these unlikely possibilities suggests that the system described presents the diverter with a new set of ground rules and that the diversion analysis will have to be reiterated. How long do you go on?

T.A. SELLERS: The purpose of the study was to identify the problem, the potential concepts for dealing with the problem, and the technology that is required to implement the options. No attempt was made to decide to what level international safeguards should be implemented.

SOME ASPECTS OF IMPLEMENTING SAFEGUARDS IN A REPROCESSING PLANT

S.V. KUMAR
Bhabha Atomic Research Centre,
Trombay, Bombay,
India

Abstract

SOME ASPECTS OF IMPLEMENTING SAFEGUARDS IN A REPROCESSING PLANT.

The success of the effective application of safeguards procedures in a reprocessing plant depends upon achieving accurate and reliable material balances in the relevant areas of the plant. This paper describes the criteria for fixing the various material balance areas in the plant and indicates procedures that could be adopted in each area. To improve the effectiveness of the safeguards measures, suggestions are made for further research and development efforts.

INTRODUCTION

The methods of implementation of safeguards have been a subject of serious consideration in IAEA and elsewhere for the past several years. Even though considerable international safeguards developments have taken place, efforts need to be continued to establish a fully effective and comprehensive system. The main consideration for effective safeguards implementation would be the need to establish technical and political confidence in the safeguards procedures and this should not retard the development of nuclear energy for peaceful purposes.

Reprocessing plants play an important role in the total safeguards efforts in a fuel cycle. The primary principle in the safeguards efforts is to ensure containment. A reactor building is by its very nature already a containment. A careful check on the inlet and outlet of fuel elements would be adequate. However, reprocessing plants require extrapolation of the containment principle since the fissile material in the form of aqueous solutions and powders is handled loose and it is necessary to control the flow through discrete material balance areas and key measuring points. The procedures adopted for striking the material balance should be so devised that they do not interfere with the regular operation of the plant and do not involve the knowledge of some of the features of the plant which may be of proprietary nature. A safeguards procedure which lays emphasis on certain strategic points will serve the purpose of maintaining its effectiveness and will also provide sufficient flexibility for the plant operator.

1. CRITERIA FOR MATERIAL BALANCE AREAS (MBAs)

The main objective in fixing the MBAs is that they should be well defined both physically and with respect to the state and form of the fissile materials involved. This obviously depends on the type of facility. It is necessary that adequate attention should be given at the design stage to ensure effective and unobtrusive implementation of safeguards. The defining of the MBAs can be governed by the basic operations carried out in a plant. They are:

Handling of the spent fuel
Head-end treatment
Solvent extraction cycles
Product conversion from solution to oxide
Product storage.

2. SPECIAL FEATURES OF THE MATERIAL BALANCE AREAS

2.1. Spent fuel handling

The spent fuel from the reactor site arrives in sealed transport containers with the relevant documentation regarding the contents and is stored in known locations in the storage racks in the fuel storage pool. It is necessary to ensure that the fuel elements are identifiable and this should be maintained till the time of charging to the dissolver. A suitable surveillance device could be utilized for this purpose.

2.2. Head-end treatment

The fissile material loses its discrete form in this area and the output from this area will be the dissolver solution and the hull. The dissolver solution is transferred to the input accountability tank under strict control on the movement of this solution. Immediately after the transfer the tank is sampled to establish the account of nuclear material.

Monitoring of fissile material content in the hull by non-destructive techniques still has certain limitations despite extensive development effort.

2.3. Solvent extraction cycles

The fissile material in this material balance area is handled entirely in the form of solutions. The input point for this area is the input accountability tank. The output streams are the uranium and plutonium product solutions and waste streams which are accounted for in key measuring points. The success of safeguards implementation in this area depends upon accurate estimation of the quantity of fissile material. This in turn depends upon the reliable measurement of volume, representative sampling and accurate chemical analysis.

Satisfactory volume measurement could be achieved by elaborate calibration procedures and statistically evolved calibration curves.

The procedures concerned with obtaining a representative accounting sample can be established after a series of preliminary trials. It is necessary to take proper care for homogenizing the contents before samples are drawn from the tanks.

The analytical techniques for estimating the fissile material concentrations to acceptable accuracies are obtainable by the coulometric methods, both in the control potential and the constant current modes.

Another interesting method for measuring volume and the total fissile material in an accountability tank could be the use of tracer techniques. Initial experiments have been quite encouraging but the plant-scale experience of this technique is still to be established.

2.4. Product conversion from solution to oxide

After conversion from solution into oxide form the uranium and plutonium products are sent out to the storage in convenient batches. The safeguards requirement in this area will be accurate weighing techniques and proper assay of the oxide samples.

2.5. Product storage

The emphasis in this area is more on the containment and surveillance.

3. MATERIAL UNACCOUNTED FOR (MUF)

Material Unaccounted For (MUF) is an unavoidable reality and values can always be expected to be positive though the aim is to have this value as low as possible. The significance of MUF would depend on whether it is due to a loss of control in the measurement system, or to a hidden or unmeasured inventory, or to identifiable or unidentifiable unmeasured losses. The quantification of the MUF could be attempted by systems analysis. It would be economically undesirable to go to extreme lengths to reduce the levels of MUF much beyond what is required for safe operation. The trends in the variation of MUF in a particular facility could be the important criteria for judging the effectiveness of safeguards applications. The problem is really complex and even a quantitative approach would call for a good deal of pragmatism.

The frequency of physical inventory should be decided in such a way that it does not unduly reduce the availability of the plant, thereby affecting production and, at the same time, keeps the safeguards requirements in view. This is very important as it has cost repercussions.

To arrive at effective safeguards measures without adversely affecting plant operations various factors should be identified and suitable development efforts should be intensified to achieve the desired objective.

In conclusion it could be said that the success of safeguards implementation largely depends on how judiciously the combination of the three basic aspects of safeguards – Accounting, Containment and Surveillance – is adopted.

DEVELOPMENT OF THE SAFEGUARDS APPROACH FOR REPROCESSING PLANTS

V. SUKHORUCHKIN, L. THORNE, D. PERRICOS,
D. TOLCHENKOV, J.E. LOVETT, T. SHEA
Department of Safeguards,
International Atomic Energy Agency, Vienna

Abstract

DEVELOPMENT OF THE SAFEGUARDS APPROACH FOR REPROCESSING PLANTS.

The present status of development in the Agency of the approach to safeguarding reprocessing facilities is briefly reviewed and some problems to be solved are indicated. In this approach the Agency should take into account the diversion hazards existing at such facilities, diversion assumptions, safeguards criteria to be satisfied and safeguards measures available, as well as the strategic value of material handled by facilities and its place in the nuclear fuel cycle.

I. INTRODUCTION

A. General

The reprocessing facility represents the point in the nuclear fuel cycle where the conversion from item accountancy, for fuel assemblies, to quantity accounting, for grams or kilograms of uranium and plutonium, occurs. The measured plutonium input to the reprocessing facility establishes the plutonium "to be accounted for" for the remainder of the fuel cycle, until the material has been re-fabricated into fuel assemblies - identifiable items - for re-use in nuclear power reactors. Hence the importance attached to reprocessing facilities, and in particular to input measurement.

Reprocessing facilities handle significant quantities of plutonium and the product is in a form which can be converted to the components of nuclear explosive devices in a relatively short period of time. A medium size plant with a throughput of 300 tons of spent fuel per year separates about 2000 kg of plutonium per year. There are also unique and serious problems of verification. Nonaccessability of equipment, questionable representativeness of samples, difficulty of independent calibration of measurement equipment; these and other difficulties make the application of safeguards very complex and provide a would-be divertor numerous opportunities for concealment.

In developing the safeguards approach for any particular type of facility, the Agency should take into account the recognized diversion hazards, diversion assumptions, the safeguards criteria to be satisfied and the safeguards measures available, as well as the strategic value of material handled by the facility. Considering the complex nature of reprocessing facilities, the recognized diversion possibilities and the quantity of sensitive material involved, the Agency will have to mobilize all possible measures for safeguarding such plants. First of all a continuous inspection regime with corresponding 100% verification of all declared input and output measurements can be foreseen. The choice of a proper combination of material accountancy, containment and surveillance (both human and instrumental) measures seems to be the central question in the development of a safeguards strategy for reprocessing plants.

B. Safeguards Criteria

The Agency's safeguards criteria with respect to significant quantities of nuclear material to be detected if diverted and the timeliness of such detection are described in [1] in detail. As applied to reprocessing facilities those criteria specify a significant quantity on the order of 8 kg of total plutonium, with a detection time of on the order of several days.

C. Diversion Hazards

States having chemical reprocessing facilities may be expected also to have conversion facilities, fabrication plants for fresh fuels, power reactors, and re-fabrication plants for processing the recovered plutonium (or U-233) into reactor fuels. Waste treatment and storage facilities will also be provided and the State may have enrichment facilities. It is necessary to consider these facilities as sources of material and/or as possible collusion centers in a State-directed programme to obtain nuclear explosives through the diversion of nuclear materials or the illicit use of facilities committed to peaceful purposes.

There exists a definite relationship between the amounts of material flowing through a fuel cycle and the power production. Accurate knowledge of the fuel content of fresh fuel assemblies, of nuclear production and loss (burn-up) during irradiation and of nuclear loss attributable to radioactive decay can establish the basis for determining what the feed to a reprocessing facility should be. It is conceivable that a State could alter these cross-linked indications to conceal an underestimate of the declared input to the reprocessing facility.

In a State having the types of facilities indicated, similar or identical forms of material will be encountered at a number of facilities. For example, spent fuel assemblies will be found at the power reactors and the reprocessing facility; plutonium nitrate will be found at the reprocessing facilty, at nitrate-to-oxide conversion plants, at chemical scrap recovery facilities operated in conjunction with mixed-oxide fuel fabrication plants, and at designated storage facilities. The possibility must be considered that materials may be diverted from a number of facilities, or, more likely that materials might be moved from facility to facility to mask diversion.

When focussing on the facility, the principal concerns are:

1) that plutonium may be diverted from the declared throughput;

2) that the facility may be used to separate plutonium from undeclared input; and

3) that uranium may be diverted from the declared throughput.

D. Safeguards Approach

Two different strategies for potential diversion can be assumed, namely abrupt and protracted diversions. For the strategy of abrupt diversion a significant amount of nuclear material is diverted in a single action or over a very short period of time. Inventory shuffling, misrepresentation and concealment might be used by the divertor to avoid or delay detection that such diversion has occurred, but it is usually assumed that an abrupt diversion could and would be detected at the next inventory verification. Thus, the safeguards system should be designed to detect the abrupt diversion of a certain amount (on the order of one significant quantity) of nuclear material from a single nuclear facility in a time frame consistent with the estimated conversion time for that material [1]. To achieve timeliness of detection under an assumption of abrupt diversion, the Agency will need to employ procedures which enable it to reach conclusions as to the amount which could be unaccounted for on a more frequent basis than can be provided by the verified clean-out physical inventories.

In the strategy of protracted diversion, quantities of nuclear material are diverted systematically at a relatively low rate over a relatively long period of time, therefore the safeguards system should be designed to detect as a minimum, a diversion rate at each facility of a certain amount (again on the order of one significant quantity) of each category of nuclear material during one year.

By its inspection activity, the Agency must maintain sufficient continuity of knowledge of the flow and inventory of nuclear material subject to safeguards and the technical conclusion of the IAEA's verification activities will be "a statement, in respect of each material balance area, of the amount of material unaccounted for over a specific period, giving the limits of accuracy of the amounts stated" [2].

In order to permit the Agency to apply safeguards effectively, access by IAEA inspectors to all strategic points at the boundaries and inside material balance areas is essential. In addition to the flow key measurement points at the boundaries of material balance areas, a sufficient number of the strategic points inside material balance areas might be chosen by the Agency to obtain necessary process data including instrument readings and measurement results relevant to the verification of source data. The Agency's inspectors should continuously observe the process, collect results of operator's measurements, take samples, make independent measurements and analyse the data collected in order to draw conclusions on possible diversion of nuclear material from the facility within a short time period.

Where material is in sealed storage, timeliness can be achieved by verifying the integrity and identity of seals at time intervals appropriate for different types of materials. More complex systems of containment and surveillance, both instrumental and human, can be used for material in storage, where simple seals alone are inadequate.

Conclusions are needed covering two time stages. The first stage conclusion refers to the observation of the process during that interval. This conclusion will be a combination of C/S measures, including human surveillance, and material accountancy based on chemical analysis of IAEA samples when these results are available on a sufficiently timely basis, on inspector NDA and other in-plant measurements and on inspector verification of certain process data measured by the operator. These quantitative tests should be characterized by timely data. The second stage conclusion complements the first by incorporating the results of other measurements (e.g. chemical analysis of samples). The second stage also permits corrections to be made to the first stage conclusions based on accountability analyses.

II. PAST ACTIVITIES

Prior to 1977 the development of a safeguards approach to reprocessing plants was conditioned by the fact that no operational plants were under safeguards. Two campaigns at the NFS West Valley plant had been attended by Agency inspectors in 1967 and 1969 but these constituted distinct self-contained campaigns and attendance was by invitation. In few senses could they be regarded as typical of conditions as they are met today or will be encountered in the future. For most of the period since the NFS campaigns the only guidance existing lay in the safeguards Document INFCIRC/66 developed prior to NPT which merely states that 'continuous inspection is foreseen'.

Clearly with the advent of NPT and the announced intention in several countries to develop reprocessing plants attention was needed to the safeguards requirement. Internally within the Agency a task force attempted to analyze the problems and establish procedures for a realistic safeguards regime. At an early stage it was seen that considerable difference existed between the idealised concepts which had always been tacitly accepted and the realities of currently developing plants. For example, it was often assumed to be axiomatic that inspectors would study the construction of each plant and become intimately familiar with every facet of the pipework and internal layout. In practice this is a situation denied even to most plant operators and any safeguards approach must recognise that total manpower, length of service, recruitment and training alone preclude these conditions being met. In addition and even more importantly the earlier safeguards agreements did not legally permit attendance of inspectors at these early stages. Similarly such essentials as verified calibration of accountancy vessels was not guaranteed by these agreements. The internal working group attempted to take into account these factors and to develop an approach that would ascribe meaningful tasks to the inspectorate. It was clear that continuous or 'on call' presence was necessary but it was also recognized that even so the approach was 'ad hoc' and would require improvement in the light of experience gained when active inspection started.

In June of 1978 the Agency called for the Advisory Group Meeting on safeguarding of Reprocessing Plants. Shortly before the meeting two groups of consultants together with the Agency's staff prepared a working paper for the Advisory Group. The Working paper [3] consisted of a detailed description of the procedures which were recognized as a possible approach to be followed by the Inspectorate in order to create a capability to detect possible diversion at reprocessing facilities and to meet safeguards objectives and criteria in respect of timeliness and quantities.

The Advisory Group Meeting was held in Vienna during the week of the 12th June and was attended by 44 experts. The working paper supplied by the Secretariat served generally as a useful basis for discussions of the Advisory Gourp, however, the Group was not able, because of the large volume of material presented and the complex matters involved, to discuss all parts of the working paper in detail, and accordingly did not attempt to prepare a revised paper.

During the Advisory Group Meeting it was recognized that there are many fields of development related to the documentation of a safeguards system for reprocessing facilities, which on the one hand should give a full possibility to satisfy the objectives and criteria of effective safeguards and on the other hand to be only an acceptable hindrance to an operator as well as having a reasonable cost.

It was also recognized that each particular plant would require a specific combination of safeguards measures to be implemented in conformity with its design features, capacity and the mode of operations.

First and foremost, the Advisory Group recommended that the Agency undertake, with the strong support and close collaboration of the Member States, a comprehensive study of safeguards systems and techniques for reprocessing facilities in order to improve the cost effectiveness of such safeguards, with the aim of developing a methodology for assessing various combinations of materials accountancy and containment-surveillance (C-S) techniques. As examples, this comprehensive study should include:

a) the development of means for quantifying the assurances given by both materials accountancy techniques and C-S and combinations thereof;

b) an analysis of the degree to which alternative safeguards approaches would meet Agency objectives and criteria;

c) the development of feasible safeguards techniques and approaches in order to enable the Agency to maintain continuity of knowledge of the flow and inventory of nuclear materials during intervals between clean-out physical inventories;

d) examination of the cost effectiveness of various alternate safeguards strategies, approaches and techniques;

e) application of the results of the generalized study in order to design optimum safeguards strategies and approaches for reprocessing plants;

f) in due course examination of the applicability of safeguards approaches developed for small or medium-sized facilities to large facilities, and development of adaptations or new approaches where necessary; and

g) indication of priorities in development of different safeguards methods and techniques.

Further research and development effort should also relate to the study of sampling techniques, sample preparation methods and problems associated with sample storaging and transportation as well as to the development of necessary NDA and analytical measurement technologies aimed at the verification of nuclear material on a rapid time scale, and the development of adequate tamper-resistant and reliable containment and surveillance devices.

III. CURRENT ACTIVITIES

A. Operational Experience

Up to now the Agency has very limited experience in applying safeguards to reprocessing facilities. The first reprocessing plants to come under Agency's safeguards have now been so operating for just over one year. During this time they have been under continuous inspection. This means that inspectors have been continuously present in the plants or in its vicinity ready to be called in to witness important operations or measurements. Much has been learned of the effectiveness of planned techniques and the areas of weakness have also become more clear. Some of this experience was described earlier in this symposium [4].

B. Development of Timely Detection Procedures

As was mentioned, the Agency should detect a would-be diversion of nuclear material at a reprocessing plant in rather a short period of time. One way to make detection time shorter is to increase the frequency of physical inventory taking and material balance closing. Because the usual clean-out physical inventory taking is very costly and often scheduled and performed by an operator with a frequency which is not sufficient for the Agency's safeguards objectives, some alternative modes of material balance verification to be implemented in intervals between two clean-out physical inventory takings are now being considered with the aim of choosing proper combinations of different methods which can satisfy safeguards criteria.

Among those alternative modes are: draindown physical inventory taking, running book inventory taking, in-process physical inventory determination, flow follow up, etc. All of these concepts have, at one time or another, been used or discussed as means of providing operational information for process engineers. In some cases there have also been attempts to generate data useful for materials accountancy, with only limited success. With improved measurement methods and other improvements possible now but not available at the time, these concepts may find an application in international safeguards.

The possibilities for application of these methods will vary considerably from plant to plant depending on a number of factors such as:

- plant size;
- availability of accurate intermediate flow measurement points;
- availability of volume, flow and density instruments at inventory locations;
- availability of sample points at inventory locations; and
- quantity of flow relative to inventory quantity.

Dynamic materials accountancy is another possibility which is being studied, and which appears to have a significant potential. Paper studies of large facilities look extremely promising for the long term future, but even in the near term it appears possible to design a semi-dynamic system which is capable of meeting the specified goals of detection sensitivity and timeliness.

TABLE I. TASTEX PROJECT

Tasks	
T-A	Evaluation of Performance and Application of Surveillance Devices in the Spent Fuel Receiving Areas.
T-B	Collection and Analysis of Gamma Spectra of Irradiated Fuel Assemblies Measured at the Storage Pond.
T-C	Demonstration of Hull Monitoring System.
T-D	Demonstration of the Loadcell Technique for Measurement of Solution Weight in the Accountability Vessel.
T-E	Demonstration of the Electromanometer for Measurement of Solution Volume in Accountability Vessel.
T-F	Study of Application of DYMAC Principals to Safeguarding Spent Fuel Reprocessing Plants.
T-G	K-edge Densitometer for Measuring Plutonium Product Concentrations.
T-H	High Resolution Gamma Spectrometer for Plutonium Isotopic Analysis.
T-I	Monitoring the Plutonium Product Area.
T-J	Resin Bead Sampling and Analytical Technique.
T-K	Isotope Safeguards Techniques.
T-L	Gravimetric Method for Input Measurements.
T-M	Tracer Methods for Input Measurements.

C. Isotopic Correlations Technique

A clearer picture is now emerging on the role of application of ICT at a chemical reprocessing plant for the LWR fuel cycle, as a result of research and development carried out at various centres, including the IAEA. Because of the potential of these techniques for verification of input and for checking of consistency and reliability of reported data, IAEA is actively interested, and participates [5,6], in these developments.

The actual procedures of applying ICT to a reprocessing campaign and how ICT would fit in with the overall scheme of inspection procedures at a reprocessing plant, are also becoming more clearly understood. It is hoped that a clear-cut enunciation of the procedures, together with quantitative indications of accuracies to be expected, based on actual inspection experience will be available in the near future. Intensive efforts are also directed towards adapting the IAEA isotopic correlations data bank to the needs of the safeguarding authority for its verification work [6].

D. TASTEX

In the Spring of 1978 agreement was reached between the governments of France, Japan and the United States of America, together with the IAEA, on a series of thirteen tasks under the acronym TASTEX, for Tokai Advanced Safeguards Technology Exercise. Although obviously largely related to the safeguarding of the Japanese PNC-Tokai reprocessing facility, many of the thirteen tasks have been worded, and will be undertaken, as studies relative to the applicability of advanced safeguards technology to model facilities. The tasks are listed in Table 1. They cover a broad range of activities, including the use of surveillance equipment, the demonstration of a number of destructive and non-destructive measurement techniques, and the potential application of dynamic materials control.

During the first two years of operation of the Tokai reprocessing facility instruments are expected to be installed in the plant that will facilitate the measurement of the plutonium concentration and isotopic abundance on the product solution, as well as monitor movement of the product material. In addition the input tank will be monitored by an electromonometer that will provide hard copy for inspection. These four hardware tasks will be demonstrated in the laboratory before installation and evaluation at the Tokai facility in the Spring and Summer of 1979.

No definite decision regarding the possible application of a dynamic materials control system is likely before the Spring 1979, but it is clear that the feasibility of such a system depends primarily on the development of NDA equipment foreseen under other tasks, notably the K-edge densitometer being developed under task T-G. In order for the dynamic materials control system to be feasible the densitometer, (or some other NDA instrument) must be extendable not only to product solutions but also to various intermediate solutions and ideally even to input solution.

E. Comprehensive Study

Each particular reprocessing plant represents its own problem for effective safeguarding and should be carefully studied in order to choose the proper combination of applicable safeguards measures. Choice of that combination can be facilitated by development of a model safeguards approach for reprocessing facilities. To develop such an approach, a comprehensive study of safeguards systems and techniques for reprocessing facilities was recommended by the Advisory Group Meeting on Safeguarding of Reprocessing Plants (Vienna, 12-16 June 1978) in order to improve the cost effectiveness of such safeguards. This study can be fulfilled only under conditions of the strong support and close collaboration of Member States and with effective coordination of R and D work in this field by Member States and by the Agency.

The idea of creating a special international group of experts to conduct this comprehensive study is now under consideration in the Agency. The study should be finished when it can demonstrate the validity of developed safeguards concepts.

F. Large Scale Reprocessing Facilities

The Agency is not expected to be faced with safeguarding large-scale commercial facilities for some years to come. Nevertheless, some preliminary concepts are under examination now. Complex combinations of material accountancy, containment and surveillance measures will definitely be required for the safeguarding of large plants. It is clear now that such plants should be designed and constructed with "safeguards in mind". Some recommendations for designers of such plants in respect of facilitation of safeguarding are expected as results of the comprehensive study above, as well as of the International Fuel Cycle Evaluation (INFCE) Programme.

VI. SUMMARY

It is clear that the development of an effective safeguards approach for reprocessing facilities is a complex problem. Historically the Agency has little experience, and even that little has no direct bearing on today's problems. Nevertheless, a broad spectrum of development efforts are underway, and the early indications are that the problems can be resolved and reprocessing facilities will prove to be safeguardable.

REFERENCES

[1] HOUGH, G., SHEA, T., TOLCHENKOV, D., "Technical criteria for the application of IAEA Safeguards", IAEA-SM-231/112, these Proceedings, Vol.I.

[2] INTERNATIONAL ATOMIC ENERGY AGENCY, The Structure and Content of Agreements between the Agency and States Required in Connection with the Treaty on the Non-Proliferation of Nuclear Weapons, INFCIRC/153 (corrected), IAEA, Vienna (1971).

[3] INTERNATIONAL ATOMIC ENERGY AGENCY, Advisory Group Meeting on Safeguarding of Reprocessing Plants, Vienna, 12–16 June 1978, Secretariat Working Paper (May 1978) (unpublished).

[4] HAGINOYA, T., FERRARIS, M., FRENZEL, W., KISS, I., KLIK, F., POROYKOV, V., THORNE, L., THORSTENSEN, S., "Experience in the application of IAEA inspection practice", IAEA-SM-231/105, these Proceedings, Vol. I.

[5] SUKHORUCHKIN, V., Verification of Plutonium Input to Reprocessing Plants (using Isotopic Correlation Safeguards Techniques) IAEA/STR-52, IAEA, Vienna (1976).

[6] SANATANI, S., SIWY, P., "Utilization of a data bank for safeguards application of isotopic correlations", ESARDA Symp. on Isotopic Correlation and its Application to the Nuclear Fuel Cycle, Stresa, Italy, 9–11 May 1978.

DISCUSSION

J. REGNIER: I should like to draw attention to the fourth paragraph in Section D of the paper. This paragraph is taken almost word for word from a document presented by the Agency at the Advisory Group meeting in June 1978, a document which was not well received. The paragraph in question goes far beyond the recommendations of INFCIRC/153 in regard to the key measuring points in the material balance areas. These measuring points are of use only during an inventory period. It seems to me, moreover, that continuous observation of the process, as referred to in the paragraph mentioned, also goes far beyond the recommendations of INFCIRC/153 and is therefore not acceptable.

V. SUKHORUCHKIN: For various reasons, the Secretariat working paper for the Advisory Group you mentioned was not discussed in detail during the meeting in June. The paper we have just presented describes the present status in the development of Agency safeguards strategy and technology for reprocessing facilities, and this development is still under way. Of course, the Agency's approach should comply with INFCIRC/153 and other properly accepted basic safeguards documents. I do not think that the approach described here went beyond the "blue book". As far as the particular case you mentioned is concerned, I would like to refer to paragraphs 108 and 116 of INFCIRC/153, which do not limit the choice of flow KMPs (key measurement points) to inputs and outputs in material balance areas.

IAEA-SM-231/57

A STATISTICAL EXAMINATION OF THE PRACTICAL PROBLEMS OF MEASUREMENT IN ACCOUNTANCY TANKS

W. DAVIES
DNPDE, Dounreay, Thurso

P.T. GOOD, A.G. HAMLIN
NMACT, Harwell,
United Kingdom

Abstract

A STATISTICAL EXAMINATION OF THE PRACTICAL PROBLEMS OF MEASUREMENT IN ACCOUNTANCY TANKS.

In the first part of the part of the paper the general problems of measurement in large accountancy tanks are considered. The generalized tank is assumed to have an extended geometry for the avoidance of criticality, to be fitted with pneumatic level indicating devices and with temperature sensors, and to contain liquid to be accounted, such as that derived from irradiated fuel elements, which is sufficiently active to generate appreciable heat and also radiolytic gases. Possible uncertainties contributed to the final measurement of fissile material contained in or discharged from the tank by the effects of hydrostatic heads, temperature, radiolysis, surface tension, and drainage are considered in detail. The magnitude of these is established from practical data and the errors combined in order to estimate the best possible performance, which, under the specified conditions, appears to be about ±0.3% (1σ). The implications for the design of large accountancy tanks are considered, with particular reference to the design of accountancy tanks for future plants where the above precision may not be adequate. The second part of the paper considers practical approaches to the problem of ensuring that actual performance of the measuring system approaches the best possible as closely as possible. In particular, a system of operation in which the accountancy tank is utilized essentially as a fixed volume, with the measuring systems restricted to determining small variations from this nominal volume, offers considerable promise.

1. INTRODUCTION

1.1 Accountancy tanks are of particular importance in reprocessing plants.

1.2 The feed to such plants consists of irradiated fuel elements containing fissile material that has been initially adequately accounted for for Safeguards purposes but which has been changed by the irradiation process to a degree which can currently be estimated on the basis of reactor data but not described precisely element by element. Accountancy of the input is therefore on an item basis, and it is not until the elements have been broken down and their fissile material extracted and dissolved for reprocessing that control over the irradiated fissile material can be re-established. This point is reached when the resultant solution is measured in the accountancy tank prior to reprocessing.

1.3 The basic measurement will be of the contents of the tank as a product of the contained volume or mass and of the analytically determined concentration of plutonium and uranium in the contained volume or mass. This basic measurement will then have to be related to the quantity of plutonium and uranium actually discharged for processing.

1.4 This paper attempts to analyse the significance of the errors arising in this process as comprehensively as possible.

1.5 It is assumed in the assessment of the various errors that calibration for the contained volume has been carried out with water at 15°C, and that operationally the tank will contain a solution of heavy metals in 3<u>M</u> nitric acid, $1{\cdot}3.10^3$ kg/m^3, at 50°C.

2. THE NATURE OF ACCOUNTANCY TANKS

2.1 The main factors which are essential in the design of accountancy tanks for substantial amounts of nuclear material are:-

2.1.1 The shape of the tank must make it impossible to assemble a critical mass of fissile material within it.

2.1.2 It must be possible to mix the liquor in the tank efficiently in order to obtain representative samples.

2.1.3 It must be possible to measure the volume or mass of the liquor with high accuracy. Where measurement techniques based on liquor height are to be employed:-

(a) The liquor height should be accurately proportional to liquor volume over practically the whole usable volume range of the tank.

(b) The tank should be tall and think, rather than short and squat.

(c) The below-liquor volumes of any dip-pipes or external pipes should be kept as small as possible, and arrangements must be made for expelling the liquor in them into the main liquor bulk at the liquor-mixing step, and (excepting in the case of pneumercator pipes) for venting the upper ends of the pipes to tank atmosphere for the liquor-measurement operation.

2.1.4 Filling and emptying arrangements must make it very difficult to over-fill or over-empty the tank in routine operation; nevertheless the tank should contain less than 1% of the "full" volume when emptied routinely.

2.1.5 Tank maintenance requirements should be minimal.

2.1.6 The tank should not be unduly difficult or expensive to fabricate.

2.2 It can be seen that many of these requirements are in conflict with each other and with the requirements of good accountancy. They lead to two types of accountancy tank which have found general acceptance - slab tanks and harp tanks.

2.3 Slab tanks have been used mainly for aqueous uranyl nitrate solutions containing up to 250 g/litre of enriched uranium. They are rectangular in profile, a typical nominal capacity of 120 litres, corresponding to a liquor height of 2 metres for a tank 2 metres wide containing a slab of liquor 30 to 40 millimetres thick. The geometry requires the tanks to be strengthened by rows of internal struts (which make calibration and liquor-mixing operations difficult.

2.4 Harp tanks have been used mainly for aqueous plutonium nitrate solutions at concentrations up to around 300 g/litre of plutonium. They are typically constructed from lengths of piping of about 150 or 230 mm bore arranged in a gate- or harp-like configuration. Tanks of nominal capacities up to 575 litres are currently being made. A typical tank consists of two verifical pipes about 4 metres high connected by three parallel pipes at about 30° to the horizontal, each of about 5 metres in length (see Figure 1). The configuration of these tanks imposes restrictions on the liquor measurement performance achievable and makes it difficult to obtain adequate liquor-mixing.

2.5 Harp tanks are generally regarded as more secure than slab tanks and can be modified under suitable conditions to the form of a "b" to give improved accountancy, measurements being confined to the upright stem and the loop being used merely to contain a fixed volume in a suitably extended geometry. An idealised tank of the type is considered in this paper. The original was about 180 litres capacity, consisting of a vertical pipe about 4 metres high, to the lower half of which is attached another pipe of about 5 metres in length which is bent into a loop. The bores of the pipes forming the main portion of the tank are partly about 150 mm and partly about 230 mm while the drainage sump at the base consists of a short length of 76 mm bore pipe closed at one end. An undesirable but unavoidable feature in practice is the relatively large liquor volume - about 3% of the total "full" volume - which can be enclosed in the various filling, emptying and air sparge mixing pipes that are within the tank. The idealised layout is shown in Figure 1.

3. ERRORS IN THE ACCOUNTANCY MEASUREMENTS

3.1 Predictable errors inherent in the Measurement of Contained Volume or Mass

3.1.1 The basic measurement will be the determination of the height to which the liquor fills the accountancy tank, coupled with a determination of the density of the liquor. Measurement systems for highly active liquors have to be capable of operating remotely without maintenance. Consequently the choice is limited and this paper considers the case in which the height of the liquor is measured by measurement of the pressure difference between the surface of the liquid and the foot of a dip tube reaching almost to the bottom of the tank. The latter allows the pressure reading to be converted either into a height, and thus through the calibration results to a volume, or directly into contained mass.

3.1.2 The basic measurement system is illustrated in Figure 1 and the general assumptions are stated in Table I. Height h_1 is measured as a function of the pressure gauge reading at M. Hydrostatic pressure at bottom of long dip tube, P_1 is:-

TABLE I. ASSUMPTIONS

Accountancy Tank	The notional tank is constructed from stainless steel tube (N.B. 6 in.) to the approximate dimensions shown in Figure 1. Pneumercator lines are stainless steel (N.B. 0·25) square ended. Other connections are ignored.
Temperature	Calibration temperature 15°C Operating temperature 50°C
Calibration fluid	Water at 15°C
Process liquor	Approx. 300 g/l of uranium plus plutonium in 3 M nitric acid at 50°C Sp.gr.1·3.

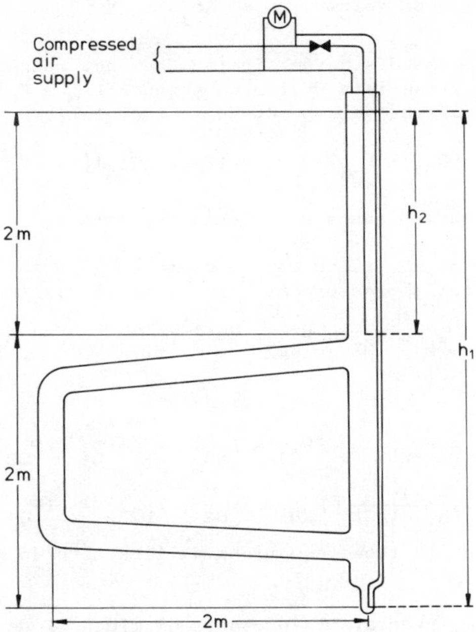

FIG.1. Schematic of accountancy tank.

$P_1 = P_{M1} - P_{D1} - P_B + P_{G1}$ where P_{M1} is pressure at gauge

P_{D1} is pressure drop in line to bottom of dip leg

P_B is mean pressure of bubble formation

P_{G1} is hydrostatic pressure of gas

Hydrostatic pressure at surface, P_s is:-

$P_S = P_{MS} - P_{DS} + P_{GS}$ where P_{MS} is pressure at gauge

P_{DS} is pressure drop in line to head space

P_{GS} is hydrostatic pressure of gas

Then

$$h_1\rho = P_1 - P_s \quad \text{where } \rho \text{ is the density of the fluid}$$

$$= (P_{M1} - P_{MS}) - (P_{D1} - P_{DS}) - P_B + (P_{G1} - P_{GS})$$

3.1.3 Provided measurements are made only in the cylindrical area, cross sectional area A, above the upper side arm, the system will behave as a uniform cylinder, height h_1 with a constant volume V attached to its base.

Thus the measured volume $V_M = Ah_1 + V$

(In practice the discharge dip-leg does not reach the bottom of the tank and therefore V consists of two components $(V_D + V_H)$ where V_D is the dischargeable part of the volume and V_H is the heel).

Thus $$V_M = \frac{A}{\rho}\left[P_G - (P_{D1} - P_{DS}) - P_B + (P_{G1} - P_{GS})\right] + (V_D + V_H) \quad (1)$$

where P_G is the observed gauge differential pressure.

3.1.4 The density of the fluid may be measured in situ by using the short dip tube. Hydrostatic pressure at base of short dip tube, P_2, is

$$P_2 = P_{M2} - P_{D2} - P_B + P_{G2}$$

Thus

$$(h_1 - h_2)\rho = (P_{M1} - P_{M2}) - (P_{D1} - P_{D2}) + (P_{G1} - P_{G2})$$

$$\rho = \frac{1}{(h_1 - h_2)}\left[P_{GD} - (P_{D1} - P_{D2}) + (P_{G1} - P_{G2})\right] \quad (2)$$

The variables in these equations will be altered by external factors as follows:-

A - temperature, hydrostatic pressure of fluid if density is different from calibration fluid, errors of measurement

ρ - temperature, composition of fluid

P_{G1} P_{GD} - instrumental and observational errors

P_{D1}, P_{DS}, P_{D2} - temperature, gas flow rates

P_B - surface tension of fluid, density of fluid

P_{G1}, P_{GS}, P_{G2} - temperature

$(V_D + V_H)$ - temperature, hydrostatic pressure if working fluid is different from calibration fluid, errors of measurement

h_1, h_2 - temperature (linear expansion of tank)

3.1.5 Appendix 1 analyses some uncertainties arising in the measurement of h_1 and h_2 and shows that these may safely be neglected as regards uncertainties in the working conditions, although gross changes e.g. from calibration fluid and temperature to process liquor and temperature may merit minor corrections.

3.1.6 As a result, equations (1) and (2) may be simplified to:-

$$V_M = \frac{AP_G}{\rho} + (V_D + V_H) \quad (3)$$

$$\rho = \frac{P_{GD}}{h_1 - h_2} \quad (4)$$

3.1.7 Equation (3) can be stated as

$$\rho V_M = AP_G + \rho(V_D + V_H) \quad (5)$$

Whence

$$M = AP_G + \frac{P_{GD}}{h_1-h_2}(V_D + V_H) \quad (6)$$

3.1.8 Equation (6) states that the relationship between P_G in the linear part of the tank and the <u>mass</u> of material present is a family of straight lines of constant slope but variable intercept, the latter being proportional to the density of the process liquor. The latter can be measured in the laboratory and related to P_G by appropriate calculation. However the use of equation (4) has the advantages that the measurement can be contemporaneous with the measurement of P_G and that no temperature correction is involved, thus removing a source of error, and that the measurement is made over a substantial fraction of the input batch thus reducing sampling error. If calibration is made against water at 15°C and liquor is measured at 50°C:-

$$\frac{L}{W} = \frac{P_{GDL}\ (h_1 - h_2)}{(h_1-h_2)\ (1+35\cdot 17.10^{-6})P_{GDW}}$$

P_{GDL}, P_{GDW} are the gauge readings with liquor and water respectively.

$$L = 0.9994\ W \frac{P_{GDL}}{P_{GDW}}$$

17.10^{-6} is coefficient of linear expansion of stainless steel

Whence it can be shown approximately that

$$L = \frac{1\cdot 6\ o_{PGL}}{P_{GDW}}$$

where o is the coefficient of variation

If the gauge can be read with a coefficient of variation of 0.08% (approx. 1 mm in 1000) this is approaching the accuracy of laboratory determination.

3.1.9 It should be noted that the expansion of the fluid within the contained volumes is corrected for by the density term whether equation (3) or (6) is used, i.e. whether volume or mass is the measured parameter, provided that density is measured contemporaneously (equation (4)). If this is not done, and if subsequently the density is measured under laboratory conditions, or analysis for plutonium or uranium is made on a mass/volume rather than a mass/mass basis, or the contents of the tank are discharged as a volume measured at a different temperature, an accurate knowledge of the temperatures of measurement (in plant and in laboratory) and of discharge, and of the coefficient of cubical expansion of the process liquor is necessary. Determination of all of these factors introduces apparently unnecessary errors and uncertainties.

3.1.10 A and $(V_D + V_H)$ can be established by regression techniques from calibration data and the associated random errors can be found. Practically determined values are quoted in Table II.

3.1.11 It remains to determine the effect of change of temperature upon A and $(V_D + V_H)$.

$$\frac{\Delta A}{A} = 2\alpha(T - T_o)$$

where T, T_o are operating and working temperatures $(T - T_o) = 35°C$

α is the coefficient of thermal expansion of the tank material - $17.10^{-6}/°C$

TABLE II. SUMMARY OF PREDICTABLE ERRORS

% of observed quantity

Uncertainty	Operating Variable	Major change*	Operating Uncertainty*	Correction for* Major change	Correction for* Operating Uncertainty
Hydrostatic pressure of gas	Liquor density	- 0.03	< ± 0.003/10% concn.	Nil	Nil
	Temperature	+ 0.02	± 0.0001/°C	M	Nil
Bubble pressure	Liquor density	Nil	Nil	Nil	Nil
	Liquor surface tension	+0.0025	+ 0.0005/10% concn.	Nil	Nil
	Temperature	- 0.03	± 0.001/°C	M	Nil
Hydrodynamic drop in pneumercator line	Gas Flow	-	± 3.10^{-5}/cc/min	-	Nil
	Temperature	+ 1.10^{-4}	± 5.10^{-6}/°C	Nil	Nil
	Liquor density	- 1.10^{-4}	± 1.10^{-5}/10% concn.	Nil	Nil
Linear expansion	Long dip leg	- 3.10^{-3}	± 7.10^{-5}/°C	Nil	Nil
	Long-short dip leg	+ 0.06	± 0.002/°C	M	Nil
Volume expansion	Change in A				
	Temperature	< -0.1	± 0.003/°C	P	Nil
	Hydrostatic pressure	<< -0.02	< - 0.002/10% concn	Nil	Nil
	Change in $V_D + V_H$				
	Temperature	+ 0.1	± 0.005/°C	P	Nil
	Hydrostatic pressure	<< -0.02	< ± 0.002/10% concn.	Nil	Nil
Meter reading			± 0.05(3σ)(Estimated)		
Determination of A			± 0.5 (3σ)		
Determination of $(V_D + V_H)$			± 0.5 (3σ)		

*Major change = change from 15°C to 50°C or change from water to process liquor but not both

Operating uncertainty = Variation due to changes in nominal operating values but excluding the effect of a major change

M = Correction marginally necessary

P = Correction probably necessary

% of observation involved + indicates that observation would be too high - that it would be too low

At an operating temperature of 50°C as opposed to a calibration temperature of 15°C, therefore

$$\frac{\Delta A}{A} = 1 \cdot 2 . 10^{-3}$$

or the change in value of A will cause the apparent mass to be low by rather less than 0.1% (dependent on the value of the second term in equation (6)). This may just merit correction but uncertainties in operating temperature can be neglected.

3.1.12 $(V_D + V_H)$ in equations (1) and (6) is a nominal volume but can be treated as if it were a physical entity enclosed in stainless steel. Thus, similarly to the above

$$\frac{\Delta(V_D + V_H)}{(V_D + V_H)} = 3\alpha(T - T_o)$$

$$= 5 \cdot 2 . 10^{-5}/^{o}C$$

On changing the temperature from 15°C to 50°C the change in the value of $(V_D + V_H)$ will cause the apparent mass to be low by about 0.1% dependent on the value of the first term in equation (5). This would merit correction but uncertainties in operating temperature can be neglected.

3.1.13 The uncertainties attached to measurement of the content of the accountancy tank are summarised in Table II. A positive value attributed to a major change indicates that the apparent content of the tank as deduced from calibration data will be too high. The table indicates that if an uncertainty of ± 0.1% is taken as criterion, then with the possible exception of the change in $V_D + V_H$, no corrections need to be made for operating uncertainties. The dominating errors are the random errors associated with meter readings and with the derivation of A and $(V_D + V_H)$. As regards major changes from the calibration conditions, initial corrections for major changes need to be made in respect of A and $(V_D + V_H)$ but are only marginally necessary in the case of a few other uncertainties.

3.2 Unpredictable Errors Inherent in the Measurement of Contained Volume or Mass

3.2.1 There are some possible errors, the significance of which can be revealed only by operational experience. These are discussed below.

3.2.2 Temperature Errors. The contents of the tank will need to be mixed, e.g. by sparging, but little can be predicted of the ultimate uniformity of temperature or its accuracy of measurement. Thus even if the density is measured in situ, which will average over an appreciable volume of the tank, there could be appreciable differences in temperature in the side arm and in the upper limb of the tank. These could manifest themselves in two ways - actual variation of the mean density from the measured density due to the temperature differences in the other parts of the tank and erroneous density measurements due to hydrodynamic pressure changes arising from circulation within the tank.

3.2.3 If the lower part of the tank is regarded as an approximately rectangular loop, a temperature differential of 1°C between the vertical arms will generate a circulatory head of

$H_c = h\rho - \frac{h\rho}{1+\alpha}$ where h is the height of the vertical limbs, approx. 1m

$\doteq h\rho\alpha$ ρ the fluid density at the lower temperature

$= 4{\cdot}4.10^{-4}$ α the coefficient of cubical expansion of the fluid

$= 4{\cdot}4.10^{-4}$ cm WG/°C for process liquor

In a steady state this head will be dissipated around the other three sides of the rectangle which, in the model accountancy tank, are in the proportions of 1:2:1:2. Thus the head appearing along the other vertical limb will be $0.2H_c$ and this will reduce the observed value of $(h_1 - h_2)$ if the side limb is hotter than the main limb and increase it if it is cooler. The error will be approximately

$$\pm \frac{0{\cdot}2H_c 100}{1{\cdot}3.200} = \pm 3{\cdot}4.10^{-5}\%$$

of the observed density per °C. For reasonable differences in internal temperature, this effect may be ignored.

3.2.4 In addition, errors will arise from temperature differences distributed around the tank. If the temperature difference in the lower arms of the tank is maintained, secondary circulation will cause the upper limb of the tank to approximate to the hotter lower limb in temperature. Taking the horizontal limbs as intermediate in temperature, if the side limb is hotter, since the upper limb is approximately equal to the lower limb, then, approximately

$$\frac{(1 + 2\ (0{\cdot}5.2) + 1\ )}{(1 + 1 + 2.2 + 1\ )} .100 = 57\%$$

of the tank contents are 1°C higher in temperature than the portion in which the density is measured. This gives a true density that is

$$\frac{(57\ \rho/(1+\alpha)) + 43\rho}{\rho} = 99.98\% \text{ of the observed density/°C}$$

Similarly, if the main limb is hotter

$$\frac{(1 + 2\ (0{\cdot}5.2)\ \ )}{(1 + 1 + 2.2 + 1)} .100 = 43\%$$

of the tank contents are 1°C lower in temperature than the portion in which the density is measured. This gives a true density that is

$$\frac{43\rho\ (1+\alpha) + 57\rho}{\rho} = 100.01\% \text{ of the observed density/°C}$$

3.2.5 These errors of 0.01 - 0.02%/°C could be significant if variations of internal temperature are much more than 1°C. If they are random over a series of runs, the cumulative error will be small, but problems will arise if there is a stable pattern of difference, which would establish a systematic error.

3.2.6 There remains a stable case where the temperature of the tank contents increases uniformly with height. If the overall gradient is 1°C then roughly

$$\frac{2 + 2.2}{1 + 1 + 2.2 + 1} .100 = 86\%$$ of the contents will be at the mean temperature of the measured portion and the residue will be approximately 0.5°C higher. Thus the true density will be

$$\frac{14\ \rho/(1 + 0.5\alpha) + 86\rho}{\rho} = 99.99\% \text{ of the observed density}/^{o}C$$

The error could be significant if the internal temperature gradient were much greater than 1°C, and, if regularly established as a stable state, could result in a cumulative systematic error.

Void Errors

3.2.7 The presence of voids would give an apparent volume larger than the true volume. They could arise in two ways - by generation of gas uniformly throughout the solution by radiolysis, or by nucleation of adherent bubbles from this or other sources.

3.2.8 Some theoretical studies have been made on radiolysis but its significance will only be revealed by experience. Provided generation is uniform and equation (6) with contemporaneous measurement of density is used to obtain contained mass, problems should be small. However, if the gas nucleates as adherent bubbles on the walls, or such bubbles arise from other sources the errors could be appreciable and extremely unpredictable. Finally even if generation is uniform a density gradient decreasing with height will be established over some part of the tank if the bubbles are large enough to rise at an appreciable velocity. Thus even a contemporaneous density measurement may be erroneous if applied to the whole contents of the tank.

3.2.9 If a column of liquid, of cross section A, is divided into n horizontal layers, a steady state is achieved when bubbles leave a layer at the rate at which they are generated within it plus the rate at which they enter it from below. If V is the volume of a layer and v the total volume of gas generated within it per second, which is small compared with V, then in the mth layer at equilibrium the volume of gas entering per second is $(m - 1)v$, the gas generated internally per second is v.

3.2.10 The hold up of the average bubble in the section, t, is

$$t = \frac{h}{a}$$ where h is the height of the layer and a is the terminal rise velocity of the bubble

The average bubble generated in the section will travel only half the height of the layer, therefore its residence time will be h/2a. Thus the hold up of the mth section is:-

$$v_m = (m - 1)v\,\frac{h}{a} + v\,\frac{h}{2a}$$

$$= \frac{hv}{2a}\,(2m - 1)$$

The mass of liquid in the mth section, neglecting the density of the air is thus:-

$$M = (V - v_m)\rho$$

$$= (V - \frac{hv}{2a}(2m - 1))\rho$$

If this is summed over n layers

$$M_n = (nV - \frac{hv}{2a}n^2)\rho$$

3.2.11 In the notional accountancy tank the density is measured over approximately one half of the total height of the column. If this is h_T, then

$$\frac{\rho act}{\rho obs} = \frac{n\rho\,(V - \frac{hvn}{2a})}{n\rho\,(V - \frac{hvn}{4a})} = 1 - \frac{h_T v}{4aV}$$ (plus higher powers of the last term, which is required to be small)

$$a = \frac{2}{9}\left(\frac{3v}{4\pi m}\right)^{\frac{2}{3}} \frac{g\rho}{\eta}$$ where m is the number of bubbles into which v is divided

$$= 1.53.10^6 \left(\frac{v}{m}\right)^{\frac{2}{3}} \text{ m/sec}$$ where it is assumed that the viscosity of the process liquor is 1.3 times that of water ($0.548\ 10^{-3}$ N·sec·m^{-2}) at 50°C

Thus, since h = 4m,

$$\frac{act}{obs} = 1 - \frac{m^{2/3}\ v^{1/3}}{1.5.10^6 V}$$

3.2.12 Since $v^{1/3}/V$ is <1 it appears unlikely that errors in the density measurement due to bubbles leading to the formation of density gradients will be significant (> 0.1%) unless very finely divided bubbles are formed - m of the order of 10^4.

4. ANALYTICAL ERRORS

4.1 Various titrimetric redox techniques of chemical analysis appear to offer the best prospects of achieving adequate performance at reasonable cost for the routine measurement of both plutonium and uranium concentration in the liquor. It seems likely that, in the case of fast reactor fuel, the chemical complexity of the feed solution will limit the measurement performance attainable.

4.2 For fast reactor fuel it is expected that, by making 2 to 6 replicate determinations of plutonium or uranium per batch, it will be possible to produce a final concentration result for each feed liquor batch which will have a total uncertainty (one-sigma limits, including random and systematic components) in the range ± 0.1% to ± 0.3%.

4.3 It should be possible to demonstrate, for each batch analysed, that negligible errors are arising from non-representative sampling by arranging matters so that the above-mentioned replicate determinations are carried out on several samples, each sample being taken after a liquor-mixing operation in the tank.

5. ERRORS INHERENT IN TRANSFER OPERATIONS

5.1 The accountancy tank is conveniently calibrated "to contain", but it will be used in a "to deliver" mode. This will introduce two errors - a systematic error representing the difference between the delivered and contained volume, and a random error representing the degree to which operational variables affect this difference.

5.2 The systematic error represents material adhering to the internal surfaces of the system after discharge. Its magnitude will depend upon physical retention of solution in undrainable spaces, on the viscosity, and on the time allowed for drainage before any heel is measured. Experimental measurement of this with solutions of appropriate composition will be necessary in order to correct for the mean systematic error, but even then errors introduced by liquor trapped in undrainable or slowly drainable spaces will not be detected except by wash-out procedures. Material trapped in this way will appear as an apparent loss on each batch processed and the total apparent loss will not be recovered by wash-out procedures - only that appropriate to a single batch.

5.3 Material adhering to the internal surfaces will need to be kept to a standard figure by an appropriate draining routine before the heel is measured. If, for instance, drainage is considered complete when the level of the heel does not rise more than 1 mm/min, the arrival of liquid at the base of the tank will be

$$\frac{\pi D^2 . 10^{-3}}{4.60} \ m^3/\text{sec}$$ where D is the diameter of the tank in m

5.4 The average velocity of a falling film, a, is

$$a = \frac{g\rho m^2}{3\eta} \ \text{metres/sec.}$$

m = thickness in m
ρ = density kg/m³
η = viscosity
g = 9.81

$$a = 6.07.10^6 m^2 \ \text{metres/sec for process liquor}$$

From above

$$\frac{D^2 . 10^{-3}}{4.60} = \pi D.\ m.\ a.$$

Thus $$a = \frac{D}{2\cdot 4.10^5 m}$$, and so

$$m^3 = 6\cdot 9\ D\ 10^{-12} \qquad (D \simeq 1\cdot 5.10^{-1} m)$$

$$m = 1.10^{-4}\ m$$

5.5 Assuming that the tank approximates to two vertical tubes, diameter D of height 4.5 m and that the film tapers uniformly to zero at the top, total hold-up is

$$H = \frac{9\pi\ Dm}{2} \ \text{cubic metres}$$

Total capacity is $\frac{9\pi D^2}{4}$ m^3 so the hold-up is approximately

$$\frac{2m}{D} 100\% = 0.1\% \text{ of capacity}$$

5.6 This calculation is very approximate - the hold-up will be increased by the vertical tubes included in the tank and by the behaviour of the horizontal sections which may drain more slowly - but suffices to indicate that the error may be among the more substantial. The random component will vary as the one third power of the actual drainage rate relative to the nominal and should not be critical.

5.7 A random transfer error is introduced by the expansion or contraction of the heel resulting from any difference in the discharge temperature from that of the previous batch. If the temperature is 1^oC degree different, then αV_H will be added to or withdrawn from the charge added to the tank when it is discharged. Under normal conditions this will represent

$$\frac{\alpha V_H}{V_T} .100 = \frac{3 \cdot 4.10^{-4} V_H .100}{167}$$

$$= 0.0002 \, V_H \% \text{ of the volume transferred/}^oC$$

and may probably be neglected under reasonably stable operating conditions if V_H is not more than about 5 litres.

5.8 A further error is introduced by the actual measurement of the heel, and may be established from the calibration data as for the full volume measurement. The equation

$$M^1 = A^1 P_G + V^1$$

may be applied if the final level lies in the narrow part of the vertical limb. From the calibration the relevant variances are known. The significance of errors in this case is discussed in the next section.

If the final level does not lie in this narrow section the relation between M^1 and P_G cannot be described mathematically and will have to be determined from calibration data, with a greater uncertainty.

6. METHODS OF OPERATING USING FULL TANK RANGE

6.1 The amount of material contained in a single discharge from the accountancy tank will be subject to systematic and random errors.

6.2 Systematic errors arise because the measurements of the capacity of the accountancy tank and of the relation between this and the mass discharged cannot be made under conditions identical to those occurring in the ultimate process. However, Table II indicates that these errors are likely to be small (0.1% or usually much less) even considering the gross change from water at 15°C to process liquor at 50°C, and that reasonable calculated corrections can be applied in respect of them with the possible exception of void errors. The significance or otherwise of the last will be revealed only by experience, but for the moment it will be assumed that systematic errors in measurement of mass transferred are negligible. Section 4 suggests that they are also negligible for analysis.

6.3 Thus only the combination of random errors needs to be considered. The amount of nuclear material discharged from the accountancy tank is given by the equation

$Q = R\,(M_1 - M_2 - H)$ where Q is the total quantity discharged.
R is the weight/weight concentration of nuclear material in the tank.
M_1 is the mass contained in the full tank.
M_2 is the measured mass contained in the empty tank.
H is the mass distributed over internal surfaces or otherwise retained but not included in M_2.

Since the variables are mutually independent

$$V(Q) = \left(\frac{\delta Q}{\delta R}\right)^2 V(R) + \left(\frac{\delta Q}{\delta M_1}\right)^2 V(M_1) + \left(\frac{\delta Q}{\delta M_2}\right)^2 V(M_2) + \left(\frac{\delta Q}{\delta H}\right)^2 V(H)$$

where V(Q) etc. are the variances of the respective quantities

or $V(Q) = (M_1 - M_2 - h)^2\, V(R) + R^2\, (V(M_1) + V(M_2) + V(H))$

$(M_1 - M_2 - H)$ is expected to be approximately $2{\cdot}15.10^2$ kg on average

R is expected to be $2{\cdot}1.10^{-2}$ for plutonium, $1{\cdot}2.10^{-1}$ for uranium

V(R) has been assessed in Section 4 as approximately $1{\cdot}8.10^{-9}$ for plutonium.

6.4 It remains to assess the remainder.

From equation (6) $M_1 = AP_G + \frac{P_{GD}}{h_1 - h_2}(V_D + V_H)$

The variables are again independent, therefore as before

$$V(M_1) = P_G^2\, V(A) + A^2 V(P_G) + \frac{(V_D + V_H)^2}{(h_1 - h_2)^2}\, V(P_{GD})$$

$$+ \left(\frac{P_{GD}\,(V_D + V_H)}{(h_1 - h_2)^2}\right)^2 V(h_1 - h_2) + \left(\frac{P_{GD}}{h_1 - h_2}\right)^2 V(V_D + V_H)$$

6.5 The terms may be assessed as follows:-

P_G If the level is measured in the upper cylindrical part of the tank then P_G will have an average value of, say 390 cm W.G.

V(A) The variance of A (A is equivalent to the free cross-sectional area of the upper part of the tank) arises from the uncertainty of its determination by calibration. A practical value has been found to be $9.0.10^{-8}$. Operational uncertainties do not contribute appreciably to this.

A A practical value of A is $1.74.10^{-1}$.

$V(P_G)$ The most obvious error here is the uncertainty of the gauge reading. gauges of adequate sensitivity have a range of 300 cm W.G. and can be read to 1 mm. Because of the restricted range, the measurement for the upper part of the tank is made from the second dip-leg. It has been shown that uncertainty in operating temperature will not introduce a significant error into the spacing of these dip-legs, nor will process uncertainties introduce significant errors due to other variables in the gauge systems.

The composite value P_G may be written

$$P_G = P_G' + \rho(h_1 - h_2)$$

$$= P_G' + P_{GD}\left(\frac{h_1 - h_2}{h_1 - h_2}\right)$$

Whence

$$V(P_G) = V(P_G') + V(P_{GD}) = 2V(P)$$

If the gauge can be read to $\pm$ 0.05 cm W.G. with certainty this can be regarded as a 95% confidence limit. At mid scale, therefore

$$\sigma = \frac{0.05}{3} = 0.017 \text{ and } V(P) = 3.10^{-4}$$

$\left(\frac{V_D + V_H}{h_1 - h_2}\right)^2$ has a practically determined value of 116 l

$V(P_{GD})$ Since this is a single measurement it is assumed equal to V(P) above.

P_{GD} is approximately 260 cm W.G.

$V(h_1 - h_2)$ could be established from engineering measurement giving a negligible variance or by a pressure difference. In the latter case its variance is 2 VP.

$V(V_D + V_H)$ has a practically determined value of $4.15.10^{-2}$. Operational uncertainties do not contribute appreciably to this.

6.6 Thus $V(M_1) = 3{\cdot}9^2.9{\cdot}9.10^{-4} + 1{\cdot}74^2.6.10^{-4} + 1{\cdot}16^2.3.10^{-4}$

$$+ (1{\cdot}3.1{\cdot}16)^2\ 6.10^{-4} + 1{\cdot}3^2.4{\cdot}2.10^{-2}$$

$$= 1{\cdot}37.10^{-2} + 1{\cdot}8.10^{-4} + 4{\cdot}04.10^{-4}$$

$$+ 1{\cdot}36.10^{-3} + 7{\cdot}10.10^{-2}$$

$$\simeq 2{\cdot}10.10^{-1}$$

since the second, third and fourth terms do not contribute significantly.

6.7 M_2 may be determined by an equation similar to (6)

$$M_2 = A^1 P_G^1 + \frac{P_{GD}}{h_1 - h_2} V_H$$

Thus $V(M_2) = {P^1_G}^2 \; V(A^1) \; + \; {A^1}^2 \; V(P^1_G) \; + \left(\frac{V_H}{h_1-h_2}\right)^2 V(P_{GD})$

$$+ \left(\frac{P_{GD}}{(h_1-h_2)^2}\right)^2 V(h_1-h_2) \; + \; \left(\frac{P_{GD}}{h_1-h_2}\right)^2 \; V(V_H)$$

Here the quantities not given earlier are:-

P^1_G assuming the heel is contained within the lowest cylindrical part of the tank, an average value would be 20 cm W.G.

$V(A^1)$ A practically determined value is $3{\cdot}9.10^{-6}$

A^1 A practically determined value is $4{\cdot}2.10^{-2}$

$V(P^1_G) = V(P)$

$V_H = 3.10^{-1}$ l.

$V(V_H)$ a practically determined value is $1{\cdot}56.10^{-4}$

Thus

$$V(M_2) = 2^2.3{\cdot}9.10^{-4} + 4{\cdot}2^2.3.10^{-8} + 3^2.3.10^{-10}$$

$$+ 1{\cdot}3^2.2{\cdot}3.10^{-6} + 1{\cdot}3^2.1{\cdot}6.10^{-4}$$

$$= 1{\cdot}56.10^{-3} + 5{\cdot}29.10^{-7} + 2{\cdot}70.10^{-9}$$

$$+ 1{\cdot}01.10^{-5} + 2{\cdot}7.10^{-4}.$$

$$= 1{\cdot}83.10^{-3}$$

since the second, third and fourth terms do not contribute significantly.

6.8 H is one of the transfer errors dealt with in Section 4 and its variance can only be established by experiments that have not yet been made. However, Section 4 suggests that its value will not exceed 0.1% of the discharge, i.e. 0.2 kg, and if we assume that the coefficient of variation is 10% then $V(H) = 4.10^{-4}$. Compared with $V(M_1)$ this is hardly significant and its value is not critical.

Thus for plutonium:-

$$V(Q) = 2{\cdot}15^2.1{\cdot}8.10^{-5} + 2{\cdot}1^2 \; (2100 + 18 + 4) \; 10^{-8}$$

$$= 8{\cdot}32.10^{-5} + 9{\cdot}26.10^{-5}$$

$$= 1{\cdot}758.10^{-4}$$

$$Q = R(M_1 - M_2 - H) \simeq 2{\cdot}1.2{\cdot}15 = 4.5 \text{ kg}$$

Therefore the coefficient of variation of Q is

$$\frac{(1{\cdot}76.10^{-4})^{\frac{1}{2}}}{4{\cdot}5} . 10^2 = 0.29\%$$

The uncertainties derived above for Q may be understatements due to the effects of the unpredictable errors discussed in Section 2.2, which have not, of course, been taken into account.

7. METHODS OF OPERATION USING RESTRICTED TANK RANGE

7.1 For this procedure, the tank is treated as an accurately calibrated volumetric vessel with one pneumercator level gauge covering the upper "full" region in the neck and another pneumercator level gauge covering the lower "empty" region in the sump. The sump region extends over only 1% of the tank volume, and although the neck region can contain up to about 16% of the "full" volume, only the lower part of the neck region corresponding to about 1% of the "full" volume is used. The advantages of this mode of operation are that the pneumercator systems are operated well within their performance capabilities, and that the required liquor density value can be inferred sufficiently accurately from liquor composition, so that no actual density measurements by gauge or liquid-weighing techniques are needed.

7.2 The amount of plutonium in a single batch of liquor transferred out of the tank is given by

$$W = (Y)\left\{(L)\ (F) + \frac{(K_1)(P_1)}{D} - \frac{(K_2)(P_2)}{D} - \frac{(K_2)(P_2)}{D} - R\right\}$$

where W = mass of plutonium transferred in grams

Y = plutonium concentration in the liquors, expressed as grams of plutonium/litre of liquor at laboratory temperature. (Obtained by chemical analysis of a sample in the laboratory).

L = calibrated volume of the tank, up to the lower end of the "neck" pneumercator dip pipe at tank operating temperature.

F = volume decrease factor applicable, taking into account the liquor composition, for the thermal contraction of the liquor from tank operating temperature to laboratory temperature. (F is expected to lie in the range 0.97 to 1.00).

K_1 & K_2 = constants, in units of litres/cm, applicable to neck and sump regions of the tank respectively.

P_1 & P_2 = differential pressures, in cm of water, indicated by the pneumercator gauges in the neck and sump respectively.

D = density of the liquor in kg/litre at laboratory temperature, to the nearest 0.01 kg/litre.

R = volume, in litres, below the sump pneumercator gauge dip pipe. (Temperature corrections for R will be negligible since the value of R is approx. 0.3% of the volume of a liquor batch only).

7.3 Experience has indicated that, ignoring possible measurement uncertainties arising from void formation and temperature gradients in the liquor, then the volume of each liquor batch at laboratory temperature i.e. the quantity with the large brackets at the right-hand side of the above expression, should be measurable with an attached total uncertainty within the range of ± 0.1% to ± 0.3%. (Each value given is a one-sigma confidence limit which includes both random and systematic components). This derived almost entirely from the calibration error (mainly systematic) associated with L, together with a random component representing the reproducibility with which the calibrated volume can be dispensed. The uncertainties attached to the measurements of the small neck and sump volumes can both be shown to be negligible in comparison. In practice, it is expected that, for each liquor batch, replicate measurements of the "full" volume will be made after each liquor-mixing and sampling operation; this will provide some information on the reproducibility of the "full" volume measurement.

7.4 Hence combining the volume measurement and chemical analysis uncertainties - both expected to be in the range $\pm$ 0.1% to $\pm$ 0.3% - gives an expected total uncertainty in the range of about $\pm$ 0.2% to $\pm$ 0.4% for one-sigma limits which include random and systematic components.

8. RESULTS OBTAINED ON AN ACTUAL TANK

8.1 A tank of the shape shown in Figure 1 has been calibrated using the usual technique of adding known volumes of water and observing the pressure differences produced as cm of water on Wallace and Tiernan gauges and as volts by use of a transducer and a digital voltmeter.

8.2 The calibration equations for the top section of the tank (152 - 185 litres) and for the sump section (0.2 - 1.5 litres) have been derived using a least squares treatment. The errors, it has been assumed, only occur in the pressure readings and the variance of the pressure is assumed to be independent of its absolute value in the region under consideration. For the 152 - 185 litre section the equation is

$$V = 0.1741P + 116.333$$

and the variance of an unknown volume derived from a measured pressure is 10.02×10^{-3}, i.e. a standard deviation of $\pm$ 0.10 litre, or a relative standard deviation of 0.06%.

For the 0.2 - 1.5 litre section the equation was

$$V = 0.04222\,P + 0.2315$$

The variance of V is 6.70×10^{-5}, a standard deviation of $\pm$ 0.008 litre.

8.3 A number of experiments were carried out where the volume of water delivered was collected and measured and compared with the volume as calculated from the pressure readings on the tank before and after delivery. Good agreement was found between the two methods.

8.4 The least squares analysis produces a figure for the variance of the pressure readings of 0.32 at 400 cm H_2O, 0.03 at 16 cm H_2O. At 400 cm this corresponds to a standard deviation of $\pm$ 0.57 cm. This is accepted as being in reasonable agreement with values deduced in 6.6.

8.5 Using the values for the variance of the pressure readings it is possible to estimate the uncertainty in estimating the volume of dissolver solution in the tank at 50°C when the density of the solution is determined using the fixed length dip tubes installed in the tank. For 170 litres of dissolver solution at 50°C the standard deviation of the volume is $\pm$ 0.28 litre, a relative standard deviation of 0.16%. The increase in uncertainty over that for water, 0.10 litre (1σ), arises from the fact that for dissolver solution the uncertainty in the pressure measurements required for the density determination must be taken into account.

9. CONCLUSIONS

9.1 Analysis of predictable errors suggests that discharges of nuclear material from a typical type of accountancy tank of "b" configuration should be accountable with a coefficient of variation of about 0.3% on a single batch. This view is substantiated by experimental measurements. Summation of discharges over a campaign should substantially reduce the final uncertainty.

9.2 There are some unpredictable errors, notably void formation and inhomogeneity of temperature which may impair performance and will have to be checked in practice.

9.3 With the conventional system here considered, attempts to improve precision would rapidly require consideration of many sources of error which now can be ignored. It appears that the performance reported is likely to be near the limit for this system.

9.4 In the construction of new accountancy tanks, more attention should be given to accountancy requirements at the design stage.

9.5 Better nuclear material measurement capability in accountancy tanks should be striven for, to meet the increasingly stringent accountancy demands. It may be necessary to discard measurement systems based on liquor height or volume and to replace them with direct liquor weighing techniques in order to:-

(a) overcome void and temperature gradient difficulties.

(b) enable total measurement uncertainties of better than $\pm$ 0.1% (one sigma) to be reached.

9.6 Ignoring the unpredicted errors (para. 9.2), the volume or mass measurement error and the analytical error are comparable in magnitude. Any significant improvement in measurement will require improvement in performance of both volume or mass measurement and analysis. The only analytical techniques capable of sufficiently good performance at present are those based on sampling-plus-chemical analysis, but further analytical development work for feed solution analysis is desirable.

APPENDIX I

ANALYSIS OF PREDICTABLE UNCERTAINTIES AFFECTING THE MEASUREMENT OF HYDROSTATIC HEAD IN THE ACCOUNTANCY TANK

1. <u>Hydrostatic pressure of gas</u>

In equation (1)

$$(P_{G1} - P_{GS}) = \rho_G h_1 - \rho'_G (h_1 - h)$$

where h is the actual height of liquor above the bottom of dip tube h_1, ρ'_G is the density of the gas above the liquid surface

The observed hydrostatic pressure for water

$$= \rho_W h$$

The observed hydrostatic pressure for process liquor

$$= \rho_L H$$

Max value of $h_1 \doteq 400$ cm

$\rho_W \doteq 1$ g/cm^3

$\rho_L \doteq 1.3$ g/cm^3

ρ_G 1.2.10^{-3} g/cm^3 at S.T.P. for air

The maximum operating pressure of the gas is approximately

$$\left(760 + \frac{\rho.5000}{13.6}\right) \text{ mm Hg}$$ where ρ is the density of the liquid

giving for water 1054 mm and for liquor 1142 mm of Hg.

The minimum operating pressure of the gas is approximately

$$\left(760 + \frac{\rho.2000}{13.6}\right) \text{ mm Hg}$$

giving for water 907 mm and for liquor 951 mm of Hg.

The working density of the gas is:-

$$\rho_G = \frac{273}{273+t} \cdot \frac{P}{760} \cdot 1{\cdot}2.10^{-3}$$ where t is the operating temperature
P is the operating pressure

For water at 15°C $\rho_G$max. = 1·58.10^{-3}, ρ_{Gmin} = 1·36.10^{-3} g/cm^3

For liquor at 15oC ρ_{Gmax} = 1·71 ρ_{Gmin} 1·42.10^{-3} g/cm^3

Meter readings will therefore be low by:-

$$\frac{\rho_G h_1 - \rho'_G(h_1-h)}{\rho_{W,L} h}$$ where ρ_G is the density of the gas in the dip tube.

ρ' is the density of the gas over the surface of the liquor
(1·14.10^{-3} g/cm^3 at 760 mm and 15°C)

$$= \frac{h_1(\rho_G - \rho'_G)}{_{W,L}h} + \frac{\rho'G}{W,L}$$ $\rho_{W,L}$ the density of water or liquor
h the actual height of the liquid column

This corresponds to:-

For water at 15°C 1·58.10^{-1} %

For liquor at 15°C 1·31.10^{-1}) %

At 50°C ρ'_G = 1·01.10^{-3} g/cm^3

For water at $50^{o}C$ $\rho_{Gmax} = 1\cdot41.10^{-3}$ $\rho_{Gmin} = 1\cdot21\ g/cm^{2}$

For liquor at $50^{o}C$ $\rho_{Gmax} = 1\cdot52.10^{-3}$ $\rho_{Gmin} = 1\cdot27\ g/cm^{2}$

These result in meter readings that are low by:-

For water at $50^{o}C$ $1\cdot41.10^{-1}$ %

For liquor at $50^{o}C$ $1\cdot17.10^{-1}$ %

Combining these changes in the table below gives the expected error in the meter reading compared with the calibration with water at $15^{o}C$

Tank Level	Water at $50^{o}C$	Liquor $15^{o}C$	Liquor $50^{o}C$
Full	$+ 0\cdot17.10^{-1}$	$+ 0\cdot29.10^{-1}$	$+ 0\cdot41.10^{-1}$
Lower	$+ 0\cdot17.10^{-1}$	$+ 0\cdot29.10^{-1}$	$+ 0.41.10^{-1}$

The combined change from calibration with water at $15^{o}C$ to measurement of concentrated product at $50^{o}C$ may marginally merit corrections, but other errors may be ignored, as may be the uncertainties in operating variables.

In equation (2)

$$(P_{G1} - P_{G2}) = \rho_{G1}\, h_1 - \rho_{G2}\, h_2$$

where ρ_{G1}, ρ_{G2} are the gas densities in dip tubes h_1, h_2

The pressure at the foot of the dip tube h_1 is

$$\left(760 + \frac{\rho_{w,L} h}{13\cdot6}\right) \text{ mm of Hg}$$

where $\rho_{W,L}$ is the density of water or process liquor
h is the actual height of the liquid above the foot of the dip tube in mm

$$\rho_{G1} = \frac{273}{(273 + t)} \left(1 + \frac{\rho_{W,L} h}{13\cdot6.760}\right) 1\cdot2\cdot10^{-3}$$

where t is the temperature

$$\rho_{G_2} = \frac{273}{(273 + t)} \left(1 + \frac{\rho_{WL}\ (h-(h_1-h_2))}{13\cdot6.760}\right) 1\cdot2.10^{-3}$$

$$\text{Thus } (P_{G1} - P_{G2})\ 10 = \frac{273.1\cdot2.10^{-3}}{273 + t} (h_1-h_2) \left(1 + \frac{\rho_{WL}}{13\cdot6.760} (h-h_2)\right)$$

where h_1, h_2 are in mm

$(h_1 - h_2) = 2000$ mm, $h_1 = 4000$ mm, $h_2 = 2000$ mm, $h = 4000$ to 2000 mm

For liquor at $15^{o}C$ $(P_{G1} = P_{G2}) = 0\cdot28 - 0\cdot23$ cm W.G.

For liquor at $50^{o}C$ $= 0\cdot25 - 0\cdot20$ cm W.G.

The observed pressure difference $(h_1 - h_2)\rho$ (Equation 2) is 260 cm W.G. for liquor.

Thus the observed reading will be 0·09 - 0·11% too low. This will need correcting for, but the change in temperature from 15 to 50°C introduces an error of only 0.01% and may be neglected.

2. Bubble pressure

At the point of maximum pressure indication, the bubble is hemispherical and is subject to excess pressures with regard to the lower lip of the tube due to the hydrostatic pressure of the fluid and the surface tension of the fluid.

Assuming that the bubble is hemispherical hydrostatic excess pressure due to bubble formation at lower lip of tube is

$$\frac{2\pi r^3 \rho g}{3\pi r^2} = \frac{2}{3} r\rho$$

Surface tension excess pressure due to bubble formation is

$$\frac{2\pi r\gamma}{\pi r^2} = \frac{2\gamma}{r} \qquad \text{where } \rho = \rho_L - \rho_G$$

$$\simeq \rho_L$$

r = radius of tube

γ = surface tension of liquid

Therefore total excess pressure P_B is

$$P_B = 2 \quad \frac{r\rho g}{3} + \frac{\gamma}{r}$$

For air and water at 15°C

$r = 0{\cdot}003$ m

$\rho \simeq 10^3$ kg/m^3

$\gamma = 73{\cdot}5\ 10^{-3}$ N/m

$$P_B = 2\left(\frac{3.10^{-3}.10^3.9{\cdot}81}{3} + \frac{73{\cdot}5.10^{-3}}{3.10^{-3}}\right)$$

$$= 2\ (9{\cdot}81 + 24{\cdot}50)$$

$$= (19{\cdot}6 + 49{\cdot}0)\ \text{N/m}^2$$

$$= (0{\cdot}20 + 0{\cdot}50)\ \text{cm W.G.}$$

The minimum reading in the linear section of the tank is 200 cm W.G. Thus the meter reading will be at most 0.1% too high for all liquids due to density, but correction for this is implicit in the calibration curve.

The reading will be a further 0.25% too high due to the surface tension effect and correction for this will not be maintained if conditions are changed. The surface tension is a function of temperature and the composition of the process liquor.

Treating these separately:-

Surface tension varies with temperature as:-

$$\gamma_T = \gamma T_o - \alpha (T - T_o)$$

Values of α are not readily available but appear to be approximately 2.10^{-4} for a variety of liquids. Using this value a change in working temperature from 15-50°C will reduce the surface tension by 7.10^{-3} N m^{-1} or approximately 10%. In the absence of better data, the correction to be made to the meter reading (0.025%) is not substantially greater than its uncertainty.

For aqueous solutions of solutes having no particular surface activity

$$\gamma_L = \gamma_W + M\Delta\gamma$$

where γ_L is the surface tension of the liquor
γ_W is the surface tension of water
M is concentration of solute in moles per litre
Δ_γ = -0.6 for nitric acid
= 2.7 for tri-ionic salts

The liquor is 3M in HNO_3, and approximately 0.8M in heavy metals (calculated as uranyl nitrate).

The change from water to process liquor will therefore increase surface tension by approximately

$$-3.0{\cdot}6 + 0{\cdot}8.2{\cdot}7 = 0{\cdot}4.10^{-3} \text{ N/m}$$

which is equivalent to approximately 0.5%. Change from water to process liquor at 15°C will therefore cause the meter reading to be too high by 0.0025%.

The effects of change of temperature and working solution may therefore be safely neglected.

3. Pressure Drop in Pneumercator Lines

It is assumed that flow resistances in the two lines to the head of the accountancy tank are equal. The differential pressure drop will therefore occur in approximately h_1 of the dip pipe.

Assume:-

Flow rate 10 cm^3 S.T.P. per minute

Bore (1/4 in. NB) 0.6 cm

ρ_G 1.5 10^{-3} g/cm^3

μ_G (viscosity) 18. 10^{-6} Ns/m^2

$$R_e = \frac{d\rho}{\mu} = \frac{10{\cdot}10^{-6}}{\pi 0{\cdot}3^2.10^{-4}.60} \cdot \frac{0{\cdot}6.10^{-2}.1{\cdot}5}{18{\cdot}10^{-6}} = 3$$

Flow is therefore laminar and unlikely to be disturbed by broad variations of temperature and fluid density.

Pressure drop in the length h_1 of tube and the changes in pressure due to gas hydrostatic head are likely to be small and Poiseuille's formula can be used

$$P_{D1} - P_{DS} = \frac{128 \mu h_1 v}{\pi d^4}$$

v = volume flow m^3/sec,
h_1 = metres
$\mu^1 = 18{\cdot}2\ 10^{-6}$ N sec/m^2 for air at 20^oC
d = diameter, m

$$= \frac{128.\ 18{\cdot}1{\cdot}10^{-6}\ 4.}{\pi 6^4 .10^{-12}} \cdot \frac{10.10^{-6}}{60}$$

$$= 0.04 N\ m^{-2}$$

$$= 4.\ 10^{-4}\ \text{cm W.G.}$$

Minimum hydrostatic head with water at 15^oC in the linear part of the tank is roughly 200 cm.

The gauge will therefore read 2.10^{-4}% higher than it should at the minimum level of the linear part of the tank.

The error will vary proportionally to gas flow rates, and inversely to gas density resulting from liquor changes, and with temperature as

$$\frac{P_{T+t}}{P_T} = \left(\frac{T+t}{T}\right)^{\frac{1}{2}} \left(\frac{T+t}{T}\right) = \left(1 + \frac{t}{T}\right)^{3/2}$$

$$\simeq \left(1 + \frac{3t}{2T}\right)$$

if t is small (T is in degrees absolute) or about $0.5\%/^oC$.

This term ($P_{D1} - P_{DS}$) in equation (1) and ($P_{D1}\ \mu\ P_{D2}$) in equation (2) can safely be neglected.

4. Linear expansion

The longer leg (h_1) approximates to the total length of the tank and will therefore be approximately invariate. Coefficient of linear expansion of stainless steel is $17.2\ 10^{-6}/^oC$. If the lower lip is 8 cm from the bottom of the tank, change in working temperature from 15 to 50^oC will increase the gap, i.e. decrease the effective length by

$$8.35.17\ 10^{-6} = 5.10^{-3}\ \text{cm}$$

This would cause the minimum reading in the linear part of the tank to be too low by approximately $2.5\ 10^{-3}$%, which is negligible

Taking the difference $h_1 - h_2$ as 200 cm the effect of changing from 15 - 50^oC will increase the gap by

$$200.35.17.10^{-6} = 1{\cdot}2.10^{-1}\ \text{cm}$$

or the observed reading will be 0.06% too high which may just merit correction although uncertainties in the working temperature will have a negligible effect on specific gravity measurements.

For operational reasons it is necessary to transfer the reference point for measurement of contained volume or mass from the longer dip leg to one of approximately half the length. Changing the operating temperature from 15 to 50° will cause the lip of this tube to move 1·2 10^{-1} cm further from the base of the tank. Thus if equation (6) is established at 15°C, at 50°C it will become

$$M = A^1P_G + (V_D + V_H + 1{\cdot}2\ 10^{-4}A^1)$$ where A^1 is in cm^2 and volumes in litres

$$1{\cdot}2\ 10^{-4}A^1 \triangleq 2{\cdot}0\ 10^{-2}\ 1$$

$$V_D + V_H \triangleq 116\ 1$$

Thus the maximum error in M when A^1P_G is small will be

$$\frac{2{\cdot}0\cdot10^{-2}}{116}\cdot 10^2 = 0.02\%$$

below the true value. This may just merit correction but the effect of uncertainties in operating temperature will be negligible.

5. Hydrostatic Pressure of Process Liquor

For simplicity assume that the tank consists of a single length of cylindrical pipe, cross section A, submitted to a uniform pressure increase equivalent to that at the bottom of the tank when the contained fluid is changed from water to process liquor

Length of pipe, $L = \frac{V\ max}{A} \triangleq \frac{0.164}{\pi_r 2}$ metres (R is in m^2)

Pressure increase, $\Delta P = 11{,}800\ N/m^2$

Wall thickness, $T = 7.1.\ 10^{-3}m$

E (stainless steel) $= 200.\ 10^9\ N/m^2$

Radius R $= 0{\cdot}077$ m

The increase in radius due to pressure increase ΔP is

$$\Delta R = \frac{R^2 \Delta P}{T^2\ E}$$

Increase in area $= \pi\left[(T + \Delta R)^2 - R^2\right] \triangleq 2\pi R\Delta R\ (\equiv 1{\cdot}8.10^{-2}\%)$

Increase in volume $= \frac{2\pi R^3 L \Delta P}{T^2 E}$

This is equivalent to $\frac{2\pi R^3 L \Delta P}{T^2 E} \cdot \frac{100}{\pi R^2 L}$

$$= \frac{2R\Delta P}{T^2 E} \; 100$$

$$= 0.018\% \text{ of the maximum volume}$$

This will be a considerable overestimate because the pressure increase on average will approach half that assumed and the errors can be neglected

DISCUSSION

D. JUNG: You stated that material or isotopic mass in fuel-bundle or assembly items could not be verified before the accountability tank stage. Do you then believe that NDA methods cannot be used for verification prior to this stage?

A.G. HAMLIN: I do not personally know of any NDA technique which gives the precision and accuracy required for nuclear accountancy, even on unirradiated fuel elements, let alone irradiated ones. Better results might be obtainable on pins.

R.J. DIETZ: The assay of the solutions in the accountability tank at the head-end of a reprocessing plant is, of course, based on the product of volume and concentration. While the "dee"-tank configuration appears to offer advantages in the volume determination, I wonder if its unusual shape might not introduce mixing difficulties and lead to concentration inhomogeneities that would cancel the improvements gained in the volume determination. Mixing could be extremely important in the hot, radioactive solutions for which the tank is intended.

A.G. HAMLIN: We have tested the mixing procedure as far as possible with inert solutions, and the air-mixing process employed appears to be satisfactory. Further experiments will be performed in due course with solutions closely resembling process liquid.

J.M. COUROUBLE: You described tanks with sub-critical geometry needed, for example, for high plutonium concentrations. Would you agree that the input tanks in reprocessing plants for LWR fuels could be of conventional geometry, bearing in mind the volume concentration? This would give better homogeneity and greater precision.

A.G. HAMLIN: I agree. In my presentation I said that I was considering tanks required for the reprocessing of highly enriched uranium and plutonium. Perhaps I should have made it clearer that I was considering plutonium concentrations appropriate to fast reactor fuel and not those normally bred into uranium-based fuels.

POSSIBLE WAYS OF VERIFYING THE INPUT OF A REPROCESSING FACILITY

L. KOCH
CEC Joint Research Centre,
Karlsruhe Establishment,
European Institute for
Transuranium Elements, Karlsruhe,
Federal Republic of Germany

Abstract

POSSIBLE WAYS OF VERIFYING THE INPUT OF A REPROCESSING FACILITY.

At the reprocessing input point the identity, integrity and input amount of fissile material have to be verified. This can be achieved by combining volume/concentration, Pu/U ratio method and isotope correlation techniques. A procedure is described. The sources of information, accuracy, tamper-resistance and inspection effort are discussed.

When nuclear fuel passes through the input point of a reprocessing facility, three pieces of information have to be verified:

Identity of the individual fuel element;
Integrity of the fuel material; and
The input amount of fissile material.

The first two pieces of information are verified by containment and surveillance measures. The change in the amount of fissile material of the spent fuel compared to the fresh material is determined by accountancy measures. During the analysis of the fuel composition, information is obtained which can be used to establish the identity of the fuel element and to verify its integrity.

1. IDENTITY VERIFICATION

The nuclide abundances in a fuel are unique and for this reason it has been proposed that they could be used as fingerprints for identification purposes [1]. If to these observations the isotope correlation technique is added, one cannot only distinguish between individual fuel elements but deduce from the isotopic composition the history and the origin of the fuel. Parameters such as neutron flux level, cooling time, reactor type, initial fuel composition etc., can be determined from the post-irradiation measurements [2].

2. INTEGRITY OF FUEL MATERIAL

Any accidental or operational losses of fuel material after the fabrication of the fuel element can be observed by a simple material balance given in Eq.(1).

$$U(O, F) = 100\, V\rho d\, ([U(R)] + [TU(R)])/(100\text{-}FT) \qquad (1)$$

V	= volume of dissolver solution
ρ	= density of dissolver solution
d	= dilution factor of sample
[(R)]	= concentration in sample
FT	= burn-up in at.%

The initial number of fuel atoms, U(O, F), must be the same as the sum of remaining uranium U(R) and transuranium atoms TU(R) plus the number of fissioned atoms, so that any substantial losses occurring in the section of the cycle, starting with the fabrication and ending with the analysis of the dissolved fuel, can be detected.

3. INPUT OF FISSILE MATERIAL

The burnup of fissile material and the conversion of fertile material cannot be accurately predicted by reactor physics calculations. Only destructive analysis of the fuel determines the amount of fissile material in the spent fuel. The three methods described in Sections 3.1 to 3.3 can be used to measure the uranium and plutonium isotopic content. The first method is widely used by plant operators.

3.1. Volume/concentration method

The volume/concentration method has already been described [3]. From the volume and density of the dissolver solution, the dilution factor of the sample, and from its isotopic concentrations the amount of the isotope X, can be calculated:

$$X(R) = [X(R)]\, V\rho d \qquad (2)$$

For explanation of symbols see Eq.(1).

This equation can be transformed to

$$X(R) = U(O, R)\, X(R)IMA \qquad (3)$$
$$U(O, R) = 100\, V\rho d\, ([U(R)] + [TU(R)])/(100 - FT)$$
$$X(R)\, IMA = X(R)/U(O, R)$$

For explanation of symbols see Section 3.2.

If this method is used by the plant operator, the safeguards authorities verify his measurement by observation, except for the determination of the concentration, which is done on a separate sample by the safeguards laboratory.

3.2. Pu/U ratio method

The so-called Pu/U ratio method makes use of the fabrication data concerning the fuel amount, U (O, F). Thus, the measurement of the mass of the dissolver solution, as well as the concentration of the nuclides in it, are eliminated. Instead, the abundance of each uranium and plutonium nuclide relative to the initial fuel amount, X(R) IMA, is measured, which implies a burnup determination. The latter can be accomplished either by determining the ^{148}Nd content or by using the isotope correlation technique.

In contrast to the method described above, the nuclide X is deduced from U(O, F) instead of from the fuel amount determined later in the plant, U (O, R). The difference between these was discussed above.

$$X(F) = U(O, F)\, X(R)\, IMA \qquad (4)$$

If this method is used by the plant operator the safeguards inspector need merely observe the sampling of the dissolver solution for an isotope analysis, which is carried out in the safeguards laboratory, to verify the nuclide abundances as stated by the plant operator.

3.3. Isotope correlation technique

Several isotope correlations have been described, by which the nuclide abundance in a spent fuel, X(cor) IMA, can be deduced from simple isotope ratio measurements. The mass of X(cor) is obtained similarly to the Pu/U ratio method, by using the initial fuel amount as stated by the fabricator of the fresh fuel.

$$X(cor) = U(O, F)\, X(cor)\, IMA \qquad (5)$$

TABLE I. FUEL PARAMETERS OBTAINED FROM CORRELATION ax+b = Xn(cor) IMA with X1: $^{242/240}Pu$ = 0.186, X2: $^{242/241}Pu$ = 0.350, X3: $^{132/131}Xe$ = 2.37 AND COMPARED WITH EXPERIMENTAL VALUE X(R) IMA. (*Note:* D5 DEFINED AS (U-235 (O) – U-235(F))/U-235 (O). F_T given in at.% F IMA.)

Parameter	a	b	X1 (cor) IMA	a	b	X2 (cor) IMA	a	b	X3 (cor) IMA	X (R) IMA
D5	2.02 E+O	2.88 E-1	6.65 E-1	1.35 E+O	1.96 E-1	6.69 E-1	4.38 E-1	3.86 E-1	6.52 E-1	6.63 E-1
F_T	1.37 E-1	4.06 E-1	2.96 E+O	8.96 E+O	1.43 E-1	2.99 E+O	2.88 E+O	3.97 E+O	2.88 E+O	3.04 E+O
^{234}U	1.57 E-3	1.19 E-4	1.75 E-4	9.38 E-4	1.51 E-4	1.77 E-4	1.21 E-4	1.14 E-4	1.73 E-4	1.69 E-4
^{235}U	5.21 E-2	1.97 E-2	9.96 E-3	8.49 E-2	2.21 E-2	9.85 E-3	1.13 E-2	3.72 E-2	1.04 E-2	1.01 E-2
^{236}U	1.22 E-2	1.38 E-3	3.66 E-3	8.14 E-3	8.31 E-4	3.68 E-3	2.56 E-3	2.48 E-3	3.58 E-3	3.72 E-3
^{238}U	1.31 E-1	9.73 E-1	9.48 E-1	8.76 E-2	9.79 E-1	9.48 E-1	2.41 E-2	1.01 E+O	9.49 E-1	9.47 E-1
^{238}Pu	7.39 E-4	3.79 E-5	9.98 E-5	4.94 E-4	7.18 E-5	1.01 E-4	1.54 E-4	2.71 E-4	9.42 E-5	9.44 E-5
^{239}Pu	7.48 E-3	3.52 E-3	4.91 E-3	4.97 E-3	3.18 E-3	4.92 E-3	1.66 E-3	9.08 E-4	4.86 E-3	4.90 E-3
^{240}Pu	8.47 E-3	3.43 E-4	1.92 E-3	5.66 E-3	4.16 E-5	1.94 E-3	1.79 E-3	2.40 E-3	1.86 E-3	1.92 E-3
^{241}Pu	5.46 E-3	1.54 E-5	1.03 E-3	3.64 E-3	2.32 E-4	1.04 E-3	1.09 E-3	1.61 E-3	9.96 E-4	1.02 E-3
^{242}Pu	2.72 E-3	1.48 E-4	3.58 E-4	1.81 E-3	2.71 E-4	3.63 E-4	5.69 E-4	1.01 E-3	3.41 E-4	3.59 E-4

4. PROCEDURE

A procedure has been developed which allows the results of the three methods mentioned to be obtained and compared. This evaluation has been integrated into the existing software used by the European Institute for Transuranium Elements in analysing reprocessing safeguards samples [4], and is being tested in the current isotope correlation experiment (ICE).

In applying these methods, four sources of information can be identified – fuel fabricator, reprocessor, safeguards laboratory and historical data. The amount of information needed depends on the method to be employed.

The *fuel* can be *identified* by comparing its post-irradiation nuclide abundances, either with historical data of similar reactors or with earlier fuel batches of the same reactor. In addition to the isotope correlations mentioned below, the rather unique correlations between ^{235}U depletion, D5, and various isotope ratios such as $^{242/240}Pu$, $^{242/241}Pu$ and $^{132/131}Xe$ can be used to verify the initial ^{235}U enrichment (Table I).

Balancing the pre- and post-irradiation quantities of fuel leads to a check of the *fuel integrity*. Information comes from the fuel fabricator, reprocessor and the safeguards laboratory. If the burnup was not measured experimentally using ^{148}Nd, historical data are needed to obtain burnup correlations. In the example given below, the same type of correlation has been applied as was given for D5. To obtain the *fuel input*, the sources of information are different for the three methods mentioned.

For the direct method (see Section 3.1) information comes from the reprocessor and safeguards laboratory, whereas for the Pu/U and the isotope correlation method (see Sections 3.2 and 3.3) no information from the reprocessor is required.

An example is given in Table II. The data refer to the reprocessing input of a KWO fuel. The last two columns show the amount of each nuclide in the spent fuel given in atoms as obtained by methods 1 and 2. For method 3 the figures were produced by three different correlations. Since the reprocessed fuel had a higher initial ^{235}U enrichment than the historical data of this reactor in the data-bank of the European Institute for Transuranium Elements, the amount of ^{235}U and ^{236}U could not be obtained from the $^{242/240}Pu$ and $^{242/241}Pu$ correlations (see columns 1 and 2). The $^{132/131}Xe$ correlation is, however, less sensitive to the initial ^{235}U enrichment. The results from this correlation are listed in column 3.

5. DISCUSSION

From the safeguards aspect the methods mentioned above have to be examined in terms of accuracy; tamper resistance; and inspection effort.

TABLE II. PRINTOUT OF AMOUNTS (IN ATOMS) OF INDIVIDUAL ISOTOPES PER INPUT BATCH AS OBTAINED BY ISOTOPE CORRELATION TECHNIQUE (COLUMNS 1–3), Pu/U RATIO-(X1 (U, O, F)) AND VOLUME CONCENTRATION METHOD X1 (U(O,R)).*

```
************************
*PW0B0105BS   *KW0VI95*        MEMBERS:  DATABANK = 158 UNULBANK = 10
************************

DATA CARDS INPUT VALUES:                        COMPUTED VALUES:

VOLUME               = 0.569835E+03 L           U (O, R) = 0.346724E+27 AT
DENSITY              = 0.143850E+01 KG/L
DILUTION FAC.        = 0.337350E+03
TOT. H. NUCLIDES     = 0.121570E+19 AT/G
U (O, F)             = 0.347170E+27 AT
```

NUCLIDE XJ	XJ (COR.) AT (1)	XJ (COR.) AT (2)	XJ (COR.) AT (3)	XJ (COR.) AT (4)	XJ (U(O,F)) AT	XJ (U(O,R)) AT
U-234	**********	**********	.60090D+23	**********	.5879673E+23	.5872127E+23
U-235	**********	**********	.36109D+25	**********	.3495308E+25	.3490821E+25
U-236	**********	**********	.12441D+25	**********	.1291576E+25	.1289919E+25
U-238	.32912D+27	.32919D+27	.32940D+27	**********	.3288671E+27	.3284450E+27
PU-238	.35104D+23	.34633D+23	.32709D+23	**********	.3277945E+23	.3273737E+23
PU-239	.17092D+25	.17051D+25	.16857D+25	**********	.1702002E+25	.1699817E+25
PU-240	.67256D+24	.66719D+24	.64588D+24	**********	.6683718E+24	.6675140E+24
PU-241	.36184D+24	.35846D+24	.34591D+24	**********	.3558840E+24	.3554271E+24
PU-242	.12611D+24	.12442D+24	.11824D+24	**********	.1244744E+24	.1243146E+24
AM-241	**********	**********	**********	**********	***********	**********
AM-243	**********	***********	**********	**********	***********	**********
CM-242	**********	***********	**********	**********	.6865630E+22	.6856821E+22
CM-244	**********	***********	**********	**********	.4807957E+22	.4801787E+22

* (This print-out has been retyped for publication.)

The reproducibility of the measurements needed for methods 1, 2 and 3 under routine operation can easily reach 0.3% and better. Deviations between operator analyses and re-measurements are caused mainly by sampling errors or wrong calibrations. Such biases occur more frequently in method 1 than in the others, where a larger number of measurements are involved. The Pu/U method eliminates the measurement of the dissolver mass by using the initial fabrication data of the fuel and seems more reliable. For this method, as well as for the isotope correlation technique, all losses during dissolution have to be measured to accuracies of about 10%. The last method requires only isotopic abundance measurements, which are less sensitive to systematic errors.

The tamper-resistance of all measurements involved for method 1 is lower than for the other methods because they are all subject to plant operations. Eliminating the determination of dissolver solution mass by the initial fuel weight (or by relating the fuel to an added tracer [5]) does not substantially increase this tamper-resistance. Any clandestine addition of depleted uranium (or the tracer) to the dissolver would lower (according to Eq.(4)) the amount of Pu in the input. Such a diversion strategy could be detected by the isotope correlation technique which would reveal the change in the consistency of the nuclide abundance of an identified fuel. Any accidental or intended cross-contamination of the fuel sample with fuel of lower exposure (having the effect of lowering the Pu amount in the input) could be detected by using fission gas correlations. The analysis of the fission gases is made *simultaneously* with the dissolution of the reprocessed fuel batch.

The greatest inspection effort is needed for method 1, the method of choice for plant operators; therefore, the two other methods can be applied without additional effort (with the exception of fission gas analysis) by using redundant information obtained during routine inspection.

6. CONCLUSION

The accuracies obtained at present for the methods mentioned have not yet reached a level which can be achieved by using existing analytical techniques. A combination of the methods seems to provide the tamper-resistance required for nuclear material safeguards without increasing the inspection effort too much.

To assess the potential of the methods, integral experiments of the type at present performed by the ESARDA working group at WAK (1st Isotope Correlation Experiment) should be carried out.

REFERENCES

[1] KOCH, L., BRAUN, H., CRICCHIO, A., in Safeguards Techniques (Proc. Symp. Karlsruhe,1970) 2, IAEA, Vienna (1970) 539.

[2] BRANDALISE, B., KOCH, K., RIJKEBOER, C., ROMKOWSKI, D., in ESARDA Symp. Isotopic Correlations and their Application to the Nuclear Fuel Cycle, Stresa, Italy, 9–11 May, 1978.

[3] KOCH, L., COTTONE, G., Tagungsber. Reaktortagung des Dtsch Atomforums (1973) 287.

[4] BRANDALISE, B., COTTONE, G., CRICCHIO, A., GERIN, F., KOCH, L., EUR-5669 (1977) (1977).

[5] MATHEWS, C.K., JAIN, H.C., KAVIMANDAN, V.D., AGGARWAL, S.G., in Safeguarding Nuclear Materials (Proc. Symp. Vienna,1975) 2, IAEA, Vienna (1976) 485.

DISCUSSION

J. BOUCHARD: You described three ways of determining the input balance – three independent, technical methods. Are these methods going to be developed from the point of view of redundancy or simply in order to select the best one? I think this is important, not only for the operator of a reprocessing plant but also for those who are developing such methods.

L. KOCH: We are comparing the potential of each method. A combination of the methods seems promising for safeguards purposes and would be easy to achieve if the plant operator were using the volume/concentration method as this would give the redundancy required for the application of the other two methods.

S. SANATANI: In safeguarding a reprocessing plant, what other problems, apart from input verification, are most important in your opinion? We need to consider, for example, losses or diversion before the fuel reaches the input tank, the mixing of fuel from different reactors, the presence of recycle acid and hull measurement.

L. KOCH: Operational losses in the head-end process are often only estimated and declared as shipper/receiver differences. If equipment is installed in the future to measure, say, hull losses, the material balance will be more accurate.

S. SANATANI: What do you mean by *identification* of dissolved fuel, considering that individual fuel elements lose their identity on dissolution?

L. KOCH: The nuclide abundance, burnup etc., are unique to a particular spent fuel and can be used to identify the latter's origin and history.

J.M. COUROUBLE: In connection with Mr. Sanatani's question I would like to point out that there is a discrepancy between the quantity of fissile material contained in the fuel assembly and that contained in the clarified dissolver solution. The discrepancy springs from very small amounts of material contained in the hulls and the insoluble material. These amounts can be estimated by appropriate measurements, so that no "loss" occurs.

TRACER TECHNIQUES IN THE INPUT ACCOUNTABILITY OF PLUTONIUM IN REPROCESSING PLANTS

C.K. MATHEWS, H.C. JAIN,
V.D. KAVIMANDAN, S.K. AGGARWAL
Radiochemistry Division,
B.A.R.C., Trombay,
Bombay, India

Abstract

TRACER TECHNIQUES IN THE INPUT ACCOUNTABILITY OF PLUTONIUM IN REPROCESSING PLANTS.

The input end of a reprocessing plant is the first point in the fuel cycle where plutonium production can be accurately measured. The volume-concentration method is the most widely used method and it involves sampling, measurement of the concentration of plutonium in the sample and the measurement of the volume, density and temperature of the solution in the tank. Other possible methods are the Pu/U ratio method and isotope correlation technique, which depend on data from the fuel fabrication plant and the reactor. Two tracer techniques have been developed at Trombay, the Magnesium Tracer Technique for the input Accountability of Plutonium (MAGTRAP) and the corresponding lead tracer technique code named LEADTRAP. The tracer method uses a suitable tracer (magnesium), a known quantity of which is added to the solution in the input accountability tank and then the tracer-to-plutonium ratio is measured in the samples withdrawn from the accountability tank using a double-spike isotope dilution technique. Knowing the amount of tracer added and the tracer to plutonium ratio, one can readily obtain the total amount of plutonium in the accountability tank. The volume of the solution in the tank and the size of the aliquot need not be known. The validity of this technique has been established by a series of experiments at a fuel reprocessing plant. Data is presented to demonstrate that accuracies of better than 1% are attainable in the measurement of the total plutonium as well as uranium for even small reprocessing batches.

INTRODUCTION

In a nuclear material accounting system covering the fuel cycle, one of the most critical measurements is that of the plutonium produced in the reactor. The input end of a reprocessing plant is the first point in the fuel cycle where an accurate assessment of the total plutonium in the irradiated fuel can be made. It is at this point that the fuel elements discharged from a reactor lose their identity, and their material content becomes openly accessible for measurement. Apart from the obvious importance of this measurement from the aspects of national and international material accounting systems, for the reprocessing plant itself an accurate measurement of the plutonium input is necessary in order to assess the processing losses. This loss is particularly important in the case of plutonium in view of the criticality and other safety problems associated with it and also its strategic importance.

From the point of view of an agency responsible for independent nuclear material accounting, the two major criteria for choosing a measurement method are accuracy and independence from data coming from other sources. From this point of view, the currently available methods for input accountability at reprocessing plants are inadequate. The most common method for measuring the plutonium input to a reprocessing plant is to determine the total amount of this element in each batch in the accountability tank by the volume concentration method [1]. This involves measuring the total volume of the solution in the tank as well as the concentration of plutonium. The volume is measured by manometers, which are precalibrated for this purpose, and the concentration is usually determined by isotope dilution mass-spectrometry. Thus, the error in the measurement of plutonium in a dissolver batch arises from errors in the values of both the total volume and the concentration, the latter implying a determination of the sample size. For the volume, one has to depend on the plant operator's data.

Another approach is to measure the plutonium-to-uranium ratio [2] in the accountability tank and then calculate the amount of plutonium from the known quantity of uranium in the fuel as obtained from the fabrication data corrected for the burnup loss in the reactor. Thus, in this method, one must depend on the data from the fabrication plant for the initial amount of uranium present in the fuel and on the reactor data for fuel burnup. Further, one must correct the input data for losses through the undissolved fuel.

A third method that has been discussed in recent years is the isotope correlation technique [3]. Here, use is made of the correlation existing for a particular reactor between the plutonium-to-uranium ratio and various heavy-element or fission-product isotope ratios. A large number of isotope correlations involving the depletion of ^{235}U as well as the build-up of plutonium isotopes and radioactive fission products have been recommended for determining the Pu/U ratio. However, each correlation is influenced to a certain extent by the main characteristics of the core and the type of reactor. The evolution of the isotopic composition of the fuel is influenced by various parameters like initial enrichment of the fuel, moderation ratio, cladding material, reactivity control method and, in case of BWRs, the voids in the coolant. Though this method is independent of any volume or concentration measurement at the plant, its accuracy is poorer than that of the other techniques mentioned above.

The use of tracer techniques, as developed at the Bhabha Atomic Research Centre, Trombay, for input accountability does not suffer from this disadvantage [4–7]. A tracer is added to the input accountability tank and the ratio of plutonium to tracer is measured in a sample of the solution using a double-spike isotope-dilution technique. Knowing the amount of the tracer element added, the total amount of plutonium in the input accountability tank can be obtained. This method is simpler and more accurate than all the other methods and it is independent of the data from the reprocessing plant as well as from other facilities in the fuel cycle. Two tracers, magnesium and lead, have been identified for this purpose. When magnesium is used as a tracer, the method is called MAGTRAP (MAGTRAP – Magnesium Tracer Technique for the Accountability of Plutonium) and when lead is used, the method is called LEADTRAP (LEADTRAP – Lead Tracer Technique for the Accountability of Plutonium).

I. CHOICE OF TRACER

To be considered as a tracer for input accountability measurements, an element must meet the following conditions:

(i) The tracer element should have at least two stable (or very long-lived) isotopes, one of which must be available in a highly enriched form to be used as spike for isotope dilution;
(ii) It should be possible to measure its isotopic abundances (at least the ratio of an abundant natural isotope to the spike isotope) very accurately;
(iii) It should, preferably, not have different oxidation states between which the isotopic exchange is slow;
(iv) It should be chemically inert so as not to be lost by adsorption, plating and other reactions;
(v) It should be inexpensive and readily available;
(vi) A chemical assay standard must be available, i.e. it must have a compound which is available in a pure and stoichiometric form;
(vii) It should preferably not be a fission product with a high fission yield; and
(viii) It should not interfere with plant operation, especially in product purity, decontamination factor, extraction efficiency etc.

Only isotope-dilution mass-spectrometry is expected to give the necessary accuracy for measurement. Outside the fission-product range the only elements that merit consideration are Li, K, Mg, Ca and Pb. Lithium, being a very light element, presents many problems in accurate isotopic composition measurement because of isotopic fractionation problems. Potassium and calcium are very abundant, and therefore problems arising from natural contamination, as well as large blank values, are expected to be serious. This narrowed our choice down to magnesium and lead. The relative isotopic abundances of these elements can be measured mass-spectrometrically with a precision of about 0.2%. They are not extractable in TBP and have a very low neutron absorption cross-section which makes it easy to meet product purity requirements, as impurity levels in product plutonium and uranium are specified in "ppm boron equivalent".

The availability of two tracer elements has the advantage that they can be alternated. This is very advantageous when the accountability tank cannot be fully emptied, leading to the build-up of blank levels if the same element is repeatedly used.

2. PRINCIPLE OF THE METHOD

Let us consider the case of MAGTRAP where magnesium is used as a tracer.

A known quantity (W_{Mg}) of natural magnesium is added to the tank containing plutonium. After thorough mixing, small aliquots of the solution are withdrawn and spiked with a mixture containing ^{242}Pu and ^{26}Mg in an accurately known ratio, R_t ($R_t = {}^{242}Pu/{}^{26}Mg$). The aliquot size of either the sample or the spike need to be known. The isotope ratios $^{242}Pu/^{239}Pu$ ($R^M_{2/9}$) and $^{26}Mg/^{24}Mg$ ($R^M_{6/4}$) are measured in the spiked solution. The total amount of plutonium, W_{Pu} in the tank is then given by the equation

$$W_{Pu} = W_{Mg} \cdot R_t \cdot \frac{R^M_{6/4} - R^O_{6/4}}{R^M_{2/9} - R^S_{2/9}} \cdot \frac{\sum_i R^S_{i/j}\,(Pu)\, M^S_i(Pu)}{\sum_i R^S_{i/j}\,(Mg)\, M^S_i(Mg)} \quad (1)$$

where $R_{i/j}$ refers to the ratio of the abundance of the ith isotope to that of the jth isotope (i, j etc. are numerically given by the last digit of the mass number and the reference isotope j is 24 in the case of Mg and 239 in the case of Pu) and the superscripts S, M and O refer to the unspiked sample, spiked mixture and natural magnesium, respectively. The summation is over all the isotopes of the element given in the bracket and the $M_i(X)$ refers to the mass of the ith isotope of the element X. It may be noted that

$$\sum_i R^S_i(Pu)\, M^S_i(Pu) = \langle A_t \cdot W_t \rangle^S_{Pu} / (AF_9)^S$$

i.e. the average atomic weight of plutonium divided by the atomic fraction of ^{239}Pu in the unspiked sample.

Eq.(1) implies (a) that the spike solution does not contain ^{239}Pu and ^{24}Mg and (b) that magnesium is absent in the tank before tracer magnesium is added (i.e. no blank). These conditions are, in general, not satisfied and hence corrections have to be made. The first of the above corrections can be taken care of by multiplying Eq.(1) with the factor

$$\frac{1 - R^{M}_{2/9}/R^{Sp}_{2/9}}{1 - R^{M}_{6/4}/R^{Sp}_{6/4}}$$

where the superscript Sp refers to the spike solution. Eq.(1) now becomes

$$W_{Pu} = W_{Mg} \cdot R_t \cdot \frac{R^{M}_{6/4} - R^{O}_{6/4}}{R^{M}_{2/9} - R^{S}_{2/9}} \cdot \frac{\sum_i R^{S}_{i/j}(Pu)\, M^{S}_{i}(Pu)}{\sum_i R^{S}_{ij}(Mg)\, M^{S}_{i}(Mg)} \cdot \frac{1 - R^{M}_{2/9}/R^{Sp}_{2/9}}{1 - R^{M}_{6/4}/R^{Sp}_{6/4}} \qquad (2)$$

To correct for the second, i.e. magnesium blank, an additional measurement is necessary. There are two alternatives:

(i) An accurate correction for blank can be made by carrying out the double spike ($^{26}Mg + {}^{242}Pu$) isotope dilution in a (blank) sample of the tank solution taken before the addition of the tracer and analysing it in the same way as the sample collected after tracer addition. If Eq.(2) is rewritten as

$$W_{Pu} = W_{Mg} \cdot Z$$

where Z stands for all the remaining factors, then it can be shown that blank correction can be accompanied by multiplying Eq.(2) by

$$1 + \frac{Z^T}{Z^b - Z^T}$$

where the superscripts b and T refer to the values of Z before and after addition of magnesium tracer. The amount of plutonium in the tank is then given by

$$W_{Pu} = W_{Mg} \cdot Z^T \left[1 + \frac{Z^T}{Z^b - Z^T}\right]$$

This method of blank correction involves double spiking followed by the measurement of both magnesium and plutonium isotope ratios. MAGTRAP using this alternative for blank correction, is referred to as MAGTRAP-I (or LEADTRAP-I).

(ii) However, if the blank is small, then an approximate knowledge of the magnesium concentration both in the blank and in the sample aliquots (C^b and C^T), respectively, would be sufficient for correction. Equation (2) in that case is multiplied by the factor

$$1 + \frac{C^b}{C^T - C^b}$$

Here, the correction factor involves only magnesium concentrations; plutonium isotope ratio measurement for the blank sample is not required as in alternative (i). The amount of plutonium in the tank is now given by

$$W_{Pu} = W_{Mg} \cdot Z^T \left[1 + \frac{C^b}{C^T - C^b} \right]$$

Knowledge of concentrations does imply that of aliquot sizes but these need be known with much less accuracy than is otherwise required. For example, if the blank is 10% of the tracer added, then the aliquot sizes need be known with ten times poorer accuracy than is demanded in plutonium measurement. In other words, volume measurement by pipetting would be adequate and this relaxation makes the handling of dissolver solution easier. When this method is applied for blank correction, the technique is referred to as MAGTRAP-II (or LEADTRAP-II).

3. VOLUME MEASUREMENT

The tracer technique can also be used to measure the total volume (or weight) of the solution in the accountability tank. After the addition of a known amount of tracer (W_{tr}), a measured aliquot (W_{al}) of the solution is mixed with a known quantity of a suitable spike (say ^{26}Mg in case of MAGTRAP). From the isotope ratio measured in the spiked mixture (for example, $^{26}Mg/^{24}Mg$ in MAGTRAP), one calculates the quantity of tracer (X) in the aliquot. The weight (or volume) of the total solution in the tank is then given by $W_T = (W_{tr}/X) \cdot W_{al}$, the units of W_{al} determining that of W_T. This variation of the tracer technique (limited to volume measurement) is called MAGTRAP-V when using magnesium as a tracer, and LEADTRAP-V when using lead as a tracer. It can be used either to measure the volume of the solution in the tank as required in the volume-concentration method, or for verifying the tank calibration. A similar technique using lithium tracer was investigated by Bokelund [8] with inconclusive results.

4. EXPERIMENTAL

General description of the experiments

The experiments consisted of adding a known amount of the tracer stock solution with accurately known concentration to the tank through a half-inch stainless-steel line. After different durations of air sparging, replicate samples were withdrawn. Weighed aliquots of the sample were spiked with known quantities of a spike solution containing the respective spikes. The spike mixture consisted of ^{26}Mg, ^{233}U and ^{242}Pu in the case of MAGTRAP, and of ^{206}Pb, ^{233}U and ^{242}Pu in the case of LEADTRAP. Magnesium/lead, uranium and plutonium were chemically separated and the relevant isotope ratios were measured.

Chemical separation

The detailed procedures for the separation of magnesium/lead, uranium and plutonium from the irradiated fuel solution are given elsewhere [5, 9]. In MAGTRAP experiments, the procedure involved the absorption of Pu (IV) and U (VI) on an anion exchange column and their later elution with dilute nitric acid. Magnesium, which appeared in the initial washings along with alkali metals, alkaline earths and rare-earths, was further separated from rare-earths using an anion exchange separation procedure in methanol-nitric acid medium. It was finally purified from fission products by using a cation exchange method. In LEADTRAP experiments lead, along with the rare-earths and fission products, was separated from uranium and plutonium using an anion exchange column and nitric acid medium. Finally, it was separated from fission products by an anion exchange method using methanol-nitric acid medium.

Isotopic composition measurement

The isotope ratios were measured using a Varian MAT CH-5 mass spectrometer equipped with a thermionic source. Uranium and plutonium isotopic compositions were measured using a double rhenium filament assembly. Details of this are given elsewhere [10].

A good deal of investigation was required to arrive at the optimum conditions for magnesium isotope ratio measurements because of the possibility of isotope fractionation. The procedure finally implemented involved the loading of a drop of a solution of $MgCl_2$ (containing about 10 μg Mg) on the sample filament of a pre-heated double-filament rhenium assembly and a well-defined schedule for heating the filament [11]. Measurements were carried out at filament temperatures of 5.5–6.0 A (ionizing filament) and 1.4–1.8 A (sample filament).

TABLE I. WEIGHT/VOLUME MEASUREMENT BY TRACER TECHNIQUE

Experiment No.	Weight of solution in the tank (kg)		Error (%)
	By DP measurement or by weighing	By tracer technique	
1.	1012.0	1016.3	+ 0.42
2.	1055.0	1061.0	+ 0.57
3.	1510.3	1504.3	− 0.40
4.	1505.5	1509.5	+ 0.26
5.	2248.3	2264.1	+ 0.70
6.	1903.7	1898.4	− 0.28
7.	1503.3	1513.9	+ 0.70
8.	1) 1383.2	1381.4	− 0.13
	2) 1383.2	1375.3	− 0.57

Mean error = 0.49%.

TABLE II. TOTAL URANIUM AND PLUTONIUM DETERMINED IN THE INPUT ACCOUNTABILITY TANK

Element	By MAGTRAP I	By MAGTRAP II	Volume-concentration method	
			MAGTRAP-V	DP measurement
Uranium (kg)	493.61 ± 3.82 (0.77%)	493.31 ± 3.57 (0.72%)	493.94 ± 1.94 (0.39%)	496.14 ± 1.95 (0.39%)
Plutonium (g)	2.839 ± 0.012 (0.42%)	2.837 ± 0.009 (0.32%)	2.841 ± 0.020 (0.70%)	2.854 ± 0.020 (0.70%)

TABLE III. TOTAL URANIUM (kg) DETERMINED IN THE INPUT ACCOUNTABILITY TANK BY MAGTRAP AND LEADTRAP

Technique	I	II	V
MAGTRAP	254.80 ± 0.37 (0.14%)	254.78 ± 0.36 (0.14%)	254.82 ± 0.13 (0.05%)
LEADTRAP	252.59 ± 1.32 (0.52%)	252.58 ± 1.30 (0.51%)	252.10 ± 0.27 (0.11%)
DP measurement			255.15 ± 0.53 (0.21%)

Isotopic composition measurements for lead were carried out by using the silica gel-phosphoric acid technique with a single rhenium filament assembly [12]. Measurements were taken at a filament temperature of about 2 A.

5. RESULTS

The results of the experiments conducted in the input accountability tank of a reprocessing plant are given in Tables I, II and III. Table I gives the results of the measurement of the volume of the solution in the tank by using the tracer technique. In some of these experiments, the accountability tank was filled with known quantities of demineralized water, while in others the differential pressure (DP) data was used as the tanks were freshly calibrated. As is clear from a comparison made in Table I, the tracer technique can be used to measure the volume of the solution in the tank with an accuracy of ± 0.5%.

Table II summarizes the results of an experiment for determining total uranium and plutonium in the input accountability tank. Here the results obtained by using tracer techniques are compared with those obtained by the volume-concentration method. In the volume-concentration method, the concentrations were measured by the isotope dilution technique with aliquots taken on weight basis. The total volume was obtained by two different methods: (1) using MAGTRAP-V, and (2) using the differential pressure method. Here, the reliability of the volume measurement by the DP method is very high as the tanks had been recently calibrated. The concentration measurement is also very accurate as the aliquot measurements were carried out by using the weight burette technique. The fact that both these variations of volume-concentration method agreed within 0.5%, as seen from Table II, enhances the confidence in the value. The total uranium and plutonium has been measured with a precision of better than 1% using the tracer technique and these values are within 1% of the value obtained by the volume-concentration method. It may therefore be concluded that the total uranium and plutonium in the input accountability tank can be determined with a precision and accuracy of better than 1% by using the tracer techniques. In this experiment, small traces of plutonium and fission products were added to the uranium and it is remarkable that the method gives better than 0.5% accuracy in the determination of total plutonium even at very low levels. If one computes the overall error from the various error components in Eq.(1), which involves only isotope ratios, one arrives at a value of about 1% for precision and accuracy, which is comparable with that achieved in these experiments.

Table III gives the results of an experiment conducted in the input accountability tank of a reprocessing plant to determine the total amount of uranium by using both lead and magnesium as tracers. As is clear, the values obtained for the total amount of uranium using these two tracers compare well with each other, thus demonstrating the suitability of lead as an alternate tracer.

TABLE IV. MEASUREMENT AND DATA REQUIREMENTS OF DIFFERENT METHODS

Operations	Volume measurement	Density measurement	Aliquotation	Isotope dilution	MS analysis	Fuel fabricator data	Reactor data	Historical data (isotope correlations)	Analysis of recycled materials
1. Tracer technique (MAGTRAP LEADTRAP)				*	*				*
2. Volume concentration	*	*	*	*	*				*
3. Pu/U ratio				*	*	*	*		*
4. Heavy isotopes correlations					*	*	*	*	*
5. Fission gas isotopes correlations					*	*		*	

* Denotes measurement/data required.

6. DISCUSSION

The tracer techniques fully meet the twin requirements of accuracy and of independence from operators' data. The operations involved for carrying out this measurement are:

(1) Addition of tracer solution (this can be brought by the experimenter)
(2) Sampling
(3) Addition of an (unmeasured) aliquot to the spike vial brought by the experimenter
(4) Chemical separation and mass spectrometric measurements (which can be carried out in an independent laboratory).

Thus, there is a minimum of intrusion in the plant and no dependence on the operators' data. It is also seen that accuracies of better than 1% are readily obtainable.

Table IV lists a comparison of the measurement and data requirements of the various methods that may be considered for input accountability. From accuracy considerations one would consider only volume-concentration, Pu/U ratio and tracer methods; others are more useful for verification purposes. From the point of view of dependence on reactor and fabrication plant data the volume-concentration and tracer techniques would be preferred. But the disadvantages of the volume-concentration method are the dependence on operators' measurements (such as differential pressure, density etc.), tank calibration data and aliquot measurement for spiking (which has to be done at the plant itself). The tracer techniques MAGTRAP and LEADTRAP do not suffer from these drawbacks.

ACKNOWLEDGEMENTS

The authors are grateful to Dr. M.V. Ramaniah, Head, Radiochemistry Division and Shri N. Srinivasan, Project Director, Reactor Research Centre for their encouragement. Thanks are due to Shri A.N. Prasad and Shri M.K.T. Nair of the Fuel Reprocessing Division.

REFERENCES

[1] SRINIVASAN, N., "Nuclear material control in a reprocessing plant", Safeguards Techniques (Proc. Symp. Karlsruhe, 1970) I, IAEA, Vienna (1970) 155.

[2] STEWARD, K.B., SCHNEIDER, R.A., "Properties of the plutonium estimate based on weighed Pu/U values", ibid. I, 583.

[3] CHRISTENSEN, D.E., SCHNEIDER, R.A., "Summary of experience with heavy element isotopic correlations", Safeguarding Nuclear Materials (Proc. Symp. Vienna, 1975) 2, IAEA, Vienna (1976) 377.

[4] MATHEWS, C.K., "Mass spectrometric techniques in fissile material accounting and management", Accountability and Management of Fissile Materials, Proc. Seminar BARC, Trombay, April 1971, BARC/I-165 (1972) 17.

[5] MATHEWS, C.K., et al., An Independent Method for Input Accountability in Reprocessing Plants: Magnesium Tracer Technique for the Accountability of Plutonium (MAGTRAP), BARC-809, BARC (1975).

[6] MATHEWS, C.K., et al., "An independent method for input accountability in reprocessing plants (MAGTRAP)", Safeguarding Nuclear Materials (Proc. Symp. Vienna, 1975) 2, IAEA, Vienna (1976) 485.

[7] AGGARWAL, S.K., et al., Lead Tracer Technique for the Accountability of Plutonium (LEADTRAP), BARC-904, BARC (1976).

[8] BOKELUND, H., Investigation of Reprocessing Input Measurement Using Tracer Technique, ETR-266, Eurochemie, Belgium (1970).

[9] RAMAKUMAR, K.L., et al., Ion Exchange Separation of Lead and Neodymium from a Solution of Irradiated Nuclear Fuel, BARC-939, BARC (1977).

[10] CHITAMBAR, S.A., et al., Mass Spectrometric Analysis of Uranium and Plutonium, BARC-865, BARC (1976).

[11] AGGARWAL, S.K., KAVIMANDAN, V.D., MATHEWS, C.K., Isotopic Analysis of Magnesium, BARC-830, BARC (1975).

[12] BARNES, I.L., Lead separation by anodic deposition and isotope ratio mass spectrometry of microgram and smaller samples, Anal. Chem. 45 (1973) 1881.

CONCEPTION D'UN SYSTEME DE MESURE DE LA QUANTITE DE MATIERES NUCLEAIRES ENTRANT DANS UNE USINE DE RETRAITEMENT

J.M. COUROUBLE, A. DENIS
Commissariat à l'énergie atomique,
Paris

J. HURE, P. PLATZER
CEA, Centre d'études nucléaires
de Fontenay-aux-Roses,
Fontenay-aux-Roses,
France

Abstract–Résumé

DESIGN OF A SYSTEM FOR MEASURING THE QUANTITY OF NUCLEAR MATERIALS ENTERING A REPROCESSING PLANT.

The accounting for fissile materials entering a reprocessing plant can be carried out only in a system especially designed for this purpose. An attempt is made to define a system totally independent of data from outside the reprocessing plant. The arrangement described should be considered a measuring and transfer device. Its basic components are buffer tanks, an accounting tank proper and its connecting tank. Instruments are used for the measurement of volume and specific gravity, and also of the content by weight of U and Pu (carried out in the laboratory). Emphasis is placed on calibration and on maintaining the reproducibility of the measurements with time. The analysis and evaluation of sources of uncertainty in the final result are discussed.

CONCEPTION D'UN SYSTEME DE MESURE DE LA QUANTITE DE MATIERES NUCLEAIRES ENTRANT DANS UNE USINE DE RETRAITEMENT.

La comptabilité des matières fissiles entrant dans une usine de retraitement ne peut se faire que dans un dispositif spécialement conçu pour l'établissement de cette comptabilité. On a cherché à définir un système totalement indépendant d'informations extérieures à l'usine de retraitement. L'installation ainsi décrite doit être considérée comme un appareil de mesure et de transit. Elle se compose essentiellement de cuves tampon, d'une cuve bilan proprement dite et sa cuve annexe. Des dispositifs permettent la mesure des volumes, des masses volumiques ainsi que des teneurs pondérales en U et Pu (effectuées en laboratoire). L'accent est porté sur l'étalonnage, le maintien de la reproductibilité des mesures dans le temps. L'analyse et l'estimation des causes d'incertitude sur le résultat final sont évoqués.

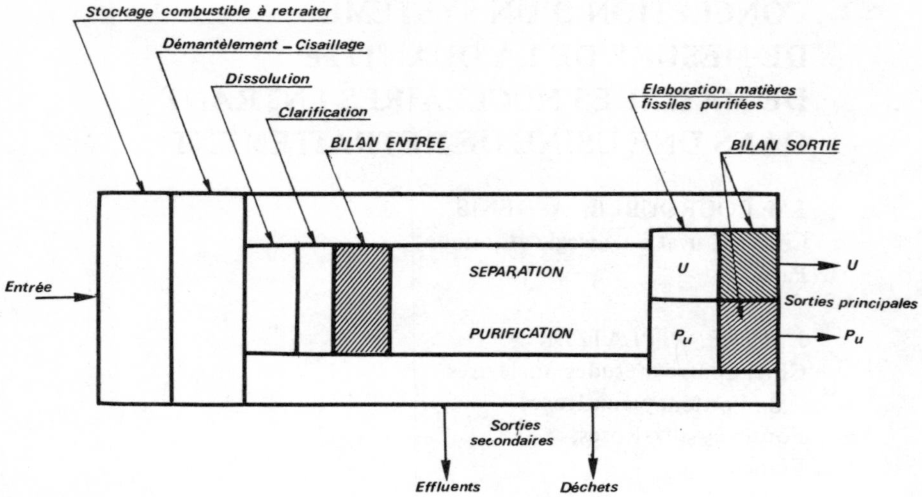

FIG.1. Situation de l'unité bilan.

1. CONCEPT ET SITUATION DU DISPOSITIF DE MESURE

La connaissance du bilan quantitatif des matières fissiles entrant et sortant d'une usine de retraitement est le premier objet du contrôle comptable. Cette comptabilité ne peut se faire que dans un dispositif *spécialement conçu* pour l'établissement de cette comptabilité. Dans une usine utilisant un procédé par voie aqueuse on dispose, après dissolution et clarification, d'un point de passage obligé de la matière fissile en phase aqueuse homogène où la mesure peut apporter le maximum de garantie de précision.

On y installera donc une unité spécifique conçue pour le transit des matières, à *passage obligé* et à *sens unique amont-aval*. Ce bilan est effectué dans les conditions suivantes:

- sur une solution exempte de particules solides, donc après clarification des solutions de dissolution (voir fig.1);
- par une méthode dite volumique qui permet de s'affranchir de toutes données extérieures pour n'utiliser que des mesures faites dans l'usine: concentration, masse volumique et volume de la solution;
- dans une unité de capacité déterminée en fonction des contraintes suivantes: *fractionnement* du combustible à l'entrée en lots correspondant à une décharge de réacteur, *temps de réponse* des dispositifs de mesure et des laboratoires d'analyse.

Ce concept de transit et d'appareil de mesure aboutit à la recherche d'une optimisation conciliant ces deux principes.

Le transit impose l'existence d'une cuve de passage dite cuve bilan, de volume suffisant pour permettre de diminuer le nombre de mesures pour un même lot de combustible.

L'appareil de mesure que représente cette unité nécessite des accessoires métrologiques indispensables au but poursuivi.

Les bases de choix concernant la conception générale de cette installation doivent aussi satisfaire à des préoccupations de sûreté nucléaire. Les critères de sûreté tiennent compte:

- d'une part des caractéristiques générales du combustible à retraiter: taux d'enrichissement maximal, taux de combustion, temps de refroidissement, quantités traitées, rapport U/Pu,
- d'autre part des caractéristiques particulières des solutions elles-mêmes: concentrations en U et Pu; acidité libre.

En ce qui concerne la filière à eau légère, l'examen de ces caractéristiques permet l'utilisation de cuves de forme conventionnelle et de grand volume. Elles ne présentent aucun risque de criticité à aucun moment.

2. ETABLISSEMENT DE LA METHODE DE MESURE

La méthode volumique employée consiste à déterminer les valeurs des paramètres suivants:

- à partir de mesures de volumes et de masses volumiques, la masse totale de solution présente (M_P) à un moment donné dans la cuve bilan,
- la teneur pondérale en U (C_U) et Pu (C_{Pu}) dans une prise d'échantillon représentative faite au même moment dans la même cuve bilan.

Le produit de ces deux paramètres fournit les masses de U et Pu présentes à un moment donné dans la cuve. Les masses *entrantes* $M_{E(U)}$ et $M_{E(Pu)}$ seront obtenues par différence entre les masses présentes avant (M_R) et après remplissage de la cuve bilan (M_P). On obtient ainsi:

$$M_{E(U)} = M_P \times C_{(U)} - M_{R(U)}$$

$$M_{E(Pu)} = M_P \times C_{(Pu)} - M_{R(Pu)}$$

La mesure précise des volumes dans la cuve bilan est assurée par un ensemble de siphons équipant cette cuve à laquelle est associée une cuve annexe (voir description ci-après et fig.2). La mesure précise des masses volumiques est assurée par la technique dite du bullage dans les cuves bilan et annexe.

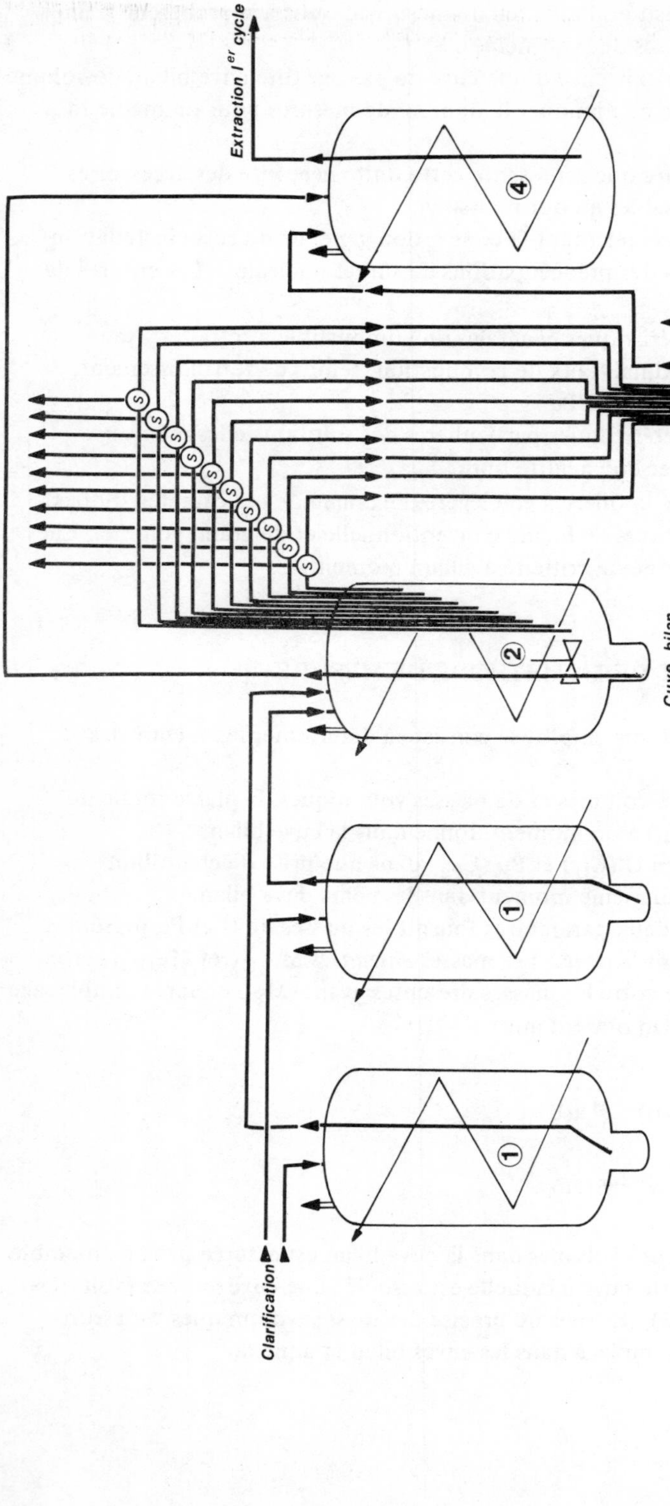

FIG.2. Unité bilan UP3, schéma des cuves.

La mesure des teneurs pondérales en U et Pu dans la prise d'échantillon est assurée, au laboratoire, par dilution isotopique et mesure par spectrométrie de masse.

3. DESCRIPTION DE L'UNITE (fig.2)

3.1. Organisation des cuves

L'analyse séquentielle du mode de remplissage de la cuve bilan en fonction de la cadence de dissolution, de la taille des lots de combustibles et des délais de réponse du laboratoire, montre qu'il est nécessaire de mettre en place le système suivant:

- 2 cuves tampon amont ①,
- 1 cuve bilan ②,
- 1 petite cuve annexe ③,
- 1 système de cuves aval ④ (voir fig.2).

Les volumes des cuves sont déterminés en fonction des contraintes suivantes: la cuve bilan ne peut être vidangée que lorsque l'échantillon a été exploité par le laboratoire; le temps de réponse de l'analyse (24 h) doit être respecté; le nombre de mesures par lot traité doit être optimisé.

Le nombre de cuves amont permet de ne pas bloquer la cadence de dissolution dans les cas particuliers du traitement des queues de lot.

Le nombre de cuves aval doit être fixé en fonction du régime de travail envisagé pour les cycles d'extraction. Les cuves bilan et annexe sont conçues pour remplir la fonction mesure qui leur incombe.

3.2. Mesure des masses de solution (M_P)

3.2.1. Mesure des volumes

La cuve bilan et sa cuve annexe sont reliées entre elles par une série de siphons assurant le transfert de l'une vers l'autre. Les tuyauteries de siphonnage sont étagées régulièrement de manière à laisser dans la cuve un volume égal à 9/10 V, 8/10 V, etc.

La détermination précise des volumes correspondant à chaque hauteur de siphon est obtenue par étalonnage préalable. Le volume excédentaire obtenu par écrêtage de la solution dans la cuve principale est déterminé par la mesure de niveau dans la cuve annexe par la technique dite du bullage. La cuve annexe a été également étalonnée au préalable.

3.2.2. Mesure des masses volumiques

Cette détermination s'effectue par une mesure de pression différentielle entre deux cannes de bullage dont le positionnement dans la cuve est réalisé avec une grande précision. Si les deux cannes A et B sont distantes d'une hauteur H la pression différentielle

$$P_A - P_B = H \times d \times g$$

d étant la masse volumique de la solution.

Une optimisation du positionnement des cannes permet d'obtenir le maximum de précision dans le cas de grands volumes ($> 20\ m^3$) et d'effectuer également la mesure dans des petits volumes (cas des queues de lot). Ce dispositif de mesure est identique pour les cuves bilan et annexe.

On obtient alors la masse de solution par les relations suivantes:

$$M_P = M_i + M_a$$

avec M_i = masse de solution après écrêtage de la cuve bilan
M_a = masse de solution dans la cuve annexe

ou

$$M_P = V_i \times d_i + V_a \times d_a - M_r$$

avec V_i = volume après écrêtage de la cuve bilan
d_i = masse volumique de V_i
V_a = volume présent dans la cuve annexe
d_a = masse volumique de V_a

M_r, masse au fond de la cuve annexe, est connu, soit par étalonnage préalable du volume résiduel V_r et mesure de d_a d'une opération précédente, soit par mesure de la masse d'acide nitrique introduite. V_i, d_i, V_a, d_a sont mesurés par les méthodes décrites ci-dessus.

3.3. Détermination des teneurs pondérales en U et Pu

Afin de pouvoir déterminer les teneurs pondérales en U et Pu nécessaires à l'établissement du bilan la cuve bilan est pourvue d'un dispositif de prise d'échantillon qui permet un prélèvement *représentatif* de la solution. Ce prélèvement est effectué après homogénéisation de la solution et avant siphonnage. Il est ensuite envoyé au laboratoire aux fins d'analyse (voir paragr. 5.2).

4. DISPOSITIFS ANNEXES

4.1. Mesure, homogénéisation et stabilisation de la température

Le produit (volume × masse volumique) n'a de sens que si ces deux grandeurs sont mesurées à la même température et si la température de la solution est homogène dans toute sa masse. Les cuves bilan et annexe sont équipées de serpentins alimentés en eau froide et d'une série de capteurs, judicieusement disposés pour vérifier une bonne homogénéité des températures au sein de la solution.

4.2. Système d'homogénéisation des solutions

Un système d'agitation permet d'assurer l'homogénéisation en un temps minimum déterminé par des essais, en tous points des cuves bilan et annexe, y compris les points bas.

4.3. Dispositif d'introduction de réactifs

Un dispositif est conçu pour l'introduction de l'acide nécessaire à un ajustage éventuel ainsi que l'introduction d'un élément marqueur pour réétalonnage des cuves.

5. DESCRIPTION DES OPERATIONS

5.1. Séquence des opérations

a) La solution clarifiée en provenance des cuves tampon est transférée dans la cuve bilan par volume unitaire, par exemple 20 m^3 (cuvée).

b) L'homogénéisation et la stabilisation de la température sont effectuées pendant le temps défini au cours des essais préalables.

c) L'échantillonnage est effectué dans les conditions prescrites pour obtenir une prise d'échantillon représentative.

d) Le siphon immergé situé juste au-dessous du niveau de la solution à mesurer est amorcé pour transférer dans la cuve annexe le volume excédentaire. Le volume à mesurer V_i est connu par le numéro du siphon amorcé et l'étalonnage de la cuve bilan.

e) La masse volumique d_i de la solution est mesurée. V_a et la masse volumique d_a sont mesurés par bullage dans la cuve annexe.

5.2. Opérations de laboratoire

Le laboratoire dispose d'installations et d'une instrumentation spécialement réservées aux analyses bilan. Ces installations sont constituées d'une ou plusieurs enceintes blindées susceptibles de recevoir, stocker et traiter des échantillons de très haute activité. Elles comprennent notamment:

- un dispositif permettant une prise d'essai dans l'échantillon,
- une balance permettant la mesure de la masse a de la prise d'essai et la mesure de la masse d'une solution diluée de cette prise d'essai (facteur de dilution D),
- un dispositif pour les opérations d'ajout de traceur et de préparation à la mesure des teneurs en U et Pu (séparation des produits de fission),
- un équipement complet de spectrométrie de masse avec traitement des données associé pour la mesure des teneurs en U et Pu ainsi que des compositions isotopiques dans la prise d'essai.

On peut donc considérer que, bien qu'étant souvent géographiquement indépendantes de l'unité bilan décrite, les installations de laboratoire destinées aux analyses bilan font partie intégrante de cette unité.

Les techniques analytiques spécialisées de mise en œuvre et de détermination par la spectrométrie de masse ont été décrites par de nombreux auteurs. Dans les usines, elles font appel à des méthodes normalisées. Retenons surtout qu'elles permettent d'obtenir par dilution isotopique les teneurs pondérales en U et Pu, et par spectrométrie de masse à thermo-ionisation, les teneurs isotopiques de chacun des constituants.

Si $C'_{(U)S}$ et $C'_{(Pu)S}$ sont les teneurs pondérales (en $g \cdot g^{-1}$) de la prise d'essai de la solution diluée nécessaire à l'analyse, les teneurs pondérales dans l'échantillon sont données par les relations:

$$C_{(U)} = C'_{(U)S} \times D$$

et

$$C_{(Pu)} = C'_{(Pu)S} \times D$$

6. EXPRESSION DES RESULTATS D'UNE OPERATION BILAN

Les masses d'uranium et de plutonium entrant, par cuvée, dans la cuve bilan sont données par les formules suivantes:

$$M_{E(U)} = [(V_i \times d_i + V_a \times d_a - M_r) \times C_{(U)}] - M_{R(U)}$$

$$M_{E(Pu)} = [(V_i \times d_i + V_a \times d_a - M_r) \times C_{(Pu)}] - M_{R(Pu)}$$

7. ETALONNAGE PREALABLE DE L'INSTALLATION

7.1. Rôle et but

L'existence de l'installation bilan est liée à la longévité de l'usine elle-même. C'est un instrument de mesure qui doit conserver de grandes qualités de reproductibilité et de justesse, à un niveau si possible inférieur à 1%, ce qui implique:
- une conception et une construction qui prennent en compte ce caractère d'instrument de mesure,
- *un étalonnage rigoureux* avant la mise en exploitation, qui doit être considéré comme une opération précise, délicate et longue, nécessitant des appareils de mesure très précis et un plan rigoureux de mise en oeuvre.

L'objectif des opérations d'étalonnage est double: d'une part établir expérimentalement, pour chaque mesure concourant au résultat final, la relation entre le signal lu et la grandeur mesurée, d'autre part évaluer l'incertitude des résultats pour chacune des mesures.

7.2. Opérations d'étalonnage

L'étalonnage portera au moins:
- dans la cuve bilan, sur les volumes à chaque siphon, le volume résiduel et la mesure de la masse volumique par bullage,
- dans la cuve annexe, sur la détermination des volumes et de la masse volumique par bullage, des volumes résiduel et de réamorçage des siphons,
- sur le dispositif de prise d'échantillon, pour vérifier que les opérations d'échantillonnage n'entraînent pas une variation significative de concentration.

Ces étalonnages seront effectués après que les dispositifs d'homogénéisation de la solution et de stabilisation des températures auront été éprouvés, testés, et que les incidences de variations éventuelles auront été chiffrées.

8. OPERATIONS DE REETALONNAGE PERIODIQUE

En cours d'exploitation, des vérifications périodiques de l'étalonnage sont indispensables pour contrôler la stabilité des caractéristiques des dispositifs de mesure. Elles peuvent se faire, soit en effectuant à nouveau les opérations de l'étalonnage préalable (au cours des arrêts de l'installation), soit en exploitation normale en utilisant la méthode dite massique.

Cette méthode consiste à faire intervenir un paramètre extérieur à l'unité, à savoir un élément marqueur.

Cet élément marqueur, de masse M_E, est introduit dans la solution en cours d'homogénéisation et sert d'élément de base à la détermination des masses totales de U et Pu.

Ces masses sont connues par les relations suivantes:

$$M_{(U)} = \frac{M_E}{m_E} \times m_{(U)}$$

et

$$M_{(Pu)} = \frac{M_E}{m_E} \times m_{(Pu)}$$

dans lesquelles m_E = masse de l'élément marqueur dans la prise d'échantillon $m_{(U)}$ et $m_{(Pu)}$ = masses respectives de U et Pu dans le même échantillon.

9. ANALYSE ET ESTIMATION DES CAUSES D'INCERTITUDE

9.1. Définition de l'estimation

L'incertitude sur le résultat final est évaluée par la composition des incertitudes partielles de chacune des mesures intervenant dans l'établissement de ce résultat.

Un examen approfondi des conditions d'application des méthodes de mesure est nécessaire pour déterminer l'origine d'erreurs systématiques éventuelles. De même, un examen périodique systématique permettra de mettre en évidence une dérive éventuelle de la justesse.

Dans la composition de ces incertitudes, il y a lieu de faire intervenir d'une part les incertitudes inhérentes à *l'étalonnage* préalable et d'autre part les incertitudes inhérentes *aux mesures* effectuées sur chaque cuvée ou opération bilan.

Pour l'établissement de la variance sur le résultat final, les premières (étalonnage) se composent différemment suivant qu'il s'agit d'une cuvée ou d'un lot de plusieurs cuvées. Pour une cuvée, les différentes incertitudes inhérentes aux *étalonnages* sont à considérer comme des variables aléatoires (on ne connaît pas le signe) et leur composition s'effectue par la somme des variances; pour un lot, le signe étant le même pour toutes les cuvées, elles sont à considérer comme des incertitudes à caractère systématique; leur composition s'effectue par la somme des variances, chacune d'elles étant affectée du coefficient K^2 (K: nombre de cuvées constituant le lot):

– pour une cuvée:

$$\sigma'^2_C = \sum \sigma'^2_i$$

– pour un lot:

$$\sigma'^2_L = K^2 \sum \sigma'^2_i$$

Les variances correspondant aux incertitudes inhérentes *aux mesures* se composent par leur somme (incertitude de caractère aléatoire):

– pour une cuvée:

$$\sigma^2_C = \sum \sigma^2_i$$

– pour un lot:

$$\sigma^2_L = K \sum \sigma^2_i$$

L'incertitude globale se ramène ainsi:

– pour une cuvée, à:

$$S^2_C = \sum \sigma'^2_i + \sum \sigma^2_i$$

– pour un lot, à:

$$S^2_L = K^2 \sum \sigma'^2_i + K \sum \sigma^2_i$$

9.2. Analyse des causes d'incertitude

L'examen doit porter sur toutes les causes possibles, notamment sur l'intervalle de confiance à apporter à:

– la mesure des volumes,
– la mesure des masses volumiques,

– la représentativité de l'échantillon (concentration ou dilution possible par le dispositif de prélèvement),
– l'analyse de l'échantillon.

Ceci suppose une bonne maîtrise des dispositifs annexes concourant à l'établissement du bilan: homogénéisation de la solution et stabilité de la température pendant les mesures.

9.3. Approximation des incertitudes

Différents cas de figure ont été estimés en fonction d'expériences antérieures.

Les résultats obtenus par ces estimations donnent une précision relative, au niveau de probabilité de 95%, de ± 0,40% pour une cuvée dans des conditions normales.

En tout état de cause, la précision espérée restera de l'ordre de quelques dixièmes pour cent et inférieure à 1%.

10. CONCLUSION

L'unité bilan est une installation indispensable dans une usine de retraitement des combustibles irradiés. Elle s'insère justement après la dissolution-clarification et avant le procédé chimique proprement dit. Elle doit être considérée comme un appareil de mesure, avec les impératifs de justesse, reproductibilité et fiabilité que cela comporte.

APPLICATION OF AN IMPROVED VOLUME CALIBRATION SYSTEM TO THE CALIBRATION OF ACCOUNTABILITY TANKS

F.E. JONES
National Bureau of Standards,
Washington, DC,
United States of America

Abstract

APPLICATION OF AN IMPROVED VOLUME CALIBRATION SYSTEM TO THE CALIBRATION OF ACCOUNTABILITY TANKS.

This paper describes a very significantly improved system for the volume calibration of nuclear materials accountability tanks. The system involves the transfer of the current technology of liquid volume measurement and differential pressure measurement to the field, enabling an improvement of tank volume calibration by one to two orders of magnitude, and a consequent improvement in process solution volume measurement, leading to significantly improved accountability of nuclear materials for nuclear safeguards purposes. The system has been used in a very successful calibration of an input accountability tank at the Savannah River Plant operated for the US Department of Energy.

INTRODUCTION

Conventionally [1], chemical process tanks have been calibrated by adding accurately weighed increments of water to the tanks and inferring the liquid depth in the tank from differential pressure measurements made using oil manometers or differential pressure gages. Calibration equations or tables relating volume of liquid to differential pressure are then developed for the particular tank.

A review of the procedures used for the gravimetric calibration of volumetric test measures [2], devices used to contain or to deliver known volumes of water, has established the equivalence of gravimetric and volumetric test measure calibration. As a consequence of this established equivalence, the volumetric test measure is an excellent device to use for calibration of process tanks and, due to practical considerations including convenience, is preferred to weigh tanks.

In a laboratory experiment [3] using a specially designed tank, the problem of calibrating a chemical process tank was explored and as a consequence the use of volumetric calibration procedures, as opposed to gravimetric procedures, was specifically recommended. In a later laboratory study [4], the same tank was used to evaluate liquid level instrumentation, three commercially available pressure gages and a sight gage, in tank volume calibration. The results of eight calibration runs indicated that the pressure gages were of approximately equal precision and that they were suitable for use in accountability tank calibration.

The present paper describes the application of volumetric calibration procedures in the very successful calibration of an input accountability tank at the Savannah River Plant near Aiken, South Carolina, operated for the U.S. Department of Energy by E.I. duPont de Nemours & Co., Inc. It is not the intent of the present paper to describe in detail the calibration arrangement and procedures and the treatment of the data, but rather to outline the calibration effort and give an example of the results.

CALIBRATION ARRANGEMENT

The input accountability tank which was calibrated at the Savannah River Plant is illustrated in Fig.1. It is essentially a right circular cylinder with a capacity of approximately 13,600 ℓ (liter) (3,600 gal). The height is approximately 3.4 m (11 ft) and the diameter is approximately 2.4 m (8 ft). The major details are shown in the Fig.1; it should be noted that the cooling coils are constructed of stainless steel tubing of 51 mm (2 in) outside diameter and are located in the interior of the tank.

The volumetric test measures used to introduce known increments of water into the accountability tank are illustrated in Fig.2. The sight gage mounted on the neck of the test measure is used to determine the quantity of of water in the test measure in excess of, or less than, the calibrated volume. Two test measures of this type, one of nominal capacity 378 ℓ (100 gal) and one of nominal capacity

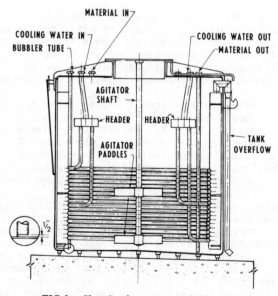

FIG.1. Sketch of an accountability tank.

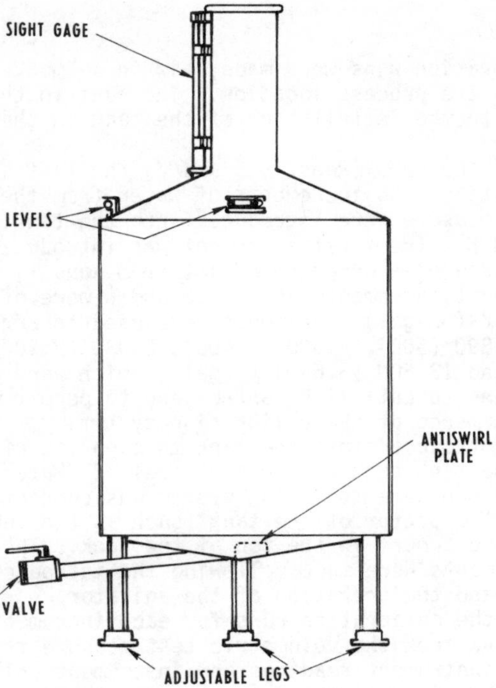

FIG.2. Volumetric test measure.

189 ℓ (50 gal), were used in the tank calibration. In addition, a volumetric standard of nominal capacity 19 ℓ (5 gal) was used to introduce smaller increments.

The gage used to make the differential pressure measurements was a null-operated quartz bourdon-type differential pressure gage. A flow of 8 mℓ s^{-1} (1 ft^3 h^{-1}) of dry nitrogen was maintained through each of the two lines communicating the differential pressure from the tank to the gage.

The temperature of the water in the test measure was measured using either a quartz crystal thermometer or a thermistor. The temperature of the water in the tank was measured using the nickel resistance thermometer to be used in the operation of the tank, and using an array of three thermistors.

The measurements of pressure, temperature and humidity to be used in the calculation of air density were made using an aneroid barometer, either a quartz crystal thermometer or a thermistor, and an electric hygrometer.

CALIBRATION RUNS

Seven calibration runs were made, six in a "mock-up" area of the plant and one in the process location. The runs in the "mock-up" area were made prior to the installation of the tank in the process location in the canyon.

In three of the runs (Runs 1, 5 and 7, the last of these being in the process location), 36 increments of water from the 378-ℓ (100-gal) volumetric test measure were introduced into the tank. In two of the runs (Runs 2 and 6), the first increment was introduced from the 189-ℓ (50-gal) test measure in order to offset from Runs 1, 5 and 7 by this amount. Subsequent increments in Runs 2 and 6 were of 378 ℓ (100 gal) except that 189-ℓ (50-gal) increments were used to arrive at the five check points, 1,890 (500), 5,300 (1,400), 8,330 (2,200) 10,200 (2,700), 12,100 (3,200) and 13,600 (3,600) ℓ (gal), which were used throughout the series of runs to establish control and to permit monitoring of the run-to-run performance of the calibration system.

Water was introduced into the tank through one of the open ports in the top of the tank such as the one labelled "Material In" in Fig.1. The differential pressure measuring system was connected to a "bubbler" tube leading to the bottom of the tank (such as the tube labelled "Bubbler Tube") and to a port in the top of the tank. The differential pressure measurements were made following the introduction of each increment of water and the operation of the agitator.

In each of the calibration runs for each increment of water transferred to the tank from the volumetric test measure the following were determined from instrument readings and instrument calibration data: temperature of the water in the test measure, temperature of the water in the tank, differential pressure, ambient air pressure and ambient air relative humidity. The volume of water introduced in each transfer was determined from the calibration volume of the test measure, the test measure sight gage reading and the temperature of the water in the test measure.

TREATMENT OF THE DATA

The volumetric test measures had been calibrated gravimetrically. The accuracy of a standard test measure as determined by such a calibration is considered to be about 3 parts in 10^5 [4]. This number is the estimate of the relative standard deviation determined from repeated calibration of a test measure.

The calibration of the differential pressure gage was determined by comparison against an air-lubricated dead weight tester in the Pressure Section of NBS. The comparison was made in the pressure range 10,340 Pa (1.5 lbf/in^2) to 39,470 Pa (5.0 lbf/in^2) at 3,947-Pa (0.5-lbf/in^2) intervals and at zero. The calibration data were fitted by linear least squares to provide a linear equation relating gage indication to differential pressure. The estimate of the standard deviation of the calibration points about the linear relationship was 0.48 Pa.

The calibration volume for the volumetric test measures is the volume at 15.56°C, therefore it was necessary to adjust this volume for the expansion of the test measure to arrive at the volume at the temperature of the water in the test measure. The accumulated mass

TABLE I. COMPARISON OF VALUES OF V_{25} FOR RUN 7 WITH VALUES, V_c, CALCULATED FROM CALIBRATION EQUATIONS

Region	V_{25} (Liters)	$V_c - V_{25}$ (Liters)	$\frac{V_c - V_{25}}{V_{25}}$ (%)
I	1515.23	-0.20	-0.013
	1894.01	-0.24	-0.013
	2272.72	+1.12	+0.049
	2651.54	+0.31	+0.012
	3030.39	-0.72	-0.024
	3409.19	+0.44	+0.013
	3788.05	+0.07	+0.002
	4166.93	-0.79	-0.019
III	6819.13	+0.50	+0.007
	7197.98	+0.07	+0.001
	7576.84	0.00	0.000
	7955.65	-0.21	-0.003
	8334.53	-0.62	-0.007
	8713.40	-0.43	-0.005
	9092.41	+0.15	+0.002
	9471.35	+0.46	+0.005
	9850.20	+0.07	+0.0007
IV	10986.76	+0.17	+0.0016
	11365.62	+0.53	+0.0047
	11744.46	-0.40	-0.003
	12123.43	-0.28	-0.0023
	12502.38	-0.01	-0.00008

of water in the accountability tank was calculated by summing the product of the volume and density of each water increment at the temperature of the water increment. The volume which the water in the tank would occupy at the reference temperature, 25°C, was calculated by dividing the accumulated mass by the density of water at the reference temperature.

The differential pressure as measured was also "reduced" to the reference temperature by adjusting for air density, the density of water at the temperature of the water in the tank and for the expansion of the tank and the "bubbler" tube. It is not asserted here that the method used for the reduction of the differential pressure is optimum, there is at present an effort at NBS to develop and test optimum algorithms. However, it is not expected that the choice of the method of reduction will substantially affect the conclusions drawn in this paper.

The calibration runs provided sets of number pairs, (V_{25}, P_{25}), the subscript referring to the reference temperature, 25°C. From these sets, calibration relationships between V_{25} and P_{25} were developed. For this development, the accountability tank was divided into four regions: I) the region occupied by the coils, II) the transition region near and immediately above the top of the coils, III) the region above the transition region but below the headers (Fig.1) and IV) the region above the headers.

In the use of the accountability tank, linear calibration equations relating volume to "depth of immersion" are used. To provide equations in this form, the calibration data, the sets of number pairs, were combined for Runs 1, 2, 5, 6 and 7 and fitted by linear least squares for each of the four regions. The data for Runs 1, 2, 5 and 6 were first "normalized" to Run 7 in an attempt to compensate for pressure drops in the mock-up area pressure installation due to the use of tubing which was smaller in inside diameter than that in the canyon pressure installation. It is clear from physical considerations that a linear equation is not appropriate for the transition region, this region could, consequently, not be used for accountability purposes. It is not asserted here that linear equations are the best representation of the data, the data are currently being analyzed at NBS to ascertain the optimum representation of the data. In the present paper, the linear equations are used to give an indication of the precision attainable in the field with the calibration procedure using volumetric test measures and a state-of-the-art differential pressure gage.

The data for Runs 3 and 4 were not used in the development of the calibration equations. Run 3 was terminated when a major discrepancy was discovered at the 1890-ℓ (500-gal) check point. The probable cause of the discrepany is operator error in the dispensing of one of the water increments. In Run 4 there was a discrepancy in volume of water in the tank which increased uniformly above about the 5300-ℓ (1400-gal) check point. The probable cause of this discrepancy is the flow into the tank of condensate from steam passing through leaking steam valves. Remedial action directed at the probable causes of the discrepancies was taken and subsequent runs (Runs 5, 6 and 7) were apparently free of both discrepancies. Also, Run 4 was apparently free of the discrepancy of Run 3.

The fit of the individual calibration points to the linear calibration equations is illustrated in Table 1, in which the calibration points for Run 7 are compared with the values of volume calculated using the equations developed using the data for Runs 1, 2, 5, 6 and 7 in regions I, III and IV. It is to be emphasized that the comparison is illustrative rather than definitive. The maximum deviation is seen to be +1.12 liters, corresponding to a relative deviation of +0.049%. Based on the results of a calibration in 1971 of a similar tank using a weigh tank and an oil manometer, the results in Table 1 represent an improvement in tank volume calibration by between 1 and 2 orders of magnitude.

CONCLUSIONS

The calibration of an accountability tank using volumetric test measures and a null-operated quartz bourdon-type differential pressure gage has been briefly described. The use of these devices has resulted in an improvement in calibration by between 1 and 2 orders of magnitude for the particular type of tank. This result represents an improvement in process solution volume measurement leading to significantly improved accountability of nuclear materials for Nuclear Safeguards purposes.

ACKNOWLEDGEMENTS

It is a pleasure to acknowledge the contributions of J.F. Houser and R.M. Schoonover, who participated in the calibration effort; of C.L. Carroll, who did many of the computations; of various members of the DOE Security and Safeguards Division at the Savannah River Plant; and of various members of several duPont groups at the Savannah River Plant.

REFERENCES

[1] RODDEN, C.J., Selected Measurement Methods for Plutonium and Uranium in the Nuclear Fuel Cycle, Office of Measurement Services, U.S. Atomic Energy Commission (1972) 61-65.

[2] SCHOONOVER, R.M., The Equivalence of Gravimetric and Volumetric Test Measure Calibration, NBSIR 74-454, National Bureau of Standards, Washington, D.C. (1974).

[3] SCHOONOVER, R.M., HOUSER, J.F., Pressure Type Liquid Level Gages, NBS Report 10396, National Bureau of Standards, Washington, D.C. (1970).

[4] SCHOONOVER, R.M., KU, H.H., WHETSTONE, J.R., HOUSER, J.F., Liquid Level Instrumentation in Volume Calibration, NBSIR 75-900, National Bureau of Standards, Washington, D.C., (1975).

VERIFICATION OF REPROCESSING PLANT INPUT AND OUTPUT ANALYSES

Practical experience in the reprocessing of light-water reactor fuels with burnups of up to 40 GW · d/t

R. BERG
Gesellschaft zur Wiederaufarbeitung
von Kernbrennstoffen mbH,
Leopoldshafen,
Federal Republic of Germany

Abstract

VERIFICATION OF REPROCESSING PLANT INPUT AND OUTPUT ANALYSES: PRACTICAL EXPERIENCE IN THE REPROCESSING OF LIGHT-WATER REACTOR FUELS WITH BURNUPS OF UP TO 40 GW · d/t.

Verification procedures for reprocessing plant input and output analyses are routinely used at the WAK reprocessing plant near Karlsruhe. Input analyses are verified by isotope correlation techniques using the measured heavy isotopes of uranium and plutonium and the radioactive fission products; and correlating these with the measured Pu/U ratios. Highly accurate and precise density measurements used to establish the homogeneity of input and output solutions, and to correlate with the heavy metal concentration and the acidity of the solutions. Verification results from recent reprocessing campaigns are presented.

1. INTRODUCTION

The precision and accuracy of the analysis of uranium, plutonium and their isotopic composition in reprocessing plant solutions are influenced by remote handling problems; complex analytical procedures, sample instability and cross-contamination problems. Due to the importance of the input analyses where the highest possible accuracy and precision are wanted for reasons of safety, Safeguards and accountancy, means for rapid and independent verification of measurements have been sought, established and implemented.

In the early stages of development; verification procedures using the existing mass spectrometric measurement data were needed by the reprocessing analyst, mainly as a tool to identify outliers in a series of measurements. Today, established correlations are considered to be important tools for Safeguards and extensive investigations, experimental as well as theoretical, are underway [1, 2].

The verification of heavy metal concentration by means of highly accurate and precise density determinations; routinely performed in all accountability solutions to establish homogeneity, has for many years been used by the author on pure uranyl nitrate solutions. The success experienced with uranium led to investigations on input- and plutonium nitrate final product solutions.

2. INPUT ANALYSIS VERIFICATION

2.1 Isotope correlation techniques

Of the large number of possible correlations existing, we have found the plot of Pu/U versus U-235 particularly useful.

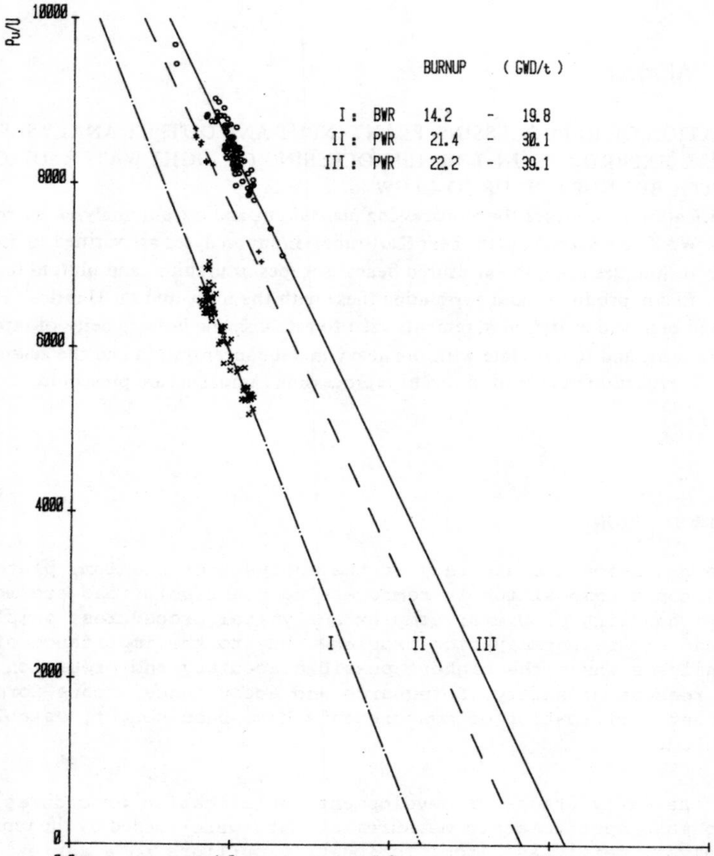

FIG.1. Pu/U ratio versus final ^{235}U content.

TABLE I. VALUES OF $\frac{Pu/U}{D_5}$

Reprocessing-campaigns	Initial enrichment wt% U-235	Reactor-type	$\frac{Pu/U}{D_5}$	rel.Std. dev.	Burn up range $MW \cdot d/t \times 10^{-3}$	number of Input-charges
A	2,2	BWR	4883	2,2%	14.2. - 19.8.	72
B	2,48	PWR	5006	1,6%	23.9. - 24.5.	12
B	2,49	PWR	4841	2,3%	24.2. - 24.5.	4
C	2,50	PWR	4836	1,93%	24.2.	2
B	2,51	PWR	4907	0,4%	24.1. - 24.5.	8
B	2,79	PWR	4551	1,8%	22.6. - 26.2.	26
B	2,80	PWR	4509	1,4%	22.9. - 26.3.	16
B	2,81	PWR	4459	2,2%	22.7. - 25.5.	6
C	2,83	PWR	4172	1,3%	21.4. - 3o.2.	1o
C	3,00	PWR	3966	0,68%	22.2. - 39.1.	6
C	3,10	PWR	3983	1,8%	22.2. - 39.1.	77

Figure 1 presents data from recent campaigns for three initial enrichments and two different reactor types. For reasons of clarity other initial enrichments were omitted. Table I gives a complete set of recent data, here the measured ratio $\frac{Pu/U}{D_5}$[1] is listed.

As can be seen from both figure 1 and table I, a linear relationship exists between Pu/U and the final enrichment for elements with the same initial enrichment coming from the same core. The linearity is maintained over a surprisingly large burn-up range (14 to about 30 GW·d/t).

The slight indications of nonlinearity observed at 39 GW·d/t (Curve III) is not unexpected, it could stem from the fact that plutonium is burnt at a rate increasing with the burn-up.

Figure 2 shows the correlation between the U-235 and the Pu-239 content in the spent fuel. Here linearity is maintained over the whole burn-up range (20 to 39 GW·d/t).

Outliers in correlations can be caused by faulty analysis, the use of incorrect initial enrichment for the element (bookkeeping error) or substitution of elements prior to dissolution. Indeed, we have in the past been able to detect and correct occasional mistakes of all the above types using correlation techniques.

We start an extensive check of results when replicate analyses give inconsistent results or when agreeing replicates give outliers in established correlations.

Additional informations we use in check procedures are Shipper's information about burn up, pre- and postirradiation uranium data, plutonium concentration and isotopic content.

---

[1] D_5, called depletion, is the difference between the initial and final ^{235}U content.

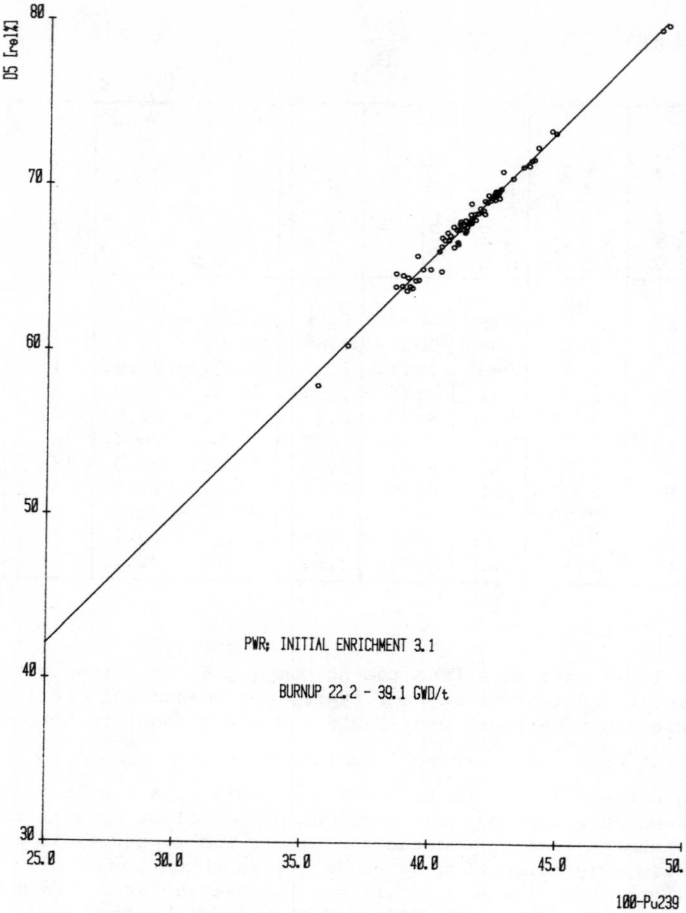

FIG.2. Relative depletion versus ^{239}Pu content.

As one PWR input charge at WAK consists of exactly one half element, comparisons between element halves have been possible. Here we have in many cases detected significant burn-up differences between the halves; as evidenced by different Pu/U-ratios; different final enrichments, different plutonium isotopic ratios and different amounts of radioactive fission products. In most of the cases where we have detected burn-up differences between two element halves the correlations are maintained.

Out of 23 dissolved elements 3 elements produced possible outliers in some of the correlations we use in the verification procedures. Since it is conceivable that extreme differences in irradiation history could cause outliers we plan in coming PWR compaigns to study selected fuel elements in close collaboration with the reactor operator.

TABLE II. REPLICATE INPUT ANALYSES IN C-CAMPAIGN

No. of input	Element analysed (concentration)	Difference between replicates		
		Average	Minimum	Maximum
40	Pu	0,65%	0,00%	1,41%
40	U	0,59%	0,06%	1,63%

TABLE III. COMPARISON BETWEEN CALCULATED (Eq.(1)) AND MEASURED URANIUM CONTENT IN REPROCESSING PLANT INPUT SOLUTIONS

Number of charges:	40
Burn-up range:	21.4. - 39.0 MW·d/kg
Acidity range:	1,71 - 4,38 M
Density range:	1,3937 - 1,4998 g/ml
Uranium concentration range:	157,6 - 203,6 g/kg
Mean difference between calculated and measured uranium concentration:	- 0,06 g/kg
Standard deviation of difference:	1,9 g/kg
Maximum differences observed:	- 4,0 and + 3,18 g/kg
Number of outliers in data set:	2

TABLE IV. COMPARISON BETWEEN CALCULATED (Eq.(1)) AND MEASURED URANIUM CONTENT IN PURE URANYL NITRATE SOLUTIONS

Number of charges:	34
Acidity range:	0,07 - 0,21 M
Density range:	1,6063 - 1,6596 g/ml
Uranium concentration range:	277,3 - 298,0 g/kg
Mean difference between calculated and measured uranium content:	0,13 g/kg
Standard deviation of difference:	0,34 g/kg
Maximum differences observed:	- 0,35 - + 0,81 g/kg
Number of outliers in data set:	0

TABLE V.: COMPARISON BETWEEN CALCULATED (Eq.(2)) AND MEASURED PLUTONIUM CONTENT IN PLUTONIUM NITRATE SOLUTIONS

Number of charges	16	
Acidity range:	3,87 - 5,6 M	
Density range:	1.3233 - 1.4998	g/ml
Plutonium concentration range:	94 - 154	g/kg
Mean difference between calculated and measured plutonium concentration:	o,72	g/kg
Standard deviation of mean:	2,1	g/kg
Maximum difference observed:	- 3,7 and 5,4	g/kg
Number of outliers in data set:	3	

The C-campaign in Table I is an example of input analytical performance under near optimum conditions. Here instrumentation and highly qualified personnel with long experience were continuously available and permitted almost immediate analysis. Table II summarizes the results from replicate concentration determinations on input solutions.

It is important to realize that biased analytical determinations can also produce correlating results. Therefore the extreeme care we apply today when calibrating the input analysis system will have to be maintained.

The isotopic correlation technique cannot yet replace the complex and difficult input analysis, has however, turned out to be an important quality control tool for the reprocessor and the Safeguards authorities.

2.2 Verification of uranium content by density-acidity measurements

At WAK highly precise and accurate density determinations are performed for homogeneity control reasons immediately after sampling the input solution. The density is determined by measuring the resonance frequency of a vibrating capillary, filled with the solution and held at a constant temperature of 20.0°C. The method is capable of delivering density results free of bias with a relative standard deviation of 0,04% for a single measurement. The equipment can be easily installed, operated and maintained under remote conditions.

Eq. (1) wich is a modified version of Moeken's formula *[3]*, can be used to calculate the uranium content in an input solution when density, acidity and burn-up is known:

$$U = 749{,}2 - 0{,}31 \cdot BU - \frac{1}{D_{20}} \cdot (752 + 23{,}28 \cdot (H^{+})) \quad (1)$$

U is the uranium concentration in g/kg,
BU is Shipper's burn-up in MW·d/kg,
D_{20} is the density measured at 20.0°C and
(H^{+}) is the acidity (HNO_3) in Mol/l.

The burn-up correction term was determined empirically using the mass spectrometric uranium results from 160 input charges. Table III summarizes results from a recent campaign.

Applying Eq. (1) to input solutions, shipper-receiver comparisons on uranium can be performed shortly after sampling; mass spectrometric results can be checked for outliers and the more tedious and less precise uranium determinations required for process control (colorimetry or potentiometry) need not to be performed.

3. OUTPUT ANALYSIS VERIFICATION

3.1 Uranyl nitrate solutions

For pure uranyl nitrate solutions we have for years used Eq. (1) with BU = 0 for the verification of the gravimetrically determined uranium content. Table IV summarizes the performance characteristics of the method.

3.2 Plutonium nitrate solutions

Since the density-uranium-acid correlation turned out to be extremely useful, similar investigations were undertaken for plutonium nitrate solutions.

Our final product has a Pu-concentration of 100 - 150 g Pu/kg and an acidity between 4 and 6 M.

We have for these concentration ranges empirically established the following equation

$$Pu = 425.17 \cdot \left[D_{20} - 1.0902 - 0.00414 \cdot ((H^+) - 4) \cdot 10 \right] \qquad (2)$$

where

Pu is the plutonium concentration in g/kg,

D_{20} is the density at 20.0°C in g/ml and

(H^+) is the acidity of the solution in mol/l, determined by potentiometric titration after precipitating the plutonium as iodate.

Table V summarizes the results from the latest plutonium nitrate product charges. The verification procedure for plutonium has recently been developed. Further investigations to improve the method are underway; in particular we plan to study and optimise the acidity determination; up to now only of interest for process reasons. The valency state distribution of plutonium might influence the density determination, we plan to study this effect in coming campaigns.

REFERENCES

[1] CHRISTENSEN, D.E., SCHNEIDER, R.A., "Summary of experience with heavy-element isotopic correlations", Safeguarding Nuclear Materials (Proc. Symp. Vienna, 1975) 2, IAEA, Vienna (1976) 377.

[2] BERG, R., FOGGI, C., KOCH, L., KRAEMER, R., WOODMAN, F.J., "Value and use of isotopic correlations in irradiated fuels", Proc. Symp. Practical Aspects of R and D in the Field of Safeguards, ESARDA,Rome (1974).

[3] MOEKEN, H.H., Short communication, Anal. Chim. Acta. **44** (1969) 225–28.

DISCUSSION

E. VAN DER STIJL: The formula for the uranium content includes a burnup value. Do you actually measure the burnup, or do you take the value given by the reactor operator?

R. BERG: For the time being no measurements are performed in connection with the application of this formula. This could be done later however. We use the shipper's burnup values, which are quite good for this purpose.

H.T. YOLKEN: Your work is to be commended. It clearly illustrates that careful measurements plus good quality control can be of benefit in regard to plant operation, commercial exchange of fuel, and nuclear safeguards. The paper further illustrates that high-quality data can be used to develop very useful correlations with other parameters for nuclear safeguards use.

S. SANATANI: What are the main headaches of a plant accountant? From my point of view the delay in getting final results appears to be a bottleneck.

R. BERG: That is true. However, the methods I have described in my paper can provide faster results for accountancy and safeguards. The new high-capacity mass-spectrometers will enable us to deliver input results within days.

S. SANATANI: Does the continuous presence of inspectors cause excessive interference to plant operators in your opinion?

R. BERG: I believe that inspector interference is acceptable and we hope it will stay like that in the future.

EXPERIENCE ACQUISE A LA HAGUE SUR LE CONTROLE DES COQUES

J. CONSTANT*, D. HEBERT**, J.P. MACAREZ+,
G. MALET*, J. REGNIER+

*CEA, Centre d'études nucléaires de Saclay,
Gif-sur-Yvette

**CEA, Centre d'études nucléaires de Cadarache,
Saint-Paul-lez-Durance

+ Compagnie générale des matières nucléaires,
La Hague,
France

Abstract–Résumé

EXPERIENCE GAINED AT LA HAGUE IN THE INSPECTION OF HULLS.

Measurement of the amount of residual fissile material in hulls after dissolution is an important problem in safeguarding reprocessing plants. Three methods have been tried out at La Hague, two of them at the AT 1 plant, a prototype for reprocessing fast neutron reactor fuel, and the third at the HAO head-end of the UP 2 plant for the processing of water-reactor oxide fuel. Experience gained in the use of these three techniques now makes it possible to review the situation and draw conclusions regarding their possible use in future facilities. The two methods used at AT 1 are cyclic neutron irradiation with counting of the delayed neutrons, and measurement of the ratio between the gamma activity of ^{144}Pr and that of ^{60}Co or ^{54}Mn before and after dissolution. The accuracies attained by means of these two methods are comparable – around 20% for roughly 1% of residual fuel – and intercomparisons have been made by applying the two procedures to the same series of containers. The technique applied at the HAO plant, which is the new oxide head-end of the UP 2 plant, is also based on measurement of the gamma activity of ^{144}Pr, but in this case using a single measurement following dissolution. The results presented relate to the first two campaigns, which involved BWR fuel from the Mühleberg reactor and PWR fuel from the Stade reactor. The conclusions drawn from t the trial application of these three techniques are being put to use in connection with future projects.

EXPERIENCE ACQUISE A LA HAGUE SUR LE CONTROLE DES COQUES.

La mesure de la quantité de matière fissile résiduelle dans les coques après dissolution est un problème important dans le contrôle des usines de retraitement. A La Hague, trois méthodes ont été expérimentées, deux d'entre elles sur l'atelier AT 1, prototype pour le retraitement des combustibles de réacteurs à neutrons rapides, la troisième sur la tête HAO de l'usine UP 2 destinée au traitement des combustibles oxydes de réacteurs à eau. L'expérience acquise dans l'utilisation de ces trois méthodes permet d'effectuer actuellement un premier bilan et d'en tirer des enseignements pour les possibilités d'application à de futures installations. Les deux méthodes appliquées à AT 1 sont l'irradiation cyclique aux neutrons avec comptage des neutrons différés et les mesures de rapport de l'activité gamma du ^{144}Pr à celle du ^{60}Co ou du

^{54}Mn avant et après dissolution. Les précisions atteintes à l'aide de ces deux méthodes sont comparables, de l'ordre de 20% pour environ 1% de combustible résiduel, et des comparaisons ont été effectuées en appliquant les deux procédés à une même série de conteneurs. La technique appliquée à l'atelier HAO, nouvelle tête oxyde de l'usine UP 2, est également fondée sur la mesure de l'activité gamma du ^{144}Pr, mais cette fois à l'aide d'une seule mesure après dissolution. Les résultats présentés concernent les deux premières campagnes, qui ont porté sur des combustibles du réacteur à eau bouillante de Mühleberg et des combustibles du réacteur à eau sous pression de Stade. Les conclusions de cette expérience d'utilisation des trois techniques sont mises à profit pour la préparation des projets futurs.

1. INTRODUCTION

Le contrôle des résidus éventuels de matières fissiles dans les coques, ces morceaux de gaines rejetés après dissolution du combustible, est important pour l'exploitant (fonctionnement correct du procédé et bilan de l'usine), pour la sûreté (criticité des stockages et contenu des déchets) et pour l'application des garanties (connaissance des flux de matières fissiles sortant du procédé). Techniquement plusieurs méthodes sont applicables et nous rappelons brièvement leurs principales caractéristiques, ainsi que les contraintes ou limitations qu'elles comportent. Pratiquement les problèmes de réalisation industrielle conduisent à rechercher des méthodes simples et efficaces. Dans le cas de l'usine de La Hague, deux méthodes, généralement considérées comme les plus intéressantes, l'interrogation neutronique active et le comptage gamma passif ont été appliquées simultanément dans l'atelier prototype AT1 utilisé pour le retraitement de plusieurs centaines de kg de combustible de réacteurs à neutrons rapides.

L'expérience correspondante, dont les conditions et des exemples de résultats sont présentés ici, permet de conclure sur la validité de ces deux méthodes.

Pour la tête oxyde de l'usine HAO, dont la capacité est prévue pour 400 t/an de combustibles de réacteurs à eau, le comptage gamma passif a été seul retenu. Les deux campagnes déjà effectuées, en 1976 sur 15 tonnes de combustible BWR de MUHLEBERG et en 1978 sur 54 tonnes de combustible PWR de STADE permettent d'effectuer un premier bilan.

2. METHODES APPLICABLES POUR LE CONTROLE DES COQUES

Avant d'examiner les différentes méthodes possibles, il est nécessaire de bien définir l'objet à contrôler et de résumer les hypothèses qui permettraient d'expliquer une présence plus ou moins importante de résidus de combustibles dans ces coques.

2.1 <u>Définition des coques</u>

Les combustibles nucléaires que nous considérons ici sont à base d'oxyde d'uranium ou d'oxyde mixte uranium-plutonium. Le gainage utilisé est soit de l'acier inoxydable, soit du zircaloy. La dissolution industrielle de cet oxyde se fait avec de l'acide nitrique dans des cuves en acier inoxydable. Il est donc exclu de dissoudre le matériau de gainage qui doit être tronçonné mécaniquement.

TABLEAU I. ACTIVITE DES COQUES (1, 2)
calculée pour le contenu d'un panier HAO et des assemblages refroidis 600 jours

ASSEMBLAGE	ACTIVITES EN CURIES		
	^{95}Zr	^{54}Mn	^{60}Co
BWR (Type MUHLEBERG)	30	10	200
PWR (Type FESSENHEIM)	30	35	410
PWR (Type STADE)	30	280	1420
Rapide (Type PHENIX)	-	1430	310

Théoriquement les coques sont ces morceaux de gaine supposés alors lavés de toute contamination par les produits de fission ou tout résidu de combustible.

Dans la réalité la situation est plus complexe, d'une part puisqu'il n'est pas exclu que de faibles quantités de combustible accompagnent ces morceaux de gaine, ce qui justifie le contrôle, et d'autre part du fait des conditions pratiques de réalisation du cisaillage mécanique qui conduisent dans certains cas à conserver avec les gaines d'autres éléments de structure des assemblages. Il en résulte alors un surcroît d'activité qui complique certaines mesures. Pour les exemples étudiés ici, la situation est la suivante :

Dans le cas de l'atelier AT1 le contenu du panier est constitué uniquement de morceaux d'acier inoxydable provenant des aiguilles. On a donc du matériau de gainage pur.

Dans le cas du HAO, on retrouve dans les paniers la totalité des matériaux de structure d'un assemblage, sauf les deux embouts. Il y a donc le zircaloy des tubes de gainage, mais aussi les grilles en inconel, et chez certains constructeurs, de l'acier inoxydable provenant des tirants de maintien.

Pour chiffrer les conséquences de ceci on a calculé dans différents cas les activités gamma pour un panier de coques semblable à ceux qui sont utilisés à l'atelier HAO. (1, 2). Les données de ce calcul et les résultats sont présentés dans le tableau I. On voit que dans tous les cas l'essentiel de l'activité provient du ^{60}Co et que cette activité, très importante, varie notablement d'un cas à l'autre.

De plus, comme le cobalt apparaît en tant qu'impureté dans la fabrication des matériaux de structure, il faut s'attendre à de très fortes variations de l'activité du ^{60}Co d'un réacteur à un autre pour une même composition théorique.

Enfin il faut noter que le problème des embouts d'assemblage qui correspondent à des masses de métaux beaucoup plus importantes n'est pas abordé ici et que sa solution n'est pas nécessairement obtenue par les mêmes procédés que le contrôle des coques.

2.2 Hypothèses pour expliquer la présence de combustible dans les coques

Après dissolution du combustible, les coques sont rincées par une solution acide, puis par de l'eau.

On retient deux causes principales d'entraînement éventuel de combustible par les coques :

- effet "berlingot" : par suite d'un défaut de cisaillage, le morceau de gaine est fermé à ses deux extrémités, de sorte qu'il ne peut y avoir contact suffisant entre le combustible et l'acide

- dissolution incomplète : les conditions de dissolution totale du combustible n'ont pas été réunies et il reste du combustible dans des coques parfaitement ouvertes.

On peut également évoquer d'autres causes qui nous semblent plus secondaires par rapport à ce que l'on cherche, soit :

- un dépôt de la solution sur les parois des coques
- un entraînement par les coques de nodules métalliques insolubles, de composition chimique inconnue, sous forme de très fines particules, cet effet étant particulièrement valable pour le cas de très fortes irradiations.

2.3 Méthodes de mesure

Parmi les méthodes qui ont été envisagées pour le contrôle du contenu des coques en matière fissile résiduelle, il y a lieu de distinguer deux catégories, les méthodes "actives" ou les méthodes "passives".

Les premières consistent à provoquer des fissions dans le combustible et à compter des évènements accompagnant la décroissance des produits de fission de courte période. Ce sont théoriquement les mieux adaptées au contrôle recherché, puisque les phénomènes observés sont directement liés à la présence de matière fissile et que l'interprétation du résultat ne nécessite pas d'hypothèse importante sinon sur la composition approchée du combustible. En pratique, nous n'avons considéré que les méthodes d'interrogation neutronique, c'est-à-dire que les fissions sont provoquées par une irradiation en neutrons.

Compte tenu de l'activité gamma importante des coques, la méthode de détection la plus appropriée consiste à compter les neutrons différés résultant des fissions en réalisant des cycles irradiation-comptage avec une période inférieure à quelques minutes (exemple de AT 1).

Très séduisante sur le plan théorique, cette méthode comporte un inconvénient principal lié à la propagation des neutrons. En effet, la grande sensibilité à l'hydrogène nécessite de se trouver clairement dans une des situations totalement à sec ou totalement en eau, et si c'est cette dernière hypothèse qui est retenue, les faibles libres parcours des neutrons dans l'eau entraînent une limitation sur le diamètre des conteneurs examinés, avec un maximum qui ne peut excéder 25 ou 30 cm. Par ailleurs, cette méthode nécessite un appareillage dont la mise en oeuvre est plus complexe que pour les méthodes passives.

Pour ces dernières, le contrôle est fondé sur un comptage direct de l'activité gamma ou neutrons associée au combustible ; on fait alors l'hypothèse que les éventuels résidus de combustible ont conservé une composition moyenne identique à celle du combustible irradié qui était contenu précédemment dans les gaines. Si la technique de détection utilisée est un comptage de neutrons, les curium sont les principaux émetteurs et on retrouve la limitation indiquée précédemment pour le diamètre des conteneurs. Les comptages de gammas offrent plus de possibilités. Dans tous les cas il s'agit de rayonnements accompagnant la décroissance de produits de fission et le choix se porte généralement sur le ^{144}Ce de période intéressante, 285 jours,et dont l'émission du descendant, le ^{144}Pr, à 2,18 MeV, bien que de faible rendement, se détache bien dans le spectre gamma du combustible et permet de travailler sur des conteneurs de relativement grand diamètre (maximum de 50 à 60 cm).

Le résultat de ces comptages neutron ou gamma peut être interprété en pseudo absolu ou en relatif. Nous appellerons ici, par convention, pseudo absolue une méthode ne faisant intervenir qu'une seule mesure par opposition à une méthode relative qui fait intervenir le rapport de deux mesures où la deuxième est alors "relative" à la première. Dans tous les cas il est nécessaire de procéder à un étalonnage complet de l'installation avec une source connue. Dans le cas des mesures en pseudo absolu, il est nécessaire en plus de disposer, pour tout combustible traité, de calculs corrects des activités spécifiques pour les émetteurs considérés. Compte tenu des très fortes variations des concentrations en curium avec le taux de combustion, le temps de refroidissement et même certaines conditions d'irradiation, l'interprétation directe du comptage neutron passif est pratiquement exclue, sauf à disposer de résultats complets d'analyse sur chaque combustible avant le contrôle des coques correspondantes. L'utilisation directe des comptages gamma du ^{144}Pr est beaucoup plus simple, le seul paramètre important dans ce cas étant le temps de refroidissement, et des abaques sont suffisantes pour l'exploitation des résultats.

L'application des méthodes en relatif supprime cette nécessité de disposer d'éléments calculés pour l'interprétation du résultat, par contre, elle impose une mesure similaire sur le combustible avant dissolution et un mode d'exploitation qui permet de relier clairement les gaines contrôlées avec les combustibles.

En résumé, la méthode qui sur le plan théorique apporte le plus de garanties est l'interrogation neutronique active. Toutefois, les contraintes liées à sa mise en oeuvre peuvent conduire sur le plan industriel à retenir des méthodes passives plus simples et plus efficaces, en particulier le comptage gamma associé au ^{144}Pr. Nous allons voir dans ce qui suit que la comparaison de ces méthodes ne fait pas apparaître de différence notable, même dans le cas de combustible de réacteurs à neutrons rapides.

3. CONTROLE DES COQUES A L'ATELIER AT 1

Le Centre de La Hague dispose, en plus de l'ensemble UP 2 construit pour le retraitement des combustibles UNGG,[1] d'un atelier prototype, AT 1,

---

[1] UNGG: filière uranium naturel-graphite-gaz.

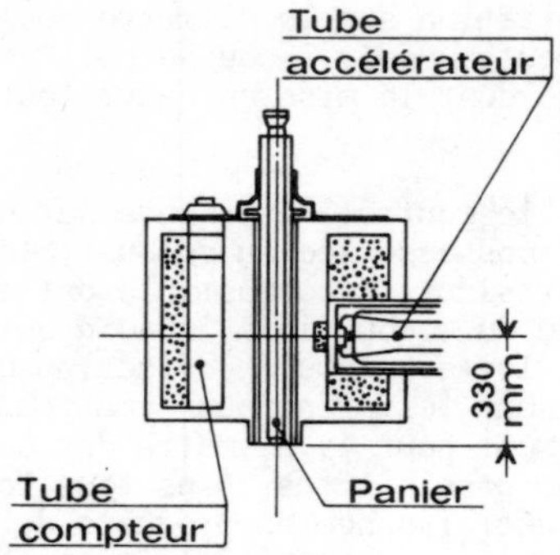

FIG.1. Bloc ralentisseur pour le comptage des neutrons retardés.

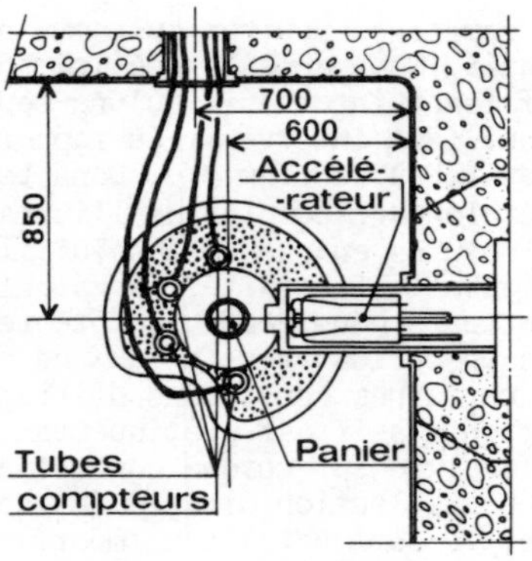

FIG.2. Implantation du bloc ralentisseur dans la cellule (dimensions en mm).

pour le retraitement par voie aqueuse de 200 kg/an de combustible de réacteurs à neutrons rapides (gainage en acier inoxydable).

Cet atelier a été principalement utilisé pour retraiter les combustibles du réacteur expérimental RAPSODIE, mais a également retraité du combustible PHENIX.

Deux techniques de contrôle des coques y sont appliquées, simultanément ou non suivant les campagnes.

3.1 Contrôle par interrogation neutronique (3, 4, 5)

Le principe de cette mesure est bien connu : on irradie cycliquement la matière fissile par une source de neutrons.

On compte l'émission de neutrons différés entre deux cycles. Lorsque le régime permanent est atteint, le comptage des neutrons retardés est stable et représentatif de la matière fissile présente, à condition que l'on connaisse sa composition isotopique.

TABLEAU II. COMPARAISON DES RESULTATS OBTENUS PAR INTERROGATION NEUTRONIQUE ET PAR SPECTROMETRIE GAMMA

Campagne	% de matière contenu dans les coques par rapport à la masse de combustible initial (± 1 écart type)			
	Neutrons	^{60}Co	^{54}Mn	Moyenne générale
RAPSODIE 74 A	1.244 ± 0.12	0.675 ± 0.21	0.657 ± 0.20	
	1.172 ± 0.12	0.770 ± 0.33	0.752 ± 0.24	
	1.422 ± 0.14	1.298 ± 0.28	1.138 ± 0.21	
	0.943 ± 0.12	0.890 ± 0.30	1.257 ± 0.21	
	0.581 ± 0.09	1.091 ± 0.70	0.915 ± 0.62	
	1.274 ± 0.14	1.501 ± 0.65	1.640 ± 0.52	
	0.951 ± 0.11	0.669 ± 0.40	0.616 ± 0.28	
	1.528 ± 0.18	1.265 ± 0.90	1.564 ± 0.61	
Moyenne	1.139	1.020	1.067	1.076
écart type	0.304	0.315	0.398	
RAPSODIE 78	0.843 ± 0.08		0.671 ± 0.19	
	1.911 ± 0.33		2.276 ± 0.69	
	0.886 ± 0.07		–	
	0.894 ± 0.16		1.138 ± 0.33	
	0.854 ± 0.12		–	
	–		1.382 ± 0.24	
	1.016 ± 0.20		0.976 ± 0.20	
	0.813 ± 0.20		1.098 ± 0.24	
Moyenne	1.031		1.257	1.144
écart type	0.393		0.551	
PHENIX 77 A	0.522 ± 0.37		0.522 ± 0.16	
	0.626 ± 0.37		0.208 ± 0.05	
	0.365 ± 0.18		0.444 ± 0.13	
	0.391 ± 0.18		0.418 ± 0.13	
	0.261 ± 0.31		1.357 ± 0.37	
	0.626 ± 0.37		0.939 ± 0.26	
	0.678 ± 0.37		0.418 ± 0.47	
	0.522 ± 0.37		1.148 ± 0.37	
	0.522 ± 0.37		1.044 ± 0.31	
	0.313 ± 0.18		0.391 ± 0.10	
	0.391 ± 0.18		0.313 ± 0.13	
Moyenne	0.474		0.655	0.565
écart type	0.138		0.391	

Ecarts relatifs exprimés en % par rapport à la valeur moyenne de la quantité de matière mesurée

Campagne	Comparaison	Ecart moyen $\overline{D}$	Ecart type sur D
RAPSODIE 74 A	Neutrons Co	11.1	32.4
	Neutrons Mn	6.7	35.4
	Co - Mn	4.6	18.7
RAPSODIE 78	Neutrons Mn	11.9	20.1

3.1.1 Description de l'appareillage

Source de neutrons : accélérateur
(10^{10} n s^{-1} en 4 π à 14 MeV)

Détecteurs : 4 compteurs à dépôt de bore

Ensemble ralentisseur (Fig. 1 et 2) : un gainage en plomb du bloc de polyéthylène assure l'atténuation des flux gamma issus du panier d'une part, et de la cellule d'autre part.

Electronique : elle comporte :

. 3 échelles de comptage
. 1 sous-ensemble horloge + programme
. 1 châssis de commande du générateur de neutrons
. 1 châssis de commande d'impression des résultats.

La résolution de la chaîne de comptage est de 50 n s.

3.1.2 Réalisation du contrôle

Le cycle de mesure se décompose de la manière suivante :

TI = temps d'irradiation : 180 s
TA = retard à l'analyse : 2 s
TC = temps de comptage des neutrons retardés : 180 s
TF = temps de comptage du bruit de fond : 180 s

Le panier contenant les coques a un diamètre assez faible (environ 100 mm), mais une grande hauteur (950 mm de remplissage maximum). Il est donc nécessaire de faire plusieurs mesures en plaçant le panier à différentes cotes par rapport à la source considérée comme quasi ponctuelle. En pratique on fait cinq mesures avec un pas de 200 mm.

Pour éviter des hypothèses sur la composition isotopique, divers étalonnages ont été préalablement effectués en introduisant dans le panier des masses connues des combustibles du même type que ceux qui seront retraités. On élimine aussi de cette façon les problèmes de géométrie de détection.

L'efficacité de détection de l'installation est de l'ordre de 50 coups comptés par gramme de combustible.

Des exemples de résultats sont présentés dans le tableau II.

Compte tenu de l'expérience acquise, on considère que :

- le seuil de détection est de l'ordre de 0,05% de la quantité initiale de matière fissile ;
- la précision est de l'ordre de 20% de la quantité mesurée, lorsque celle-ci est supérieure à 0,5% de la quantité initiale.

3.2 Contrôle par spectrométrie gamma

L'émission à 2,18 MeV du ^{144}Pr (T = 17 minutes) est particulièrement intéressante pour son énergie élevée, ainsi qu'à cause de l'abondance et de l'activité du produit de fission père, le ^{144}Ce (T = 285 jours).

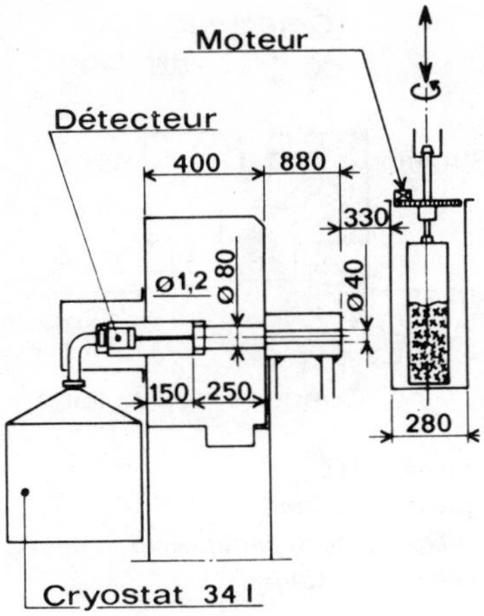

FIG.3. Installation de spectrométrie γ à AT 1 (dimensions en mm).

Dans le cas d'AT 1 on mesure le rapport des activités du ^{144}Pr et du ^{60}Co (ou du ^{54}Mn) avant et après cisaillage. On en déduit la quantité de combustible non dissoute, la quantité de matériau de structure étant supposée invariante.

3.2.1 Description de l'appareillage

- Détecteur Ge(Li)
  volume 15 cm^3 - résolution 2,4 KeV à 1,33 MeV
- Collimation (Fig. 3)
  1° longueur 880 mm diamètre 40 mm
  2° longueur 250 mm diamètre 80 mm
  3° longueur 150 mm diamètre variable 1.6 - 3 - 4 et 6 mm
- Electronique
  - Préampli
  - Amplificateur
  - Codeur et Bloc mémoire 4 000 canaux
  - Horloge
  - Imprimante

3.2.2 Réalisation du contrôle

La durée des mesures est très variable, de quelques minutes à quelques dizaines de minutes, suivant le temps de refroidissement du combustible.

Pour avoir une valeur moyenne de l'activité contenue dans le panier on fait tourner celui-ci sur son axe (40 tours/heure), et on le fait défiler axialement devant le trou de collimation.

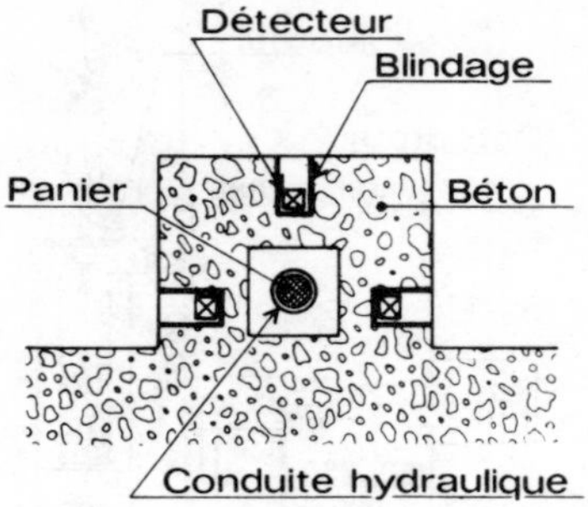

FIG.4. Disposition des détecteurs à HAO.
Panier: diamètre 400 mm, longueur 1000 mm.
Il y a deux fois trois détecteurs placés comme indiqué sur la figure, à 700 mm l'un de l'autre, symétriquement par rapport au centre du panier.

Le contenu des canaux dans la zone intéressante du spectre gamma est imprimé. Le dépouillement et le calcul de l'aire des pics sont effectués manuellement.

Le contenu du bloc mémoire peut aussi être sorti sur ruban perforé qui est alors traité par un ordinateur PDP 8.

Le seuil de détection de cette installation est de l'ordre de 0,2% de la quantité initiale de combustible, et la précision des résultats est d'environ 20% pour plus de 0,5% de matière résiduelle.

3.3 Comparaison des deux méthodes

Le tableau II donne une comparaison des résultats obtenus par les deux méthodes, lors des dernières campagnes.

Une analyse statistique de ces résultats (analyse de variance, comparaison des résultats appariés) ne permet pas de mettre en évidence un écart systématique entre les méthodes au niveau de confiance de 95 %, bien que cet écart varie entre 5 et plus de 30 % en valeur relative.

L'utilisation de la méthode de GRUBBS (6) sur les résultats obtenus sur la campagne RAPSODIE 74 A permet d'estimer pour l'erreur aléatoire un écart type relatif (coefficient de variation) voisin de 20 % pour les trois méthodes.

Les principales causes d'incertitude pour les différents résultats sont les suivantes :

. Spectrométrie gamma avec le ^{60}Co : le nombre de coups comptés à 1,33 MeV (seul pic utilisable) est faible lors de la mesure avant dissolution.

. Spectrométrie gamma avec le ^{54}Mn : l'absorption dans les gaines des photons de 835 KeV devient non négligeable.

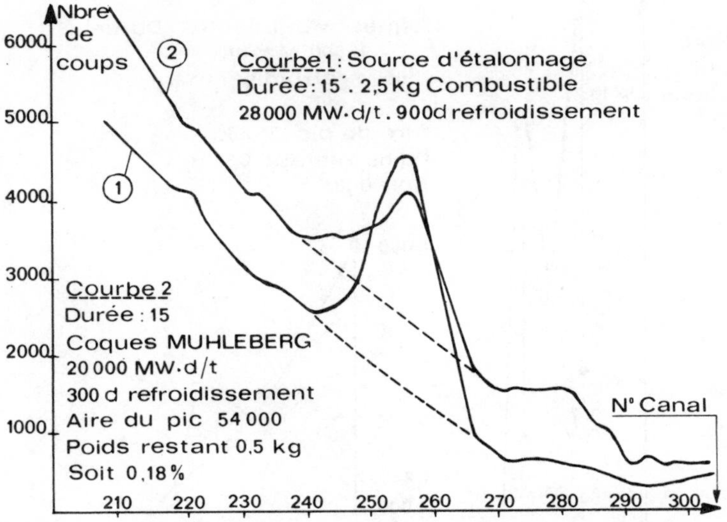

FIG.5. Comparaison entre l'étalonnage et la mesure sur un conteneur.

Dans les deux cas le comptage à 2,18 MeV reste faible pour les coques.

. Activation neutronique : la méthode est sensible au bruit de fond de la cellule (stockage d'autres éléments) et à la possibilité de fissions sur d'autres noyaux que les isotopes fissiles considérés dans l'interprétation des résultats.

4. CONTROLE DES COQUES A L'ATELIER HAO

Cet atelier a été ajouté en tête de l'usine UP 2 pour permettre le traitement de 400 tonnes par an de combustibles oxydes, issus principalement des réacteurs à eau.

En raison des contraintes liées au volume de coques à contrôler et à la réalisation de l'installation de rejet et de stockage de ces coques, la méthode retenue est un comptage gamma de l'activité liée à la décroissance du ^{144}Ce.

4.1 Description de l'appareillage

- Détecteurs : Six détecteurs NaI (Tl) avec source américium incorporée pour chaque détecteur :
  Un ampli + deux sélecteurs monocanaux
- Douze échelles binaires
- Une horloge
- Un interface vers l'ordinateur
- Blindage (Fig. 4) : la géométrie est très ouverte
  Il n'y a pas collimation, mais atténuation.

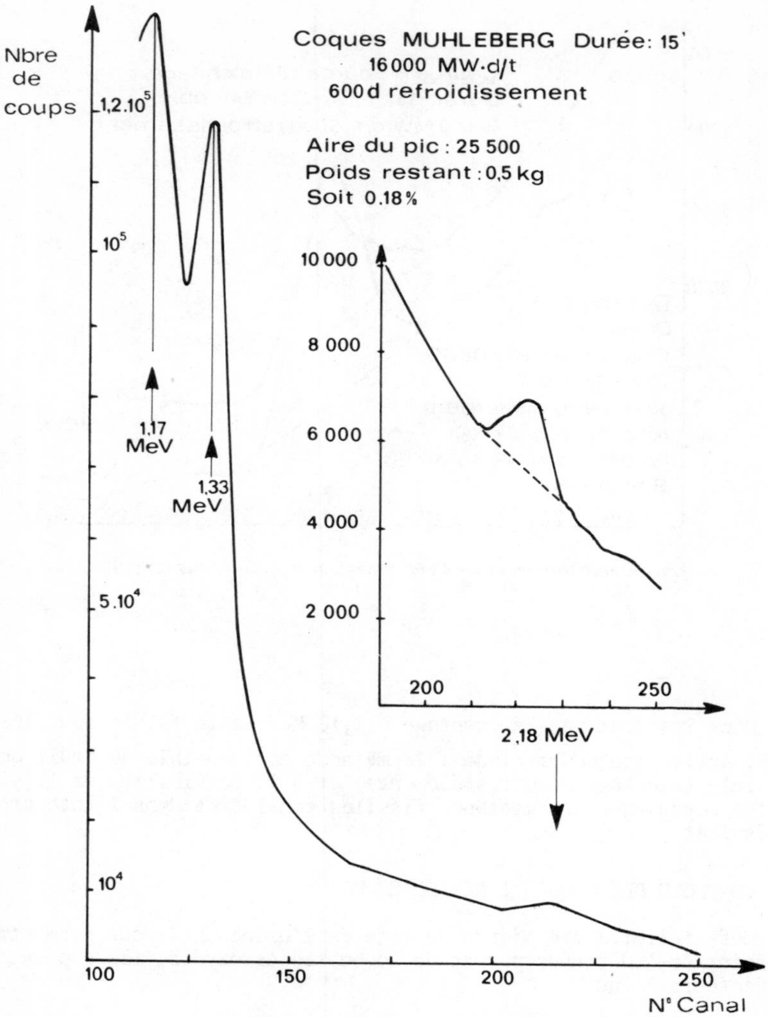

FIG.6. Mesure d'un conteneur BWR.

4.2 Réalisation des mesures

La méthode de dépouillement mise en oeuvre est la méthode SP 2 testée et mise au point à Saclay (7).

Les impulsions issues de chaque détecteur sont, après amplification, envoyées à l'entrée de deux sélecteurs monocanaux de largeurs de bande d'énergie différentes mais centrées sur l'énergie 2,18 MeV du ^{144}Pr. Les impulsions issues de ces deux sélecteurs sont stockées dans des échelles binaires.

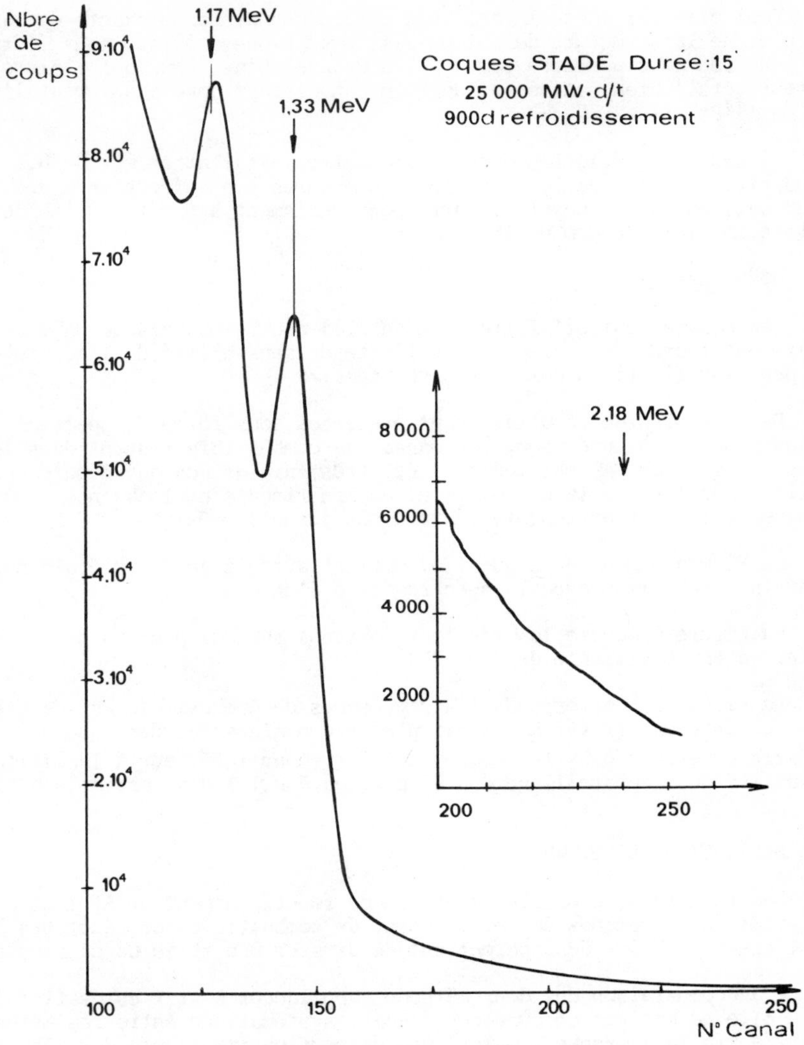

FIG. 7. Mesure d'un conteneur PWR.

L'hypothèse de la linéarité du bruit de fond dans la zone étudiée permet de déterminer l'aire du pic à 2,18 MeV à l'aide de ces deux contenus d'échelle. Pendant toute la durée de la mesure, le panier est immobile. Les six détecteurs travaillent simultanément.

La géométrie est telle que l'on est sûr de détecter du combustible quel que soit son emplacement dans le panier.

La valeur moyenne des six détecteurs est représentative de l'activité contenue dans le panier.

Pour éliminer toute incertitude sur la connaissance exacte de la géométrie et de la structure des blindages, nous avons utilisé, pour l'étalonnage, une source constituée par 0,5 % d'un assemblage PWR dont l'histoire en réacteur était bien connue, et dont on connaissait donc exactement l'activité en ^{144}Pr.

Le seuil de détection obtenu avec cette installation est de 0,1 % de la quantité initiale de combustible. Les mesures sont affectées d'une incertitude qui, en valeur absolue, correspond également à environ 0,1 % de la quantité initiale de combustible.

4.3 Résultats

Les mesures ont été faites pendant les deux campagnes du HAO. Les résultats ont fourni des valeurs à la limite de sensibilité de l'appareillage, donc peu significatives des masses rejetées.

Des exemples de résultats sont présentés sous forme de spectres plutôt que sous forme numérique. Les masses de combustible restant dans les coques à l'atelier HAO ont toujours été très faibles, ce qui dénote une excellente qualité de la dissolution, un bon rinçage et l'absence à peu près certaine de l'effet berlingot, en tout cas jusqu'à présent.

La Figure 5 montre ce que l'on obtient avec la source d'étalonnage et permet la comparaison avec la spectrométrie d'un panier.

La Figure 6 montre les résultats obtenus sur des paniers de coques du réacteur à eau bouillante de MUHLEBERG.

La Figure 7, correspondant à des coques de combustible PWR de STADE, montre qu'on peut arriver à n'avoir plus aucun signal et donc que le combustible restant dans les coques est largement inférieur à la limite de sensibilité de l'appareillage que l'on situe à 0,1 % d'un assemblage cisaillé.

5. BILAN ET PERSPECTIVES

Les résultats de plusieurs années de fonctionnement de AT 1 et des deux premières campagnes de retraitement de combustibles de réacteurs à eau dans l'ensemble HAO - UP 2 permettent de dresser une série de conclusions :

- La comparaison des deux méthodes appliquées à AT 1 est satisfaisante puisqu'elle ne met pas en évidence d'écart systématique entre les méthodes. L'écart entre les moyennes, malgré une dispersion importante des résultats, reste inférieur à 0,1 % de la masse contenue dans le combustible initial. Ceci renforce les hypothèses, indispensables pour l'application des comptages gamma, suivant lesquelles il n'y a pas de matière fissile résiduelle, en quantité non négligeable, sous une forme différente du mélange constituant le combustible irradié présent avant dissolution. Le fait que cette conclusion soit obtenue avec du combustible en oxyde mixte de réacteurs à neutrons rapides à taux de combustion élevé pour lequel le risque de présence de nodules insolubles à la dissolution est le plus élevé est particulièrement encourageant.

- L'application industrielle de la méthode de comptage gamma passif dans le cas d'une usine de forte capacité, telle que l'ensemble HAO - UP 2, est bien adaptée, ne nécessite qu'un appareillage relativement simple et permet un contrôle systématique de grandes quantités de coques en un temps limité. Les spectres représentés sur les Figures 5, 6 et 7 montrent ce que l'on peut attendre de cette méthode.

- Quelle que soit la méthode appliquée, le seuil de détection dans des applications pratiques semble bien se situer au voisinage de 0,1 % de la matière fissile contenue initialement dans les gaines correspondantes. Cette même valeur de 0,1 % est également un bon ordre de grandeur de l'incertitude en valeur absolue qui affecte les résultats de mesure lorsque la quantité décelée est supérieure à ce seuil. S'il s'avérait un jour nécessaire de réduire ces limites, il apparaît clairement qu'il faudrait un effort considérable afin non seulement de réduire le seuil de détection par des mesures plus complètes sur les coques dans des conteneurs de très petit diamètre, mais aussi de garantir des incertitudes plus faibles en limitant les corrections liées à la propagation des rayonnements et aux étalonnages.

Ces premières conclusions, jointes à l'analyse d'un certain nombre de difficultés techniques qui ont été rencontrées dans la mise en oeuvre de ces contrôles, sont à la base des études que nous poursuivons actuellement pour les réalisations projetées dans le cadre des usines futures.

REFERENCES

[1] DARROUZET, M., MARTIN-DEIDIER, L., Calcul de l'activité des matériaux de structure des assemblages, Communication privée.

[2] COSTA, L., Rayonnement émis par un assemblage moyen du réacteur Phénix, Communication privée.

[3] BELIARD, L., JANOT, P., Détermination de la quantité de matière fissile présente dans un échantillon par irradiation au moyen d'une source de neutrons et comptage des neutrons retardés, Rapport CEA-R-3272 (1967).

[4] JOVER, P., JANOT, P., GEY, G., MALET, G., Contrôle des pertes de matière fissile dans le traitement du combustible de Rapsodie, Rapport CEA-R-3961 (1970).

[5] CONSTANT, J., Le contrôle des déchets de gaine après dissolution: l'expérience AT 1, à paraître.

[6] JAECH, J.L., Errors of measurement with more than two methods, J. Inst. Nucl. Mater. Manage. **4** 4 (1976).

[7] GUERY, M., Méthode SP 2, Communication privée.

DISCUSSION

E. VAN DER STIJL: There is a considerable difference in the amount of fuel left behind in the leached hulls of fast-breeder fuel and LWR fuel. Is this due to the difference in burnup or to the different type of facility (pilot plant or industrial-scale)?

D. HEBERT: Perhaps Mr. Régnier from Cogema would like to answer.

J. REGNIER: Dissolution of high burnup PuO_2 might be one reason; less efficient leaching of the hulls might be another. For future industrial breeder fuels low values are expected.

TESTING AND DEMONSTRATING THE AUTOMATION OF THE NUCLEAR MATERIAL ACCOUNTABILITY CONTROL AT AN IRRADIATED FUEL REPROCESSING FACILITY

F. POZZI, G. OSSOLA
Comitato Nazionale Energia Nucleare,
Fuel Cycle Department,
EUREX Plant, Saluggia

G. BARDONE
Comitato Nazionale Energia Nucleare,
Fuel Cycle Department,
Casaccia, Rome,
Italy

Abstract

TESTING AND DEMONSTRATING THE AUTOMATION OF THE NUCLEAR MATERIAL ACCOUNTABILITY CONTROL AT AN IRRADIATED FUEL REPROCESSING FACILITY.

The paper summarizes some preliminary evaluations and results obtained during work carried out at the CNEN-EUREX reprocessing plant within a programme in co-operation with the IAEA concerning: (1) TDR (Time Domain Reflectometry) system installation in the plutonium final-product tank and in three cold make-up tanks; also a comparison with the conventional dip-tube pneumatic system; (2) installation, testing and demonstration of an X-ray absorption system (MAX-1) for continuous Pu determination in the second extraction cycle process solutions. Installation of a second MAX-1 absorptiometer system and of a dual-energy gamma-absorptiometer system for measuring Pu in the final product solution. This is to collect sufficient data in order to evaluate both systems; and (3) testing of a technique for measuring the residual nuclear material in the hulls by weighing and comparison with a gamma spectrometric system using a NaI detector.

INTRODUCTION

This paper summarizes some preliminary evaluations and results obtained during the pre-operational tests for the CANDU reprocessing campaign at the CNEN-EUREX plant, within a programme of co-operation with the IAEA concerning testing systems and techniques useful for safeguarding reprocessing plants.

The EUREX plant is a multipurpose pilot plant, located in northern Italy, for reprocessing, under industrial plant conditions, MTR or low enriched fuel elements. It became operational in October 1970 and up to now 506 MTR fuel elements have been reprocessed, using both tertiary amine (TCA) and TBP as extractants [1]. During 1975 – 1977 modifications were made to permit the reprocessing of spent fuel elements from power reactors. The activities of the joint CNEN-IAEA programme and the related work in progress are as follows:

(1) Testing was carried out of four complete TDR systems and a comparison made with the conventional dip-tube pneumatic system. The TDR systems were each installed in the plutonium final-product tank and in three cold make-up tanks. Calibrations of the vessels were carried out using both the TDR and the dip-tube pneumatic systems and preliminary data on the accuracy and precision of the TDR were obtained. An approach to the automatic treatment of the TDR system data was also made.

(2) Testing was done of two in-line X-ray absorptiometer systems for measuring Pu in process streams and a comparison was made with a dual-energy gamma-absorptiometer system for measuring Pu in a high concentration range (Pu final product). The systems were installed, and calibrated using U and Pu solutions, and the hydraulic circuit tested in order to control the flow-rates and circuits' hold-up.

(3) A weighing system and a gamma spectrometric system were installed to measure NM losses in the leached hulls. A weighing system, using a strain gauge, was installed and some data collected during the cold dissolution tests.

The gamma spectrometric system was also installed and calibrations, using a specially assembled standard source in simulated plant operating conditions, are in progress.

1. TDR (TIME DOMAIN REFLECTOMETRY)

1.1. Generalities

The TDR (Time Domain Reflectometry) system is a technique by which mismatches can be determined in a circuit, or in a signal transmission line [2]. The TDR analysis starts with the propagation of an incident signal (voltage pulse) in a circuit or in a cable being examined, followed by observation of the reflected signal returning from the controlled system.

All conditions causing discontinuities can be physically located by recording, with respect to time, the position of the reflected wave form.

When a coaxial probe, matched to the characteristic impedance of the system (50 ohm), with air as dielectric medium, is put into a liquid phase, the change of the dielectric constant at the interface between the two fluids will give a signal reflection which will be displayed on the CRT oscilloscope associated with the TDR system. It is then possible to measure the dielectric constant of

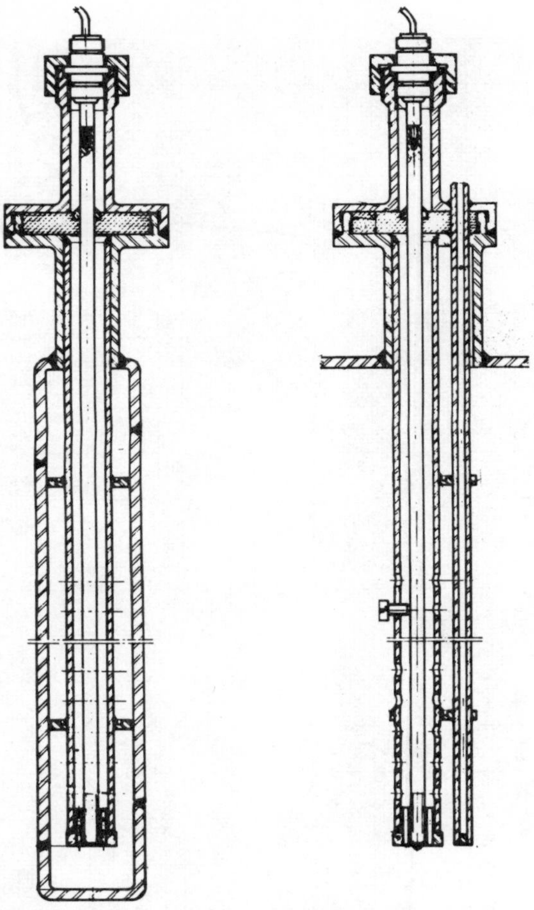

FIG.1. EUREX plant. TDR probe for Pu tank.

liquids and the liquid level in stationary or mobile conditions, even in the presence of foam or emulsion.

1.2. EUREX TDR probes

The TDR probes are corrosion-proof stainless-steel probes. Their coaxial configuration, with air as the dielectric medium, is the basic feature, as described in Ref.[2].

The diameter of the coaxial probe conductors has been calculated in order to yield an impedance as close as possible to 50 ohm. A series of regular holes in the outer conductor allows penetration of the liquid between the two conductors.

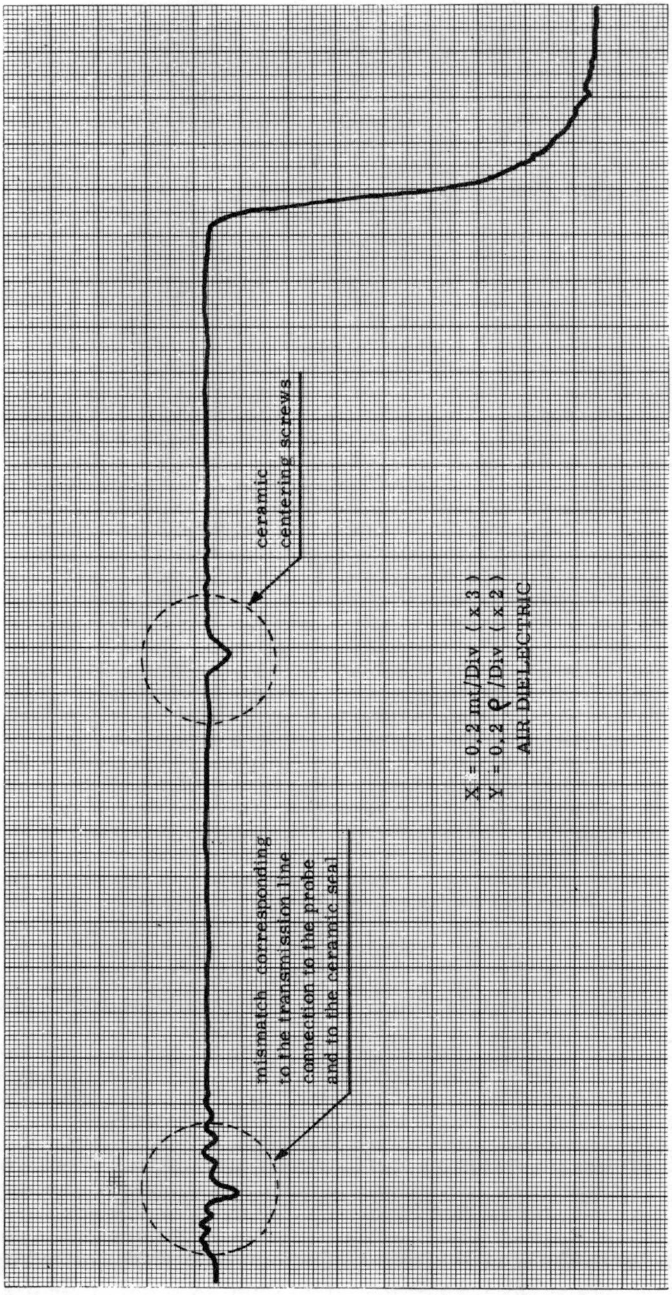

FIG.2. Typical signature of a probe installed at the EUREX plant.

The material thickness was chosen as sufficient for safety and good lite expectancy. A thermoresistor, mounted in a stainless-steel tube, is installed in close proximity to the probe's outer conductor. The probes, designed and constructed for the EUREX vessels (see Fig.1), are of three different heights – one of 1.5 m, one of 2 m and one of 2.5 m, and in the two higher probes the inner conductor is centred, at half its height, by a series of ceramic screws.

A typical recorder signature of the EUREX type probe is shown in Fig.2.

1.3. Transmission line

The signal from the probe is transferred to the electronic unit by a coaxial polyethylene polyfoam cable, with low capacitance and low dielectric losses, protected by a stainless-steel tube.

In the case of the make-up probes the coaxial cable, before reaching the TDR unit, passes through a commutation unit which can be manually or automatically operated. This unit is planned to control many probes with a single TDR electronic unit.

1.4. Data acquisition system

The signal (behaviour) is transferred from the oscilloscope to an X-Y recorder and the TDR signatures are manually elaborated.

A series of experimental tests were performed in order to allow the automatic collection and elaboration of the TDR data by interfacing the TDR electronic unit with an IBM System/7 computer. The interface provides the analog-to-digital conversion of the X-Y signals, which are then sent to the computer at a rate of 100 data/second to find on the TDR signature the inflexion point corresponding to the liquid-air interface (liquid level).

Furthermore, a microprocessor system has been almost completed. This will supply continuous information to the control room operators, with a permanent display of all TDR probe signatures and parameters of interest (liquid level, volume, temperature, etc.). The block diagram of the system is shown in Fig.3. This last development could be, in our opinion, a very useful feature for safeguard purposes.

1.5. Experimental

The calibrations of all vessels were carried out using the standard procedure for accountability vessel calibration, recording both the TDR and dip-tube manometric system data.

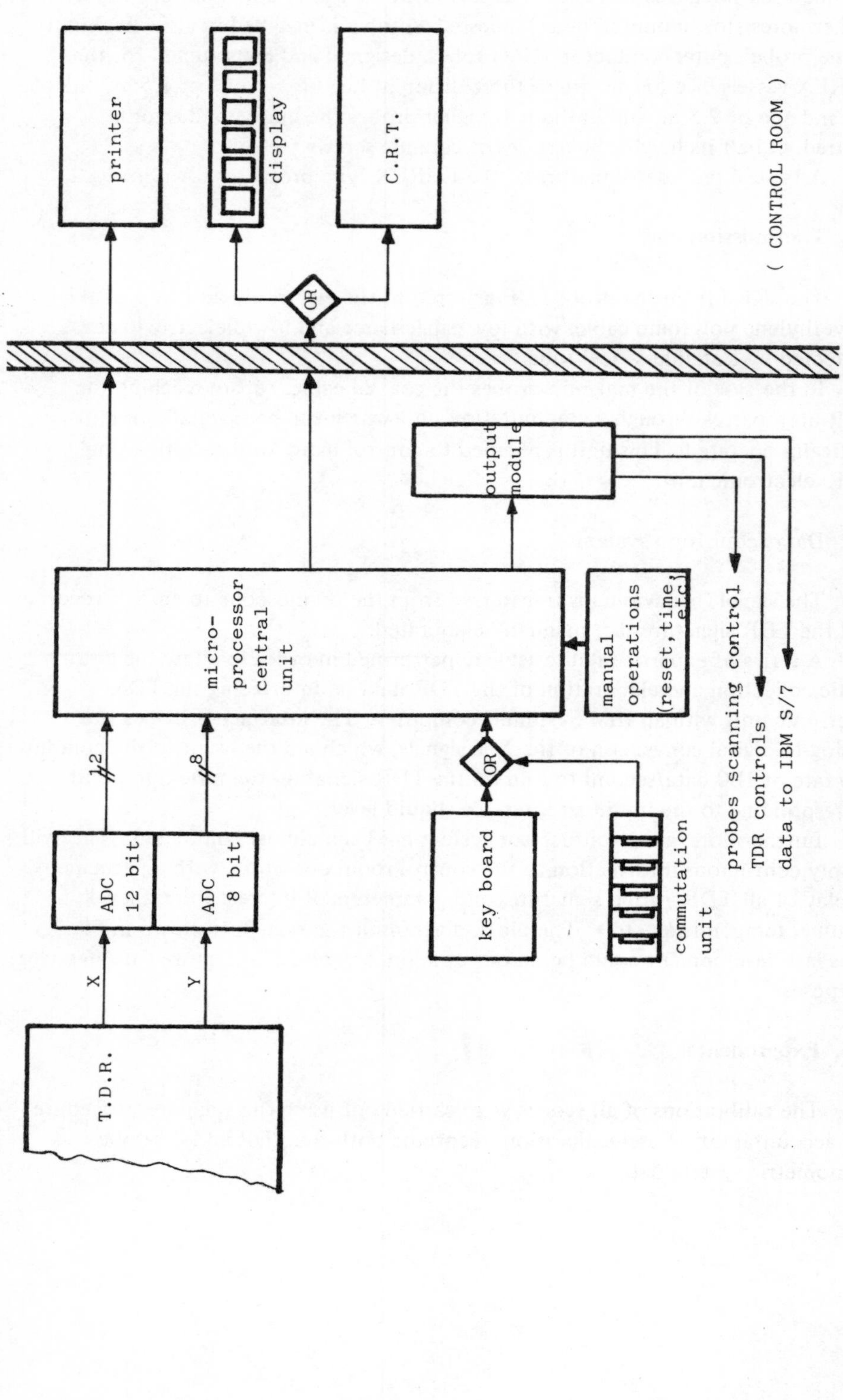

FIG.3. Block diagram of the microprocessor system for TDR automation.

FIG.4. MAX-1 absorptiometer system installed on the second extraction cycle stream.

FIG. 5. MAX-1 and dual-energy beam gamma absorptionmeters in-line, installed for measuring the plutonium final-product concentration.

Applying the cumulative regression model [3] for the treatment of the data, the average error value obtained with seven calibration runs was 2% for TDR (at 95% C.L.), and 0.7% for the manometric system. The factors limiting the calibration precision for TDR are, in our opinion, the following:

(a) Low reading sensitivity of the TDR and of the associated recording system compared with the manometer reading sensitivity; and
(b) The ceramic centring screws which cause, when present, a discontinuity in the calibration curve.

2. MAX-1 AND DUAL-ENERGY GAMMA-ABSORPTIOMETER SYSTEMS

2.1. General description of the two systems

The MAX-1 absorptiometer [4] is an apparatus developed by CNEN based on the X-ray absorption technique. It is a double-beam system with static compensating of the instrumental drift, using two ^{241}Am sources as primary sources, which excite two different targets. The apparatus has already been applied for in-line measuring of uranium during the MTR campaign, using tin and europium as targets, yielding 25 and 41 keV X-ray fluorescence energies. The two targets were changed for Pu concentration measurement in order to eliminate the Pu X-ray interference. Ba and Ho, which yield 32 and 47.5 keV, respectively, as X-ray fluorescence energies, were selected. The two beams are alternatively used as analysis or reference beams. Both energies, detected by an X-ray NaI (Tl) scintillation detector, are then shaped, discriminated by two single-channel analysers and finally sent to a ratio counter. A view of the MAX-1 absorptiometer system installed in-line is given in Fig.4.

The dual-energy gamma absorptiometer is part of a system designed by General Electric (GE) in order to provide accountability and safeguards controls at the GE Midwest Fuel Recovery Plant. Because of the GE plant delay, after a loan arrangement between IAEA and ERDA, the absorptiometer was transferred to the EUREX plant.

It is an in-line instrument [5] for high-level plutonium concentration measurements, using two gamma rays – the first of low energy (^{241}Am), whose absorption is primarily dependent on the plutonium concentration, and the second of high energy (^{137}Cs), which is primarily dependent on the solution density. Consequently it can be used to give plutonium concentrations independent of the nitric acid present.

To compensate for any interfering gamma activity present, the instrument provides remote-controlled shuttering on both gamma beams. It can also move the source and the detector for measuring standard absorbers, providing a reference

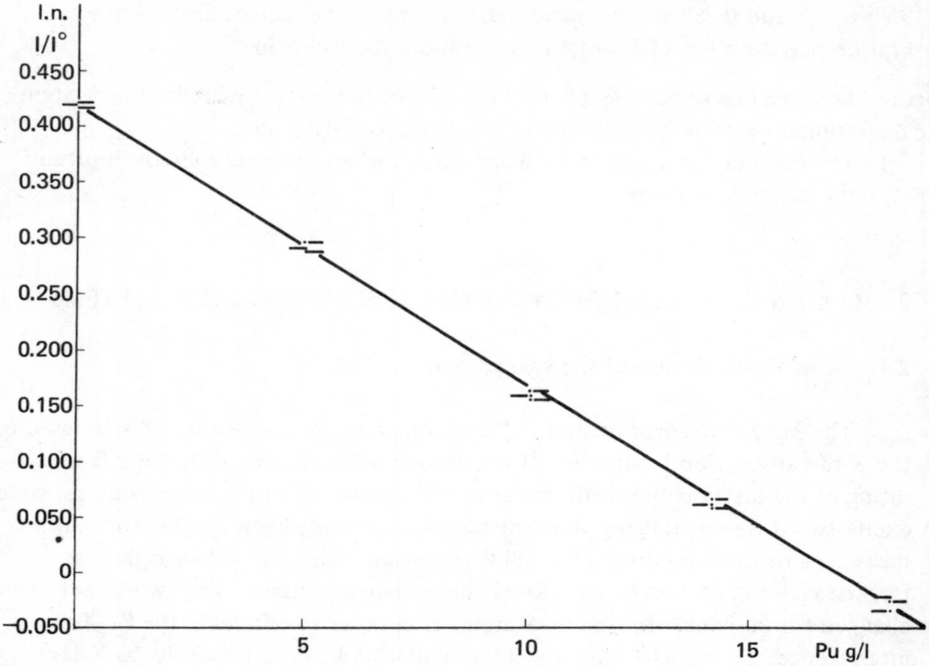

Sample	I/I°	l.n. I/I°	I/I°	l.n. I/I°	I/I°	l.n. I/I°
HNO_3 2M	1.517598	0.417	1.520941	0.4193	1.521599	0.4198
Pu 5.06 g/l	1.350047	0.3001	1.337373	0.2907	1.337246	0.2906
Pu 10.16 g/l	1.169382	0.1565	1.176993	0.1629	1.172524	0.1592
Pu 14.17 g/l	1.072123	0.0696	1.057656	0.0560	1.061082	0.0592
Pu 18.22 g/l	0.979191	−0.0210	0.982777	−0.0174	0.966722	−0.0338
Temp.	21.4 – 21.8°C		22 – 22.4°C		23.2 – 23.1°C	

FIG.6. MAX-1 calibration curve for Pu solutions.
Date: 2.4.1976 Conditions: HV 1257
Measuring beam: Ba SCA 1 E 3.72, E 2.72
Cell width: 5 mm SCA 2 E 7.02, E 3.20
Matrix: HNO_3 2M See operating manual
Collimators: Ba (1) 4 mm, (2) 4 mm; 4650 counts (HNO_3 2M)
Ho (1) 3.8 mm, (2) 3.3 mm; 3030 counts (reference)

and a rapid check of the system operation. Figure 5 shows the two absorptiometer systems installed in the plutonium final-product sampling box.

2.2. Equipment installation and preliminary tests

2.2.1. MAX-1 installed for Pu concentration measurements in the solution from the second extraction cycle

Continuous measurements can be carried out by only using the sampling station. Modifications were then carried out on the second-cycle sampling station in order to allow a circuit to be installed for both continuous measurements and routine sampling for laboratory analysis. A calibration curve for plutonium standard solutions, ranging from 0.5 to 20 g/l, has been made (see Fig.6).

2.2.2. MAX-1 and dual-energy gamma absorptiometers installed for Pu final-product concentration measurements

The final-product section of the EUREX plant has been modified in order to adapt it for reprocessing spent fuel elements coming from power reactors. Two new vessels for Pu final-product storage have been installed. Plutonium will be sampled and analysed by the two absorptiometers installed in an area near the process cell. The Pu solution is transferred by lift from the vessel to the sampling pot, then by gravity to the absorptiometer cells and finally back to the vessel.

The absorptiometers are mounted in a stainless-steel box (see Fig.5), where a conventional sampling station of the two Pu final-product vessels is also installed for laboratory sample analysis. The tests performed up to now have been mainly directed towards checking the correct functioning of the circulation system, using liquids at different densities to determine both the circuit hold-up and the amount of liquid remaining in the circuit when the circulation is stopped.

Calibration curves with uranyl nitrate solutions at different nitric acid concentrations were prepared with satisfactory results. Particularly on the dual-energy beam gamma absorptiometer, a series of eight calibration runs using uranium solutions were performed, obtaining an average precision of 0.5% (1 σ).

3. LEACHED HULLS MONITORING SYSTEM

3.1. Description of weighing system and preliminary test

The weighing system is installed in the maintenance cell which is located on the chopping cell and applies a strain gauge connected to a transducer indicator installed in front of the cell (see Fig.7). The basket containing the hulls, when

FIG. 7. Leached hulls weighing system installation.

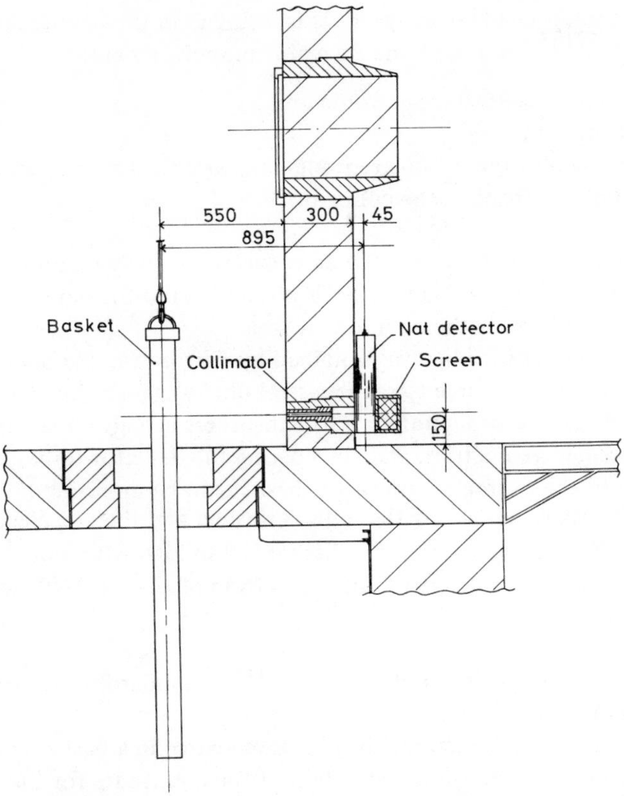

FIG. 8. Scheme of the leached hulls gamma monitoring system.

extracted from the dissolver, is drained and is then transferred in a special container installed on the load cell. The weighing is carried out by pneumatic actuation of the load cell.

After a calibration curve has been obtained, to establish the accuracy and precision of the weighing system a series of measurements were performed using two well-calibrated standards, basket-shaped, of 10 and 20 kg. The precision obtained was within 0.05% (1σ) and a bias of +0.2% was noted.

3.2. Description of the gamma monitoring system and preliminary calibration data

Like the dual-energy beam gamma absorptiometer, the gamma monitoring system was transferred from the GE MRF to the EUREX plant. The system uses a NaI (Tl) 2 in. × 3 in. scintillation detector linked either to a single-channel system or to a multi-channel analyser. It is installed in the maintenance cell near the weighing system (see Fig.8) and its main components are:

(a) Measuring head with the electronic unit
(b) Collimator plug
(c) Transfer system to move in a reproducible way the basket containing the leached hulls in front of the collimator.

(a) The measuring head, i.e. the detector, is vertically mounted for reasons of limited space (see Fig.8). A lead screen is placed behind the detector to provide the necessary shielding for the operational area.

(b) The collimating plug was obtained by perforating the maintenance cell wall. In the 125-mm-dia. hole a stainless-steel plug was placed and a 2 × 2 cm removable lead collimator mounted. Lead absorbers can also be inserted. With the above-described geometries, and considering the detector-to-basket distance, a 13-cm section of the overall length of the basket can be monitored.

(c) The basket containing the hulls is washed and then displaced by 13 cm step by step using a bridge crane. The basket is 2 m high with a diameter of 125 mm; it can contain hulls from the dissolution of three CANDU-type fuel elements.

The whole system was tested by using a ^{144}Ce standard source of 0.5 Ci, especially assembled.

Calibrations are in progress with the above source in a basket containing cold hulls in order to establish the variation of the count-rate for the different positions in the basket, and the absorption factor.

ACKNOWLEDGEMENTS

Appreciation is expressed by the authors to Mr. R. Di Bona, Mr. G. Gasso, Mr. G.P. Godio, Mr. B. Mattia and Mr. V. Pagliai, for their helpful co-operation during all the work.

REFERENCES

[1] NOTIZIARIO CNEN, **23** 4 (April 1977).
[2] DE CAROLIS, M., BARDONE, G., TDR Methods and Apparatus for Measurement of Levels and Physical Characteristics of Moving or Static Fluids in Pipe-lines or Tanks, CNEN Rep. RT/CHI (1974) 7.
[3] HOUGH, C.G., Statistical Analysis. Accuracy of Volume Measurements in a Large Process Vessel, HW-62177 (Oct. 1959).

[4] DE CAROLIS, M., BARDONE, G., Methods and Apparatus for Measuring the Concentration of Elements in Solution and the Thickness of Solid Samples, CNEN Patent No.53367A/72.
[5] CARTAN, F., Loadout Accountability Safeguards Monitoring System for GE-MFRP, communication to the IAEA.

DISCUSSION

H.T. YOLKEN: Have you considered in-line liquid flow measurements to provide continuous monitoring of bulk quantities? These real-time measurement results could be used in conjunction with real-time NDA results.

G. BARDONE: Your suggestion is very interesting. At CNEN we have carried out preliminary evaluations and studies in this connection and some systems have in fact been installed on cold pipes. However, since we have a large number of systems in operation which still have to be fully evaluated, we have preferred to concentrate on them.

D. ÖNER: What is the enrichment of the MTR fuel elements that are reprocessed and where do they come from?

G. BARDONE: The MTR fuel elements reprocessed at the CNEN EUREX plant to date have had enrichments of 20 and 90%. They all came from the CEC reactor in Italy.

J. REGNIER: I don't quite understand why you weigh the hulls. Could you give a few details in this connection?

G. BARDONE: The techniques used for measuring nuclear material losses in leached hulls at the EUREX plant include both gamma spectrometry and the weighing method. Here are some details of the weighing method: (1) The weight of the fuel element or of the fuel cladding must be known with adequate reliability. For the CANDU-type fuels, AECL supplies the cladding weight to EUREX. If the above information is not available, the fuel element can be weighted; (2) All materials (including fuel head, etc.) must be collected during the chopping operation; (3) the precision of the weighing system must be within 0.5%. The EUREX weighing system precision is better than this; and (4) during pre-operational tests the water remaining in the leached hulls weighed about 100 g. This value was more or less constant, so that it can be assumed as a systematic error.

Mathematical evaluations lead us to believe that with the EUREX weighing system we will be able to detect a nuclear material loss of 1% or more. However, the weighing technique is still in the experimental stage and a reliable evaluation will be available only after a significant period of hot operations.

DEVELOPMENT AND DEMONSTRATION OF SAFEGUARDS TECHNIQUES IN THE TOKAI FUEL REPROCESSING PLANT

K. NAKAJIMA, T. KOIZUMI,
T. YAMANOUCHI, S. WATANABE,
N. SUYAMA
Power Reactor and Nuclear Fuel
Development Corporation,
Tokai-mura, Ibaraki-ken,
Japan

Abstract

DEVELOPMENT AND DEMONSTRATION OF SAFEGUARDS TECHNIQUES IN THE TOKAI FUEL REPROCESSING PLANT.

During commissioning tests since 1974 of the Tokai fuel reprocessing plant some efforts have been concentrated on establishing the accountability procedures and demonstrating the safeguards techniques and equipment. Examples of nuclear material balance in the plant and tentative installation of surveillance equipment in the fuel receiving and storage area are presented, and experimental results on the hull-monitoring system and the direct weighing devices, using strain-gauge for the solution in the accountability vessels, are discussed. TASTEX, an international project on the development and demonstration of advanced safeguards techniques for the Tokai plant is also introduced.

1. INTRODUCTION

1.1. Plant description

The Tokai fuel reprocessing plant is a facility which uses the Purex process, and has a capacity of 0.7 t/d U. The main type of fuel to be reprocessed is zircaloy or stainless-steel-clad, low-enriched uranium for light water reactor fuel. The final products are uranium trioxide and plutonium nitrate solution.

The facility is located at Tokai-mura, about 120 km north-east of Tokyo. The site opens to the Pacific Ocean and is 220 000 m^2 in area.

1.1.1. Fuel receiving, storage and head-end process

After initial cooling in the reactor pond, the fuel is transferred to the plant storage pond and stored until ready for reprocessing. The head-end system consists of equipment to chop by mechanical means the fuel assemblies into

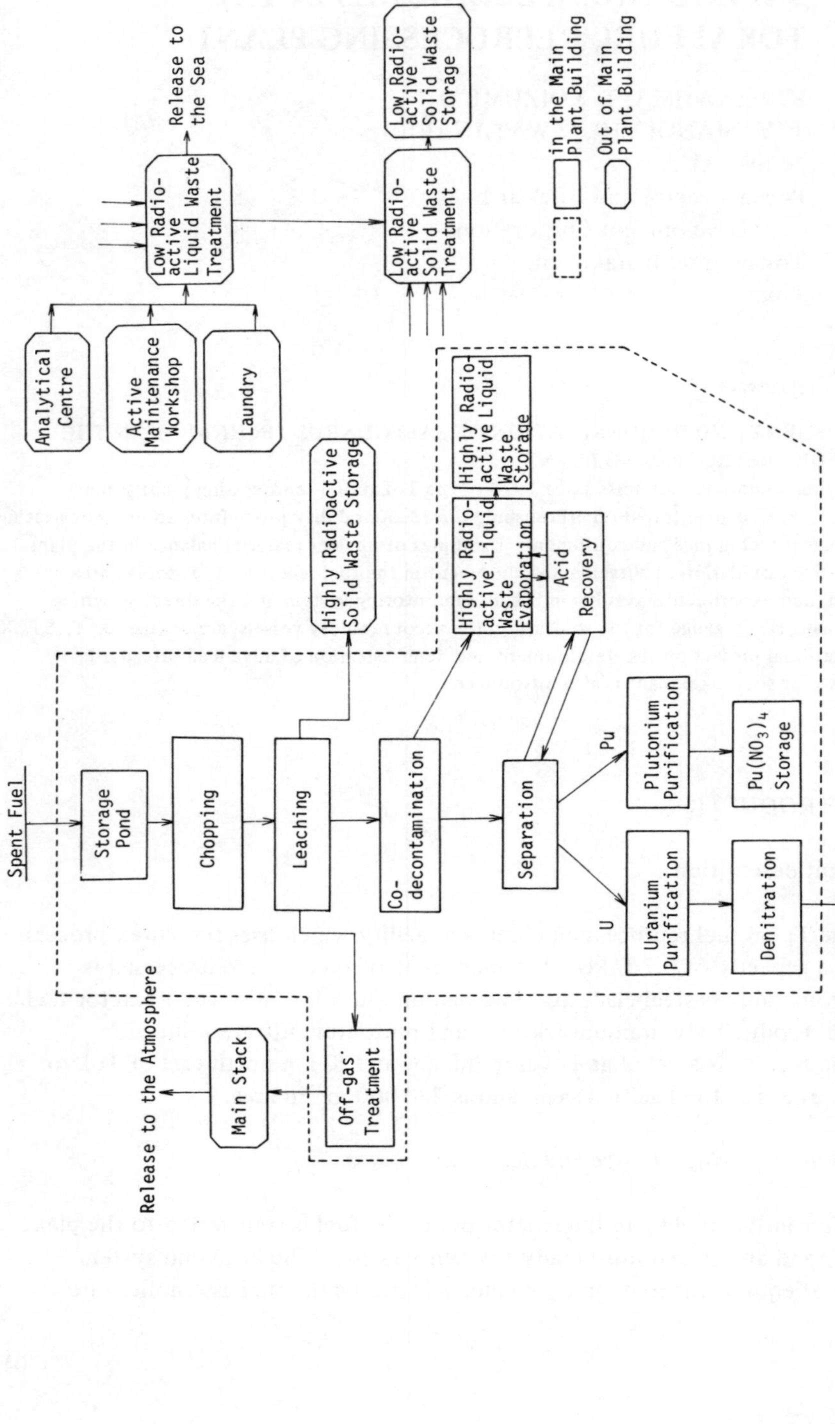

FIG.1. Schematic flow diagram of the process.

short pieces and to leach the "meat" with nitric acid from the chopped fuel in a batch-wise operation in the dissolvers. The dissolver solution is transferred to an accountability vessel. The leached hulls are transferred to the high-level radioactive solid-waste storage.

1.1.2. Solvent extraction

The recovery and purification of the uranium and plutonium products are accomplished using the Purex solvent extraction process with 30% TBP in normal dodecane in mixer-settler contactors. The dissolver solution is first co-decontaminated from fission products in the first cycle, and then the uranium and plutonium are separated from each other in the second partition cycle. After partitioning, uranium and plutonium streams are fed to an additional extraction cycle to be further purified from fission products and from each other.

1.1.3. Product

The plutonium nitrate product, which is concentrated to 250 g/l by evaporation, is stored in annular vessels of criticality safe design in the main process building.

The uranyl nitrate solution is concentrated by evaporation, converted by thermal denitration into UO_3 powder and contained in bottles with "bird-cages" for criticality safety. The bottles are transported to a separate UO_3 storage facility.

1.1.4. Waste treatment and storage

Highly active liquid waste, mainly from the first extraction cycle, is concentrated and stored in large vessels, which are equipped with continuous cooling, ventilating and stirring devices, in the main process building.

Medium-active liquid wastes from the second and third extraction cycles etc. are concentrated in the acid recovery evaporator. The concentrate is transferred to the highly active waste evaporator. Nitric acid is recovered from the condensate in a distillation column and recycled by appropriate steps in the process.

A low-active liquid-waste treatment system consists of the evaporation and flocculation process. The treated effluent is discharged to the sea after being monitored.

Highly active solid wastes from the chop and leach operation are monitored by the hull monitoring system, then transferred and stored under water in the highly active solid-waste storage facility.

TABLE I. TIME SCHEDULE OF THE HOT TEST AT PNC REPROCESSING PLANT

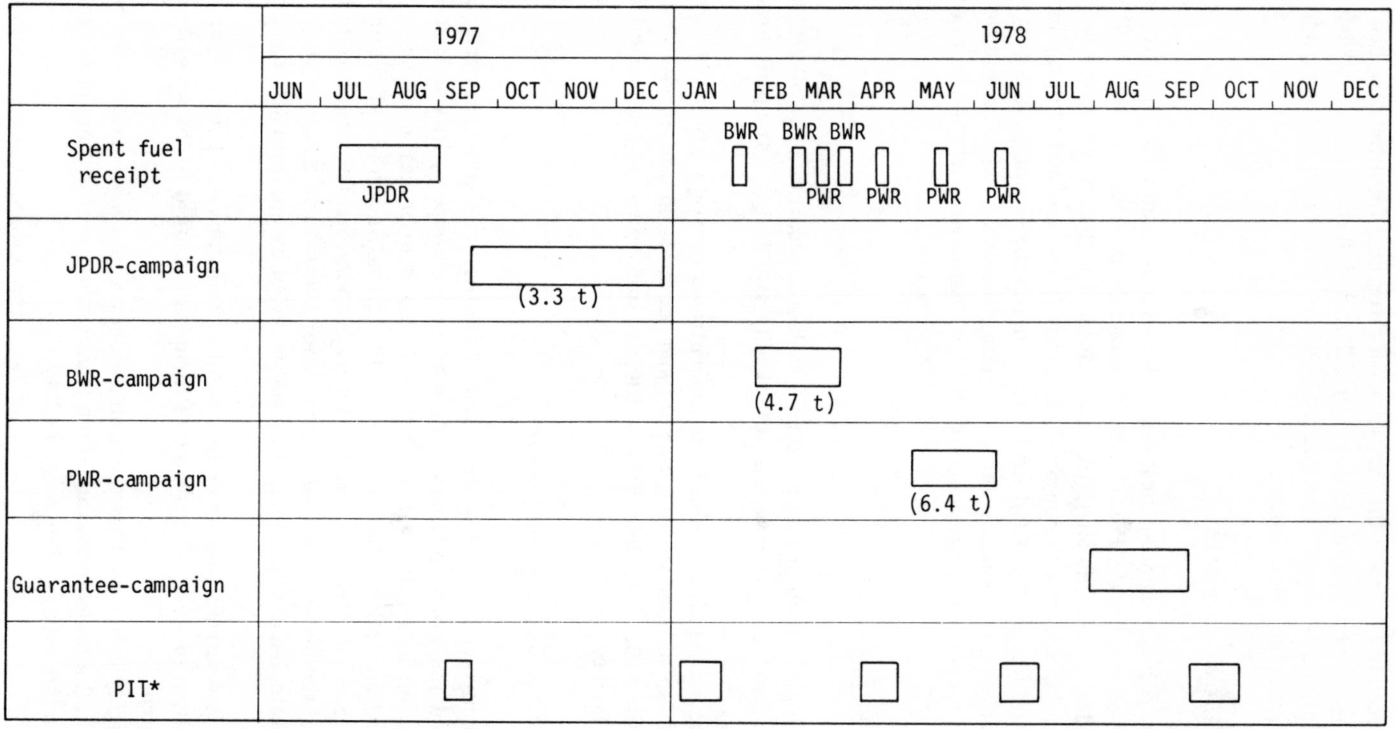
1977
1978
JUN
JUL
AUG
SEP
OCT
NOV
DEC
JAN
FEB
MAR
APR
MAY
JUN
JUL
AUG
SEP
OCT
NOV
DEC
Spent fuel receipt
JPDR
BWR
BWR
BWR
PWR
PWR
PWR
PWR
JPDR-campaign
(3.3 t)
BWR-campaign
(4.7 t)
PWR-campaign
(6.4 t)
Guarantee-campaign
PIT*

* PIT=Physical Inventory Taking.

TABLE II. CHARACTERISTICS OF SPENT FUELS TO BE PROCESSED DURING THE HOT TEST

	JPDR	BWR	PWR
Name of Reactor	Japan Power Demonstration Reactor (JAERI)	Fukushima No.1 Reactor (TEPCO)	Mihama No.2 Reactor (KEPCO)
Burn up (MW·d/t)	110 ∿ 5,640 Av. 4,030	6,640 ∿ 11,940 Av. 10,330	10,840 ∿ 29,850 Av. 18,750
Initial enrichment (%)	2.6	2.09	2.27 or 3.03
Cooling time at Sep. 1977 (days)	>2,875	1,754	508 ∿ 1,635
Cladding Material	Zircaloy-2	Zircaloy-2	Zircaloy-4
Number of received and processed assemblies	71 (57)[a]	72 (24)[a]	40 (16)[a]
Amount of Uranium (t)	∿4.1 (3.6)[a]	∿14.1 (4.7)[a]	∿16.0 (6.4)[a]
Period of Processing	Sep. ∿ Dec. 1977	Feb.∿Apr. 1978 Aug.∿Sept. 1978	May.∿Jun. 1978 Aug.∿Sept. 1978

JAERI : Japan Atomic Energy Research Institute.
TEPCO : Tokyo Electric Power Co.
KEPCO : Kansai Electric Power Co.

[a] The number of assemblies and the amounts of uranium which have been processed before guarantee tests (GT) are shown in brackets.

Burnable low-active solid wastes are incinerated, and non-burnable wastes are incorporated into concrete in steel drums, which are stored in the low-active solid-waste storage facility. The schematic flow diagram is shown in Fig.1.

1.2. Operation mode of the hot test

The hot test consists of four campaigns, the JPDR, BWR, PWR and guarantee campaigns.

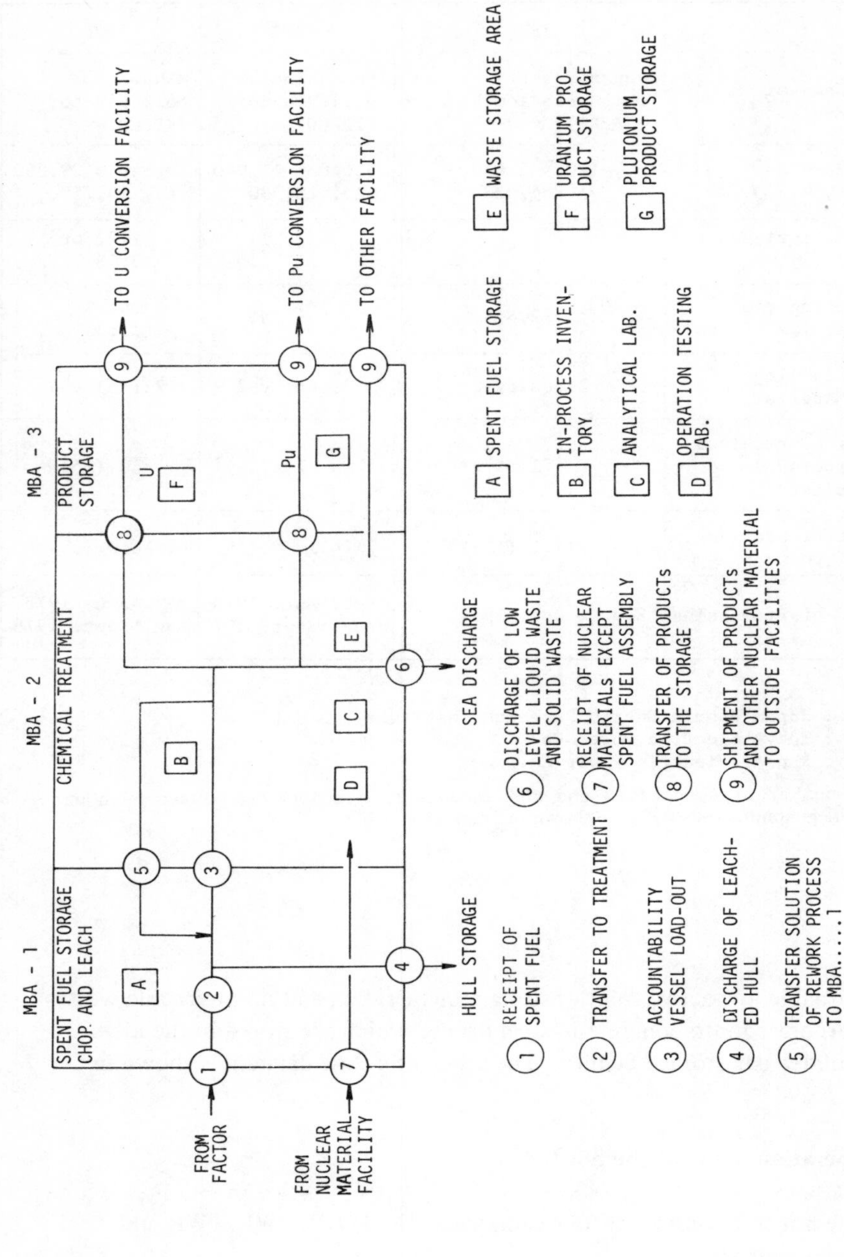

FIG.2. MBAs and KMPs in the plant.

A time schedule is shown in Table I and characteristics of the spent fuels to be processed during the hot test are shown in Table II. During each campaign the following items were examined.

(a) Plutonium and uranium accountabilities
(b) Shielding capability
(c) Equipment performance such as decontamination factors of the different cycles, of the evaporators etc.
(d) Losses at different cycles, and total losses
(e) Instrumentation capability
(f) Behaviour of iodine, tritium, krypton and transuranic elements
(g) Solid contents in dissolved solution
(h) Degree of solvent degradation etc.

The results from the JPDR to PWR campaigns show that these items were examined satisfactorily. The physical inventory-taking of nuclear material was also carried out at the end of each campaign on a clear-out basis.

2. SAFEGUARDS MEASURES

The safeguards measures of the reprocessing plant consist of three internationally agreed measures of containment, surveillance and material accountancy.

2.1. MBAs and KMPs in the plant

The plant is divided into three material balance areas (MBAs):

(a) MBA-1, which is a shipper-receiver difference (SRD) area comprising the receiving and storage facility of irradiated fuel elements, the mechanical treatment cells, the dissolver, the clarification unit for dissolver solution, and the input accountability vessel;
(b) MBA-2, which is a MUF area comprising the chemical process area, the waste treatment area, the analytical laboratory, and the operation testing laboratory (OTL);
(c) MBA-3, which is a product storage area comprising the plutonium product storage vessels and the uranium product storage facility.

Strategic points, which are key measurement points (KMPs) in each MBA, are divided into two categories for determining nuclear material flow and physical inventory. MBAs and KMPs are shown in Fig.2.

TABLE III. MEASUREMENT METHODS FOR THE MAIN KMPs

KMP	Measurement point	Chemical analysis		Sampling	Volume or weight measurement
		Uranium	Plutonium		
FKMP-3	Input accountability vessel	Isotopic dilution -Mass spectrometry	Isotopic dilution -Mass spectrometry	Circulation using air lift and vacuum line	Pneumatic bubble system using dip tube
FKMP-5 and 6	Waste solution	Solvent-extraction -DBM spectrophotometry	Solvent-extraction -α counting	Circulation using air lift and vacuum line	Pneumatic bubble system using dip tube
FKMP -8,9	UO_3 product	$K_2Cr_2O_7$ titration	-	Proposional sampler	Weighing on a large scale
	Plutonium product accountability vessel	-	$Ce(SO_4)_2$ titration	Using vacuum	Pneumatic bubble system using diptube
IKMP	In-process inventory vessels	Case by case, methods same as FKMPs	Case by case, methods same as FKMPs	Case by case, methods same as FKMPs	Case by case, methods same as FKMPs

In general, the application of the accountability measurement is rather difficult for the MBA-1 because inputs from the reactor are not measured directly; then, these amounts are only estimates from reactor data, so that the uncertainty becomes rather high. Therefore, the safeguards measure taken in this MBA is chiefly the containment and surveillance method.

In the MBA-2, the accountability method can be applied because most of the nuclear materials are in solution form so that they are able to be measured easily.

In MBA-3, the product storage area, the removal of nuclear materials is not so frequent that containment and sealing are useful or necessary.

2.2. Accountability measurements

2.2.1. Material flow measurements

The flow measurement of nuclear materials in the KMPs is carried out batch-wise by volume measurement, sampling, and chemical analysis in the accountability vessels, which are equipped with the conventional dip-tube system for measuring liquid level and density. The signal is pneumatically transmitted to a recorder in the control room and to a manometer.

The samples are obtained from the sampling stations. Before sampling, the contents of the vessels are mixed by vigorous sparging. Sampling is started by inserting the needles of the sampling block through the rubber cap of a sampling bottle. Liquid from the vessel is drawn into a sampling bottle using an air lift and a vacuum line. After sampling, the sampling bottle is transferred to the analytical laboratory by the pneumatic transfer system. The accountability samples are analysed in the analytical laboratory according to procedures of the analytical manual. The measurement and analytical methods for each flow KMP are shown in Table III.

2.2.2. Physical inventory-taking (PIT)

At the end of each campaign, PIT is carried out. Just after completing the feeding of active solution to the extraction cycle, uranium solution, free from fission product and plutonium, is put into the feeding vessel and fed to the series of mixer-settlers. After the concentration of plutonium in the partition cycle and the plutonium purification cycle is reduced to less than 0.5 mg/l, the feed is replaced with nitric acid for rinsing. The acid rinsing is carried out continuously until the concentration of uranium drops below 0.1 g/l in the whole extraction cycle.

Solutions containing a certain amount of nuclear materials are gathered into the in-process vessels which are provided with volume measurement and sampling equipment. After the preparation of PIT, measurements of volume and sampling of the inventory vessels are carried out at the same time.

Samples are analysed by using the same methods for flow measurement.

2.3. Containment and surveillance method

A surveillance system, using cameras and closed-circuit television (CCTV) to the fuel receiving area and the storage pond in the MBA-1, is used. The provisions by PNC for surveillance are described in Section 4.2. Also, the IAEA has installed a CCTV system, which consists of two TV cameras, a control console which includes a video recorder, a monitor, and a control unit. The shearing, dissolving, clarification of dissolver solution and extraction feed adjustment are performed in the cells as rigid containment. The transfer of liquid samples to OTL is done by a special shielded container under the supervision of the inspector. A sealing system to ensure containment of the mechanical treatment cells was discussed by PNC and the IAEA. For transfer of chopped fuel pieces to OTL, a hatch through which chopped fuel can be removed is provided in the side wall of the dissolver loading cell (DLC). The hatch cover should be attached by the normal wire seal.

The uranium product – UO_3 powder – is contained in a bottle sealed by the inspector; and the plutonium product, supposedly most vulnerable to diversion, is stored in vessels in the storage cells, but the plutonium load-out facility is now physically closed. The area is under personnel access control by the use of electromagnetic locking devices. An automatic surveillance system over the area is being investigated in TASTEX – see Section 4.1.

3. ACCOUNTABILITY RESULTS AND MUF EVALUATION

3.1. Material balance during JPDR campaign

3.1.1. Preparation before the starting of the campaign

The receipt of JPDR fuel assemblies from the JAERI site started on 15 July 1977, and finished at the end of August. The receipt was 71 spent fuel assemblies, which were stored in the spent fuel storage pond in the MBA-1.

The other receipt was a small amount of plutonium and uranium as standard reference material for analysis and training for the analytical laboratory operation.

3.1.2. Material balance of uranium

The material balance of uranium during the JPDR campaign, from 1 September 1977 to 9 January 1978, is shown in Fig.3.

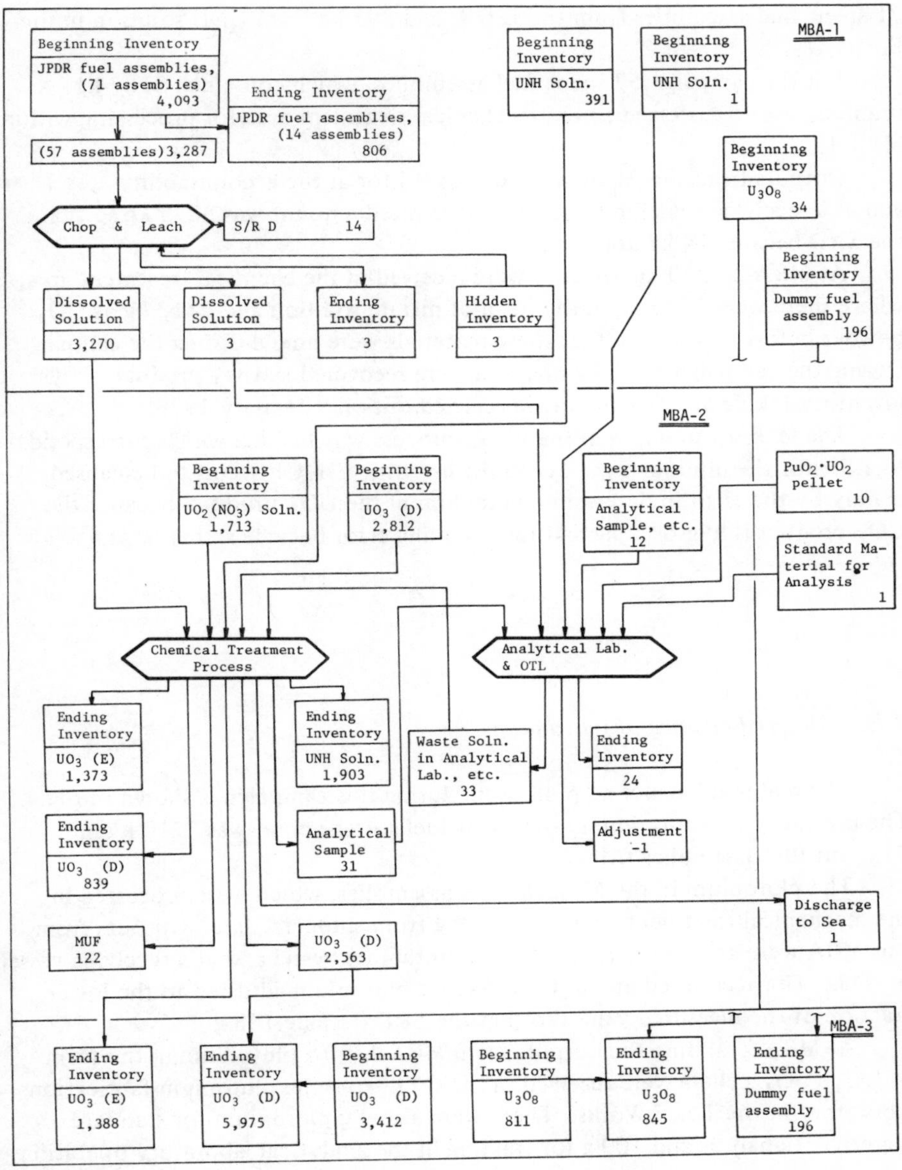

Unit : kgU, (D) : Depleted Uranium, (E) : Enriched Uranium, OTL : Operation Testing Lab.

FIG. 3. Material balance of uranium during the JPDR campaign.

The inventory of uranium in the spent fuel storage pond was 4093 kg in 71 spent fuel assemblies from the JPDR, and 391 kg from UNH solution in the buffer vessels.

For the campaign 57 spent fuel assemblies, which contained 3287 kg uranium were transferred to the mechanical treatment area for processing within MBA-1.

Outputs from this MBA were accounted for at the accountability vessel and at a receiving vessel in OTL. The accounted amount was 3273 kg so that the SRD became 14 kg uranium.

At MBA-2 3270 kg uranium were received in the chemical treatment area, where there were 1713 kg in the form of nitrate solution and 2812 kg as UO_3 powder before starting, so that these materials were mixed during the campaign. During the campaign 1388 kg uranium were recovered as UO_3 product. Physical inventory-taking after clean-out was carried out on 9 January 1978.

The MUF of uranium in the whole process was 122 kg, which corresponds to 1.5% for the input and the beginning inventory. MUF seems to be caused mainly by the error in determining uranium in the UO_3 product, because the UO_3 product is hygroscopic and the sampling error became rather large.

3.1.3. Material balance of plutonium

The material balance of plutonium during this campaign is shown in Fig.4. The inventory of plutonium in the spent fuel storage pond was 7210 kg in 71 spent fuel assemblies from JPDR.

The plutonium in the 57 spent fuel assemblies, which were processed in the mechanical treatment area, was 5589 g from shippers' data. Outputs from this MBA were accounted for at the accountability vessel and at a receiving vessel in OTL. The accounted amount was 5401 g plutonium adjusted to the lower fraction of the measured values so that the SRD became 188 g.

At MBA-2, adding the outputs from MBA-1, 90 g plutonium in the form of UO_2-PuO_2 pellets were received in the OTL from the plutonium fabrication laboratory in the Tokai Works. There were also 4 g plutonium for standard material of analysis and 109 g for training in the analytical laboratory operation as starting inventory, so that the total starting inventory became 5605 g. During this campaign, there was nothing to transfer to another MBA. From the result of PIT, the MUF of plutonium was minus 36 g, which corresponds to 0.64% for the input and the starting inventory. MUF agrees with the book inventory within an allowance for measurement error.

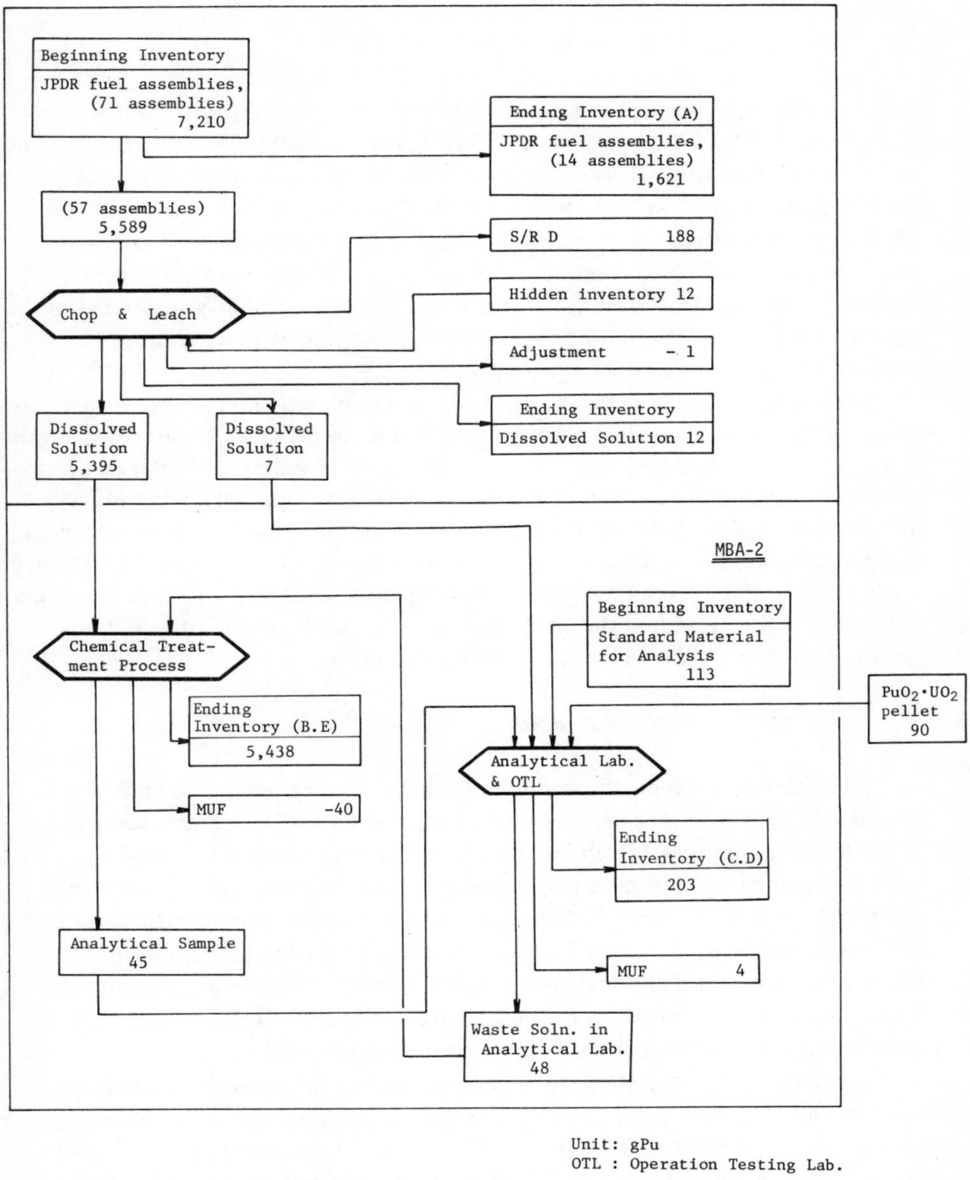

FIG.4. Material balance of plutonium during the JPDR campaign.

3.2. Material balance during BWR campaign

3.2.1. Material balance of uranium

The material balance of uranium from 9 January to 5 April 1978 is shown in Fig.5. At MBA-1, 72 assemblies of BWR fuel from Fukushima and 10 assemblies of PWR fuel from Mihama were received in the spent fuel storage pond during the campaign; 13 850 kg and 3935 kg of uranium were contained in BWR and PWR spent fuel, respectively. For processing, 24 assemblies of BWR spent fuel, which contained 4623 kg uranium, were transferred to the mechanical area. Outputs from this MBA were accounted for at the accountability vessel and in the OTL. The accounted amount was 5445 kg uranium so that the shipper-receiver difference became 27 kg.

At MBA-2, during the campaign, 5442 kg uranium were received in the chemical treatment area where there were 1903 kg in the form of nitrate solution and 2212 kg as UO_3 powder, which was used as a seed material for the denitrator operation. Also, 214 kg uranium as UO_3 powder was transferred from MBA-3 to MBA-2 as seed material for the denitrator operation; 8004 kg were recovered as product-formed UO_3 powder and transferred into the uranium product storage facility. The MUF of uranium in the whole process was 18 kg which corresponds to 0.18% for the input and the starting inventory. MUF agrees with the book inventory within an allowance for measurement error.

3.2.2. Material balance of plutonium

The material balance of plutonium during this campaign is shown in Fig.6. At MBA-1, 72 BWR fuel assemblies, which contained 66725 g plutonium, and 10 PWR fuel assemblies, which included 21412 g plutonium, were received from Fukushima and Mihama, respectively, during the campaign. Twenty-four BWR fuel assemblies, which included 20801 g plutonium, were transferred to the mechanical treatment area for processing and 17 g were accounted for in the remaining solution and recycled samples. The amount of outputs from this MBA which was accounted for at the accountability vessel and in the OTL was 20392 g so that the shipper-receiver difference became 426 g.

At MBA-2, 20392 g plutonium were received in the chemical treatment area. In MBA-2, there were 5438 g plutonium as nitrate solution which had been produced in the JPDR campaign, 23 772 g were recovered in the form of nitrate solution, and transferred into the plutonium product storage located in the basement of the main process building.

After PIT, the MUF of plutonium was estimated at minus 587 g in the chemical treatment process area; minus 9 g in the analytical laboratory and OTL; and 1037 g in the plutonium product storage of MBA-3 so that the total MUF

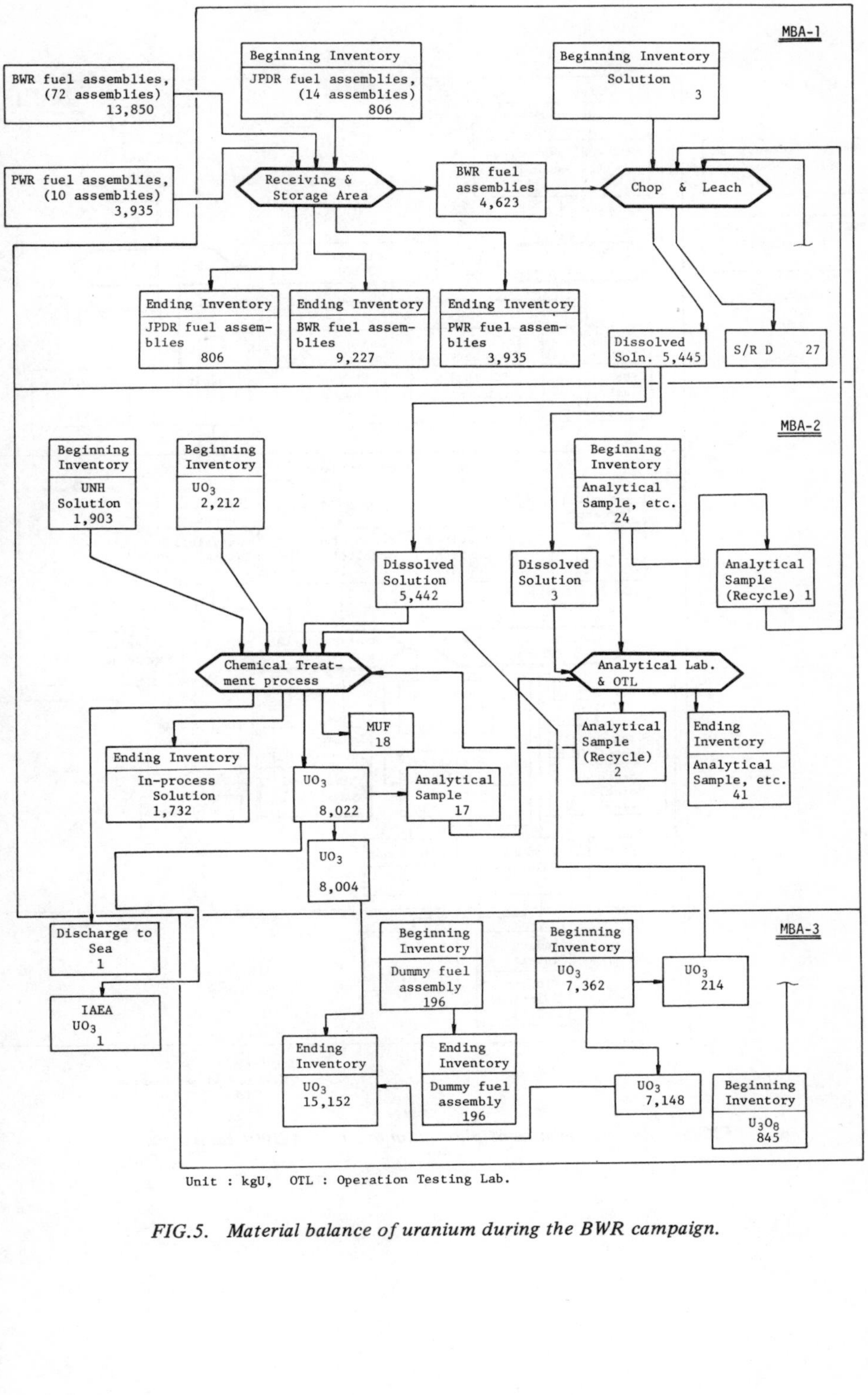

FIG.5. Material balance of uranium during the BWR campaign.

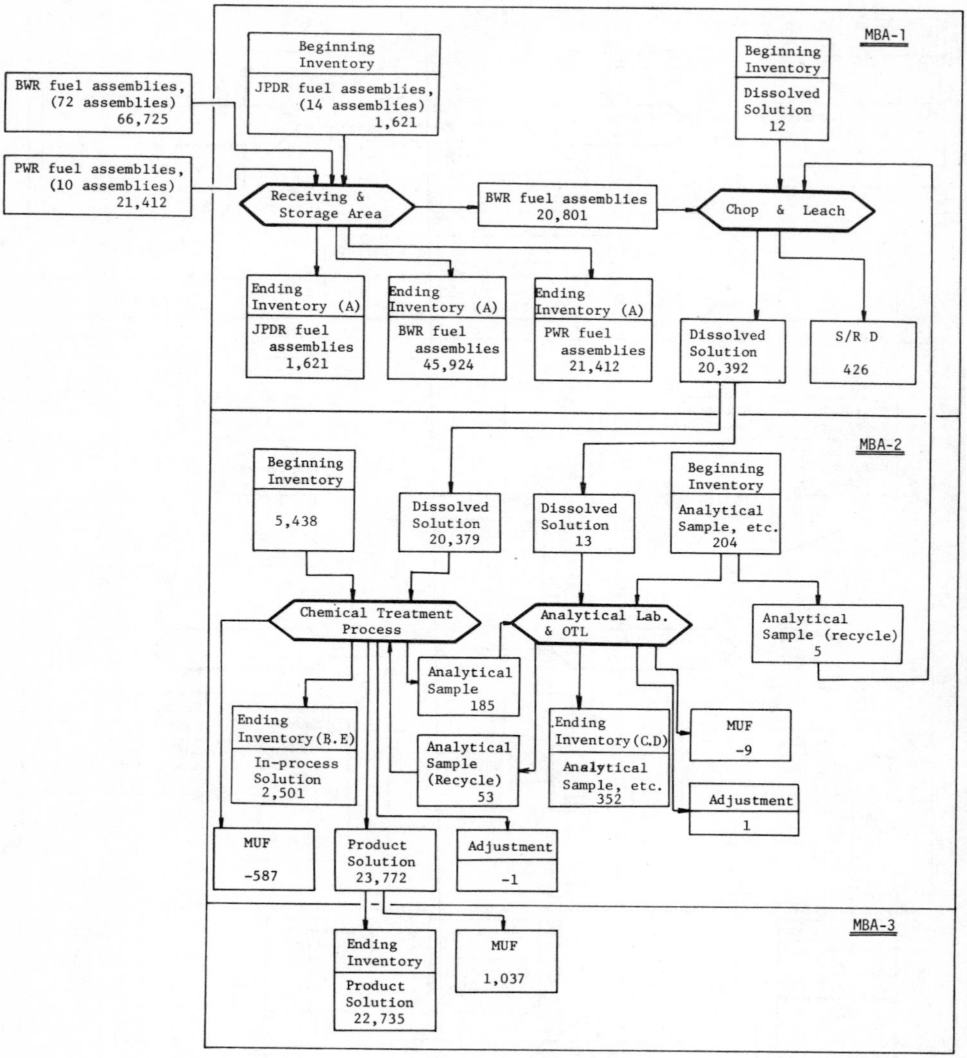

Unit: gPu
OTL : Operation Testing Lab.

FIG. 6. Material balance of plutonium during the BWR campaign.

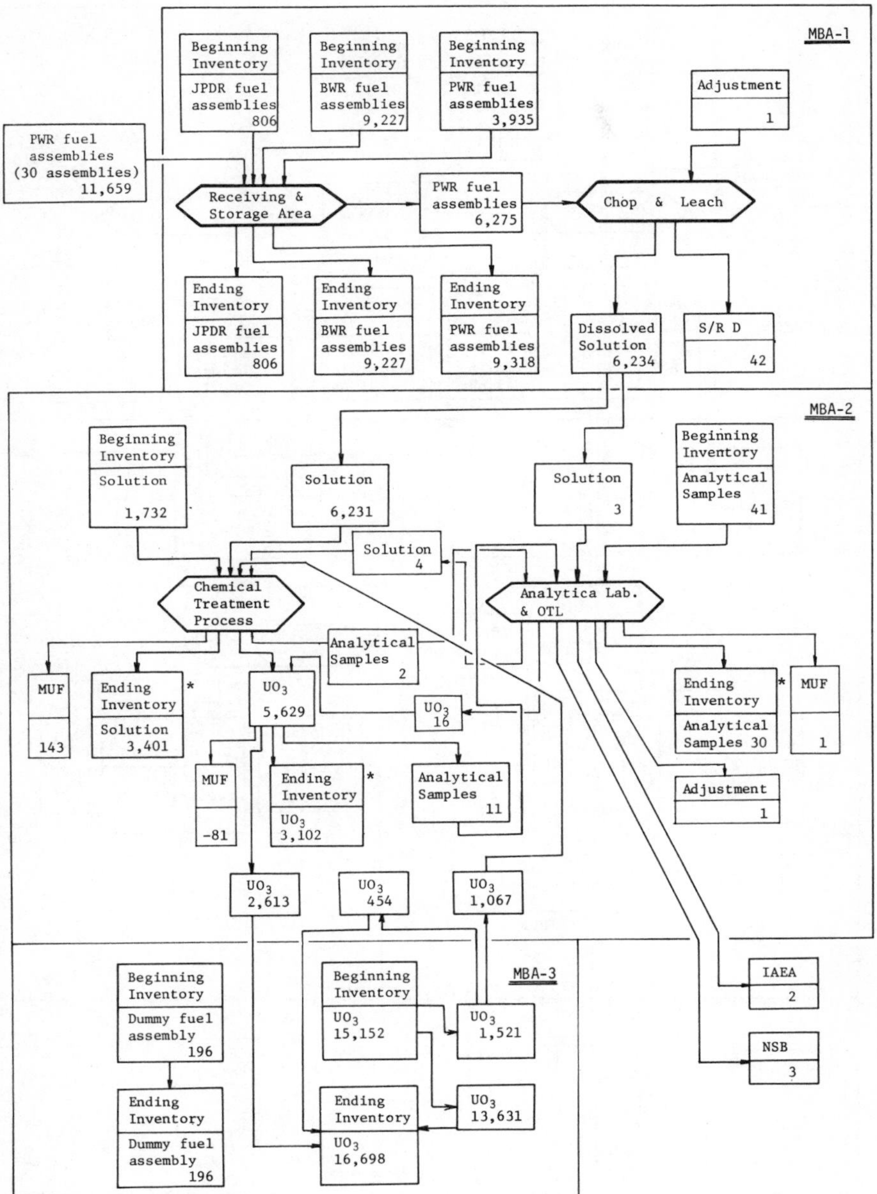

Unit : kgU

FIG.7. Material balance of uranium during thr PWR campaign.

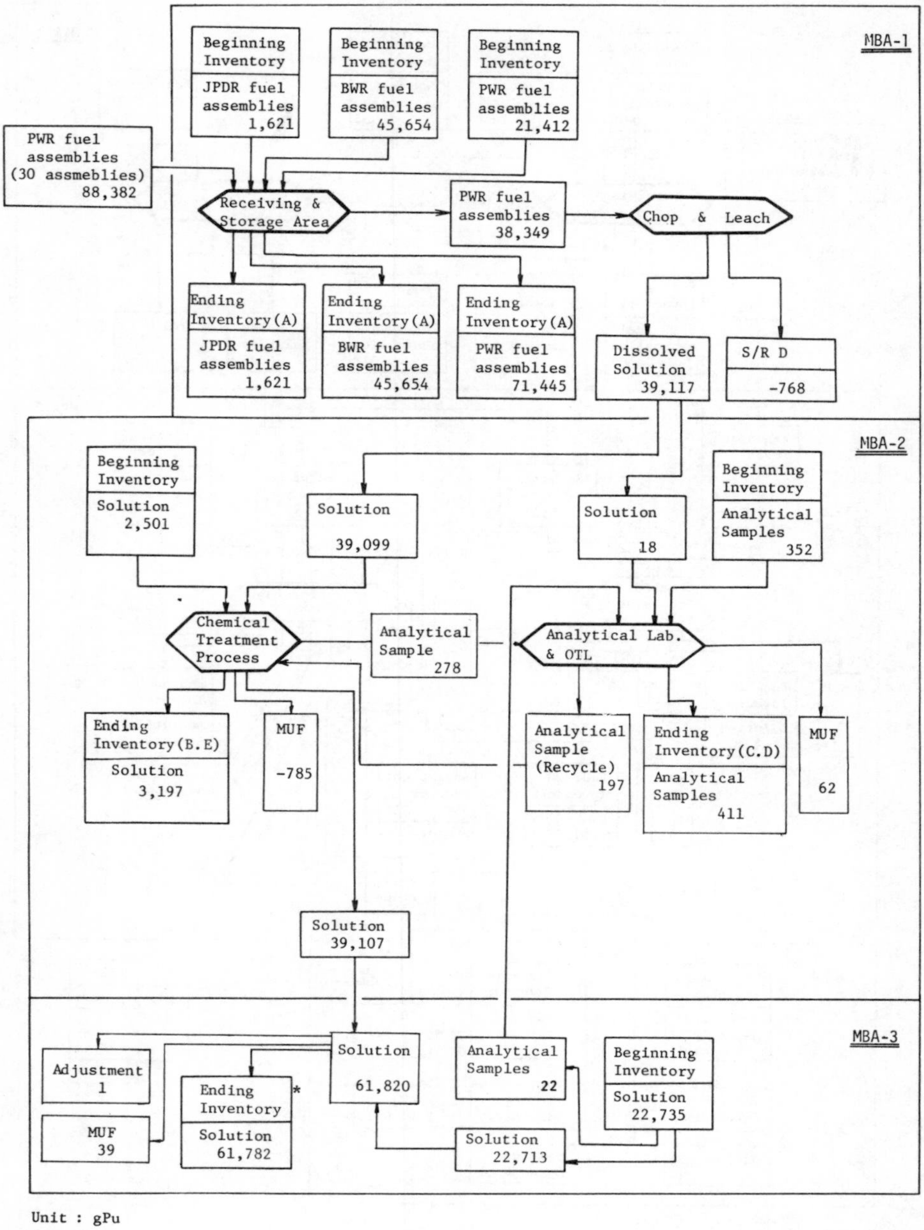

FIG.8. Material balance of plutonium during the PWR campaign.

in whole process was 441 g plutonium. MUF in MBA-3 seems to be caused by the hold-up on the pipe surface because this pipe-line was used for the first time for plutonium solution.

3.3. Material balance during PWR campaign

3.3.3. Material balance of uranium

The material balance of uranium from 6 April to 15 June 1978 is shown in Fig.7.

At MBA-1, 30 assemblies of the PWR fuel which included 11 659 kg uranium from Mihama, were received in the spent fuel storage pond during the campaign; 16 PWR fuel assemblies, which had 6275 kg uranium, were treated in chop and leach processes. After accounting, the dissolver solution, which included 6234 kg uranium, was transferred into the chemical process area and OTL, so that the SRD became 42 kg uranium. At MBA-2, during the campaign 6231 kg uranium were received in the chemical treatment area, where 1732 kg uranium in the form of nitrate solution and 1067 kg in the form of UO_3 powder were transferred from MBA-3 for the seed material of the denitration; 5629 kg were recovered as product-formed UO_3 powder and 2613 kg uranium product was transferred into the uranium product storage facility. After PIT, the MUF of uranium was estimated at 64 kg, which corresponds to 0.70% for the input and the starting inventory.

3.3.4. Material balance of plutonium

The material balance of plutonium from 6 April to 15 June 1978 is shown in Fig.8. At MBA-1, 30 assemblies of the PWR fuel that included 88 382 kg plutonium from Mihama were received in the spent fuel storage pond during the campaign; 16 PWR fuel assemblies that included 38 349 g of plutonium were treated in chop and leach processes. After accounting, the dissolver solution, which included 39 117 g plutonium, was transferred to the chemical process area and OTL, so that SRD became minus 768 g. At MBA-2, during the campaign 39099 g plutonium were received in the chemical treatment area, where there were 2501 g in the form of nitrate solution; 39 107 g were recovered as product in the form of nitrate solution, and transferred into the plutonium storage vessel located in the basement of the main process building. After PIT, the MUF of plutonium was estimated at minus 785 g in the chemical treatment process area; 62 g in the analytical laboratory and OTL; and 39 g were estimated as MUF in MBA-3.

TABLE IV. ESTIMATED σ_{MUFe} OF PLUTONIUM DURING BWR CAMPAIGN
Unit (g)

KMP	Batch	Throughput	Measurement error (RSD %)	σ_i
Input	FU-1- 1	1,092	o Volume measurement : 0.5	13
	2	1,448	o Plutonium concentration by Mass Spec. : 1.0	18
	3	1,428		18
	4	1,521	o Sampling : 0.5	19
	5	1,665	Total : 1.22	21
	6	1,323		16
	7	1,760		22
	8	1,978		25
	9	1,733		21
	10	1,964		24
	11	1,774		22
	12	2,693		33
	Total	20,379		σ_1 = 75
Product output	1	4,000	o Volume measurement : 0.4	48
	2	6,144	o Plutonium concentration : 1.0	74
	3	6,503	o Sampling : 0.5	78
	4	7,125	Total : 1.20	86
	Total	23,772		σ_2=146
In-process Inventory	B	2,466	Measurement errors depending upon each inventory vessel	24
	C	272		2
	D	80		~
	E	35		0
	Total			σ_3=24

$$\sigma_{MUFe} = \sqrt{\sigma_1{}^2+\sigma_2{}^2+\sigma_3{}^2} = \sqrt{75^2+146^2+24^2} = 166$$

TABLE V. ACTUAL MUF AND VARIANCE MUF OF PLUTONIUM DURING THE HOT TEST
Unit (g)

Campaign	Actual MUF	σ_{MUF} [a]	$1.645\times\sigma_{MUF}$ [b]
JPDR	-36	68	112
BWR	-596	166	274
PWR	-723	231	381

[a] σ_{MUF} indicates that the standard deviation of the variance of MUF is caused by measurement errors at main KMPs.

[b] One-side test for MUF at Type I error of 5% in such a manner that the critical region=$1.645\times\sigma_{MUF}$.

3.4. MUF evaluation

It is well known that the MUF evaluation technique which uses the variance of MUF from the measurement errors at KMPs is useful as a safeguards measure. The actual MUF and MUF variance of plutonium during the JPDR, BWR and PWR campaigns are shown as Tables IV and V. In MBA-2 all actual MUF of plutonium was evaluated as minus, so that no diversion occurred; nevertheless, it seems to be necessary to check measurement systems, especially volume measurements of input and plutonium product accountability vessels. The results of a one-sided test for MUF at a Type I error of 5% show that:

For JPDR campaign: no significance
For BWR campaign: significant
For PWR campaign: significant

It is necessary to investigate the reasons in the next campaign.

4. APPLICATION OF NEW SAFEGUARDS TECHNIQUES FOR THE PLANT

4.1. The circumstances of TASTEX project

4.1.1. Before starting the hot tests of the Tokai reprocessing plant, a survey of plant safeguardability was carried out on 10 September 1977 by a United States delegation.

4.1.2. In the joint communique issued by the US and Japanese Governments on 12 September 1977, it was stated that:

"Japan is willing to improve the safeguardability and physical security at the facility, and for this purpose is prepared to cooperate with the IAEA in the testing of advanced safeguards instrumentation, and to make timely preparations to facilitate the use of such instrumentation in the initial period.

The United States is prepared to participate in this safeguards testing through agreed means.

Japan and the United States will promptly consult with the IAEA to facilitate implementation of this testing program."

4.1.3. A joint Japan-US-IAEA programme for testing advanced safeguards instrumentation at the Tokai reprocessing plant (TASTEX) was proposed by the United States of America on 12 January 1978.

The eleven items of the proposal are as follows:

- (T-A) Evaluation of performance and application of surveillance devices in the spent fuel receiving areas
- (T-B) Collection and analysis of gamma spectra of irradiated fuel assemblies measured at the storage pool
- (T-C) Demonstration of hull monitoring system
- (T-D) Demonstration of the load cell technique for measurement of solution weight in accountability vessels
- (T-E) Demonstration of the electromanometer for measurement of solution volume in accountability vessels
- (T-F) Study of application of DYMAC principles to safeguarding spent fuel reprocessing plants
- (T-G) K-edge densitometer for measuring plutonium product concentrations
- (T-H) High-resolution gamma spectrometer for plutonium isotopic analysis
- (T-I) Monitoring the plutonium product area
- (T-J) Resin-bead sampling and analytical technique
- (T-K) Isotope safeguards techniques

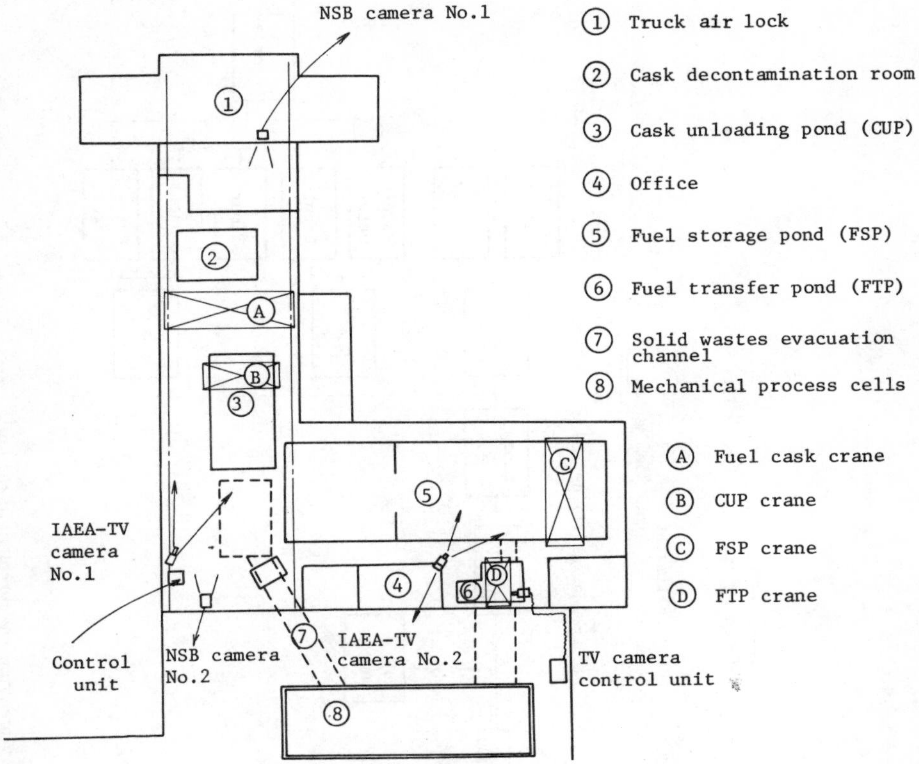

FIG.9. Spent fuel receiving and storage area of the PNC plant and location of surveillance equipment.

Meetings were held in Tokyo and at the Tokai facility, 2 February to 8 March 1978 and the agreed-upon programme included all eleven items of the initial United States proposal and two additional items proposed by France as follows:

(T-L) Gravimetric method for input measurement
(T-M) Tracer method for input measurement

Investigations on these thirteen items are to be carried out now, so that the results will be available in the near future.

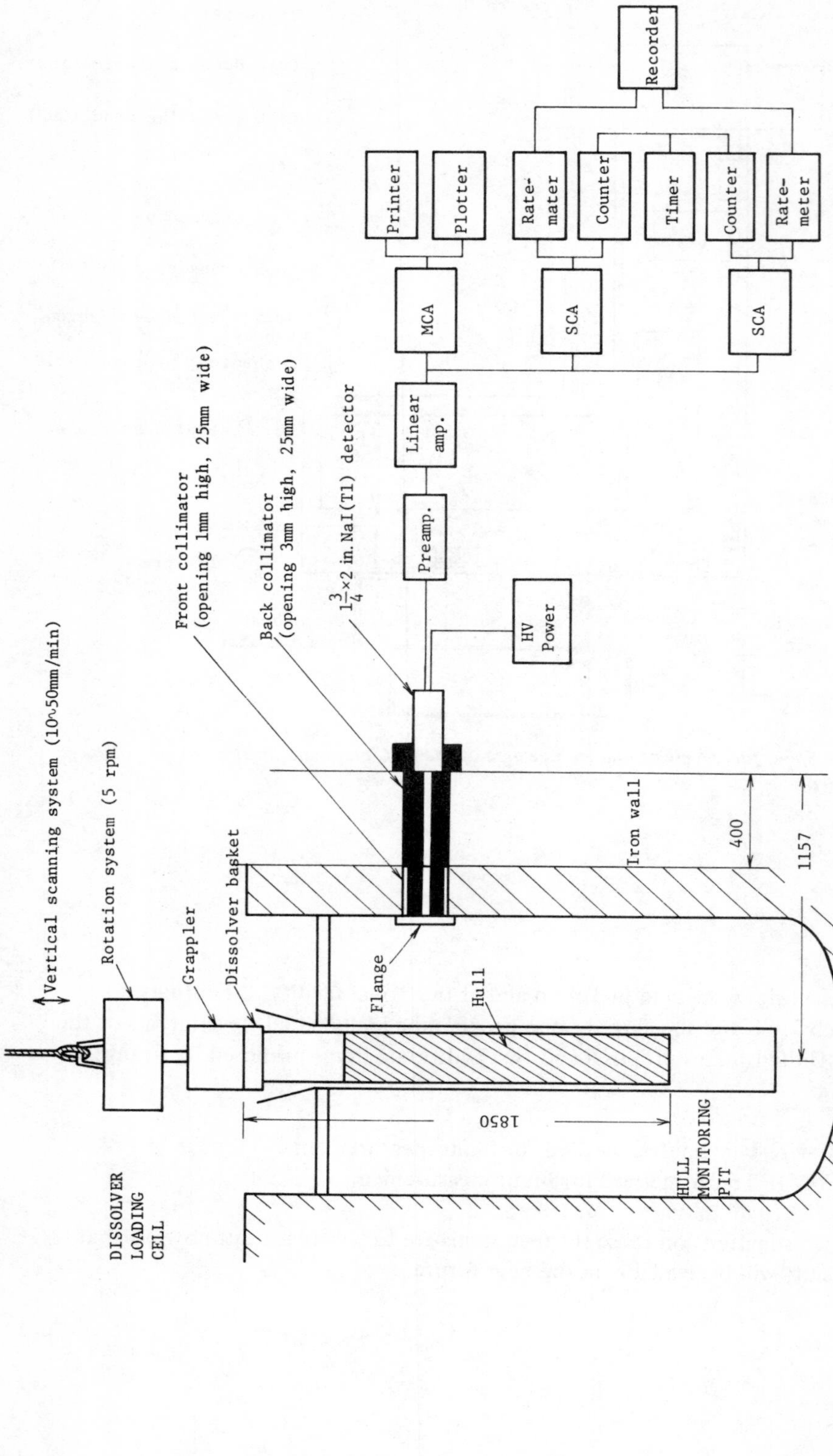

FIG.10. Gamma-ray spectrum measurement of the hull.

4.2. Application of surveillance devices in the spent fuel receiving area

Since nuclear materials are handled solely by an individual fuel assembly in the fuel receiving and storage area, the material control will be performed by following up the movement of the assemblies and by grasping the fuel assembly number in the fuel storage pond, after the individual fuel assembly has been identified and registered.

Some equipment and components, shown in Fig.9 for surveillance of the area, are being demonstrated and evaluated for their performance and adaptability.

4.2.1. Crane operating monitor

In the plant, a 110-tonne overhead crane is provided for handling fuel casks whose weight is generally over 70 t and a specially tailored shielding container of 40 t, which is used for evacuating solid wastes, i.e. leached hulls, from the mechanical process cells.

Also, for handling the fuel assemblies removed from the cask and the fuel storage baskets, specially designed cranes are installed over the cask unloading pond (CUP), fuel storage pond (FSP) and fuel transfer pond (FTP).

Correlation between the movements of the fuel-handling devices and the actual flow of fuel assemblies is being investigated by having operation monitors installed on the cask crane and the pond cranes.

The monitors can record operating time of the cranes on recording paper disks and discriminate the movement into X (travelling), Y (traversing) and Z (lifting) directions. An analysis on the various patterns of the records resulting from the sequential procedures of fuel handling in routine and non-routine modes, will be conducted in order to characterize the flow of fuel assemblies through the fuel reception and transfer.

Transmittance of the signals of operation monitors into a computer analyser will be discussed.

4.2.2. Underwater surveillance CCTV system

A system composed of a TV camera and video monitor and recorder with a superimposer of date-time has been installed at the FTP.

The camera is located over the FTP and views underwater through a tube one end of which dips in the water. The camera view covers the fuel assembly receiving position of a fuel conveyor to the mechanical process cell.

The stability and operability of the system will be investigated first, and a further combination of material control systems and operation control systems will be discussed.

TABLE VI. THE REMAINING RATIOS OF ^{106}Ru-^{106}Rh AND ^{137}Cs AND THE ESTIMATED NUCLEAR FUEL MATERIALS

Basket No.	^{106}Ru-^{106}Rh (%)	^{137}Cs (%)	U (g)	Pu (g)	Basket No.	^{106}Ru-^{106}Rh (%)	^{137}Cs (%)	U (g)	Pu (g)
B- 1	0.28	0.19	370	1.3	P- 5	0.47	0.18	370	1.9
" 2	0.08	0.11	220	0.8	" 6	0.30	0.15	300	1.5
" 3	1.1	0.15	300	1.2	" 7	0.13	0.46	930	1.9
" 4	1.0	0.13	340	1.0	" 8	0.24	0.13	260	1.4
" 5	0.72	0.11	220	0.9	" 9	0.84	0.20	400	2.2
" 6	0.45	0.12	230	1.0	" 10	0.94	0.16	320	1.8
" 7	1.1	0.17	340	1.4	" 11	0.07	0.18	360	2.0
" 8	0.77	0.19	370	1.6	" 12	0.34	0.17	330	1.8
" 9	1.9	0.34	660	2.9	" 13	0.48	0.19	370	2.1
" 10	2.0	0.17	330	1.5	" 14	1.1	0.16	310	1.7
" 11	2.0	0.22	430	2.0	" 15	0.13	0.12	240	1.3
" 12	2.0	0.16	310	1.5	" 16	0.11	0.14	280	1.6
" 13	2.3	0.53	1,000	4.6	" 17	0.28	0.18	350	2.5
" 14	2.2	0.38	750	3.4	" 18	0.33	0.13	260	1.8
" 15	0.50	0.32	630	2.9	" 19	0.76	0.20	410	2.9
" 16	0.19	0.26	520	2.4	" 20	0.53	0.15	310	2.2
" 17	1.1	0.32	620	2.9	" 21	0.47	0.24	470	3.3
" 18	0.60	0.38	730	3.5	" 22	0.27	0.13	260	1.8
" 19	0.17	0.39	760	3.6	" 23	0.14	0.17	330	2.3
" 20	0.31	0.36	700	3.3	" 24	0.21	0.17	330	2.4
" 21	0.43	0.42	810	3.8	" 25	0.34	0.16	320	2.2
" 22	0.36	0.43	840	4.8	" 26	0.36	0.12	250	1.8
" 23	0.44	0.33	640	3.0	" 27	0.10	0.19	380	2.7
" 24	0.68	0.27	530	2.5	" 28	0.10	0.14	270	2.9
BWR Total			12650	57.8	" 29	1.0	0.20	400	2.2
P- 1	0	0.15	290	1.4	" 30	0.77	0.16	310	1.8
"- 2	0.15	0.12	250	1.2	" 31	1.4	0.20	410	2.3
"- 3	0.12	0.16	330	1.6	" 32	1.3	0.17	330	1.8
"- 4	0.11	0.16	310	1.6	PWR Total			11,040	63.9

B-1, B-2 burn-up : ∿7000 MW · d/t U
B-3∿B-24 burn-up : 8600∿10200 MW · d/t U
P-1∿P-16, P-29∿P-32 burn-up : 10800∿13000 MW · d/t U
P-17∿P-28 burn-up : 19200∿19500 MW · d/t U

The other CCTV system, scheduled to be installed in the CUP, is now being procured. The camera will be contained in a water-tight casing with remote-control mechanism.

4.2.3. Surveillance cameras

Two sets of motor-driven cameras (ROBOT, 36 CE) using 35-mm roll film (200 ft) have been delivered.

The camera has a motorized film-winding mechanism, a random shutter controller from one shot per one second to one per 20 min, and a superimposer of date-time and serial number. And the camera is mounted in a water-proof housing with a window electrically heated.

Operability and reliability will be demonstrated.

The Safeguards Division, Nuclear Safety Bureau, Science and Technology Agency, are also demonstrating their compact surveillance cameras originating from an 8-mm movie camera in the area.

4.3. Demonstration of a hull-monitoring system

By means of γ-ray spectrometry, fission products remaining in the hulls were quantitatively analysed. From the obtained values, the nuclear fuel materials (U, Pu) remaining in the hulls were estimated. Twenty-four baskets filled with the hulls from BWR fuel and thirty-two baskets from PWR fuel were measured. The hulls from one assembly of BWR fuel or one half assembly of PWR fuel were put in one basket. Figure 10 illustrates a hull-monitoring system which is composed of a γ-ray spectrum measuring system using a 1.75 in. × 2 in. NaI(Tl) detector and a basket-scanning system.

Energy-resolving power of the detector was sufficient for the γ-rays of ^{125}Sb, ^{106}Ru-^{106}Rh and ^{134}Cs to be mixed with the γ-ray peak (662 keV) of ^{137}Cs. And then a correction of the mixed γ-rays was necessary. The basket-scanning speed was set up at 10 mm/min vertical with rotation of 5 rpm. The measured γ-ray counts from the hulls were transformed into the activity of the isotope using the following equations

$$A = C/\eta F G(\rho) \frac{V_m}{V_n} \Omega \epsilon K T$$

$$\rho = \frac{W}{V_h}, \quad V_h = \pi R^2 K, \quad \Omega = \frac{a}{4\pi L^2}$$

where A is total activity of a specific isotope in a dissolver basket, L is the distance from the detector to the basket centre axis, C is the measured peak counts for the isotope, η is γ-ray emission rate per disintegration, F is γ-ray transmission in the flange and the basket wall, $G(\rho)$ is γ-ray transmission in the hulls which is a function of the hull density ρ, V_m is volume of the hulls in the viewer of the collimator, V_h is volume of the total hulls in the basket, Ω is a solid angle of the collimated detector, ϵ is detection efficiency of the measured γ-ray, K is a coefficient transforming dps to curie, T is measuring time, ρ is the measured hull density, W is weight of the hulls, R is inter-radius of the basket, H is height of the hulls and a is section area of the collimator.

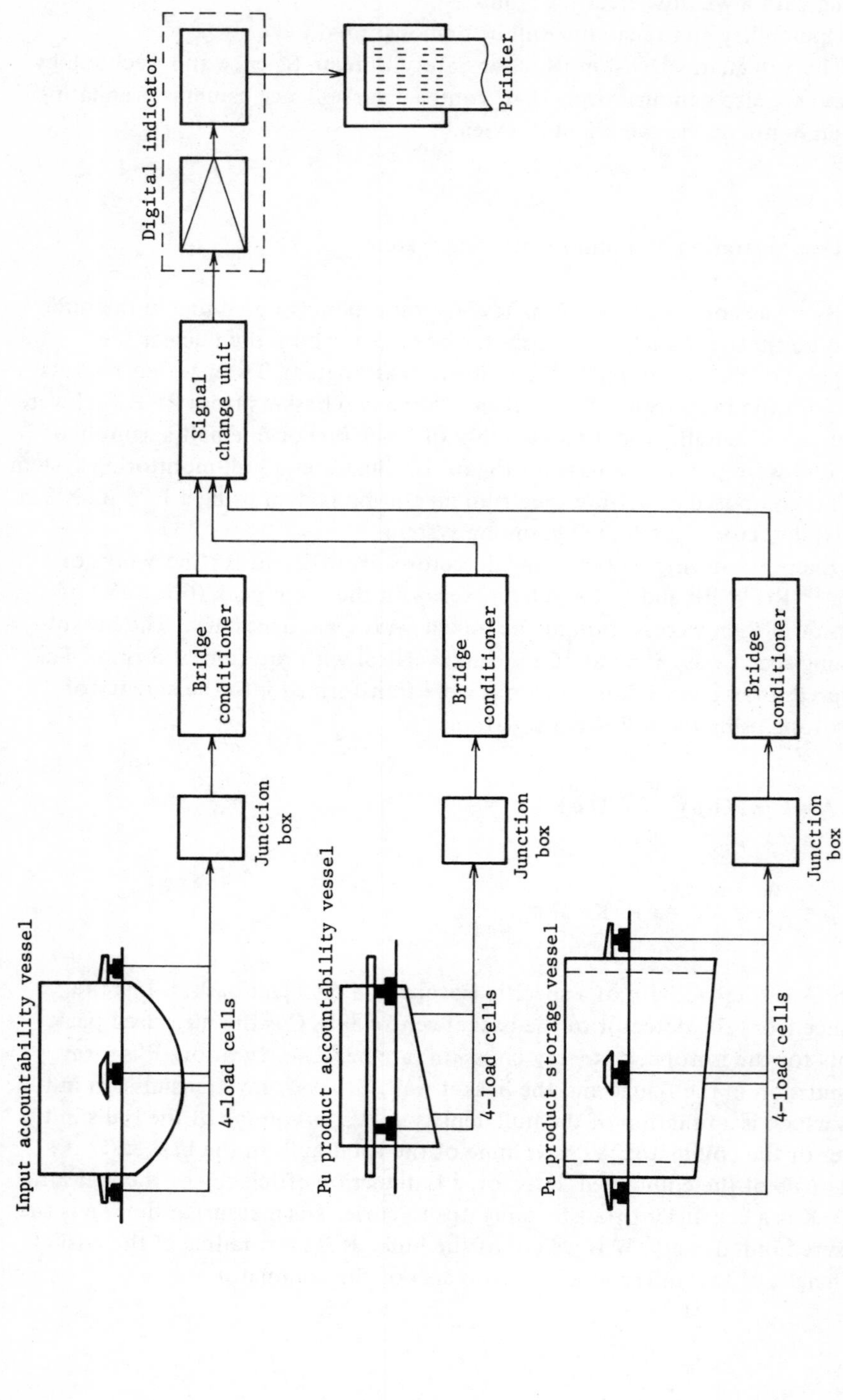

FIG.11. Block diagram of the installed load cell system.

Since the cooling time of the fuels processed was as long as three or five years, the fission products with the γ-ray peaks clearly detected were ^{137}Cs (half-life: 30 a), ^{106}Ru-^{106}Rh (369 d) and ^{125}Sb (2.53 a). The high-energy (2.18 MeV) γ-ray of ^{144}Ce-^{144}Pr, which was not affected by the scattering of the other γ-rays, but had a low γ-ray emission rate (0.74%), could be scarcely detected. The isotope was originally selected as a representative indicator.

The measured activities of ^{137}Cs and ^{106}Ru-^{106}Rh remaining in the hulls in each basket were shown in Table VI. The accuracy of measurement might be ± 40%. The results for ^{106}Ru-^{106}Rh were very variable, but for ^{137}Cs were comparatively uniform. And the distribution of ^{137}Cs and ^{106}Ru-^{106}Rh along the vertical axis of basket was observed as not being similar to each other.

For the time being, it is difficult to clarify which of the measured isotopes, ^{137}Cs and ^{106}Ru-^{105}Rh, represents more closely the residual uranium and plutonium.

Assuming that a ratio between the activity of ^{137}Cs, derived from the burnup calculation using the ORIGEN code, contained in the chopped fuels in the basket, and the measured one remaining in the hulls, was the same at a ratio between the initial charge of uranium and plutonium in the basket and their residue after leaching; the estimate of uranium and plutonium remaining in the leached hulls is given in Table V. Referring to the SRD values shown in Section 3, the estimate seems to be acceptable.

A further investigation will be conducted into a more accurate measurement using Ge (pure) detector and establishing the correlation between the activities of the indicator isotope, i.e. ^{137}Cs and ^{144}Pr, remaining in the leached hulls, and the residual uranium and plutonium through the accumulated plant operation data.

4.4. Demonstration of the weighing technique of the accountability vessel by load cell system

Reprocessing of nuclear fuel requires the accurate accounting of the nuclear material in the solution. The conventional volume and concentration method have been employed, but if the accurate accounting of the nuclear material is achieved directly, many advantages will be gained. So PNC started to accept a strain-gauge-type load cell system for the input accountability vessel (251V10), for the Pu product accountability vessel (266V23), and for the Pu product storage vessel (267V10). A block diagram of the installed load cell system is shown in Fig.11.

From January to September 1972, a feasibility study was performed, after which the load cells to be installed in the Tokai plant were manufactured.

Then the strain-gauge load-cell systems were mounted on the vessels in the plant. From November 1973 to March 1977, a demonstration of the weighing

TABLE VII. STANDARD DEVIATION AND BIAS OF X_D/X_L DURING HOT RUN

	251V10 (600∿3,700kg)		266V23 (40∿50kg)		(70∿400kg)	
	σ	δ	σ	δ	σ	δ
September∿December 1977 JPDR campaign	1.4% (N=9)	-0.03%	3.6% (N=5)	-2.6%		
February∿April 1978 BWR campaign	2.2% (N=27) 0.87% (N=24) [a]	-0.9% -0.3% [a]	2.1% (N=5)	-2.6%	5.2% (N=4)	+17.2%
May∿June 1978 PWR campaign	0.68% (N=16) 0.75% (N=16) [a] 0.74% (N=16) [b]	-0.01% ±0 % [a] -0.01% [b]	5.9% (N=6) 4.0% (N=6) [a] 2.1% (N=6) [b]	+12.9% + 3.4% [a] + 2.5% [b]	0.83% (N=6) 0.67% (N=6) [a] 0.55% (N=6) [b]	-18.2% -12.5% [a] -12.2% [b]

[a] Initial value correction (IVC). [b] IVC & average of periodic digital indication

Standard deviation $\sigma = \sqrt{\dfrac{\Sigma(Xi - \bar{X})^2}{N - 1}}$

$Xi = X_D/X_L$

X_D = measured weight by diptube system

X_L = measured weight by load cell system

$\bar{X}$ = average of Xi

N = number

$\delta = \bar{X} - 1$

technique during the blank test, chemical test and uranium cold test, was performed and the results are as follows. According to the calibration of mass measurement, using demineralized water for each vessel, the standard deviations for 251V10 and 267V10 were 0.3%; these values were satisfactory, but the standard deviation for 266V23 at 1.9% was unsatisfactory.

Using non-irradiated $UO_2(NO_3)_2$, the standard deviation of the ratio of measured weight by the load-cell system and the measured weight by the dip-tube system, was 0.4% in summer and 0.7% in winter, and these values were a little more than expected.

Using active $UO_2(NO_3)_2$ and $Pu(NO_3)_4$, the standard deviation of the ratio of measured weight by the dip-tube system and measured weight by the load-cell system obtained from September 1977 to June 1978 during the JPDR, BWR and PWR campaigns, are shown in Table VII.

The accumulated exposure of γ-radiation on the load cell was 7.4×10^5 R (251V10), 1.6×10^4 R (266V23) and 1.9×10^4 R (267V10).

These levels of γ-radiation were far below the designed level of 10^9 R. The drop of insulation resistance and the change of initial value caused by radiation damage to the strain-gauge that leads to the destruction of the load cell system did not occur. But the temperature change of the solution in the vessels would possibly affect load cell indication. To clarify the reason for the deviation we will investigate the data by analysis during the guarantee test campaign.

DISCUSSION

A.G. HAMLIN: I note that surveillance cameras were used in these experiments. Did the information they provided add anything not revealed by other control methods tested in the experiment?

T. KOIZUMI: PNC surveillance cameras were not used to control the reception of spent fuels for the commissioning test, as they had only just been delivered. The performance of the individual cameras is currently being checked on site. The next stage will be to use a combination of cameras and crane monitors. The application of a C/S concept like that described by Mr. Sellers in his paper[1] at the Tokai Plant is also being considered in connection with TASTEX.

W.C. BARTELS: The technical content of your paper constitutes a major contribution to nuclear material accounting at chemical reprocessing plants. Those of us who have had to work with the limited data provided by campaigns of the past are particularly grateful for your results.

---

[1] SELLERS, T.A., IAEA-SM-231/81, those Proceedings, Vol.II.

With reference to the TASTEX programme, I wish to report on the very effective co-operation that has been developed with Japan. During a recent meeting of United States laboratory co-ordinators, examples of such co-operation were abundant. The Tokai plant operator had recently agreed, for instance, to additional floor space for electronics associated with the K-edge plutonium densitometer and had agreed to a separate glove-box for the HRG system for plutonium isotopic measurements.

I should also like to ask a question. You showed us the load cells on the input accountability tank, and I should like to ask how well they performed. We tried them on the accountability tank of a plant in the United States of America but we encountered difficulties due to temperature-induced strains in piping connected to the tank.

T. KOIZUMI: We realize that changes in the temperature of the solution in the vessels can induce strains in the piping connected to the vessels and thus affect the indication of the load cells. Such phenomena are more significant on the plutonium product output accountability vessel, where the accounting batch size is small and the connecting pipes are relatively inflexible; this can be seen from Table VII showing the standard deviation and bias of the measurement.

INVESTIGATION INTO THE POSSIBILITIES OF DETERMINING THE URANIUM AND PLUTONIUM CONTENT OF VVER-TYPE FUEL FOR SAFEGUARDS IN CONNECTION WITH ISOTOPE CORRELATION TECHNIQUES

A. HERMANN, H.-C. MEHNER
Academy of Sciences of the GDR,
Central Institute of Nuclear Research,
Rossendorf, Dresden,
German Democratic Republic

Abstract

INVESTIGATION INTO THE POSSIBILITIES OF DETERMINING THE URANIUM AND PLUTONIUM CONTENT OF VVER-TYPE FUEL FOR SAFEGUARDS IN CONNECTION WITH ISOTOPE CORRELATION TECHNIQUES.

Correlations between burnup, Pu/U ratio and U/Uo ratio on the one hand and the FP ratios $^{134}Cs/^{137}Cs$ and $^{154}Eu/^{137}Cs$ on the other hand have been investigated for the VVER-70-type power reactor. A method is proposed that allows the uranium and plutonium content of closed fuel containments (fuel assemblies, fuel rods) to be determined on the basis of proved correlations between fuel parameters by measuring γ-active FP outside the fuel containment. In the case of correlations with the $^{134}Cs/^{137}Cs$ ratio a dependence on the measuring position along the fuel rod (assembly) has been noted (influence of the irradiation history). The experimental results are compared with theoretical calculations using the COFIP and COHN codes.

INTRODUCTION

Realization of nuclear safeguards in all stages of the nuclear fuel cycle requires all known methods or those being developed to be carefully examined in order to discover a simple and sufficiently precise technique of determining fuel components, which could obtain general acceptance. It is at the stage of reactor irradiation due to the closed fuel containments (fuel assemblies, fuel rods), that nuclear safeguards by experimental methods are especially complicated. Isotope correlations are a promising means for obtaining valuable information about the content of fissionable isotopes within the above-mentioned closed containments of irradiated fuel. This paper proposes a method that allows the uranium and plutonium content to be determined on the basis of proved correlations between fuel parameters (content of U and Pu isotopes, of fission products (FP) and fuel burnup) by measuring γ-active FP outside the fuel containment.

1. CHARACTERIZATION OF EXPERIMENTAL DATA

Experimental data obtained on fuel rods and a whole fuel assembly of the VVER-70 power reactor at Rheinsberg (GDR) were used in order to discover relations between non-destructive measurable fuel parameters (γ-active FP), content of uranium and plutonium isotopes (measured by destructive methods [1]) and fuel burnup. The following treatments were applied to solutions of fuel rod samples, 10 mm long:

γ-spectrometry for γ-active FP;
Spectrophotometry and coulometry for measuring U content [1];
Spectrophotometry for measuring Pu content [2];
Mass spectrometry for determining isotope ratios of U isotopes [1].

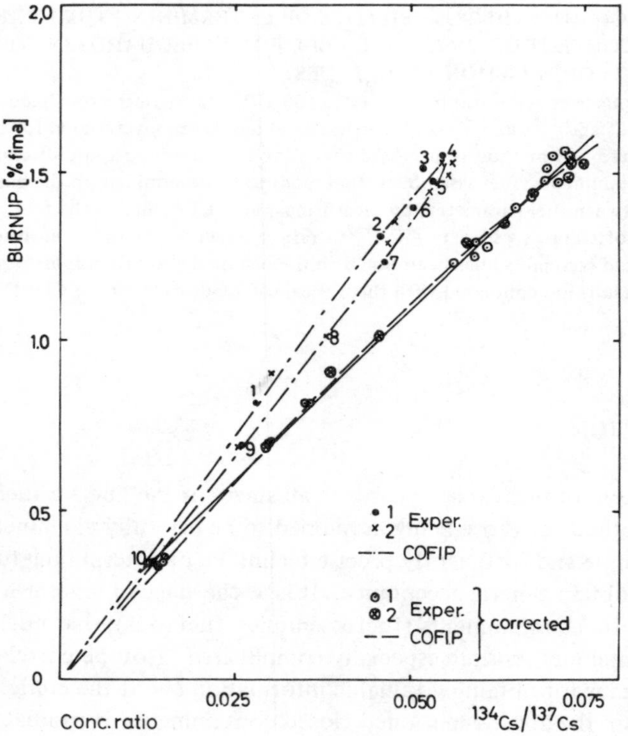

FIG.1. Correlation burnup – $^{134}Cs/^{137}Cs$ concentration ratio for VVER-70 fuel, 2% initial ^{235}U enrichment. (1: measurements on solutions of fuel element segments; 2: measurements on a whole fuel assembly; the figures at the measuring points denote the position of measurement along the element or assembly.)

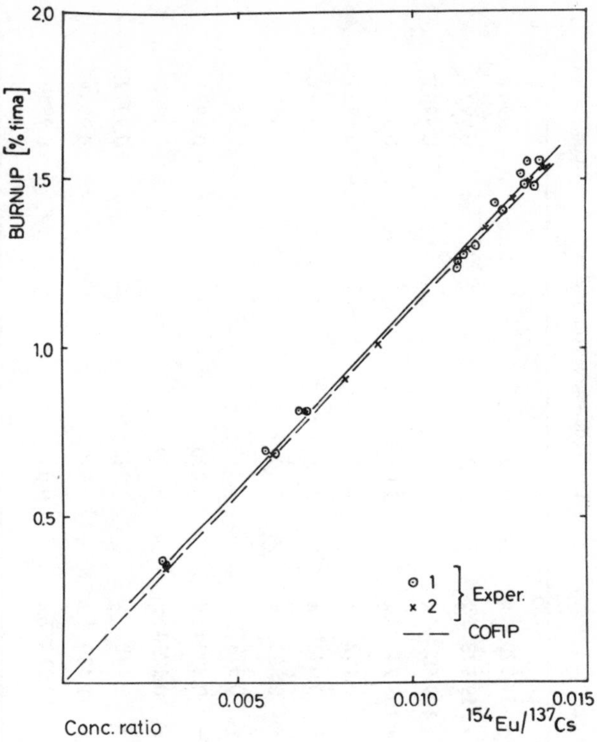

FIG.2. Correlation burnup – $^{154}Eu/^{137}Cs$ concentration ratio for VVER-70 fuel, 2% initial ^{235}U enrichment. (1: measurements on solutions of fuel element segments; 2: measurements on a whole fuel assembly.)

The γ-spectrometric measurements on whole fuel assemblies were performed with the aid of a container, a collimator system and a Ge(Li)-detector in the reactor hall of the VVER-70 power station [3].

Correlations between fuel parameters were found by differential methods, i.e. with discrete measuring positions of fuel rods (or fuel assemblies), but they can also be used for averaged data of whole fuel rods or fuel assemblies. Measurements were performed on 10 points (always in the middle between two distance lattices [4]) and the mean of the 10 measurements results in an averaged value of the measured parameter for the fuel rod (fuel assembly), which is suitable for practical use.

TABLE I. RESULTS OF THE REGRESSION ANALYSIS OF THE CORRELATION CURVES
(2% initial ^{235}U enrichment; burnup range: 0.3 – 1.6% fima)

Correlation y	Correlation x	Equation of the regression line	Relative deviation (%) from the regression line[b]	Correlation coefficient
Burnup	$^{134}Cs/^{137}Cs$	y = − 0.00442 + 27.979 x	3.79	0.9944
Burnup	$^{134}Cs/^{137}Cs$ [a]	y = 0.08988 + 19.973 x	2.72	0.9971
Burnup	$^{154}Eu/^{137}Cs$	y = 0.040067 + 109.50 x	2.98	0.9965
Burnup	$^{154}Eu/^{137}Cs$ [a]	y = 0.04411 + 108.24 x	2.95	0.9966
Pu/U	$^{134}Cs/^{137}Cs$	y = 0.000684 + 0.1027 x	4.40	0.9923
Pu/U	$^{134}Cs/^{137}Cs$ [a]	y = 0.001029 + 0.0729 x	3.15	0.9961
Pu/U	$^{154}Eu/^{137}Cs$	y = 0.000975 + 0.3830 x	2.30	0.9979
Pu/U	$^{154}Eu/^{137}Cs$ [a]	y = 0.000993 + 0.3781 x	2.27	0.9979
U/Uo	$^{134}Cs/^{137}Cs$	y = 0.9998 − 0.3934 x	0.070	−0.9943
U/Uo	$^{134}Cs/^{137}Cs$ [a]	y = 0.9985 − 0.2794 x	0.029	−0.9990
U/Uo	$^{154}Eu/^{137}Cs$	y = 0.9985 − 1.448 x	0.042	−0.9979
U/Uo	$^{154}Eu/^{137}Cs$ [a]	y = 0.9985 − 1.430 x	0.040	−0.9981
U/Uo	Burnup	y = 0.9992 − 0.01337 x	0.037	−0.9989

[a] Ratios corrected for irradiation history

[b] $$\sqrt{\frac{1}{n-2}\sum_{i=1}^{n}(y_i - Y_i)^2 / \frac{1}{n}\sum_{i=1}^{n} y_i}$$

2. ISOTOPE CORRELATIONS VALUABLE FOR SAFEGUARDS

A number of correlations between fuel parameters (content of U and Pu isotopes, of FP and fuel burnup) have been found to be useful for safeguards [5, 6], but there is a lack of data concerning the VVER-type PWR, especially where the connection between non-destructive and destructive measurable isotopes is concerned. Points the data should be used to evaluate are:

Confirmation of isotope correlations, which have been found in other PWRs or BWRs, for the VVER-type reactor;
Making practical use of correlations;
Investigation of the applicability limits of correlations depending on irradiation conditions, fuel parameters etc.

The results we have achieved allow only the first two problems to be dealt with.

2.1. Correlations between the burnup and γ-active fission product ratios

The long-lived FP which are suitable for safeguard measurements are ^{137}Cs, ^{134}Cs, ^{106}Ru and ^{154}Eu. By defining useful correlations as those being linear with best linear indices [5] – after investigating all possible ratios between these isotopes – the correlations between the burnup and the $^{134}Cs/^{137}Cs$ atom ratio as well as the $^{154}Eu/^{137}Cs$ atom ratio must therefore be considered to be the most suitable ones.

Figure 1 demonstrates the correlation between the burnup and the $^{134}Cs/^{137}Cs$ ratio for the case of 2% initial ^{235}U enrichment. In general a linear correlation between burnup and the $^{134}Cs/^{137}Cs$ atom ratio at the end of reactor irradiation does exist, but nevertheless a notable dependence on the measuring position can be noticed (measuring points 1 to 10 equally distributed from the bottom to the top of the fuel assembly or fuel element). This dependence is caused by changes in the axial neutron flux density as a result of moving the control rods during irradiation and can be abolished by correcting the $^{134}Cs/^{137}Cs$ ratio for irradiation history. Figure 1 shows the good agreement between experimental data measured on the whole fuel assembly as well as on two single fuel elements (from 90) of the same assembly (including one element with a corner position).

When considering a great number of fuel element measurements with this correlation, the picture seems to be more complicated [7]. However, the correlation between burnup and the $^{134}Cs/^{137}Cs$ ratio is well expressed by a straight line with a relative deviation of the experimental points of 1–2% when the atom ratios are corrected for irradiation history. A dependence of the slope of the correlation straight line on the initial ^{238}U enrichment has been proved to exist [8].

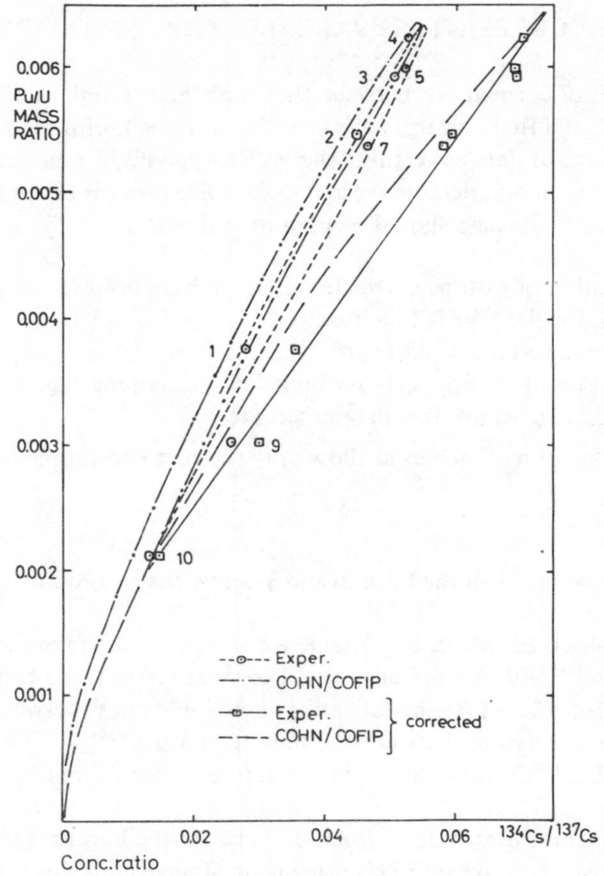

FIG.3. Correlation Pu/U – $^{134}Cs/^{137}Cs$ concentration ratio for VVER-70 fuel, 2% initial ^{235}U enrichment. (The figures at the measuring points denote the position of measurement along the fuel element.)

The reliability of the experimental results has been emphasized by results of theoretical calculations using the COFIP code [9], which calculates the FP concentrations of ^{106}Ru, ^{134}Cs, ^{137}Cs, ^{144}Ce and ^{154}Eu and also the burnup on the basis of a one-group formalism for cross-sections assuming a known irradiation history.

In Fig.2 the experimental data and the results of COFIP calculations for the correlation ^{154}Eu/^{137}Cs burnup are collected together. Besides a good linear correlation this relationship is notable for not exhibiting the above-mentioned dependence on the measuring position and it can be used without correcting for irradiation history. Other advantages of this correlation are the greater linear range with increasing burnup and a diminished dependence on the neutron spectrum as compared with the ^{134}Cs/^{137}Cs ratio.

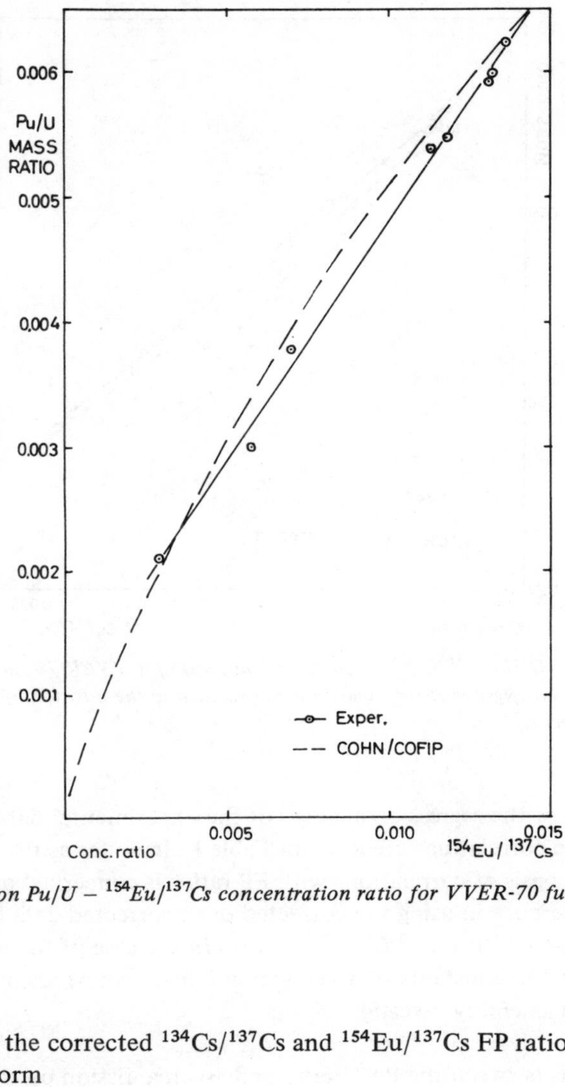

FIG.4. Correlation Pu/U – ^{154}Eu/^{137}Cs concentration ratio for VVER-70 fuel, 2% initial ^{235}U enrichment.

Regarding the corrected ^{134}Cs/^{137}Cs and ^{154}Eu/^{137}Cs FP ratios an approximation of the form

$$\beta \approx 2 \frac{\alpha}{\sigma_i^a} \frac{Y_{j\,25}}{Y_{i\,25}} \frac{n_{i+1}^*}{n_j^*} \qquad (1)$$

(β-burnup; α-proportionality coefficient between β and the fluence; σ_i^a = neutron absorption cross-section of FP i; Y = fission yield; n* = corrected number of atoms) has been derived, i and j denoting primary FP. Equation (1) again illustrates the theoretical background of the linear correlation found between burnup and the ^{134}Cs/^{137}Cs and ^{154}Eu/^{137}Cs ratios.

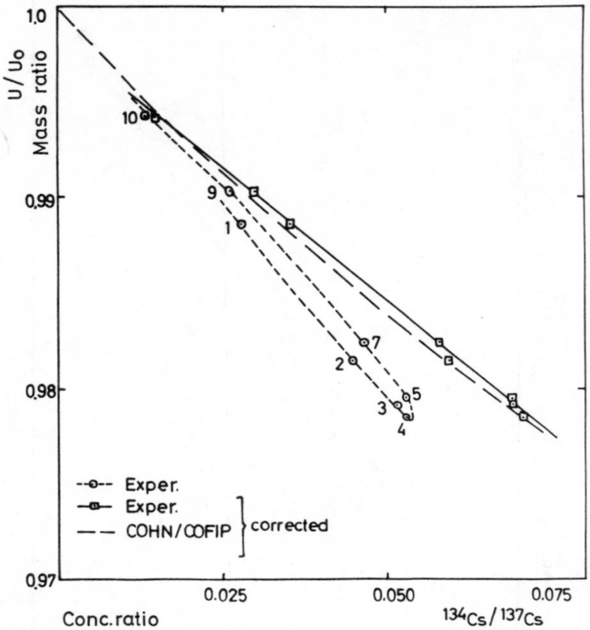

FIG.5. Correlation U/Uo – $^{134}Cs/^{137}Cs$ concentration ratio for VVER-70 fuel, 2% initial ^{235}U enrichment. (The figures at the measuring points denote the position of measurement along the fuel element.)

The results of the regression analysis of the experimental data representing the most suitable correlations are given in Table I. In so far as the burnup determination on the basis of correlations with FP ratios is concerned one should notice the minimal difference in using the corrected or uncorrected data for the $^{154}Eu/^{137}Cs$ ratio in comparison with the $^{134}Cs/^{137}Cs$ ratio. In the case of the uncorrected $^{134}Cs/^{137}Cs$ ratio the constants of the regression lines can be assumed to be valid only for the fuel assembly investigated.

2.2. Correlations between the Pu/U ratio and γ-active fission product ratios

A linear correlation between Pu/U and $^{134}Cs/^{137}Cs$ has been reported to exist [10, 11]. In the case of the VVER-70-type PWR our data confirm the existence of such a correlation within the investigated burnup region, as can be seen from Fig.3. Figure 4 demonstrates the correlation between Pu/U and the $^{154}Eu/^{137}Cs$ ratio, and again this correlation must be considered preferable because, as a result of the long half-life of ^{154}Eu, the irradiation history hardly influences the correlation curve. This conclusion can be drawn from the data of Table I and Fig.4.

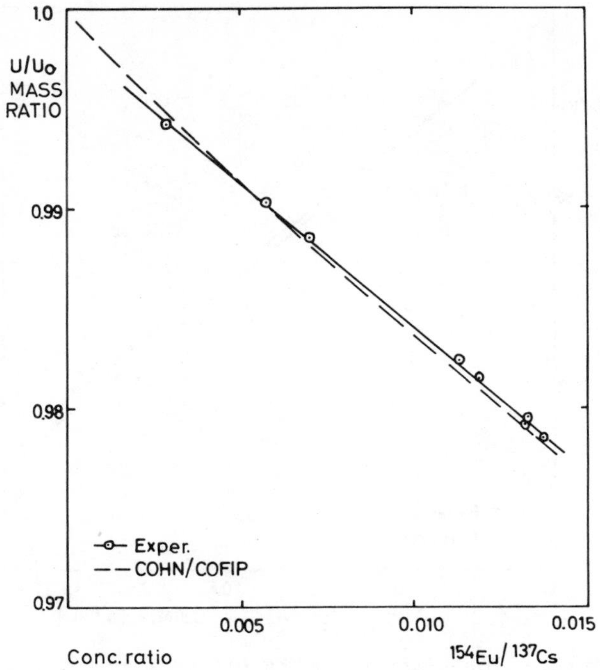

FIG.6. Correlation U/Uo – $^{154}Eu/^{137}Cs$ concentration ratio for VVER-70 fuel, 2% initial ^{235}U enrichment.

In Figs 3 and 4 the experimental data are compared with results of theoretical calculations, using the COHN and COFIP codes for concentrations of heavy nuclides and FP. The COHN code describes the build-up and depletion of heavy nuclides by solving the coupled differential equations using the simplification of dividing the whole process into reaction chains of successive nuclides [9]. The calculated curves cannot be fitted completely to the experimental data, but nevertheless the agreement is sufficient.

The correlations Pu/U – $^{134}Cs/^{137}Cs$ and Pu/U – $^{154}Eu/^{137}Cs$ are not exactly linear, but a straight line is seen to be a good approximation within the investigated burnup range (0.3–1.6% fima). In the case of $^{134}Cs/^{137}Cs$, taking into account the irradiation history, a correction is necessary.

A few data concerning the correlation Pu/U – $^{134}Cs/^{137}Cs$ should be mentioned [12]. These data were obtained on fuel of the VVER-365 reactor (Novo-Voronezh, USSR, 3% initial ^{235}U enrichment) and agree fairly well with our results.

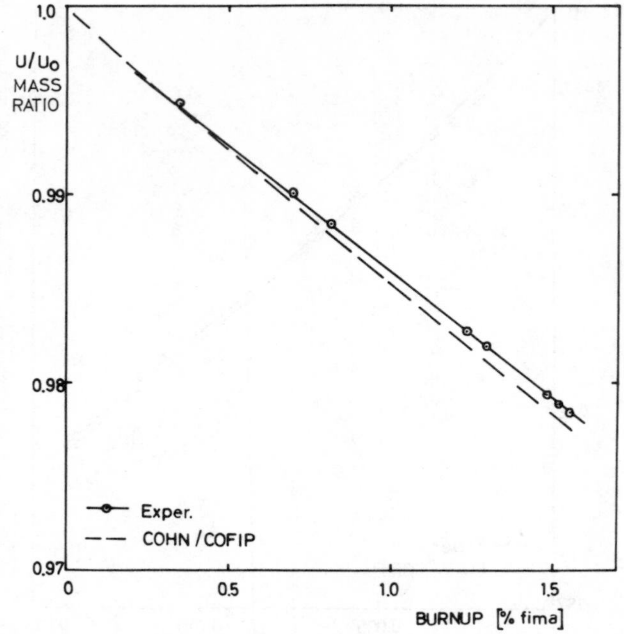

FIG. 7. Correlation U/Uo – burnup for VVER-70 fuel, 2% initial ^{235}U enrichment.

2.3. Correlations between the U/Uo ratio and γ-active fission-product ratios and the burnup

A relation between remaining and initial uranium content (U/Uo) and some other easily measurable information is important for determining uranium content in fuel rods (or assemblies) after irradiation. The correlations between U/Uo and the $^{134}Cs/^{137}Cs$ and $^{154}Eu/^{137}Cs$ ratios are demonstrated in Figs 5 and 6. Again a dependence of the results for the $^{134}Cs/^{137}Cs$ ratio on the measuring position has to be noted, which will disappear by correcting the values by allowing for irradiation history. In the case of $^{154}Eu/^{137}Cs$ such a correction is not necessary. The experimental data are compared with the results of COFIP and COHN calculations, which are illustrated by the dashed lines of Figs 5 and 6. In general, a linear correlation between U/Uo and $^{134}Cs/^{137}Cs$ or $^{154}Eu/^{137}Cs$ can be assumed for practical use within the burnup range investigated. The correlation coefficients given in Table I underline this conclusion. However, as can be seen from the theoretical curve, the dependence between the investigated ratios is not exactly linear.

TABLE II. DETERMINATION OF OF U AND CONTENT OF IRRADIATED FUEL (2% initial ^{235}U enrichment) FROM MEASURED FISSION PRODUCT RATIOS

Object of measurement	Measured value		Determined by correlation techniques			Uranium content (kg)		Plutonium content (g)	
	Isotope	Measured mean fission ratio product	Burnup	Pu/U	U/Uo	By correlation techn.	Calculated	By correlation techn.	Calculated
A1	$^{154}Eu/^{137}Cs$	0.01035	1.173	0.00491	0.98354 0.98354[a]	110.047 110.047[a]	110.006[b] 110.005[c]	540.2 540.2[a]	562.5[b] 532.5[c]
	($^{154}Eu/^{137}Cs$) corrected	0.0104	1.170	0.00490	0.98359 0.98359[a]	110.052 110.052[a]		538.7 538.7[a]	
	$^{134}Cs/^{137}Cs$	0.0420	1.170	0.00497	0.98329 0.98359[a]	110.018 110.052[a]		546.3 546.5[a]	
	($^{134}Cs/^{137}Cs$) corrected	0.0537	1.163	0.00491	0.98347 0.98368[a]	110.039 110.062[a]		541.0 541.1[a]	
A2	$^{134}Cs/^{137}Cs$	0.0376	1.048	0.00452	0.98499 0.98522[a]	110.013 110.039[a]	109.952[c]	497.4 497.5[a]	500.3[c]
	($^{134}Cs/^{137}Cs$) corrected	0.0483	1.055	0.00452	0.98499 0.98513[a]	110.013 110.028[a]		497.7 497.8[a]	
E1	($^{134}Cs/^{137}Cs$) corrected	0.0517	1.122	0.00477	0.98405 0.98422[a]	1.223 1.224[a]	1.222[b] 1.222[c]	5.835 5.836[a]	6.250[b] 5.917[c]
	$^{154}Eu/^{137}Cs$	0.0100	1.138	0.00479	0.98401	1.223		5.860	

A: Fuel assembly; E: Fuel element.
[a] U/Uo by the correlation U/Uo-burnup (burnup determined by correlation techniques).
[b] Calculation code COHN.
[c] Calculation code GRUPA [13].

An exclusively good linear correlation becomes evident between U/Uo and the measured burnup (Fig.7). Therefore, the determination of U/Uo using the U/Uo-burnup correlation has been assumed to be more advantageous than the other possibility, which goes directly from FP ratios to U/Uo. As will be proved later in the case of the $^{134}Cs/^{137}Cs$ ratio, a difference between the two possibilities does exist, but nevertheless it is still difficult to decide which way of determining U/Uo would turn out to be the better one.

3. DETERMINATION BY CORRELATION TECHNIQUES OF URANIUM AND PLUTONIUM IN FUEL ASSEMBLIES

With regard to the data evaluated it has been possible to propose a method to determine the uranium and plutonium content of fuel assemblies (fuel rods) based on measurements of the mean $^{134}Cs/^{137}Cs$ or $^{154}Eu/^{137}Cs$ ratios, as well as isotope correlations verified by our experimental investigations. Naturally, this is only possible provided that an initial value for the uranium content exists (certificate of the fuel assembly). Consequently, the following equations for the uraniumccontent U or the plutonium content Pu can be used:

$$U = Uo\,(U/Uo) \tag{2}$$

$$Pu = U\,(Pu/U) \tag{3}$$

The results of two fuel assemblies and one fuel element are given in Table II. The determined values could not be compared with experimental data; therefore, a comparison with calculated data was performed. The experimental plutonium content (based on correlations) was found to agree in the range of 1 to 4% – i.e. in dependence on the correlation used – with the calculated one, the uranium content in the range of 0.01 to 0.07%.

Judging from the results of Table II there seems to be no difference whether the $^{134}Cs/^{137}Cs$ – or the $^{154}Eu/^{137}Cs$ – correlations are used. However, it is the disadvantages of the $^{134}Cs/^{137}Cs$ ratio, mentioned above, that must be borne in mind. The different possibilities of determining U/Uo (on the basis of the correlations U/Uo-$^{134}Cs/^{137}Cs$, U/Uo-$^{154}Eu/^{137}Cs$ or U/Uo-burnup) bring the same results within the range of this method's uncertainty. With a standard deviation of the FP ratio measurements of ± 2.5%, the uranium content is determined by correlation techniques with a possible error quota of ± 2.5%, the plutonium content with ± 4.2% or ± 4.7% using the Pu/U-$^{154}Eu/^{137}Cs$ or Pu/U-$^{134}Cs/^{137}Cs$ correlations, respectively.

The agreement of experimental and calculated results within that range of uncertainty provides evidence of the suitability of our method. In the case of fuel assembly A1 this is emphasized by results of two different calculation codes (COHN and GRUPA) of different groups of scientists.

4. CONCLUSIONS

Useful correlations and their practical application have been pointed out between burnup, Pu/U and U/Uo, on the one hand, and the $^{134}Cs/^{137}Cs$ and $^{154}Eu/^{137}Cs$ FP ratios on the other. Correlations using the $^{154}Eu/^{137}Cs$ ratio are preferable because they do not, or only slightly, depend on irradiation history and neutron spectrum and a better accuracy will be obtained.

In further investigations the range of applicability of the correlation lines will have to be determined. In spite of the similarity of the correlation lines for different types of PWR and BWR [6, 11, 12], for practical use it will be necessary to establish (or to prove) the correlations for each type of reactor and fuel characteristics (initial enrichment etc.).

REFERENCES

[1] HERMANN, A., STEPHAN, H., HÜBENER, S., NEBEL, D., TREBELJAHR, S., NIESE, U., SUS, F. in Proc. IV Comecon Symp. Investigations in the Field of Irradiated Fuel Reprocessing, 28 March – 1 April 1977, Karlovy Vary, CSSR, **3** (1974) 144.

[2] NEBEL, D., HERMANN, A., TREBELJAHR, S., J. Radioanalyt. Chem. **28** (1975) 133.

[3] GRABER, H., HERMANN, A., BARANIAK, L., HOFMANN, G., HÜBENER, S., MEHNER, H.-C., NAGEL, S., GÜNTHER, H., in Proc. IV Comecon Symp. Investigations in the Field of Irradiated Fuel Reprocessing, 28 March – 1 April 1977, Karlovy Vary, CSSR, **3** (1974) 80.

[4] GRABER, H., HERMANN, A., HÜTTIG, W., GÜNTHER, H., SCHIFF, B., Kernenergie **20** (1977) 98.

[5] CHRISTENSEN, D.E., SCHNEIDER, R.A., in Safeguarding Nuclear Materials (Proc. Symp. Vienna, 1975) **2**, IAEA, Vienna (1976) 377.

[6] GUPTA, D., KFK-2400 (1976).

[7] SYKOV, K.I., MILLER, O.A., At. Ehnerg. **39** (1975) 206.

[8] MEHNER, H.-C., GRABER, H., HOFMANN, G., in Proc. IV Comecon Symp. Investigations in the Field of Irradiated Fuel Reprocessing, 28 March – 1 April 1977, Karlovy Vary, CSSR, **3** (1974) 252.

[9] MEHNER, H.-C., HOFMANN, G., GRABER, H., to be published.

[10] BANNELLA, R., GIURIATO, A., PAOLETTI GUALANDI, M., BRESESTI, A.M., BRESESTI, M., D'ADAMO, D., Trans. Am. Nucl. Soc. **15** (1972) 681.

[11] PAOLETTI GUALANDI, M., PERONI, P., BRESESTI, M., CUYPERS, M., D'ADAMO, D., LEZZOLI, L., Safeguarding Nuclear Materials (Proc. Symp. Vienna, 1975) **2**, IAEA, Vienna (1976) 613.

[12] NOVIKOV, Yu.B., GABESKIRIYA, V.Ya., MASLENNIKOVA, M.N., At. Ehnerg. **43** (1977) 278.

[13] MÖLLER, W., VEB KKAB Berlin, priv. communication.

PRECISE ISOTOPIC CORRELATIONS IN PLUTONIUM RECYCLING OF LIGHT-WATER REACTOR FUEL

P. DE REGGE, R. BODEN
Studiecentrum voor Kernenergie,
Centre d'étude nucléaire,
Mol, Belgium

Abstract

PRECISE ISOTOPIC CORRELATIONS IN PLUTONIUM RECYCLING OF LIGHT-WATER REACTOR FUEL.

A series of measurements on pellets obtained from a plutonium recycling assembly with a burnup ranging between 12 and 45 GW·d/t indicates the existence of precise correlations. The relationships between the different isotopes, the burnup and the actinide concentrations are described together with some of the underlying analytical equations. On the basis of selected correlations the consistency of the isotopic data can be verified within narrow limits and the Pu/U ratio can be predicted within one per cent.

INTRODUCTION

Isotopic correlation techniques are emerging as reliable tools for the verification of declared values at various stages of the nuclear fuel cycle. They are based on regularities observed between concentrations of heavy elements, fission products and isotopic compositions in irradiated fuels, providing the possibility of checking analytical data for internal consistency. A large number of data have been collected in different data banks [1–3] and hundreds of correlations have been tested, many of them yielding correlation coefficients in excess of 0.95 or 0.99. Most of the correlations aim at the inspection of analytical measurements at the input accountability tank of a reprocessing plant, but they can provide a means for the verification of other external informations such as reactor type, initial enrichment and composition, and fuel exposure. Isotopic correlations developed at present cover essentially heavy-water, boiling-water and pressurized-water reactors fuelled by natural or low-enriched uranium. The development of isotopic correlations for PWR recycling plutonium fuel has been recommended on several data sheets of the IAEA Safeguards Manual [4], but no chemical processing data have been made available for this type of fuel, mainly because the recycling of Pu in thermal reactors is still in the demonstration stage. Data stemming from post-irradiation research on prototype fuel are reported for the

Saxton reactor [5] and were recently presented for the Garigliano reactor [6], while preliminary data for the BR3 reactor were given in Ref.[7]. In this paper additional data for the BR3 reactor are presented.

1. FUEL CHARACTERISTICS

The assembly IF 70/4 was the first plutonium pin assembly ever to be introduced in the BR3 reactor at Mol. It was irradiated in the Vulcain core from 29 November 1966 to 18 November 1968. After examinations and leak-testing it was reloaded in core 2b from 1 August 1969 to 21 December 1970. The irradiation corresponds to 890 equivalent fuel power days. The assembly was cooled in the BR3 storage pool until the post-irradiation examination was started in August 1973. The examinations were completed in 1975 [8]. The correlation

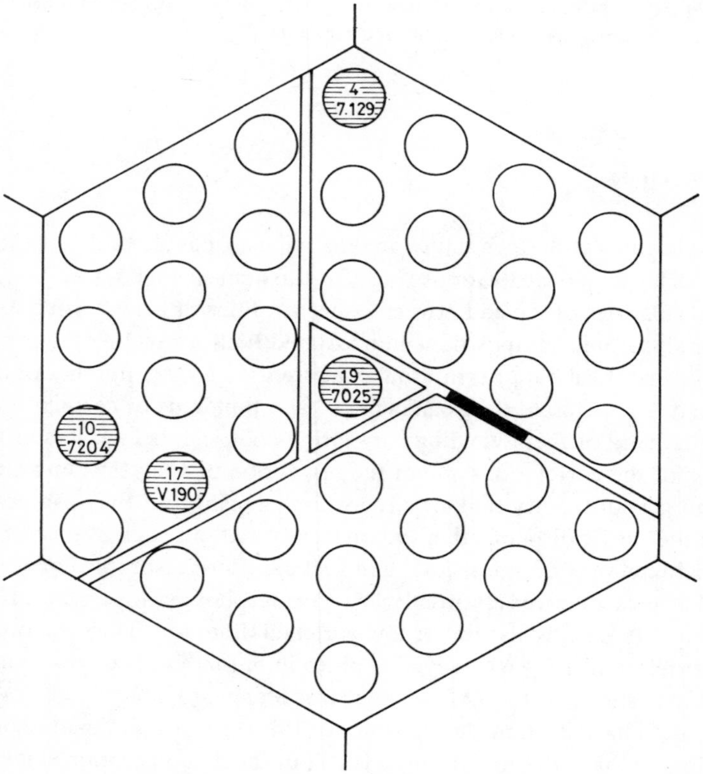

FIG.1. Pin location in the IF 70/4 assembly. (Dimensions in mm.)

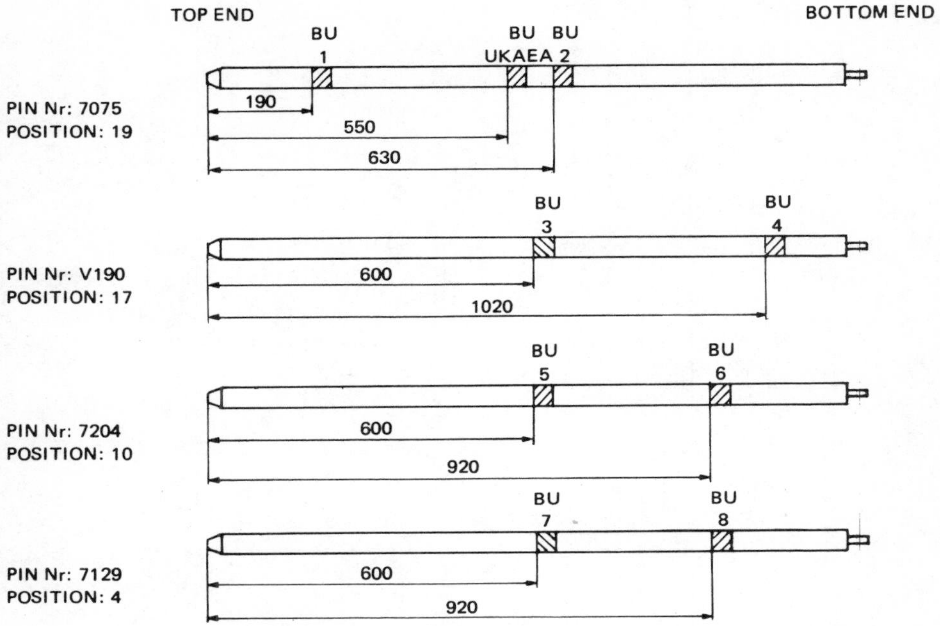

FIG.2. Burnup samples location in the BR3-IF 70/4 fuel rods.

studies involve both the initial fuel characteristics and the results of the post-irradiation examinations. The assembly fuel pins were of two types, resulting from two different fabrication techniques, 18 pins being manufactured by pelletizing and 19 pins by vibratory compaction. The samples were taken from the four pins shown in Fig.1. The sampling positions on the pins are given in Fig.2. The initial fuel composition and isotopic data are given in Table I. There is some uncertainty on the data of the isotopic analysis and therefore also on the isotopic composition because of the decay of ^{241}Pu. Therefore, initial fuel characteristics are indicated in the figures but not included in the calculation of the regression lines.

2. POST-IRRADIATION DATA

Eight samples were taken to determine the achieved burnup. The isotopic composition of uranium and plutonium was determined by thermal ionization mass spectrometry and the concentration of the different actinides was measured by mass-spectrometric isotopic dilution and alpha-ray spectrometry. The burnup determinations involve measurement of the neodymium isotopic composition and

TABLE I - SUMMARY OF THE DATA USED FOR ISOTOPIC CORRELATION STUDIES ASSEMBLY IF 70/4

Pellet	Date	Burnup MW·d/t	10^3Pu/U	^{234}U %	^{235}U %	^{236}U %	^{238}U %	^{238}Pu %	^{239}Pu %	^{240}Pu %	^{241}Pu %	^{242}Pu %
1	211270	12950	59.85	.004	.609	.027	99.360	.129	77.806	17.152	4.520	0.393
2	"	27900	54.84	.005	.521	.050	99.424	.294	68.091	22.669	7.891	1.056
3	"	33050	51.15	.004	.455	.060	99.481	.318	61.051	28.503	8.537	1.591
4	"	19350	48.67	.004	.545	.040	99.411	.150	72.263	21.472	5.474	0.641
5	"	43950	38.54	.005	.345	.076	99.574	.661	41.162	40.852	12.592	4.733
6	"	36350	42.55	.001	.408	.062	99.529	.455	49.518	36.712	10.366	2.948
7	"	33500	50.06	.003	.470	.056	99.471	.431	59.793	27.745	10.160	1.871
8	"	25700	52.58	.002	.512	.046	99.440	.309	64.944	25.148	8.336	1.263
A	?	0	(66.50?)	.006	.725	.000	99.269	?	88.2	10.3	1.3	0.2
B	?	0	69.53	.006	.725	.000	99.269	.018	91.3	7.87	0.79	0.041

A is the initial fuel composition of samples 1, 2, 5, 6, 7 and 8

B is the initial fuel composition of samples 3 and 4

concentration as well as the activities of a few selected radioisotopes. Although interesting correlations may be drawn from the measurement of fission products, they are of limited interest from the viewpoint of international safeguards since they require additional measurements which are usually not required for the operation of the reprocessing plant. The correlations described in the following sections are therefore based only on uranium and plutonium concentrations and their isotopic composition. The data for uranium and plutonium are given in Table I.

3. ISOTOPIC CORRELATIONS

The use of isotopic correlation techniques is expected to be centred mainly on the input accountability tank of a reprocessing plant [9]. The primary quantity of interest for safeguards purposes at the reprocessing of LWR fuel is the amount of plutonium entering the plant. Among the different methods used in present reprocessing facilities to measure the plutonium input, the Pu/U ratio method is best suited for independent verification by isotopic correlation techniques. The method relies on the equation

$$Pu = U_f \left(\frac{Pu}{U_f}\right) \qquad (1)$$

and $$U_f = U_0 + Pu_0 - F - Pu \qquad (2)$$

or $$Pu = \left(\frac{Pu}{U_f}\right) (U_0 + Pu_0) \left(1 - B - \left(\frac{Pu}{U_f}\right) \left(\frac{U_0}{Pu_0 + U_0}\right)\right) \qquad (3)$$

$U_0 + Pu_0$: weight of initial heavy elements

B : burnup in FIMA

In the correction term U_f has been equated with U_0.

The independent verification of Pu by isotopic correlation necessitates the estimation of both the burnup and the Pu/U_f ratio. The burnup, however, is only involved in the correction term and a relatively poor estimate has only a restricted effect on the overall precision. Nevertheless, careful selection of the correlation is essential to obtain a precision which is fitted to the verification of an analytical measurement rather than to the detection of an obvious outlier. Therefore, any relationship with a correlation coefficient less than |0.99| should be discarded as being nothing more than an interesting trend but unsuitable for

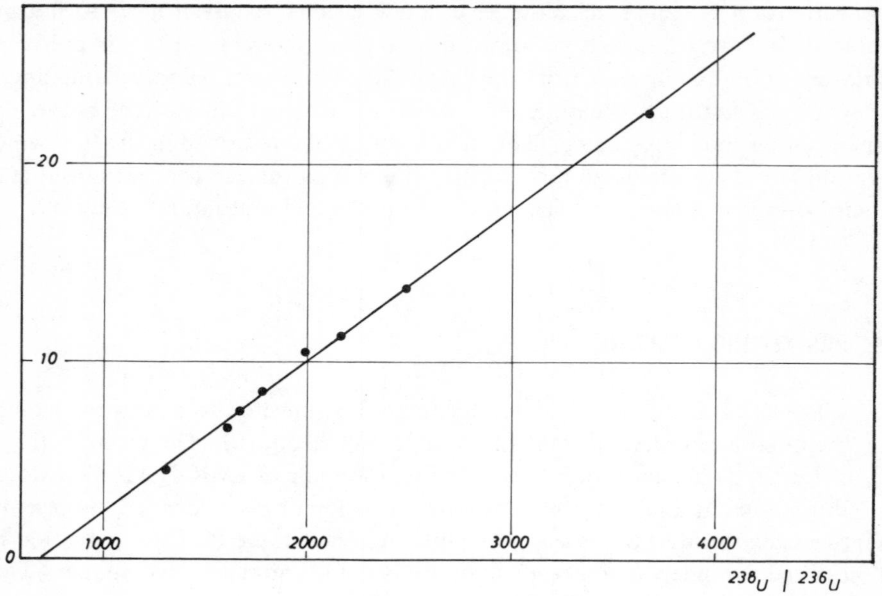

FIG.3. Correlation between $^{235}U/^{236}U$ *and* $^{238}U/^{236}U$.

any verification purpose. This is especially true for data obtained from post-irradiation research since the sampling and the measurements are usually produced in better conditions than in a reprocessing plant.

3.1. Consistency test of isotopic data

The purpose of these correlations is to verify the consistency of a given isotopic composition with the initial fuel data. For uranium isotopes a family of curves has been published [5] to estimate the initial enrichment by plotting the $^{236}U/^{235}U$ ratio versus the $^{236}U/^{238}U$ ratio. The curves can be linearized by taking the reciprocal values of the ratios. As shown in Fig.3 the slope of the line is related to the initial enrichment. Both ways of plotting the data, however, lack sensitivity to errors in the ^{236}U content, which means that deviations up to about 5% on this isotope will remain undetected. Since ^{238}U is constant within 0.3% in this type of fuel the relation can be transformed to

$$\frac{^{235}U}{^{236}U} = 5.146 + 0.00756 \ \frac{99.4}{^{236}U} \qquad (4)$$

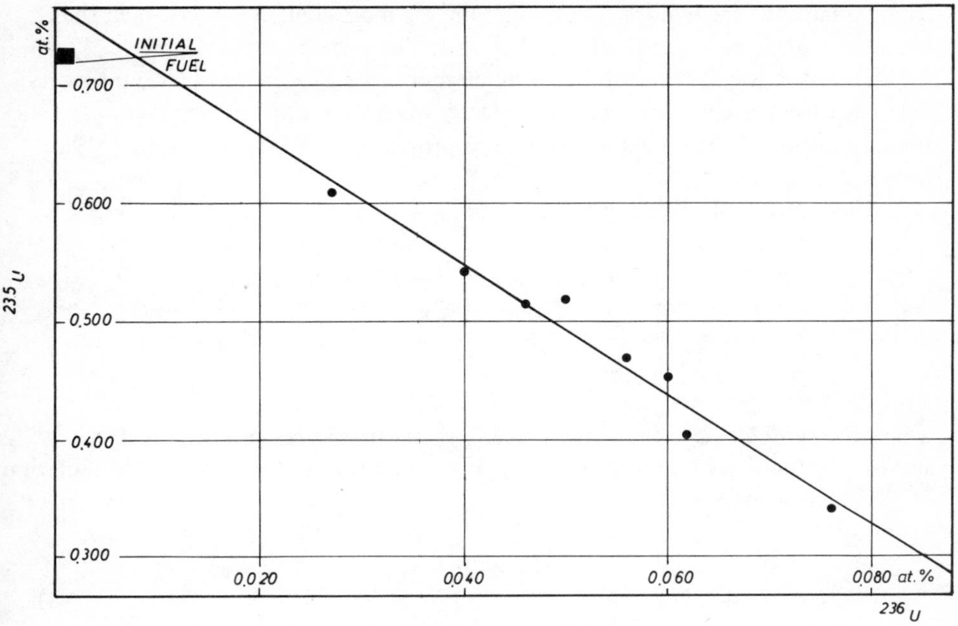

FIG.4. Correlation between ^{235}U and ^{236}U.

Multiplying both sides by ^{236}U yields

$$^{235}U = - 5.146\ ^{236}U + 0.751 \quad (5)$$

Eq.(5) shows the relationship between ^{235}U and ^{236}U. Putting ^{236}U equal to zero yields an estimate of the initial enrichment. Rearrangement of Eq.(5) yields

$$(0.751 - {}^{235}U) = 5.146\ ^{236}U \quad (6)$$

which relates the depletion in the isotope ^{235}U with the amount of ^{236}U. This equation is also the clue to the physical basis for the correlation, the coefficient of ^{236}U being simply the ratio of the total to the capture cross-section of ^{235}U in the particular neutron spectrum averaged over the irradiation time. In addition, Eqs (4) to (6) show that a number of empirically determined correlations can be found which are in fact only transformations of the same basic physical relation.

Using the same data, Eq.(5) has a correlation coefficient of 0.9825, whereas Eq.(4) has one of 0.9987. This is due to the different sensitivity of the correlations to the quality of the isotopic data, Eq.(5) being more sensitive to errors in the measurement of the minor isotope ^{236}U, as can be seen from Fig.4.

The situation for Pu isotopes is much more complex and, in fact, no theoretical basis has yet been formulated for many of the correlations observed. By analogy to Eq.(6) the depletion of ^{239}Pu is proportional to the amount of ^{240}Pu formed.

The amount of ^{239}Pu is given by

$$N_{49} = \frac{N^{\circ}_{28}\,\sigma_{28}\,\phi}{(\sigma_{49} - \sigma_{28})\,\phi}\, e^{-\sigma_{28}\phi t} + \left(N^{\circ}_{49} - \frac{N^{\circ}_{28}\,\sigma_{28}\,\phi}{(\sigma_{49} - \sigma_{28})}\right) e^{-\sigma_{49}\phi t} \qquad (7)$$

where the symbols have their usual meaning, the first index referring to the Z and the second indicating the atomic mass of the isotope concerned. The depletion of ^{239}Pu can then be written as

$$N^{\circ}_{49} - N_{49} = \left(N^{\circ}_{49} - \frac{N^{\circ}_{28}\,\sigma_{28}\,\phi}{(\sigma_{49} - \sigma_{28})\,\phi}\right)\left(1 - e^{-\sigma_{49}\phi t}\right) \qquad (8)$$

where $e^{-\sigma_{28}\phi t} \approx 1$

Using the same conventions the amount of ^{240}Pu is given by

$$N_{40} = \frac{N^{\circ}_{28}\,\sigma_{28}\,\sigma_{c\,49}\,\phi^2}{(\sigma_{49} - \sigma_{28})\,(\sigma_{40} - \sigma_{28})\,\phi^2}\, e^{-\sigma_{28}\phi t} +$$

$$\left(N^{\circ}_{49} - \frac{N^{\circ}_{28}\,\sigma_{28}\,\phi}{(\sigma_{49} - \sigma_{28})\,\phi}\right) \frac{\sigma_{c\,49}\,\phi}{(\sigma_{40} - \sigma_{49})\,\phi}\, e^{-\sigma_{49}\phi t} +$$

$$\left[N^{\circ}_{40} - \frac{N^{\circ}_{28}\,\sigma_{28}\,\sigma_{c\,49}\,\phi^2}{(\sigma_{49} - \sigma_{28})\,(\sigma_{40} - \sigma_{28})\,\phi^2} - \left(N^{\circ}_{49} - \frac{N^{\circ}_{28}\,\sigma_{28}\,\phi}{(\sigma_{49} - \sigma_{28})\,\phi}\right) \frac{\sigma_{c\,49}\,\phi}{(\sigma_{40} - \sigma_{49})\,\phi}\right] e^{-\sigma_{40}\phi t} \qquad (9)$$

where σ_{c49} refers to the capture cross-section of ^{239}Pu while σ_{49} refers to the total cross-section.

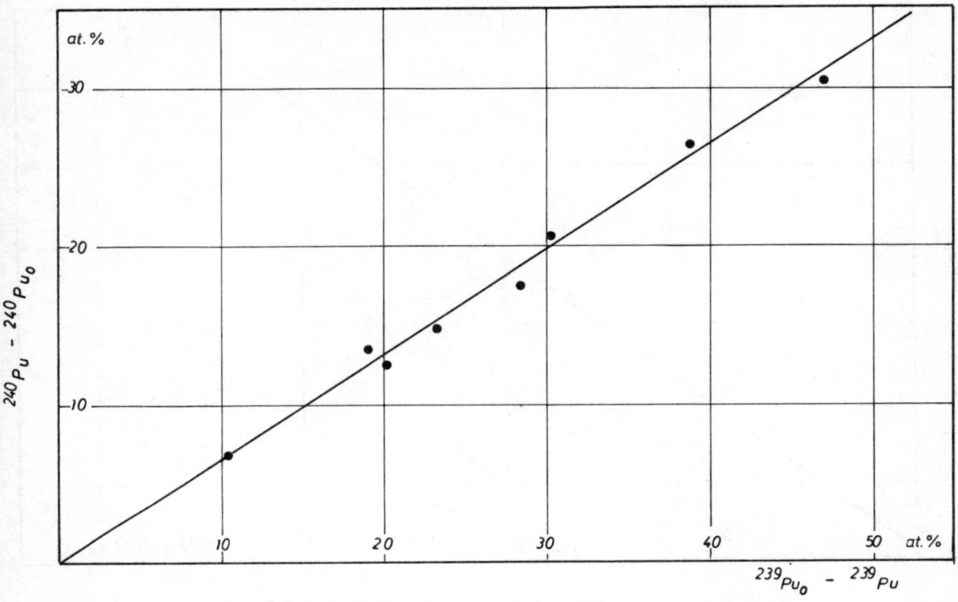

FIG.5. The formation of ^{240}Pu versus the depletion of ^{239}Pu.

The production of ^{240}Pu is then given by

$$N_{40} - N^{\circ}_{40} = \left[\left(N^{\circ}_{49} - \frac{N^{\circ}_{28}\,\sigma_{28}\,\phi}{(\sigma_{49} - \sigma_{28})\,\phi} \right) \frac{\sigma_{c\,49}\,\phi}{(\sigma_{40} - \sigma_{49})\,\phi} + \frac{N^{\circ}_{28}\,\sigma_{28}\,\sigma_{c\,49}\,\phi^{2}}{(\sigma_{49} - \sigma_{28})\,(\sigma_{40} - \sigma_{28})\,\phi^{2}} - N^{\circ}_{40} \right] \times \left(1 - e^{-\sigma_{40}\,\phi t} \right) - \frac{\sigma_{c\,49}\,\phi}{(\sigma_{40} - \sigma_{49})\,\phi} \left(N^{\circ}_{49} - \frac{N^{\circ}_{28}\,\sigma_{28}\,\phi}{(\sigma_{49} - \sigma_{28})\,\phi} \right) \left(1 - e^{-\sigma_{49}\,\phi t} \right) \quad (10)$$

Substitution of Eq.(8) in (10) yields

$$N_{40} - N^{\circ}_{40} = C \left(1 - e^{-\sigma_{40}\,\phi t} \right) - \frac{\sigma_{c49}\,\phi}{(\sigma_{40} - \sigma_{49})\,\phi} \left(N^{\circ}_{49} - N_{49} \right) \quad (11)$$

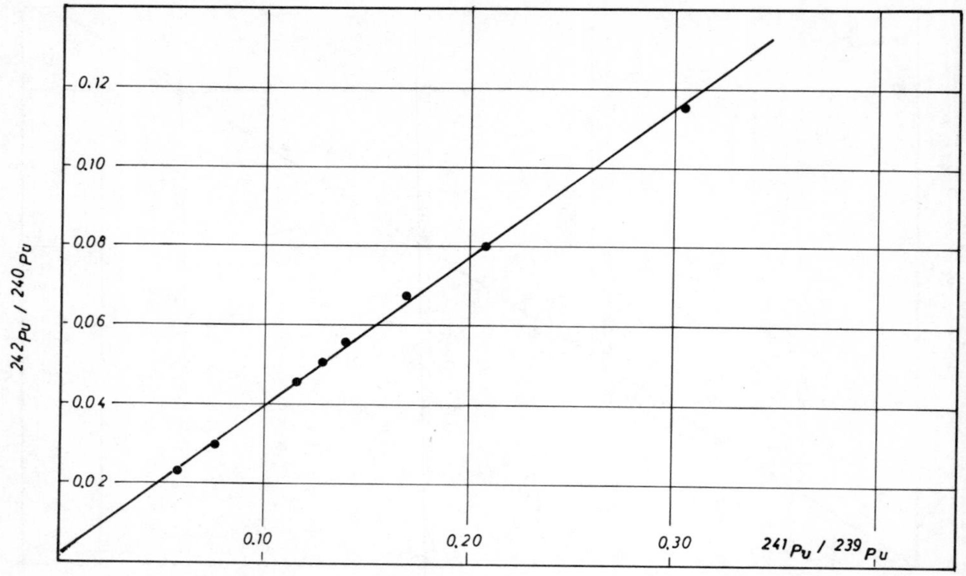

FIG.6. The correlation between $^{242}Pu/^{240}Pu$ and $^{241}Pu/^{239}Pu$.

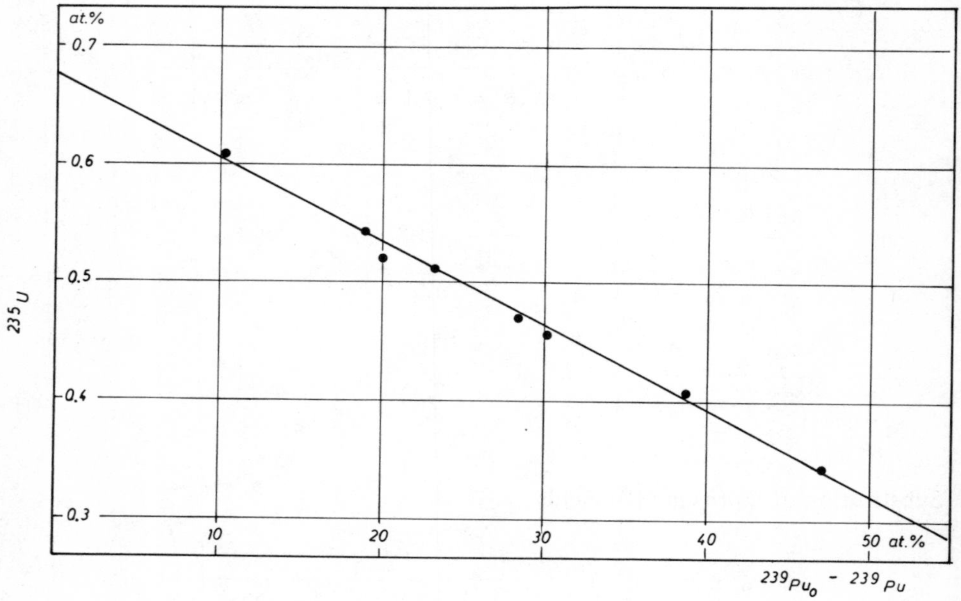

FIG.7. The correlation between at.% of ^{235}U and the depletion of ^{239}Pu.

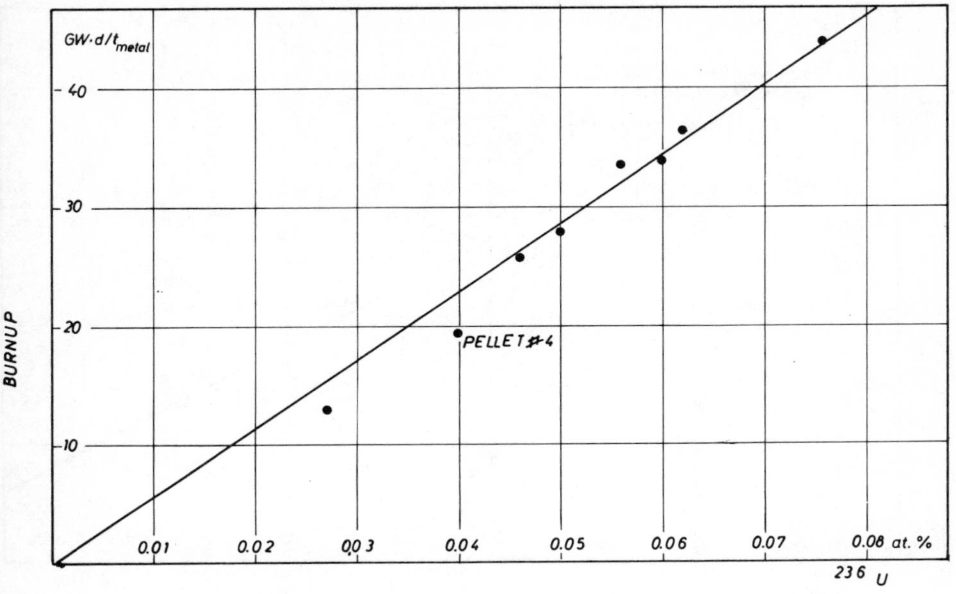

FIG.8. The correlation between burnup and at.% of ^{236}U.

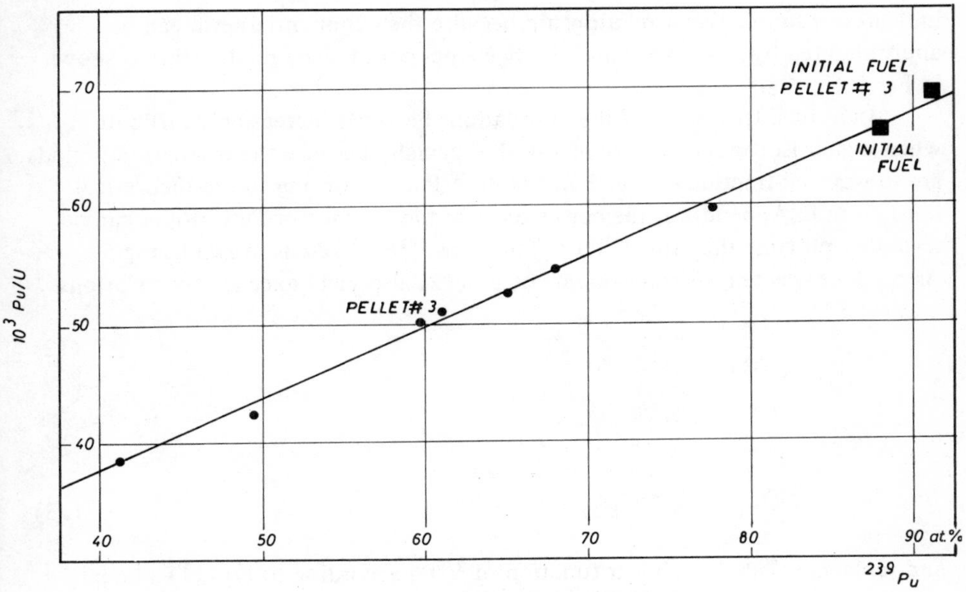

FIG.9. The correlation between Pu/U and at.% of ^{239}Pu.

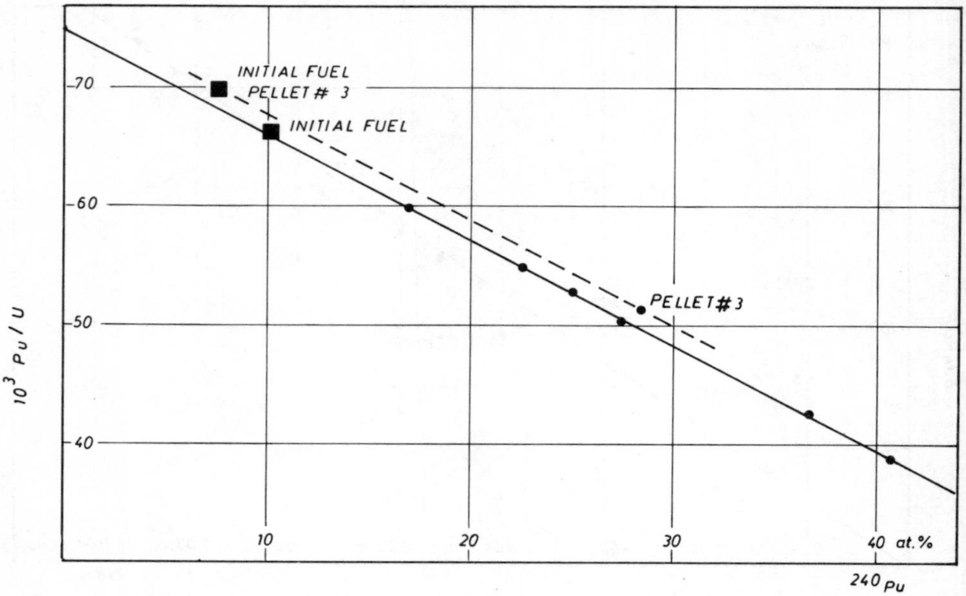

FIG.10. The correlation between Pu/U and at.% of ^{240}Pu.

indicating a simple linear relationship because the exponential term can be approximated by a series expansion. See Appendix I. The relationship is shown in Fig.5.

Analytical treatment of the correlations becomes increasingly difficult when higher isotopes of Pu are involved, especially because the irradiation periods are substantial fractions of the half-life of ^{241}Pu. According to the theoretical basis given in Appendix I, the consistency of the four major Pu isotopes can be tested by plotting the ratio ^{242}Pu/^{240}Pu versus ^{241}Pu/^{239}Pu as shown in Fig.6. As can be expected, rearrangements of Eq.(12) also yield excellent correlations.

$$^{242}\mathrm{Pu}/^{240}\mathrm{Pu} = a(^{241}\mathrm{Pu}/^{239}\mathrm{Pu}) \tag{12}$$

for instance

$$^{242}\mathrm{Pu}/^{241}\mathrm{Pu} = a(^{240}\mathrm{Pu}/^{239}\mathrm{Pu}) \tag{13}$$

and replacing ^{240}Pu by a linear function of ^{239}Pu according to Eq.(11) a linear relation is obtained between ^{242}Pu/^{241}Pu and the reciprocal of ^{239}Pu. The relation between ^{235}U and $^{239}Pu^{\circ}$-^{239}Pu as shown in Fig.7 has a correlation coefficient of 0.996.

Plotting ^{235}U versus ^{239}Pu shows clearly discrepant values for data points 3 and 4 which are due to the difference in ^{239}Pu content of the initial fuel. This correlation thus has the ability to indicate differences as small as 3% in the initial Pu isotopic composition.

The existence of this relationship can again be inferred from the analytical expressions for the individual isotopes. This relation opens a realm of correlations between uranium and plutonium isotopes or their combinations.

3.2. Estimation of the fuel exposure

None of the Pu isotopes nor their ratios give a reasonable correlation with the fuel burnup because a substantial part of the fissions originated in ^{239}Pu formed during irradiation from ^{238}U and in ^{241}Pu. The ^{236}U content, however, can be used as an estimator for the fuel burnup with good confidence as is shown in Fig.8. The regression line, calculated from the data points without any constraints includes the point (0.0) within its probable error limits, confirming thus the initial fuel composition. In principle, burnup should also correlate well with the ^{235}U depletion, but the correlation deteriorated by two data points. This emphasizes the dependence of the quality of the correlations on the ability of the mass spectrometry to perform accurate measurements on the minor isotopes. From the safeguards viewpoint, however, the burnup value is only involved in the correction factor in Eq.(3) and the estimate based on Fig.8 is sufficiently accurate.

3.3. The actinide concentration ratio

Most important for the verification of declared material balances is the existence of accurate and reliable correlations between the concentration ratio Pu/U and parameters which are directly accessible to the safeguards inspectors. Since an isotopic composition measurement is less subject to error than a concentration determination by spiking, the demonstration of correlations between isotopic data and concentrations is of interest, provided that the required accuracy can be obtained and the theoretical basis is available to allow extrapolations. The accuracy required for Pu in irradiated fuel processing has been stated as ±1% [10]. Two correlations have been found to fulfil this requirement, they are shown in Figs 9 and 10 and can be expressed as

$$10^3\ Pu/U = 13.778 + 0.5993\ {}^{239}Pu \qquad (14)$$

and

$$10^3\ Pu/U = 75.237 - 0.8909\ {}^{240}Pu \qquad (15)$$

TABLE II. COMPARISON OF THE PU/U RATIO CALCULATED FROM THE CORRELATION WITH THE ANALYTICAL DETERMINATION

Sample	(1) Pu/U x 10^3	(2) Pu/U from Eq.(14) x 10^3	Ratio (2)/(1)	(3) Pu/U from Eq.(15) x 10^3	Ratio (3)/(1)
1	59.85	60.41	1.0093	60.05	1.0033
2	54.84	54.58	0.9954	55.04	1.0037
3	51.15	50.37	1.0156	49.84	0.9745 [a]
5	38.54	38.45	0.9976	38.84	1.0079
6	42.55	43.45	1.0212	42.53	0.9996
7	50.06	49.61	0.9911	50.52	1.0092
8	52.58	52.70	1.0023	52.83	1.0048

$$\frac{1}{7}\Sigma\ (|1-(2)/(1)|) = 0.92\ \%$$

$$\frac{1}{7}\Sigma\ (|1-(3)/(1)|) = 0.78\ \%$$

[a] Although sample 3 has been included in the regression analysis its different initial composition is indicated by this ratio.

both with correlation coefficients in excess of 0.996. Data point 4 has been omitted from the regression analysis because its Pu/U ratio is obviously incompatible with the other data. Careful chemical analysis by different methods, however, has confirmed the Pu/U ratio. It should be remembered, however, that Eqs(14) and (15) are rearrangements of the physical relationship which involves the change of the ratio Pu/U as a function of the depletion in ^{239}Pu or the formation of ^{240}Pu. This is reflected by the fact that the omission of data point 3, which belongs to a fuel rod with a slightly different initial composition, increases the correlation coefficient of Eq.(15) to > 0.9995. Lack of precision on the data referring to the unirradiated fuel, however, prevents the use of differential data, illustrating once more that the precision displayed in many correlations is frequently due to poor analytical data rather than to an intrinsic limitation of the relationship. The discrepancy of data point 4 with respect to Eqs (14) and (5) can be due to a local inhomogeneity of the Vipac fuel [11]. To illustrate the precision of both correlations the Pu/U ratios have been calculated from the isotopic data using the regression lines. The results are given in Table II. The average error is ±0.92% and ±0.78%, respectively, which is certainly comparable to the precision obtained by common analytical methods on highly radioactive solutions.

4. CONCLUSIONS

Although isotopic correlations have been studied and demonstrated mainly on an empirical basis, it is essential to define the underlying physical phenomena and to provide the analytical proof for their existence before they can be used as an *a priori* requirement for verification purposes rather than as an interesting *a posteriori* phenomenon. It has further been shown that many empirical correlations are in fact algebraic rearrangements of a few basic relationships describing the nuclear transformations occurring in the reactor core. It has been demonstrated that the precision attainable by isotopic correlation techniques is comparable to that obtained by common analytical methods if the correlations are based on accurate analytical data. It should be remembered that correlations calculated from historical plant data reflect the precision of those data. This leads to the usual picture of a number of points scattered within a few per cent of a straight relationship. The foregoing paragraphs have shown that the scatter can be reduced to a fraction of one per cent, which is a primary requirement for the use of isotopic correlation techniques as a verification tool for safeguards purposes. Once the existence of precise linear relationships has been demonstrated by careful analysis of data obtained in post-irradiation work, it can be expected that dissolution of fuel batches, being essentially a linear combination of a large number of samples, will show similar relationships. The correlations found in

PWR recycling plutonium fuel are promising in this respect and it will be interesting to compare the results with similar irradiations and with reprocessing data for mixed-oxide fuel when they are available.

REFERENCES

[1] CHRISTENSEN, D.E., SCHNEIDER, R.A., STEWART, K.B., Rep. BNWL-SA-4273 (1972).

[2] SANATANI, S., SIWY, D., in Safeguarding Nuclear Materials (Proc. Symp. Vienna, 1975), IAEA, Vienna (1976).

[3] BEETS, C., BEMELMANS, P., in ESARDA Conf. Isotopic Correlations Techniques, May 1978, Stresa, ICT-24 (1978).

[4] INTERNATIONAL ATOMIC ENERGY AGENCY, Safeguards Technical Manual, Part E, IAEA Vienna (1975) 161–219.

[5] CHRISTENSEN, D.E., PREZBINDOWSKI, D.L., J. Inst. Nucl. Mater. Manage. 2 1 (1973) 821–54.

[6] SCHOOF, S., STEINERT, H., KOCH, L., in ESARDA Conf. Isotopic Correlation Techniques, May 1978, Stresa, ICT-6 (1978).

[7] DE REGGE, P., BODEN, R., ibid. ICT-33.

[8] TRAUWAERT, E., VAN OUTRYVE D'YDEWALLE, B., Rep. BN-7512-02 (1975).

[9] GUPTA, D., Rep. KFK-2400 (1976).

[10] INTERNATIONAL ATOMIC ENERGY AGENCY, Safeguards Technical Manual, Part A, IAEA, Vienna (1975) 36.

[11] TRAUWAERT, E., VAN OUTRYVE D'YDEWALLE, B., Rep. BN-7405-01 (1974).

Appendix I

Equation (8) can be written as

$$(N^{\circ}_{49}-N_{49}) = B\,(1 - e^{-\sigma_{49}\,\phi t})$$

$$\text{or } \phi t = -\frac{1}{\sigma_{49}}\ \ln\left(1 - \frac{N^{\circ}_{49}-N_{49}}{B}\right)$$

and $(1 - e^{-\sigma_{40}\,\phi t})$ can be written as

$$1 - \left(1 - \frac{N^{\circ}_{49}-N_{49}}{B}\right)^{(\sigma_{40}/\sigma_{49})}$$

which can be written as its series expansion

$$1 - (1 - (\frac{\sigma_{40}}{\sigma_{49}}) (\frac{N^{\circ}_{49} - N_{49}}{B}) + (\frac{N^{\circ}_{49} - N_{49}}{2B^2})^2 (\frac{\sigma_{40}}{\sigma_{49}}) (\frac{\sigma_{40}}{\sigma_{49}} - 1) \; \ldots\ldots)$$

$$= (\frac{N^{\circ}_{49} - N_{49}}{B}) (\frac{\sigma_{40}}{\sigma_{49}}) (1 - \frac{N^{\circ}_{49} - N_{49}}{2B} (\frac{\sigma_{40}}{\sigma_{49}} - 1) \; \ldots\ldots$$

Substituting this into Eq. (10) yields

$$(N_{40} - N^{\circ}_{40}) = \left[\frac{\sigma_{C49} \, \phi}{(\sigma_{40} - \sigma_{49}) \, \phi} + \frac{N^{\circ}_{28} \, \sigma_{28} \, \sigma_{C49} \, \phi^2}{B(\sigma_{49} - \sigma_{28})(\sigma_{40} - \sigma_{28})\phi^2} - \frac{N^{\circ}_{40}}{B} \right] X$$

$$(N^{\circ}_{49} - W_{49}) \; (\frac{\sigma_{40}}{\sigma_{49}}) \; (1 - \frac{N^{\circ}_{49} - N_{49}}{2B} (\frac{\sigma_{40}}{\sigma_{49}} - 1) \; \ldots\ldots)$$

$$- \frac{\sigma_{C49} \, \phi}{(\sigma_{40} - \sigma_{49}) \, \phi} \quad (N^{\circ}_{49} - N_{49}) \qquad (10a)$$

Taking the irradiation characteristics into account it can be seen that the first term is a correction term to the second and that higher order corrections to the first term can be neglected, which means that the formation of ^{240}Pu is essentially proportional to the depletion of ^{239}Pu.

Equations (12) and (13) are the extension of the above procedure to the major isotopes.

The analytical expressions for the isotopes are given as follows:

$$N_{49} = \frac{N^{\circ}_{28} \, \sigma_{28} \, \phi}{(\sigma_{49} - \sigma_{28}) \, \phi} \, e^{-\sigma_{28} \phi t} + (N^{\circ}_{49} - \frac{N^{\circ}_{28} \, \sigma_{28} \, \phi}{(\sigma_{49} - \sigma_{28}) \, \phi}) \, e^{-\sigma_{49} \phi t}$$

$$\text{or } N_{49} = A \, e^{-\sigma_{28} \phi t} + B \, e^{-\sigma_{49} \phi t}$$

$$N_{40} = \frac{A \, \sigma_{C49} \, \phi}{(\sigma_{40} - \sigma_{28}) \, \phi} \, e^{-\sigma_{28} \phi t} + \frac{B \, \sigma_{C49} \, \phi}{(\sigma_{40} - \sigma_{49}) \, \phi} \, e^{-\sigma_{49} \phi t}$$

$$+ \left[N^{\circ}_{40} - \frac{A \, \sigma_{C49} \, \phi}{(\sigma_{40} - \sigma_{28}) \, \phi} - \frac{B \, \sigma_{C49}}{(\sigma_{40} - \sigma_{49}) \, \phi} \right] e^{-\sigma_{40} \phi t}$$

$$\text{or } N_{40} = A' e^{-\sigma_{28}\phi t} + B' e^{-\sigma_{49}\phi t} + C e^{-\sigma_{40}\phi t}$$

$$N_{41} = \frac{A'\sigma_{40}\phi e^{-\sigma_{28}\phi t}}{(\lambda+\sigma_{41}\phi-\sigma_{28}\phi)} + \frac{B'\sigma_{40}\phi e^{-\sigma_{49}\phi t}}{(\lambda+\sigma_{41}\phi-\sigma_{49}\phi)}$$

$$+ \frac{C\sigma_{40}\phi\, e^{-\sigma_{40}\phi t}}{(\lambda+\sigma_{41}\phi-\sigma_{40}\phi)} + \Bigg[N^{o}_{41} - \frac{A'\sigma_{40}\phi}{(\lambda+\sigma_{41}\phi-\sigma_{28}\phi)}$$

$$- \frac{B'\sigma_{40}\phi}{(\lambda+\sigma_{41}\phi-\sigma_{49}\phi)} - \frac{C\,\sigma_{40}\phi}{(\lambda+\sigma_{41}\phi-\sigma_{40}\phi)}\Bigg] e^{-(\sigma_{41}\phi+\lambda)t}$$

$$\text{or } N_{41} = A''e^{-\sigma_{28}\phi t} + B''e^{-\sigma_{49}\phi t} + C'e^{-\sigma_{40}\phi t} + D\, e^{-(\lambda+\sigma_{41}\phi)t}$$

Similarly

$$N_{42} = A''' e^{-\sigma_{28}\phi t} + B''' e^{-\sigma_{49}\phi t} + C'' e^{-\sigma_{40}\phi t}$$

$$+ D' e^{-(\lambda+\sigma_{41}\phi)t} + E\, e^{-\sigma_{42}\phi t}$$

where $A''' = \dfrac{A''\,\sigma_{C41}\,\phi}{(\sigma_{42}-\sigma_{28})\,\phi}$ and the meaning of the other terms can be inferred from the foregoing equations. Starting with Eq.(10) now it can be seen that

$$N^{o}_{40}-N_{40} = C(1-e^{-\sigma_{40}\phi t}) + \frac{\sigma_{C49}\,\phi}{(\sigma_{40}-\sigma_{49})\,\phi}\,(N^{o}_{49}-N_{49})$$

$$\text{and } N^{o}_{41} - N_{41} = N^{o}_{41}(1 - e^{-(\lambda+\sigma_{41}\phi)t}) - A''\,(1 - e^{-(\lambda+\sigma_{41}\phi)t})$$

$$- B''\,(1 - e^{-(\lambda+\sigma_{41}\phi)t}) - C'\,(1 - e^{-(\lambda+\sigma_{41}\phi)t})$$

$$+ B''\,(1 - e^{-\sigma_{49}\phi t}) + C'\,(1 - e^{-\sigma_{40}\phi t})$$

$$\text{or } N^{o}_{41} - N_{41} = D\,(1 - e^{-(\lambda+\sigma_{41}\phi)t})$$

$$+ (N^{o}_{40}-N_{40})\,\frac{\sigma_{40}\,\phi}{(\lambda+\sigma_{41}\phi-\sigma_{40}\phi)}$$

$$+ \frac{\sigma_{C49}\,\phi}{(\sigma_{40}-\sigma_{49})\,\phi}\left[\frac{\sigma_{40}\phi}{(\lambda+\sigma_{41}\phi-\sigma_{49}\phi)} - \frac{\sigma_{40}\phi}{(\lambda+\sigma_{41}\phi-\sigma_{40}\phi)}\right](N^{o}_{49}-N_{49})$$

The similar rearrangement for ^{242}Pu yields

$$N^{\circ}_{42}-N_{42} = E(1 - e^{-\sigma_{42}\phi t}) + (N^{\circ}_{41}-N_{41}) \left(\frac{\sigma_{C41}\phi}{\sigma_{42}\phi-\sigma_{41}\phi-\lambda}\right)$$

$$+ (N^{\circ}_{40}-N_{40}) \left[\frac{\sigma_{40}\phi}{(\lambda+\sigma_{41}\phi-\sigma_{40}\phi)} \cdot \frac{\sigma_{C41}\phi}{(\sigma_{42}-\sigma_{40})\phi}\right.$$

$$\left. - \frac{\sigma_{40}\phi}{(\lambda+\sigma_{41}\phi-\sigma_{40}\phi)} \cdot \frac{\sigma_{C41}\phi}{(\sigma_{42}\phi-\lambda\cdot\sigma_{41}\phi)}\right]$$

$$+ (N^{\circ}_{49}-N_{49}) \left(\frac{\sigma_{C49}\phi}{(\sigma_{40}-\sigma_{49})\phi}\right) X$$

$$\left[\frac{\sigma_{40}\phi}{(\sigma_{41}\phi+\lambda-\sigma_{49}\phi)} \cdot \frac{\sigma_{C41}\phi}{(\sigma_{42}-\sigma_{49})\phi} - \frac{\sigma_{40}\phi}{(\lambda+\sigma_{41}\phi-\sigma_{40}\phi)} \cdot \frac{\sigma_{C41}\phi}{(\sigma_{42}-\sigma_{40})\phi}\right.$$

$$\left. - \frac{\sigma_{40}\phi}{(\lambda+\sigma_{41}\phi-\sigma_{49}\phi)} \cdot \frac{\sigma_{C41}\phi}{(\sigma_{42}-\sigma_{41}\phi-\lambda)} + \frac{\sigma_{40}\phi}{(\lambda+\sigma_{41}\phi-\sigma_{40}\phi)} \frac{\sigma_{C\,41}\phi}{(\sigma_{42}\phi-\sigma_{41}\phi-\lambda)}\right]$$

or remembering that $(1- e^{-\sigma_{42}\phi t}) \approx 0$

$$N^{\circ}_{42}-N_{42} = K_1 (N^{\circ}_{49}-N_{49}) + K_2 (N^{\circ}_{40}-N_{40}) + K_3 (N^{\circ}_{41}-N_{41})$$

Division by $(N^{\circ}_{40}-N_{40})$ and substitution by Eq.(10a) yields

$$\frac{(N^{\circ}_{42}-N_{42})}{(N^{\circ}_{40}-N_{40})} = \frac{K_1}{K_4} + K_2 + \frac{K_3}{K_4} \frac{(N^{\circ}_{41}-N_{41})}{(N^{\circ}_{49}-N_{49})} \qquad (12a)$$

$$\text{with} \quad K_4 = \frac{\sigma_{C49}\phi}{(\sigma_{40}-\sigma_{49})\phi} \left[\frac{\sigma_{40}}{\sigma_{49}} \left(1 + \frac{A'}{B} - \frac{N^{\circ}_{40}}{B}\right) - 1\right]$$

Equation (12a) shows a general linear relationship between the formation and depletion of four major Pu isotopes. The particular initial conditions and irradiation characteristics of this assembly where such that

$$\frac{K_1}{K_4} \approx - K_2$$

which leaves

$$\frac{(N^\circ_{42}-N_{42})}{(N^\circ_{40}-N_{40})} = \frac{K_3}{K_4}\,\frac{(N^\circ_{41}-N_{41})}{(N^\circ_{49}-N_{49})} \tag{12b}$$

and because

$$\frac{N^\circ_{42}}{N^\circ_{40}} = \frac{K_3}{K_4}\,\frac{N^\circ_{41}}{N^\circ_{49}}$$ it is also true that

$$\frac{N_{42}}{N_{40}} = \frac{K_3}{K_4}\,\frac{N_{41}}{N_{49}}$$ as shown in Fig.6.

ACTIVITIES OF THE EUROPEAN SAFEGUARDS RESEARCH AND DEVELOPMENT ASSOCIATION (ESARDA) IN THE FIELD OF ISOTOPIC CORRELATIONS

C. FOGGI
CEC Joint Research Centre,
Ispra, Italy

Abstract

ACTIVITIES OF THE EUROPEAN SAFEGUARDS RESEARCH AND DEVELOPMENT ASSOCIATION (ESARDA) IN THE FIELD OF ISOTOPIC CORRELATIONS.

Isotopic correlation techniques have long been one of the themes of the collaboration among ESARDA partners; a Working Group has been set up, which includes specialists from the relevant countries as well as observers from outside the ESARDA area and from the IAEA. The Working Group serves as a forum for exchange of information, for co-ordination of research work and promotion of collaborative research programmes. This paper reviews the activity of the group and describes some recent collaborative programmes, namely: the ICE (Isotopic Correlation Experiment) which is being carried out at the WAK reprocessing plant; the data banks of isotopic compositions of irradiated nuclear fuels; and the Mol-IV experiment. The paper also reviews the findings of the Symposium on The Isotopic Correlation and its Application to the Nuclear Fuel Cycle, which was held in Stresa from 9–11 May 1978, under the sponsorship of ESARDA.

INTRODUCTION

The European Safeguards Research and Development Association (ESARDA)[1] has engaged in studies of the Isotope Correlation Technique (ICT) since its beginning. In 1974 a Working Group for this topic was established with representatives from industry, research organizations and the CEC Directorate of Safeguards. The motivation for their action lies in the potential application of ICT to the verification of nuclear material accountancy at reprocessing input and, moreover, in support of nuclear material management such as fuel management, consistency checks of experimental post-irradiation data, and reactor physics calculations. To assess the verification potential of ICT with regard to the conventional reprocessing input

---

[1] ESARDA is an association of European Organizations created for the purpose of collaboration on research work on the safeguards of nuclear materials. The Organizations party to the Association are: CEN/SCK (Belgium), CNEN (Italy), the Commission of the European Communities (CEC), ECN (Netherlands), ENS (Denmark), KfK (Federal Republic of Germany), UKAEA (United Kingdom).

analytical tools, the scope of the working was enlarged to include Reprocessing Input Analysis.

The activities of the Group cover the exchange of information and experience in the field of ICT, co-ordination of the activities, and collaboration in specific tasks which may include carrying out "common projects". Observers from other organizations may be invited: at present, the International Atomic Energy Agency and the US Brookhaven Northwest Laboratories (BNWL) have permanent observers in the Working Group.

This paper reviews recent progress. In particular, advances made in an Isotope Correlation Experiment – ICE, at present being conducted at the WAK plant in Karlsruhe, are reported, the status of the Data Bank of Isotopic Compositions is described, and the scope and the structure of the former Joint Safeguards Experiment MOL-IV are briefly summarized.

With regard to the exchange of information, the highlights of a recent symposium on The Isotope Correlation and its Application to the Nuclear Fuel Cycle are given.

1. COMMON PROJECTS

Two major common projects are currently in progress, for which the organizations involved are contributing with their expertise or their resources. The status of these projects are described later on. In the past, a Joint Safeguards Experiment (Mol-IV) was carried out within the framework of ESARDA. This is briefly outlined.

1.1. Progress of the Isotope Correlation Experiment (ICE)

The first isotope correlation experiment, ICE, was planned in detail by the Working Group in September 1977. The agreed aims were:

To determine the accuracy of ICT (and other related techniques) under routine conditions;

To evaluate the additional effort for inspection and analysis;

To prove its benefit for safeguards and other fuel management purposes;

To describe the additional information required in order to employ this technique;

To check the applicability of proposed ICT procedures (data banks etc.).

The experiment consisted of three phases:

(1) Pre-experimental phase, defining the scope of the experiments and the information required;

(2) Experimental phase, during which the reprocessing input batches were analysed under routine conditions (main analyses) and in which special analyses were performed to supplement the experiment;
(3) Post-experimental phase: evaluation of the data and experiences accumulated during the experiment.

For the purpose of this experiment five fuel assemblies discharged from the BWR reactor of Kernkraftwerk Obrigheim (KWO) were chosen because they were reprocessed sequentially during a routine campaign at WAK at the beginning of 1978. Detailed information on the initial fuel weight of the assemblies under study was supplied from KWU. In addition to the shipper estimates of the spent fuel the reactor operator, KWO, supplied the irradiation history of each assembly which might be used for in-pile correction of short-lived radio-nuclides such as ^{241}Pu, ^{134}Cs.

The initial enrichment of the fuel was 3.10 wt.% ^{235}U and the burnup achieved about 30000 MW·d/t.

During the campaign, exactly half a fuel assembly was dissolved for each input batch. The 10 batches thus obtained were sampled according to the standard procedure of the plant and diluted, and samples were distributed to four laboratories:

WAK plant laboratory,
IAEA Safeguards laboratory, SALE,
Euratom safeguards laboratory (European Institute for Transuranium Elements),
Referee laboratory (IRCH-KfK).

The samples were analysed for the isotopic concentration of uranium and plutonium isotopes by all these laboratories. In addition the burnup, trans-plutonides and fission gases were measured by the European Institute of Transuranium Elements. The results obtained are now under evaluation by the Working Group as described previously for the third phase (3).

1.2. Status of the ESARDA data-bank of fuel isotopic compositions

During recent years, the Working Group has taken the initiative, on a European scale, of collecting data on the isotopic composition of irradiated nuclear fuels, generated during a number of reprocessing campaigns or post-irradiation fuel analyses. These data are now generally available for research purposes and for application. The task of gathering these data into a bank has been entrusted to the Joint Research Centre of the CEE (Ispra Establishment). Experimental data have been supplied by the Karlsruhe Establishment of the JRC, the CEN/SCK (Mol), the ECN (Petten), and the BNFL Windscale. The ECN has also provided the

TABLE I. DATA BANKS OF ISOTOPIC COMPOSITIONS OF IRRADIATED FUELS REPORTED AT THE SYMPOSIUM

Owner	No. of data records (measured data)	No. of data records (calculated data)	No. of reactors	Purpose of the bank	Notes
ESARDA	264	–	7	Research and development of ICT	Uses the ADABAS data management system on the IBM 370/175
Karlsruhe Establishment of the JRC	279	–	9	Research; verification of analyses	
CEN/SCK	275	–	10	Research; verification of the Pu input at the reprocessing	Uses the computer HP 9815 A
IAEA	450 (includes calculated data)		6	Safeguards	Uses the ADABAS data management system on the IBM 370/158
Safeguards Directorate of the CEC				Safeguards	
Battelle Northwest Laboratories	917	39		Research; safeguards verification of technical models	Uses the data control system MISTY

statistical methods (and supplied the relevant computer software) for data correlation. The bank is now operational [1] and undergoing further development. It contains a total of 350 measured or re-measured data sets, half of which are relevant to fuel pellet samples and the other half to reprocessing dissolution samples. The data are relevant to nine different reactors.

The data management is performed by means of the System ADABAS; the statistical treatment is based on linear data correlation which takes into account errors in the X and Y directions [2]; incorporation of routines for polynomial and exponential data correlation is in progress [3, 4]. Each set of stored data includes (whenever available): isotopes of U, Pu, Am, Cm, Kr, Xe, Nd, Cs and Eu; fuel burnup; fresh fuel data and general information on the reactor and the experiment.

1.3. The joint safeguards experiment – Mol-IV

The experiment was carried out during 1972–73 [5, 6] in collaboration with a number of other organizations[2], with the aim of gaining more practical experience in the application of isotopic correlations to obtain independent estimates of the total plutonium at the dissolver tank of a reprocessing plant.

While most of the effort was devoted to heavy isotopes and destructive analyses, correlations were also investigated among fission products and by non-destructive measurements. Fission-product activities measured, either in the input solution or before dissolution (non-destructive assay of spent fuel assembly), and dissolver off-gas activities, provided redundant Pu input estimates.

The experiment covered several reprocessing campaigns carried out at the Eurochemic plant, on the fuel of the following reactors: SENA (PWR), TRINO VERCELLESE (PWR), KRB (BWR), DODEWAARD (BWR), NPD (HWR). The estimates obtained for the Pu/U ratio exhibit a relative standard deviation of around 1%. The experiment generated 54 verified data sets for use in the data banks of isotopic compositions.

---

[2] Co-operation with the following organizations:

ACDA	:	United States Arms Control and Disarmament Agency as contractor for Battelle Memorial Institute in Richland (PNL) as well as the Oak Ridge National Laboratory (ORNL).
SENA	:	Société d'Energie Nucléaire Franco-Belge.
ENEL	:	Ente Nazionale per l'Energia Elettrica, Rome, Italy.
Eurochemic	:	Mol, Belgium.
General Electric	:	San José, California, USA.
IAEA	:	International Atomic Energy Agency, Vienna.

2. INFORMATION EXCHANGE AND CO-ORDINATION

During regular meetings of the Working Group, the results obtained in the field of ICT are discussed, current actions are reported and future studies are co-ordinated. To extend the exchange of information, in 1977 the Working Group took the initiative of organizing a topical symposium, which was held in Stresa from 9 to 11 May 1978.

2.1. Highlights of the symposium on The Isotopic Correlation and its Application to the Nuclear Fuel Cycle

The symposium was organized to summarize past experience, to exchange information on the new experimental data available and on theoretical advances, to illustrate the benefits to be gained by the application of ICTs, and to highlight the areas in which further research may be needed. A total of 29 papers were presented, 22 from the EEC countries, three from the USA, two from Japan, and two from the IAEA. Twenty-two papers had been prepared by research institutes, three by Safeguards Authorities, one by an electrical utility, and three jointly by nuclear companies or utilities and research institutes. A panel discussion followed the presentation of the papers.

Valuable information was obtained from the various sessions on a number of fuel types, including new ones. Three papers dealt with Pu-U mixed-oxide fuels: correlations between burnup and radioactive fission products were investigated by non-destructive techniques using a fuel rod irradiated in the Halden BHWR reactor; a systematic correlation analysis was made on the isotopic data of post-irradiation samples from a Garigliano reactor fuel assembly [7] (BWR) and from a BR-3 reactor fuel assembly [8]. In all cases interesting correlations were found, but more investigation is still needed. Several papers [9–15] reported experimental data and results of theoretical studies concerning uranium-oxide fuels. The potential of the heavy isotope and fission-product correlations for reactor operator problems was illustrated by some of these papers [9, 11, 13], which described applications to the validation of computing codes or to the determination of various fuel parameters, such as burnup or Pu build-up. An evaluation of correlations based on fission products in a highly enriched uranium fuel was also reported [16].

Computerized data banks of irradiated fuel isotopic compositions have been created by various laboratories and organizations [7–21]; they are complemented by appropriate statistical routines for data correlation. The data are derived from post-irradiation experiments (measurement of pellet samples) or from reprocessing input analyses (batch samples); in some cases also calculated data sets are collected. The situation reported at the symposium is shown in Table I.

Owing to some overlapping of the data collected by the various banks, the accumulated historical information involves a total of 39 reactors, subdivided into the following types: 13 PWRs, 13 BWRs, 6 HWRs, 6 graphite-moderated, 1 FBR.

Statistical models used for data correlation include linear [17-20, 20a] polynomial and exponential [3, 4] regression (with possible consideration of errors in X and Y directions). A method of correlation which also allows the determination of the proper functional relationship between the data has been proposed in Ref.[22]. Use of data banks to solve various problems has been illustrated in Refs [19, 21, 23–25]. Contacts and exchanges of information between the various bank owners take place within the framework of the ESARDA Working Group and of an IAEA Research Coordination Group.

Two sessions dealing with the applications of the ICT to the nuclear fuel cycle revealed an unexpectedly wide use of these techniques. Apart from the first attempts which are being performed by Safeguards Authorities [24, 25], broad application has been described in various fuel management areas such as reactor operation problems related to the validation of computing codes and to the determination of Pu build-up and burnup in the fuel; waste management problems; consistency checks of post-irradiation analyses [9, 13, 23, 26–29]. Experiments for application to safeguards at the reprocessing plant have already been performed [6]. The application of a particular type of correlation to a uranium enrichment plant, using the centrifuge process, is described in Ref.[30].

The panel concluded the symposium by pointing out that the many applications already made of the ICT testify to its usefulness as a technique to supplement and verify calculations and other measurement techniques; in the area of safeguards applications further experiments under routine conditions still need to be carried out.

3. OUTLOOK

The potential of ICTs has been recognized not only for safeguards but also in the various fields of nuclear material management such as consistency checks of post-irradiation analysis; waste management; validation of computer codes; and reactor fuel management. However, integral tests are still needed to assess the benefits to be gained and the resources to be deployed with the use of this technique, and to compare it with other methods. This goal will be achieved by performing experiments under routine conditions with the co-operation of plant operators and fuel owners. By this means operational procedures will be obtained, not only for safeguards, but also for other applications to the nuclear fuel cycle. The ESARDA working group is working toward this end, and is now considering the feasibility of another isotope correlation experiment, possibly extended to include the NDA of complete assemblies.

REFERENCES

(The papers which were presented at the symposium on The Isotopic Correlation and its Application to the Nuclear Fuel Cycle (Stresa 9–11 May 1978) are quoted with the abbreviation STRESA/ICT followed by a progressive number; the proceedings will be published shortly.)

[1] TSURUTA, H., MATSUURA, S., SUZAKI, T., OKASHITA, H., UMEZAWA, H., "Correlation between burnup and fission-product ratios obtained from non-destructive measurement on a mixed oxide fuel", Stresa/ICT-3.

[2] FOGGI, C., ZIJP, W.L., "Data treatment for the isotopic correlation technique", Safeguarding Nuclear Materials (Proc. Symp. Vienna, 1975) **2**, IAEA, Vienna (1976) 405.

[3] ZIJP, W.L., AALDIJK, A.K., "Fitting of polynomial and exponential functions to experimental data points with errors in two directions", Stresa/ICT-27.

[4] ZIJP, W.L., AALDIJK, J.K., FOGGI, C., "Fitting functions to isotopic composition data present in the ESARDA data bank", Stresa/ICT-28.

[5] BEETS, C., Contributions to the Joint Safeguards Experimental Mol-IV, at the Eurochemic Reprocessing Plant, Mol-Belgium, CEN/SCK-BLG 486.

[6] BEETS, C., "The joint safeguards experiment MOL IV at the Eurochemic reprocessing plant Mol, Belgium", Stresa/ICT-30.

[7] SCHOOF, S., STEINERT, H., KOCH, L., "The influence of plutonium recycling in BWRs on isotope correlations", Stresa/ICT-6.

[8] De REGGE, P., "Isotopic correlations in Pu recycling of LWR fuel", Stresa/ICT-33.

[9] ROUSSET, P., "Isotopic correlations of fission gas applied to light water reactors", Stresa/ICT-16.

[10] UMEZAWA, H., OKASHITA, H., MATSUURA, S., "Studies on correlation among heavy isotopes in irradiated nuclear fuels", Stresa/ICT-4.

[11] DAWSON, J.T., EDENS, D.J., HUGHES, R.P., "A comparison between predicted and measured concentrations of plutonium isotopes", Stresa/ICT-1.

[12] WIESE, H.W., MARZO, M., "Heavy isotopes correlations: A theoretical investigation on correlations between actinide isotopes in PWRs", Stresa/ICT-10.

[13] BOUCHARD, J., DARRUZET, M., ROBIN, M., "Utilisation des corrélations isotopiques sur les combustibles irradiés. Application pratique au cas des mesures sur assemblages de réacteurs à eau", Stresa/ICT-2.

[14] FOGGI, C., FRENQUELLUCCI, "Fission product nuclear data requirements for the calculation of isotopic correlations in LWR fuels", Stresa/ICT-23.

[15] FOGGI, C., FRENQUELLUCCI, F., "Isotopic correlations based on radioactive fission product nuclides in power reactor irradiated fuels", Stresa/ICT-21.

[16] MAECK, W.J., EMEL, W.A., "Isotope correlations for highly enriched ^{235}U thermal reactor fuels", Stresa/ICT-9.

[17] FOGGI, C., PAGLIARI, V., "The ESARDA data bank of isotopic compositions of irradiated nuclear fuels", Stresa/ICT-22.

[18] BEMELMANS, C., BEETS, C., FRANSSEN, F., PETENYI, A., "A portable isotopic data bank", Stresa/ICT-24.

[19] SIWY, P., SANATANI, S., "Utilization of a data bank for safeguards application of isotopic correlations", Stresa/ICT-18.

[20] NAPIER, B., TIMMERMAN, C.L., "Developing isotopic functions", Stresa/ICT-11.

[20a] STEWART, K., TIMMERMAN, C.L., "Isotopic safeguards statistics", Stresa/ICT-12.

[21] WELLUM, R., DE MEESTER, R., KAMMERICHS, K., KOCH, L., "The categorisation of binary isotope correlations from the data bank of the Transuranium Institute, Karlsruhe", Stresa/ICT-5.

[22] CHALLE, J., "The application of quantiles to isotopic correlations", Stresa/ICT-26.

[23] BEETS, C., BEMELMANS, P., FRANSSEN, F., BAIRIOT, H., "Application of the CEN-SCK data bank to the reprocessed fuel", Stresa/ICT-25.

[24] Van der STIJL, H., ARENZ, H.J., "Verification of input measurements at reprocessing plants using isotopic correlation techniques", Stresa/ICT-20.

[25] SANATANI, S., "Application of isotopic correlation techniques (ICT) to international safeguards" Stresa/ICT-19.

[26] VENCHIARUTTI, R., COCCHI, A., ZAFFIRO, B., "Potential application of the isotopic correlations to safeguards and the fuel cycle", Stresa/ICT-32.

[27] BRANDALISE, B., KOCH, L., RIJKEBOER, C., ROMKOWSKI, D., "The application of isotope correlations in verifying automatic analysis of irradiated fuels", Stresa/ICT-7.

[28] GUARDINI, S., GUZZI, G., "The use of isotopic correlation techniques as consistency check and elaboration of post irradiation examination data", Stresa/ICT-17.

[29] ERNSTBERGER, R., KOCH, L., WELLUM, R., "Isotope correlations used in fuel and waste management", Stresa/ICT-8.

[30] BAHM, W., FUHSE, W., "Contribution to the application of isotopic correlation techniques in centrifuge enrichment plants", Stresa/ICT-31.

LA PRISE EN COMPTE DE L'HISTOIRE DU COMBUSTIBLE DANS LE CONTROLE A L'ENTREE DES USINES DE RETRAITEMENT

J. BOUCHARD*, G. DEAN**, P. PATIGNY+, M. ROBIN*

*CEA, Centre d'études nucléaires de Cadarache, Saint-Paul-lez-Durance

**CEA, Centre d'études nucléaires de Fontenay-aux-Roses, Fontenay-aux-Roses

+Compagnie générale des matières nucléaires, La Hague, France

Abstract–Résumé

POSSIBILITIES OFFERED BY THE ANALYSIS OF FUEL HISTORY FOR SAFEGUARDS SURVEILLANCE AT THE INPUT END OF REPROCESSING PLANTS.

Simplified calculations with a limited amount of data on the history of the fuel and its irradiation are a means of predicting all the final compositions of the irradiated fuel to a high degree of accuracy. The combination of these values with the results of destructive or non-destructive measurements carried out on the fuel when it enters the reprocessing plant provides several possible ways of verifying the accuracy of the results. Among the applications which have been discussed, the one which involves the determination of uranium and plutonium masses entering the plant by what is called the "gravimetric balance" method has been the subject of detailed studies during the first reprocessing campaigns on water-reactor fuel at La Hague. The accuracy attained in the results and the confidence that can be placed in them in view of the large number of cross-checks clearly demonstrates the importance of these methods which also have the advantage that they make it possible to reduce the number of complex analyses (such as neodymium measurements) and thus the cost of determining the input balance. Since they supply an explanation for the results of indirect measurements such as the gamma emission from certain fission or activation products in terms of burnup, the cooling time or the fuel masses, these methods provide means for non-destructive surveillance. Two important examples are the identification of irradiated assemblies and monitoring of the quantities of fissile material present in hulls.

LA PRISE EN COMPTE DE L'HISTOIRE DU COMBUSTIBLE DANS LE CONTROLE A L'ENTREE DES USINES DE RETRAITEMENT.

L'application de calculs d'évolution simplifiés avec un nombre limité de données historiques sur le combustible et son irradiation permet de prévoir avec une très bonne approximation toutes les compositions finales de ce combustible irradié. En combinant ces valeurs avec les résultats de mesures destructives ou non destructives effectuées sur le combustible à l'entrée de l'usine de retraitement on obtient un grand nombre de possibilités pour vérifier l'exactitude des résultats. Parmi les applications présentées, celle qui concerne la détermination des masses d'uranium et de plutonium entrant dans l'usine par la méthode dite de *bilan gravimétrique* a donné lieu à de larges investigations lors des premières campagnes de retraitement de combustibles de réacteurs à eau à La Hague. La précision atteinte sur les résultats et la confiance que l'on peut lui attribuer compte tenu du grand nombre de recoupements montrent tout l'intérêt de méthodes qui, par ailleurs, offrent la possibilité de réduire le nombre d'analyses complexes comme celles du néodyme et donc le coût des déterminations du bilan d'entrée. En permettant d'interpréter des résultats de mesures indirectes, telles que l'émission gamma de certains produits de fission ou d'activation, en termes de taux de combustion, de temps de refroidissement ou de masses de combustibles, ces méthodes apportent des solutions de contrôles non destructifs. L'identification des assemblages irradiés et le contrôle des quantités de matières fissiles présentes dans les coques en sont deux exemples importants.

INTRODUCTION

Un combustible irradié est par définition un combustible qui a séjourné un certain temps dans un réacteur et a de ce fait subi des modifications de composition liées aux réactions nucléaires, principalement la capture et la fission.

Lorsque ce combustible arrive dans une usine de retraitement on pourrait pour déterminer sa composition adopter à l'extrême l'une des deux démarches suivantes :

- Ignorer son passé, considérer que l'on reçoit un produit dont les caractéristiques sont en majeure partie inconnues et donc mesurer tous les paramètres qui sont indispensables pour l'établissement du bilan et plus généralement l'ensemble des opérations associées au retraitement de ce combustible.
- Supposer au contraire que ce combustible est parfaitement identifié et reconstituer par le calcul toutes les caractéristiques à partir de sa composition initiale et son historique d'irradiation.

La première démarche a l'avantage de la simplicité puisqu'elle revient à déconnecter complètement le retraitement des réacteurs. Elle serait par contre extrêmement lourde dans son application en raison du nombre et de la complexité des mesures à réaliser. La seconde, strictement appliquée, serait irréaliste pour deux raisons essentielles : une précision insuffisante des estimations calculées de certaines compositions et la possibilité, jamais exclue, d'une erreur d'identification.

Dans la pratique il est nécessaire de rechercher une voie intermédiaire entre ces deux extrêmes. La solution actuellement retenue à LA HAGUE consiste à mesurer l'essentiel (par exemple, le rapport Pu/U et les teneurs isotopiques de ces deux éléments), et à tenir compte de l'histoire du combustible pour assurer des recoupements et déterminer des paramètres secondaires (comme les teneurs ou activités des produits de fission et des transplutoniens).

Les études qui sont présentées ici, tout en servant de fondement à l'utilisation actuelle de relations calculées entre certains paramètres, tendent également à montrer que l'on pourra progressivement accroître la prise en compte de l'histoire du combustible et par là même soit réduire le nombre de mesures à effectuer soit bénéficier d'informations supplémentaires intéressantes pour le fonctionnement et le contrôle de l'usine.

Après une présentation théorique appuyée sur des confirmations expérimentales nous examinons trois domaines d'application, la détermination du bilan d'entrée, l'identification non destructive des combustibles et la détermination d'autres données nécessaires pour le contrôle des usines de retraitement.

1. OBJECTIFS

1.1. Intérêt pour les contrôles de garanties

Le bilan d'entrée des usines de retraitement est l'un des points importants pour la comptabilité des matières nucléaires dans le cycle du combustible. En effet les dernières mesures sont normalement effectuées à l'occasion de la fabrication du combustible, l'estimation de la quantité de Pu et d'U résiduel étant ensuite obtenue par calcul à la sortie du réacteur.

La prise en compte de l'histoire du combustible, c'est-à-dire de données concernant sa composition initiale et les conditions de son irradiation en réacteur ainsi que des compositions finales, qui peuvent en être déduites grâce à des calculs appropriés, présente l'intérêt pour le contrôle des matières nucléaires de permettre :

- Des recoupements entre les quantités de matières estimées par l'exploitant du réacteur et mesurées à l'entrée de l'usine de retraitement.
- La vérification de l'exactitude des résultats obtenus à partir de méthodes d'analyse indépendantes.
- Une réduction du délai de réponse et du coût par la mise au point de nouvelles techniques d'établissement de bilan et l'utilisation de méthodes d'analyses rapides.
- L'identification du combustible par analyse non destructive à son arrivée à l'usine ou au départ du réacteur.
- La détermination des quantités de matières nucléaires contenues dans les coques.

Sur ces différents points les intérêts du contrôle des garanties ne sont pas fondamentalement différents de ceux de l'exploitant.

1.2. Autres applications

Elles se situeront au niveau du fonctionnement de l'usine et de la sûreté de l'installation. Pour l'essentiel les problèmes de contrôle se résument à une bonne connaissance du combustible qui est nécessaire pour :

- Réaliser des bilans précis,
- Optimiser la conception ou le fonctionnement de l'unité en fonction des combustibles traités,
- Assurer la sûreté de l'installation sans prendre des marges trop importantes pour les problèmes de criticité ou de protection biologique,
- Avoir une gestion précise des rejets.

Pour tous ces problèmes une identification non destructive des assemblages et l'interprétation correcte de quelques mesures effectuées après dissolution doivent permettre , avec le support de calculs prenant en compte les données "historiques" du combustible, de disposer de réponses complètes et sûres.

La connaissance des compositions offre des possibilités pour :

- Suivre le combustible,
- Vérifier l'absence de matières fissiles dans les sous-produits comme les coques, ou déterminer les faibles quantités éventuellement présentes,
- Prévenir les risques d'accumulation dans certaines parties du procédé.

On peut obtenir ces données par une simple interprétation des résultats de mesures mettant en oeuvre des calculs fondés sur l'histoire du combustible.

2. PRESENTATION DES METHODES

L'évolution du combustible en cours d'irradiation est régie par des lois physiques bien connues, même si elles ne peuvent pas toujours être parfaitement représentées dans les calculs.

TABLEAU I. PRECISION DES CALCULS *REACTEUR* POUR UN ASSEMBLAGE AVEC UNE METHODE AJUSTEE

Paramètres \ Précision de connaissance du taux de combustion	5 %	1 %
Appauvrissement en ^{235}U	5 %	1 à 1,5 %
Formation de ^{236}U	6 à 7 %	2 %
Rapport Pu/U	3 à 5 %	2 %
Teneurs isotopiques Pu (240,241,242)	5 à 10 %	3 à 5 %

Des codes de neutronique très complets permettent d'effectuer un bilan détaillé des réactions nucléaires intervenant en particulier dans le combustible et donc de déterminer les compositions finales après irradiation, à partir des valeurs initiales provenant de la fabrication du combustible.

2.1. Précision des calculs de composition effectués avec les codes de réacteur

Pour le cas de réacteurs à eau pressurisée de grande puissance et avec les méthodes actuellement disponibles au CEA, l'estimation des incertitudes affectant les valeurs calculées des compositions moyennes pour un assemblage est donnée dans le tableau I. Les valeurs finales dépendent du taux de combustion, paramètre qui peut être indiqué par l'exploitant du réacteur à partir de ses mesures de bilan thermique ou déterminé par des mesures destructives ou non sur le combustible. La précision sur la détermination des compositions est évidemment liée à l'incertitude affectant la connaissance du taux de combustion et deux exemples en sont donnés dans le tableau I.

2.2. Méthode simplifiée pour les calculs de compositions

Les méthodes de calcul de réacteur sont complexes et d'une utilisation assez lourde dans la mesure où elles nécessitent une représentation détaillée de toute la configuration à chaque étape de l'irradiation. En fait, cette complexité est essentiellement liée à la recherche d'autres informations que les modifications de composition du combustible, en particulier l'évolution en réactivité et les distributions de puissance. Il était donc naturel de chercher à simplifier ces méthodes dans le cadre d'applications strictement limitées à la détermination des compositions finales du combustible.

C'est ce qui a conduit à developper au CEA une procédure de calcul dont les principes sont les suivants :

- Prise en compte des principales caractéristiques du réacteur et du combustible pour permettre des calculs neutroniques "quasi ponctuels" précis.
- Utilisation de codes de cellules qualifiés pour la détermination au niveau de l'assemblage ou de zones d'assemblage des sections efficaces moyennes et de leur variation avec l'irradiation. Dans le cas des réacteurs à eau le code utilisé est APOLLO (1).
- Réalisation des calculs d'évolution complets avec un code sans dimension d'espace en tenant compte d'un historique de puissance approximatif. Ce code, EVOGENE, est décrit par ailleurs (2).

TABLEAU II. ORDRES DE GRANDEUR DES ECARTS EXPERIENCE-CALCUL AVANT AJUSTEMENT
Echantillon de combustible de la Centrale nucléaire des Ardennes (CNA)

Taux de combustion moyen	12000 MW·d/t	20000 MW·d/t	25000 MW·d/t
Δ (U^{235})	+ 1.5	+ 0.8	+ 2.5
U^{236}/U^{238}	- 2.0	- 0.8	- 0.5
Pu^{239}/U^{238}	- 5.2	- 4.2	- 7.1
Pu^{241}/Pu^{239}	+ 3.5	+ 2.3	+ 1.7
Pu^{242}/Pu^{239}	+ 1.5	+ 1.0	+ 2.1

(Ecarts en pour-cent)

TABLEAU III. ECARTS EXPERIENCE-CALCUL APRES AJUSTEMENT

	C.N. A.*	STADE **
Δ (U^{235})	+ 0.7	- 0.4
U^{236}/U^{238}	- 1.0	+ 0.1
Pu^{239}/U^{238}	- 0.6	- 0.4
Pu^{240}/Pu^{239}	+ 1.3	+ 0.8

* Résultats de mesures sur échantillons prélevés dans les crayons

** Résultats de mesures effectuées sur prélévements dans le dissolveur à LA HAGUE

Ce schéma nous permet d'obtenir pour chaque combustible des tables de composition en fonction du taux de combustion et du temps de refroidissement ainsi que des formulaires rendant compte des relations entre les principaux paramètres caractéristiques des compositions finales.

La simplification adoptée, qui est considérable en termes de moyens à mettre en oeuvre pour parvenir aux valeurs recherchées, revient pour l'essentiel à négliger l'influence sur les sections efficaces des perturbations pouvant avoir affecté le combustible pendant son irradiation et dues par exemple à son environnement immédiat, aux mouvements de barres de contrôle, à des variations de température du modérateur, etc....

L'importance de ces approximations n'est pas du même ordre de grandeur pour tous les paramètres et nous examinons ce point dans le paragraphe suivant (2.3.).

La qualification de cette méthode simplifiée est obtenue par deux voies:
- Le bénéfice de la qualification du code de cellule et des bibliothèques neutroniques associées qui résulte des travaux de physique des réacteurs.
- Des comparaisons globales effectuées entre les valeurs calculées par la méthode simplifiée et des résultats expérimentaux.

TABLEAU IV. SENSIBILITE DE DIFFERENTES CORRELATIONS AUX PERTURBATIONS DE SPECTRE PENDANT L'IRRADIATION ET AUX INCERTITUDES SUR LA CONNAISSANCE DES DONNEES HISTORIQUES

Influence sur la relation \ PERTURBATION			A	B	C	D
TCF	⟷	Δ (U5)	3 %	5 %	2 %	2 %
TCF	⟷	Δ (U6)	1 %	5 %	0	1 %
TCF	⟷	Pu/U	0	5 %	6 %	8 %
Pu/U	⟷	Pu9/Pu0	0,5 %	10 %	3 %	5 %
Pu/U	⟷	$\frac{Pu2/Pu0}{Pu0/Pu9}$	0	1 %	2 %	4 %
Pu/U	⟷	$\frac{Pu - Pu9}{Pu}$	0,3 %	0	7 %	11 %

Référence : PWR Zr type 17 × 17; enrichissement initial = 3,2%; taux de combustion = 24000 MW·d/t.

Perturbation :

A : Erreur de 1 % sur l'enrichissement initial.
B : Erreur de 10 % sur le taux de combustion annoncé (sans itération). En pratique les conséquences de cette incertitude sont négligeables dès lors que l'on itère sur le taux de combustion.
C : Empoisonnement partiel de l'assemblage pendant irradiation (Pyrex).
D : Variation de 5 % du rapport de modération moyen.

Nous ne reviendrons pas sur le premier point qui dans le cas du code APOLLO par exemple a déjà fait l'objet de publication (3).

Pour le second nous donnons quelques exemples de résultats dans les tableaux II et III. Le premier contient des ordres de grandeur d'écarts expérience-calcul avant ajustement du formulaire et le second les écarts moyens après ajustement pour des combustibles du réacteur de la Centrale Nucléaire des Ardennes et ceux du réacteur de STADE qui ont été retraités à LA HAGUE. Il est particulièrement intéressant de constater la bonne cohérence obtenue pour ces derniers alors que l'ajustement n'avait pas pris en compte leurs résultats.

2.3. Influence des approximations sur différents paramètres

Afin de juger de l'importance des approximations de la méthode simplifiée pour les différents paramètres, compositions ou relations entre compositions, qui sont plus particulièrement intéressants pour le contrôle à l'entrée des usines de retraitement, nous avons réalisé une étude d'effets différentiels en créant artificiellement des perturbations nettement plus importantes que celles qui peuvent affecter des assemblages d'un même réacteur. Les résultats sont

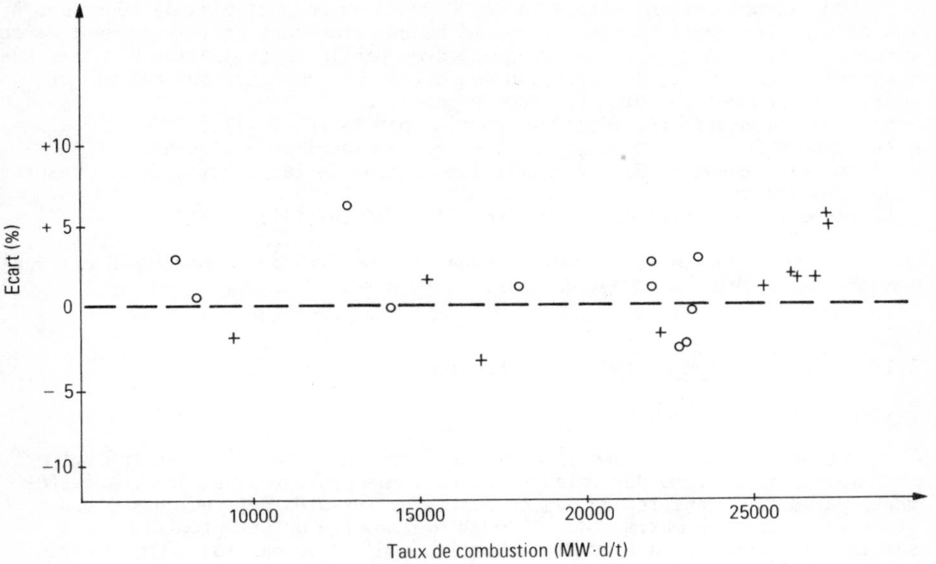

FIG.1. Ecart entre les taux de combustion déduits de $^{134}Cs/^{137}Cs$ et de la teneur en Nd.
○ CNA deuxième cycle; + CNA troisième cycle.

présentés dans le tableau IV. De ces quelques exemples nous pouvons déduire une série de conclusions :

- Les relations entre taux de combustion et variation des compositions isotopiques de l'uranium sont sensibles à la composition initiale (cas A) ce qui est évident et nécessite d'avoir toujours un recoupement avec d'autres déterminations indépendantes.
- La sensibilité des différentes relations au taux de combustion est plus ou moins importante mais ici ne constitue pas un critère de choix puisque le taux de combustion est un des résultats et que par itération on peut réduire très facilement les conséquences d'une mauvaise évaluation initiale.
- Les perturbations de spectre en cours d'irradiation, et nous avons ici deux exemples: introduction d'absorbants (cas C) et variation du rapport de modération (cas D), sont particulièrement sensibles au niveau des relations impliquant le rapport Pu/U. Les valeurs indiquées font ressortir la nécessité d'être prudent dans l'utilisation de ces relations et en particulier de ne pas juger de leur intérêt sur la faible sensibilité aux erreurs éventuelles sur la composition initiale du combustible ou sur le taux de combustion.

3. APPLICATION POUR LES BILANS D'ENTREE

La méthode classique d'établissement du "bilan gravimétrique" uranium à l'entrée de l'usine de retraitement nécessite une technique analytique longue et coûteuse dont la phase la plus lourde consiste en la détermination par spectrométrie de masse du néodyme de fission.

Dans l'application faite à LA HAGUE pour les combustibles de réacteurs à eau on cherche à mettre au point des méthodes permettant progressivement de remplacer en grande partie la mesure du néodyme par la détermination d'autres paramètres. Par exemple, à l'occasion des dernières campagnes des calculs de masses fissionnées ont été effectués à partir de :
- L'irradiation intégrée (données fournies par le réacteur),
- Le rapport Pu/U déterminé par analyse au laboratoire de l'usine,
- L'appauvrissement en U_{235} à partir des données de fabrication et des mesures du laboratoire,
- Le rapport des activités des césium 134 et césium 137.

Plusieurs possibilités pour recouper les mesures du rapport Pu/U ont également fait l'objet d'investigations.

3.1. Détermination du taux de combustion

3.1.1. Précision requise

Une étude détaillée des incertitudes affectant les bilans en uranium et plutonium, compte tenu des valeurs extrêmes que peuvent atteindre les différents paramètres dans le cas des combustibles irradiés de réacteurs à eau (taux de combustion maximum de 40000 MW·d/t), montre qu'une précision de 5 % sur la détermination du taux de combustion conduit au maximum à une contribution à l'incertitude de 0,2 % sur les quantités d'uranium et de plutonium. Si dans cette détermination de taux de combustion la part d'incertitude liée aux erreurs systématiques n'excède pas 2 % on peut réduire l'influence des erreurs aléatoires par un nombre suffisant de mesures de telle sorte que globalement sur un lot important de combustible, correspondant par exemple à une recharge de réacteur à eau c'est-à-dire un tiers de coeur, la contribution de cette détermination à l'incertitude finale sur le bilan en masses d'U et Pu n'excède pas 0,1 %. Ceci est un objectif raisonnable et il est important de noter que ces ordres de grandeur sont ceux que l'on peut atteindre sur les valeurs annoncées par l'exploitant du réacteur et que l'on dispose ainsi d'une possibilité de recoupement.

3.1.2. Méthode de référence

Les travaux effectués depuis de nombreuses années sur les mesures de taux de combustion ont conduit à retenir la méthode du néodyme, la teneur de l'isotope 148 de cet élément étant mesurée par spectrométrie de masse et rapportée à la quantité d'uranium 238 (4). Cette mesure est complexe et sa mise en oeuvre systèmatique nécessite des moyens importants. La précision d'une détermination est de 2 à 3 % avec une contribution de l'erreur systématique liée en particulier à la connaissance absolue des rendements de fission du ^{148}Nd qui se situe entre 1 et 2 %. Nous la considérons donc comme méthode de référence. Elle a été appliquée à LA HAGUE sur les 17 dissolutions de la première campagne (combustible du réacteur BWR de MUHLEBERG) et sur 15 des 81 dissolutions de la seconde campagne (combustible du réacteur PWR de STADE).

3.1.3. Méthode du rapport césium

L'intérêt du rapport d'activités des ^{134}Cs et ^{137}Cs pour la détermination du taux de combustion a été montré par ailleurs (5).

Des études de laboratoire ont été réalisées au CEA sur des solutions de combustibles irradiés du réacteur PWR de la Centrale Nucléaire des Ardennes (CNA) avec une large gamme de taux de combustion et de temps de refroidissement. Des exemples de résultats sont donnés sur la Figure 1. L'interprétation des résultats est effectuée par la méthode simplifiée présentée au paragraphe 2.2.

TABLEAU V. DETERMINATIONS DE TAUX DE COMBUSTION
Comparaison des méthodes (test de Student) [8]

METHODES	|D̄|			s D			Test de STUDENT		
	CNA	MUL	STA	CNA	MUL	STA	CNA	MUL	STA
R-Δ(U5)	-	0.13	0.33	-	1.52	1.46	0	-	0.33
R-Δ(U6)	-	0.41	0.31	-	3.24	2.02	0	-	-
R-Pu/U	-	-	0.30	-	-	2.85	0	0	-
Δ(U5)-Δ(U6)	0.02	0.27	0.26	1.96	3.51	2.08	-	-	-
Δ(U5)-Pu/U	-	-	0.63	-	-	2.88	0	0	-
Δ(U6)-Pu/U	-	-	0.61	-	-	2.56	0	0	0.61
Nd-Δ(U5)	0.68	1.05	0.04	1.27	2.15	1.75	0.68	-	-
Nd-Δ(U6)	0.66	1.72	0.92	1.37	3.00	1.91	0.66	-	-
Nd-Pu/U	-	-	0.69	-	-	3.35	0	0	-
Nd - R	-	0.91	0.14	-	1.98	1.64	0	-	-
Nd - Cs	-	0.59	-	-	2.95	-	0	-	-

Légendes

CNA = Réacteur PWR de la Centrale Nucléaire des Ardennes (33 mesures sur prélèvements de crayons)
MUL = Réacteur BWR de MUHLEBERG (17 mesures sur dissolutions à LA HAGUE)
STA = Réacteur PWR de STADE (81 mesures sur dissolutions à LA HAGUE, excepté pour Nd, seulement 15 valeurs)

D̄ = Ecart moyen entre méthodes } exprimés en %
sD = Ecart type sur D̄ }

Test de STUDENT = 0 pas de test,
- : pas d'écart systématique,
ou la valeur la plus probable de l'écart systématique.

Méthodes :

R : Taux de combustion annoncé par l'exploitant du réacteur
Δ(U5) : Taux de combustion déduit de l'appauvrissement en ^{235}U
Δ(U6) : Taux de combustion déduit de la formation de ^{236}U
Pu/U : Taux de combustion déduit de la formation de plutonium
Nd : Taux de combustion déduit de la mesure du néodyme.

Pour les deux campagnes de retraitement considérées des mesures systématiques, par spectrométrie gamma, ont été effectuées sur les prélèvements de solutions dans le dissolveur utilisés pour les autres analyses. Une interprétation complète de ces résultats n'a pu être réalisée, certaines informations sur l'historique d'irradiation étant incomplètes. Néanmoins l'utilisation directe des résultats de mesure en valeurs relatives montre pour la campagne MUHLEBERG une très bonne cohérence avec les valeurs néodyme alors que pour la campagne STADE la dispersion est beaucoup plus importante.

TABLEAU VI. DETERMINATIONS DE TAUX DE COMBUSTION
Estimation des erreurs aléatoires [9]

METHODES	ECART TYPE (%)			
	CNA 33 mesures	MUL 14 mesures	STADE 15 mesures	STADE 81 mesures
Réacteur	-	0.4	1.1	1.3
Néodyme	0.0	1.4	1.3	-
Δ (U5)	1.3	1.3	1.0	1.3
Δ (U6)	1.6	2.9	1.1	1.5
Pu / U	-	-	3.1	2.5
Césium	-	2.3	-	-

Légendes : Voir tableau V

3.1.4. Utilisation de mesures isotopiques sur l'Uranium et le Plutonium

La recherche du taux de combustion à partir des teneurs isotopiques de l'Uranium ou du Plutonium mesurées sur des échantillons prélevés au dissolveur a été systématiquement entreprise. Cette recherche a été guidée par les conclusions des études théoriques et expérimentales présentées par ailleurs(2) et dont les points principaux sont rappelés en 2.3.

Le passage des variations de compositions isotopiques (appauvrissement en ^{235}U, formation de ^{236}U, rapport Pu/U) aux valeurs de taux de combustion a été réalisé grâce aux formulaires présentés en 2.2.

Une analyse statistique des résultats obtenus pour les campagnes MUHLEBERG et STADE est présentée dans les tableaux V et VI où sont rappelées également les valeurs obtenues sur des échantillons de C.N.A.

On peut constater que cette analyse portant sur un grand nombre de résultats conduit pour l'estimation des erreurs aléatoires à un écart type par mesure inférieur à 3 % et dans la plupart des cas nettement meilleur. Le test de STUDENT, pour la détermination d'erreurs systématiques éventuelles,est généralement négatif, ceci malgré le grand nombre de mesures et donc l'erreur aléatoire réduite sur l'écart moyen entre méthodes, et dans les quelques cas où il est positif l'estimation d'erreur systématique entre méthodes est inférieure à 1 %.

3.2. Détermination du rapport Pu/U

3.2.1. Précision requise

Dans le cas de combustibles Uranium de réacteur à eau la mesure du rapport Pu/U est l'élément principal d'incertitude sur le bilan plutonium déduit de la méthode gravimétrique. La précision recherchée est meilleure que 1 %, l'objectif raisonnable étant l'incertitude sur une mesure directe de ce rapport par spectrométrie de masse avec la technique de double dilution isotopique, soit environ 0,5 %.

TABLEAU VII. DETERMINATIONS DU RAPPORT Pu/U POUR STADE
(81 dissolutions)

COMPARAISON DES METHODES			ESTIMATION DES ERREURS ALEATOIRES	
METHODES	$\bar{D}$ (%)	sD (%)	METHODE	σ (%)
Mesure - Réacteur	- 1.85	2.46	Mesure	0.4
Mesure - TCF	0.37	1.50	Réacteur	2.9
Mesure - Pu9 / Pu0	1.17	1.79	TCF	1.1
Mesure - $\frac{Pu2 / Pu0}{Pu0 / Pu9}$	5.16	1.98	Pu9/Pu0	1.7
Réacteur - TCF	2.25	3.10	$\frac{Pu2/Pu0}{Pu0/Pu9}$	1.0
Réacteur - Pu9 / Pu0	3.02	3.42		
Réacteur - $\frac{Pu2 / Pu0}{Pu0 / Pu9}$	6.86	2.12		
T C F - Pu9 / Pu0	0.80	0.85		
T C F - $\frac{Pu2 / Pu0}{Pu0 / Pu9}$	4.64	1.80		
Pu9/Pu0 - $\frac{Pu2 / Pu0}{Pu0 / Pu9}$	3.84	2.16		

Légendes :

$\bar{D}$ = Ecart moyen entre méthodes
sD = Ecart type sur $\bar{D}$
σ = Ecart type estimé par méthode
Mesure = Détermination directe du rapport $\frac{Pu}{U}$ par spectrométrie de masse
Réacteur = Estimation du rapport Pu/U fournie par l'exploitant
T C F = Valeur de Pu/U déduite du taux de combustion mesuré
Pu9/Pu0 = Valeur de Pu/U déduite du rapport isotopique mesuré Pu9/Pu0
$\frac{Pu2/Pu0}{Pu0/Pu9}$ = Valeur de Pu/U déduite du rapport isotopique mesuré $\frac{^{242}Pu/^{240}Pu}{^{240}Pu/^{239}Pu}$

3.2.2. Précision du calcul

Nous avons indiqué au paragraphe 2.1. les précisions que l'on peut attendre du calcul "réacteur", compte tenu de l'incertitude sur la connaissance du taux de combustion. On voit que même en supposant ce dernier très bien connu grâce à une exploitation complète de toutes les méthodes indiquées en 3.1., la précision sur une détermination individuelle du rapport Pu/U par le calcul sera insuffisante pour atteindre l'objectif visé. Par contre au niveau d'un lot de combustible correspondant à la charge d'un réacteur ou à une fraction importante de cette charge il n'est pas exclu d'aboutir à une précision suffisante à partir des calculs décrits en 2.2. et d'une détermination précise du taux de combustion. Les exemples montrés dans le tableau VII font nettement ressortir cette tendance.

TABLEAU VIII. DETERMINATIONS DES MASSES D'URANIUM POUR STADE

a) Comparaison des méthodes

METHODES DE DETERMINATION DU T C F	ECARTS SUR LES MASSES U en %					
	$\|\bar{D}\|$		s D		TEST	
	15 Mes.	81 Mes.	15 Mes.	81 Mes.	15 Mes.	81 Mes.
Nd - Réacteur	0.02	-	0.11	-	-	0
Nd - Δ (U5)	0.01	-	0.12	-	-	0
Nd - Pu/U	0.02	-	0.15	-	-	0
Réacteur - Δ(U5)	0.03	0.03	0.05	0.07	-	0.03
Réacteur - Pu/U	0.00	0.02	0.11	0.09	-	-
Δ (U5) - Pu/U	0.03	0.05	0.12	0.11	-	0.05

b) Estimation des erreurs aléatoires

M E T H O D E	ECART TYPE (%)	
	15 Mesures	81 Mesures
Néodyme	0.03	-
Réacteur	0.03	0.04
Δ (U5)	0.04	0.06
Pu/U	0.10	0.09

Légendes : Voir tableau V

3.2.3. Utilisation de mesures isotopiques

Plusieurs "indicateurs" du rapport Pu/U, autres que le taux de combustion ont été proposés par divers auteurs. Là encore une analyse systématique a été effectuée sur les résultats des campagnes de LA HAGUE en partant des conclusions des études déjà citées (2).

Les résultats présentés dans le tableau VII montrent que toutes ces déterminations sont plutôt moins précises que celle qui résulte du taux de combustion. Les écarts notés sur la moyenne des lots peuvent être réduits par une qualification plus poussée du calcul des isotopes correspondants. Les combinaisons de rapports isotopiques sont certainement susceptibles d'amélioration en jouant sur des compensations d'effets résultant de perturbations éventuelles mais on risque alors de se heurter au problème de limite de précision des résultats d'analyses.

3.3. Détermination des masses d'Uranium et de Plutonium

A partir des différentes déterminations exposées précédemment pour les taux de combustion, le bilan en masses d'Uranium et de Plutonium a été déterminé pour les deux campagnes de retraitement déjà citées.

Le tableau VIII qui contient les résultats définitifs de la campagne STADE montre que l'on dispose d'un assez large éventail de moyens pour réduire l'importance des mesures à effectuer dans l'application de la méthode gravimétrique ou pour contrôler les résultats à partir de mesures indépendantes.

On peut noter que l'écart type sur une détermination de masse d'Uranium est inférieur à 0,1 % et qu'il n'apparaît pas d'écart systématique dû à des différences entre méthodes de détermination du taux de combustion qui dépasse 0,05 %. Pour le bilan en masses de Plutonium il est clair qu'il faut ajouter à ces valeurs l'incertitude sur la détermination du rapport Pu/U.

4. APPLICATION A L'IDENTIFICATION NON DESTRUCTIVE DES COMBUSTIBLES IRRADIES

C'est un second exemple de domaine d'application des études présentées ici. Il est particulièrement intéressant puisque sa conception n'a été possible qu'avec le développement de ces études.

Des études préliminaires ont été effectuées au CEA à partir de l'expérience acquise dans les mesures de spectrométrie gamma sur des assemblages de réacteurs à eau et des travaux sur l'interprétation des mesures de compositions de combustibles irradiés.

La technique de mesures, ses conditions pratiques de mise en oeuvre et les principes du contrôle d'identification sont présentés dans une autre communication à ce colloque (6). Nous ne traiterons ici que de l'apport de la prise en compte de l'histoire du combustible pour un tel contrôle.

4.1. Mesure du taux de combustion

Les études que nous avons effectuées nous ont conduit à ne retenir, dans le cas de mesures non destructives sur des combustibles de réacteur à eau que les deux solutions suivantes :

- La mesure du rapport des activités des césium 134 et 137
- La mesure de la teneur absolue en 137 Cs.

Dans l'un et l'autre cas on ne peut pas envisager d'interpréter les résultats sans connaître un minimum d'informations sur l'histoire du combustible, l'importance, en quantité, étant toutefois plus grande dans le premier que dans le second. La solution proposée au terme des études préliminaires déjà citées consiste à exploiter en ligne ces informations avec des formulaires dérivés de la méthode de calcul présentée en 2.2.

Les principales données qui doivent être fournies pour cette exploitation sont :

- Géométrie et composition initiale du combustible,
- Type de réacteur dans lequel a eu lieu l'irradiation,
- Historique approché du flux et date de déchargement.

Les deux premières catégories sont nécessaires pour le calcul neutronique des sections efficaces, de la même manière que dans les autres applications. La troisième est d'une importance particulière dans le cas du rapport Césium en raison de la période radioactive du ^{134}Cs, 2, 1 ans. Comme il s'agit d'une information précise sur le fonctionnement du réacteur il est intéressant de fixer le degré de précision avec lequel elle doit être connue. L'étude de différents schémas de fonctionnement tels que ceux qui sont résumés dans le tableau IX permet de conclure que l'idéal est la connaissance d'un historique mensuel de puissance du réacteur et des taux de combustion annoncés à la fin de chaque cycle. A défaut la simple prise en compte avec ces mêmes valeurs des

TABLEAU IX. EXEMPLES DE SENSIBILITE DES RAPPORTS $^{134}Cs/^{137}Cs$ ET $^{144}Ce/^{137}Cs$ A L'HISTOIRE D'IRRADIATION

HISTOIRE	PUISSANCE DE FONCTIONNEMENT EN % DE LA PUISSANCE NOMINALE				
	Schéma de fonctionnement				
	Référence	1	2	3	4
0	75%	75%	75%	75%	75%
1 an	0*	0	0	0	0
	75%	75%	75%	50% 100% 50%	75%
2 ans	0	0	0	0	0
	75%	100% 50%	50% 100% 50%	75%	75%
3 ans	Décharg^t	Décharg^t	Décharg^t	Décharg^t	Décharg^t

Ecarts en % avec la référence pour les rapports à la date de déchargement:

$^{134}Cs/^{137}Cs$	0	−1.8	−0.1	0	− 3.6
$^{144}Ce/^{137}Cs$	0	−4.6	−0.5	+ 0.2	−10.4

* 0 = arrêt

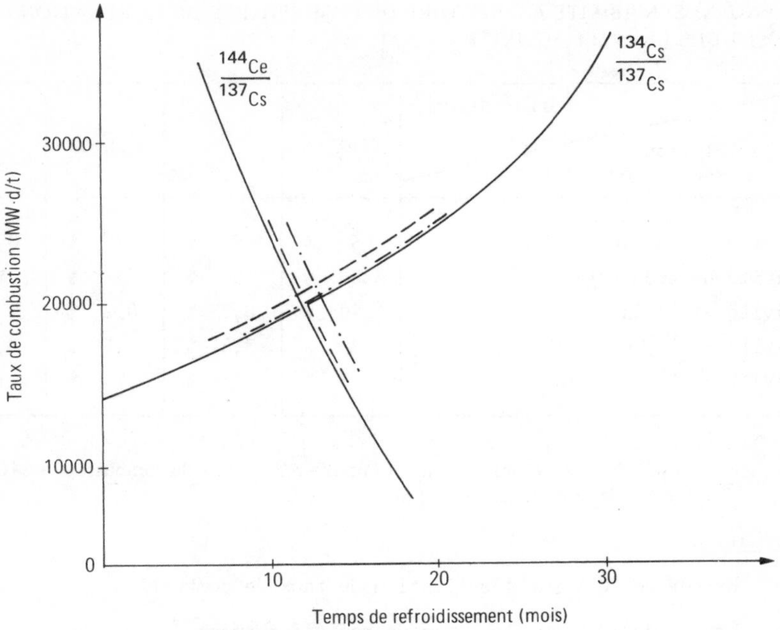

FIG.2. Détermination du taux de combustion et du temps de refroidissement pour des rapports d'activités donnés. ——— courbes de référence; – – – – écart de 10% sur l'enrichissement initial; —·—·— écart de 10% sur la durée d'irradiation.

taux de combustion et des dates de début et de fin de cycles est suffisante dans la grande majorité des cas ; l'approximation qui en résulte ne peut entraîner d'erreurs supérieures à 5 % sur le taux de combustion.

4.2. Validité du contrôle

Puisque la mesure du taux de combustion nécessite la prise en compte de données sur l'histoire du combustible elle ne peut, seule, constituer une réelle identification de l'élément combustible. C'est pourquoi la solution proposée comprend en outre une mesure du temps de refroidissement, fondée sur des produits de fission émetteurs γ de périodes très différentes (^{95}Zr, ^{144}Ce, ^{106}Ru,....), et un contrôle d'identification par comparaison des valeurs obtenues pour les deux paramètres, taux de combustion et temps de refroidissement, avec les valeurs annoncées par la fiche suiveuse du combustible. Deux questions peuvent alors se poser :

- Quelle est la qualité de l'identification réalisée à partir de ces deux paramètres?
- Que se passe-t-il si l'une des données non contrôlées est fausse ?

Le schéma de la Figure 2 apporte des éléments de réponse à ces questions. On constate que les rapports d'activité des ^{134}Cs et ^{137}Cs d'une part et des ^{144}Ce et ^{137}Cs d'autre part permettent une discrimination précise des deux paramètres recherchés et qu'une erreur sur l'enrichissement initial est décelable par ce contrôle si elle est importante.

TABLEAU X. SENSIBILITE A L'HISTOIRE DU COMBUSTIBLE DE LA RELATION ENTRE MASSE ET DIFFERENTES ACTIVITES

Perturbation / Effet sur la relation entre masse de combustible et	A	B	C	D
Activité γ totale	5 %	5 %	5 %	0,5 %
Activité neutron totale	50 %	3 %	1 %	16 %
Activité γ ^{137}Cs	10 %	0,2 %	0,2 %	0
Activité γ $^{136}Cs/^{137}Cs$	8 %	3,5 %	2,5 %	3,5 %
Activité γ ^{144}Ce	1,5 %	7 %	7 %	1 %

Références : PWR Zr 17 × 17; enrichissement initial = 3,2%; taux de combustion = 24000 MW·d/t temps de refroidissement = 1 an.

Perturbations :

A = Erreur de 10 % sur l'estimation du taux de combustion

B = Erreur de 10 % sur le temps de refroidissement

C = Erreur de 10 % sur la durée d'irradiation à puissance moyenne constante

D = Erreur de 10 % sur l'enrichissement initial.

5. APPLICATION POUR D'AUTRES CONTROLES NUCLEAIRES DANS L'USINE DE RETRAITEMENT

Les activités gamma et neutron des combustibles irradiés permettent de réaliser un certain nombre de contrôles nucléaires intéressant le fonctionnement du procédé, la sûreté de l'installation et l'application des garanties sur les matières nucléaires. Dans la plupart des cas, que l'on s'intéresse aux activités totales ou à l'activité spécifique de tel ou tel isotope, l'interprétation du résultat nécessite l'utilisation de valeurs calculées à partir de l'histoire du combustible. Là encore pour que ces calculs puissent être réalisés de manière relativement simple et ne nécessitent qu'un ensemble limité de données "historiques" il est indispensable de choisir judicieusement les paramètres mesurés compte tenu de l'objectif poursuivi.

5.1. Sensibilité de divers paramètres aux données et à la méthode de calcul

A titre d'exemple nous avons examiné la sensibilité à la précision de connaissance des données "historiques" et aux approximations de la méthode de calcul dite simplifiée (§ 2.2.) de la relation entre diverses activités et la quantité de combustible correspondante, en supposant que les problèmes de condition de mesure et d'interprétation de la propagation des rayonnements sont résolus par ailleurs. Les résultats sont présentés dans le tableau X.

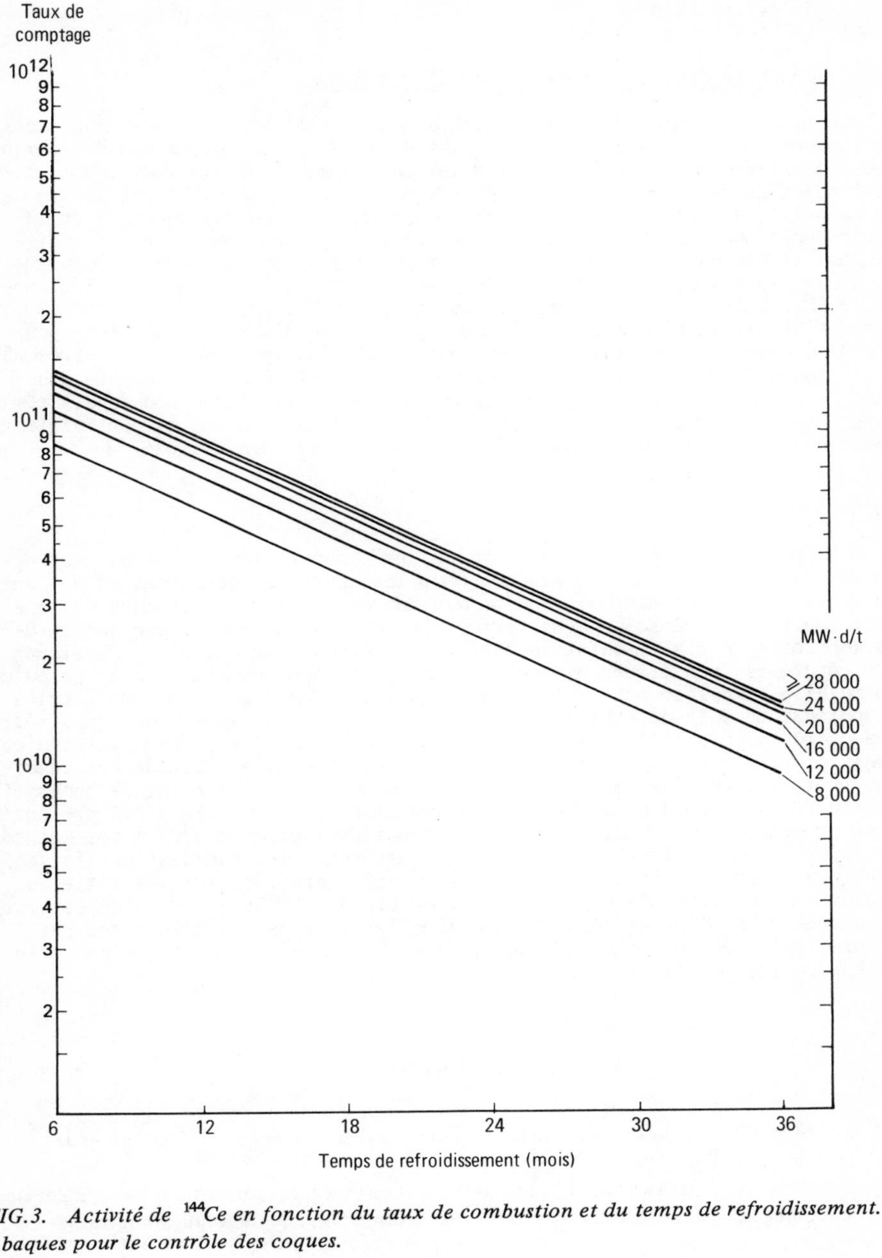

FIG.3. Activité de ^{144}Ce en fonction du taux de combustion et du temps de refroidissement. Abaques pour le contrôle des coques.

On constate aisément que l'utilisation dans ce but d'une mesure d'activité neutronique totale est beaucoup plus sensible, et donc plus délicate, que celle de l'activité gamma d'un produit de fission de longue période. Il ne s'agit là bien sûr que d'un exemple.

5.2. Contrôle des coques par spectrométrie gamma

Parmi les nombreuses applications à des contrôles nucléaires nous avons retenu une de celles qui concernent plus directement l'application des garanties, c'est-à-dire le contrôle des quantités de matières fissiles éventuellement rejetées avec les résidus de gaines après dissolution. L'application de ce contrôle fait l'objet d'une autre communication à ce colloque (7) et là encore nous nous limiterons au problème de la prise en compte de l'histoire du combustible dans l'exploitation des résultats de ce contrôle.

Dans le cas des combustibles de réacteurs à eau la méthode adoptée consiste à mesurer la quantité absolue de ^{144}Ce et à en déduire la quantité de combustible résiduel. La Figure 3 donne un exemple d'abaques calculées pour l'exploitation des résultats correspondants et dans le tableau X on a résumé l'influence des principaux paramètres sur cette relation. On constate que dans ce cas si le calcul est nécessaire pour une utilisation sensée des résultats, les approximations sur la méthode ou les données n'ont que de faibles conséquences à l'exception toutefois du temps de refroidissement.

CONCLUSION

La prise en compte de l'historique des combustibles retraités peut apporter des informations importantes pour les contrôles de garanties à l'entrée de l'usine. Elle ne nécessite qu'un minimum de données historiques mais ne peut fournir des renseignements précis et sûrs que moyennant une interprétation correcte des résultats à partir des méthodes de calcul de physique des réacteurs. Les études présentées dans cette communication montrent qu'un formulaire simplifié mais suffisamment qualifié donne de très bons résultats. L'application à la détermination du bilan d'entrée par la méthode gravimétrique offre la possibilité de réduire le nombre d'analyses du néodyme et d'obtenir plusieurs valeurs indépendantes pour vérifier l'exactitude des résultats. Les renseignements des deux premières campagnes de retraitement de combustibles de réacteurs à eau à LA HAGUE dont la synthèse des résultats a été présentée sont prometteurs et confirment l'intérêt qui peut être accordé à ces méthodes. Les autres applications, dans le domaine des contrôles non destructifs, doivent également faciliter la mise en oeuvre des garanties sur les matières nucléaires. Le contrôle des coques par mesure du ^{144}Ce est déjà en service, notamment à LA HAGUE et l'identification d'assemblages si elle n'est pas encore appliquée sur un plan industriel ne semble pas présenter d'obstacle technique insurmontable.

REFERENCES

[1] KAVENOKY, A., Un système modulaire pour le calcul des réacteurs à eau ordinaire, Bull. Inf. Sci. Tech. (Paris) n° 212 (1976).

[2] BOUCHARD, J., DARROUZET, M., ROBIN, M., «Etude des corrélations isotopiques sur les combustibles irradiés», Symp. on Isotopic Correlation and its Application to the Nuclear Fuel Cycle, ICT/2, Stresa, 9–11 mai 1978.

[3] BOUCHARD, J., KAVENOKY, A., REUSS, P., «NEPTUNE, Un système modulaire pour le calcul des réacteurs à eau ordinaire», Conf. Nucl. Europ., Paris, avril 1975.

[4] ROBIN, M., «Specific burn-up determination and qualification of fuel irradiation dosimetry calculations based on post irradiation measurements», ASTM-Euratom, Palo Alto, oct. 1977).
[5] BOUCHARD, J., FREJAVILLE, G., ROBIN, M., Mesures de distribution de puissance et de taux de combustion par spectrométrie γ sur les combustibles irradiés de réacteurs à eau, Note CEA-N-1982 (1977).
[6] BERAHA, R., et al., «Contrôle des combustibles irradiés des réacteurs à eau par gammamétrie sur les sites des centrales», Présents Comptes rendus, IAEA-SM-231/47.
[7] CONSTANT, J., et al., «Expérience acquise à La Hague sur le contrôle des coques», Présents Comptes rendus, IAEA-SM-231/46.
[8] AFNOR, Recueil de normes de la statistique (1974).
[9] JAECH, J.L., Errors of measurement with more than two measurement methods, J. Inst. Nucl. Materials Management **4** 4 (1976).

DISCUSSION

C.N. BEETS: I should like to make a comment. The accuracy and precision achieved with the isotopic correlation technique (ICT) must be compared with those obtained using other techniques. When the correlation techniques were first used, they compared favourably with reference techniques such as mass-spectrometry (IDA 72).[1] Since considerable progress has been made both in measuring techniques and in theoretical predictions, it can be assumed that the correlations will become more exact.

L. KOCH: First, I should like to make a comment. Mr. Bouchard, you say that the correlation between the Pu/U ratio and the burnup is less precise than, say, the ^{236}U depletion. Since, however, this correlation is less sensitive to the initial enrichment, it could be used for other enrichments.

I also have a question. As we mentioned in our paper IAEA-SM-231/26[2], the correlations existing between burnup and plutonium isotopic ratios, e.g. $^{242}Pu/^{240}Pu$, are precise enough to meet the requirements of the Pu/U ratio method. Did you consider these correlations in your work?

J. BOUCHARD: We did not investigate every conceivable correlation in our perturbation studies, but if there are any other interesting possibilities we shall not fail to consider them. However, we are convinced that it is always possible to fault an empirical correlation, and for this reason we believe that it is essential to leave out this stage for precise applications and to concentrate on relations which are solidly based on reactor calculations. We then require a minimum of information in regard to the fuel history, e.g. initial enrichment, but then I do not

[1] BEYRICH, W., DROSSELMEYER, E. (Eds.), The Interlaboratory Experiment IDA-72 on Mass Spectrometric Isotope Dilution Analysis, Kernforschungszentrum Karlsruhe, Fed.Rep. Germany, Rep. KFK 1905 (two volumes). Also identified as European Community Euratom Rep EUR 5203e (July 1975).

[2] These Proceedings, Vol.II.

think it is possible to develop precise control applications using isotopic correlations without this minimum amount of information, whatever method is chosen.

A.G. HAMLIN: As a representative of the operators of production plants, who have no interest in isotope correlation in their daily business, I should like to strongly support your conclusion that in this field we should proceed only on a basis of sound theory.

Frankly, the structure of isotopic correlation as a mass of individual pairs extracted from this or that fuel or from a few chance pellets terrified me. Sooner or later, isotopic correlation would inevitably have picked on one of the outliers that always seemed to occur among the half dozen ratios presented, or the plant would have been processing fuel whose irradiation history differed from that from which the correlation was derived so that the latter was invalidated.

I strongly maintain that if operators are to be accused it should be on a basis of sound theory.

S. SANATANI: Since the Agency did not present a paper on isotopic correlations during this symposium, I should like to mention that we do have a continuing programme on this subject in the form of a Research Co-ordination Programme with participants from Belgium, Euratom, Japan and the United States of America. We have had three meetings at yearly intervals and the next meeting will be in Vienna from 20 to 23 November 1978. Both data banks and ICT procedures are discussed under this programme and the Agency might invite one or two more Member States to join the programme at a later stage.

My other comment is that the interest of ICT varies according to whether one is a physicist, a chemist, or a safeguards inspector, and we should consider the different viewpoints in order to get a complete picture.

J. BOUCHARD: I am not sure that I made myself clear. I did not intend to imply that the work which has been carried out in the past was of no use. A great many experimental results have been examined and the empirical correlations deduced from them can certainly be used to make an approximate check on measurement results, for example in cases where no information on the history of the fuel is available.

What I said was that these empirical correlations may not be adequate for precise applications so that it becomes necessary to resort to calculated relations based on reactor physics and confirmed by experimental results. This is particularly important in the case of international safeguards for which, in my opinion, one cannot rely on empirical relations or historical data the validity of which can never be absolutely guaranteed in any particular case.

ИЗОТОПНЫЕ КОРРЕЛЯЦИИ В ОБЛУЧЕННОМ ТОПЛИВЕ РЕАКТОРОВ ТИПА ВВЭР

О.А. МИЛЛЕР, С.В. ПИРОЖКОВ, Ю.Ф. РОДИОНОВ,
В.М. ШАТИНСКИЙ, В.П. ТАРАСЕВИЧ
Институт атомной энергии им. И.В. Курчатова
Государственного комитета
по использованию атомной энергии СССР,
Москва

Г.Н. ЯКОВЛЕВ, А.А. ЗАЙЦЕВ, Г.Н. РОБУЛЕЦ
Научно-производственное объединение "Энергия"
Министерства энергетики и электрификации СССР,
Москва

В.П. КРУГЛОВ, В.М. ИЛЯСОВ, Ю.Б. НОВИКОВ
Нововоронежская атомная электростанция им. 50-летия СССР,
Союз Советских Социалистических Республик

Abstract–Аннотация

ISOTOPIC CORRELATIONS IN IRRADIATED FUEL FROM VVER POWER REACTORS.

The paper describes studies on the isotopic composition of irradiated VVER reactor fuel carried out for the purpose of checking calculational programs for burnup and deriving isotopic correlations for uranium and plutonium. The authors present experimental data on the isotopic composition of uranium and plutonium in irradiated fuel from the VVER-365 and VVER-440 reactors at the Novovoronezh Nuclear Power Station. The discrepancies between these data and data calculated using the ROR and UNIRASOS programs are: for ^{238}U, less than 1%; for ^{235}U, 2–3%; for plutonium, 5–7%. Correlations were obtained for the plutonium isotopes N_{239Pu}/N_{Pu} from the burnup, N_{Pu}/N_U from the content of ^{235}U, N_{Pu}/N_U and N_{240Pu}/N_{239Pu} from the ^{235}U and ^{238}U isotopic ratio (N_5/N_8), and $(100\text{-}N_{239Pu})/D_5$ from the ^{235}U content. The discrepancy between the experimental and calculated values was usually not more than 3–5%. These experimental data on isotopic correlations can be of use for the purposes of IAEA safeguards.

ИЗОТОПНЫЕ КОРРЕЛЯЦИИ В ОБЛУЧЕННОМ ТОПЛИВЕ РЕАКТОРОВ ТИПА ВВЭР.

Описаны результаты исследований изотопного состава облученного топлива реакторов типа ВВЭР, проведенных с целью проверки расчетных программ выгорания и получения корреляционных отношений изотопов урана и плутония. Приведены экспериментальные данные по изотопному составу урана и плутония в облученном топливе ВВЭР-365 и ВВЭР-440 Нововоронежской АЭС. Расхождение их с расчетами по программам РОР и УНИРАСОС составило: для ^{238}U – менее 1%, для ^{235}U – 2-3%, для Pu – 5-7 %. Получены корреляционные отношения для изотопов плутония: $N_{Pu\text{-}239}/N_{Pu}$ в зависимости от выгорания, N_{Pu}/N_U – от содержания ^{235}U, N_{Pu}/N_U и $N_{Pu\text{-}240}/N_{Pu\text{-}239}$ – от отношения изотопов ^{235}U и ^{238}U ($N_{U\text{-}235}/N_{U\text{-}238}$), а также $(100 - N_{Pu\text{-}239})/D_{U\text{-}235}$ – от содержания ^{235}U. Расхождение экспериментальных и расчетных значений, как правило не превышало 3-5 %. Описанные экспериментальные данные по изотопным корреляциям могут оказаться полезными для целей гарантий МАГАТЭ.

ВВЕДЕНИЕ

Изотопные корреляции в облученном топливе реакторов используются для целей гарантий МАГАТЭ, главным образом, в качестве одного из дополнительных способов контроля ядерных материалов при переработке облученного топлива.

Как отмечается в ряде работ [1-9], корреляционные зависимости некоторых изотопов урана и плутония в принципе позволяют определять с высокой точностью их концентрации в растворах облученного топлива.

Такие экспериментальные данные были получены для ряда энергетических реакторов с водой под давлением и с кипящей водой [10].

С целью получения экспериментальных данных по изотопным корреляциям урана и плутония для реакторов типа ВВЭР, а также с целью экспериментальной проверки расчетных программ выгорания РОР и УНИРАСОС [11, 12] в Советском Союзе были проведены исследования изотопного состава урана и плутония в топливе этих реакторов.

В докладе приведены результаты исследований изотопного состава урана и плутония в облученном топливе реакторов типа ВВЭР-365 и ВВЭР-440 Нововоронежской АЭС. Описаны методы исследований, результаты экспериментальной проверки расчетных программ выгорания, а также экспериментальные корреляционные соотношения изотопов плутония: $N_{Pu\text{-}239} / N_{Pu}$ в зависимости от выгорания, N_{Pu} / N_U – от содержания ^{235}U, $N_{Pu\text{-}240} / N_{Pu\text{-}239}$ и $N_{Pu\text{-}239} / N_{Pu}$ – от отношения изотопов ^{235}U и ^{238}U ($N_{U\text{-}235} / N_{U\text{-}238}$), а также $\frac{(100 - N_{Pu\text{-}239})}{D_{U\text{-}235}}$ – от содержания ^{235}U (N_i – процентное содержание элемента i; $D_{U\text{-}235}$ – разность начального и конечного процентного содержания ^{235}U).

Перечисленные изотопные корреляции, по нашему мнению, наиболее характерны для топлива реакторов типа ВВЭР.

МЕТОДЫ ИССЛЕДОВАНИЙ

В целях получения более полной информации о режимах облучения, спектрах, интегральных нейтронных потоках и о состоянии топлива после облучения исследования проводились на отдельных образцах, взятых из определенных мест некоторых твэлов в кассете. Как правило, вырезались образцы из концов, середины твэла, а также из мест с максимальным выгоранием. Нам представляется, что полученные из таких образцов величины концентраций изотопов урана и плутония более точно характеризуют изотопные корреляции, чем, например, экспериментальные данные, полученные при анализе образцов, взятых из различных кассет, т.к. в последнем случае влияние условий облучения учесть более сложно.

Дело в том, что кассеты в реакторах типа ВВЭР переставляются в активной зоне от края к середине в течение трех лет облучения по мере выгорания и на количество урана и плутония в них оказывает влияние боковой и торцевой отражатели, окружающие кассеты и элементы системы регулирования [13].

Изотопный состав урана и плутония, заключенных в облучаемых твэлах, зависит от расположения кассет по отношению к краю активной зоны. Эта зависимость в большей степени сказывается для твэлов, расположенных в периферийных частях кассеты. Поэтому для данных исследований оказался необходимым описанный выше выбор образцов.

Из вырезанных образцов изымалась таблетка или часть ее, которая затем измельчалась, перемешивалась, и из нее отбиралась навеска 0,3-0,5 г, которая затем растворялась в крепкой азотной кислоте. Далее, выделение урана и плутония проводилось либо по схеме ионного разделения, либо путем соосаждения с карбонатом лантана и очистки урана и плутония экстрагенным хроматографическим методом с использованием ТБФ.

Полученные растворы урана и плутония анализировались методом изотопного разбавления с последующим масс-спектрометрическим анализом. Использовалась также методика альфа-спектрометрии [14].

Погрешности измерений изотопов ^{235}U и ^{238}U составили около 0,5%, ^{236}U – 3%, изотопов ^{239}Pu и ^{240}Pu – 0,5-1 %, ^{241}Pu и ^{242}Pu – 3-5%.

ИЗОТОПНЫЙ СОСТАВ УРАНА И ПЛУТОНИЯ В ОБЛУЧЕННОМ ТОПЛИВЕ ВВЭР

По описанной методике были выполнены исследования изотопного состава урана и плутония в образцах облученного топлива реактора ВВЭР-365 (начальное обогащение ^{235}U – 3%) и реактора ВВЭР-440 (начальное обогащение – 3,3%) II и III блоков НВ АЭС соответственно, как наиболее близко соответствующих серийному реактору ВВЭР-440 (начальное обогащение – 3,6%).

Экспериментальные значения отношений содержания изотопов урана и плутония в зависимости от величины выгорания или от содержания ^{235}U сравнивались с расчетными данными, полученными по программам РОР (для топлива ВВЭР-365) и УНИРАСОС (для топлива ВВЭР-440).

На рис. 1 приведена зависимость отношения содержания изотопов плутония $N_{Pu\text{-}239}/N_{Pu}$ от выгорания для топлива ВВЭР-365 и ВВЭР-440. Из сравнения приведенных кривых можно сделать вывод о слабом влиянии начального обогащения урана на это отношение. Экспериментальные точки имеют разброс по сравнению с соответствующими значениями на расчетных кривых в пределах ±5%. Таким образом, данное отношение может быть использовано для определения общего выгорания топлива по относительному содержанию изотопа ^{239}Pu.

На рис. 2 приведена зависимость отношения N_{Pu}/N_U от содержания ^{235}U для топлива реактора типа ВВЭР-365. Разброс экспериментальных точек по сравнению с соответствующими значениями на расчетной кривой оказался равным 2-3%, за исключением образца с содержанием ^{235}U, равным 1,72%, который тоже был вырезан из верхнего конца твэла и облучался в более мягкой области нейтронного спектра. Ранее [15] отмечалось, что расхождение между экспериментальными и расчетными величинами концентраций составляют для ^{238}U – менее 1%, для ^{235}U – 2-3%, для Pu – 5-7%, для выгорания – 2-3%.

Для приведенных на рис. 1-2 отношений содержания изотопов плутония в зависимости от выгорания или концентрации ^{235}U расхождение между расчетными и экспериментальными величинами составили 3-5%.

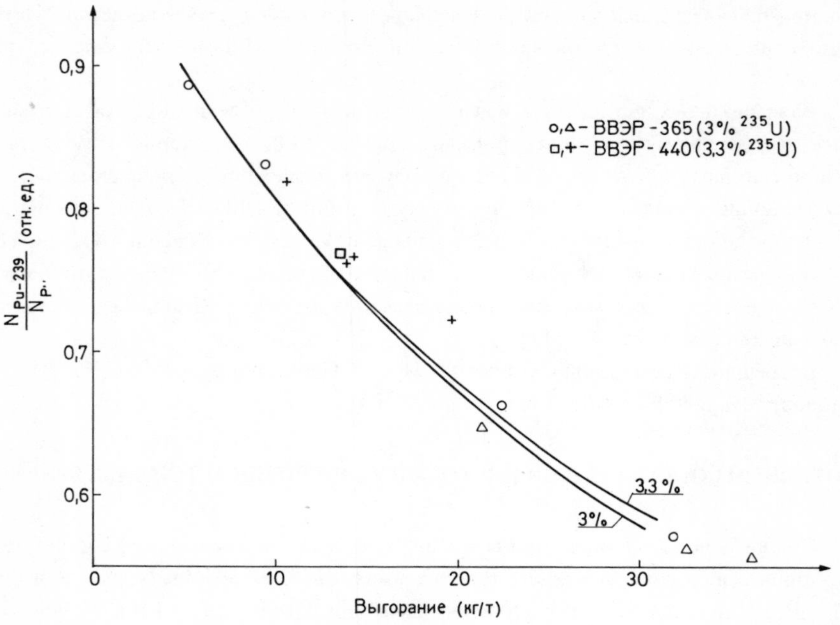

Рис. 1. Зависимость $\frac{N_{Pu-239}}{N_{Pu}}$ от выгорания для топлива ВВЭР-365 и ВВЭР-440.

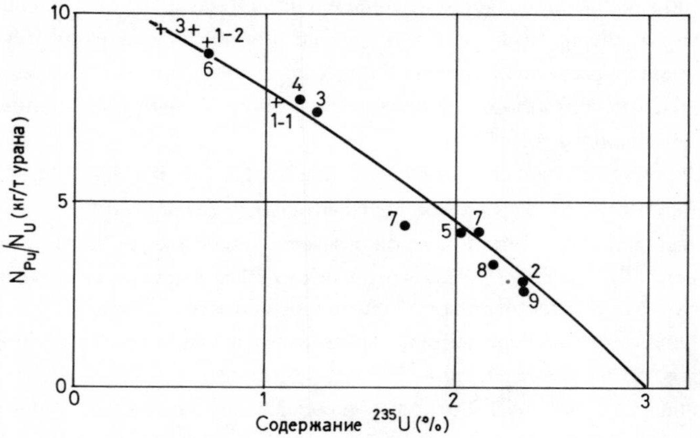

Рис. 2. Содержание плутония в зависимости от содержания ^{235}U для топлива ВВЭР-365.

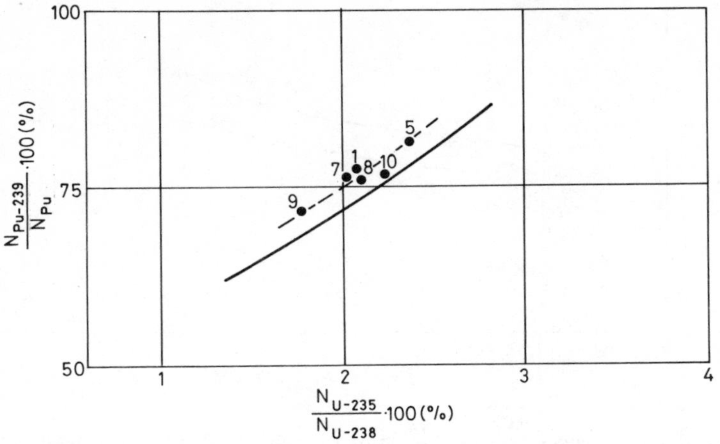

Рис. 3. Относительное содержание изотопов урана и плутония в топливе ВВЭР-440 (начальное обогащение – 3,3 %).

Из приведенных сравнений видно, что расчетные программы выгорания топлива в реакторах типа ВВЭР – РОР и УНИРАСОС – обеспечивают совпадение расчетных и экспериментальных данных по изотопным корреляциям плутония в пределах 3-5%.

ИЗОТОПНЫЕ КОРРЕЛЯЦИИ В ОБЛУЧЕННОМ ТОПЛИВЕ ВВЭР

Как отмечалось выше, наиболее полезными для целей гарантий являются корреляционные зависимости между отношениями содержания плутония и его изотопов и содержанием ^{235}U. Эти зависимости могут быть описаны в виде гладких кривых [14]. В настоящем докладе приведены зависимости отношения $\frac{100 - N_{Pu-239}}{D_{U-235}}$ от содержания $^{235}U\ (N_{U-235})$, а также отношения содержания изотопов плутония N_{Pu-239}/N_{Pu} и N_{Pu-240}/N_{Pu-239} от отношения изотопов урана N_{U-235}/N_{U-238}. Перечисленные корреляционные соотношения позволяют путем измерений изотопного состава урана и плутония проверить содержание плутония и ^{239}Pu в перерабатываемых растворах облученного топлива реакторов типа ВВЭР.

На рис. 3-5 приведены упомянутые корреляционные зависимости, полученные расчетным путем по программам РОР и УНИРАСОС, а также экспериментальные точки, полученные путем анализа изотопного состава урана и плутония в образцах облученного топлива реакторов типа ВВЭР-365 и ВВЭР-440.

На рис. 3 приведена зависимость N_{Pu-239}/N_{Pu} от отношения N_{U-235}/N_{U-238} в топливе ВВЭР-440. Расхождение расчетной кривой с экспериментальными данными лежит в пределах 5%, причем наблюдалось смещение экспериментальных точек в сторо-

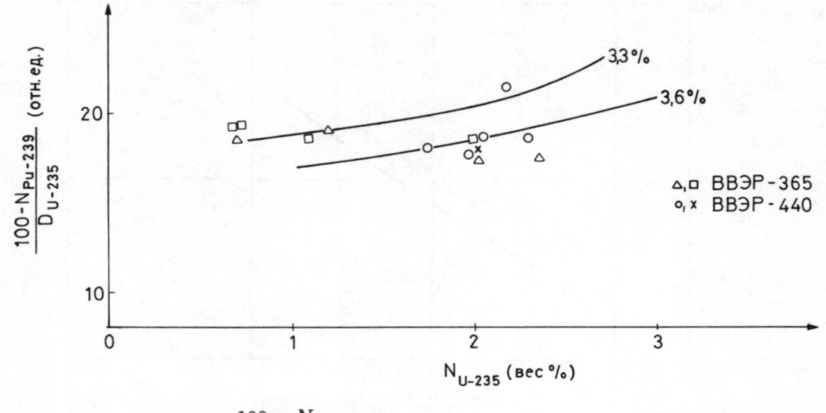

Рис. 4. Зависимость $\frac{100 - N_{Pu\text{-}239}}{D_{U\text{-}235}}$ от содержания ^{235}U для ВВЭР-365 и ВВЭР-440.

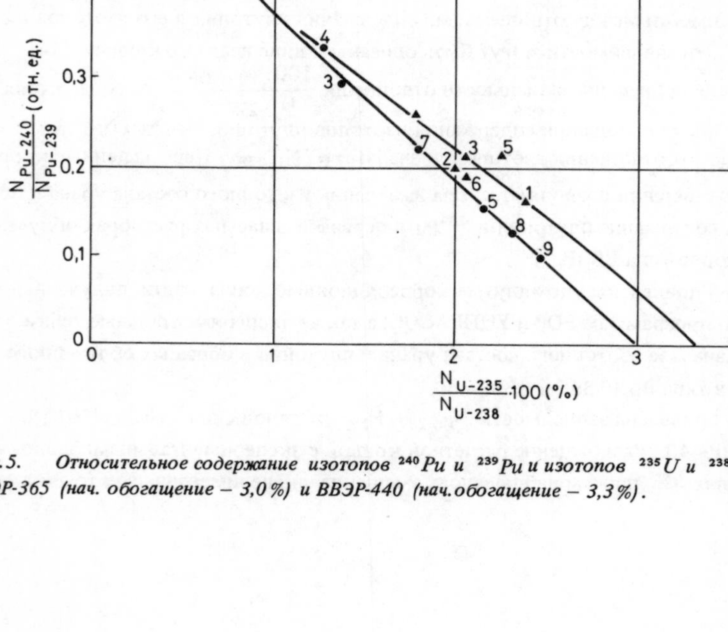

Рис. 5. Относительное содержание изотопов ^{240}Pu и ^{239}Pu и изотопов ^{235}U и ^{238}U в топливе ВВЭР-365 (нач. обогащение – 3,0%) и ВВЭР-440 (нач. обогащение – 3,3%).

ну более глубоких выгораний. Это может быть объяснено тем обстоятельством, что данная кассета облучалась на краю активной зоны вблизи бокового отражателя, где спектр нейтронов был более мягким по сравнению с расчетным.

Зависимость отношения $\frac{100 - N_{Pu\text{-}239}}{D_{U\text{-}235}}$ от содержания ^{235}U приведена на рис. 4. Расчетные кривые для начального обогащения 3,3 и 3,6% имеют наклон в сторону больших выгораний. Однако, экспериментальные точки расположились вблизи значения 18,5 ± 0,9 (т.е. в пределах ±5%) во всем диапазоне выгорания топлива. Из графика четко видна зависимость данного отношения от начального обогащения.

На рис. 5 приведены зависимости отношения $N_{Pu\text{-}240}/N_{Pu\text{-}239}$ от $N_{U\text{-}235}/N_{U\text{-}238}$ для начальных обогащений 3-3,3%. Экспериментальные данные удовлетворительно согласовались с расчетными, в пределах ±5% (для топлива ВВЭР-440). Очевидна зависимость этого отношения от начального обогащения топлива.

Из приведенных графиков видно, что расчетные и экспериментальные величины корреляционных соотношений для облученного топлива реакторов типа ВВЭР-365 и ВВЭР-440 согласуются между собой с погрешностями 3-5%.

Следует отметить при этом влияние на эти корреляционные соотношения начального обогащения топлива, а также спектра интегрального нейтронного потока.

КРАТКИЕ ВЫВОДЫ

Исследования изотопного состава урана и плутония в облученном топливе реакторов типа ВВЭР позволило определить изотопные корреляции, а также оценить погрешность (расхождение с расчетными данными), которая оказалась равной 3-5%.

Исследования проводились на образцах, вырезанных из твэлов одной кассеты, что позволило выявить влияние на эти корреляции начального обогащения и спектров интегральных нейтронных потоков. Полученные корреляционные соотношения в облученном топливе реакторов типа ВВЭР-440 могут оказаться полезными для целей гарантий МАГАТЭ.

ЛИТЕРАТУРА

[1] CHRISTENSEN, D.E., EMING, R.A., GAINES, E.P., KRAEMER, R., SCHNEIDER, R.A., STIEFF, R.L., WINTER, H.A., "A Summary of Results Obtained from the First MIST Experiment of Nuclear Fuels Services", Safeguards Techniques, v. I (Proc. Symp. Karlsruhe, 1970), IAEA, Vienna (1970) 563.

[2] STEWART, K.B., SCHNEIDER, R.A., "Properties of the Pu Estimate Based on Weighted Pu/U Values", Safeguards Techniques, v. I (Proc. Symp. Karlsruhe, 1970) IAEA, Vienna (1970) 583.

[3] WEITKAMP, C., BAECKMAN, A.V., BÖHNEL, K., KÜCHLE, M., KECH, L., "The Role of Nuclear Data in Nuclear Material Safeguards", Nuclear Data in Science and Technology, v. I (Proc. Symp. Paris, 1973) IAEA, Vienna (1973) 197.

[4] ARENZ, H.-J., VAN DER STIJL, E., "Euratom Experience of Verification Methods in Reprocessing Facilities", Safeguarding Nuclear Materials, v. II (Proc. Symp. Vienna, 1975) IAEA, Vienna (1976) 361.

[5] CHRISTENSEN, D.E., SCHNEIDER, R.A., "Summary of Experience with Heavy Element Isotopic Correlations", Safeguarding Nuclear Materials, v. II (Proc. Symp. Vienna, 1975) IAEA, Vienna (1976) 377.

[6] FOGGI, C., ZIJP, W.L., "Data Treatment for the Isotopic Correlation Technique", Safeguarding Nuclear Materials, v. II (Proc. Symp. Vienna, 1975) IAEA, Vienna (1976) 405.

[7] FOGGI,.C., FRENQUELLUCCI, F., PEROLISA, G., "Isotope Correlations Based on Fission-product Nuclides in LWR Irradiated Fuels: a Theoretical Evaluation", Safeguarding Nuclear Materials, v. II (Proc. Symp. Vienna, 1975) IAEA, Vienna (1976) 425.

[8] SANAFANI, S., SIWY, P., "IAEA Bank of Correlated Isotopic Composition Data", Safeguarding Nuclar Materials, v. II (Proc. Symp. Vienna, 1975) IAEA, Vienna (1976) 439.

[9] KOCH, L., JRC 42 (1978) 359.

[10] SUKHORUCHKIN, V.,"Verification of Plutonium Input to Reprocessing Plants", IAEA/STR-52 (1976).

[11] ОВЧИННИКОВ, Ф.Я. и др., Эксплуатация Реакторных Установок Нововоронежской АЭС, М., Атомиздат, 1972.

[12] ОВЧИННИКОВ, Ф.Я. и др., Эксплуатационные Режимы АЭС с Реакторами Типа ВВЭР, М., Атомиздат, 1977.

[13] ЗЫКОВ, К.И., МИЛЛЕР, О.А., Ат. Энерг. 39 (1975) 208.

[14] РОДИОНОВ, Ю.Ф., ШАТИНСКИЙ, В.М., ЗЕЛЕНКОВ, А.Г., Isotopenprascis 12 7/8 (1976) 271.

[15] МИЛЛЕР, О.А., и др., "Некоторые технические аспекты учета и контроля ядерных материалов на атомных установках топливного цикла", Nuclear Power and its Fuel Cycle, 7 (Proc. Symp. Vienna, 1977) IAEA, Vienna (1977) 525.

[16] НОВИКОВ, Ю.Б. и др., Ат. Энерг. 43 (1977) 240.

CHAIRMEN OF SESSIONS

Session I	H.W. SCHLEICHER	Commission of the European Communities (CEC)
Session II	A. BURTSCHER	Austria
Session III	D. GUPTA	Federal Republic of Germany
Session IV	C. CASTILLO-CRUZ	Mexico
Session V	A.G. HAMLIN	United Kingdom
Session VI	V.M. GRYAZEV	Union of Soviet Socialist Republics
Session VII	H. KRINNINGER	Federal Republic of Germany
Session VIII	G. Robert KEEPIN	United States of America
Session IX	A. PETIT	France
Session X	A. VON BAECKMANN	International Atomic Energy Agency (IAEA)

SECRETARIAT

Scientific Secretary	J.E. LOVETT	Department of Safeguards, IAEA
Administrative Secretary	Edith PILLER	Division of External Relations, IAEA
Editor	Monica KRIPPNER	Division of Publications, IAEA
Records Officer	D.J. MITCHELL	Division of Languages, IAEA

LIST OF PARTICIPANTS

AUSTRALIA

McDonald, N.R. — Australian Embassy,
Mattiellistrasse 2/4,
A-1040 Vienna, Austria

Quealy, K.J. — Australian Atomic Energy Commission,
Research Establishment,
Private Mail Bag,
Sutherland 2232, N.S.W.

AUSTRIA

Burtscher, A. — Österreichische Studiengesellschaft
für Atomenergie GmbH,
Lenaugasse 10, A-1082 Vienna

Müller, H.K. — Institut für Theoretische Physik und Reaktorphysik,
Technische Universität Graz,
Kopernikusgasse 24, A-8010 Graz

Putz, F. — Österreichische Studiengesellschaft
für Atomenergie GmbH,
Lenaugasse 10, A-1082 Vienna

Schmidt, F.W. — Federal Chancellery,
Austrian Authority for Accounting and
Control of Nuclear Material (AAAC),
Hohenstaufengasse 3, A-1010 Vienna

BELGIUM

Beets, C. — Centre d'étude de l'énergie nucléaire (SCK/CEN),
Boeretang 200, B-2400 Mol

Bemelmans, P.J. — Centre d'étude de l'énergie nucléaire (SCK/CEN),
Boeretang 200, B-2400 Mol

BELGIUM (cont.)

De Canck, H.	Belgonucléaire, Europalaan 20, B-2480 Dessel
De Regge, P.	Centre d'étude de l'énergie nucléaire (SCK/CEN), Boeretang 200, B-2400 Mol
Ingels, R.	Belgonucléaire, Europalaan 20, B-2480 Dessel
Verstappen, G.	Centre d'étude de l'énergie nucléaire (SCK/CEN), Boeretang 200, B-2400 Mol

BRAZIL

Braga Mello, E.	Commissão Nacional de Energia Nuclear, Rua General Severiano 90, Botafogo, Rio de Janeiro
Rodrigues, C.	Institute of Atomic Energy, C.P. 11049, São Paulo

BULGARIA

Dragnev, T.N.	Institute of Nuclear Research and Nuclear Energy, Bulgarian Academy of Sciences, Boulevard Lenin 72, Sofia

CANADA

Amundrud, D.L.	Atomic Energy of Canada Limited, Whiteshell Nuclear Research Establishment, Pinawa, Manitoba R0E 1L0
Dennys, R.G.	Atomic Energy of Canada Limited, Power Projects, Sheridan Park, Mississauga, Ontario L5K 1B2
Head, D.A.	Atomic Energy Control Board, Box 1046, Ottawa, Ontario K1P 5S9
Hewitt, J.S.	University of Toronto, Department of Chemical Engineering, Toronto, Ontario M5S 1A4

Stirling, A.J. — Atomic Energy of Canada Limited,
Chalk River Nuclear Laboratories,
Chalk River, Ontario K0J 1J0

DENMARK

Frederiksen, P. — Risø National Laboratory,
Safeguards Office,
DK-4000 Roskilde

EGYPT

Sultan, M.A. — Atomic Energy Establishment,
101 Kasr El-Eini Street, Cairo

FINLAND

Heinonen, O.J. — University of Helsinki,
Department of Radiochemistry,
Unioninkatu 35, SF-00170 Helsinki 17

Rautjärvi, J.S. — Institute of Radiation Protection,
P.O. Box 268, SF-00101 Helsinki 10

FRANCE

Aycoberry, C.F. — Cogéma (Compagnie générale des matières nucléaires),
La Boursidière, RN 186,
F-92357 Le Plessis-Robinson Cedex

Bouchard, J. — CEA, Centre d'études nucléaires de Cadarache,
B.P. 1, F-13115 Saint-Paul-lez-Durance

Busquet, P.R. — Commissariat à l'énergie atomique,
31-33 rue de la Fédération, F-75752 Paris Cedex 15

Courouble, J.-M. — Commissariat à l'énergie atomique,
Département des procédés industriels,
Division de chimie,
31–33 rue de la Fédération, F-75752 Paris Cedex 15

Daste, B.L. — Société franco-belge de fabrication de combustibles (FBFC),
6 avenue Bertie Albrecht, F-75008 Paris

FRANCE (cont.)

Déan, G. — CEA, Centre d'études nucléaires de Fontenay-aux-Roses, B.P. 6, F-92260 Fontenay-aux-Roses

Denis, A. — Commissariat à l'énergie atomique, Département des procédés industriels, Division de chimie, 31–33 rue de la Fédération, F-75752 Paris Cedex 15

Desneiges, P. — CEA, Centre d'études nucléaires de Fontenay-aux-Roses, B.P. 6, F-92260 Fontenay-aux-Roses

Doumerc, J. — Compagnie pour l'étude et la réalisation de combustibles atomiques (CERCA), 41, avenue Montaigne, F-75008 Paris

Dugas, R. — Secrétariat général du comité interministériel de la sécurité nucléaire, 27 rue Oudinot, F-75007 Paris

Dumesnil, P. — CEA, Centre d'études nucléaires de Saclay, SES-SAI, B.P. 2, F-91190 Gif-sur-Yvette

Geraud de Galassus, B. — Commissariat à l'énergie atomique, 31–33 rue de la Fédération, F-75752 Paris Cedex 15

Goens, J.R. — Eurodif, 116 avenue Aristide Briand, F-92220 Bagneux

Guay, P. — Commissariat à l'énergie atomique, 31–33 rue de la Fédération, F-75752 Paris Cedex 15

Hébert, D. — CEA, Centre d'études nucléaires de Cadarache, B.P. 1, F-13115 Saint-Paul-lez-Durance

Herpin, A. — Cogéma (Compagnie générale des matières nucléaires), La Boursidière, RN 186, Bâtiments G.H.I., F-92357 Le Plessis-Robinson Cedex

Huré, J. — CEA, Centre d'études nucléaires de Fontenay-aux-Roses, B.P. 6, F-92260 Fontenay-aux-Roses

Jeanpierre, G.	CEA, Centre d'études nucléaires de Fontenay-aux-Roses, B.P. 6, F-92260 Fontenay-aux-Roses
Lachoviez, S.	CEA, Centre d'études nucléaires de Fontenay-aux-Roses, B.P. 6, F-92260 Fontenay-aux-Roses
Lamorlette, G.	Cogéma (Compagnie générale des matières nucléaires), La Boursidière, RN 186, Bâtiment J, F-92357 Le Plessis-Robinson Cedex
Lecomte, F.	CEA, Centre d'études nucléaires de Fontenay-aux-Roses, B.P. 6, F-92260 Fontenay-aux-Roses
Malet, G.	CEA, Centre d'études nucléaires de Saclay, B.P. 2, F-91190 Gif-sur-Yvette
Masson, H.	Société générale pour les techniques nouvelles, 23 boulevard Georges Clémenceau, F-92400 Courbevoie
Masson, P.	Délégation générale à l'armement, Mission Atome, 10 rue Saint-Dominique, F-75007 Paris
Petit, A.	Commissariat à l'énergie atomique, 31 – 33 rue de la Fédération, F-75752 Paris Cedex 15
Régnier, J.	Cogéma (Compagnie générale des matières nucléaires), La Boursidière, RN 186, Bâtiment J, F-92357 Le Plessis-Robinson Cedex
Robin, M.	CEA, Centre d'études nucléaires de Cadarache, DRE/SEN, B.P. 1, F-13115 Saint-Paul-lez-Durance
Sauzay, G.	CEA, Centre d'études nucléaires de Saclay, DGI/SGC, B.P. 2, F-91190 Gif-sur-Yvette
Teboul, A.	Société de fabrication d'éléments catalytiques (SFEC), B.P. 33, F-84500 Bollène
Zečević, V.	Framatome, Tour Fiat, Cedex 16, F-92084 Paris La Défense

GERMAN DEMOCRATIC REPUBLIC

Graber, H.	Zentralinstitut für Kernforschung Rossendorf, Postfach 19-, DDR-8051 Dresden
Siebert, H.U.	Staatliches Amt für Atomsicherheit und Strahlenschutz der DDR, Waldowallee 117, DDR-1157 Berlin-Karlshorst

GERMANY, FEDERAL REPUBLIC OF

Avenhaus, R.	Kernforschungszentrum Karlsruhe GmbH, Institut für Datenverarbeitung in der Technik, B.P. 3640, D-7500 Karlsruhe
Berg, R.	Gesellschaft zur Wiederaufarbeitung von Kernbrennstoffen mbH, D-7514 Eggenstein-Leopoldshafen 2
Beyrich, W.	Kernforschungszentrum Karlsruhe GmbH, Entwicklungsgruppe Kernmaterialsicherung, Postfach 3640, D-7500 Karlsruhe
Bödege, R.	Deutsche Gesellschaft für Wiederaufarbeitung von Kernbrennstoffen, Bünteweg 2, D-3000 Hanover
Brückner, Chr.	Kernforschungszentrum Karlsruhe GmbH, Postfach 3640, D-7500 Karlsruhe
Büker, H.	Kernforschungsanlage Jülich GmbH, Postfach 1913, D-5170 Jülich
Duerr, R.	Bundesvermögensverwaltung, Josef-Danzer-Strasse 33, D-8033 Planegg
Filss, P.	Institut für Chemische Technologie der Kernforschungsanlage Jülich GmbH, Postfach 1913, D-5170 Jülich 1
Fischer, P.	Kernforschungszentrum Karlsruhe GmbH, Weberstrasse 5, D-7500 Karlsruhe 1
Fuchs, R.	Badenwerk AG, Postfach 1680, D-7500 Karlsruhe 1
Gupta, D.	Kernforschungszentrum Karlsruhe GmbH, Postfach 3640, D-7500 Karlsruhe

Hagenberg, W.	ALKEM GmbH, Industriegelände, D-6450 Hanau 11
Heger, H.	Vereinigung Deutscher Elektrizitätswerke, c/o RWE-KW, Am Guten Mann, D-5403 Mülheim-Kärlich
Hein, H.	Gesellschaft zur Wiederaufarbeitung von Kernbrennstoffen mbH, D-7514 Eggenstein-Leopoldshafen 2
Heinzelmann, M.	Kernforschungsanlage Jülich GmbH, Postfach 1913, D-5170 Jülich
Hoffmann, D.	Gesellschaft für Reaktorsicherheit mbH, Glockengasse 2, D-5000 Cologne 1
Jarsch, V.	Kernforschungszentrum Karlsruhe GmbH, Postfach 3640, D-7500 Karlsruhe
Krinninger, H.	Internationale Natrium-Brutreaktor-Bau GmbH (INB), Friedrich-Ebert-Strasse, D-5060 Bergisch-Gladbach 1 (Bernsberg)
Lauer, P.	Babcock-Brown Boveri Reaktor GmbH (BBR), Heppenheimer Strasse 27 – 29, D-6800 Mannheim 31
Matussek, P.	Kernforschungszentrum Karlsruhe GmbH, Institut für Angewandte Kernphysik, Postfach 3640, D-7500 Karlsruhe
Münch, E.	Kernforschungsanlage Jülich, Postfach 1913, D-5170 Jülich
Onnen, S.	Kernforschungszentrum Karlsruhe GmbH, Postfach 3640, D-7500 Karlsruhe
Ottmar, H.	Kernforschungszentrum Karlsruhe GmbH, Institut für Angewandte Kernphysik, Postfach 3640, D-7500 Karlsruhe
Rezniczek, A.	Unternehmens-Beratung Aachen (UBA), Roermonder Strasse 600, D-5100 Aachen
Ruppert, E.	Interatom, Friedrich-Ebert-Strasse, D-5060 Bergisch-Gladbach 1 (Bernberg)

GERMANY, FEDERAL REPUBLIC OF (cont.)

Scheller, S. — Permanent Mission of the Federal Republic of Germany to the IAEA,
Metternichgasse 3, A-1030 Vienna

Schinzer, F. — NUKEM GmbH,
Postfach 11 00 80, D-6450 Hanau 11

Schnell, C. — Dornier System GmbH,
Postfach 1360, D-7990 Friedrichshafen

Stein, G. — Kernforschungsanlage Jülich GmbH,
Postfach 1913, D-5170 Jülich

Vallée, J. — Unternehmensberatung Aachen (UBA),
Roermonder Strasse 600, D-5100 Aachen

von Eyss, H. — Gesellschaft für Reaktorsicherheit mbH,
Glockengasse 2, D-5000 Cologne 1

von Osten, W. — Bundesministerium für Forschung und Technologie,
Heinemannstrasse 2, D-5300 Bonn 2

Voss, F. — Kernforschungszentrum Karlsruhe GmbH,
Postfach 3640, D-7500 Karlsruhe

Weppner, J. — Kernforschungszentrum Karlsruhe GmbH,
Postfach 3640, D-7500 Karlsruhe

GREECE

Mitsonias, K. — Greek Atomic Energy Commission,
Nuclear Research Center "Demokritos",
Aghia Paraskevi, Athens

HUNGARY

Biro, T. — Institute of Isotopes,
P.O. Box 77, H-1525 Budapest

Szabó, E. — Central Institute of Physics,
P.O. Box 49, H-1525 Budapest

INDIA

Iyer, M.R.	Bhabha Atomic Research Centre, Health Physics Division, Trombay, Bombay 400 085
Kumar, S.V.	Bhabha Atomic Research Centre, Fuel Reprocessing Division, Trombay, Bombay 400 085

IRAN

Ghods, A.	Atomic Energy Organization of Iran, P.O. Box 12–1198, Teheran
Somehea, M.H.	Atomic Energy Organization of Iran, Safeguards and Physical Protection Division, P.O. Box 12–1198, Teheran

ITALY

Bardone, G.	Comitato Nazionale per l'Energia Nucleare (CNEN), Dip. Ciclo del Combustibile, Laboratorio Ricerche Retrattamento, CSN Casaccia, S.P. Anguillarese Km1 + 300, I-00100 Rome
Bazzan, A.	Comitato Nazionale per l'Energia Nucleare (CNEN), Viale Regina Margherita 125, I-00198 Rome
Cocchi, A.	Ente Nazionale per l'Energia Elettrica (ENEL), Via G.B. Martini 3, I-00198 Rome
Frazzoli, F.	Istituto di Fisica, Facoltà di Ingegneria, Università degli Studi, Piazzale delle Scienze 5, I-00100 Rome
Leonardi, Valeria	S.F. Combustibili per Reattori Veloci, Corso Porta Romana 68, I-20122 Milan
Pozzi, F.	Comitato Nazionale per l'Energia Nucleare (CNEN), I-13040 Saluggia (Vercelli)
Sansone, C.	AGIP Nucleare, Corso di Porta Romana 68, CP 1629, I-20122 Milan

ITALY (cont.)

Sarno, G.	Centro Applicazioni Mitare Energia Nucleare, San Piero Agrado, Pisa
Tabet, E.	Istituto Superiore di Sanità, Viale Regina Elena 299, Rome
Vanni, P.	Comitato Nazionale per l'Energia Nucleare (CNEN), Viale Regina Margherita 125, I-00198 Rome
Venchiarutti, R.	Comitato Nazionale per l'Energia Nucleare (CNEN), Direzione Centrale, Relazioni Esterne, Viale Regina Nargherita 125, I-00198 Rome

JAPAN

Ikawa, K.	Japan Atomic Energy Research Institute, Tokai Research Establishment, Tokai-mura, Naka-gun, Ibaraki-ken 319–11
Koizumi, T.	Power Reactor & Nuclear Fuel Development Corp., Muramatsu, Tokai-mura, Ibaraki-ken 319–11
Matsuda, Y.	Nippon Atomic Industry Group Co., Ltd, Nuclear Research Laboratory, 4–1 Ukishima-cho, Kawasaki-ku, Kawasaki-shi
Minato, T.	Power Reactor and Nuclear Fuel Development Corp., 9–13, 1-chome, Akasaka, Minato-ku, Tokyo 107
Tsujimura, S.	Japan Atomic Energy Research Institute, Tokai-mura, Naka-gun, Ibaraki-ken 311–19

MALAYSIA

Lau, H.M.	Tun Ismail Atomic Research Centre, 14th Floor, Bangunan Oriental Plaza, Jalan Parry, Kuala Lumpur 04–01
Salikin, S.	Tun Ismail Atomic Research Centre, 14th Floor, Bangunan Oriental Plaza, Jalan Parry, Kuala Lumpur 04–01

MEXICO

Castillo-Cruz, C.	Instituto Nacional de Energía Nuclear, Subgerencia de Seguridad de Plantas Nucleares, Av. Insurgentes Sur 1079, Apdo Postal 27–190, Mexico 18, D.F.
Ramirez de Alba, F.	Instituto Nacional de Energía Nuclear, Subgerencia de Seguridad de Plantas Nucleares, Av. Insurgentes Sur 1079, Apdo Postal 27–190, Mexico 18, D.F.

NETHERLANDS

Baas, J.	Ministry of Health and Environmental Protection, Dokter Reijersstraat 10, Leidschendam/The Hague
Harry, R.J.S.	Netherlands Energy Research Foundation ECN, NL-1755 ZG Petten (N.H.)
Nienhuys, K.	Groningen State University, Chemistry Laboratories, Nijenborgh 16, NL-9747 AG Groningen
Van der Hulst, P.	N.V. Kema, Utrechtseweg 310, NL-6812 AR Arnhem
Van der Meijden, D.	Groningen State University, Chemistry Laboratories, Nijenborgh 16, NL-9747 AG Groningen
Van Raaphorst, J.G.	Netherlands Energy Research Foundation ECN, Westerduinweg 3, NL-1755 ZG Petten (N.H.)
Zijp, W.L.	Netherlands Energy Research Foundation ECN, Westerduinweg 3, NL-1755 ZG Petten (N.H.)

NORWAY

Feyling, R.	Institutt for Atomenergi, P.O. Box 40, N-2007 Kjeller

PAKISTAN

Ullah, H.	Pakistan Atomic Energy Commission, Directorate of Nuclear Fuels and Materials, Jan Chambers College Road, P.O. Box 1114, Islamabad

SPAIN

López-Menchero y Ordonez, E.M.	Agregaduria de Industria y Energía, Embasada de España en Austria y Delegaciones permanentes de España ante el O.I.E.A., Prinz Eugen Strasse 18/2/1/20, A-1040 Vienna, Austria
Sevilla, B.A.	Junta de Energía Nuclear, Ciudad Universitaria, Madrid-3

SWEDEN

Ek, P.I.	Swedish Nuclear Power Inspectorate, P.O. Box 27106, S-10252 Stockholm
Ekecrantz, L.	Swedish Nuclear Power Inspectorate, Sehlstedsgatan 11, S-10252 Stockholm
Sjöborg, L.	National Defense Research Institute, Fack, S-104 50 Stockholm
Waak, I.	South Swedish Power Company Ltd, Fack, Carl Gustafs Väg 1, S-200 70 Malmö 5

TURKEY

Öner, D.	Cekmece Nuclear Research and Training Center, P.K. 1, Hava-Alam-Yesilköy, Istanbul

UNION OF SOVIET SOCIALIST REPUBLICS

Babaev, N.S.	I.V. Kurchatov Institute of Atomic Energy, Moscow
Gryazev, V.M.	Institute of Nuclear Reactors, Dimitrovgrad

UNITED KINGDOM

Brown, F.	Safeguards Office, Atomic Energy Division, Department of Energy, Thames House South, Millbank, London SW1P 4QS
Clegg, J.C.	British Nuclear Fuels Ltd, Risley, Warrington WA3 6AT
Davies, W.	UKAEA, Dounreay Nuclear Power Development Establishment, Thurso, Caithness KW14 7TZ, Scotland
Ellingsen, J.E.	Safeguards Office, Atomic Energy Division, Department of Energy, Thames House South, Millbank, London SW1P 4QJ
England, C.J.	UKAEA, Atomic Energy Research Establishment, Harwell, Didcot, Oxon OX11 ORA
Griffith, J.	Springfields Nuclear Power Development Laboratories, Salwick, Preston PR4 OXJ
Hamlin, A.G.	Nuclear Material Accounting and Control Team (NMACT), UKAEA, Harwell, Didcot, Oxon OX11 ORA
Leake, J.W.	UKAEA, Atomic Energy Research Establishment, Harwell, Didcot, Oxon OX11 ORA
Lees, E.W.	UKAEA, Atomic Energy Research Establishment, Nuclear Physics Division, Harwell, Didcot, Oxon OX11 ORA
Mummery, G.B.	Central Electricity Generating Board, Walden House, 24 Cathedral Place, London EC4P 4EB
Phillips, G.	UKAEA, Atomic Energy Research Establishment, Chemistry Division, Harwell, Didcot, Oxon OX11 ORA
Rawson, D.F.	UKAEA, Risley Nuclear Power Development Establishment, Risley, Warrington, Cheshire WA3 6AT
Sturman, H.G.	British Nuclear Fuels Ltd, Head Office, Risley, Warrington, Cheshire WA3 6AT

UNITED STATES OF AMERICA

Allen, W.B.	Pacific Gas & Electric Co., 77 Beale Street, San Francisco, California
Augustson, R.H.	Los Alamos Scientific Laboratory, Group Q-3, MS 539, P.O. Box 1663, Los Alamos, NM 87545
Baron, N.	Los Alamos Scientific Laboratory, Group Q-3, MS 539, P.O. Box 1663, Los Alamos, NM 87545
Bartels, W.C.	US Department of Energy, Room A2-1016, Washington, DC 20545
Bieber, A.M., Jr.	Brookhaven National Laboratory, Building 197 C, Upton, NY 11973
Burnett, R.F.	US Nuclear Regulatory Commission, Office of Standards Development, Washington, DC 20555
Cohen, K.D.	US Nuclear Regulatory Commission, Washington, DC 20555
DeMerschman, A.W.	Westinghouse Hanford Company, P.O. Box 1970, Richland, WA 99352
De Vito, V.J.	Institute of Nuclear Materials Management, P.O. Box 628, Piketon, OH 45661
Dietz, R.J.	Los Alamos Scientific Laboratory, P.O. Box 1663, MS 541, Los Alamos, NM 87545
Doher, L.W.	Rockwell International, Rocky Flats Plant, P.O. Box 464, Golden, CO 80401
Dowdy, E.J.	Los Alamos Scientific Laboratory, P.O. Box 1663, Los Alamos, NM 87545
Glancy, J.E.	Science Applications, Inc., 1200 Prospect Avenue, P.O. Box 2351, La Jolla, CA 92038
Houck, F.S.	US Arms Control and Disarmament Agency, Washington, DC 20451

Howard, E.M.	US Nuclear Regulatory Commission, Office of Standards Development, Washington, DC 20555
Johnson, M.E.	Los Alamos Scientific Laboratory, P.O. Box 1663, S-1, MS 606, Los Alamos, NM 87545
Jones, F.E.	National Bureau of Standards, Washington, DC 20234
Keepin, G. Robert	Los Alamos Scientific Laboratory, P.O. Box 1663, Ms 541, Los Alamos, NM 87545
Krick, M.S.	Los Alamos Scientific Laboratory, P.O. Box 1663, Ms 540, Los Alamos, NM 87545
Levin, S.A.	Union Carbide Corporation, Oak Ridge Gaseous Diffusion Plant, P.O. Box P, Oak Ridge, TN 37830
Mahy, J.F.	Permanent Mission of the United States of America to the IAEA, Schmidgasse 14, A-1082 Vienna, Austria
McGee, W.F.	US General Accounting Office, 441G Street, NW, Washington, DC 20548
Miller, H.	National Nuclear Corporation, 3150 Spring Street, Redwood City, CA 94063
Murphey, W.M.	US Arms Control and Disarmament Agency, Washington, DC 20451
Murrell, J.S.	Goodyear Atomic Corporation, P.O. Box 628, Piketon, OH 45661
Nilson, R.	Exxon Nuclear, 2955 Gea Wash Way, Richland, WA 99352
Olson, R.G.	University of California, Lawrence Livermore Laboratory, P.O. Box 808, Livermore, CA 94550
Pasternak, T.	Science Applications Inc., P.O. Box 2351, 1200 Prospect Avenue, La Jolla, CA 92038

UNITED STATES OF AMERICA (cont.)

Perry, R.B.	Argonne National Laboratory, Nondestructive Assay Section, Building 16, 9700 South Cass Avenue, Argonne, IL 60439
Pike, D.H.	Oak Ridge National Laboratory, P.O. Box X, Oak Ridge, TN 37830
Prokoski, F.J.	US ACDA, Washington, DC 20451
Reed, W.P.	National Bureau of Standards, Office of Standard Reference Materials, Washington, DC 20234
Scarborough, J.M.	US Department of Energy, New Brunswick Laboratory, 9800 South Cass Avenue, Chicago, IL 60439
Schrack, R.A.	National Bureau of Standards, Washington, DC 20234
Seabaugh, P.W.	Monsanto Research Corporation,Mound Laboratory, P.O. Box 32, Miamisburg, OH 45342
Sellers, T.A.	Sandia Laboratories, P.O. Box 5800, Albuquerque, NM 87185
Sonnier, C.S.	Permanent Mission of the United States of America to the IAEA, Schmidgasse 14, A-1080 Vienna, Austria
Sorenson, R.J.	Battelle Pacific Northwest Laboratories, P.O. Box 999, Richland, WA 99342
Strohm , W.W.	Monsanto Research Corporation, Mound Laboratory, P.O. Box 32, Miamisburg, OH 45342
Ting, P.	US Nuclear Regulatory Commission, Washington, DC 20555
Todd. J.L.	Sandia Laboratories, Albuquerque, NM 87115
Turel, S.P.	US Nuclear Regulatory Commission, Office of Standards Development, Washington, DC 20555
Weber, H.J.	IRT Corporation, P.O. Box 80817, San Diego, CA 92138

Woods, E.G.	US General Accounting Office, 441G Street, NW, Room 4821, Washington, DC 20548
Yolken, H.T.	US National Bureau of Standards, Building 221-Room B318, Washington, DC 20234
Zucker, M.S.	Brookhaven National Laboratory, Upton, Long Island, NY 11973

VENEZUELA

Fernandez Zarraga, A.A.	Consejo Nacional para el Desarrollo de la Industria Nuclear (CONAN), Apartado 68233, Caracas 106

YUGOSLAVIA

Bulović, V.	Boris Kidrič Institute of Nuclear Sciences, Vinča, P.O. Box 522, YU-11001 Belgrade
Jurčević, M.	Nuclear Power Plant Krško, ul. 4, Julija 38, YU-68270 Krško
Martinc, R.	Boris Kidrič Institute of Nuclear Sciences, Vinča, P.O. Box 522, YU-11001 Belgrade

ORGANIZATIONS

ATOMIC INDUSTRIAL FORUM INC. (AIF)

Graham, F.	7101 Wisconsin Avenue, Washington, DC 20014 (also: AERE Harwell, Didcot, Oxon OX11 ORA, United Kingdom)

COMMISSION OF THE EUROPEAN COMMUNITIES (CEC)

Busca, G.	Jean Monnet Building, B4-003, B.P. 1907, Kirchberg, Luxembourg

COMMISSION OF THE EUROPEAN COMMUNITIES (cont.)

Crutzen, S.J.	Euratom, CCR, Material Science Division, I-21020 Ispra (VA), Italy
Cuypers, M.	Joint Research Centre, Ispra Establishment, I-21020 Ispra (VA), Italy
De Bièvre, P.	Euratom, Central Bureau for Nuclear Measurements, B-2440 Geel, Belgium
Foggi, C.	Euratom, CCR, I-21020 Ispra (VA), Italy
Haas, R.	Jean Monnet Building, B4-003, B.P. 1907, Kirchberg, Luxembourg
Kloeckner, W.F.	Euratom Safeguards – DG XVII-E, Jean Monnet Building, B4-003, B.P. 1907, Kirchberg, Luxembourg
Koch, L.	European Institute for Transuranium Elements, Postfach 2266, D-7500 Karlsruhe, Federal Republic of Germany
Kschwendt, H,	Euratom Safeguards Directorate, Jean Monnet Building, B4-003, B.P. 1907, Kirchberg, Luxembourg
Ley, J.	Euratom, CCR, I-21020 Ispra (VA), Italy
Meister, H.	Euratom, CCR, I-21020 Ispra (VA), Italy
Meyer, H.	Euratom, Central Bureau for Nuclear Measurements, B-2440 Geel, Belgium
Miranda, U.	Jean Monnet Building, B4-003, B.P. 1907, Kirchberg, Luxembourg
Müller, K.H.	Euratom, CCR, I-21020 Ispra (VA), Italy

Neu, H.	Euratom, CCR, I-21020 Ispra (VA), Italy
Prosdocimi, A.	Joint Research Centre, Ispra Establishment, I-21020 Ispra (VA), Italy
Schleicher, H.W.	Jean Monnet Building, B4-003, B.P. 1907, Kirchberg, Luxembourg
Sharpe, B.W.	Jean Monnet Building, B4-003, B.P. 1907, Kirchberg, Luxembourg
Stanchi, L.	Joint Research Centre, Ispra Establishment, I-21020 Ispra (VA), Italy
Stanners, W.	Jean Monnet Building, B4-003, B.P. 1907, Kirchberg, Luxembourg
Van der Stijl, E.	Directorate of Safeguards, Jean Monnet Building, Kirchberg, Luxembourg
Van der Stricht, E.	Jean Monnet Building, B4-003, B.P. 1907, Kirchberg, Luxembourg
Vinche, C.J.	Euratom, CCR, I-21020 Ispra (VA), Italy

INTERNATIONAL ATOMIC ENERGY AGENCY (IAEA)

All the participants below are from the Agency's Department of Safeguards

Alston, W.
Bahm, W.
Beetle, T.M.
Beyer, N.
Brandão, F.
De Carolis, M.
Dermendjiev, E.
Dornow, V.A.
Ermakov, S.
Ferraris, M.
Ferris, Y.
Frenzel, W.
Ghods, A.
Gmelin, W.
Haginoya, T.
Heaysman, B.K.

INTERNATIONAL ATOMIC ENERGY AGENCY (cont.)

Hough, G.
Jung, D.
Keddar, A.
Kiss, I.
Klik, F.
Konnov, Y.
Kotte, E.
Kuhn, E.
McGovern, D.E.
Mecheski, D.E.
Nägele, G.G.
Nardi, J.
Nelson, G.
Oudejans, L.
Parsick, R..
Perricos, D..
Petrunin, D.
Poroykov, V.
Rehak, W.W.
Remagen, H.H.
Rundquist, D.E.
Sanatani, S.
Sanders, K.E.
Sapir, J.
Schnaible, M.
Shea, T.
Sher, R.
Siller, H.
Smith, G.D.
Sukhoruchkin, V.
Terrey, D.R.
Thompson, S.
Thorstensen, S.
Tsutsumi, M.
von Baeckmann, A.
Werner, H.V.
Woelfl, E.

INTERNATIONAL ORGANIZATION FOR STANDARDIZATION (ISO)

Braatz, U. — Vereinigung Deutscher Elektrizitätswerke (VDEW)
Avenue de Tervuren 267, B-1150 Brussels, Belgium

LEAGUE OF ARAB STATES

Khidair, S.S. — Délégation permanente auprès de l'ONU,
9 rue du Valais, CH-1202 Geneva, Switzerland

AUTHOR INDEX

Roman numerals are volume numbers.
Italic Arabic numerals refer to the first page of a paper.
Upright Arabic numerals denote questions and comments in discussions.

TRANSLITERATION INDEX

Алексеев, И.Н.	Alekseev, I.N.
Бабаев, Н.С.	Babaev, N.S.
Бибичев, Б.А.	Bibichev, B.A.
Виллун, К.	Villun, K.
Воронков, А.А.	Voronkov, A.A.
Гаджиев, Г.И.	Gadzhiev, G.I.
Галкин, Б.Я.	Galkin, B.Ya.
Голубев, Л.И.	Golubev, L.I.
Горобец, А.К.	Gorobets, A.K.
Грунер, В.	Gruner, V.
Грязев, В.М.	Gryazev, V.M.
Демидов, Л.И.	Demidov, L.I.
Зайцев, А.А.	Zaitsev, A.A.
Зиберт, Х.-У.	Siebert, H.-U.
Илясов, В.М.	Ilyasov, V.M.
Казаринов, Н.М.	Kazarinov, N.M.
Круглов, В.П.	Kruglov, V.P.
Лещенко, Ю.И.*	Leshchenko, Yu.I.
Майоров, В.П.	Majorov, V.P.
Миллер, О.А.	Miller, O.A.
Нигматуллин, Н.Р.	Nigmatullin, N.R.
Новиков, Ю.Б.	Novikov, Yu.B.
Пирожков, С.В.	Pirozhkov, S.V.
Протасенко, Ю.М.	Protasenko, Yu.M.
Робулец, Г.Н.	Robulets, G.N.
Родионов, Ю.Ф.	Rodionov, Yu.F.
Седов, В.К.	Sedov, V.K.
Семенов, А.Л.	Semenov, A.L.
Сунчугашев, М.А.	Sunchugashev, M.A.
Тарасевич, В.П.	Tarasevich, V.P.
Федотов, П.И.	Fedotov, P.I.
Фортаков, В.Б.	Fortakov, V.B.
Хиленко, С.Н.	Khilenko, S.N.
Хоффман, Д.	Hoffmann, D.
Шатинский, В.М.	Shatinskij, V.M.
Яковлев, Г.Н.	Yakovlev, G.N.
Яковлева, И.В.	Yakovleva, I.V.

* На стр. 521 и 531 тома I эта фамилия ошибочно напечатана как Ю.И. Лешенко.

* On pp. 521 and 531, Vol.I, this name appears wrongly as Ю.И. Лешенко. The transliteration is correct throughout.

INDEX OF PAPER SYMBOLS

The following conversion table is provided for the convenience of readers and to encourage the use of SI units.

FACTORS FOR CONVERTING SOME OT THE MORE COMMON UNITS TO INTERNATIONAL SYSTEM OF UNITS (SI) EQUIVALENTS

NOTES:

(1) SI base units are the metre (m), kilogram (kg), second (s), ampere (A), kelvin (K), candela (cd) and mole (mol).

(2) ▶ indicates SI derived units and those accepted for use with SI;
▷ indicates additional units accepted for use with SI for a limited time.
[*For further information see The International System of Units (SI), 1977 ed., published in English by HMSO, London, and National Bureau of Standards, Washington, DC, and International Standards ISO-1000 and the several parts of ISO-31 published by ISO, Geneva.*]

(3) The correct abbreviation for the unit in column 1 is given in column 2.

(4) ✱ indicates conversion factors given exactly; other factors are given rounded, mostly to 4 significant figures.
≡ indicates a definition of an SI derived unit: [ ] in column 3+4 enclose factors given for the sake of completeness.

Column 1 *Multiply data given in:*	Column 2	Column 3 *by:*	Column 4 *to obtain data in:*	
Radiation units				
▶ becquerel	1 Bq	(has dimensions of s^{-1})		
disintegrations per second (= dis/s)	1 s^{-1}	≡ 1.00×10^{0}	Bq	✱
▷ curie	1 Ci	= 3.70×10^{10}	Bq	✱
▷ roentgen	1 R	[= 2.58×10^{-4}	C/kg]	✱
▶ gray	1 Gy	[≡ 1.00×10^{0}	J/kg]	✱
▷ rad	1 rad	= 1.00×10^{-2}	Gy	✱
sievert *(radiation protection only)*	1 Sv	[= 1.00×10^{0}	J/kg]	✱
rem *(radiation protection only)*	1 rem	[= 1.00×10^{-2}	J/kg]	✱
Mass				
▶ unified atomic mass unit ($\frac{1}{12}$ of the mass of ^{12}C)	1 u	[= 1.66057×10^{-27}	kg, approx.]	
▶ tonne (= metric ton)	1 t	[= 1.00×10^{3}	kg]	✱
pound mass (avoirdupois)	1 lbm	= 4.536×10^{-1}	kg	
ounce mass (avoirdupois)	1 ozm	= 2.835×10^{1}	g	
ton (long) (= 2240 lbm)	1 ton	= 1.016×10^{3}	kg	
ton (short) (= 2000 lbm)	1 short ton	= 9.072×10^{2}	kg	
Length				
statute mile	1 mile	= 1.609×10^{0}	km	
nautical mile (international)	1 n mile	= 1.852×10^{0}	km	✱
yard	1 yd	= 9.144×10^{-1}	m	✱
foot	1 ft	= 3.048×10^{-1}	m	✱
inch	1 in	= 2.54×10^{1}	mm	✱
mil (= 10^{-3} in)	1 mil	= 2.54×10^{-2}	mm	✱
Area				
▷ hectare	1 ha	[= 1.00×10^{4}	m^2]	✱
▷ barn *(effective cross-section, nuclear physics)*	1 b	[= 1.00×10^{-28}	m^2]	✱
square mile, (statute mile)^2	1 $mile^2$	= 2.590×10^{0}	km^2	
acre	1 acre	= 4.047×10^{3}	m^2	
square yard	1 yd^2	= 8.361×10^{-1}	m^2	
square foot	1 ft^2	= 9.290×10^{-2}	m^2	
square inch	1 in^2	= 6.452×10^{2}	mm^2	
Volume				
▶ litre	1 l *or* 1 ltr	[= 1.00×10^{-3}	m^3]	✱
cubic yard	1 yd^3	= 7.646×10^{-1}	m^3	
cubic foot	1 ft^3	= 2.832×10^{-2}	m^3	
cubic inch	1 in^3	= 1.639×10^{4}	mm^3	
gallon (imperial)	1 gal (UK)	= 4.546×10^{-3}	m^3	
gallon (US liquid)	1 gal (US)	= 3.785×10^{-3}	m^3	
Velocity, acceleration				
foot per second (= fps)	1 ft/s	= 3.048×10^{-1}	m/s	✱
foot per minute	1 ft/min	= 5.08×10^{-3}	m/s	✱
mile per hour (= mph)	1 mile/h	= { 4.470×10^{-1}	m/s	
		1.609×10^{0}	km/h	
▷ knot (international)	1 knot	= 1.852×10^{0}	km/h	✱
free fall, standard, g		= 9.807×10^{0}	m/s^2	
foot per second squared	1 ft/s^2	= 3.048×10^{-1}	m/s^2	✱

This table has been prepared by E.R.A. Beck for use by the Division of Publications of the IAEA. While every effort has been made to ensure accuracy, the Agency cannot be held responsible for errors arising from the use of this table.

Column 1 *Multiply data given in:*	Column 2	Column 3 *by:*	Column 4 *to obtain data in:*	
Density, volumetric rate				
pound mass per cubic inch	1 lbm/in^3	= 2.768×10^4	kg/m^3	
pound mass per cubic foot	1 lbm/ft^3	= 1.602×10^1	kg/m^3	
cubic feet per second	1 ft^3/s	= 2.832×10^{-2}	m^3/s	
cubic feet per minute	1 ft^3/min	= 4.719×10^{-4}	m^3/s	
Force				
▶ newton	1 N	[≡ 1.00×10^0	$m \cdot kg \cdot s^{-2}$]	*
dyne	1 dyn	= 1.00×10^{-5}	N	*
kilogram force (= kilopond (kp))	1 kgf	= 9.807×10^0	N	
poundal	1 pdl	= 1.383×10^{-1}	N	
pound force (avoirdupois)	1 lbf	= 4.448×10^0	N	
ounce force (avoirdupois)	1 ozf	= 2.780×10^{-1}	N	
Pressure, stress				
▶ pascal	1 Pa	[≡ 1.00×10^0	N/m^2]	*
▷ atmosphere[a], standard	1 atm	= $1.013\,25 \times 10^5$	Pa	*
▷ bar	1 bar	= 1.00×10^5	Pa	*
centimetres of mercury (0°C)	1 cmHg	= 1.333×10^3	Pa	
dyne per square centimetre	1 dyn/cm^2	= 1.00×10^{-1}	Pa	*
feet of water (4°C)	1 ftH_2O	= 2.989×10^3	Pa	
inches of mercury (0°C)	1 inHg	= 3.386×10^3	Pa	
inches of water (4°C)	1 inH_2O	= 2.491×10^2	Pa	
kilogram force per square centimetre	1 kgf/cm^2	= 9.807×10^4	Pa	
pound force per square foot	1 lbf/ft^2	= 4.788×10^1	Pa	
pound force per square inch (= psi)[b]	1 lbf/in^2	= 6.895×10^3	Pa	
torr (0°C) (= mmHg)	1 torr	= 1.333×10^2	Pa	
Energy, work, quantity of heat				
▶ joule (≡ W·s)	1 J	[≡ 1.00×10^0	N·m]	*
▶ electronvolt	1 eV	[= $1.602\,19 \times 10^{-19}$	J, approx.]	
British thermal unit (International Table)	1 Btu	= 1.055×10^3	J	
calorie (thermochemical)	1 cal	= 4.184×10^0	J	*
calorie (International Table)	1 cal_{IT}	= 4.187×10^0	J	
erg	1 erg	= 1.00×10^{-7}	J	*
foot-pound force	1 ft·lbf	= 1.356×10^0	J	
kilowatt-hour	1 kW·h	= 3.60×10^6	J	*
kiloton explosive yield (PNE) (≡ 10^{12} g-cal)	1 kt yield	≃ 4.2×10^{12}	J	
Power, radiant flux				
▶ watt	1 W	[≡ 1.00×10^0	J/s]	*
British thermal unit (International Table) per second	1 Btu/s	= 1.055×10^3	W	
calorie (International Table) per second	1 cal_{IT}/s	= 4.187×10^0	W	
foot-pound force/second	1 ft·lbf/s	= 1.356×10^0	W	
horsepower (electric)	1 hp	= 7.46×10^2	W	*
horsepower (metric) (= ps)	1 ps	= 7.355×10^2	W	
horsepower (550 ft·lbf/s)	1 hp	= 7.457×10^2	W	
Temperature				
▶ temperature in degrees Celsius, t where T is the thermodynamic temperature in kelvin and T_0 is defined as 273.15 K	$t = T - T_0$			
degree Fahrenheit	$t_{°F} - 32$	$\times \left(\frac{5}{9}\right)$ gives	t *(in degrees Celsius)*	*
degree Rankine	$T_{°R}$	$\times \left(\frac{5}{9}\right)$ gives	T *(in kelvin)*	*
degrees of temperature difference[c]	$\Delta T_{°R}$ (= $\Delta t_{°F}$)	$\times \left(\frac{5}{9}\right)$ gives	ΔT (= Δt)	*
Thermal conductivity[c]				
1 $Btu \cdot in/(ft^2 \cdot s \cdot °F)$	*(International Table Btu)*	= 5.192×10^2	$W \cdot m^{-1} \cdot K^{-1}$	
1 $Btu/(ft \cdot s \cdot °F)$	*(International Table Btu)*	= 6.231×10^3	$W \cdot m^{-1} \cdot K^{-1}$	
1 $cal_{IT}/(cm \cdot s \cdot °C)$		= 4.187×10^2	$W \cdot m^{-1} \cdot K^{-1}$	

[a] atm abs, ata: atmospheres absolute;
atm (g), atü: atmospheres gauge.

[b] lbf/in^2 (g) (= psig): gauge pressure;
lbf/in^2 abs (= psia): absolute pressure.

[c] The abbreviation for temperature difference, deg (= degK = degC), is no longer acceptable as an SI unit.

HOW TO ORDER IAEA PUBLICATIONS

An exclusive sales agent for IAEA publications, to whom all orders and inquiries should be addressed, has been appointed in the following country:

UNITED STATES OF AMERICA	UNIPUB, 345 Park Avenue South, New York, NY 10010

In the following countries IAEA publications may be purchased from the sales agents or booksellers listed or through your major local booksellers. Payment can be made in local currency or with UNESCO coupons.

ARGENTINA	Comisión Nacional de Energía Atomica, Avenida del Libertador 8250, RA-1429 Buenos Aires
AUSTRALIA	Hunter Publications, 58 A Gipps Street, Collingwood, Victoria 3066
BELGIUM	Service du Courrier de l'UNESCO, 202, Avenue du Roi, B-1060 Brussels
CZECHOSLOVAKIA	S.N.T.L., Spálená 51, CS-113 02 Prague 1 Alfa, Publishers, Hurbanovo námestie 6, CS-893 31 Bratislava
FRANCE	Office International de Documentation et Librairie, 48, rue Gay-Lussac, F-75240 Paris Cedex 05
HUNGARY	Kultura, Hungarian Trading Company for Books and Newspapers, P.O. Box 149, H-1389 Budapest 62
INDIA	Oxford Book and Stationery Co., 17, Park Street, Calcutta-700 016 Oxford Book and Stationery Co., Scindia House, New Delhi-110 001
ISRAEL	Heiliger and Co., Ltd., Scientific and Medical Books, 3, Nathan Strauss Street, Jerusalem
ITALY	Libreria Scientifica, Dott. Lucio de Biasio "aeiou", Via Meravigli 16, I-20123 Milan
JAPAN	Maruzen Company, Ltd., P.O. Box 5050, 100-31 Tokyo International
NETHERLANDS	Martinus Nijhoff B.V., Booksellers, Lange Voorhout 9-11, P.O. Box 269, NL-2501 The Hague
PAKISTAN	Mirza Book Agency, 65, Shahrah Quaid-e-Azam, P.O. Box 729, Lahore 3
POLAND	Ars Polona-Ruch, Centrala Handlu Zagranicznego, Krakowskie Przedmiescie 7, PL-00-068 Warsaw
ROMANIA	Ilexim, P.O. Box 136-137, Bucarest
SOUTH AFRICA	Van Schaik's Bookstore (Pty) Ltd., Libri Building, Church Street, P.O. Box 724, Pretoria 0001
SPAIN	Diaz de Santos, Lagasca 95, Madrid-6 Diaz de Santos, Balmes 417, Barcelona-6
SWEDEN	AB C.E. Fritzes Kungl. Hovbokhandel, Fredsgatan 2, P.O. Box 16356, S-103 27 Stockholm
UNITED KINGDOM	Her Majesty's Stationery Office, P.O. Box 569, London SE 1 9NH
U.S.S.R.	Mezhdunarodnaya Kniga, Smolenskaya-Sennaya 32-34, Moscow G-200
YUGOSLAVIA	Jugoslovenska Knjiga, Terazije 27, P.O. Box 36, YU-11001 Belgrade

---

Orders from countries where sales agents have not yet been appointed and requests for information should be addressed directly to:

Division of Publications
International Atomic Energy Agency
Kärntner Ring 11, P.O. Box 590, A-1011 Vienna, Austria

HOW TO ORDER IAEA PUBLICATIONS

An exclusive sales agent for IAEA publications, to whom all orders and inquiries should be addressed, has been appointed in the following country:

[illegible]

In the following countries IAEA publications may be purchased from the sales agents or booksellers listed or through your major local booksellers. Payment can be made in local currency or with UNESCO coupons.

[illegible]

Orders from countries where sales agents have not yet been appointed and requests for information should be addressed directly to:

Division of Publications
International Atomic Energy Agency
Kärntner Ring 11, P.O. Box 590, A-1011 Vienna, Austria

79-6377